Engineering Science

Engineering Science

For Foundation Degree and Higher National

Mike Tooley and Lloyd Dingle

 Routledge
Taylor & Francis Group

LONDON AND NEW YORK

First published 2012
by Routledge
2 Park Square, Milton Park, Abingdon, Oxon OX14 4RN

Simultaneously published in the USA and Canada
by Routledge
711 Third Avenue, New York, NY 10017

Routledge is an imprint of the Taylor & Francis Group, an informa business

British Library Cataloguing in Publication Data
A catalogue record for this book is available from the British Library

Library of Congress Cataloging-in-Publication Data
Tooley, Michael H.
Engineering science : for foundation degree and higher national / Mike Tooley and Lloyd Dingle. – 1st ed.
p. cm.
Includes index.
1. Engineering. 2. Engineering mathematics. I. Dingle, Lloyd. II. Title.
TA147.T659 2012
620–dc23   2011050168

ISBN: 978-1-85617-775-7 (pbk)
ISBN: 978-0-203-11496-4 (ebk)

Typeset in Times
by Cenveo Publisher Services

Printed and bound in India by Replika Press Pvt. Ltd.

Contents

vi **Contents**

Preface

Introduction

This book has been designed to provide you with a thorough introduction to the core engineering science that you will study as part of a Foundation Degree, First Degree or Higher National engineering course. The two authors have many years of experience of teaching engineering at this level and their principal aim has been that of capturing, within a single volume, the core knowledge required of all engineering students studying courses beyond level 3.

What is engineering science?

Engineering covers a vast field of disciplines ranging from aeronautical engineering to shipbuilding and from biomedical engineering to telecommunications. Although these sectors may be diverse, engineers need a common core of scientific knowledge in order to be able to work effectively in them. No matter what branch of engineering you choose to study (and eventually work in), you will need a thorough grounding in the underpinning scientific principles. This common core of knowledge, based on physics and applied mathematics, is what we refer to as 'engineering science'.

Engineering science provides a thorough grounding in applied physics for all engineering specialisms including aeronautical engineering, electronics, civil engineering, marine engineering, nanotechnology, mechanical engineering, control engineering, etc. Engineering science is *the* key to understanding and solving many everyday problems, such as why some structures are stronger than others, some materials conduct electricity more readily than others and why an airframe is able to withstand the stresses imposed on it during high-speed flight. Engineering science provides us with a means of bridging the gap between different disciplines by applying scientific principles to the analysis and solution of real-world problems.

About this book

We have organised this book into seven main parts:

Part I: Mechanics of materials

Part II: Dynamics

Part III: Thermodynamics

Part IV: Electrostatics and electromagnetism

Part V: Direct current

Part VI: Transients

Part VII: Alternating current

Each part has been divided into chapters and each chapter is devoted to a major topic. Each chapter is then further sub-divided into decimal-numbered sections.

Throughout this book we have provided worked examples that show how the ideas introduced in the text can be put into practice. We have also included problems and questions at various stages in the text. Depending on the nature of the topic, these questions take a variety of forms, from simple *Test your knowledge* problems requiring short numerical answers to those that may require some additional research or the use of an analytical software package in their solution. Your tutor may well ask you to provide answers to these questions as coursework or homework but they can also be used to help you with revision for course assessments. In addition, a set of *Review questions* will be found at the end of each chapter. *The answers to all of the questions and problems will be found on the book's website.* Some additional information is also available for downloading from the site (see page xii).

Part I provides a study of the mechanics of materials and the behaviour of solid bodies under the influence of loads. It examines the way in which a body behaves as a result of the loads that are imposed, the stresses that they endure and the subsequent strains and deflections. This is of utmost importance to engineers involved with the design and application of engineering structures, such as beams, columns, plates and catenaries and their use in applications such as airframes, boilers, bridges, pressure vessels and the hulls of ships.

Chapter 1 provides an introduction to fundamental concepts for readers who may not have studied mechanical engineering before. This chapter explains basic concepts such as the resolution of forces, coplanar forces, and the relationship between stress and strain. It provides essential background reading for the chapters that follow.

Chapter 2 is devoted to an analysis of beams and bending, whilst Chapter 3 considers the concept of torsion in shafts used as power transmission. The forces and stresses acting on pressure vessels are considered in Chapter 4, and Chapter 5 deals with concentrically loaded columns and struts.

Chapter 6 provides an introduction to strain energy and an analysis of the deflection of beams from externally loaded components. The last chapter in Part I provides an introduction to the analysis of complex stress and strain. The chapter finishes with a section on the use of various types of gauge used for strain measurement.

Part II is concerned with the motion of bodies under the action of forces. We begin our study of dynamics in Chapter 8 by looking at fundamental concepts of linear and angular motion and the forces that create such motion. We next consider momentum and inertia, together with the nature and effects of friction that acts on linear and angular motion machines and systems.

In Chapter 9 we consider the motion of one or two single and multilink mechanisms before briefly considering the geometry and output motions of various types of cam. Chapter 10 continues by looking at power transmission systems and their application in the form of belt drives, friction clutches, gears, screw drives, dynamic balancing, rotors and flywheels. Chapter 11 concludes our short study of engineering dynamics with a brief look at the nature and effects of mechanical vibration and oscillatory motion.

Part III provides an introduction to thermodynamics and the relationship between heat and mechanical energy and the conversion of one to another. Our study of thermodynamics in this part of the book is constrained to gases and liquids as the working fluid.

A companion chapter available for downloading from the book's website covers vapours, vapour processes and systems.

Chapter 12 reviews thermodynamic fundamentals including topics such as pressure, specific heat, latent heat and the gas laws. Once again, this introductory chapter provides an essential foundation for the chapters that follow. Chapter 13 is concerned with closed and open thermodynamic systems and an introduction to the first and second laws of thermodynamics, whilst Chapter 14 looks at the perfect gas processes including constant pressure, constant volume, isothermal, isentropic and polytropic processes. The concepts of reversibility and reversible work are also covered.

The concept of entropy and its use in determining end states for thermodynamic cycles is introduced in Chapter 15. The Carnot cycle, Otto cycle, diesel cycle and constant pressure cycles are covered, together with their isentropic efficiencies. Chapter 16 is concerned with the practical cycles and efficiencies of combustion engines, whilst Chapter 17 provides an introduction to the separate subject of heat transfer by conduction, convection and radiation.

Finally in Part V, Chapter 18 is devoted to fluid mechanics with an introduction to fluid statics (such as thrust forces on immersed bodies and the buoyancy of immersed and floating bodies) and fluid dynamics (such as fluid momentum, Bernoulli theory and applications, fluid flow instruments, viscosity and energy losses in a piped system).

Part IV provides an introduction to electrostatics and electromagnetism. These two chapters are essential background reading for the chapters that follow. Chapter 19 begins by explaining the nature of electric charge and how it is quantified before describing the ways that it can be concentrated and stored in capacitors. The chapter also introduces the important concept of the electric dipole that exists where two charges of equal magnitude but opposite polarity are placed in close proximity to one another. Chapter 20 examines the comparable phenomenon that occurs when a magnetic dipole is created by a bar magnet having north and south poles at its opposite extremities. The chapter also explains the concepts of magnetic flux and how it is quantified as well as describing ways in which it can be shaped and concentrated.

Part V provides an introduction to direct current and the application of various theorems that can be used to solve electric and magnetic circuits. In Chapter 21 we introduce direct current electricity and the fundamental nature of electric current, potential difference (voltage) and opposition to current flow (resistance). We also compare electric and magnetic circuits, showing that magnetic flux is analogous to electric current, magnetomotive force is analogous to electromotive force, and reluctance of a magnetic circuit is analogous to resistance in an electric circuit.

Finally in Part V, Chapter 22 explains, with the aid of a number of worked examples, the use of circuit theorems as an aid to the solution of even the most complex of series/parallel direct current circuits containing multiple sources of electromotive force and current.

In Part VI we examine the behaviour of circuits in which transient conditions exist when there is a change from one steady-state condition to another, for example when a switch is opened or closed, when a supply voltage is first applied to a circuit or when it is disconnected. In Chapter 23 we investigate the growth and decay of voltage and current in first-order circuits where resistance and capacitance or resistance and inductance are present. Our analysis of transients continues in Chapter 24 when we develop an understanding of the behaviour of first- and second-order systems, using the Laplace transform.

Part VII is dedicated to a study of alternating current. Here again, an introductory chapter is provided for readers who may not have studied this topic before. Chapter 25 begins by explaining fundamental terms and concepts. This chapter also explains the important concepts of reactance (both capacitive and inductive) and impedance.

Chapter 26 provides an introduction to the use of a simple yet powerful method of solving even the most complex of a.c. circuits by using complex numbers and j-notation. The phenomenon of resonance is explained in Chapter 27. We examine the effects of series and parallel resonance and introduce the relationship between Q-factor and bandwidth. We also explain the effects of loading and damping on the performance of a resonant circuit.

Mutual inductance and the transformer principles are discussed in Chapter 28. This chapter also provides a discussion of transformer losses, regulation, efficiency and the effect of loads that are not purely resistive. Chapter 29 takes this a stage further by providing an introduction to power and power factor in a.c. circuits.

Chapter 30 is devoted to waveforms that are not purely sinusoidal. This chapter includes an introduction to Fourier analysis, a powerful technique based on the concept that all waveforms, whether continuous or discontinuous, can be expressed in terms of a convergent series. Finally, Chapter 31 concludes Part VII with an investigation of the power contained in a complex waveform.

Essential mathematics

The authors recognise that, for many engineering students, mathematics can be challenging. In order to help with this we have provided three useful introductions to key mathematical topics on the book's companion website: *Vectors and vector operations*, *Introduction to the calculus* and *an Introduction to differential equations*. In addition, two further *Essential mathematics* topics can be downloaded from the book's website. These include *Algebraic fundamentals* and *Trigonometric identities*.

It is important to note that these brief *Essential mathematics* sections are not designed to replace conventional mathematics textbooks. Instead, they are designed to provide a succinct and easily digestible reference at the point at which they are needed to support the main text.

About SCILAB

In order to simplify the mathematical content of Parts IV to VII we have included a number of examples that make use of open-source mathematical modelling software developed by the *Institut national de recherche en informatique et en automatique* (INRIA). INRIA is a French research institute dedicated to technology transfer and is active in the computer science, control theory and applied mathematics sectors. INRIA is organised in five key sectors:

- Aeronautics, defence and aerospace
- Software publishing and embedded systems
- Energy, transport and sustainable development
- Health, life sciences and biotechnology
- Telecommunications, networks and multimedia

INRIA's mathematical modelling package, SCILAB, is well documented and supported. The software is applicable to a very wide range of scientific, control and engineering applications and can be freely downloaded from www.scilab.org. SCILAB is ideal for student use and has many features that are comparable with the immensely popular (but not freely distributed) MATLAB package.

A few words of advice and encouragement

This book has been designed to provide you with a thorough introduction to each of the main topics that you will study as part of your engineering science course. Despite this, you are advised to make use of other reference books and materials wherever and whenever possible. You should also get into the habit of using all of the resources that are available to you. These include your tutor, your college or university library, computer centre, engineering laboratories, and other learning resources including the Internet (see the book's website for some useful web links). You should also become familiar with selecting materials that are appropriate to the topics that you are studying. In particular, you may find it useful to refer to materials that will provide you with several different views of a particular topic.

Finally, we would like to offer a few words of advice and encouragement to students. At the beginning of your engineering course you will undoubtedly find that some topics appear to be more difficult than others. Sometimes you may find the basic concepts difficult to grasp (perhaps you have not met them before), you may find the analytical methods daunting, or you might have difficulty with things that you cannot immediately visualise.

No matter what the cause of your temporary learning block, it is important to remember two things: you would not be the first person to encounter the problem, and there is plenty of material available to you that will help you overcome it. All that you need to do is to recognise that it *is* a problem and then set about doing something about it. A regular study pattern and a clearly defined set of learning goals will help you get started. In any event, do not give up – engineering is a challenging and demanding career and your first challenge along the road to becoming a practising engineer is to master the core scientific knowledge that engineers use in their everyday work. And that is what you will find in this book.

May we wish you every success with your engineering studies!

Mike Tooley and Lloyd Dingle

Acknowledgements

The authors would like to thank a number of people who have helped in producing this book. In particular, we would like to thank Gavin Fidler, Emma Gadsden and all members of the team at Taylor & Francis for their patience and perseverance. Last, but by no means least, we would like to say a big 'thank you' to Wendy and Yvonne. But for your support and understanding this book would never have been finished!

Website

More information and resources, *together with the answers to questions and problems*, are available from the book's companion website: www.key2engineeringscience.com

Part I

Mechanics of materials

The study of the mechanics of materials is concerned with the behaviour of solid bodies under the influence of loads. The ways in which these bodies behave as a result of the loads imposed on them, the stresses they endure, their subsequent strains and deflections, together with their internal reaction to these externally imposed loads, are of the utmost importance to engineers, particularly with respect to the design and in-service endurance of engineering structures.

In Chapter 1 some *fundamental concepts* are covered that may or may not be familiar to the reader but which are designed to provide essential background for the topics that follow. These topics include forces, resolution of forces, coplanar force systems, simple stress and strain, and thermal stress and strain. Chapter 2 is concerned with the analysis of *beams* and includes topics on shear force and bending moment, engineers' theory of bending, centroid and second moment of area, beam selection and the slope and deflection of beams. In Chapter 3 we consider the concept of *torsion in shafts* used as power transmitters; topics covered include engineers' theory of torsion, polar second moments of area and the power transmitted by shafts. Chapter 4

is concerned with the forces and stresses created by and acting on *pressure vessels*, where both thick-walled and thin-walled pressure vessels are considered, together with an application of the theory to the stress design of pressure vessels for specific functions. In Chapter 5 concentrically loaded *columns and struts* are considered; topics covered include the determination of parameters such as slenderness ratio, radius of gyration and effective length, Euler theory and the Rankine–Gordon relationship. Chapter 6 provides an introduction to *strain energy*, where strain energy is considered as a result of direct stress, shear stress, torsion and bending. Castigliano's theorem is introduced and its use for analysing deflection of beams from externally loaded components is covered. Finally, in Chapter 7, we analyse complex stress and strain, starting with the analysis of stresses on oblique planes that result from direct tensile loading. Two-dimensional stresses acting both directly and in shear are analysed, together with the use of Mohr's circle. Complex strain is then analysed and principal strains determined. The chapter finishes with a section on strain gauging and the use of strain gauge rosettes to determine principal strains.

Chapter 1

Fundamentals

We begin our study in this chapter with a reminder of the concepts of force, vectors, the analysis of coplanar force systems and compound bars, followed by a study of one- and two-dimensional stress and strain. This lays the foundation for the study of beams, shafts, pressure vessels, columns, struts, strain energy and complex stress and strain, that follow in the subsequent chapters of this, the first part of *Engineering Science*.

Some of the fundamental concepts covered in this first chapter may be unfamiliar to all readers, so readers with a particular area of weakness should ensure that they attempt and successfully complete the 'Test your knowledge' (TYK) questions at the end of each section. For this particular chapter, these TYK exercises have been designed not only to enable immediate revision and consolidation but also to act as review questions at the end of each section, rather than at the end of the chapter which, you will find, is the normal format for many of the subsequent chapters.

1.1 Force

In its simplest sense a force is a push or pull exerted by one object on another. In a member in a static structure, a push causes compression and a pull causes tension. Members subject to compressive and tensile forces have special names. A member of a structure that is in *compression* is known as a *strut* and a member in *tension* is called a *tie*.

Only rigid members of a structure have the capacity to act as both a strut and a tie. Flexible members, such as ropes, wires or chains, can only act as ties.

Force cannot exist without opposition, as you will know from your previous study of Newton's laws. An applied force is called an *action* and the opposing force it produces is called a *reaction*.

> **Key Point.** The action of a force always causes an opposite reaction

The effects of any force depend on its three characteristics: *magnitude*, *direction* and *point of application*.

In general, *force* is that which changes, or tends to change, the state of rest or uniform motion of a body.

$$F = ma = \frac{m(v - u)}{t} \tag{1.1}$$

This formula is a consequence of Newton's second law, which you should already be familiar with. If not, you should refer to Part II of this book, where Newton's laws are revised at the beginning of Chapter 8, Fundamentals of Dynamics.

The SI unit of force is the *Newton* (N). The Newton is defined as follows: *1 Newton is the force that gives a mass of 1 kg an acceleration of 1 m s^{-2}.*

> **Key Point.** Force = mass (*m*) × acceleration (*a*) and is a vector quantity

Note that weight force is a special case where the acceleration acting on the mass is that due to gravity, so *weight force* may be defined as $F = mg$. On the surface of the earth the gravitational acceleration is taken as

$g = 9.81 \, \mathrm{m \, s^{-2}}$. This means, for example, that the weight of a body (with the same mass) will vary from that on earth if it is taken to a place such as the moon, where it is subject to the influence of the moon's much lower gravity.

> **Key Point.** Weight = mass × acceleration due to gravity and it is a vector quantity

Since forces have magnitude, direction and a point of application and are therefore vector quantities, they can be represented graphically in two dimensions by scale drawing. The *resultant* and/or *equilibrant* of coplanar force systems may also be determined by resolving forces using trigonometry or, in the case of three-dimensional force systems, by use of *position vectors* and *vector arithmetic* (see *Essential Mathematics 1 – Vector Operations* on the book's companion website).

1.2   Vector representation and combination of forces

You will be aware that a *force* may be represented on paper as a *vector* quantity, provided an arrow is drawn to a convenient scale representing its magnitude and the arrow is offset at some angle θ representing its direction (Figure 1.1). Please note that throughout this book all vector quantities will be identified using emboldened text.

In addition to possessing the properties of magnitude and direction from a given reference (Figure 1.1), vectors must obey the *parallelogram law* of combination. This law requires that two vectors $\mathbf{v_1}$ and $\mathbf{v_2}$ may be replaced by their equivalent vector $\mathbf{v_T}$ which is the diagonal of the parallelogram formed by $\mathbf{v_1}$ and $\mathbf{v_2}$, as shown in Figure 1.2a.

This vector sum is represented by the vector equation: $\mathbf{V_T} = \mathbf{V_1} + \mathbf{V_2}$.

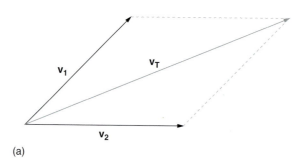

(a)

Figure 1.2 (a) Parallelogram law

Note: the plus sign in this equation refers to the addition of two vectors, and should not be confused with ordinary scalar addition, which is simply the sum of the magnitudes of these two vectors and is written as $V_T = V_1 + V_2$ in the normal way *without* emboldening (see also *Essential Mathematics 1 – Dot products*).

Vectors may also be added head-to-tail using the *triangle law* as shown in Figure 1.2b. It can also be seen from Figure 1.2c that the order in which vectors are added does not affect their sum.

> **Key Point.** Two vectors may be added using the parallelogram rule or triangle rule

The vector difference $\mathbf{V_1} - \mathbf{V_2}$ is obtained by *adding* $-\mathbf{V_2}$ to $\mathbf{V_1}$. The effect of the minus sign is to reverse the direction of the vector $\mathbf{v_2}$ (Figure 1.2d).

The vectors $\mathbf{v_1}$ and $\mathbf{v_2}$ are known as the components of the vector $\mathbf{V_T}$.

In the two examples that follow we are asked to find the resultant of the force systems, that is, the single equivalent force that can replace all the forces in the system.

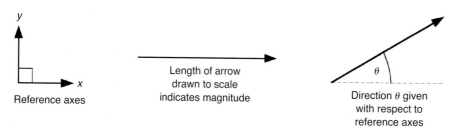

Figure 1.1 Graphical representation of force

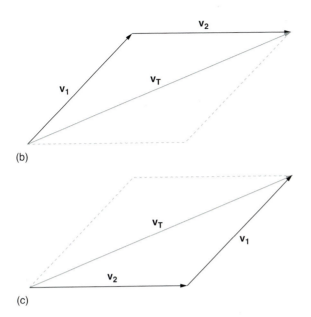

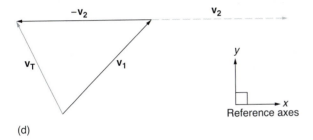

Figure 1.2 (b) Triangle law and (c) Reverse order

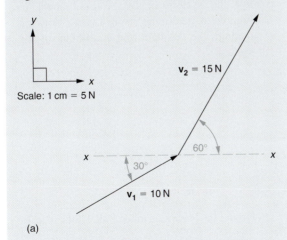

Figure 1.2 (d) Vector subtraction

Example 1.1 Two forces act at a point as shown in Figure 1.3. Find by vector addition their resultant.

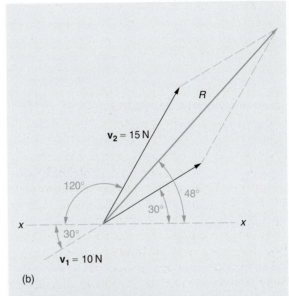

(b)

Figure 1.3 (b) Vector diagram

From the vector diagram the resultant vector **R** is 4.8 cm in magnitude and, from the scale, this gives the magnitude of the resultant $R = 24$ N at an angle of 48° from the *x*-axis.

Note that a *space diagram* is first drawn to indicate the orientation of the forces with respect to the reference axes; these axes should always be shown. Also note that the *line of action* of vector $\mathbf{v_1}$, passing through the point 0, is shown in the space diagram and can lie anywhere on this line, as indicated on the vector diagram.

Key Point. The resultant is the single equivalent force that replaces all the forces in a force system

Example 1.2 Find the result of the system of forces shown in Figure 1.4, using vector addition.

From the diagram, the resultant = 6.5 cm = 6.5 × 10 N = 65 N acting at an angle of 54° from the *x*-reference axis. This answer may be written mathematically as: resultant = 65 N∠54°

(a)

Figure 1.3 Vector addition using the parallelogram law (a) Space diagram

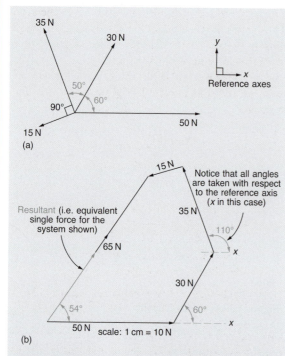

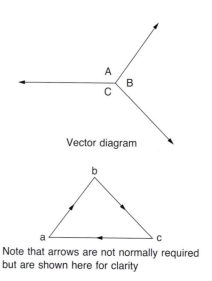

Figure 1.4 caption: **Figure 1.4** Vector addition using polygon of forces method (a) Space diagram (b) Vector diagram

that lie in the adjacent spaces either side of the vector arrow representing that force.

Figure 1.5 Bow's notation

Key Point. Vector addition of two or more forces acting at a point (concurrent coplanar forces) can be achieved using the head-to-tail rule

Note that for the force system in Example 1.2 vector addition has produced a *polygon*. Any number of forces may be added vectorially in any order, provided the *head-to-tail rule* is observed. In this example, if we were to add the vectors in reverse order, the same result would be achieved. Also note that for any concurrent coplanar force system the single force, the *equilibrant*, that brings the system into static equilibrium is equal in magnitude and opposite in direction to the resultant.

Key Point. The *equilibrant* is opposite in direction and equal in magnitude to the resultant and maintains the system of *concurrent* coplanar forces in static equilibrium.

Bow's notation is a convenient system of labelling the forces for ease of reference when there are three or more forces to be considered. Capital letters are placed in the space between forces in a clockwise direction, as shown in Figure 1.5. Any force is then referred to by the letters

The vectors representing the forces are then given the corresponding *lower case* letters. Thus the forces AB, BC and CA are represented by the vector quantities *ab*, *bc* and *ca*, respectively. This method of labelling applies to any number of forces and their corresponding vectors.

Test your knowledge 1.1

1. For the system of two forces shown in Figure 1.6, determine, using a vector drawing method, the resultant and equilibrant of the system.

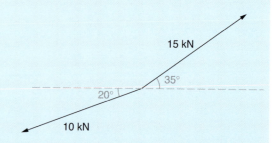

Figure 1.6 Force system for TYK 1.1 question 1

2. For the system of forces shown in Figure 1.7, determine, using the polygon of forces

method, the magnitude and direction of the force that will put the system into equilibrium.

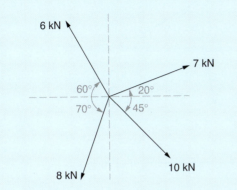

Figure 1.7 Force system for TYK 1.1 question 2

1.3   Coplanar force systems

Forces that act within a two-dimensional plane, such as the plane of this paper, are referred to as *coplanar forces*. If all the *lines of action* of these forces pass through the same point, known as *the point of concurrence* (Figure 1.8), then we have a *concurrent coplanar force system*.

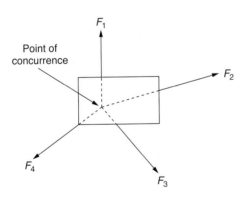

Figure 1.8 Concurrent coplanar force system

When dealing with forces that act on rigid structures and other rigid bodies, we ignore deformation within these bodies and concentrate only on the effects of external forces that act on them. Under these circumstances it is not necessary to restrict the application of these forces to one point. In fact, any

external force acting on a body will produce the same effect, provided it acts anywhere along its line of action. Thus, in Figure 1.9, the external effects of the force F on the two securing bolts (A, B) that rigidly support the solid metal bracket will be the same, no matter whether this force is applied at point P or point Q, along its line of action.

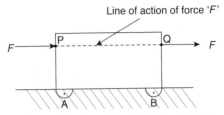

Same external reaction at A + B, no matter whether force F is applied at P or at Q

Figure 1.9 Illustration of the principle of transmissibility

We may summarise and generalise the above situation by the *principle of transmissibility*, that states:

A force may be applied at any point on its line of action to a rigid body without altering the resultant effects of the force external to the rigid body on which it acts.

The usefulness of this principle will become clearer later when we consider force systems in which forces act other than at a point.

Key Point.   A force may be applied at any point along its line of action to a rigid body without altering the resultant effects of the force external to the rigid body on which it acts. This is known as *the principle of transmissibility*

All the force systems that we have considered so far, that have been solved using vectors, have involved only concurrent coplanar forces.

We now need to consider another system of coplanar forces, where the lines of action of the forces in the system do not pass through the same point. This system (Figure 1.10a) is known as a *non-concurrent coplanar force system*. In this system there is not only a tendency for the force to move the body in a certain linear direction, but also to make it rotate. This system may again be represented by a single resultant force but, in addition, because the forces do not act through the same point there must also be a *turning moment* created between the line of action of the resultant and the point about which the rotation of the body takes place. This idea is illustrated in Figure 1.10b.

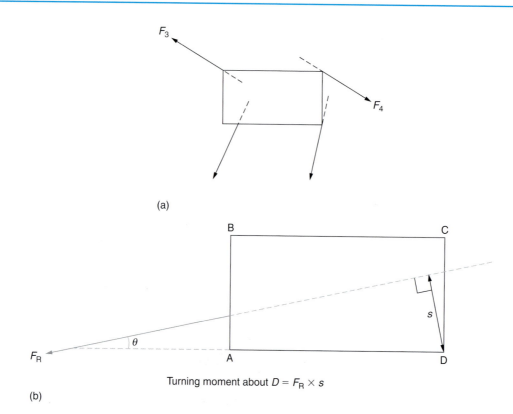

(a)

(b)

Turning moment about $D = F_R \times s$

Figure 1.10 (a) Non-concurrent coplanar force system (b) A resultant force and turning moment

The concept of turning moments and equilibrium is central to an understanding of non-concurrent coplanar force systems and their resolution. For this reason, we next revise the key concepts concerned with turning moments, equilibrium and couples.

1.3.1 Turning moments and equilibrium

A moment is a turning force, producing a turning effect. The magnitude of this turning force depends on the magnitude of the *force* applied and the *perpendicular distance* from the pivot or axis to the line of action of the force (Figure 1.11a).

The moment of a force (M) is defined as the product of the magnitude of force (F) and its perpendicular distance (s) from the pivot or axis to the line of action of the force. This may be written mathematically as: $M = Fs$, where the SI unit for a moment is the *Newton-metre* (N m).

Moments are always concerned with *perpendicular* distances. From Figure 1.11a, you should note that moments can be clockwise (*CWM*) or anticlockwise (*ACWM*). Conventionally, we consider *clockwise moments to be positive* and *anticlockwise moments to be negative*.

If the line of action of the force passes through the turning point it has no turning effect and so no moment. Figure 1.11b illustrates this important point.

The *resulting moment* is the difference in magnitude between the total clockwise moment and the total anticlockwise moment. Note that if the body is in *static equilibrium* this *resultant will be zero*.

Key Point. For static equilibrium the algebraic sum of the moments is zero

When a body is in equilibrium there can be no *resultant force* acting on it. However, reference to Figure 1.12 shows that a body subject to two equal and opposite non-concurrent coplanar forces is not necessarily in equilibrium even when there is no resultant force acting on it.

The resultant force on the body is zero but the two forces would cause the body to rotate, as indicated. Therefore, in the case illustrated, a clockwise restoring moment would be necessary to bring the system into equilibrium. This leads us to a second condition that must be satisfied to ensure that a body is in static equilibrium. This is known as *the principle of moments*, which states that *when a body is in static equilibrium*

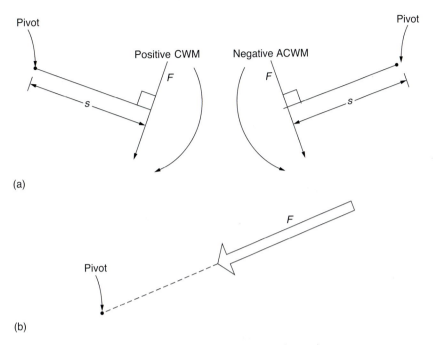

(a)

(b)

Figure 1.11 (a) Definition of a moment (b) Line of action passing through a pivot point

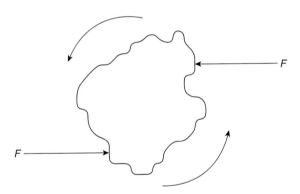

Figure 1.12 Non-equilibrium condition for equal and opposite non-concurrent coplanar forces acting on a body

under the action of a number of forces, the total CWM about any point is equal to the total ACWM about the same point. This principle may be represented algebraically by the formula:

$$\sum CWM = \sum ACWM \qquad (1.2)$$

One other further necessary condition for static equilibrium is that *upward forces = downward forces.* This further condition is necessary, for example, when considering the static equilibrium of a beam with simple supports and a number of point loads, where the reactions at the supports must balance the forces acting down on the beam.

1.3.2 Couples

So far, with respect to force systems, we have been restricted to the turning effect of forces taken one at a time. *A couple occurs when two equal forces acting in opposite directions have their lines of action parallel.*

Example 1.3 Figure 1.13 shows the turning effect of a couple on a beam of regular cross-section. Determine the moment (turning effect) of the couple.

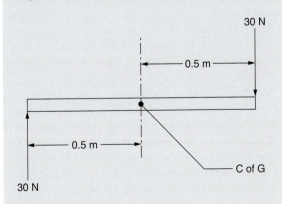

Figure 1.13 Turning effect of a couple

Taking moments about the centre of gravity (C of G), that is the point at which all the weight of the beam is deemed to act, we get:

$$(30 \times 0.5) + (30 \times 0.5) = \text{turning effect}$$

So *moment of couple* = 30 N m.

Example 1.4   Figure 1.14 shows the turning effect of a couple on a beam of irregular cross-section that we will again try to resolve about its centre of gravity

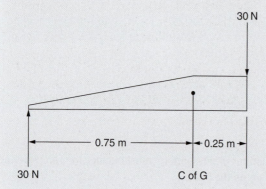

Figure 1.14 Turning effect of a couple with irregular cross-section beam

Again, taking moments about the C of G gives:

$$(30 \times 0.75) + (30 \times 0.25) = \text{turning effect}$$

So *moment of couple* = 30 N m

It can be seen from the above two examples that the moment is the same in both cases and is independent of the position of the fulcrum. Therefore, if the fulcrum is assumed to be located at the point of application of one of the forces, *the moment of a couple is equal to one of the forces multiplied by the perpendicular distance between them.* Thus in both cases, shown in Examples 1.3 and 1.4, the moment of the couple = (30 N × 1 m) = 30 N m, as before.

Another important application of the couple is its *turning moment* or *torque*. The definition of torque is as follows: *torque is the turning moment of a couple and is measured in Newton-metres* (N m), *that is, torque T = force F × radius r*. The turning moment of the couple given above in Example 1.3 is = $F \times r$ = (30 N × 0.5 m) = 15 N m. Thus the formula for torque is:

$$T = Fr \qquad (1.3)$$

Key Point.   The moment (turning effect) of a couple = force × distance between forces, and the turning moment (torque) = force × radius

1.4   Resolution of forces for coplanar systems

Graphical solutions to problems involving vector forces are sufficiently accurate for many engineering problems and are invaluable for estimating approximate solutions to more complicated force problems. However, it is sometimes necessary to provide more accurate results, in which case a mathematical method will be required. One such mathematical method is known as the *resolution of forces*. We look at a number of examples of this method, initially for *concurrent* coplanar force systems and then for *non-concurrent* coplanar force systems.

Consider a force F acting on a bolt A (Figure 1.15). The force F may be replaced by two forces P and Q, acting at right angles to each other, which together have the same effect on the bolt.

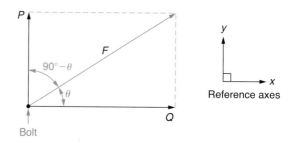

Figure 1.15   Resolving force F into its components

From your previous knowledge of the trigonometric ratios you will know that:

$$\frac{Q}{F} = \cos\theta \quad \text{and so} \quad Q = F\cos\theta$$

Also, $\dfrac{P}{F} = \cos(90 - \theta)$ and we know that

$$\cos(90 - \theta) = \sin\theta$$

therefore $P = F\sin\theta$

So, from Figure 1.15, $P = F\sin\theta$ and $Q = F\cos\theta$ and the single force F has been resolved or split into its two equivalent forces of magnitude $F\cos\theta$ and $F\sin\theta$

acting at right angles to one another (they are said to be *orthogonal* to each other). $F \cos \theta$ is known as the *horizontal component of F* and $F \sin \theta$ is the *vertical component* of F.

> **Key Point.**   Forces or loads that are *orthogonal* act at *right angles* to one another

Example 1.5   Three coplanar forces, *A*, *B* and *C*, are all applied to a pin joint (Figure 1.16a). Determine the magnitude and direction of the equilibrant for the system.

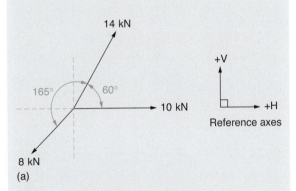

Figure 1.16 (a) Space diagram for force system

Each force needs to be resolved into its two orthogonal components, which act along the vertical and horizontal axes, respectively. Using the normal algebraic sign convention with our axes, above the origin *V* is positive and below it is negative. Similarly, *H* is positive to the right of the origin and negative to the left. Using this convention, we need only consider acute angles for the sine and cosine functions; these are tabulated below.

Magnitude of force	Horizontal component	Vertical component
10 kN	+ 10 kN (→)	0
14 kN	+ 14 cos 60 kN (→)	+ 14 sin 60 kN (↑)
8 kN	− 8 cos 45 kN (←)	− 8 sin 45 kN (↓)

Then total horizontal component

$$= 10 + 7 - 5.66 \text{ kN} = 11.34 \text{ kN}(\rightarrow)$$

and total vertical component

$$= 0 + 12.22 - 5.66 \text{ kN} = 6.46 \text{ kN}(\uparrow)$$

Since both the horizontal and vertical components are positive, the resultant force will act upwards to the right of the origin. The three original forces have now been reduced to two that act orthogonally. The magnitude of the resultant $\mathbf{F}_R$ or the equilibrant may now be obtained using Pythagoras's theorem on the right-angled triangle obtained from the orthogonal vectors, as shown in Figure 1.16b.

From Pythagoras we get $R^2 = 6.46^2 + 11.34^2 = 170.33$ and so resultant $\mathbf{F}_R = 13.05$ kN, so the *magnitude* of the equilibrant also $= 13.05$ kN.

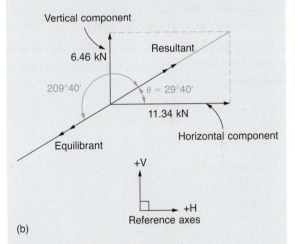

Figure 1.16 (b) Resolution method

From the right-angled triangle shown in Figure 1.16b, the angle θ that the resultant $\mathbf{F}_R$ makes with the given axes may be calculated using the trigonometric ratios.

Then $\tan \theta = \dfrac{6.46}{11.34} = 0.5697$ and $\theta = 29.67^\circ$

therefore the resultant $\mathbf{F}_R = 13.05$ kN $\angle 29.67^\circ$

The *equilibrant* will act in the opposite sense and therefore $= 13.05$ kN $\angle 209.67^\circ$.

Key Point. Pythagoras's theorem may be used to find the single resultant force of two orthogonal forces

We consider one final example on concurrent coplanar forces, concerned with *equilibrium on a smooth plane*. Smooth in this case implies that the effects of friction may be ignored. A body is kept in equilibrium on a plane by the action of three forces, as shown in Figure 1.17. These are the:

1. *weight W* of the body acting vertically down;

2. *reaction R* of the plane to the weight of the body. *R* is known as the *normal reaction*, normal in this sense meaning at right angles to the plane in this case; and

3. *force P* acting in some suitable direction to prevent the body sliding down the plane.

Forces *P* and *R* are dependent on the:

*   angle of inclination of the plane,
*   magnitude of *W*, and
*   inclination of the force *P* to the plane.

It is therefore possible to express the magnitude of both *P* and *R* in terms of *W* and the trigonometric ratios connecting the angle θ.

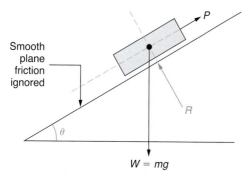

Figure 1.17 Equilibrium on a smooth plane

In the example that follows we consider the case when the body remains in equilibrium as a result of the force *P* being applied parallel to the plane.

Example 1.6   A crate of mass 80 kg is held in equilibrium by a force *P* acting parallel to the plane as indicated in Figure 1.18a. Determine, using the

resolution method, the magnitude of the force *P* and the normal reaction *R*, ignoring the effects of friction.

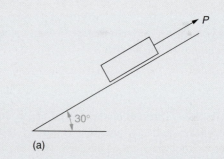

(a)

Figure 1.18 (a) Crate acted on by force *P*

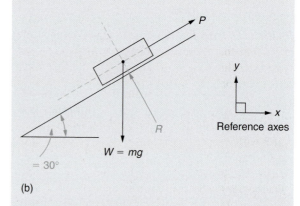

(b)

Figure 1.18 (b) Space diagram

Figure 1.18b shows the space diagram for the problem, clearly indicating the nature of the forces acting on the body.

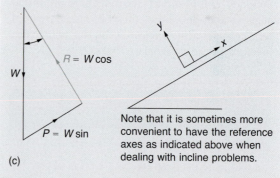

Note that it is sometimes more convenient to have the reference axes as indicated above when dealing with incline problems.

(c)

Figure 1.18 (c) Vector components of force acting on crate

W may therefore be resolved into the two forces P and R, since the force component at right angles to the plane $= W \cos \theta$ and the force component parallel to the plane $= W \sin \theta$ (Figure 1.18c).

Equating forces gives:

$$W \cos \theta = R \quad \text{and} \quad W \sin \theta = P$$

So, remembering the mass/weight relationship, we have:

Then, $\quad W = mg = (80)(9.81) = 784.8 \text{ N.}$

$$R = 784.8 \cos 30° = 679.7 \text{ N and}$$

$$P = 784.8 \sin 30° = 392.4 \text{ N.}$$

We now consider examples involving *non-concurrent* coplanar force systems and see the subtle differences in techniques that we must apply in order to solve such systems. In this first simple example, we consider forces acting horizontally and vertically on a body.

Example 1.7 A flat metal plate is pivoted at its geometric centre and is acted upon by horizontal and vertical forces as shown in Figure 1.19.

If the plate is free to rotate about the pivot, determine the magnitude and direction of the resultant of the coplanar forces acting on the system and also find the perpendicular distance of its line of action from the pivot.

Now, we remember that for non-coplanar systems there may not only be a resultant force but also a resulting turning moment that acts on the system. Therefore, we must not only resolve forces but also

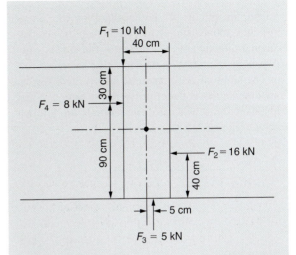

Figure 1.19 Pivoted metal plate

find the resultant moment of the system by applying the principle of moments and so, in this case, find the perpendicular distance of the line of action of the resultant from the pivot. Thus a table of values of forces and moments can be set up and these values calculated, as shown below.

Please note:

1) The use of the *sign convention*: positive forces are to the right and upwards, clockwise moments are positive, and vice versa.

2) That for the *moment of each component* in the table *we ignore the sign of the component as the moment itself is only dependent on whether it is clockwise (positive) or anticlockwise (negative)*, so in row one of the table the anticlockwise moment of the vertical component of the force F_1 about the pivot is negative, as required.

Force F (kN)	Horizontal component F_H	Vertical component F_V	Moment of F_H about pivot (kN m)	Moment of F_V about pivot (kN m)
10	0	-10	0	$-(-10 \times 0.2) = -2$
16	-16	0	$(-16 \times 0.2) = +3.2$	0
5	0	5	0	$-(5 \times 0.05) = -0.25$
8	8	0	$(8 \times 0.3) = 2.4$	0
Totals	-8	-5	$+5.6$ kN m	-2.25 kN m

14 Part I

The magnitude of the resultant of the force system is found using Pythagoras's theorem and the angular direction of the resultant is found using the trigonometric ratios, in the same way as before.

The sums of the horizontal and vertical forces are both negative, therefore the line of action of the resultant force will act down towards the left, as shown (Figure 1.20).

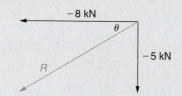

Figure 1.20 Resultant magnitude

Its magnitude will be: $R = \sqrt{(-8)^2 + (-5)^2} = \sqrt{89} = 9.43$ kN and the angle θ is found using the tangent ratio, $\tan\theta = \frac{-5}{-8} = 0.625$, therefore $\theta = \tan^{-1} 0.625 = 32°$.

Now, from the table the sum of the moments $= 5.6 - 2.25$ N m $= 3.35$ N m. Then, calling the perpendicular distance from the line of action of the resultant to the pivot point d, we get the sum of the moments $\sum M = R \times d$ or $3.35 = 9.43 \times d$ and $d = \frac{3.35}{9.43} = 0.355$ m. Now, because the resulting turning moment is positive the line of action of the resultant must be below the pivot point, as illustrated in Figure 1.21.

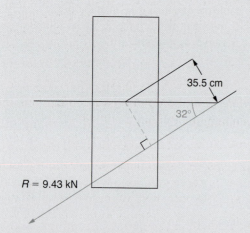

Figure 1.21 Line of action of the resultant with respect to the pivot

Key Point. Non-concurrent coplanar force systems can be reduced to a resultant force and a turning moment

Key Point. The sign convention for non-concurrent coplanar force systems dictates that upward forces and clockwise turning moments are positive, and vice versa

In the next example we include two forces that are not horizontal or perpendicular. We resolve these forces using the techniques you are already familiar with. Note also that in Example 1.8 all angles are measured from the horizontal, so that all the horizontal components of the forces will then be given by $F_H = F\cos\theta$ and all vertical components are given by $F_V = F\sin\theta$. Do remember that this is only applicable if the components of the forces are measured in this way.

Example 1.8 Determine the magnitude and direction of the resultant force and turning moment about point A for the force system shown.

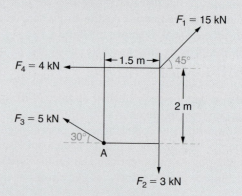

Figure 1.22 Non-concurrent force system

From Figure 1.22 we tabulate the values as we did before, noting that we have resolved F_1 and F_3 into their horizontal and vertical components, giving each their appropriate sign.

Note: When taking moments of the components about A, take only the positive value (modulus) of the components. The signs of the moments are determined only by whether they are clockwise (positive) or anticlockwise (negative). This rule can be seen in operation for the vertical moment of the 3 kN force and the horizontal moment of the 4 kN force.

Force F (kN)	Horizontal component F_H	Vertical component F_V	Moment of F_H about point A (kN m)	Moment of F_V about point A (kN m)
15	$15\cos 45 = 10.6$	$15\sin 45 = 10.6$	$(10.6 \times 2) = 21.2$	$-(10.6 \times 1.5) = -15.9$
3	0	-3	0	$(-3 \times 1.5) = 4.5$
5	$-5\cos 30 = -4.3$	$5\sin 30 = 2.5$	0	0
4	-4	0	$-(4 \times 2) = -8$	0
Totals	$+2.3$	$+10.1$	$+13.2$ kN m	-11.4 kN m

Then, using Pythagoras as before, we have $R = \sqrt{(2.3)^2 + (10.1)^2} = \sqrt{107.3} = 10.36$ kN and the direction is found using the tangent ratio. Then $\tan \theta = \dfrac{10.1}{2.3} = 4.391$ and $\theta = \tan^{-1} 4.391 = 77.17°$.

The turning moment about A is the sum of the clockwise and anticlockwise turning moments given in the table; these are $\sum M = 13.2 - 11.4 = 1.8$ kN m (clockwise). The line of action is therefore upwards towards the right.

In Example 1.8 we could have gone on to find the perpendicular distance from the line of action of the resultant to point A if we so wished, in a similar manner to that given in Example 1.7.

In the next two examples you will meet a simple engineering application of the resolution of forces method for non-concurrent coplanar force systems and a final example where we introduce the use of *Varignon's theorem*.

Example 1.9 A pulley holds a cable that laps it by an angle of 130 degrees as shown in Figure 1.23. The tensions in either side of the cable are 6 kN and 4 kN as indicated. What is the magnitude of the resultant force on the pulley?

All that is required to solve this problem is to resolve the tensile forces in the cable into their horizontal and vertical components. The diagram shows these components for the 4 kN force. I hope you can see from the geometry of the situation that the angle $A = 50°$. Then the horizontal and vertical components of F_2 are $F_{2H} = 4\cos 50 = 2.57$ kN and $F_{2V} = -4\sin 50 = -3.06$ kN, respectively.

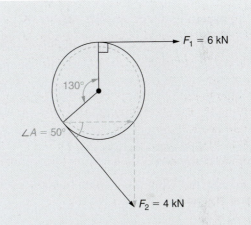

Figure 1.23 Pulley and cable assembly

Since the 6 kN force is horizontal, then the total horizontal force components $= 2.57 + 6.0 = 8.57$ kN and the total vertical force components $= -3.06$ kN. Therefore, by Pythagoras, the magnitude of the resultant force on the pulley is $= \sqrt{(8.57)^2 + (3.06)^2} = 9.1$ kN.

Its direction, although not asked for, may be easily found using the tangent ratio as before. Its line of action will act down towards the right.

In the following final example we apply a very useful theorem. In some text books it is given the name *Varignon's theorem*, and it states that: *the moment of a force about any point is equal to the sum of the moments of the concurrent components of that force about the same point*. This theorem or principle may be represented by the moment of a force about a point in terms of a *vector cross product*.

You already know from your previous study, and from what has been said in this section, that the moment of a force is given by the force F multiplied by the perpendicular distance of the line of action of the force from the point of reference; in symbols, $M = Fd$. By considering the situation in Figure 1.24, where we know that the moment of the force F about A is given by $M = Fd$, then we may also represent this moment by the vector cross-product expression:

$$\mathbf{M} = \mathbf{r} \times \mathbf{F} \qquad (1.4)$$

where $\mathbf{r}$ is a positive vector that runs from the point A to *any point* on the line of action of the vector force $\mathbf{F}$.

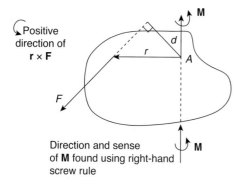

Figure 1.24 The vector cross-product situation for the moment $\mathbf{M}$ about A

The *magnitude* of this cross product is given by:

$$M = Fr \sin \theta = Fd,$$

noting that $d = r \sin \theta$.

Now the direction and sense of the moment $\mathbf{M}$ may be found using the *right-hand screw rule*, which you may have already met and used in your study of magnetic fields around current-carrying conductors. If the fingers of the right hand are curled towards the positive direction of $\mathbf{r}$ and the positive direction of $\mathbf{F}$, then the thumb points in the positive sense of $\mathbf{M}$ (see Figure 1.24). Note also that the vector cross product $\mathbf{r} \times \mathbf{F}$ is not the same as $\mathbf{F} \times \mathbf{r}$, where in the latter case the positive direction and sense of $\mathbf{M}$ are direct opposites. For a fuller explanation of vector cross products and their applications, including Varignon's theorem, you should refer to *Essential Mathematics 1 – Vector Operations*, on the book's website.

Key Point. Varignon's theorem states that: the moment of a force about any point is equal to the sum of the moments of the concurrent components of that force about the same point

You may be wondering why we bother with a vector approach to the solution of moments, where for simple two-dimensional problems in which the vertical distance d is easily calculated the conventional scalar approach is simpler. However, for some two-dimensional problems where $\mathbf{F}$ and $\mathbf{r}$ are not perpendicular but are easily expressed in vector notation, and for more complex three-dimensional problems such as mechanisms (in Part II), the vector approach is often best.

Example 1.10   Figure 1.25 shows a flat metal signpost acted upon by a coplanar force of 1500 N at point A.

a)   Calculate the magnitude of the moment about the base B, using Varignon's theorem.

b)   Then calculate the magnitude of the moment, using the principle of transmissibility.

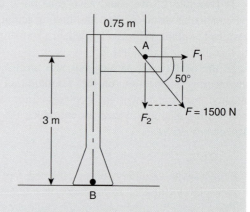

Figure 1.25 Signpost acted upon by a coplanar force

a)   In this example you will note that the 1500 N force F is non-concurrent with base B of the signpost, but, from Varignon's theorem, if we find the horizontal and vertical concurrent components of the force (F_1, F_2) about A, we will then be able to sum the moments of these forces about the base B, using vector cross products, as required. Then the *magnitude* of the moment M_B:

$$M_B = r_1 \times F_1 + r_2 \times F_2 \quad \text{or}$$

$$M_B = (3)(1500 \cos 50) + (0.75)(1500 \sin 50)$$

$$= 3754.3 \, \text{N m}.$$

b) By moving force F along its line of action, the moment about base B may be calculated by first eliminating the moment of one of the components F_1 or F_2 (see Figure 1.26), where the force F has been moved back along its line of action to align vertically (at point P) with the base B, so eliminating the moment of the force F_2.

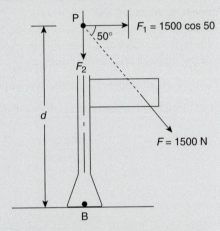

Figure 1.26 Situation for finding the magnitude of the moment M_B using transmissibility

Then $d = 3 + 0.75 \tan 50° = 3.8938$ m and the magnitude of moment $M_B = (3.8938)(1500 \cos 50) = 3754.33$ N m, as before. *Note*: The force F could just as easily have been moved down (transmitted) until the force F_1 aligned horizontally with B and the moment found in a similar manner to the above.

Test your knowledge 1.2

1. With respect to the line of action of forces, what does the term orthogonal mean?

2. What is the sign convention for forces, and for moments?

3. Find the magnitude and direction of the resultant and the equilibrant for the concurrent force system shown in Figure 1.27.

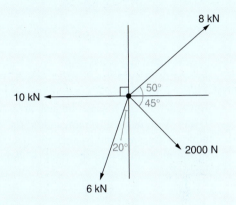

Figure 1.27 Concurrent force system

4. A flat metal plate is acted upon by a system of forces and is free to pivot about A, as shown in Figure 1.28. Determine the magnitude and direction of the resultant of the force system and also find the perpendicular distance of its line of action from the pivot.

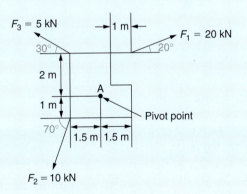

Figure 1.28 Plate force system

5. A drive belt is wrapped around a pulley assembly forming an angle of lap of 220° as shown in Figure 1.29. If the tensions in the belt are 10 kN and 8 kN, find the magnitude of the resultant force.

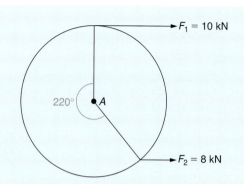

Figure 1.29 Drive belt system

6. Figure 1.30 shows the situation for a coplanar force acting on a plate hinged at point P. Using Varignon's theorem and/or the principle of transmissibility, find the magnitude of the moment M_P of the force F about the hinge P.

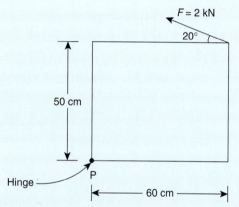

Figure 1.30 Hinged plate acted on by a coplanar force

1.5 Simple stress and strain

As you have already seen when you revised the concept of force, components that are subject to tensile loads in engineering structures are known as *ties*, while those subject to compressive loads are known as *struts*. In both cases the external loads cause internal extension (tensile loads) or compression (compressive loads) of the atomic bonds of the material from which the component is made. The intensity of the loads causing the atomic bonds to extend or compress is known as *stress* and any resulting deformation is known as *strain*. We will

now revise the nature of stress and strain in engineering components that are under load.

> **Key Point.** Structure components designed to carry tensile loads are known as *ties*, while components designed to carry compressive loads are known as *struts*.

1.5.1 Stress

If a solid, such as a metal bar, is subjected to an external force (or load), a resisting force is set up within the bar and the material is said to be in a state of stress. There are three basic types of stress that are described next:

- *Tensile stress* – which is set-up by forces tending to pull the material apart;
- *Compressive stress* – produced by forces tending to crush the material; and
- *Shear stress* – produced by forces tending to cut through the material, i.e. tending to make one part of the material slide over the other.

Figure 1.31 illustrates these three types of stress.

Stress is defined as force per unit area, that is:

$$\text{stress}, \sigma = \frac{\text{force}, F}{\text{area}, A} \qquad (1.5)$$

Also,

$$\text{shear stress}, \tau = \frac{\text{shear force } (F_s)}{\text{area resisting shear } (A)} \qquad (1.6)$$

Note: the Greek letter σ is pronounced *sigma* and the letter τ is pronounced *tau*.

The basic SI unit of stress is the N m^{-2} or pascal (Pa). The megapascal (MPa) is the same as the MN m^{-2} or the N mm^{-2}, which are commonly used.

> **Key Point.** $1\,\text{MN m}^{-2} = 1\,\text{N mm}^{-2} = 1\,\text{MPa}$

1.5.2 Strain

A material that is altered in shape due to the action of a force that acts on it is *strained*.

This may also mean that a body is strained internally even though there may be little measurable difference in its dimensions, just a stretching of the bonds at the

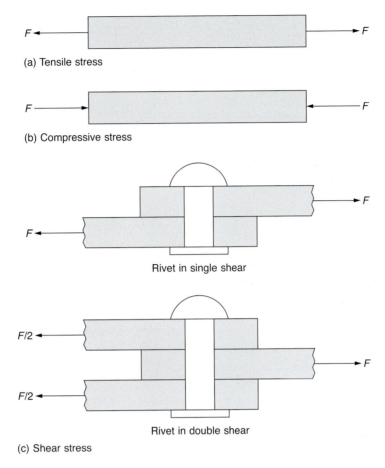

(a) Tensile stress

(b) Compressive stress

Rivet in single shear

Rivet in double shear

(c) Shear stress

Figure 1.31 Basic types of stress

atomic level. Figure 1.32 illustrates three common types of strain resulting from the application of external forces (loads) that are described next:

- *Direct tensile strain* resulting from an axial tensile load being applied;
- *Compressive strain* resulting from an axial compressive load being applied; and
- *Shear strain* resulting from equal and opposite cutting forces being applied.

Direct strain may be defined as: *the ratio of change in dimension (deformation) over the original dimension,* that is:

$$\text{direct strain, } \varepsilon = \frac{\text{deformation } (x)}{\text{original length } (l)} \quad (1.7)$$

(where both x and l are in metres). The symbol ε is the Greek lower case letter *epsilon*. Note also that the deformation for tensile strain will be an extension and for compressive strain it will be a reduction.

Note that shear strain γ (gamma), the angle of deflection resulting from the shear stress, is measured in radian and is therefore dimensionless, as is direct strain.

Key Point.   Since strain is a ratio of dimensions it has no units

1.5.3   Hooke's law and the elastic modulus

Hooke's law states that: within the elastic limit of a material the change in shape is directly proportional to the applied force producing it.

A good example of the application of Hooke's law is the *spring*. A spring balance is used for measuring weight force, where an increase in weight will cause a corresponding extension (see Figure 1.33).

The *stiffness* (k) of a spring is the force required to cause a certain (unit deflection), that is:

$$\text{spring stiffness } (k) = \frac{\text{force}}{\text{unit deflection}} \quad (1.8)$$

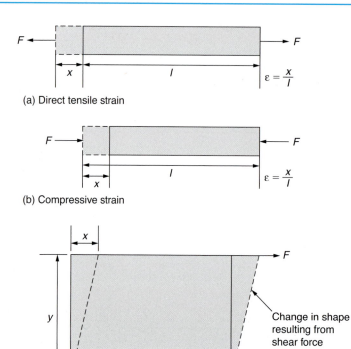

(a) Direct tensile strain

(b) Compressive strain

(c) Shear strain

Figure 1.32 Types of strain

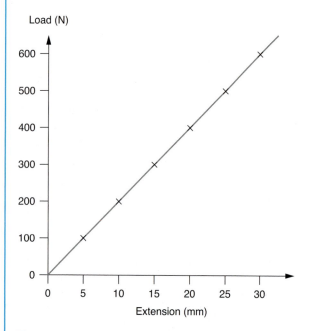

Figure 1.33 Load/extension graph for a spring

The SI unit of spring stiffness is the $N\,m^{-1}$. By considering Hooke's law, we can show that stress is directly proportional to strain while the material remains *elastic*. That is, while the external forces acting on the material are only sufficient to stretch the atomic bonds, without fracture, so that the material may return to its original shape after the external forces have been removed.

Hooke's law states that the extension is directly proportional to the applied load, provided the material remains within its elastic range. Since stress is load per unit area, the load applied to any cross-sectional area of material can be converted into stress provided the cross-sectional area of the material is known. The corresponding extension of the stressed material divided by its original length will in fact provide a value of its strain. Therefore we can quote Hooke's law in terms of stress and strain, rather than load and extension, and say that *stress is directly proportional to strain in the elastic range*, i.e.

Stress $\propto$ strain or stress = (strain × a constant) so:

$$\frac{\text{stress}}{\text{strain}} = \text{a constant}$$

This constant of proportionality will depend on the material and is given the symbol E. Thus

$$\frac{\text{stress } (\sigma)}{\text{strain } (\varepsilon)} = E \qquad (1.9)$$

where E is known as the *modulus of elasticity*. Because strain has no units, the modulus of elasticity has the same units as stress. Modulus values tend to be very high; for this reason GN m^{-2} or GPa are the preferred SI units.

Key Point. The elastic modulus of a material may be taken as a measure of the stiffness of that material

Also, the relationship between the shear stress (τ) and shear strain (γ) is known as the *modulus of rigidity* (G), i.e.

$$\text{Modulus of rigidity } (G) = \frac{\text{shear stress } (\tau)}{\text{shear strain } (\gamma)} \qquad (1.10)$$

with units of GPa or GN m^{-2}

Example 1.11 A rectangular steel bar 10 mm × 16 mm × 200 mm long extends by 0.12 mm under a tensile force of 20 kN. Find:

a) the stress,

b) the strain and

c) the elastic modulus, of the bar material.

a) Now the tensile stress $= \dfrac{\text{tensile force}}{\text{cross-sectional area}}$

Also, tensile force = 20 kN = 20×10^3 N and cross-sectional area = 10 × 16 = 160 mm^2. Remember, tensile loads act against the cross-sectional area of the material.

Then, substituting in the above formula we have:

$$\text{tensile stress } (\sigma) = \frac{20000 \text{ N}}{160 \text{ mm}^2} \qquad \sigma = 125 \text{ N mm}^{-2}$$

b) Now, strain $\varepsilon = \dfrac{\text{deformation (extension)}}{\text{original length}}$

Also, extension = 0.12 mm and the original length = 200 mm.
Then, substituting gives

$$\varepsilon = \frac{0.12 \text{ mm}}{200 \text{ mm}} = 0.0006.$$

c) $E = \dfrac{\text{stress}}{\text{strain}} = \dfrac{125 \text{ N mm}^{-2}}{0.0006}$

$$= 208000 \text{ N mm}^{-2} \text{ or } 208 \text{ GN m}^{-2}$$

Example 1.12 A, 10 mm diameter rivet holds three sheets of metal together and is loaded as shown in Figure 1.34. Find the shear stress in the rivet.

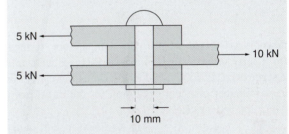

Figure 1.34 A rivet in double shear

We know that the rivet is in double shear. So the area resisting shear

$$= 2 \times \text{the cross-sectional area}$$

$$= 2\pi r^2 = 2\pi 5^2 = 157 \text{ mm}^2$$

and shear stress $= \dfrac{10000}{157} = 63.7 \text{ N mm}^{-2}$

$$= 63.7 \text{ MN m}^{-2}.$$

Note that when a rivet is in double shear, the area under shear is multiplied by 2. With respect to the load, we know from Newton's laws that to every action there is an equal and opposite reaction, thus we only use the action or the reaction of a force in our calculations, not both.

Factor of safety: engineering structures, components and materials need to be designed to cope with all normal working stresses. The *safety factor* is used in these materials to give a margin of safety and to take account of a certain *factor of ignorance*. It may be defined as the ratio of the ultimate stress to the working stress, i.e.:

$$\text{factor of safety} = \frac{\text{ultimate stress}}{\text{working stress}} \qquad (1.11)$$

Factors of safety vary in engineering design, dependent on the structural sensitivity of the member under consideration. They are often around 2.0 but can be considerably higher for joints, fittings, castings and primary loadbearing static structures and somewhat lower for aircraft structures, where weight saving is essential for efficient structural design.

1.5.4 Thermal stress and strain

If a metal bar (Figure 1.35) is rigidly held at one end and then subject to heat, causing a temperature rise, there will be an unconstrained increase in length given by:

$$\Delta L = L\alpha\Delta t \qquad (1.12)$$

where (Δt) or $(t_2 - t_1)$ is the temperature change (in $^\circ$C), ΔL is the change (increase in this case) in length (in metres), L is the original length (in metres) and α is the coefficient of linear expansion for the metal from which the bar is made.

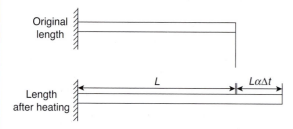

Figure 1.35 Metal bar, rigidly supported at one end, subject to heating

Now, if the bar is totally constrained and so prevented from expanding or contracting, there will be stresses set up in the bar as a result, i.e.:

Stress due to temperature change, with total constraint

$$\sigma = E\varepsilon = E\alpha\Delta t \qquad (1.13)$$

and for fractional (n) partial restraint, the stress set up is given by

$$\sigma = (1 - n)E\alpha\Delta t \qquad (1.14)$$

Example 1.13 Find the stress set up due to a temperature rise of 80°C in an aluminium rod 30 cm long, rigidly fixed at both ends. Take the linear coefficient of expansion for aluminium as $\alpha = 24 \times 10^{-6}/^\circ$C and its modulus of elasticity $E = 80\,\text{GN m}^{-2}$.

Then, first finding the unconstrained increase in length due to temperature rise and then using formula 1.13, we get:

$$\Delta L = L\alpha\Delta t = (300)(24 \times 10^{-6})(80)$$
$$= 0.576\,\text{mm}$$

and so strain

$$\varepsilon = \frac{\text{compression}}{\text{original length}} = \frac{0.576}{300} = 0.00192,$$

and from equation 1.13,

$$\sigma = E\varepsilon = (80 \times 10^3)(0.00192)$$
$$= 153.6\,\text{MN m}^{-2}.$$

Note the manipulation of units in this example!

To assist you in tackling the problems in TYK 1.3 (below) and those in TYK 1.4, that follows the next section in this chapter on compound bars, and also as a useful source of reference, Table 1.1 sets out some useful properties of engineering metals.

Test your knowledge 1.3

1) Define: a) tensile stress, b) shear stress, c) compressive stress.

2) State Hooke's law and explain its relationship to the elastic modulus.

3) Define in detail the terms: a) elastic modulus, b) shear modulus.

Table 1.1 Some typical properties of engineering metals

Material	Modulus of elasticity E (GN m^{-2})	Shear modulus G (GN m^{-2})	Tensile strength (MN m^{-2})	Linear coefficient of thermal expansion (α) ($\times 10^{-6}$/C)	Mass density (kg m^{-3})
Aluminium alloy	70	26	390	24	2780
Brass	102	38	350	20	8350
Bronze	115	45	310	18	7650
Hard-drawn copper	120	44	270	18	8900
Magnesium	45	17	378	29	1790
Mild steel	207	80	480	12	7800
Stainless steel	200	73	650	11.7	8000
Titanium	107	40	550	9.5	4500

4) What is the engineering purpose of the factor of safety?

5) If the ultimate tensile stress at which failure occurs in a material is 550 MPa and a factor of safety of 4 is to apply, what will be the allowable working stress?

6) A steel bar of length 20.0 cm and diameter 2 cm is subject to a compressive load of 40 kN, which compresses the bar by 4.5×10^{-5} m. Determine a) the modulus of elasticity of this steel and b) the extension under a tensile load of 20 kN.

7) A welded length of mild steel railway track is laid at 20°C. Determine the stress in the rails at 2°C if all contraction is prevented. Take the required values for α and the elastic modulus for mild steel from Table 1.1.

1.6 Compound bars

A compound bar consists of two or more members that are securely fastened at their ends. Some structural components formed in this way, such as concrete reinforced with steel bars, are able to share the best attributes of the individual members that go to make up the bars or columns. For example, in steel-reinforced concrete columns the steel is protected from corrosion and fire by being embedded in the concrete, while the concrete, which is very weak in tension, has pre-tensioned steel cables to keep it in compression during service and so maintain the advantage of the concrete's good compressive strength.

When a load is applied to a compound bar or compound component, it will be shared between the individual materials from which the bar is made. Under differing loading and heating conditions the individual members of the bar will extend or contract together (since they are securely fastened to one another) and so will have equal strains. However, if they have different elastic moduli, they will have different stresses. These two principles of load-sharing and equal strains can be used in various ways to solve numerical engineering problems concerned with compound bars. We will develop the techniques and, in so doing, produce one or two useful formulae for the solution of problems concerned with compound bars, through the series of examples that follow.

Example 1.14 Five steel rods each with a diameter of 30 mm reinforce a short concrete column 30 cm in diameter (Figure 1.36). If the modulus of elasticity for the steel is 15 times that of the concrete, determine the load that can be taken by the column if the stress in the column must not exceed 5 MN m^{-2}.

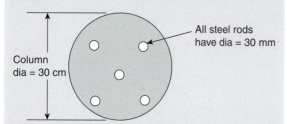

Column dia = 30 cm

All steel rods have dia = 30 mm

Figure 1.36 Cross-section of steel rod reinforced concrete column

Let the stresses in the steel and the concrete be σ_s and σ_c, respectively, and A_s and A_c be their areas.

Then, knowing that $\varepsilon = \frac{\sigma}{E}$ and that the strains for the concrete and steel rods are equal, we may write that:

$$\frac{\sigma_s}{E_s} = \frac{\sigma_c}{E_c} \quad \text{and so, } \sigma_s = \frac{E_s \sigma_c}{E_c}.$$

Also, we are given that the ratio of the moduli is 15, therefore $\sigma_s = 15\sigma_c$ and, knowing that the stress in the concrete $\sigma_c \leqslant 5 \mathrm{MN\,m^{-2}}$, then $\sigma_s = (15)(5) = 75 \mathrm{MN\,m^{-2}}$. Also, the total area of steel rods

$$A_s = \frac{5(\pi)(30)^2}{4} = 3534.29 \, \text{mm}^2$$

and total area of column

$$= \frac{\pi(300)^2}{4} = 70685.83 \, \text{mm}^2,$$

and so the area of the concrete

$$A_c = 70685.83 - 3534.29 = 67151.54 \, \text{mm}^2,$$

then:

Maximum load taken by column = load in concrete + load in steel and, from $\sigma = \frac{W}{A}$, the maximum load taken by column

$W_M = \sigma_c A_c + \sigma_s A_s = (5 \times 10^6)(67151.54 \times 10^{-6})$
$+ (75 \times 10^6)(3534.29 \times 10^{-6}) = 600829.47$ N or:

The total maximum load taken by the column $W_M = 600.8$ kN.

Note that we talk about a 'short' column to ensure the column is not subject to the added complication of bending as well as compression, as a result of the imposed loads. The question of how short is short? will be answered, when you study columns and struts in Chapter 5.

1.6.1 Equivalent modulus

A useful relationship that may be used in the solution of compound bar problems is that known as the *equivalent modulus*. This relationship may be formulated and expressed as follows.

First, by generalising the relationship for total load we used in Example 1.14, i.e., from $W_M = \sigma_c A_c + \sigma_s A_s$, we may write:

$$W_T = \sigma(A_1 + A_2 + \ldots\ldots A_n)$$
$$= \sigma_1 A_1 + \sigma_2 A_2 + \ldots\ldots + \sigma_n A_n,$$

where σ is *the stress in the equivalent single bar.* Dividing through by the strain ε, which will be common for our compound bar where all components are of equal length and securely attached to one another, we have:

$$\frac{\sigma}{\varepsilon}(A_1 + A_2 + \ldots\ldots + A_n)$$
$$= \frac{\sigma_1}{\varepsilon}A_1 + \frac{\sigma_2}{\varepsilon}A_2 + \ldots\ldots + \frac{\sigma_n}{\varepsilon}A_n$$

and

$$E_E(A_1 + A_2 + \ldots\ldots A_n)$$
$$= E_1 A_1 + E_2 A_2 + \ldots\ldots + E_n A_n,$$

where E_E is the equivalent or combined elastic modulus of the single bar. So equivalent

$$E = \frac{E_1 A_1 + E_2 A_2 + \ldots\ldots + E_n A_n}{A_1 + A_2 + \ldots\ldots + A_n}$$

or,

$$E_E = \frac{\sum EA}{\sum A} \qquad (1.15)$$

Using the relationship

$$W_T = \sigma(A_1 + A_2 + \dots\dots A_n),$$

we find that the stress in the equivalent bar, subject to an external load W_T, is $\sigma = \dfrac{W_T}{\sum A}$, and similarly from the relationship

$$\frac{\sigma}{\varepsilon}(A_1 + A_2 + \dots\dots + A_n)$$

$$= \frac{\sigma_1}{\varepsilon}A_1 + \frac{\sigma_2}{\varepsilon}A_2 + \dots\dots + \frac{\sigma_n}{\varepsilon}A_n,$$

after substitution of

$$\sigma = \frac{W_T}{\sum A}$$

and on rearrangement, we may express the strain in the equivalent single bar as

$$\varepsilon = \frac{W_T}{E_E \sum A} = \frac{x}{L}.$$

Also, from $E = \dfrac{stress}{strain}$ and formula 1.15 we get

$$\frac{\dfrac{W_T}{\sum A}}{\dfrac{x}{L}} = \frac{\sum EA}{\sum A} \quad \text{or} \quad \frac{W_T L}{\sum Ax} = \frac{\sum EA}{\sum A};$$

therefore,

$$x = \frac{W_T L}{\sum EA}$$

and, again from 1.15, we may write the relationship in terms of the common change in length (x) for the equivalent single bar as:

$$x = \frac{W_T L}{E_E \sum A} \qquad (1.16)$$

The above formulae should be considered as useful additional tools in your armoury for solving engineering problems concerned with compound bars. On the basis of the information you have available about a problem will depend which approach you adopt for its solution; in many cases this will require the application of logic rather than just the use of formulae. In the next example we will use the formulae we have just found, together with some logical thinking, to solve the problem.

Example 1.15 A heavy-duty electrical cable consists of four copper conductors, each 5 mm in diameter, securely attached to a central mild steel conductor, also 5 mm in diameter, that gives the cable additional strength. Assuming the cable can be treated as a compound bar, then:

a) Calculate the equivalent modulus for the cable, taking the values for the individual moduli for its components from Table 1.1.

b) Determine the common extension in an initial 1.5 m length of the cable when subject to a tensile load of 5 kN.

c) Determine the stresses in each of the conductors when subject to this 5 kN load.

a) From formula 1.15 and Table 1.1,

$$E_E = \frac{\sum EA}{\sum A}$$

$$= \frac{\left(\begin{array}{c} 4(120 \times 10^9 \times \pi/4 \times 5^2 \times 10^{-6}) \\ +(207 \times 10^9 \times \pi/4 \times 5^2 \times 10^{-6}) \end{array} \right)}{\pi/4(5 \times 5^2)10^{-6}}$$

$$E_E = \frac{(9424.778 + 4064.435)10^3}{(98.174)10^{-6}}$$

$$= 137.4 \times 10^9 \, \text{N m}^2,$$

so $E_E = 137.4 \, \text{GN m}^{-2}$.

b) There are a number of approaches we could adopt to find the common extension in the 1.5 m length of the cable. We will find the equivalent stress and then use $E = \dfrac{stress}{strain}$ to determine the equivalent strain and so the common extension. Then,

$$\sigma = \frac{W_T}{\sum A} = \frac{5000}{98.174 \times 10^{-6}} = 50.93 \, \text{MN m}^{-2}.$$

Therefore, strain in the equivalent bar

$$= \frac{stress}{E} = \frac{50.93 \times 10^6}{137.4 \times 10^9} = 0.371 \, \text{mm}$$

and, since all strains will be equal, the common extension = strain × original length = 0.371 × 1.5 = 0.556 mm.

c) To find the stress in the copper and the mild steel cables we again use the relationship $E = \dfrac{stress}{strain}$, so the stress in the copper = strain $\times E_c = 0.371 \times 10^{-3} \times 120 \times 10^9 = 44.52$ MN m^{-2}

and the stress in the steel = strain $\times E_s = 0.371 \times 10^{-3} \times 207 \times 10^9 = 76.8$ MN m^{-2}

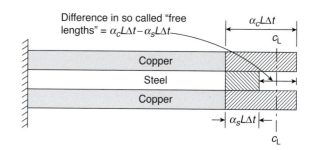

Figure 1.37 Thermal expansion of a compound bar

1.6.2 Compound bars subject to temperature change

We will now consider the effects of temperature change on compound bars. We have already considered the effects of temperature change on components made from a single material, when we considered thermal stress and strain in section 1.5.4. With the material rigidly fixed at one end (Figure 1.35) and having the other end free to expand or contract due to heating or cooling, we found that this change in length could be found using equation 1.12. We also found that if the component was held rigidly at both ends, to prevent completely any change in length due to temperature change, stresses were set up within the component that could be found using equation 1.13. Look back at Example 1.13 and see how we found the stress set up in a rigidly constrained aluminium rod that was subject to a temperature rise.

We can use the above argument and resulting formulae when dealing with problems associated with the heating or cooling of compound bars. Consider the situation shown in Figure 1.37, where two copper bars and one steel bar are rigidly joined together to form a compound bar, subject to a temperature rise.

Imagine for a moment that all the expansion takes place at the right-hand end of the bar as shown in Figure 1.37. If the copper and steel were free to expand separately, then, due to their different coefficients of linear expansion, the copper components would lengthen more than the steel, to the positions shown by the hatched lines on the diagram. However, the two different metals are rigidly secured to one another, therefore they will be affected by the movement of one on the other. The copper, with the higher coefficient of expansion, will attempt to pull the steel up to its free

length position, while the steel with the lower coefficient of expansion will try to hold back the copper. In reality an intermediate compromise position is achieved, as indicated on the diagram by the centre line C_L, whereby there is an effective compression of the copper from its free expansion position and an effective extension of the steel. From this situation and Newton's third law, two rules may be inferred:

Rule 1) *The extension of the short member + the compression of the long member = difference in the free lengths.* From the situation shown in Figure 1.37, this difference in free lengths is given as $(\alpha_c - \alpha_s)\Delta tL$.

Rule 2) From Newton's third law (to every action there is an equal and opposite reaction) we may state that: *The tensile force applied to the short member by the long member is equal to the compressive force applied to the long member by the short member.*

Then, since force = stress × area or $F = \sigma A$ and the forces in the two different materials of our compound bar are equal but opposite in magnitude, we may write that:

$$F_1 = F_2 \quad \text{or that} \quad \sigma_1 A_1 = \sigma_2 A_2,$$

where the suffices indicate the two different materials for the members of our compound bar. Note that we are equating *forces*, but the stress in each of the members may be different, dependent on their areas!

$$\sigma_1 A_1 = \sigma_2 A_2 \qquad (1.17)$$

Example 1.16 A compound bar consists of two copper plates, each having a cross-section of 25 × 10 mm, securely sandwiching a central mild steel bar having the same cross-section as each of the copper bars. Find the stresses set up in the steel and copper when they are subject to a temperature rise of 150°C. Take the values of the elastic modulus and coefficients of linear expansion for mild steel and copper from Table 1.1.

Now, from $F = \sigma A$, and because, from the data given in the question, the area of the copper is twice the area of the steel, the stress in the steel $\sigma_s = 2\sigma_c$. Knowing that the elastic modulus

$$E = \frac{\text{stress}}{\text{strain}} = \frac{\sigma}{x/L},$$

where x is the change in length, then $x = \dfrac{\sigma L}{E}$ and, using E_s and E_c for the elastic moduli, we may write:

Extension of the steel due to $\sigma_s = \dfrac{L\sigma_s}{E_s}$ and compression of the copper due to $\sigma_c = \dfrac{L\sigma_c}{E_c}$. From rule 1), *the extension of the steel + compression of copper = difference in free lengths*, and from equation 1.12 for free expansion, i.e., $\Delta L = L\alpha\Delta t$, we may write the useful relationship that:

$$\frac{L\sigma_s}{E_s} + \frac{L\sigma_c}{E_c} = (\alpha_c - \alpha_s)\Delta t L \qquad (1.18)$$

Then, knowing that $\sigma_s = 2\sigma_c$ and after substitution, we get:

$$\frac{2L\sigma_c}{E_s} + \frac{L\sigma_c}{E_c} = (\alpha_c - \alpha_s)\Delta t L$$

or

$$\frac{(2E_c\sigma_c + E_s\sigma_c)L}{E_s E_c} = (\alpha_c - \alpha_s)\Delta t L,$$

where it can clearly be seen that L is a common factor and can be eliminated from the equation. So we get

$$\frac{(2E_c\sigma_c + E_s\sigma_c)}{E_s E_c} = (\alpha_c - \alpha_s)\Delta t$$

and then, after substituting our given values, we have:

$$\frac{(240\sigma_c + 207\sigma_c)}{(207)(120) \times 10^9} = (18 - 12)10^{-6}(150)$$

or

$$447\sigma_c = (18 - 12)10^{-6}(150) \times 24840 \times 10^9.$$

Then the stress in the copper $\sigma_c = 50.013\,\text{MN m}^{-2}$ compressive.

So the stress in the steel $2\sigma_c = \sigma_s = 100.026\,\text{MN m}^{-2}$ tensile.

We conclude our short study of compound bars with one final example that involves both the stresses due to external force and the stresses due to temperature change. However, in Example 1.17, the components of the compound bar are physically separated (see Figure 1.38) and loaded axially by, in this case, the nut being tightened against the washer, creating a tensile force in the bolt and a compressive force in the sleeve. If this same system is heated, a compressive force will act on the sleeve and a tensile force will act on the bolt as a result of the temperature rise.

Example 1.17 A mild steel bolt, washers and nut assembly secure and clamp an external aluminium protective sleeve, as shown in Figure 1.38.

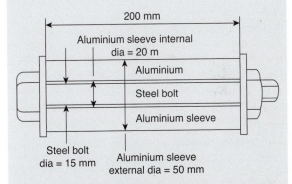

Figure 1.38 Steel bolt and aluminium sleeve assembly

The mild steel bolt and the aluminium sleeve have an active length of 200 mm, with diameters as shown. Slack in the assembly is taken up by adjusting the nut until finger tight. The pitch of the single start thread on the bolt is 1.2 mm.

a) If the nut is now tightened by one half of a complete turn, find the stresses due to tightening and the change in length of the aluminium sleeve.

b)   If, from the hand tight position, the sleeve and bolt are both heated evenly from an initial temperature of 18°C up to 60°C, find the stresses set up due to the temperature change.

Take the values for the elastic moduli and coefficients of linear expansion for mild steel and aluminium from Table 1.1, as necessary.

a)   There are a number of ways of tackling this problem. We will adapt the approach we used in Example 1.16. Then, from rule 1), we know that:

The *extension of the bolt + compression of*

the sleeve = the difference in the free lengths

where, for our example, the difference in the free lengths is given by the axial advance of the nut and for our single start thread this is simply:

Thread pitch × number of turns

$$= 1.2 \text{ mm} \times 0.5 \text{ turn} = 0.6 \times 10^{-3} \text{ m}.$$

Then, from equation 1.18, $\dfrac{L\sigma_s}{E_s} + \dfrac{L\sigma_a}{E_a} = 0.6 \times 10^{-3}$ m. We also know that the force in the aluminium sleeve equals the force in the steel bolt, as a result of the tightening of the nut, and that, from equation 1.17, we may represent this equality of the forces as $\sigma_s A_s = \sigma_a A_a$. So, finding the area of the steel bolt as $A_s = \dfrac{\pi}{4}(15 \times 10^{-3})^2 = 176.715 \times 10^{-6}$ m^2 and finding the area of the aluminium sleeve as $A_a = \dfrac{\pi}{4}((50 \times 10^{-3})^2 - (20 \times 10^{-3})^2) = \dfrac{\pi}{4}(2100 \times 10^{-6}) = 1649.34 \times 10^{-6}$ m^2, we then find that:

$$176.715\sigma_s = 1649.34\sigma_a, \text{ so } \sigma_s = 9.33\sigma_a.$$

Then, substituting these values into the above relationship that we found from equation 1.18, we have:

$$\frac{(0.2)(9.33\sigma_a)}{207 \times 10^9} + \frac{(0.2)(\sigma_a)}{70 \times 10^9} = 0.6 \times 10^{-3} \text{ m}$$

or

$$\frac{130.62\sigma_a + 41.4\sigma_a}{14490 \times 10^9} = 0.6 \times 10^{-3},$$

and so the stress in the aluminium sleeve $\sigma_a = 50.54$ MN m^{-2} and the stress in the mild steel bolt $\sigma_s = 9.33\sigma_c = 471.53$ MN m^{-2}.

b)   The ratios of the stresses for the sleeve and the bolt will remain the same as in part a) since the areas of the two metals remain the same. Thus we may again use a modified version of equation 1.18 where, in this case, we represent the stress in the steel bolt in terms of its stress ratio with that of the aluminium sleeve (as before) and where the difference in the free length, due to the greater expansion of the aluminium sleeve compared with the steel bolt, is given by the right-hand side of the equation,

i.e. $\dfrac{L9.33\sigma_a}{E_s} + \dfrac{L\sigma_a}{E_a} = (\alpha_a - \alpha_s)\Delta tL$. Again, since in this case L is a common factor, we may write:

$$\frac{9.33\sigma_a}{207 \times 10^9} + \frac{\sigma_a}{70 \times 10^9}$$
$$= (24 - 12)10^{-6}(18 - 60),$$

so the stress in the aluminium sleeve due to temperature change $\sigma_a = 8.49$ MN m^{-2} and the stress in the mild steel bolt *due to temperature change* $\sigma_s = (9.33)(8.49) = 79.21$ MN m^{-2}. Now, *if the nut had been tightened as in part a) and then heat applied*, the final stresses in the components would have been found as follows.

The total stress in the aluminium sleeve would equal the stress due to tightening the nut plus the stress due to the temperature rise, both of these stresses being compressive. So we get that:

Total stress in the aluminium sleeve σ_a

$$= 50.54 \text{ MN m}^{-2} + 8.49 \text{ MN m}^{-2}$$

$$= 59.03 \text{ MN m}^{-2} \text{ } compressive.$$

The stress in the mild steel bolt would equal the stress due to tightening the nut plus the stress due to the temperature rise. In this case both of these stresses are tensile. So we get that:

Total stress in the mild steel bolt σ_s

$$= 471.53\,\text{MN m}^{-2} + 79.21\,\text{MN m}^{-2}$$

$$= 550.74\,\text{MN m}^{-2}\ \textit{tensile}.$$

Test your knowledge 1.4

1) A cable consists of four aluminium rods, each 6 mm in diameter, all securely attached to three central stainless steel strengthening rods, each of which is also 6 mm in diameter.

   a) Calculate the equivalent modulus for the cable, taking moduli values for the individual materials of the cable from Table 1.1.

   b) Using the common modulus or otherwise, determine the common extension in an initial 2.0 m length of the cable when subject to a tensile load of 25 kN.

2) A short steel-reinforced concrete column has a cross-section of 400×300 mm and a total cross-sectional area of reinforcing steel of $6000\,\text{mm}^2$. Calculate the maximum load the column can carry if the compressive stress in the steel is not to exceed $240\,\text{MN m}^{-2}$. Take the ratio of the modulus of the steel to that of the concrete as 12:1.

3) A mild steel tube, 25 mm external diameter and 20 mm internal diameter, encloses a copper rod 18 mm in diameter and is rigidly held in position vertically between two steel-plated jaws (Figure 1.39). If, at its initial temperature, the rod and tube assembly is rigidly held by the jaws but stress free, find the stress in both the rod and the tube when the assembly is subject to an increase in temperature of $200°\text{C}$. Take the values of the moduli and coefficients of linear expansion for the mild steel and the copper from Table 1.1.

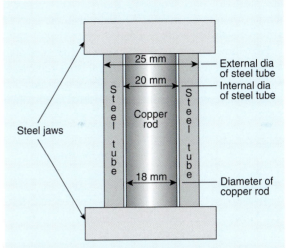

Figure 1.39 Tube and rod assembly held rigidly between steel jaws

4) Find, for the mild steel tube and copper rod assembly of question 3:

   a) the stress in the steel tube and in the copper rod resulting from the steel jaws imparting a uniaxial load of 50 kN downwards onto the rod and tube assembly;

   b) the *total stress* in the steel tube and the copper rod that results from the combined stresses due to heating by $200°\text{C}$ and due to the applied uniaxial 50 kN load.

1.7 Poisson's ratio and two-dimensional loading

You may have already, from your previous work, met the definition of *Poisson's ratio*. Here we review and define Poisson's ratio, and relate it to simple stress and strain in two dimensions. Later, in Chapter 7, when we discuss complex stress and strain, we will consider the relationship between Poisson's ratio and three-dimensional loading. Poisson's ratio gives the relationship between the lateral and axial strain for a solid under load, provided the solid is only subject to loads within its elastic range. It is expressed as:

$$\text{Poisson's ratio } (\nu) = -\frac{\text{Lateral strain}}{\text{Axial strain}} \qquad (1.19)$$

The minus sign results from the convention that compressive strains are negative and tensile strains are positive. Since Poisson's ratio always relates a tensile strain to a compressive strain, then the laws of arithmetic always produce a minus sign.

For a three-dimensional solid, such as a bar subject to a tensile load (Figure 1.40), the lateral strain represents a decrease in width (negative lateral strain) and the axial strain represents elongation (positive longitudinal strain).

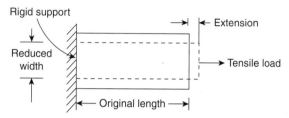

Figure 1.40 A bar subject to a tensile load, showing resulting strains

If this same bar were subject to a compressive axial load there would be a subsequent reduction in length, with a corresponding increase in width

> **Key Point.** By convention, tensile strain is considered to be positive and compressive strain negative

In tables it is normal to show only the *magnitudes* of the strains considered, so tabulated values of Poisson's ratio are positive. Table 1.2 shows typical values of Poisson's

Table 1.2 Typical values of Poisson's ratio for a variety of materials

Material	Poisson's ratio
Aluminium	0.33
Manganese bronze	0.34
Cast iron	0.2–0.3
Concrete (non-reinforced)	0.2–0.3
Marble	0.2–0.3
Nickel	0.31
Nylon	0.4
Rubber	0.4–0.5
Steel	0.27–0.3
Titanium	0.33
Wrought iron	0.3

ratio for a variety of materials. Note that for most *metals*, when there is no other available information, Poisson's ratio may be taken as 0.3.

> **Example 1.18** A flat metal plate subject to an axial force extends by 0.09 mm in the direction of the force, as shown in Figure 1.41.
>
>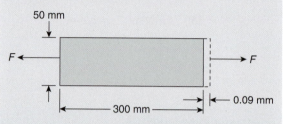
>
> **Figure 1.41** Flat plate subject to axial force
>
> Find the change in width of the metal plate, if Poisson's ratio for the metal is 0.33.
>
> We can find the axial (longitudinal) strain from the dimensions given in Figure 1.41 and the relationship Longitudinal strain $= \dfrac{\text{extension}}{\text{original length}} = \dfrac{0.09}{300} = 0.0003$; also, from equation 1.19, the lateral strain = Poisson's ratio × longitudinal strain, so lateral strain $= (0.33)(0.0003) = 0.000099$. Then, from equation 1.7, lateral strain $= \dfrac{\text{change in width (deformation)}}{\text{original width}}$, so:
>
> Change in width $= (0.000099)(50) = 0.00495$ mm.

1.7.1　Poisson's ratio in two dimensions

A two-dimensional stress system is one in which all the stresses lie within one plane, such as the plane of this paper, the *xy* plane. Consider the flat plate (Figure 1.42) under the action of two separate stresses, σ_x in the *XX* direction (Figure 1.42a) and σ_y in the *YY* (Figure 1.42b) direction.

Figure 1.42c shows the result of combining these two separate stresses. Then, using equation 1.19 for the elastic modulus and the definitions of Poisson's ratio, it can be shown that *in the XX direction:*

the strain due to σ_x (Figure 1.42a) *acting alone* $= \dfrac{\sigma_x}{E}$ (tensile) and the strain due to σ_y (Figure 1.42b) acting alone $= \nu\dfrac{\sigma_x}{E}$ (compressive). Remembering the sign convention for stresses, i.e., that tensile stresses are positive and compressive stresses are negative, we may

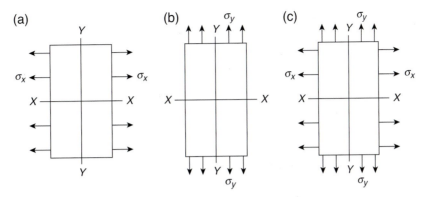

Figure 1.42 Flat plate under the action of two separate stresses and combined stresses

combine these two results (Figure 1.42c) to give the strain due to both σ_x and σ_y acting together, i.e.:

The combined strain in the XX direction

$$\varepsilon_x = \frac{\sigma_x}{E} - \nu\frac{\sigma_y}{E} \qquad (1.20)$$

A similar result may be shown for the strains due to both σ_x and σ_y acting together *in the YY direction* (Figure 1.42c), i.e.:

The combined strain in the YY direction

$$\varepsilon_y = \frac{\sigma_y}{E} - \nu\frac{\sigma_x}{E} \qquad (1.21)$$

Example 1.19 If the plate loaded as shown in Figure 1.43 is made from a steel with $\nu = 0.3$ and $E = 205$ GPa, determine the changes in dimension in both the x and y directions.

In the x direction the plate is subject to a 12 kN load acting over an area of 600 mm². Thus the stress in the x direction $= L/A = 20$ N mm⁻². Similarly, the stress in the y direction $= 9000/300$ N mm⁻² $= 30$ N mm⁻².

Then, from equation 1.20, the total combined strain in the x direction is:

$$\varepsilon_x = \frac{20 - 0.3(30)\,\text{N mm}^{-2}}{205 \times 103\,\text{N mm}^{-2}} = 5.37 \times 10^{-5},$$

and since strain = change in length (dimension)/ original length, then:

Change in length = strain × original length

$$= (5.37 \times 10 - 5) \times (100)\,\text{mm} = 0.00537\,\text{mm}.$$

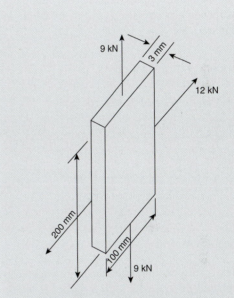

Figure 1.43 Plate loaded in two dimensions

So change in dimension in the x direction is an extension = 0.00537 mm.

Similarly, from equation 1.21, the total combined strain in the y direction is:

$$\varepsilon_y = \frac{30 - 0.3(20)\,\text{N mm}^{-2}}{205 \times 10^3\,\text{N mm}^{-2}} = 1.17 \times 10^{-4},$$

and so the change in length

$$= (1.17 \times 10^{-4}) \times (200)\,\text{mm} = 0.0234\,\text{mm}.$$

So change in dimension in the y direction is an extension = 0.0234 mm.

Test your knowledge 1.5

1. A metal bar 250 mm long has a rectangular cross-section of 60 mm × 25 mm. It is subjected to an axial tensile force of 60 kN. Find the change in dimensions if the metal has an elastic modulus $E = 200$ GPa and Poisson's ratio is 0.3.

2. A rectangular steel plate 200 mm long and 50 mm wide is subject to a tensile load along its length. If the width of the plate contracts by 0.005 mm, find the change in length. Take Poisson's ratio for the steel as 0.28.

3. A flat aluminium plate is acted on by mutually perpendicular stresses σ_1 and σ_2 as shown in Figure 1.44.

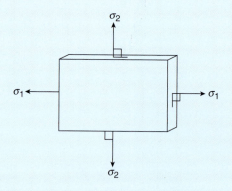

Figure 1.44 Figure for TYK 1.5 question 3

The corresponding strains resulting from these stresses are $\varepsilon_1 = 4.2 \times 10^{-4}$ and $\varepsilon_2 = 9.0 \times 10^{-5}$. Find the values of the stress if E for aluminium = 70 GPa and the value of Poisson's ratio is 0.33.

1.8 Chapter summary

The aim of this chapter has been to review and revise some of the basic concepts that will be needed to aid your understanding of the topics concerned with the mechanics of materials that follow in the remaining chapters of Part I. Your familiarity, or otherwise, with these topics will depend on your background, so in order to assess your understanding several TYK exercises have been provided, which you should be able to complete without too much difficulty.

We started by explaining the concept of "force" and how it was defined using Newton's second law. You should appreciate that *force is a vector quantity* that has both *magnitude* and *direction* as well as a point of application. We then went on to look at the vector representation and combination of forces, by first looking at how we add up two vectors using the *parallelogram and triangle laws*. We defined *concurrency*, the *equilibrant* and *resultant* of force systems and showed how to solve concurrent force systems with more than two forces by using the polygon of forces method. You were then able to check your understanding of the graphical solution of force systems by completing the questions set in TYK 1.1.

The resolution of coplanar force systems was then looked at in some detail, with both concurrent and non-concurrent systems being considered, also the fact that when resolving non-concurrent force systems, the single resultant force is always accompanied by a *turning moment*. The method of tabulating values to find total vertical and horizontal components of the forces and turning moments within the system was also explained. The principle of transmissibility and Varignon's theorem were also covered. You may not have met either of these concepts before; however, they are useful tools for solving problems concerned with non-concurrent coplanar force systems and well worth knowing. You were then able to assess your ability to tackle problems concerned with force systems by completing the questions in TYK 1.2.

In section 1.5 you were reintroduced to elements of *simple stress and strain*, many of which you will already be familiar with, including the definitions for tensile and shear stress and strain, the elastic modulus and shear modulus. A brief introduction to one-dimensional *thermal stress and strain* also formed part of this section. Again, an assessment of the knowledge and application you required at this stage was provided by the questions set in TYK 1.3.

Section 1.6 provided you with a fairly comprehensive introduction to the analysis of *compound bars* under varying conditions. In the first part of the section you considered compound bars subject to simple *axial compressive and tensile loads* and found the stresses in the components of the bar, as well as establishing common stresses and strains, using, among other techniques, the concept of the *equivalent modulus*. You next considered the stresses set up in compound bars that had been subject to temperature change, using *rule 1* and *rule 2*. You should commit these two rules to memory and fully understand their use as demonstrated in the three rather lengthy examples given in the section.

Finally, you should not only be able to determine the magnitude of the stresses set up as a result of *combined external loading and heating*, but also be able to decide on the nature of the forces that act on the individual members of the bar, i.e., whether the members tended to be compressed or extended. Being able to successfully complete the questions in TYK 1.4 will show that you have clearly understood the subject matter in this section.

Finally, in section 1.7 you were introduced to *Poisson's ratio* and its application to two-dimensional loading. You should note how the *formulae for combined*

strains vary according to the axes used in the plane. This section has been included as a precursor to the study of three-dimensional stress and strain that you will meet in Chapter 7. Again, you can assess your own understanding of the concepts discussed in this section by attempting the questions given in TYK 1.5.

The first of the *Essential Mathematics* topics (*Vectors and Vector Operations*) is given on the book's website These resumés have been included to provide the necessary mathematical background you may need to apply to specific areas of the science, as already mentioned in the introduction to the book.

Chapter 2

Simply supported beams

In this chapter we look at the simply supported beam as a structural member, and in a subsequent chapter we will look at the nature and use of columns and struts. These structural members are commonly used in the construction of bridges, buildings and pylons and, in the case of beams, they are primarily designed to take bending loads.

We start with a review of one or two fundamental concepts associated with beams including the conditions for equilibrium, which you met in Chapter 1, together with the methods we use to determine beam reactions and other unknown variables. We then consider shear force and bending moments and their determination. Next we will look at the stresses imposed on beams due to bending, including the verification and use of the engineers' theory of bending. Next we consider centroids and the second moment of area and how these concepts are used in the selection of beams for a particular engineering purpose. Finally, we spend some time considering how beams deflect under load and look at the methods used to determine the slope and deflection of these loaded beams; in particular we consider the use of the successive integration method and Macaulay's method.

In order to assist you with some of the essential background calculus, needed to study and understand some of the more difficult concepts you will meet in this chapter, you should consult *Essential Mathematics 2* on the companion website, as necessary.

2.1 Revision of fundamentals

Before we introduce the idea of the simply supported beam and show the methods we adopt to determine the shear forces and bending moments in such beams, we will review and revise the fundamental concepts of static equilibrium and the principle of moments, which you may already have met but which are needed for any study of beams.

For a *beam to remain in static equilibrium*, two conditions must apply:

1. *The upward forces must equal the downward forces*

2. *The sum of the clockwise moments (CWM) must equal the sum of the anticlockwise moments (ACWM).*

You will know from Newton's second law that for every action there is an equal and opposite reaction. Thus, for the beam shown in Figure 2.1, the actions are the points of application of the downward loads W_1 to W_4, while the reactions that balance these loads are shown acting vertically upwards through the supports (R_1 and R_2).

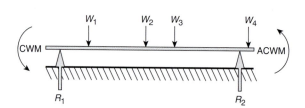

Figure 2.1 Uniform beam showing conditions for equilibrium

978-1-85617-775-7, Engineering Science, Mike Tooley & Lloyd Dingle

Also shown is the other condition for equilibrium, whereby the algebraic sum of the moments created by the forces acting along the beam must be in balance, i.e. must be equal. In other words, *the principle of moments states that: the sum of the CWM = the sum of the ACWM*.

You should also remember that a *uniform beam* is such that it has an equal cross-section along its whole length and is manufactured from a *consistent* material, that is one with an even density throughout. Under these circumstances, the *centre of gravity* of the beam acts vertically down through its geometric centre.

> **Key Point.** For static equilibrium, upward forces equal downward forces and the sum of the CWM equals the sum of the ACWM

When solving shear force and bending moment problems on beams, the first step is often to determine the reactions at the supports. If the beam is in static equilibrium (which it will be during our study of statics), the method we use to find these reactions is based on the application of the principle of moments. An example will remind you of the technique.

Example 2.1 A uniform horizontal beam is supported as shown in Figure 2.2. Determine the reactions at the supports R_A and R_B.

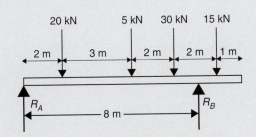

Figure 2.2 Uniform beam for Example 2.1

Now, in order to produce one equation to solve one unknown, it is necessary to take moments about one of the reactions in order to eliminate that reaction from the calculation. So, taking moments about R_A, we get:

$$(2 \times 20) + (5 \times 5) + (7 \times 30) + (9 \times 15) = 8 \times R_B$$

The above equation corresponds to the sum of the CWM = the sum of the ACWM,

and $40 + 25 + 210 + 135 = 8R_B$

then $R_B = 51.25 \text{ kN}$

and using the first condition for equilibrium, that is, upward forces = downward forces, we get:

$$R_A + 51.25 \text{ kN} = 70 \text{ kN} \quad \text{so that} \quad R_A = 18.75 \text{ kN}$$

Instead of using the first condition we can take moments about R_B to find the reaction at R_A. Then:

$(1 \times 15) + (8R_A) = (6 \times 20) + (3 \times 5) + (1 \times 30)$
and $8R_A = 120 + 15 + 30 - 15$;
then $R_A = 18.75 \text{ kN}$ as before.

Now it is not always the case that the loads on a beam are *point loads*. Beams can also be subjected to loads that are distributed for all, or part, of their length. These loads are known as, *uniformly distributed loads* (UDLs). For UDLs the whole mass of the load is assumed to act at a point through the centre of the distribution.

Example 2.2 For the beam system shown in Figure 2.3, determine the reactions at the supports R_A and R_B, taking into consideration the weight of the beam.

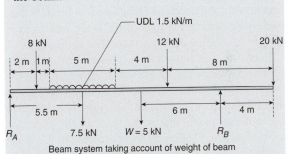

Figure 2.3 Beam system taking account of weight of beam

From what has been said, the UDL acts as a point load of magnitude (1.5 kN × 5 = 7.5 kN) at the centre of the distribution, which is 5.5 m from R_A.

In problems involved with reaction it is essential to eliminate one reaction from the calculations because only one equation is formed and only one unknown can be solved at any one time. This is achieved by taking moments about one of the reactions and then, since the distance from that reaction is zero, its moment is zero and it is eliminated from the calculations.

So, taking moments about A (thus eliminating A from the calculations), we get:

$$(2 \times 8) + (5.5 \times 7.5) + (10 \times 5) + (12 \times 12) + (20 \times 20) = 16RB$$

or $651.25 = 16RB$

so the reaction at $B = 40.7$ kN.

We could now take moments about B in order to find the reaction at A. However, at this stage it is easier to use the fact that for static equilibrium:

upward forces = downward forces,
so $R_A + R_B = 8 + 7.5 + 5 + 12 + 20$
and $R_A + 40.7 = 52.5$
and so the reaction at $A = 11.8$ kN.

2.1.1 Fundamental terminology

Before we look at shear force and bending in beams, we need to be familiar with one or two important terms and the methods we use to classify the type of beam we are considering.

The loads that act on a beam create *internal* actions in the form of *shear stresses* and *bending moments*. The lateral loads that act on a beam cause it to bend, or flex, thereby deforming the axis of the beam into a curve (Figure 2.4) called the *deflection curve* of the beam.

In this section we will only consider beams that are symmetric about the xy plane, which means that the y-axis as well as the x-axis is an *axis of symmetry* of the cross-sections. In addition, all loads are assumed to act in the xy plane, that is the plane of the paper. As a consequence, bending deflections occur in this same plane, which is known as the *plane of bending*. Thus the deflection curve AB of the beam in Figure 2.4 is a plane curve that lies within the plane of bending.

Key Point.   The loads that act laterally on a beam cause it to bend within the xy plane; this bending is known as the deflection curve

Beams are normally classified by the way in which they are supported. The *cantilever* (Figure 2.5a) is a beam that is rigidly supported at one end. The *simply supported beam* is either supported at its ends on rollers or smooth surfaces (Figure 2.5b), or supported by one of the ends being pin-jointed and the other resting on a roller or smooth surface (Figure 2.5c).

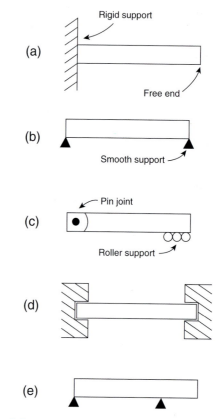

Figure 2.5 Types of beam support

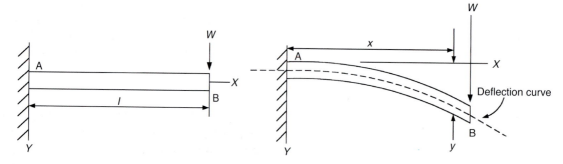

Figure 2.4 Cantilever subject to symmetric bending

Figure 2.5d shows a built-in or *encastre beam*, where both of its ends are rigidly fixed. Finally, Figure 2.5e illustrates a simply supported beam with overhang, where the supports are positioned some distance from the ends.

> **Key Point.** An encastred beam has both of its ends rigidly fixed and built-in. Bending moments are set up at either end that oppose the beam bending as a result of lateral loads

Now, as already mentioned, when a beam is loaded reactions and resisting moments occur at the supports, while shear forces and bending moments of varying magnitude and direction occur along the length of the beam. We now consider these shear forces and bending moments.

2.2 Shear force and bending moment

We have mentioned shear force and bending moment, but what are they and exactly what causes them? Consider the simply supported uniform beam shown in Figure 2.6.

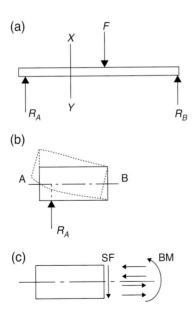

Figure 2.6 Nature of shear force and bending moment for simply supported beam

If we cut the beam at $X–Y$ and consider the section to the left of the cut (Figure 2.6a), then the effect

of the upward force from the support R_A is to try to bend the beam about the neutral axis AB, moving it upwards (Figure 2.6b). However, the portion of the beam to the right of the section $X–Y$ prevents this movement by applying an equal and opposite force downwards. The horizontal forces result from the beam being compressed above the neutral axis AB and stretched below this axis (Figure 2.6b). It is these components of force that resist the bending and set up the bending moment within the beam material (Figure 2.6c).

Remember that the shear force and resulting bending moment that we are considering is that which results from the flexing or bending of the beam created by the load.

2.2.1 Determination of shear force and bending moment

We can use the ideas presented above to determine the bending moment and shear force set up in a beam at any particular cross-section. These cross-sections can be taken wherever there exists a support, point load or distributed load.

The *bending moment* at any section of a beam just uses the principle of moments for its solution about a chosen cross-section. It is therefore *the algebraic sum of the moments of all external loads acting on the beam to one side only of the section*.

> **Key Point.** The bending moment at any section of a beam is found by calculating the algebraic sum of the moments of all external loads acting on the beam to *one side only* of the section

The shear force that acts at a section will only have a step change in value along the beam where a load normal to the beam acts. Also, the *shear force is considered positive when acting upwards on the left-hand side of a section and downwards on the right-hand side of the section*, and negative when the situation is reversed (see Example 2.5).

> **Key Point.** Shear force (SF) is considered positive when acting upwards on the left-hand side of a section and downwards on the right-hand side of the section

Example 2.3   Ignoring the weight of the beam, find the shear force and bending moment at *AB* and at *CD*, for the simply supported uniform beam shown in Figure 2.7.

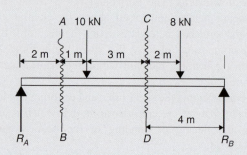

Figure 2.7 Beam for Example 2.3

We first find the reactions at the supports. Then, taking moments about R_A gives:

$(3 \times 10) + (8 \times 8) = 10R_B$

or $10R_B = 94$ and $R_B = 9.4\,\text{kN}$

Then, using first condition for equilibrium, $R_A + 9.4 = 18\,\text{kN}$ or $R_A = 8.6\,\text{kN}$

Now the shear force acting at the section *AB* is simply that due to the reaction R_A, that is $SF = +8.6\,\text{kN}$. The bending moment *to the left* of section *AB* is simply the moment due to the reaction, i.e., $BM = (2 \times 8.6) = 17.2\,\text{kN m}$

The SF to the left of section *CD* is that due to the reaction, which is $+8.6\,\text{kN}$, opposed by that due to the 10 kN load, which is acting down; then:

$SF = 8.6 - 10\,\text{kN} = -1.4\,\text{kN}$

The BM to the left of the section *CD* is the algebraic sum of the moments to the left of the section line, i.e., $(6 \times 8.6) - (3 \times 10)\,\text{kN m}$ or $BM = 21.6\,\text{kN m}$

Note that it does not matter which side of the section line we take moments, left or right, the bending moment will be the same!

So, for example, the BM to the right of *CD* is $(-2 \times 8) + (4 \times 9.4) = 21.6\,\text{kN m}$, as before.

If instead of a point load or loads we have a uniformly distributed load or a combination of point loads and UDLs, we treat the UDL as a point load with its centre at the mid-position of the UDL and then find shear force and bending moments in the same manner as that given above.

Example 2.4   Find the BM for the section *AB* shown in Figure 2.8.

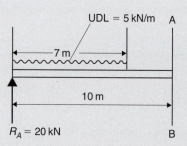

Figure 2.8 Beam section for Example 2.4

As before, to find the BM to the left of the cut *AB* we consider the algebraic sum of the moments to the left.

The UDL $= 5\,\text{kN/m}$, acting over 7 m, so the total load is $(5 \times 7) = 35\,\text{kN}$, and can be represented by a point load acting 3.5 m from the end of the beam or 6.5 m from AB.

Then, taking moments at *AB*, we get: $(10 \times 20) - (6.5 \times 35)\,\text{kN m}$

or $BM = -27.5\,\text{kN m}$

When determining bending moments, it is convenient to use a sign convention in the same way as we did for shear forces. We have in fact already used such a convention in the above example. Bending moments will be considered *positive if they compress the top section of the beam and stretch the lower section, i.e., if the beam sags* (Figure 2.9a) and *negative when the loads cause the beam to stretch on its top surface or hog* (Figure 2.9b).

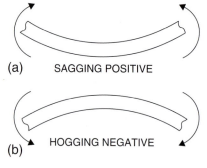

Figure 2.9 Sign convention for bending moment

Key Point. Using the convention, bending moments that tend to make the beam *sag are positive*, while bending moments that make the beam *hog are negative*

The next example generalizes the method for finding shear force and bending moment.

Example 2.5 Draw the shear force and bending moment diagrams for the beam shown in Figure 2.10.

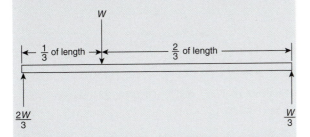

Figure 2.10 Simply supported beam with concentrated load

We have already been given the reactions at the left-hand and right-hand supports: they are $2W/3$ and $W/3$, respectively. So we may now determine the shear force and bending moment values by producing our free-body diagram, making cuts either side of the load at the arbitrary points x and r (Figure 2.11).

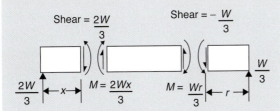

Figure 2.11 Free body diagram for cuts either side of the load

For vertical equilibrium, using our sign convention, at the left-hand cut the shear force is again $2W/3$ (positive), but this changes on the right-hand side of the load to $-W/3$. So at the load the shear force on the beam is the sum of these two shear forces, as shown in Figure 2.12. Note that the shear force only changes where a transverse load acts.

Figure 2.12(b) shows the resulting bending moment diagram, where the bending moment under the load is obtained by putting $x = \frac{1}{3}$ or $r = \frac{2}{3}$ into the equation $M = \dfrac{2Wx}{3}$ or $M = \dfrac{Wr}{3}$, respectively.

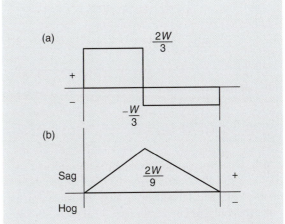

Figure 2.12 Shear force and bending moment diagram

2.2.2 Shear force and bending moment diagrams

So far we have concentrated on finding shear force and bending moments at a single cut point for a section. It is beneficial for engineers to know how the shear forces and bending moments are distributed across the whole length of the beam. To do this we construct shear force and bending moment diagrams that give the whole picture. Computer software may also be used to establish these forces and moments for more complex loading cases.

2.2.3 Shear force and bending moment diagrams for concentrated loads

When constructing these diagrams all we need to do is extend the techniques we have already mastered for finding the SF and BM at a section cut, remembering that the shear force will vary only where there is a transverse (normal) point load acting on the beam. The technique for beams subject to point loads is illustrated in Example 2.6.

Example 2.6 Draw the shear force and bending moment diagram for the beam shown in Figure 2.13.

The first step is to calculate the reactions R_A and R_B.

The reactions may be found in the normal manner by considering rotational and vertical equilibrium. Taking moments about R_A gives:

$$(9 \times 2) + (6 \times 4) = 6R_B \quad \text{and} \quad R_B = 7\,\text{kN}.$$

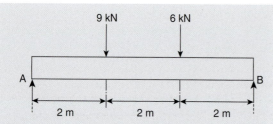

Figure 2.13 Simply supported beam with two concentrated loads

Also, for vertical equilibrium $R_A + 7 = 15$ and so $R_A = 8$ kN. Now considering the shear forces at the 9 kN load, and remembering that the shear force changes only where a transverse loads acts, on the left-hand side of the 9 kN load we have a positive shear force $R_A = +8$ kN. This is opposed at the load by the 9 kN force acting in a negative direction, so the shear force to the right of this load $= -1$ kN. This shear force continues until the next transverse load is met, in this case the 6 kN load, again acting in a negative direction, producing a net load of -7 kN which continues to R_B. This shear force distribution is clearly illustrated in Figure 2.14.

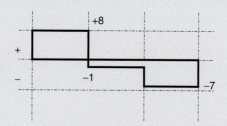

Figure 2.14 Shear force distribution for Example 2.16

The bending moment diagram can be easily determined from the shear force diagram by following the argument given in Example 2.5. If we consider the free body diagram for AC (Figure 2.15), then taking moments about C gives:

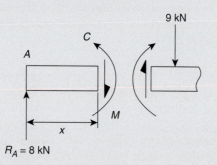

Figure 2.15 Free body diagram for section AC, Example 2.6

$RAx - M = 0$ and at the 9 kN load where $x = 2$ m we have $(8 \times 2) = M = 16$ N m (sagging).

Similarly, for section DE, consider the free body diagram shown in Figure 2.16.

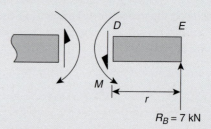

Figure 2.16 Free body diagram for section DE, Example 2.6

So, taking moments about D, when $r = 2$, then from $-R_B r + M = 0$, we have $-(7 \times 2) = -M = 14$ N m. These results can now be displayed on the bending moment diagram (Figure 2.17).

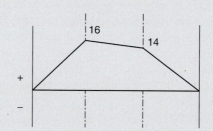

Figure 2.17 Bending moment diagram, Example 2.6

Note that the loads produce a sagging moment downwards, which intuitively you would expect.

2.2.4 Shear force and bending moment diagrams for uniformly distributed loads

We will now consider drawing shear force (SF) and bending moment (BM) diagrams for a beam subject to a UDL. These loads are normally expressed as force per unit length.

The beam shown in Figure 2.18 is subject to a UDL of 10 kN/m in addition to having a point load of 40 kN acting 2 m from A.

As usual, we first find the reaction at A and B. To do this you will remember, from your previous work, to treat the UDL as a point load acting at the centre of the uniform beam. This gives a load of 80 kN acting 4 m from A. Then the reactions may be calculated as

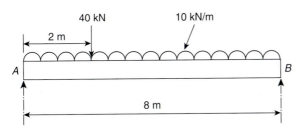

Figure 2.18 Beam subject to concentrated load and a UDL

$R_A = 70$ kN and $R_B = 50$ kN. We now cut the beam and, consider the free body on the left of the cut (Figure 2.19).

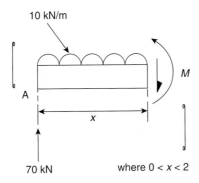

Figure 2.19 Free body diagram for beam to left of cut

Our first cut is made at a distance x from the left-hand support and to the left of the point load. We consider the left-hand part because it is the simpler of the two (you would of course get the same values for SF and BM if you considered the right-hand part). Then, from the convention used in Figure 2.19, we have:

(a)   SF $= (70 - 10x)$ kN

(b)   Taking moments about the cut gives a BM $= \left(70x - \dfrac{10x^2}{2}\right)$ kN m

Next we consider a point to the right of the cut, as shown in Figure 2.20.

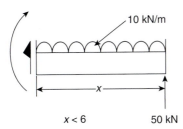

Figure 2.20 Free body diagram for beam to right of cut

Again, using the convention shown in Figure 2.20, we have:

(a)   SF $= (10r - 50)$kN

(b)   BM $= \left(50r - \dfrac{10r^2}{2}\right)$ kN m sagging

You should note that in both of the above expressions for BM there is a term where the unknown is squared. This part of the expression refers to the UDL and it will produce a parabola as the values of x and r (representing the lengths along the beam) vary. Thus, for any beam section that *only carries a UDL*, the BM plot will always be a parabola.

If we plot the SF for our beam (Figure 2.18), starting at point A and then subsequently at 1 metre intervals for the whole length of the beam, we may obtain the SF and then BM diagrams, as illustrated in Figure 2.21.

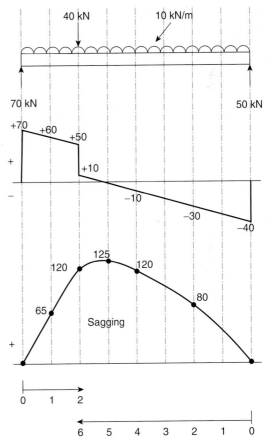

Figure 2.21 Shear force and bending moment plots for beam shown in Figure 2.18

You should ensure that you follow the above argument by verifying one or two of these SF and BM values for yourself.

Plotting diagrams point by point like this is a tedious business! In the last part of this chapter, when we study the slope and deflection of beams, you will see how we use bending theory and the calculus to determine expressions for finding the BM, slope and deflection of beams subject to different loading conditions.

Example 2.7 Consider a beam with loads as shown in Figure 2.22 and draw the SF and BM diagrams for the beam.

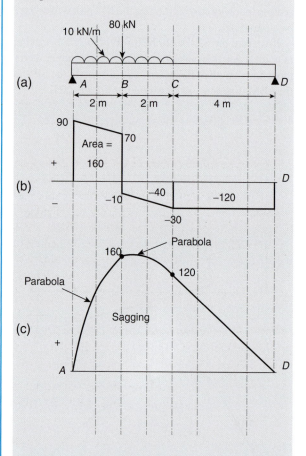

Figure 2.22 a) Loaded beam for Example 2.7; b) Resulting shear force for Example 2.7; c) Resulting bending moment for Example 2.7

The reactions may again be found in the usual way, where $R_A = 90$ kN and $R_D = 30$ kN, both acting upwards.

To obtain the SF diagram we start at the left-hand end of the beam, where:

$$R_A = +90 - [(wdx) \text{ or } (10)(2)] = 70 \text{ kN}.$$

At point B a downward force of 80 kN acts, giving $+90 - 20 - 80 = -10$ kN. At D we have a SF of -30 kN, resulting from the reaction R_D; this is constant until it meets the distributed load at C.

Now moving from C to B the SF changes constantly from -30 to -10 kN, as indicated on Figure 2.22b.

To obtain the BM plot we again start at point A where the bending moment is zero. *Adding up the areas of the shear force diagram* between A and B gives a bending moment at B of $+160$. Continuing, we may deduce that the sum of the areas between A and C is given by $\int_A^C Fdx = 160 - 40 = 120$. At D it can easily be seen, by again summing the areas on the SF diagram between A and D, that the BM $= 160 - 160 = 0$. The completed BM diagram is shown in Figure 2.22c.

Note that the SF is zero at point B, so the maximum BM occurs at B. The slope of the BM diagram is everywhere given by the ordinate of the SF diagram and, because the SF varies linearly, the BM diagram is parabolic (between A and B, B and C), whereas between C and D where the SF is constant the BM is a straight line – remember your integration!

Test your knowledge 2.1

1. Find the reactions at the supports for the beam system shown in Figure 2.23.

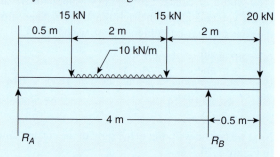

Figure 2.23 Beam system for TYK 2.1.1

2. Find the value of the BM for the section AB shown in Figure 2.24.

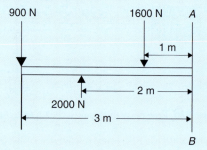

Figure 2.24 Section AB for TYK 2.1.2

3. Find the reactions at the supports and the bending moment at the section *AB* for the simply supported beam shown in Figure 2.25.

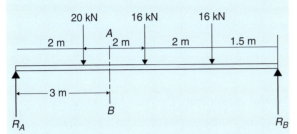

Figure 2.25 Simply supported beam for TYK 2.1.3

4. Draw the SF and BM diagrams for the simply supported beam shown in Figure 2.26.

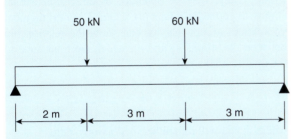

Figure 2.26 Simply supported beam for TYK 2.1.4

5. A beam is simply supported over a span of 8 m and overhangs the right-hand support by 2 m. The whole beam carries a uniformly distributed load of 10 kN/m, together with a point load of 120 kN, 2.5 m from the left-hand support. Draw the SF and BM diagrams, showing appropriate numerical values.

2.3   Engineers' theory of bending

Here we will look at the theory of bending related to beams, where the important relationship $\dfrac{M}{I} = \dfrac{\sigma}{y} = \dfrac{E}{R}$ is developed.

When a transverse load is applied to a beam, bending moments are set up which are resisted internally by the beam material. If the bending moment resulting from the load is sufficient, the beam will deform, creating tensile and compressive stresses to be set up within the beam (Figure 2.27).

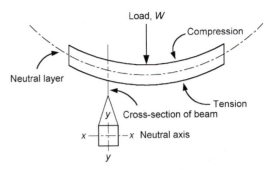

Figure 2.27 Deformation of a beam, resulting in tensile and compressive stresses

Between the areas of tension and compression there is a layer within the beam, which is unstressed, termed the *neutral layer*. Its intersection with the cross-section is termed the *neutral axis*. Engineers' theory of bending is concerned with the relationships between the stresses, the beam geometry and the curvature of the applied bending moment. In order to formulate such relationships we make several assumptions, which simplify the mathematical modelling. These assumptions include:

(a) the beam section is symmetrical across the plane of bending and remains within the plane after bending;

(b) the beam is straight prior to bending and the radius of bend is large compared with the beam cross-section;

(c) the beam material is uniform (homogeneous) and has the same elastic modulus in tension and compression (this is not the case, for example, with ceramic materials).

> **Key Point.**   The neutral layer between the areas of tension and compression in a beam that is deformed (bent) under load is known as the neutral axis

2.3.1   Deformation in pure bending

Consider a beam subject to pure bending. Figure 2.28 shows an exploded view of the beam element in which the neutral axis, discussed above, is clearly seen.

Figure 2.28a shows the general arrangement of the beam before and after bending. We consider the

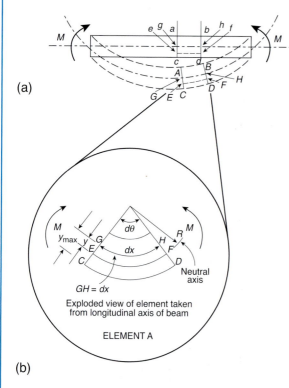

(a)

(b)

Figure 2.28 Situation for the analysis of deformation in pure bending

deformations that occur between the sections *ac* and *bd*, which are δx apart, when the beam is straight. The longitudinal fibre of the beam material *ef*, which is at a distance *y* from the neutral axis, has the same initial length as the fibre *gh* at the neutral axis, prior to bending.

Now, during bending we can see from Figure 2.28 that *ef* stretches to become *EF*. However, *gh*, which is at the neutral axis, is not strained (since stress is deemed to be zero) and so it has the same dimensions when it becomes *GH*. So, by simple trigonometry, when *R* equals the radius of curvature of *GH* we may write:

$GH = gh = dx$ where $dx = R\,d\theta\,(\theta$ in radians) and $EF = (R + y)d\theta$. We also see from the figure that the longitudinal strain of *EF* is given by:

$$\varepsilon_x = \frac{EF - ef}{ef}.$$

Now since we know that prior to bending $ef = gh$ and that after bending, because *gh* is on the neutral axis $gh = GH$, it must equal $ef = r\,d\theta$ (from above). Also, since

$EF = (R + y)d\theta$, we have:

$$\varepsilon_x = \frac{(R + y)d\theta - Rd\theta}{Rd\theta}$$

or

$$\varepsilon_x = \frac{Rd\theta + yd\theta - Rd\theta}{Rd\theta}$$

and, on division by $d\theta$,

$$\varepsilon_x = \frac{y}{R} \tag{2.1}$$

We also showed that $dx = Rd\theta$ (θ in radians) or $R = \dfrac{dx}{d\theta}$ and, on substitution into equation 2.1, we get:

$$\varepsilon_x = \frac{d\theta}{dx}y \tag{2.2}$$

From our earlier work on Poisson's ratio in section 1.7.1, we established relationships for two-dimensional strains. These may be extended to three-dimensional strains, as you will see later in Chapter 7. For the moment you should accept that we can show the relationship between the strains ε_y and ε_z, which are transverse to the longitudinal strain ε_x, as:

$$\varepsilon_y = \varepsilon_z = -\frac{v\sigma}{E}x \tag{2.3}$$

where *v*, you will remember, is Poisson's ratio.

Note: If you find the verification of these formulae difficult do not worry, just concentrate on the meaning of the results.

2.3.2   Stresses due to bending

We have already shown that $\varepsilon_x = \dfrac{y}{R}$ (equation 2.1) and we also know that $\varepsilon = \dfrac{\sigma}{E}$. Then, substituting for strain into equation 2.1 yields:

$$\frac{\sigma}{E} = \frac{y}{R} \tag{2.4}$$

We can also show, by considering the cross-section of element A in Figure 2.28, that the bending moment *M* imposed on the beam at the element is related to the stress on the element, the distance *y* from the neutral axis and the second moment of area *I* by:

$$\frac{M}{I} = \frac{\sigma}{y} \tag{2.5}$$

The second moment of area *I* is a measure of the bending efficiency of the structure, in that it measures the resistance to bending loads. The greater the distance and amount of mass of the structure away from the bending axis, the greater the resistance to bending.

If we combine equations 2.4 and 2.5, we may write:

$$\frac{M}{I} = \frac{\sigma}{y} = \frac{E}{R}. \qquad (2.6)$$

The above equation is known as the general bending formula or *the engineers' theory of bending*, where, at any given section in a beam in a state of bending:

sigma (σ) is the longitudinal stress at a layer some distance y from the neutral axis,
I is the second moment of area about the neutral axis (see section 2.4),
M is the bending moment,
E is the elastic modulus of the beam material, and
R is the radius of curvature of the beam material.

For a beam the maximum bending stress (σ_{max}) will occur where the distance from the neutral axis is a maximum (y_{max}). From equation 2.6 we may write:

$$M = \frac{I\sigma_{max}}{y_{max}} \qquad (2.7)$$

The quantity I/y_{max} depends only on the cross-sectional area of the beam under consideration and is known as the section modulus Z; therefore:

$$M = Z\sigma_{max} \qquad (2.8)$$

Key Point. The section modulus $Z = \dfrac{I}{y_{max}}$ is dependent only on the cross-sectional area of the beam

Example 2.8 Figure 2.29 shows a simplified diagram of a cantilever lifting gantry, 1.8 m long, designed to raise loads of up to 3 kN. If the maximum design stress of the cantilever beam material, is 250 MPa, determine a suitable diameter for the bar.

Now, the design load is given as 3 kN. However, in design work we should always consider a factor of safety. In view of the fact that suspended loads are involved we must ensure the complete safety of the operators, so we will choose a factor of safety of 2. Thus the maximum design load = 6 kN.

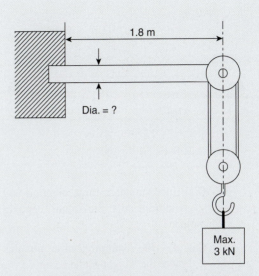

Figure 2.29 Cantilever lifting gantry for Example 2.8

The maximum bending moment for the cantilever beam is furthest from the support. In this case the maximum BM = 6 × 1.8 = 10.8 kN m.

Now the diagram shows a beam of circular cross-section. We need to calculate the second moment of area I for this beam. This is given by the formula:

$$I = \frac{\pi d^4}{64} \text{ (see Table 2.1 in the next section)}$$

We are given the maximum allowable design stress ($\sigma_{max} = 250$ MN/m^2) which on the beam occurs at the maximum distance from the neutral axis (y_{max}), in this case at distance $d/2$.

So, using engineers' theory of bending

where $\dfrac{M}{I} = \dfrac{\sigma}{y}$:

$$\frac{10800}{\pi d^4/64} = \frac{250 \times 10^6}{d/2} = 76,$$

so $d = 76$ mm.

Test your knowledge 2.2

(1) A simply supported beam (Figure 2.30) carries the UDL as shown.

The maximum allowable bending stress for the beam is 120 MPa.

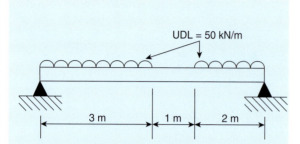

Figure 2.30 Figure for TYK 2.2.1

a) Construct the shear force diagram and so determine the maximum bending moment.

b) Calculate the required section modulus Z. Ignore the weight of the beam.

(2) A simply supported beam (Figure 2.31) carries the UDL as shown:

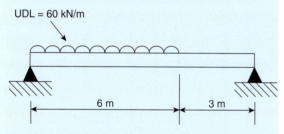

Figure 2.31 Figure for TYK 2.2.2

a) Construct the shear force diagram.

b) Sketch the bending moment diagram.

c) Determine the maximum bending moment.

d) Given that the section modulus for the beam is $3500\,cm^3$, determine the allowable bending stress. Ignore the weight of the beam.

2.4   Centroid and second moment of area

Beams have to resist the loads which act upon them, undergoing high stresses due to these loads and the resulting bending moments. The resistance of a beam to bending moments and shear forces depends on:

- the inherent properties of the beam material, such as homogeneity, ultimate tensile strength and elastic modulus, and
- the shape and size of the beam section.

Properties of a beam that can be attributed to their shape include: cross-sectional area, centroid of area (their centre of gravity), second moment of area (or moment of inertia) and their section modulus Z, mentioned earlier.

The centroid of area is, for calculation purposes, the point at which the total area is considered to act, whereas the *second moment of area* of a beam, or for that matter any shape, is a measure of the resistance to bending of that shape.

The mathematical derivation for centroids and second moments of area is covered in *Essential Mathematics 2*, on the book's website. Here, we concentrate on the practical techniques necessary to determine these properties.

Key Point.   The centroid of area is the point at which the total area is considered to act, whereas the second moment of area of a cross-section is a measure of its resistance to bending

2.4.1   Calculating centroids of area

To determine the centroid of area of a particular shaped cross-section, we adopt the following procedure:

1. Divide the shape into its component parts

2. Determine the area of each part

3. Assume the area of each part to act at its own centre of area, and

4. Take moments about a convenient point or axis to determine the centre of gravity (centroid) of the whole area.

Note: If we were finding the moment of inertia of a body, or the second moment of mass of a body, then the centre of gravity of the weight of the whole body would be determined. The moment of inertia or second moment of mass is analogous with the second moment of area for beams, where we are interested in cross-sectional areas to determine their resistance to bending. The above procedure is best illustrated by example.

Example 2.9 Determine the centroid of area of the beam section shown in Figure 2.32.

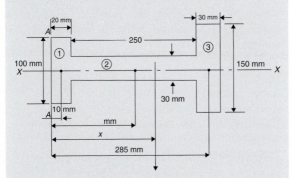

Figure 2.32 Beam section for Example 2.9

The figure is symmetrical about the centre line X–X, therefore the centroid of area ($\bar{x}$) must lie on this line, some distance x from the datum face A.

Following the procedure, we first find the individual areas of the beam cross-sections 1, 2 and 3, noting that the centroids for these individual areas occur at their geometric centres. This will be the case for homogeneous materials such as steels.

Then: Area 1 $= (20 \times 100) = 2000\,\text{mm}^2$;

Area 2 $= (30 \times 250) = 7500\,\text{mm}^2$;

Area 3 $= (30 \times 150) = 4500\,\text{mm}^2$;

Total Area $= 14000\,\text{mm}^2$.

Again, following the procedure, we take moments about the datum face A, equating the moment for total area to the moments for the individual area.

That is: $14000x = (2000 \times 10) + (7500 \times 145)$

$+ (4500 \times 285)$

or $14000x = (20000) + (1087500)$

$+ (1282500)$ and

$$x = \frac{2390000}{14000} = 170.7\,\text{mm}$$

So the centroid of area $\bar{x}$ is on the centreline, at a distance $x = 170.7$ mm from A.

Note that this procedure is similar to that which we used for finding the reaction at the supports of a uni-form beam. In this case areas have been substituted in place of loads. We may choose any convenient point or face of the cross-section to act as a datum.

Here is one more example; please make sure you can follow the argument.

Example 2.10 Find the centroid of area for the cross-section of the uniform beam shown in Figure 2.33.

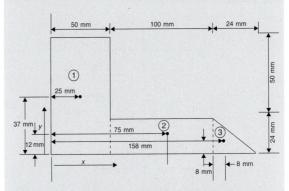

Figure 2.33 Cross-section of beam for Example 2.10

Area 1 $(50 \times 74) = 3700\,\text{mm}^2$;

Area 2 $(100 \times 24) = 2400\,\text{mm}^2$;

Area 3 $0.5(24 \times 24) = 288\,\text{mm}^2$;

Total area $= 6388\,\text{mm}^2$.

Taking moments about the left-hand vertical edge of the beam to find position of centroid $\bar{x}$, we get:

$6388x = (3700 \times 25) + (2400 \times 75)$

$+ (288 \times 158),$

which gives

$$\bar{x} = \frac{319004}{6388} = 49.94\,\text{mm}.$$

Similarly, taking moments about the base, we get:

$6388y = (3700 \times 37) + 2400 \times 12 + (288 \times 8),$
which gives

$$\bar{y} = \frac{168004}{6388} = 26.3\,\text{mm}.$$

Then the centroid is 49.94 mm to the right of the left-hand vertical edge of the beam and 26.3 mm above the base datum.

2.4.2 Second moment of area (*I*) and its determination

We have already said that *the second moment of area* is a measure of the efficiency of the beam to resist bending loads. Consider the universal 'I'-section beam shown in Figure 2.34.

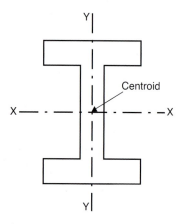

Figure 2.34 Universal beam showing principal axes

The beam is shown with its principal axes X–X and Y–Y. These axes intersect at the centroid. Normally, when considering the second moment of area, it is necessary to find this about the X–X and Y–Y axes. This is shown in a later example, where on substitution of values into the formulae for the second moment of area, it is found that when the second moment of area is taken about the X–X axis, with the greater amount of material further away from the reference axis, the higher will be the value of the second moment of area and the higher will be the resistance to bending of the beam.

In general, then, the resistance to bending of a beam will be greater as its area and depth increase.

This is easily demonstrated by considering a steel rule (Figure 2.35), where it is easily bent around

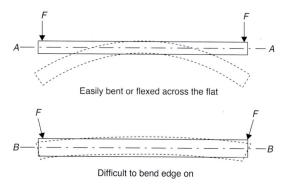

Figure 2.35 Resistance to bending of steel rule

cross-section *A–A* but it is difficult to bend around cross-section *B–B*.

The determination of the formula for the second moment of area about the principal axes for a rectangular cross-section beam is shown next, in Example 2.11.

Example 2.11   Derive the formulae for the second moment of area I_{xx} about the X–X axis for the rectangular cross-section shown in Figure 2.36a.

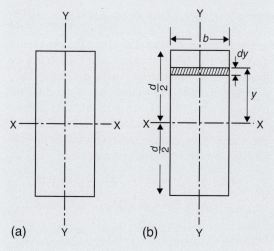

Figure 2.36 a) Rectangular section b) Rectangular section showing elemental strip, breadth and depth relationships

We are required to find the second moment of area about the X–X principal axis. Figure 2.36b shows the same rectangular section with an elemental strip that has its breadth and depth marked on it.

The first stage in determining the second moment of area of a beam cross-section is to find its centroid. This is easily found for a rectangular cross-section and is shown in Figure 2.36 as being at the intersection of the principal axes.

If we consider this elemental strip having breadth *b* and depth *dy*, at a distance *y* from the X–X axis, then the second moment of area of this strip is its area multiplied by distance $y = (b)(dy)(y^2)$ or $by^2 dy$. So the second moment of *half of the rectangular area* is the sum of all these quantities.

Then, using the integral calculus to sum these quantities, we have:

I_{xx} for half the rectangular area

$$\int_0^{d/2} by^2 dy = \left[\frac{by^3}{3}\right]_0^{\frac{d}{2}} = \frac{bd^3}{24},$$ then for the complete rectangular area $I_{xx} = \frac{bd^3}{12}$. By symmetry, the second moment of area about the Y–Y principal axis is given as, $I_{yy} = \frac{db^3}{12}$.

Example 2.12 that follows shows how we can determine the second moment of area of a rectangular cross-section when moments are taken about its base.

Example 2.12   Find the second moment of area of the rectangle shown in Figure 2.37, about its base edge.

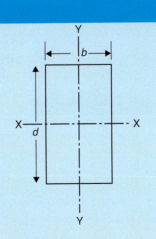

Figure 2.37 Second moment of area for rectangle

The rectangle is shown set up on suitable reference axes. It is, in this case, convenient to turn the rectangle through 90° and let the base edge lie on the y-axis (Y–Y). The diagram shows a typical elemental strip with area parallel to the reference axis (Y–Y) whose area is $b\delta x$.

Now, the second moment of area

$$\text{about the y-axis} = \sum Ax^2 = \sum_{x=0}^{x=d}(b)(\delta x)(x^2)$$
$$= \int_0^d (b)(dx)(x^2).$$

So $I_{yy} = b\int_0^d x^2 dx = b\left[\frac{x^3}{3}\right]_0^d = \frac{bd^3}{3}$

The table given below summarises the second moment of area for the more common rectangular and circular cross-sections, about their various axes.

2.4.3   Parallel axes theorem

It is sometimes necessary or more convenient to determine the second moment of area of any shape about an axis B–B, parallel to another axis A–A, *at a perpendicular distance S*. Then, the area of the shape multiplied by the distance squared (AS^2) must be added to the second moment of area about A–A, that is:

$$I_{BB} = I_{AA} + AS^2 \qquad (2.9)$$

This relationship is known as *the parallel axes theorem*.

Table 2.1   Formulae for the second moment of area for some common cross-sections

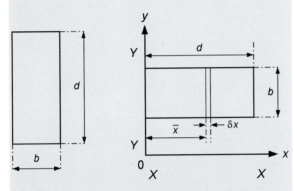

$$I_{XX} = \frac{bd^3}{12}, \quad I_{YY} = \frac{db^3}{12}$$

Figure 2.38 Rectangular section about principal axes

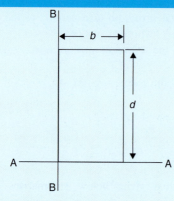

$$I_{AA} = \frac{bd^3}{3}, I_{BB} = \frac{db^3}{3}$$

Figure 2.39 Rectangular section about one edge

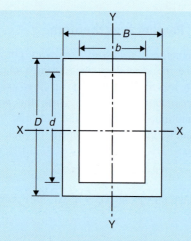

$$I_{XX} = \frac{BD^3 - bd^3}{12},$$
$$I_{YY} = \frac{DB^3 - db^3}{12}$$

Figure 2.40 Hollow rectangular section about one edge

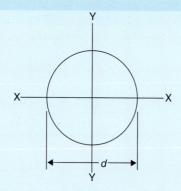

$$I_{XX} = I_{YY} = \frac{\pi d^2}{64}$$

Figure 2.41 Circular section about principal axes

Example 2.13   Consider the I-section universal beam shown in Example 2.9 (Figure 2.32). Determine the second moment of area about the principal axes I_{XX} and I_{YY} for this beam.

The beam is reproduced here (Figure 2.42), showing the I-section in its normal vertical position, with the principal axes marked on it in their correct positions about the centroid of area.

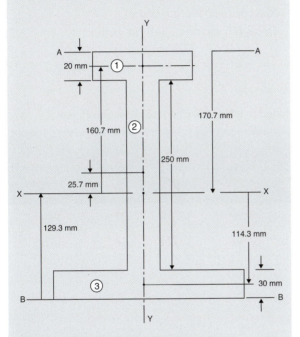

Figure 2.42 I-section beam (universal beam)

We know from Example 2.9 that the centroid of area for the whole beam acts through the centre of the web, at 170.7 mm from the A–A axis. We can verify the centroid position by taking moments about the base B–B. So:

$14000x = (4500 \times 15) + (7500 \times 145) + (2000 \times 290)$ and $14000x = 1810000$, or:

$$x = \frac{1810000}{14000} = 129.3 \text{ mm, as expected!}$$

To find I_{XX}, we first divide the I-beam into three rectangular sections, as we did when finding its centroid. Then, using the parallel axis theorem, that is equation 2.9, we find the individual values of I_{XX} for each of the rectangular areas, then sum them to find the total value of I_{XX} for the whole beam section.

So, using the parallel axes theorem and taking moments about X–X, we get for the top flange (area 1):

$$I_{XX_1} = I_{11} + A_1(S_1)^2$$
$$= \frac{(100)(20)^3}{12} + (2000)(160.7)^2 \text{ or}$$
$$I_{XX_1} = 51.716 \times 10^6 \text{ mm}^4.$$

Similarly, for the web, we get:

$$I_{XX_2} = I_{22} + A_2(S_2)^2$$
$$= \frac{(30)(250)^3}{12} + (7500)(25.7)^2 \text{ or}$$
$$I_{XX_2} = 44.016 \times 10^6 \text{ mm}^4.$$

And for the bottom flange, we get:

$$I_{XX_3} = I_{33} + A_3(S_3)^2$$
$$= \frac{(150)(30)^3}{12} + (4500)(114.3)^2 \text{ or}$$
$$I_{XX_3} = 59.128 \times 10^6 \text{ mm}^4.$$

Therefore, the total second moment of area about $X - X$ for the beam section is:

$$I_{XX} = 154.86 \times 10^6 \text{ mm}^4.$$

Now, to find I_{YY} we note that all the individual rectangular second moments of area lie on the Y–Y axis. Therefore, $S = 0$ and in all cases the total second moment of area about Y–Y reduces to the sum of the individual rectangular second moments, that is:

$$I_{YY} = I_{YY_1} + I_{YY_2} + I_{YY_3} \text{ or}$$
$$I_{YY} = \frac{(20)(100)^3}{12} + \frac{(250)(30)^3}{12}$$
$$+ \frac{(30)(150)^3}{12}, \text{ therefore}$$
$$I_{YY} = 10.67 \times 10^6 \text{ mm}^4.$$

Make sure that you are able to follow the whole argument and obtain the same numerical values.

2.5 Beam selection

Now, you will have noted that it is quite tedious to calculate values of I immediately the beam is anything other than rectangular, and yet this is only one of the criteria necessary for beam selection.

To aid us in our selection of a particular beam, tables of beams exist that may be used to select a beam for a specified purpose, to sustain a particular loading regime. These tables quote key dimensions and inherent properties such as the elastic and plastic modulus, together with the second moment of area (mass moment of inertia) and radius of gyration (which you will meet a little later in Chapter 5), of particular beam sections.

Thus a table for universal I-section beams may contain some or all of the following headings (see Table 2.2), the dimensional information being illustrated in Figure 2.43.

Overall (serial) size (mm)

Mass per metre length (kg/m)

Depth (D) and breadth (B) of section

Table 2.2 Some typical dimensions and properties found in tables for Universal I-section Beams

Serial size (mm)	Mass per metre kg/m	Second moment of area		Radius of gyration		Elastic section modulus Z		Area of section (cm²)
		X–X Axis (cm⁴)	Y–Y Axis (cm⁴)	X–X Axis (cm)	Y–Y Axis (cm)	X–X Axis (cm³)	Y–Y Axis (cm³)	
914 × 419	388	718742	45407	38.13	9.58	15616	2160	494.5
	343	625287	39150	37.81	9.46	13722	1871	437.5
406 × 178	74	27279	1448	17.0	3.91	1322	161.2	104.4
	67	24279	1269	16.9	3.85	1186	141.9	85.4
	60	21520	1108	16.8	3.82	1059	124.7	76.1
	54	18576	922	16.5	3.67	922.8	103.8	68.3
254 × 102	28	4004	174	10.5	2.19	307.6	34.13	36.2
	25	3404	144	10.3	2.11	264.9	28.23	32.1
	22	2863	116	10.0	2.02	225.4	22.84	28.4

Thickness of the web (t) and flange (T)

Root radius (r)

Depth between fillets (d)

Area of section (cm^2)

Second moment of area (cm^4)

Radius of gyration (cm)

Elastic section modulus (cm^3) (This is the section modulus $Z = I/y_{max}$ you met earlier)

Plastic section modulus (cm^3)

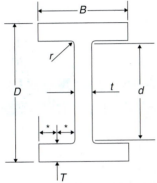

*Flange thickness measured equidistant
from web to edge of flange

Figure 2.43 Key dimensions for universal I-section beam

A slightly modified, very brief extract from such tables is shown in Table 2.2. It is given here purely for the purpose of illustrating the examples that follow.

Full unabridged versions of these tables may be obtained from the following reference source: BS 5950-1: 2000 (revised version). In addition, much useful information is now presented on the web. Particularly good sites, where tables of properties and dimensions for universal beams and columns may be found, include Corus Construction at www.corusconstruction.com and Techno Consultants Ltd at www.technouk.com. The properties and dimensions given in Table 2.2 may be used in conjunction with bending theory to select a beam for a particular function, as the following example illustrates.

Example 2.14   A rolled steel universal I-section beam with a serial size of 406 × 178 has a mass of 60 kg/m. What is the maximum safe allowable bending moment this beam can sustain, given that the maximum allowable bending stress in tension and compression must not exceed 165 N/mm^2?

This example requires us to determine the appropriate section modulus (Z) from tables and then use engineers' theory of bending to determine M_{max}.

Then, from Table 2.2 for this beam section,

$$Z_{XX} = 1059 \, \text{cm}^3 \quad \text{or} \quad 1059 \times 10^3 \, \text{mm}^3.$$

Now, from $\dfrac{M}{I} = \dfrac{\sigma}{y} = \dfrac{E}{R}$, ignoring the last term and remembering that $Z = \dfrac{I}{y}$, we get $\dfrac{M}{\sigma} = \dfrac{I}{y} = Z$ and therefore $M = \sigma Z$, which will be a *maximum* when Z is a maximum, that is when we use the beam correctly about the X–X axis.

So, $M_{max} = (165 \, \text{N/mm}^2)(1.059 \times 10^6 \, \text{mm}^3)$ or

$$M_{max} = 174.7 \times 10^6 \, \text{N mm}.$$

We finish our section on beam selection with an example that involves the use of a beam made from a non-homogeneous engineering material, i.e. wood.

Key Point.   Wood is a non-homogeneous material, that is a material whose engineering properties may vary throughout

Example 2.15   A roof support is made from a wooden rectangular section beam 100 mm wide and 200 mm deep (Figure 2.44). If the maximum allowable bending moment anywhere along the beam is 14×10^6 N mm, determine the maximum allowable bending stress.

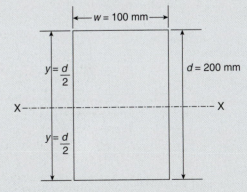

Figure 2.44 Rectangular wooden beam section

In this case we have no tabulated data for such a beam, the properties of wooden beams being dependent not only on beam geometry but also on the particular species of wood being used. So we need first to calculate Z.

Figure 2.44 shows the beam with distance y marked. This is the distance to the edge of the beam material from its neutral axis and, because the beam is symmetrical about the X–X axis, the distance y is equal in either direction from this axis.

Therefore, knowing that $Z = \dfrac{I_{XX}}{y}$, $y = \dfrac{d}{2}$ (that is, half the depth) and that $I_{XX} = \dfrac{bd^3}{12}$,

then $Z = \dfrac{I_{XX}}{y} = \dfrac{bd^3}{12} / \dfrac{d}{2} = \left(\dfrac{bd^3}{12}\right)\left(\dfrac{2}{d}\right) = \dfrac{bd^2}{6}$.

So $Z = \dfrac{(100)(200)^2}{6} = 0.667 \times 10^6 \text{ mm}^3$ and, knowing that

$$M_{\max} = Z\sigma_{\max}, \text{ then } \sigma_{\max} = \dfrac{M_{\max}}{Z}$$

$$= \dfrac{14 \times 10^6 \text{ N mm}}{0.667 \times 10^6 \text{ N mm}^3} = 21 \text{ N/mm}^2$$

Test your knowledge 2.3

1. For the beam section shown in Figure 2.45, calculate:

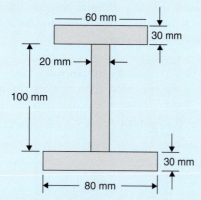

Figure 2.45 I-section beam for TYK 2.3.1

a) the centroid of area,

b) I_{XX}, the second moment of area about the X–X principal axis.

2. A rolled steel universal I-section beam with a serial size of 254 × 102 (Table 2.2) has a mass of 25 kg/m acting upon it. What is the maximum safe allowable bending moment this beam can sustain, given that the maximum allowable bending stress must not exceed 80 N/mm²?

3. A footbridge support is made from a wooden rectangular-section beam 150 mm wide and 250 mm in depth. If the beam has a maximum allowable bending stress of 35 N/mm², determine the maximum allowable bending moment along the whole length of the beam.

2.6 Slope and deflection of beams

In order that engineers may estimate the loading characteristics of beams in situations where, for example, they are used as building support lintels, supports for engineering machinery or in bridges, it is important to establish mathematical relationships that enable the appropriately dimensioned beam of the correct material to be chosen.

We have already established one very important relationship, *engineers' theory of bending*, from which other important equations may be found. From equation 2.6 we know that

$$\dfrac{M}{I} = \dfrac{E}{R}$$

and, on rearrangement, we may write:

$$\dfrac{1}{R} = \dfrac{M}{EI} \qquad (2.10)$$

Now consider Figure 2.46, which shows a cantilever beam being deflected downwards (compare with Figure 3.4).

At A the beam has been deflected (bent) by an amount y and at B the deflection is $y + dy$. The deflection of beams, in practice, is very small; therefore, from simple geometry for very small angles, we may assume that the arc distance ds is approximately equal to dx. So, as $d\theta$ approaches zero,

$$dx = Rd\theta, \text{ from which } \dfrac{1}{R} = \dfrac{d\theta}{dx}.$$

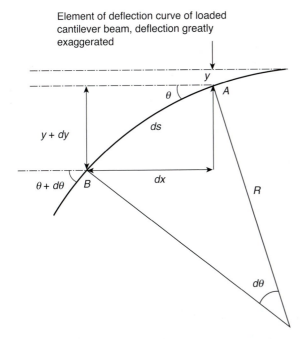

Element of deflection curve of loaded cantilever beam, deflection greatly exaggerated

Figure 2.46 Element of deflection curve of cantilever beam, greatly exaggerated

Also, since the arc distance *AB* is very small, we may approximate it to a straight line which has a gradient (slope) equal to the tangent of the angle; therefore

$$\theta = \frac{dy}{dx} \text{ and so, on differentiation, } \frac{d\theta}{dx} = \frac{d^2y}{dx^2}$$

and substituting

$$\frac{1}{R} = \frac{d\theta}{dx} \text{ into } \frac{d\theta}{dx} = \frac{d^2y}{dx^2} \text{ gives } \frac{1}{R} = \frac{d^2y}{dx^2} \quad (2.11)$$

Finally, combining equations 2.10 and 2.11 yields:

$$\frac{d^2y}{dx^2} = \frac{M}{EI} \quad (2.12)$$

2.6.1   Deflections by integration

The above bending moment differential equation provides the fundamental relationship for the deflection curve of a loaded beam. Further manipulation of this equation, using integration, produces relationships in terms of the *slope* (gradient) and *deflection* (distance away from the neutral unloaded equilibrium position) of the beam.

When the beam is of uniform cross-section and *M* is able to be expressed mathematically as a function of *x*, we may integrate the equation $d^2y/dx^2 = M/EI$ with respect to *x* and get:

$$\frac{dy}{dx} = \frac{1}{EI} \int M dx + A \quad (2.13)$$

where *A* is the constant of integration and dy/dx is the slope of the beam. Integrating a second time yields:

$$y = \frac{1}{EI} \iint M dx dx + Ax + B \quad (2.14)$$

which is an expression for the *deflection* of the beam *y*.

At this stage we introduce two relationships (without proof) that we need in order to further develop our mathematical argument. They are:

$$\frac{dF}{dx} = -\omega \quad (2.15)$$

and

$$\frac{dM}{dx} = F \quad (2.16)$$

where ω is the distributed load and *F* the shear force due to a point load. Verification of equations 2.15 and 2.16 may be found in *Essential Mathematics 2 – Introduction to the Calculus*, on the companion website.

Now, if we differentiate equation 2.12 once more, i.e. differentiate $\frac{d^2y}{dx^2} = \frac{M}{EI}$, we get

$$\frac{d^3y}{dx^3} = \frac{1}{EI} \times \frac{dM}{dx},$$

and substituting

$$\frac{dM}{dx} = F \text{ gives } \frac{d^3y}{dx^3} = \frac{F}{EI} \quad (2.17)$$

Finally, using equation 2.15, that is $\dfrac{dF}{dx} = -\omega$, and differentiating $\dfrac{d^3y}{dx^3} = \dfrac{F}{EI}$ again, we get

$$\frac{d^4y}{dx^4} = \frac{dF}{dx} \times \frac{1}{EI} \text{ and so}$$

$$\frac{d^4y}{dx^4} = -\frac{\omega}{EI} \tag{2.18}$$

Note that the *slope* and *deflection* equations we have just found all contain $1/EI$, where the modulus (E) multiplied by the second moment of area (I) provides a measure of the *flexural rigidity* of the beam. The elastic modulus is a measure of the stiffness of the beam material and the second moment of area (sometimes known as the moment of inertia) is a measure of the beam's resistance to bending about a particular axis. The signs associated with the equations we have just integrated adopt the convention that the *deflection (y) is positive upwards* and *bending moments (M) causing sagging are positive*.

> **Key Point.**   The slope and deflection of a beam under load are *inversely proportional to EI*, that is the *flexural rigidity* of the beam

The integration method can be applied to standard cases that include cantilevers and simply supported beams having a variety of combinations of concentrated loads, distributed loads, or both. Space does not permit us to consider all possible cases, so just one example of these applications is given in full below. This is followed by a table of some of the more important cases that may be solved by use of the integrating method.

When applying the integrating method to the slope and deflection of beams we will use the following symbols, which match with our previous theory: x as the distance measured from the left-hand support of the beam or from the fixed end of a cantilever; l for length; y for deflection; ω per unit length for distributed loads; and F for shear force due to point load W.

Note: In some textbooks v is used for y and P for W.

> **Example 2.16**   To determine the deflection and slope of a cantilever subject to a point load we will consider the cantilever shown previously in Figure 2.4, subject to a single point load at its free end. The load causes a 'hogging' bending moment in the beam, which, according to our convention, is *negative* so

$$-M = W(l - x) = Wl - Wx.$$

Now, using our fundamental equation 2.12 for a negative bending moment, where

$$\frac{d^2y}{dx^2} = -\frac{M}{EI}, \text{ on substitution,}$$

$$EI\frac{d^2y}{dx^2} = -W(l - x),$$

so on integration we have

$$EI\frac{dy}{dx} = -W\left(lx - \frac{x^2}{2}\right) + A.$$

The constant of integration can be found from the 'boundary conditions'. The slope of the beam at the fixed end is *zero*, since the beam is deemed to be level at this point. In other words, when $x = 0$, the slope $dy/dx = 0$. This implies that $A = 0$, so our equation becomes

$$EI\frac{dy}{dx} = -W\left(lx - \frac{x^2}{2}\right),$$

or the slope of the beam is given by:

$$\frac{dy}{dx} = -\frac{W}{EI}\left(lx - \frac{x^2}{2}\right) \tag{2.19}$$

If we integrate equation 2.19 again we get

$$y = -\frac{W}{EI}\left(\frac{lx^2}{2} - \frac{x^3}{6}\right) + B.$$

The constant of integration B may again be found from the boundary conditions. These are that $y = 0$ at $x = 0$; this infers that $B = 0$.

So our equation for deflection y may be written as:

$$y = -\frac{W}{EI}\left(\frac{lx^2}{2} - \frac{x^3}{6}\right) \tag{2.20}$$

We can see quite clearly from Figure 2.4 that the maximum deflection occurs at the free end, where $x = l$; then

$$y_{max} = -\frac{W}{EI}\left(\frac{l^3}{2} - \frac{l^3}{6}\right)$$

so $y_{max} = -\dfrac{Wl^3}{3EI}$ $\hspace{1cm}$ (2.21)

Also, the maximum slope occurs at the free end and, from equation 2.19 when $x = l$, we have:

$$\left(\frac{dy}{dx}\right)_{max} = -\frac{Wl^2}{EI} \qquad (2.22)$$

Note the use of boundary conditions to establish values for the constants. You should ensure that you follow the logic of the argument when establishing such conditions, which vary from case to case. Now let us assume that the cantilever discussed above is made from a steel with an elastic modulus $E = 210\,GPa$, and the point load $W = 40\,kN$. If the length of the beam is $l = 2.5\,m$ and the maximum deflection $y_{max} = 1.5\,mm$, determine:

a) the second moment of area for the beam,

b) the maximum slope.

a) The second moment of area is easily calculated using equation 2.21, where:

$$I = -\frac{Wl^3}{(3)(E)(y_{max})}$$

$$= -\frac{(40 \times 10^3)(2.5)^3}{(3)(210 \times 10^9)(-1.5 \times 10^{-3})}$$

$$= 661.3 \times 10^{-6}\,m^4$$

or $I = 661.3 \times 10^6\,mm^4$.
Note the careful use of units!

b) The maximum slope may now be determined using equation 2.22, where:

$$\left(\frac{dy}{dx}\right)_{max} = -\frac{Wl^2}{EI}$$

$$= -\frac{(40 \times 10^3)(2.5)^2}{(3)(210 \times 10^9)(663.3 \times 10^{-6})}$$

$$= -0.0022\,m/m,$$

or the *maximum slope downwards* is 2.2 mm/m.

The process shown in the above example, where successive integration is applied to the fundamental equations to establish expressions for the slope and deflection, may be used for many standard cases involving plane bending. Remember that the constants of integration may be found by applying the unique boundary conditions that exist for each individual case. Laid out in Table 2.3 are some standard cases for the slope and deflection associated with simply supported beams and cantilevers.

2.6.2 Principle of superposition

This principle states that if *one relationship connecting the bending moment, slope and deflection for a particular loading has been determined,* for example a cantilever subject to a concentrated load at its free end (Figure 2.47), *then we may add, algebraically, any other known relationship to it for a given loading situation.*

Table 2.3 Some standard cases for slope and deflection of beams

Situation	Slope $\left(\dfrac{dy}{dx}\right)$	Deflection (y)	$\left(\dfrac{dy}{dx}\right)$	y_{max}
Cantilever with end couple	$-\dfrac{Mx}{EI}$	$-\dfrac{Mx^2}{2EI}$	$-\dfrac{Ml}{EI}$ at B	$-\dfrac{Mx^2}{EI}$ at B
Cantilever with concentrated end load	$-\dfrac{W}{EI}\left(lx - \dfrac{x^2}{2}\right)$	$-\dfrac{W}{EI}\left(\dfrac{lx^2}{2} - \dfrac{x^3}{6}\right)$	$-\dfrac{Wl^2}{2EI}$ at B	$-\dfrac{Wl^3}{3EI}$ at B

Table 2.3 Cont'd

Cantilever with distributed load 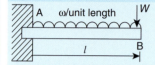	$-\dfrac{\omega}{2EI}\left(l^2 x - lx^2 + \dfrac{x^3}{3}\right)$	$-\dfrac{\omega}{2EI}\left(\dfrac{l^2 x^2}{2} - \dfrac{lx^3}{3} + \dfrac{x^4}{12}\right)$	$-\dfrac{\omega l^3}{6EI}$ at B	$-\dfrac{\omega l^4}{8EI}$ at B
Simply supported beam with point load 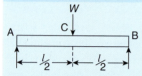	$\dfrac{W}{EI}\left(\dfrac{lx}{2} - \dfrac{x^2}{2}\right)$	$\dfrac{W}{2EI}\left(\dfrac{lx^2}{4} - \dfrac{x^3}{6} - \dfrac{l^3}{24}\right)$	$\dfrac{Wl^2}{16\,EI}$ at A and B	$-\dfrac{Wl^3}{48EI}$ at C
Simply supported beam with distributed load 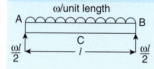	$\dfrac{\omega}{2EI}\left(\dfrac{l^2 x}{4} - \dfrac{x^3}{3}\right)$	$\dfrac{\omega}{2EI}\left(\dfrac{l^2 x^2}{8} - \dfrac{x^4}{12} - \dfrac{5l^4}{192}\right)$	$\dfrac{\omega l^3}{24EI}$ at A and B	$-\dfrac{5\omega l^4}{384EI}$ at C

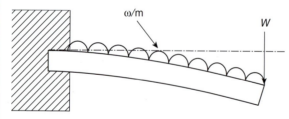

Figure 2.47 Cantilever beam subject to loads causing deflection

The cantilever shown in Figure 2.47 is subject to both a concentrated load at its free end and a distributed load over its whole length. So, for example, at a particular section, using the expressions for maximum deflections in Table 2.3 and the principle of superposition, we may write:

$$y_{\max} = -\frac{Wl^3}{3EI} + \frac{\omega l^4}{8EI}.$$

This principle enables us to combine standard loading situations, for example, obtaining the deflection at a section of a beam subject to complex loading, by summing the individual deflections caused at the section by the individual components of the loaded beam.

Example 2.17 Use the principle of superposition to determine the resultant deflection at the centre of the beam shown in Figure 2.48, given that $EI = 550\,\text{kN m}^2$.

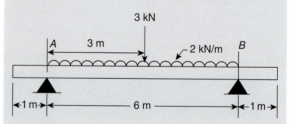

Figure 2.48 Loaded beam for Example 2.17

We may very easily split the problem into two components by recognizing that we have one standard central point load, equidistantly spaced between the supports, and one UDL running the length between the supports AB. So, from Table 2.3, we can find both deflections and then sum them using the principle of supposition to find the resultant negative deflection that is downwards (in a negative y direction), causing a sagging bending moment.

1) $y_{1max} = -\dfrac{Wl^3}{48EI} = -\dfrac{(3000)(6^3 \times 10^3)}{(48)(550 \times 10^3)}$
   $= -24.55\,\text{mm}$

2) $y_{2max} = -\dfrac{5\omega l^4}{384EI} = -\dfrac{5(2000)(6^4 \times 10^3)}{(384)(550 \times 10^3)}$
   $= -61.31\,\text{mm}$

Thus, resultant deflection at centre $= -24.55 - 61.31 = -85.91\,\text{mm}$.

Notice that the overhang at either end of the beam played no part in the determination of the resultant deflection. However, you might like to consider the effect on the beam loading if two point loads had been placed at the two extremities of the beam.

2.6.3 Macaulay's method

From the procedures and examples you have just covered, you may have realised that the method of successive integration, to determine the slope and deflection equations for each separate loading situation, is tedious and time-consuming. We are required to find values for the constants of integration by considering boundary conditions on each separate occasion and, apart from the most simple of cases, this can involve quite complex algebraic processes, as you may have noticed! All is not lost, however, since a much simpler method exists which enables us to formulate one equation which takes into account the bending moment expression for the whole beam and also enables us to find the necessary boundary conditions much more easily. This simplified process is known as *Macaulay's method*. To assist us in formulating the Macaulay expression for a given loading situation of a beam, we need to use Macaulay functions. The graph of the Macaulay unit ramp function (Figure 2.49) will help us to understand its nature.

Macaulay functions are used to represent quantities that begin at some particular point on the *x*-axis. In our case, for the unit ramp function, this is at $x = a$. To the left of the point *a* the function has the value zero; to the right of *a* the function is denoted by F_1 and has the value $(x - a)$. This relationship may be written mathematically as:

$$F^1(x) = \langle x - a \rangle^1 = \begin{cases} 0 & \text{when } x \le a \\ x - a & \text{when } x \ge a \end{cases}$$

Now, the pointed brackets indicate that the function is a *discontinuity* function. At *zero* our function can

Graph of Macaulay function F_1 (unit ramp function)

$F_1(x) = \langle x - a \rangle^1$ where F = function.

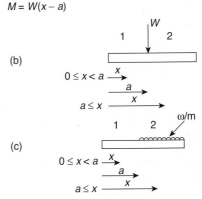

At point 1, $M = 0$
At point 2, $M = W(x - a)$
So Macaulay expression for all values of *x* is
$M = W(x - a)$

At point 1, $M = 0$
At point 2, $M = \frac{1}{2}\,\omega(x - a)^2$
So again Macaulay expression for all values of *x* is
$M = \frac{1}{2}\,\omega\langle x - a \rangle^2$

Figure 2.49 Application of Macaulay function to beams

take *two* values, zero or one. This may be expressed mathematically as:

$$F^0(x) = \langle x - a \rangle^0 = \begin{cases} 0 & \text{when } x \le a \\ 1 & \text{when } x \ge a \end{cases}$$

So what does all this mean in practice? If we apply our function to a beam where a point load W is applied some way along the beam (Figure 2.49b) at a distance *a*, we can determine the bending moments in the normal way by first considering forces to the left of *x*. This gives:

$$M = 0 \quad \text{for} \quad x \ge 0 < a.$$

Also, for forces to the right of *a*, we may write the bending moment as:

$$M = W \langle x - a \rangle.$$

Now, in using the Macaulay expressions we must remember that we are applying the unit ramp function or step function and different rules apply. In particular, *all terms which make the value inside the bracket negative are given the value zero.*

Figure 2.49c shows our beam subject to a uniformly distributed load commencing at distance a. Again, the bending moment at a distance x along the beam may be determined in the normal manner. Considering forces to the left of x at point 1, we have

$$M = 0 \quad \text{when} \quad 0 \le x \le a$$

and, at point 2, $\quad M = \frac{1}{2}\omega\,\langle x - a\rangle^2$.

Thus the Macaulay expression for the bending moment for all values of x, in this case for uniformly distributed loads, is:

$$M = \frac{1}{2}\omega\,\langle x - a\rangle^2 \qquad (2.23)$$

Do remember the constraints that apply when considering the expression inside the pointed brackets (these constraints are summarised at the end of this chapter). The method you should adopt in determining the Macaulay expression for the whole beam is illustrated by the next example, where the total procedure for obtaining the Macaulay expression is illustrated.

Example 2.18 Determine the Macaulay expressions for the slope and deflection of the simply supported beam shown in Figure 2.50.

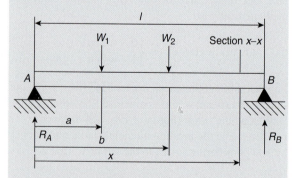

Figure 2.50 Loaded beam for example 2.18

Now, we take the origin from the extreme left-hand end, as shown in the figure. Consider a section (x–x) at the extreme right-hand end of the beam, from which we write down the Macaulay expression. Once written, the ramp function rules apply.

So, taking moments in the normal way gives

$$M = R_A x - W_1(x - a) - W_2(x - b).$$

Therefore the *Macaulay expression for the entire beam, subject to the concentrated loads as shown, is:*

$$M = R_A x - W_1\,\langle x - a\rangle - W_2\,\langle x - b\rangle \qquad (2.24)$$

and, using equation 2.12 where $\dfrac{d^2y}{dx^2} = \dfrac{M}{EI}$, gives:

$$\frac{d^2y}{dx^2} = \frac{1}{EI}\left(R_A x - W_1\,\langle x - a\rangle - W_2\,\langle x - b\rangle\right),$$

and integrating in the normal way gives as an expression for the slope:

$$\frac{dy}{dx} = \frac{1}{EI}\left(\frac{R_A x^2}{2} - W_1\frac{\langle x - a\rangle^2}{2} - W_2\frac{\langle x - b\rangle^2}{2} + A\right),$$

and, integrating a second time, gives the expression for the deflection:

$$y = \frac{1}{EI}\left(\frac{R_A x^3}{6} - W_1\frac{\langle x - a\rangle^3}{6} - W_2\frac{\langle x - b\rangle^3}{6} + Ax + B\right)$$

The constants of integration can be found by applying the boundary conditions. At the supports the deflection is zero, so when $x = 0$, $y = 0$ and when $x = l$, $y = 0$. From the deflection equation it can be seen that when $x = 0$, $y = 0$ and so $B = 0$.

Also, when $x = l$, $y = 0$; then again, from the deflection equation, the value of A can be found by evaluating the expression

$$A = \left(\frac{R_A l^2}{6} - W_1\frac{\langle l - a\rangle^3}{6l} - W_2\frac{\langle l - b\rangle^3}{6l}\right),$$

where the Macaulay terms in the pointed brackets will be positive and so valid.

We now have all the algebraic expressions necessary to determine the slope and deflection for any point along the beam.

2.6.4 Using the Macaulay method for uniformly distributed loads

In using the Macaulay ramp function for a uniformly distributed load (UDL) we started from a point a, an arbitrary distance from the zero datum. This is in order to comply with the requirements of the function. We then made the assumption that the UDL continued to the right-hand extremity of the beam. This has two consequences for use of the Macaulay method:

> **Key Point.** The *integration method* should be used for a *UDL that covers the entire span of the beam.* The Macaulay method is not appropriate in this case

(i) For a UDL that covers the complete span of the beam, the Macaulay method is not appropriate. If we wish to find the slope and deflection for UDLs that span the entire beam, then we need to use the successive integration method, discussed earlier.

(ii) For a UDL that starts at a but does not continue for the entire length of the beam (Figure 2.51a), we cannot directly apply the Macaulay terms to our bending moment equation, for the reason given above. We need to modify our method in some way to accommodate this type of loading. In effect, what we do is allow our UDL to continue to the extremity of the beam (Figure 2.51b) and counter this additional loading by introducing an equal and opposite loading into the bending moment expression.

Figure 2.51a shows a simply supported beam with a UDL positioned between points a and b. If we imagine that this load is extended to the end of the beam (Figure 2.51b), then our bending moment expression, as previously explained, would be:

$$M = R_A x - \omega \frac{\langle x - a \rangle^2}{2}$$

This of course is incorrect, but if we now counterbalance the increase in the UDL by a corresponding decrease over the distance $(x - b)$, and introduce this as an additional Macaulay term, we obtain the correct *Macaulay expression for the bending moment of a beam subject to a partial length UDL*, that is:

$$M = R_A x - \omega \frac{\langle x - a \rangle^2}{2} + \omega \left\langle \frac{(x - b)^2}{2} \right\rangle \qquad (2.25)$$

> **Example 2.19**   For the beam shown in Figure 2.52, determine the deflection of the beam at its centre if $EI = 100\,\text{MN m}^2$. We first find the reactions at the supports by taking moments about R_B, then
>
> $$10R_A = (30 \times 7) + (80 \times 5) + (50 \times 3)$$

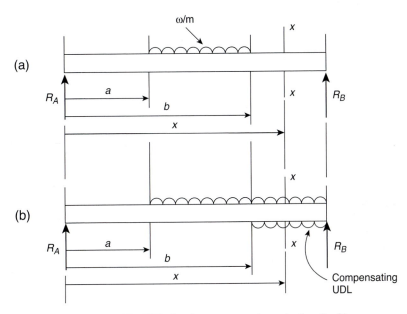

Figure 2.51 Macaulay compensation method for UDL that does not span the entire length of beam

Therefore, $R_A = 76$ kN and so $R_B = 84$ kN.

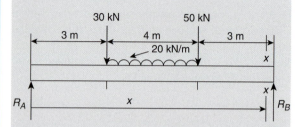

Figure 2.52 Figure for Example 2.19

Taking the origin as R_A, we apply the Macaulay method shown previously. So, by considering section $x - x$ to the extreme right of the beam and taking moments of the forces to the left of $x - x$ we get:

$$M = 10^3 \left[76x - 30 \langle x - 3 \rangle - 50 \langle x - 7 \rangle \right. $$
$$\left. - \frac{20}{3} \langle x - 3 \rangle^2 + \frac{20}{2} \langle x - 7 \rangle^2 \right]$$

where the term $+\dfrac{20}{2} \langle x - 7 \rangle^2$ is the compensating Macaulay term for the UDL. Since $\dfrac{d^2 y}{dx^2} = \dfrac{M}{EI}$, then

$$\frac{d^2 y}{dx^2} = \frac{10^3}{EI} \left[76x - 30 \langle x - 3 \rangle - 50 \langle x - 7 \rangle \right. $$
$$\left. - \frac{20}{2} \langle x - 3 \rangle^2 + \frac{20}{2} \langle x - 7 \rangle^2 \right];$$

therefore

$$\frac{dy}{dx} = \frac{10^3}{EI} \left[\frac{76x^2}{2} - \frac{30}{2} \langle x - 3 \rangle^2 - \frac{50}{2} \langle x - 7 \rangle^2 \right. $$
$$\left. - \frac{20}{6} \langle x - 3 \rangle^3 + \frac{20}{6} \langle x - 7 \rangle^3 + A \right]$$

and

$$y = \frac{10^3}{EI} \left[\frac{76x^3}{6} - \frac{30}{6} \langle x - 3 \rangle^3 - \frac{50}{6} \langle x - 7 \rangle^3 \right. $$
$$\left. - \frac{20}{24} \langle x - 3 \rangle^4 + \frac{20}{24} \langle x - 7 \rangle^4 + Ax + B \right]$$

Now, the boundary conditions yield the following values. When $x = 0$, $y = 0$, so that $B = 0$ because all the Macaulay terms are negative and equate to zero.

When $x = 10$, $y = 0$; now all the Macaulay terms are positive and substituting them into the deflection equation, with these values, we find $A = -879.3$ (you should check this value for yourself).

Now when $x = 5$ (centre of beam), then $Ax = (-879.3 \times 5) = -4396.7$ and, after eliminating the negative Macaulay terms from the expression for the deflection, we get:

$$y = \frac{10^3}{EI} \left[\frac{(76)(5)^3}{6} - \frac{30}{6}(2)^3 - \frac{20}{24}(2)^4 - 4396.7 \right],$$

giving $y = -0.0287$ m or the deflection is 28.7 mm (downwards).

1. For the situation shown in Figure 2.53, for a beam simply supported at its extremities and subject to a UDL over its whole length:

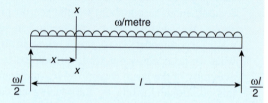

Figure 2.53 Simply supported beam subject to UDL for TYK 2.4.1

a) Using the fundamental relationship $\dfrac{d^2 y}{dx^2} = \dfrac{M}{EI}$, write down a relationship for the bending moment M_{XX}.

b) From the relationship found in part (a), determine by integration expressions for the *slope* and *deflection* of this beam.

c) Taking into consideration the boundary conditions for the situation illustrated in Figure 2.53, determine the constants of integration A and B.

d) Determine expressions for the maximum deflection $y_{\max}$ and maximum slope $\left(\dfrac{dy}{dx} \right)_{\max}$ and state where these occur.

e) Given that the beam length $l = 8\,\text{m}$, $\omega = 2\,\text{kN/m}$ and EI for the beam material is $500\,\text{kN}\,\text{m}^2$, determine values for $\left(\dfrac{dy}{dx}\right)_{\text{max}}$ and y_{max}.

2. A beam simply supported at its ends is 4 m long and carries concentrated loads of 30 kN and 60 kN at distances of 1 m and 2.5 m from the left-hand end of the beam, respectively. Given that the section modulus for the beam $EI = 16 \times 10^6\,\text{N}\,\text{m}^2$.

   a) Determine, using Macaulay's method, the slope and deflection under the 60 kN load.

   b) Determine, using an appropriate initial estimate, the position and value of the maximum deflection of the beam.

2.7   Chapter summary

In this chapter we first revised one or two fundamental concepts concerning simply supported beams, including the conditions for equilibrium and the methods used to find the reactions at their supports. We went on to discuss the nature of beam supports, where the majority of this chapter is dedicated to beams that are *simply supported*, although the *cantilever* beam has also been covered as well as mention being made of *encastred* (built-in) beams.

We then considered, in section 2.2, the nature and determination of shear forces (SF) and bending moments (BM) for a variety of beams. The SF and BM are created as a result of the flexing or bending of the beam when subject to concentrated (point load W) or uniformly distributed loads (UDL, symbol ω). The SF that *acts at a section* only step-changes (that is changes either side of the section) where a concentrated load acts normal to the beam, causing the shear. At every section in a beam carrying a concentrated load normal to the beam, there will be resultant forces either side of the section that must be equal and opposite. These *shear forces* are considered *positive when acting upwards to the left of a section and downwards to the right of a section* and *negative when reversed*. Look back at Figure 2.11 and note that the reaction force $\dfrac{2W}{3}$ is *acting*

upwards to the left of the cut, hence positive, while the reaction force $-\dfrac{W}{3}$ is *acting upwards to the right of the section* and so is *negative*.

If the beam is subject to a UDL, then the shear force changes progressively along its length until it meets the beam supports or a concentrated load, where again a step change takes place either side of the load.

The bending moment (BM or just M) at any chosen cross-section of a beam uses the principle of moments for its solution about the particular cross-section. That is, the algebraic sum of the moments of all external loads acting on the beam to *one side of the cross-section* determines the BM at that point. The sign convention for BM considers them to be *positive if they compress the top section of the beam and stretch the lower section*, i.e., they cause the beam to *sag*. Bending moments are considered *negative when the loads cause the beam to stretch on its top surface or hog*.

SF and BM diagrams may be used not only to illustrate resultant shear forces but also, by considering the area underneath SF diagrams, to determine the corresponding bending moments at particular sections or at chosen key points along the beam. Figure 2.54 shows SF and BM diagrams for some standard cases.

Figure 2.54a illustrates the situation for a beam simply supported at its ends subject to a concentrated load W at its centre. Under these circumstances the load is equally distributed between the two supports, i.e., each support takes half the load $W/2$. Note also that the area of the SF diagram at any position along the beam corresponds to the BM values at that particular position. The maximum BM (M_{max}) for this particular situation occurs at the midpoint and is $= \left(\dfrac{W}{2}\right)\left(\dfrac{l}{2}\right) = \dfrac{Wl}{4}$.

Figure 2.54b illustrates the situation for the same beam as in (a) subject this time to a UDL (ω/m) over its whole length, where in this case $M_{\text{max}} = \left(\dfrac{\omega l}{2}\right)\left(\dfrac{l}{4}\right) = \dfrac{\omega l^2}{8}$ and occurs at the centre of the beam. Note also the parabolic shape of the BM diagram for this beam, when subject to a UDL, across its whole length.

Figure 2.54c illustrates the situation for a cantilever beam subject to a concentrated load at its free end, where the SF is *upwards to the left of any section* along the beam towards the end load and so is *positive*, as indicated on the SF diagram. The BM diagram clearly shows that $M_{\text{max}} = Wl$ at the fixed end of the beam, the furthest distance (l) from the concentrated load W at the free end of the beam, tapering down to zero when the load

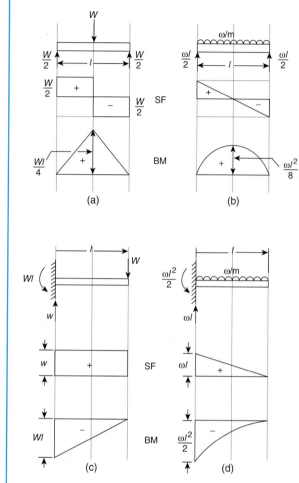

Figure 2.54 SF and BM diagrams for some standard cases

W is reached. It is also *negative* since the load causes the beam to *hog*.

Figure 2.54d illustrates the situation for the same cantilever beam subject this time to a UDL across its whole length, where again the SF is a positive maximum at the fixed support, tapering down to zero at the free end of the beam. The BM diagram again shows that it has a maximum at the fixed end of the beam, where

$$M_{\max} = \omega l \left(\frac{l}{2} \right) = \frac{\omega l^2}{2},$$

because the UDL is treated as a concentrated load that acts at the centre of the UDL. In this case the centre of the UDL is the centre of the beam. The bending moments due to the UDL cause the beam to *hog*, so they are *negative* and parabolic in nature, as shown in the diagram.

In section 3 of this chapter we looked at the development and use of engineers' simple bending

theory and established the important relationship given in equation 2.6, i.e., $\dfrac{M}{I} = \dfrac{\sigma}{y} = \dfrac{E}{R}$, where:

sigma (σ) is the longitudinal stress at a layer some distance y from the neutral axis,

y = the distance from the neutral axis,

I is the second moment of area about the neutral axis (see next section),

M is the bending moment,

E is the elastic modulus of the beam material, and

R is the radius of curvature of the beam material.

We also established, for a beam, that the maximum bending stress ($\sigma_{\max}$) occurs where the distance from the neutral axis is a maximum ($y_{\max}$) and, from equation 2.6, we found that $M_{\max} = \dfrac{I\sigma_{\max}}{y_{\max}}$. We gave the special name of *section modulus* Z to the quantity $I/y_{\max}$ that depended only on the cross-sectional area of the beam under consideration, Z being a measure of the *flexural rigidity* of the beam.

In section 4 of the chapter we established the techniques for finding the centroid (*the point where the total area is deemed to act*) and second moment of area (*I, a measure of the resistance to bending of the beam*) for some common cross-sections. Look back at Table 2.1 to see the formulae we established for finding the second moment of area about the principal axes for rectangular and circular shapes. This involved us in the use of the integral calculus to determine these second moments of area. We then found that it was sometimes useful to determine I from axes that were some perpendicular distance from the principal axes, and in order to find I about these new axes we used the parallel axis theorem $I_{BB} = I_{AA} + AS^2$, where S = the perpendicular distance from the principal axis to the axis of choice and A = the area of the shape.

Armed with the bending theory from section 3, plus the techniques needed to establish centroids and second moments of area and property tables (Table 2.2), we were able in section 5 to select a beam for a particular function, after establishing the desired parameters. In order to increase the range of beams you are able to select, you will need to use the reference sources given earlier, i.e. BS 5950-1:2000 and/or www.technouk.com, or others.

In section 6, we looked at the analytically rather difficult subject of the *slope* and *deflection* of beams. Figure 2.55 shows pictorially the slope and deflection of a beam (very much exaggerated), subject to a UDL across its whole length.

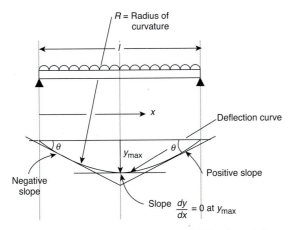

Figure 2.55 Illustration of the slope and deflection of a beam

It can clearly be seen from Figure 2.55 that the *slope* lines have a gradient equal to the tangent of the angle θ, therefore $\theta = \dfrac{dy}{dx}$ and at $y_{\max}$ this angle $\theta = 0$, so the slope at $y_{\max} = \dfrac{dy}{dx} = 0$. Note also that the slope of the beam goes from negative to positive from the left-hand side to the right-hand side of the beam, this change in sign of the slope indicating that there is a local minimum that in turn indicates a negative maximum for the deflection, as the deflection is in the negative y-direction. The above argument leads us to the very important relationship $\dfrac{d^2y}{dx^2} = \dfrac{M}{EI}$ (equation 2.12), where for beams of uniform cross-section the bending moment M and the section modulus $Z = EI$ are directly related to the differential of the slope of the beam in terms of the deflection y and function x (see argument at the beginning of Chapter 6).

Together with equation 2.12 and our knowledge of the calculus we were able to use the integration method to solve problems involving the slope and deflection of beams, where the following relationships were established:

- deflection = y,
- slope = $\dfrac{dy}{dx}$ (measured by θ in radians or by mm/m),
- bending moment $M = EI\dfrac{d^2y}{dx^2}$ and
- shear force $SF = EI\dfrac{d^3y}{dx^3}$

The integration method is all about solving the above relationships that involve differential equations, using direct integration.

We built up a number of *standard cases* using integration for the slope and deflection of beams, shown in Table 2.3, and in order to extend the use of these standard cases, we looked at the *Principle of superposition.* Look back now to remind yourself of this very useful method for finding the slope and deflection of beams subject to a variety of standard loading situations.

The remainder of section 5 was given over to Macaulay's method for determining the slope and deflection of beams by determining (using the Macaulay function) expressions for the bending moment of the entire beam, such as that given in equation 2.24. Figure 2.56 shows the nature of the Macaulay function.

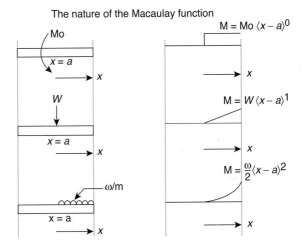

Figure 2.56 The nature of the Macaulay function

It can clearly be seen that this ramp function of x is of the form $f_n(x) = \langle x - a \rangle^n$, such that for $x \langle a, f_n(x) = 0$ and when $x \rangle a, f_n(x) = (x - a)^n$. What this mathematical shorthand is really saying is that when the value of the variable x is less than the value of a, or is equal to a, the Macaulay term equates to zero; also that when x is greater than a, then the (positive) Macaulay term *is valid* and the normal rules of algebra apply.

Note also that when determining the Macaulay expression for distributive loads *that do not extend to the end of the beam* we must use the *modified Macaulay expression* (see equation 2.25), where an extra term is introduced to counterbalance the increase in the UDL needed to reach the end of the beam.

Section 2.5 only serves as an introduction to the two basic methods (successive integration and Macaulay's method) for solving slope and deflection problems. Many cases have been omitted in the interests of space, but it is hoped that the knowledge gained here will act as a sufficient foundation for those wishing to pursue the subject further.

2.8 Review questions

1. Draw the shear force (SF) and bending moment (BM) diagrams for the beam shown in Figure 2.57 and find the position and value of the maximum bending moment.

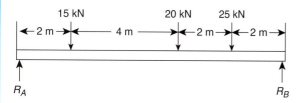

Figure 2.57 Beam with concentrated loads for review question 2.8.1

2. Draw the SF and BM diagrams for the beam shown in Figure 2.58 and find the position and value of the maximum bending moment.

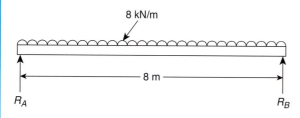

Figure 2.58 Beam with UDL for review question 2.8.2

3. Draw the shear force (SF) diagram for the beam shown in Figure 2.59 and identify significant SF values.

4. Find the position and magnitude of the maximum BM for question 3.

5. A simply supported beam (Figure 2.60) carries the concentrated load and UDL as shown.

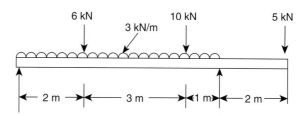

Figure 2.59 Beam system for review question 2.8.3

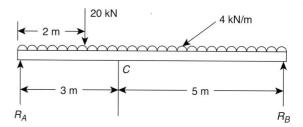

Figure 2.60 Beam system for review question 2.8.5

a) Construct the SF and BM diagrams.

b) Determine the maximum bending moment.

c) Given that the beam has a section modulus of 1200 cm³, determine the allowable bending stress, ignoring the weight of the beam.

6. For the beam section shown in Figure 2.61:

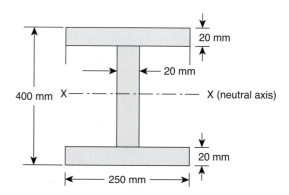

Figure 2.61 Beam section for review question 2.8.6

a) Calculate the second moment of area about the neutral X–X axis.

b) Using simple bending theory, determine the value of the section modulus Z for the beam.

7. A cantilever beam rigidly supported at one end is 3 m long and has a concentrated load of 30 kN acting downwards at its free end. If the beam is made from

steel with an elastic modulus $E = 210\,\text{GPa}$ and the maximum deflection $y_{\text{max}} = 2.0\,\text{mm}$, determine:

a) The second moment of area for the beam

b) The maximum slope.

8. For the built-in, rigidly supported (encastred) beam shown in Figure 2.62, subject to a central concentrated load W:

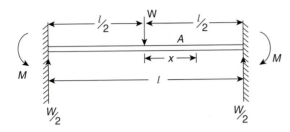

Figure 2.62 Rigidly supported beam for review question 2.8.8

a) Using the fundamental relationship (equation 2.12), produce an expression for the bending moment at point A.

b) Using successive integration and the appropriate boundary conditions, show that the maximum bending moment $M_{\text{max}} = \dfrac{Wl}{8}$ and the maximum deflection $y_{\text{max}} = -\dfrac{Wl^3}{192EI}$.

9. A beam of length 6 m is simply supported at its ends and carries a distributed load $\omega = 10\,\text{kN/m}$ that commences 1 m from the left-hand end of the beam and continues to the right-hand end of the beam. Two point loads of 30 kN and 40 kN act 1 m and 5 m, respectively, from the left-hand end. If the beam material is made of steel with $EI = 20 \times 10^6\,\text{N m}^2$, determine the deflection at the centre.

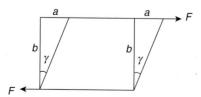

Chapter 3

Torsion and shafts

Drive shafts for pumps and motors, propeller shafts for motor vehicles and aircraft as well as pulley assemblies and drive couplings for machinery, are all subject to *torsion* or *twisting* loads. These torsion loads set up shear stresses within the shafts. Engineers need to be aware of the nature and magnitude of these torsion loads and shear stresses in order to design against premature failure and to ensure safe and reliable operation during service.

3.1 Review of shear stress and strain

In order to study the theory of torsion it is necessary to review the fundamental properties of shear stress and strain. We looked at this subject in Chapter 1 but we present it again here in just a little more depth, since shear stress and strain result when shafts are subject to torsion.

If two equal and opposite parallel forces, not acting in the same straight line, act on a body, then that body is said to be loaded in *shear* (Figure 3.1). Shear stress, as you may already have seen, is defined in a similar

way to tensile stress, except that the area in shear acts parallel to the load, so:

$$\text{Shear stress} = \frac{\text{Load causing shear } (F)}{\text{Area in shear } (A)} \quad (3.1)$$

Figure 3.1 illustrates the phenomenon known as *shear strain*. With reference to this figure, shear strain may be defined in two ways as:

- the distance through which the material has been deformed (*a*) divided by the distance between the two shearing forces (*b*), or
- by the *angle of deformation* γ in radians.

Noting also that $\tan \gamma = \dfrac{a}{b}$, the relationship between shear stress and shear strain is again similar to that of tensile stress and strain. In the case of shear the modulus is known as *the modulus of rigidity* or *shear modulus* (*G*), and is defined as:

$$G = \frac{\text{Shear stress } (\tau)}{\text{Shear strain } (\gamma)} \quad (3.2)$$

The modulus of rigidity (*G*) measures the shear stiffness of the body, in a similar way to the elastic modulus. The units are the same, *G* being measured in N/m². Riveted joints are often used where high shear loads are encountered, while bolted joints are often used where the primary loads are tensile.

Key Point. The angle of deformation γ in radians provides a measure of the shear strain

Figure 3.1 Shear strain and the angle of deformation

978-1-85617-775-7, Engineering Science, Mike Tooley & Lloyd Dingle

Example 3.1   Figure 3.2 shows a bolted coupling in which the two 12 mm securing bolts act at 45° to the axis of the load. If the pull on the coupling is 80 kN, calculate the direct and shear stresses in each bolt.

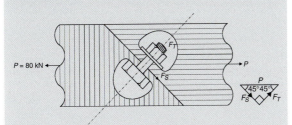

Figure 3.2 Bolted coupling subject to a tensile load

We know that the axis of the bolts is at 45° to the line of action of the load $P = 80$ kN. We will assume that the bolts are ductile and so the load is shared equally between the bolts. Then the shear area on each bolt is $= \dfrac{\pi(12)^2}{4} = 113.1$ mm^2 and the shear force (F_s) from the diagram $= P \sin 45 = (80 \times 10^3)(0.7071) = 56568.5$ N. This shear force is shared between the two bolts. Thus, shear force on each bolt $= \dfrac{56568.5}{2 \times 113.1} = 250.08$ N/mm^2.

Similarly, the direct tensile force $= P \cos 45 = (80 \times 10^3)(0.7071) = 56568.5$ N. Again, the direct tensile force on each bolt is $= \dfrac{56568.5}{2 \times 113.1} = 250.08$ N/mm^2. Thus, as would be expected by the symmetry of bolt coupling and the direction of the forces acting on it, both the direct and shear stresses have a value of 250 MN/m^2.

3.2   Engineers' theory of torsion

Shafts are the engineering components that are used to transmit torsional loads and twisting moments or torque. They may be of any cross-section but are often circular, since this cross-section is particularly suited to transmitting torque from pumps, motors and other power supplies used in engineering systems.

We have been reviewing shear stress because if a uniform circular shaft is subject to a torque (twisting moment) then it can be shown that every section of the shaft is subject to a state of pure shear. In order to help us derive the engineers' theory of torsion that relates torque, the angle of twist and shear stress, we must first make the following fundamental assumptions:

(i)   The shaft material has uniform properties throughout.

(ii)   The shaft is not stressed beyond the elastic limit, in other words it obeys Hooke's law.

(iii)   Each diameter of the shaft carries shear forces which are independent of, and do not interfere with, their neighbours'.

(iv)   Every cross-sectional diameter rotates through the same angle.

(v)   Circular sections which are radial before twisting are assumed to remain radial after twisting.

Key Point.   Engineers' theory of torsion is based on the assumption that the shaft is not stressed beyond its elastic limit

Let us first consider torsion from the point of view of the *angle of twist*. Figure 3.3 shows a circular shaft of radius R that is firmly fixed at one end and at the other is subject to a torque (twisting moment) T.

Imagine that on our shaft we have marked a line of length L which, when subjected to torque T, twists from position p to position q. The angle γ is the same angle of deformation as we identified in Figure 3.1 and is therefore a measure of the *shear strain* of our shaft.

Then, from basic trigonometry: arc length $pq =$ the radius R multiplied by θ, the *angle of twist* seen in cross-section. It also follows that the arc length $pq =$ the length of the line L multiplied by the *angle of distortion* γ (the strain angle in radians), so arc length (*the linear measure of the strain*) $pq = R\theta = L\gamma$. So, from the dimensions of this equation, the *angle of twist* θ must also be measured in *radians* while *the angle of distortion* is given by:

$$\gamma = \frac{R\theta}{L} \tag{3.3}$$

Also, from the definition of the modulus of rigidity, we know that the shear modulus $G = \dfrac{\text{Shear stress } (\tau)}{\text{Shear strain } (\gamma)}$, or that

$$\gamma = \frac{\tau}{G} \tag{3.4}$$

Then, combining equations 3.3 and 3.4 and rearranging gives the relationship:

$$\frac{\tau}{R} = \frac{G\theta}{L} \tag{3.5}$$

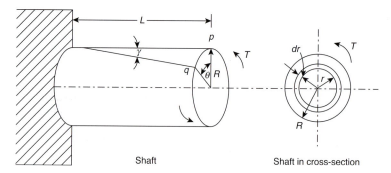

Figure 3.3 Circular shaft subject to torque

The above relationship (3.5) is independent of the value of the radius R, so that any intermediate radius emanating from the centre of the cross-section of the shaft may be considered. The related shear strain can then be determined for that radius.

With a little more algebraic manipulation we can also find expressions that relate the *shear stresses* developed in a shaft subject to pure torsion. Their values are given by equations 3.4 and 3.5. If these two equations are combined and rearranged we have:

$$G\theta = \tau = \frac{G\theta R}{L} \qquad (3.6)$$

Equation 3.6 is useful because it relates the shear stress and shear strain with the angle of twist per unit length. It can also be seen from this relationship that the shear stress is directly proportional to the radius, with a maximum value of shear stress occurring at the outside of the shaft at radius R. Obviously the shear stress at all other values of the radius will be less. Other values of the radius apart from the maximum are conventionally represented by lower case r (see Figure 3.3).

Figure 3.4 shows the cross-section of a shaft subject to torsion loads, with the corresponding shear stress distribution, which increases as the radius increases.

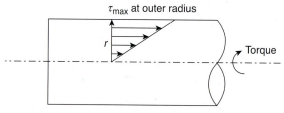

Figure 3.4 Shear stress distribution for a shaft subject to a torque

Although not part of our study at this time, it should be noted that the shear stresses shown in Figure 3.4 have complementary shears of equal value running normal to them, i.e. parallel with the longitudinal axis of the shaft.

We will now derive the relationship between the stresses and the **torque** that is imposed on a circular shaft. This will involve the simple use of the integral calculus. Figure 3.3 shows the cross-section of our shaft, which may be considered as being divided into minute parts or *elements*, these elements having radius r and thickness dr. From elementary theory we know that the force set up on each element = stress × area. The *area* of each of these cross-sectional elements of thickness dr at radius (approximately) equal to r is $2\pi r\,dr$ and, if τ is the shear stress at radius r, then the shear force on the element is $2\pi r\,dr\,\tau$. So, since the force will produce a moment about the centre axis of the shaft that contributes to the torque, the torque carried by the element will be $2\pi r^2 \tau\,dr$. The total torque (T) on the section will then be the sum of all such contributions across the section of the shaft. However, before we can find the sum of these elements, i.e. evaluate $T = \int_{0}^{R} 2\pi r^2 \tau\,dr$, we need to replace the shear stress τ with an expression in terms of the radius, because the shear stress varies with the radius across the section. Then, using equation 3.6, where $\tau = \dfrac{G\theta R}{L}$, we get that when $R = r$ $T = \int_{0}^{R} 2\pi \dfrac{G\theta}{L} r^3 dr$

or $T = \dfrac{G\theta}{L} \int_{0}^{R} 2\pi r^3 dr$. When expressed in this way,

$\int_{0}^{R} 2\pi r^3 dr = J$ the *polar second moment of area*. Thus,

$T = \dfrac{G\theta J}{L}$, so

$$\frac{T}{J} = \frac{G\theta}{L} \qquad (3.7)$$

Now, combining equations 3.5 and 3.7 gives the relationship known as *the engineers' theory of torsion*

$$\frac{T}{J} = \frac{\tau}{R} = \frac{G\theta}{L} \qquad (3.8)$$

Note that when $r = R$, that is when the radius is a maximum for a particular shaft, then, from equation 3.8, the maximum torque may be found using $T = \frac{\tau J}{R}$.

In practice, it has been shown that the engineers' theory of torsion, based on the assumptions given earlier, shows excellent correlation with experimental results. So, although the derivation of equation 3.8 has been rather arduous, you will find it very useful when dealing with problems related to torsion!

3.3   Polar second moment of area

We will now consider in a little more detail the **polar second moment of area J**, identified above, that is

$$J = \int\limits_{0}^{R} 2\pi r^3 dr.$$

One solution of this particular integral between these limits gives the *polar second moment of area for a solid shaft* as

$$J = \frac{2\pi R^4}{4} = \frac{\pi R^4}{2} \quad \text{or} \quad J = \frac{\pi D^4}{32} \qquad (3.9)$$

The polar second moment of area, defined above, is a measure of the resistance to bending of a shaft. The polar second moment for a hollow shaft (Figure 3.5) is analogous to that for a solid shaft, except that the shear area is treated as an annulus.

Then, from Figure 3.5b, the *polar second moment of area for a hollow shaft* is given by $J = \int_{r}^{R} 2\pi r^3 dr$, so that

$$J = \frac{\pi}{2}\left(R^4 - r^4\right) \quad \text{or} \quad J = \frac{\pi}{32}\left(D^4 - d^4\right) \quad (3.10)$$

When the difference between the diameters is very small, as in a very thin-walled hollow shaft, the errors encountered on subtraction of two very large numbers close together prohibits the use of equation 3.10. We then have to use an alternative relationship which measures the polar second moment of area for an individual element. For very thin-walled hollow shafts this is a

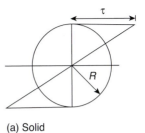

(a) Solid

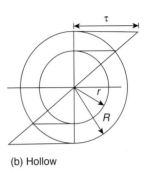

(b) Hollow

Figure 3.5 Shear stress distribution for a solid and a hollow shaft subject to torsion

much better approximation to the real case, that is, *the polar second moment of area for a very thin-walled hollow cylinder (where $t = dr$) is approximately*

$$J = 2\pi r^3 t \qquad (3.11)$$

Key Point. When establishing the polar second moment of area for a very thin-walled hollow cylinder, a good approximation is given by $J = 2\pi r^3 t$, where $t = dr$

Example 3.2   A solid circular shaft 40 mm in diameter is subjected to a torque of 800 N m.

a)   Find the maximum stress due to torsion.

b)   Find the angle of twist over a 2 m length of shaft, given that the modulus of rigidity of the shaft is 60 GN/m².

a)   The maximum stress due to torsion occurs when $r = R$, that is at the outside radius of the shaft. So in this case $R = 20$ mm. Using the standard relationship $\frac{T}{J} = \frac{\tau}{R}$, we have the values of R and T, so we only need to find the value of J for our solid shaft then we will be able to find the maximum value of the shear stress τ_{max}. For a solid circular shaft $J = \frac{\pi D^4}{32}$, so $J = \frac{\pi (40)^4}{32} = 0.251 \times 10^6$ mm^4 and, on substitution into the standard relationship,

$$\tau_{max} = \frac{(25)(800 \times 10^3)}{0.251 \times 10^6} = 79.7 \text{ N/mm}^2. \text{ This}$$

value is the maximum value of the shear stress, which occurs at the outer surface of the shaft. Notice the manipulation of the units. Care must always be taken to ensure consistency of units, especially where powers are concerned!

b) To find θ we again use the engineers' theory of torsion, which, after rearrangement, gives $\theta = \frac{LT}{GJ}$. Substituting our values for L, T, J and G we get $\theta = \frac{(2000)(800 \times 10^3)}{(60 \times 10^3)(0.251 \times 10^6)} = \frac{(\text{mm})(\text{N mm})}{(\text{N mm}^{-2})(\text{mm}^4)} = 0.106$ rad or angle of twist

$$= 6.07°.$$

Please note the careful manipulation of units.

3.4   Power transmitted by a shaft

The power transmitted by shafts is an important topic, especially when considering motor vehicle drive shafts and the power needed by rotating machines when subject to torque loads. One very useful definition of power for a rotating shaft carrying a torque relates this torque to the angular velocity of the shaft, so that the power transmitted by a shaft in Watts is given by

Power (Watts) = torque (T) $\times$ angular velocity (ω in rad/s), so that

$$\text{Power (Watts)} = T\omega \qquad (3.12)$$

Sometimes the angular velocity is given in revolutions per minute (rpm or rev/min) and the power is given in the old measure of horsepower, particularly brake horsepower when quoting performance figures for motor vehicles. Under these circumstances, for consistency of units, you should note that $T\omega = 2\pi nT$ Watts, where n is the angular velocity in revolutions per second (rev/s). To convert horsepower (hp) to Watts we multiply by 745.7, that is 1 hp = 745.7 W.

Example 3.3   Calculate the power that can be transmitted by a hollow circular propshaft, if the maximum permissible shear stress is 60 MN/m^2 and it is rotating at 100 rpm. The propshaft has an external diameter of 120 mm and an internal diameter of 60 mm.

Again, we use the engineers' theory of torsion and note that the maximum shear stress (60 MN/m^2) will be experienced on the outside surface of the propshaft, where $R = 60$ mm.

Then, using $T = \frac{\tau J}{R}$ where $J = \frac{\pi}{32}(D^4 - d^4)$,

$J = \frac{\pi}{32}(120^4 - 60^4) = 19.09 \times 10^6$ N mm, and

so torque $T = \frac{(60)(19.09 \times 10^6)}{60} = 19.09 \times 10^6$ N mm or $T = 19.09 \times 10^3$ N m. Now, the angular velocity (ω) in rad/s is $= \frac{2\pi 100}{60} = 10.47$ rad/s and we know that power $P = T\omega$, so the maximum power transmitted by the shaft $P_{max} = (19.06 \times 10^3)(10.47) = 199.6$ kW.

Example 3.4   Calculate the diameter of a shaft that will transmit 60 kW of power at 180 rpm, given that the shear stress is to be limited to 60 MN/m^2 and the twist of the shaft is not to exceed 0.02 radian per 2 metre length of shaft. The shear modulus of the shaft material $G = 80$ GN/m^2.

The torque resulting from the power to be transmitted is $T = \frac{P}{\omega} = \frac{60000}{2\pi \times 3} = 3183$ N m. Now we need to look at the size of shaft required to meet both the shear stress and angle of twist requirements. From equation 3.8, we find that the maximum shear stress is given by the relationship

$\tau_{max} = \frac{TR}{J}$, where $R =$ the full radius of the shaft and $J = \frac{\pi R^4}{2}$, so that $\tau_{max} = \frac{2TR}{\pi R^4} = \frac{2T}{\pi R^3}$.

Then $\tau_{max} = \frac{2TR}{\pi R^4} = \frac{2T}{\pi R^3}$ and $R^3 = \frac{2T}{(\pi)(\tau_{max})} = \frac{(2)(3183) \text{ N m}}{(\pi)(60 \times 10^6) \text{ N m}^{-2}} = 0.00003377$ m^3, so that $R = 0.03232$ m $= 32.32$ mm on a stress basis.

Now, radius based on the angle of twist may be found again from equation 3.8 as $\frac{T}{J} = \frac{G\theta}{L}$,

where $J = \dfrac{TL}{G\theta}$ or, $\dfrac{\pi R^4}{2} = \dfrac{TL}{G\theta}$ and so $R^4 =$
$\dfrac{2TL}{G\theta} = \dfrac{(2)(3183)}{(80 \times 10^9)(0.02)} = 0.000003979$ or $R =$
$0.04466\,\text{m} = 44.66\,\text{mm}$, based on angle of twist. The shaft must be made to this dimension to meet the criteria, so diameter of shaft must be a minimum of 8.93 cm.

We have, so far, considered solid and hollow shafts separately and made of just one material. In the final section of this chapter we consider shafts of differing diameters and differing materials along their length i.e., composite shafts.

3.5 Composite shafts

In certain composite drive shafts, due to the physical coupling of components, the complete shaft, although made from the same material, may be made up from two or more shafts of varying diameter along its length, or it may differ geometrically in some other way. Alternatively, the shaft may also differ in material as well as in geometry. Figure 3.6 illustrates the set-up for two shaft sections rigidly joined or machined together and made from the same material *and carrying the same torque*. For this arrangement we say that the shafts are joined in *series*.

In this arrangement the total torque T transmitted by each section of the shaft is equal, i.e. $T = T_1 = T_2$, and by equating these torques and using our standard equation 3.8, where $\dfrac{T}{J} = \dfrac{\tau}{R} = \dfrac{G\theta}{L}$, we obtain the very

useful relationship that

$$T = \frac{G_1 J_1 \theta_1}{L_1} = \frac{G_2 J_2 \theta_2}{L_2} \qquad (3.13)$$

From equation 3.13, we can also find relationships in terms of the torque and relative dimensions, such as the surface radii (R_1, R_2) and the surface (maximum) shear stresses (τ_1, τ_2). So, again using equation 3.8 (the engineers' theory of torsion), we may write that $T = \dfrac{\tau_1 J_1}{R_1} = \dfrac{\tau_2 J_2}{R_2}$. Then, after substitution of the formula for the polar second moment of area for a solid shaft, $\dfrac{\tau_1 \pi R_1^4}{2 R_1} = \dfrac{\tau_2 \pi R_2^4}{2 R_2}$, and on simplification, we get

$$\frac{\tau_1}{\tau_2} = \left(\frac{R_2}{R_1}\right)^3 \qquad (3.14)$$

For the set-up shown in Figure 3.6, where the sections of the shaft are in series and subject to the same torque, the total angle of twist of the shaft (θ) will be the sum of the two different twist angles for each section, i.e. $\theta = \theta_1 + \theta_2$. So we may again find some useful relationships concerning the angle of twist, dimensions and torque by using the engineers' theory of torsion. Thus from equation 3.8 we find that

$$\theta = \frac{TL_1}{GJ_1} + \frac{TL_2}{GJ_2} \text{ and so } \theta = \frac{T}{G}\left(\frac{L_1}{J_1} + \frac{L_2}{J_2}\right) \quad (3.15)$$

We also know from the theory that $\theta_1 = \dfrac{\tau_1 L_1}{R_1 G}$ and $\theta_2 = \dfrac{\tau_2 L_2}{R_2 G}$, so that $\theta = \left(\dfrac{\tau_1 L_1}{R_1 G} + \dfrac{\tau_2 L_2}{R_2 G}\right)$, or that the total twist of the shaft is given by

$$\theta = \frac{1}{G}\left(\frac{\tau_1 L_1}{R_1} + \frac{\tau_2 L_2}{R_2}\right) \qquad (3.16)$$

Key Point. The torque acting on all sections of a composite shaft joined in series is equal, i.e. $T = T_1 = T_2 = \cdots T_n$

An alternative set-up for a composite shaft is shown in Figure 3.7, where shafts consisting of two materials

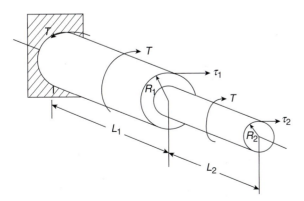

Figure 3.6 Shaft with two different diameter sections

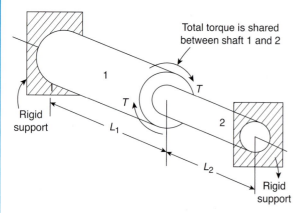

Figure 3.7 Composite shaft arrangement where the applied torque is shared

and/or two different diameters are rigidly fixed together and at each end, so that the applied torque is *shared* between them, as shown.

The composite shaft shown in Figure 3.7 has its components connected to each other in parallel between rigid supports, so this arrangement is often referred to as being just that, connected in *parallel*.

Key Point.   The torque acting on all sections of a composite shaft joined in parallel is shared i.e.
$T = T_1 + T_2 + \cdots T_n$

When connected in this way, the total torque $T = T_1 + T_2$, i.e. it is shared between components. Also, by virtue of the components being rigidly fixed together, the angle of twist of each component must be equal, so that $\theta = \theta_1 = \theta_2$. Thus, in a similar way to shafts connected in series, we can find some useful relationships connecting torque, angle of twist and dimensions for shafts connected in parallel. So, from engineers' theory of torsion, we get that

$$T = \frac{G_1 J_1 \theta_1}{L_1} + \frac{G_2 J_2 \theta_2}{L_2} \qquad (3.17)$$

In terms of the maximum shear stress, $T = \frac{\tau_1 J_1}{R_1} + \frac{\tau_2 J_2}{R_2}$.

Also, we find from $\theta = \theta_1 = \theta_2$ that

$$\theta = \frac{T_1 L_1}{GJ_1} = \frac{T_2 L_2}{GJ_2} \qquad (3.18)$$

Example 3.5   Figure 3.8 shows a composite circular shaft 2 m long rigidly fixed at both ends, consisting of one section made from a light alloy 1.2 m long with a diameter of 60 mm rigidly joined to another section made from steel 0.8 m long with a diameter of 40 mm.

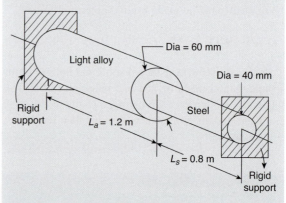

Figure 3.8 Composite circular shaft

If for the steel $\tau_{max} = 60\,MN/m^2$, determine the maximum torque that can be applied at the joint, given that the shear modulus for the steel $G_s = 80\,GN/m^2$ and for the light alloy the shear modulus is $G_a = 30\,GN/m^2$.

The set-up shown in Figure 3.8 shows that the two sections of the shaft are connected in parallel and the torque acting at the joint is shared between the two components of the shaft, so that $T = T_a + T_s$. Also, the angle of twist must be the same for each section at the joint where the applied torque acts, so $\theta_a = \theta_s$. Then, from $T_s = \frac{\tau_s J_s}{R_s}$, the torque in the steel

where $J_s = \frac{\pi R^4}{2}$ is given by $T_s = \left(\frac{\tau_s}{R_s}\right)\left(\frac{\pi R^4}{2}\right)$

or $T_s = \left(\frac{60 \times 10^6}{0.02}\right)\left(\frac{\pi (0.02)^4}{2}\right) = 754\,N\,m.$

The angle of twist for the steel is given by

$$\theta_s = \frac{T_s L_s}{G_s J_s} = \left(\frac{754}{80 \times 10^9}\right)\left(\frac{0.8}{(\pi)\left((0.02)^4 / 2\right)}\right) =$$

0.03 rad.   Therefore $\theta_a = 0.03\,rad$ and

so,   from $T_a = \dfrac{\theta_a G_a J_a}{L_a}$,   we get $T_a =$

$$\frac{(0.03)\,(30 \times 10^9)\,(\pi \times (0.03)^4)}{(1.2)(2)} = 954.3\,N\,m.$$

Therefore, the total torque that can be applied at the joint is $T = T_a + T_s = 954.3 + 754 = 1798.3\,N\,m.$

3.6   Chapter summary

In this chapter we started in section 3.1 by revising the nature of shear stress and strain as well as looking at how shear stresses may stem from components that have only been subject to longitudinal loads. Shear stress may be caused by the orientation of these components, for example the bolted coupling in Example 3.1. We look again at shear stress (τ) and strain (γ) and the shear modulus or modulus of rigidity (G) because of their importance to our study of torsion theory and its application to shafts.

Section 3.2 was given over entirely to our derivation of the *engineers' theory of torsion*, where we first made some fundamental assumptions that placed limits on the range and use of the theory. These assumptions are repeated here, as a reminder:

(i)   The shaft material has uniform properties throughout.

(ii) The shaft is not stressed beyond the elastic limit, in other words it obeys Hooke's law.

(iii) Each diameter of the shaft carries shear forces which are independent of, and do not interfere with, their neighbours'.

(iv) Every cross-sectional diameter rotates through the same angle.

(v)   Circular sections which are radial before twisting are assumed to remain radial after twisting.

Note in particular from the above assumptions that we are dealing with shafts that are only subject to stresses within their *elastic* limits. The important relationship known as engineer's theory of torsion was developed, the results being summarised as equation 3.8, i.e.

$$\frac{T}{J} = \frac{\tau}{R} = \frac{G\theta}{L}$$

where T = the torque or twisting moment (N m), J = polar second moment of area (m^4), τ = shear stress (N/m^2), R = radius of cross-section (m), G = shear modulus (GN/m^2), θ = angle of twist (radians) and L = length of shaft (m). Note also that the shear stress created by a torque acting on a shaft will be a maximum at the external surface of the shaft, i.e. τ_{max} occurs when $r = R$.

In section 3 the polar second moment of area (J) was explained in more detail. Do remember that J is *a measure of the resistance to bending* of the shaft and

that the combination GJ is referred to as the *torsional rigidity* of the shaft. We looked at the definitions of J for two particular circular shaft cross-sections, where for *solid shafts* $J = \dfrac{\pi R^4}{2}$ or $J = \dfrac{\pi D^4}{32}$ and for *hollow shafts* $J = \dfrac{\pi}{2}\left(R^4 - r^4\right)$ or $J = \dfrac{\pi}{32}\left(D^4 - d^4\right)$. You should also have noted that for very thin-walled hollow shafts the arithmetic errors created from the calculation $\left(D^4 - d^4\right)$ when D and d are very close in value prohibits its use. Under these circumstance, an alternative approximation using $J = 2\pi r^3 t$ (where $t = dr$) is recommended.

The main reason for our study of circular shafts was considered in section 4. This refers to their ability to transmit power from prime movers, such as from motors to gearboxes or generators or to other services. The important formula relating power with torque and angular velocity was introduced, i.e. *power* $= T\omega$, where the power in *Watts* is given by the product of the torque (N m) and the angular velocity (rad/s).

Also, note how we deal with angular velocities when they are given in rpm or in revs/s, i.e. $T\omega = 2\pi nT$, where n is the angular velocity in rev/s. Note also how to convert horse power (hp) into Watts using the fact that 1 hp = 745.7 W.

In section 5, *composite shafts* were introduced, treating a composite shaft as one that may be made up from parts that vary in cross-section, diameter and/or materials. Two types of composite shaft were identified, based on the way they were connected and/or secured. Those *joined in series* (Figure 3.6) *carry the same torque*, so that for a series-joined shaft with two sections $T = T_1 = T_2$. This is also true for series-joined shafts with *more than two sections*. The *total angle of twist for a series*-joined composite beam *will equal the sum of the individual sections* that go to make up the beam, i.e. $\theta = \theta_1 + \theta_2 + \cdots\cdots\theta_n$. Remember also that these individual sections may vary geometrically, or may vary physically by virtue of the material from which they have been manufactured, irrespective of the way they have been joined. Figure 3.7 illustrates a composite shaft made up from two sections that have been joined in such a way that *they share the applied torque* ($T = T_1 + T_2$), i.e. they are joined in *parallel*. Under this method of joining and application it can be seen from Figure 3.7 that *the angle of twist in each section is equal*, that is, $\theta = \theta_1 = \theta_2$ etc.

For both series- and parallel-joined composite shafts, a number of equations were formulated using engineers' theory of torsion, to enable torques, angles of twist, shear stresses and other useful parameters to be determined.

3.7   Review questions

1. Find the maximum torque that can be transmitted by a solid drive shaft 75 mm in diameter, if the maximum allowable shear stress is 80 MN/m^2.

2. Calculate the maximum power that can be transmitted by a solid shaft 150 mm in diameter when rotating at 360 rpm, if the maximum allowable shear stress is 90 MN/m^2.

3. A solid shaft rotating at 140 rpm has a diameter of 80 mm and transmits a torque of 5 kN m. If $G = 80$ GN/m^2, determine the value of the maximum shear stress and the angle of twist per metre length of the shaft.

4. A circular hollow shaft has an external diameter of 100 mm and an internal diameter of 80 mm. It transmits 750 kW of power at 1200 rpm. Find the maximum and minimum shear stress acting on the shaft and determine the angle of twist over a 2 m length. Take $G = 75$ GN/m^2.

5. A steel transmission shaft is 0.5 m long and 60 mm in diameter. For the first part of its length it is bored to a diameter of 30 mm and for the rest of its length to a diameter of 40 mm. If the shear stress is not to exceed 60 MN/m^2, find:

    a) The least value of the polar second moment of area J,

    b) The maximum allowable torque, and

    c) The maximum power that may be transmitted at a speed of 300 rpm.

Chapter 4

Pressure vessels

Cylinders and spheres that act as pressure vessels are divided into two major categories, thin-walled and thick-walled. We start our study in this chapter by considering thin-walled, mainly cylindrical pressure vessels that may be used, for example, as compressed air containers, boilers, submarine hulls, aircraft fuselages, condenser casings and hydraulic reservoirs. We then take a look at thick-walled vessels subject to pressure. These, for example, function as hydraulic linear actuators, extrusion dies, gun barrels and high-pressure gas bottles.

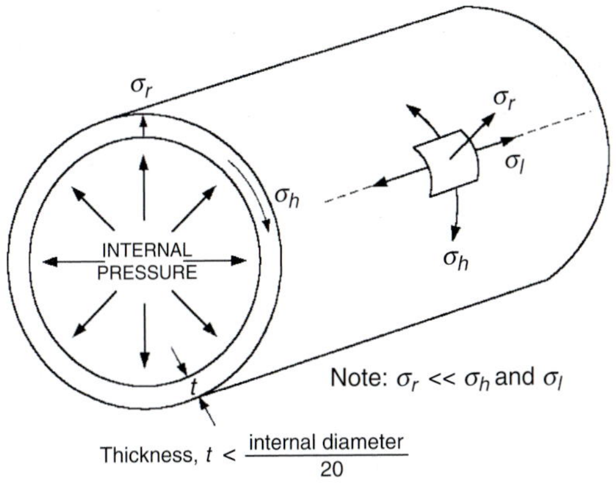

Figure 4.1 Vessel subject to internal pressure

4.1 Thin-walled pressure vessels

When a thin-walled pressure vessel is subject to internal pressure, three mutually perpendicular principal stresses are set up in the vessel material. These are the *hoop stress* (often referred to as the circumferential stress), the *radial stress* and the *longitudinal stress*. Figure 4.1 shows a vessel subject to internal pressure and the nature of the resulting hoop, longitudinal and radial stresses.

We may define a thin-walled pressure vessel as one in which the ratio of the wall thickness to that of the inside diameter of the vessel is greater than 1:20, that is less than the fraction 1/20. Under these circumstances it is reasonable to assume that the hoop and longitudinal stresses are uniform across the wall thickness and that the radial stress is so small in comparison with the hoop

and longitudinal stresses that for the purpose of our calculations it may be ignored.

> **Key Point.** A thin-walled pressure vessel is defined as one in which the ratio of the wall thickness to that of the inside diameter of the vessel is greater than 1:20

4.1.1 Hoop and longitudinal stress

Hoop or *circumferential stress* is the stress that resists the bursting effect of the applied internal pressure (Figure 4.2a).

If the internal diameter is d, the applied pressure is p, the length is L and the wall thickness is t, then the force tending to burst or separate the vessel is given by pressure × internal diameter × length, or that force = pdL. This is resisted by the hoop stress (σ_h) acting on an

978-1-85617-775-7, Engineering Science, Mike Tooley & Lloyd Dingle

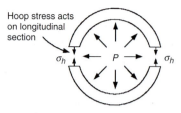

Internal pressure resisted by hoop stress

(a)

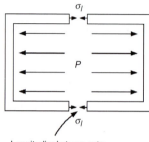

Longitudinal stress acts
on diametrical section

Internal pressure resisted by longitudinal stress

(b)

Figure 4.2 Hoop and longitudinal stresses resulting from internal pressure

area $2tL$, i.e. the area of both the top and bottom sections. For equilibrium these values may be equated, that is:

$$\text{Hoop stress} = \frac{F}{A}, \text{ so } F = \sigma_h \times A \text{ or } F = \sigma_h \times 2tL;$$

also, from above, $F = pdL$, then $pdL = 2\sigma_h tL$, so

$$\sigma_h = \frac{pd}{2t} \tag{4.1}$$

In a similar manner the force trying to separate the pressure vessel across its diameter, that is longitudinally (Figure 4.2b), is given by $pressure \times area = \frac{p\pi d^2}{4}$. This force is resisted by the longitudinal stress (σ_l) acting on the area (πdt), so, for equilibrium we have $\frac{p\pi d^2}{4} = \sigma_l \pi dt$, so

$$\sigma_l = \frac{pd}{4t} \tag{4.2}$$

Note: In the case of spherical pressure vessels the stress is the same in all tangential directions to the

inner surface, thus the derivation of the formulae for stress in spherical vessels is analogous with that for the longitudinal stress in a cylinder, since this would be unchanged if the ends of the cylinder were hemispherical. Thus the expression $\frac{pd}{4t}$ may be used for the stress in a thin-walled spherical vessel.

> **Key Point.** In spherical pressure vessels the stress is the same in all tangential directions to its inner surface

In developing the formulae for hoop and longitudinal stress we have ignored the efficiency of the joints that go to make up a cylindrical pressure vessel. You will note that the *hoop stress acts on a longitudinal section* and that the *longitudinal stress acts on a diametrical section* (Figure 4.2). It therefore follows that the efficiency of the longitudinal joints that go to make up the pressure vessel directly affects the hoop stress at the joint. Similarly, the longitudinal stress is affected by the efficiency of the diametrical joints.

The joint efficiency (η) is often quoted as the strength of the joint/strength of the undrilled plate. So, taking this into account for *hoop stress* we may write:

$$\sigma_h = \frac{pd}{2t\eta_l} \tag{4.3}$$

Similarly, for *longitudinal stress* we may write:

$$\sigma_l = \frac{pd}{4t\eta_h} \tag{4.4}$$

> **Example 4.1** A thin cylindrical vessel with a diameter of 80 cm and a wall thickness of 10 mm is subject to an internal pressure of 2 MN/m²:
>
> a) Calculate the hoop stress and longitudinal stresses in the cylinder wall.
>
> b) Find the maximum volume of a spherical pressure vessel that can withstand the same pressure, with the same maximum stress and wall thickness.
>
> Using equations 4.1 and 4.2, we get:
>
> a) $\sigma_h = \frac{pd}{2t} = \frac{(2 \times 10^6)(0.8)}{(2)(0.01)} = 80 \text{ MN/m}^2$ and
>
> $$\sigma_l = \frac{pd}{4t} = \frac{(2 \times 10^6)(0.8)}{(4)(0.01)} = 40 \text{ MN/m}^2$$

b) For the spherical vessel the stress is given by

$\sigma = \dfrac{pd}{4t}$ that acts in all tangential directions.

So, using the maximum allowable stress we found from part (a), that is $80\,MN/m^2$, the largest diameter is given by $d = \dfrac{4t\sigma}{p} = \dfrac{(4)(0.01)(80 \times 10^6)}{(2 \times 10^6)} = 1.6\,m$. The maximum volume of the sphere is then given by: $V = \dfrac{4\pi r^3}{3} = \dfrac{(4)(\pi)(0.8)^3}{3} = 2.144\,m^3$.

Example 4.2 A cylindrical compressed-air vessel has an internal diameter of $50\,cm$ and is manufactured from steel plate $4\,mm$ thick. The longitudinal plate joints have an efficiency of 90% and the circumferential joints have an efficiency of 55%. Determine the maximum air pressure that may be accommodated by the vessel if the maximum permissible tensile stress of the steel is $100\,MPa$.

Using equations (4.3 and 4.4), we have:

$$\sigma_h = \frac{pd}{2t\eta_l} = \frac{(p)(0.5)}{(2)(0.004)(0.9)} = 69.44\,p.$$

Knowing that the maximum allowable stress in the vessel $= 100\,MPa$, then the maximum air pressure that can be accommodated by the longitudinal joints is $p = \dfrac{100 \times 10^6}{69.44} = 1440092\,Pa$ or $p = 1.44\,MPa$ (remembering that $1\,Pa = 1\,N/m^2$).

Similarly: $\sigma_l = \dfrac{pd}{4t\eta_h} = \dfrac{p(0.5)}{(4)(0.004)(0.55)} = 56.818\,p$

Then the maximum air pressure that can be accommodated by the circumferential joints is $p = \dfrac{100 \times 10^6}{56.818} = 1760005\,Pa$ or $p = 1.76\,MPa$.

So the maximum air pressure is determined by the longitudinal joints and is $1.44\,MPa$.

4.1.2 Dimensional change in thin cylinder pressure vessels

The *change in length* of a thin cylinder subject to internal pressure may be determined from the longitudinal strain, provided we ignore the radial stress. The longitudinal strain may be found from equation 1.2 that you met when studying Poisson's ratio in Chapter 1, Fundamentals, that is:

$$\varepsilon_l = \frac{\sigma_l}{E} - \frac{v\sigma_h}{E}.$$

Now, since change in length = longitudinal strain × original length, then:

$$\text{Change in length } = \frac{1}{E}\,(\sigma_l - v\sigma_h)\,l.$$

So substituting equations 4.1 and 4.2 for hoop and longitudinal stress into the above equation gives the change in length $= \dfrac{1}{E}\left(\dfrac{pd}{4t} - v\dfrac{pd}{2t}\right)l$. Or:

$$\textit{Change in length } = \frac{pd}{4tE}\,(1 - 2v)\,l \tag{4.5}$$

In a similar manner to the above, the change in diameter may be determined from the *diametrical strain*, which is equal to the change in diameter divided by the original diameter. The change in diameter, in turn, is directly related to the change in circumference, where:

$$\text{Change in circumference } = \text{original circumference} \\ \times \text{ circumferential strain}$$

Now the *circumferential strain* (ε_h) is a direct result of the *circumferential stress*, that is *the hoop stress* (σ_h). So, in symbols, *the change in circumference* $= (\pi d)(\varepsilon_h)$. Now the new strained circumference is equal to the original circumference plus the change in circumference, so the strained circumference $= \pi d + \pi d\varepsilon_h = \pi d(1 + \varepsilon_h)$. However, this strained circumference has the new strained diameter $= d(1 + \varepsilon_h) = d + d\varepsilon_h$, from which it can be seen that the change in diameter $= d\varepsilon_h$, and so the diametrical strain (ε_d) $= \dfrac{\text{change in diameter}}{\text{original diameter}} = \dfrac{d\varepsilon_h}{d} = \varepsilon_h$; so we have shown that *the diametrical strain = the hoop (circumferential) strain*. Thus the *change in diameter* may be written in a similar manner to the change in length, given above, i.e. change in diameter $d\varepsilon_h = \dfrac{d}{E}(\sigma_h - v\sigma_l)$. Now, substituting for σ_h and σ_l using equations 4.1 and 4.2, we get that $d\varepsilon_h = \dfrac{d}{E}\left[\dfrac{pd}{2t} - v\dfrac{pd}{4t}\right]$, and after extracting common factors we find that the

$$\text{change in diameter} = \frac{pd^2}{4tE}(2 - v) \qquad (4.6)$$

1. Again, we can also find a relationship for the *change in internal volume of a thin-walled cylinder* by considering some of our earlier work on strain in two dimensions and extending it to three-dimensional volumetric strain. You will learn in Chapter 7, when you study complex stress and strain, that the volumetric strain is equal to the sum of the mutually perpendicular linear strains, that is:

$$\varepsilon_v = \varepsilon_x + \varepsilon_y + \varepsilon_z. \qquad (4.7)$$

Key Point.   The volumetric strain is equal to the sum of the mutually perpendicular linear strains

If we accept this volumetric strain relationship for a moment, then *for a cylinder* we have longitudinal strain (ε_l) and two perpendicular diametrical strains ($2\varepsilon_d$), so on substituting ε_l for ε_x and $2\varepsilon_d$ (the diametrical strain) for ($\varepsilon_y + \varepsilon_z$) into equation 4.6, the volumetric strain for a cylinder is:

$$\varepsilon_v = \varepsilon_l + 2\varepsilon_d \qquad (4.8)$$

Then, from the formula for change in length $= \frac{1}{E}(\sigma_l - v\sigma_h)l$, we get that the longitudinal strain $\varepsilon_l = \frac{1}{E}(\sigma_l - v\sigma_h)$, so substituting this equation into equation 4.7 gives $\varepsilon_v = \frac{1}{E}(\sigma_l - v\sigma_h) + \frac{2}{E}(\sigma_h - v\sigma_l)$. Then, substituting equations 4.1 and 4.2 into this equation for σ_h and σ_l, respectively gives:

$$\varepsilon_v = \frac{1}{E}\left(\frac{pd}{4t} - v\frac{pd}{2t}\right) + \frac{2}{E}\left(\frac{pd}{2t} - v\frac{pd}{4t}\right).$$

After a little algebra, this equation for the volumetric strain simplifies to $\varepsilon_v = \frac{pd}{4tE}(5 - 4v)$, and therefore, after multiplying by the original volume V,

$$\text{the change in internal volume} = \frac{pd}{4tE}(5 - 4v)V \qquad (4.9)$$

Example 4.3   A thin cylinder has an internal diameter of 60 mm and is 300 mm in length. The cylinder has a wall thickness of 4 mm and is subject to an internal pressure of 80 kPa. Then, assuming Poisson's ratio is 0.3 and that $E = 210\,\text{GN/m}^2$, determine the:

a)   change in length of the cylinder;

b)   hoop stress;

c)   longitudinal stress.

(a)   Using equation 4.5, the change in length

$$= \frac{pd}{4tE}(1 - 2v)l$$

$$= \left[\frac{(80 \times 10^3)(0.06)}{(4)(0.004)(210 \times 10^9)}\right]$$

$$\times (1 - 0.6)(0.3) = 0.171\,\mu\text{m}$$

(b)   Hoop stress is given by

$$\sigma_h = \frac{pd}{2t} = \left[\frac{(80 \times 10^3)(0.06)}{(2)(0.004)}\right]$$

$$= 0.6\,\text{MN/m}^2$$

(c)   Longitudinal stress is given by

$$\sigma_l = \frac{pd}{4t} = \left[\frac{(80 \times 10^3)(0.06)}{(4)(0.004)}\right]$$

$$= 0.3\,\text{MN/m}^2$$

4.2   Thick-walled pressure vessels

When dealing with the theoretical analysis of thin cylinders, we made the assumption that the hoop stress was constant across the thickness of the cylinder (Figure 4.2a) and that there was no difference in pressure across the cylinder wall. But when dealing with thick-walled pressure vessels where the wall thickness is substantial we can no longer accept either of these two assumptions. So with *thick-walled vessels*, where *the wall thickness is normally greater than one-tenth of the diameter*, both the hoop (σ_h) stress and the radial (σ_r) or diametrical stress vary across the wall. If thick-walled vessels are subject to pressure, then the value of these stresses may be determined by use of the *Lamé equations* given below:

$$\sigma_r = a - \frac{b}{r^2} \qquad (4.10)$$

$$\sigma_h = a + \frac{b}{r^2} \qquad (4.11)$$

Key Point. In thick-walled pressure vessels the radial (diametrical) stress and the hoop (circumferential) stress vary across the vessel walls

These equations may be used to determine the radial and hoop stresses at any radius *r* in terms of constants (*a*) and (*b*). For any pressure condition there will always be two known stress conditions that enable us to find values for the constants. These known stress conditions are often referred to as *boundary conditions*. Section 4.2.1 is given over to a condensed derivation of the Lamé equations. This is worth trying to follow, as it provides yet another application of the calculus and gives an insight into the true meaning of equations 4.10 and 4.11.

4.2.1 Derivation of the Lamé equations

Consider a cylinder which is long in comparison with its diameter, where the longitudinal stress and strain are assumed to be constant across the thickness of the cylinder wall.

We take an element of the cylinder wall (Figure 4.3) when the cylinder is subject to an internal and external pressure, and where the axial length is considered to be unity. Then for *radial equilibrium*, applying elementary trigonometry and remembering that for *small angles* sin θ *approximately equals* θ *(in radians)*, from Figure 4.3 we may write:

Key Point. For small angles sin θ approximately equals θ (in radians)

$(\sigma_r + d\sigma_r)(r + dr)\,d\theta = \sigma_r r d\theta + 2\sigma_h dr\dfrac{d\theta}{2}$ (make sure you can see how this equation was arrived at by studying Figure 4.3 carefully). Then from this equation, after simplification, neglecting second-order small quantities, we may write: $\sigma_r dr + r d\sigma_r = \sigma_h dr$ (again you should make sure that you can obtain this simplification). Then, on division by *dr* and multiplication throughout by (−1), we may write:

$$\sigma_h - \sigma_r = \frac{r d\sigma_r}{dr} \qquad (4.12)$$

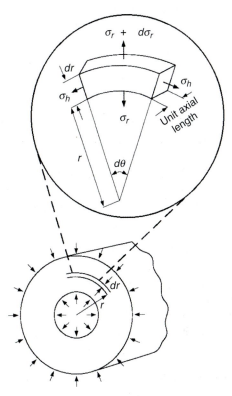

Figure 4.3 Element of cylinder wall subject to internal and external pressure

Now, if we let the longitudinal stress be (σ_l) and the longitudinal strain be (ε_l), then, following *our assumption that the longitudinal stress and strain are constant across the cylinder walls*, we may write:

$$\varepsilon_l = \frac{1}{E}\left[\sigma_l - v(\sigma_r + \sigma_h)\right] \qquad (4.13)$$

Equation 4.13 is formed on the basis of the relationship between Poisson's ratio and three-dimensional loading, which is a simple extension of Poisson's ratio and two-dimensional loading that you covered in Chapter 1. So for the moment you are asked to accept this relationship, without proof, as you were when we mentioned it earlier in deriving equation 4.9 for changes in internal volume. Also, since σ_l and ε_l are constant, as of course are Poisson's ratio *v* and the modulus *E*, then $\sigma_r + \sigma_l = $ constant (say 2*a*), or

$$\sigma_h = 2a - \sigma_r \qquad (4.14)$$

where the constant 2*a* was chosen arbitrarily to simplify the algebra.

Then, on substituting equation 4.14 for σ_h into equation 4.12 and multiplying through by r, we get:

$$2ar - 2r\sigma_r = \frac{r^2 d\sigma_r}{dr} \text{ or } 2r\sigma_r + \frac{r^2 d\sigma_r}{dr} - 2ar = 0.$$

Now we are manipulating the algebra in this particular way in order to obtain a differential equation with respect to r as the variable, so we may write:

$$\frac{\sigma_r dr^2}{dr} + \frac{r^2 d\sigma_r}{dr} - \frac{d}{dr} ar^2 = 0$$

Note that here we have integrated the first and third terms of this equation with respect to r and given their exact equivalence in differential form. You should also remember that the differential of a constant is zero so in this case $\frac{d\sigma_r}{dr}$ is zero, so the middle term will be zero. So we may now write that $\frac{\sigma_r dr^2}{dr} - \frac{d}{dr} ar^2 = 0$ or $\frac{d}{dr}\left(\sigma_r r^2 - ar^2\right) = 0$. If we now integrate this equation and take into consideration the constant of integration, we get that $\sigma_r r^2 - ar^2 = $ constant (say $-b$, chosen arbitrarily to verify the Lamé equation), so that after division by r^2 we have $\sigma_r = a - \frac{b}{r^2}$ (equation 4.10), as required.

Finally, from equation 4.13 we get that $\sigma_h = 2a - \left(a - \frac{b}{r^2}\right)$ or $\sigma_h = a + \frac{b}{r^2}$ (equation 4.11), as required.

4.2.2   Thick cylinders subject to internal pressure

Consider a thick cylinder which is only subject to internal pressure (p), the external pressure being zero.

For the situation shown in Figure 4.4 the *boundary conditions* are: $\sigma_r = -p$ when $r = r_1$ and $\sigma_r = 0$ when $r = r_2$. The internal pressure is considered as a *negative radial stress*, because it tends to produce thinning (compression) of the cylinder wall. By substituting the boundary conditions into the Lamé equations 4.10 and 4.11, and determining expressions for the constants a and b, it can be shown (after a bit of algebra) that for thick cylinders subject to internal pressure only, the:

Radial stress $\sigma_r = \dfrac{pr_1^2}{r_2^2 - r_1^2}\left(\dfrac{r^2 - r_2^2}{r^2}\right)$ (4.15)

Hoop stress $\sigma_h = \dfrac{pr_1^2}{r_2^2 - r_1^2}\left(\dfrac{r^2 + r_2^2}{r^2}\right)$ (4.16)

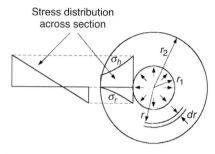

Figure 4.4 Thick cylinder subject to internal pressure

You will be asked to derive equations 4.15 and 4.16 from the Lamé equations as an exercise at the end of this chapter!

The maximum values for the radial and hoop stress across the section can be seen in Figure 4.4 and may be determined using:

$$\sigma_{r_{\max}} = -p \qquad (4.17)$$

$$\sigma_{h_{\max}} = p\left(\frac{r_1^2 + r_2^2}{r_2^2 - r_1^2}\right) \qquad (4.18)$$

4.2.3   Longitudinal and shear stress

So far we have established that the longitudinal stress in thick-walled pressure vessels may be assumed to be uniform. This *constant longitudinal stress* (σ_l) is determined by considering the longitudinal equilibrium at the end of the cylinder. Figure 4.5 shows the arrangement for the simple case where the *cylinder is subject to internal pressure*.

> **Key Point.**   In thick-walled pressure vessels the longitudinal stress may be assumed to be uniform

So, assuming equilibrium conditions and resolving forces horizontally, we have: $(p)(\pi r_1^2) = (\sigma_l)(r_2^2 - r_1^2)$, so

$$\sigma_l = \frac{pr_1^2}{r_2^2 - r_1^2} \qquad (4.19)$$

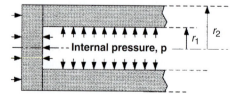

Figure 4.5 Longitudinal equilibrium at end of cylinder

In a similar manner it can be shown that for a thick-walled cylinder subject to both *internal and external pressure*, the constant longitudinal stress is given by:

$$\sigma_l = \frac{p_1 r_1^2 - p_2 r_2^2}{r_2^2 - r_1^2} \qquad (4.20)$$

The radial stress (σ_r), hoop stress (σ_h) and longitudinal stress (σ_l) are all principal stresses and are mutually perpendicular; look back at Figure 4.1. Then *the maximum shear stress (τ_{max}) will be equal to half the difference between the maximum and minimum principal stresses* (at this stage you may take this statement as fact; alternatively, if you wish to seek an explanation now, look forward to the section in Chapter 7 that deals with material subject to combined direct and shear stresses). So, accepting the above statement, if we consider the case for *internal pressure only*, we have already seen that *radial stress is compressive*, the *hoop and longitudinal stresses are tensile*.

> **Key Point.** The radial stress, hoop stress and longitudinal stress are all principal stresses and are mutually perpendicular

Therefore, remembering the sign convention, and noting the fact that the hoop stress is larger than the longitudinal stress, the maximum difference is given by ($\sigma_h - \sigma_r$) and so the maximum shear stress is:

$$\tau_{max} = \frac{\sigma_h - \sigma_r}{2}.$$

If we then substitute for σ_h and σ_r their equivalent, in terms of the Lamé coefficients, then from equations 4.10 and 4.11, we get

$$\tau_{max} = \frac{\left(a + \dfrac{b}{r^2}\right) - \left(a - \dfrac{b}{r^2}\right)}{2} = \frac{b}{r^2}.$$

Now, applying the *boundary conditions* to the Lamé equations, for a *cylinder subject to internal pressure only*, as we did in section 4.2.2 where $\sigma_r = -p$ when $r = r_1$ and $\sigma_r = 0$ when $r = r_2$, we get that

$$-p = a - \frac{b}{r_1^2} \quad \text{and} \quad 0 = a - \frac{b}{r_2^2}.$$

Then, solving these two equations simultaneously for (*b*) gives

$$b = \frac{p r_1^2 r_2^2}{r_2^2 - r_1^2},$$

and so

$$\tau_{max} = \frac{p r_1^2 r_2^2}{\left(r_2^2 - r_1^2\right) r^2} = \frac{b}{r^2}.$$

Now *the maximum shear stress (τ_{max}) occurs at the inside wall of the cylinder* where $r = r_1$, so that:

$$\tau_{max} = \frac{p r_2^2}{r_2^2 - r_1^2} \qquad (4.21)$$

Example 4.4   The cylinder of a hydraulic actuator has a bore of 100 mm and is required to operate up to a pressure of 12 MPa. Determine the required wall thickness for a limiting tensile stress of 36 MPa. In this example, the boundary conditions are that at $r = 50$ mm, $\sigma_r = -12$ MPa, and since the maximum tensile hoop stress occurs at the inner surface then, at $r = 50$ mm, $\sigma_h = 36$ MPa. In order to simplify the arithmetic when calculating the constants in the Lamé equations we will use the relationship that $1\,\text{MPa} = 1\,\text{N/mm}^2$ and work in N/mm². Therefore, using $\sigma_h = a + \dfrac{b}{r^2}$, $\sigma_r = a - \dfrac{b}{r^2}$ (inner surface) where $\sigma_r = -12\,\text{N/mm}^2$, $r = 50$ mm, $\sigma_h = 36\,\text{N/mm}^2$, we have:

$$-12 = a - \frac{b}{2500}$$

$$\text{and} \quad 36 = a + \frac{b}{2500}$$

Adding the two equations:

$$24 = 2a, \quad \text{so} \quad a = 12,$$

from which, on substitution,

$$-12 = 12 - \frac{b}{2500} \quad \text{and} \quad b = 60000.$$

At the outer surface, $\sigma_r = 0$; therefore,

$$0 = a - \frac{b}{r^2} = 12 - \frac{60000}{r^2}$$

$$\text{and} \quad r = 70.7 \text{ mm.}$$

Therefore the required wall thickness is $70.7 - 50 = 20.7$ mm.

Example 4.5   A thick-walled cylindrical pressure vessel has an internal diameter of 300 mm and is designed to take an internal pressure of 25 MPa. Calculate:

a)   the required wall thickness if the maximum shear stress $\tau_{max} = 60\,MN/m^2$, b) the longitudinal and hoop stresses under the above circumstances.

a)   As the maximum shear stress occurs at the inside wall of the cylinder, where $r = r_1$, we may use equation 4.21 where $\tau_{max} = \dfrac{pr_2^2}{r_2^2 - r_1^2}$. Knowing that $\tau_{max} = 60\,N/mm^2$ and the internal pressure $p = 25\,N/mm^2$, then,

$$60 = \frac{(25)r_2^2}{r_2^2 - (150)^2} \text{ or } 35r^2 = (60)(150)^2, \text{ so}$$

$$r_2 = \sqrt{\frac{(60)(150)^2}{35}} = 196.4\,mm \text{ and the wall}$$

thickness $t = 196.4 - 150 = 46.4\,mm$

b)   Knowing both the internal and external radii we can find the longitudinal stress and the hoop (circumferential) stress, when the pressure vessel is subject only to an *internal pressure* of $p = 25\,N/mm^2$, using equations 4.18 and 4.19. So,

$$\sigma_l = \frac{pr_1^2}{r_2^2 - r_1^2} = \frac{(25)(150)^2}{(196.4)^2 - (150)^2} = 35\,MN/m^2$$

$$\sigma_h = p\left(\frac{r_1^2 + r_2^2}{r_2^2 - r_1^2}\right) = 25\left(\frac{(150)^2 + (196.4)^2}{(196.4)^2 - (150)^2}\right)$$

$$= 95\,MN/m^2$$

4.3   Pressure vessel applications

An application of the theory of thin cylinders and other thin and thick-walled shells may be applied to the design and manufacture of pressure vessels. These vessels could be used for food or chemical processing, the storage of gases and liquids under pressure or even the fuselage of a pressurised aircraft. Whatever the industrial use, formulae can be derived from some of the relationships we have already found, to assist us in determining the required design parameters.

Let us consider some useful relationships which aid the *design of thin cylinders and spherical shells for the plant and process industry*. The governing standards for British and European unfired fusion-welded pressure vessel design are laid down in PD: 5500: 2009 and EN 13445. The following formulae are based on these standards but are not necessarily identical to those laid down in the standards.

For a thin-walled cylindrical shell, the minimum thickness required to resist internal pressure can be determined from equation 4.1, i.e. $\sigma_h = \dfrac{pd}{2t}$. We use this equation because the hoop stress in pressure vessels is greater than the longitudinal stress and so must always be considered first.

For our study of process industry pressure vessels, we let d be the internal diameter and e be the minimum thickness of shell material required. Then the mean diameter of the vessel will be given by:

$$d + \left(2 \times \frac{e}{2}\right) \text{ or simply } (d + e),$$

and substituting this into equation 4.1 gives

$$e = \frac{p(d + e)}{2\sigma_d},$$

where in this case, as before, p is the internal pressure and σ_d is the design hoop (circumferential) stress. Rearranging this formula by separating e as a common factor gives *the minimum thickness of a cylindrical pressure vessel* as:

$$e = \frac{pd}{2\sigma_d - p} \tag{4.22}$$

Equation 4.22 takes the same form as that laid down in the standards quoted above.

The relationship between the *principal stresses of a spherical shell* is similar to those we have found for the hoop and longitudinal stress (principal stresses) of a thin-walled cylinder. That is, the principal stresses for a spherical shell are given by:

$$\sigma_1 = \sigma_2 = \frac{pd}{4t} \tag{4.23}$$

Then, in a similar manner to the above, an equation for the *minimum thickness of a sphere* can be obtained from equation 4.23, that is:

$$e = \frac{pd}{4\sigma_d - p} \tag{4.24}$$

Equation 4.24 differs slightly from that given in the standards because, as you may have noticed, it is derived from thick-walled shell theory, which we considered in section 4.2. The formula for determining the minimum thickness of material required to resist internal pressure within a sphere is given in the design standards as:

$$e = \frac{pd}{4\sigma_d - 1.2p} \quad (4.25)$$

Now, assuming our pressure vessel has welded joints, then we need to take into account the integrity of these joints by using joint efficiency factors (η), which you have already met in our earlier work in section 4.1.

> **Key Point.** An efficiency factor is introduced to take into account the integrity of the vessel's fusion-welded joints

The efficiency of welded joints is dependent on the heat treatment they receive after welding, and on the amount of non-destructive examination used to ascertain their integrity. If the joint integrity can be totally relied upon as a result of heat treatment and non-destructive examination, then the joint may be considered to be 100% efficient. In practice, welded joint efficiency factors vary from about 0.65 up to 1.0. In other words, the efficiency of welded joints normally ranges from around 65% up to 100%. If we wish to consider *cylindrical* and *spherical* vessels with welded joints, then equations 4.22 and 4.25 may be written as follows:

$$e = \frac{pd}{2\eta\sigma_d - p} \quad (4.26)$$

$$e = \frac{pd}{4\eta\sigma_d - 1.2p} \quad (4.27)$$

In the design of cylindrical pressure vessels due consideration must be given to the way in which the vessels are closed at their ends. For example, flat plates or domes of various shapes could be used. There are, in fact, four principal methods used to close cylindrical pressure vessels: flat plates and formed flat heads, hemispherical heads, and torispherical and ellipsoidal heads.

Flat plates may be plain or flanged, and then bolted or welded in position (Figure 4.6).

Torispherical domed heads, which are often used to close cylindrical pressure vessels, are formed from part of a torus and part of a sphere (Figure 4.7).

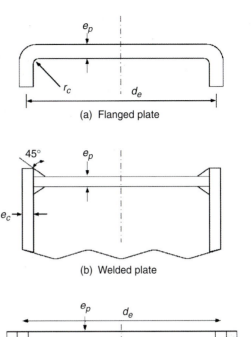

(a) Flanged plate

(b) Welded plate

(c) Bolted plate

KEY

d_e = nominal plate diameter
r_c = corner radius
e_p = minimum thickness for plate
e_c = minimum thickness for cylinder

Figure 4.6 Flat plate closures for cylindrical pressure vessel

Torispherical heads are generally preferred to their ellipsoidal counterparts (Figure 4.8) because they are easier and cheaper to fabricate. Hemispherical heads (Figure 4.9) are the strongest form of closure that can be used with cylindrical pressure vessels and are generally used for high-pressure applications. They are, however, costly, so their use is often restricted to applications where pressure containment and safety take priority over all other considerations, including costs.

Minimum thickness formulae for the methods of closure can be determined in a similar manner to those we found earlier for cylinders and spheres, by analysing their geometry, stresses and method of constraint at their periphery. Time does not permit us to derive these relationships, but three typical formulae for flat, ellipsoidal and torispherical heads are given below.

The equation for determining *the minimum thickness of flat heads* is:

$$e = c_p d_e \sqrt{\frac{p}{\sigma_d}} \qquad (4.28)$$

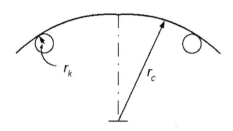

where $\dfrac{r_k}{r_c}$ not less than 0.06

and r_k = knuckle radius, r_c = crown radius

(a) Torisphere

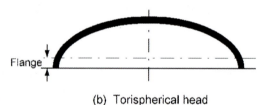

(b) Torispherical head

Figure 4.7 Geometry of torispherical domed head for cylindrical pressure vessel

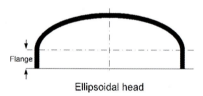

Ellipsoidal head

Figure 4.8 Ellipsoidal head for cylindrical pressure vessel

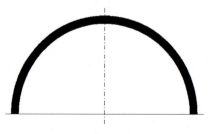

Hemispherical head

Figure 4.9 Hemispherical head for cylindrical pressure vessel

where c_p = a design constant, dependent on edge constraint, σ_d = design stress, d_e = nominal diameter.

The *minimum thickness required for ellipsoidal heads* is given by:

$$e = \frac{pd}{2\eta\sigma_d - 0.2p} \qquad (4.29)$$

where the symbols have their usual meaning.

Finally, the *minimum thickness required for torispherical heads* is given by:

$$e = \frac{pR_c C_s}{2\eta\sigma_d + p(C_s - 0.2)} \qquad (4.30)$$

where R_c = crown radius, R_k = knuckle radius and C_s = stress concentration factor for torispherical heads and is found from $C_s = \dfrac{1}{4}\left(3 + \sqrt{\dfrac{R_c}{R_k}}\right)$.

We have spent some time looking at formulae that can be used in determining minimum thickness of materials for pressure vessels. For a full account of all aspects of pressure vessel design and to ensure that the latest and most accurate criteria are being followed, those involved with the design of pressure vessels should always consult the appropriate design standards, such as PD: 5500: 2009 or EN 13445 or other relevant standards.

Example 4.6   Consider the pressure vessel shown in Figure 4.10. Estimate the thickness of material required for the cylindrical section and domed ends.

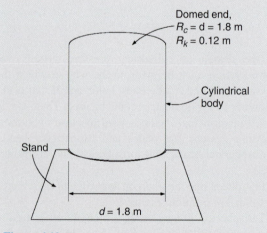

Figure 4.10 Pressure vessel set-up for Example 4.6

Assume that the pressure vessel is to be used in the food processing industry and so the material for construction is to be a stainless steel (SS). The vessel is to operate at a design temperature of 250°C and a design pressure of 1.6 MPa, so we will use a typical design stress of 115 MPa.

a) *The cylindrical section*
   Using equation 4.22, we have:

$$e = \frac{pd}{2\sigma_d - p} = \frac{(1.6 \times 10^6)(1.8)}{(2)(115 \times 10^6) - (1.6 \times 10^6)}$$

$$= 12.6\,\text{mm}$$

Now we need to consider any relevant safety factors. The design pressure will have a safety factor built-in, so it will already be set at a higher value than normal operating pressure. Therefore we need only consider an allowance for corrosion and fabrication problems. The SS is highly resistant to corrosion, but as a result of the fabrication process it may have 'locked-in' stresses which we are not aware of. Therefore it would be prudent to make some allowance for the unknown without incurring too much unnecessary expense or material wastage. In practice, further detail design will determine the necessary allowances. For the purpose of this exercise, we will assume that a further 2 mm is an appropriate allowance; then: $e = 12.6 + 2 = 14.6\,\text{mm}$. Now we are unlikely to obtain 14.6 mm SS plate as stock, so we would use 15 mm plate.

b) *The domed head*
   We shall first consider a standard dished torispherical head and assume that the head is formed by pressing. Therefore it has no welded joints and so the efficiency (η) may be taken as 1.0. Let us also assume that the torisphere has a crown radius equal to the diameter of the cylindrical section (its maximum value) and a knuckle radius of 0.12 m.
   Then, using equation 4.30, we have for a torispherical head:

$$e = \frac{pR_cC_s}{2\eta\sigma_d + p(C_s - 0.2)},$$

$$\text{where } C_s = \frac{1}{4}\left(3 + \sqrt{\frac{R_c}{R_k}}\right),$$

so $C_s = \dfrac{1}{4}\left(3 + \sqrt{\dfrac{1.8}{0.12}}\right)$

$$= 1.72 \quad \text{and in this case } \eta = 1.0$$

So $e = \dfrac{(1.6 \times 10^6)(1.8)(1.72)}{(2)(115 \times 10^6) + (1.6 \times 10^6)(1.72 - 0.2)}$

$$= 21.5\,\text{mm}.$$

Let us now try an ellipsoidal head. Then, from equation 4.29, we get:

$$e = \frac{(1.6 \times 10^6)(1.8)}{(2)(115 \times 10^6) - (0.2)(1.6 \times 10^6)}$$

$$= 12.5\,\text{mm}.$$

Since the fabrication of both the torispherical and ellipsoidal head incur similar effort, the most economical head would be *ellipsoidal with a minimum thickness of 15.0 mm*, after the normal allowances are added.

4.4   Chapter summary

In section 4.1 of this chapter, we first introduced some useful relationships concerned with *thin-walled pressure vessels*, in particular those shells that were *cylindrical* and *spherical in shape*, where for the purposes of calculation thin-walled pressure vessels were defined as those shells where the ratio of the wall thickness to that of the inside diameter is greater than 1:20.

When these shells were subject to *internal pressure*, this resulted in three mutually perpendicular stresses being set up. These were *hoop stress (circumferential stress), radial stress* and *longitudinal stress* (Figure 4.1). It was found by analysis in section 4.1.1 that the *hoop stress was always larger than the longitudinal stress* when the vessel was subject to internal pressure, and therefore was always considered first. The *hoop stress in cylindrical pressure vessels* was that which resisted the bursting effect of the internal pressure (Figure 4.2a) and was given by equation 4.1, i.e. $\sigma_h = \dfrac{pd}{2t}$. Also, the force trying to separate the *cylindrical pressure vessel* across its diameter (Figure 4.2b) was known as the *longitudinal stress* and was given by equation 4.2, i.e. $\sigma_l = \dfrac{pd}{4t}$. The *stress in spherical pressure vessels* was known to be the same in all tangential directions and therefore was analogous with the stress in a longitudinal cylinder where, in effect, the ends are hemispherical and the stress is therefore also given by $\sigma = \dfrac{pd}{4t}$. Equations 4.3 and 4.4 take into consideration the joint efficiency, that is they include an *efficiency factor* for welded joints. Dimensional changes in thin-walled pressure vessels were then looked at and relationships established for change in length, change in diameter and change in internal volume of cylindrical pressure vessels.

In section 4.2 we considered *thick-walled pressure vessels* (those where the wall thickness is generally greater than one-tenth of the diameter) and derived the *Lamé equations* (equations 4.10 and 4.11), which may be used to determine the *radial stress and hoop (circumferential) stress* in these vessels. The important difference between the analysis of thin- and thick-walled pressure vessels was emphasised, i.e. that of the *variation in radial stress and hoop stress across the vessel walls*. The idea of *boundary conditions* was also introduced, these being needed when finding the Lamé constants under varying pressure conditions.

We then considered thick cylinders subject to internal pressure and noted that the *radial stress* was a maximum when acting on the inside wall of the vessel, tending to compress the wall and so treated as a *negative stress* (equation 4.17). A relationship for the hoop (circumferential) stress for a cylinder under internal pressure was also established (equation 4.18). We then went on to look at the *longitudinal* and *shear stresses* that are set up in thick-walled pressure vessels and established two relationships for longitudinal stress when the vessel is subject solely to internal pressure (equation 4.19) and also when subject to internal and external pressure (equation 4.20). We established the relationship between the principal stresses (radial, hoop and longitudinal) and the maximum shear stress and noted that the maximum shear stress occurred at the inside wall of the cylinder and was given by equation 4.21.

In section 4.3 we considered applications of pressure vessel theory to their design, in particular to the design of thin-walled cylinders for the food processing industry, where the prime material used to manufacture these vessels is stainless steel. The design equations (4.22–4.30) that were established for the body and various heads of these vessels were primarily concerned with the *minimum thickness of the material* to be used under varying pressure, stress and dimensional considerations.

Example 4.6 illustrated the procedure we might adopt to determine the necessary design parameters, where we established that the ellipsoidal head proved (in this case) to be the most economically viable choice. Please remember that if you are involved in the design of pressure vessels, no matter what the circumstances, you should always consult the *latest* national and international design standards, to ensure you are working to current criteria.

4.5   Review questions

1. A thin cylinder has an internal diameter of 50 mm, a wall thickness of 3 mm and is 200 mm long. If the cylinder is subject to an internal pressure of 10 MPa, taking $E = 200$ GPa and Poisson's ratio $v = 0.3$, determine the change in internal diameter and the change in length.

2. A cylindrical reservoir used to store compressed gas is 1.8 m in diameter and is assembled from plates 12.5 mm thick. The longitudinal joints have an efficiency of 90% and the circumferential joints have an efficiency of 50%. If the reservoir plating is to be limited to a tensile stress of 110 MPa, determine the maximum safe pressure of the gas.

3. A thin cylinder has an internal diameter of 240 mm, a wall thickness of 1.5 mm and is 1.2 m long. The cylinder material has $E = 210$ GPa and Poisson's ratio $v = 0.3$. If the internal volume change is found to be 14×10^{-6} m^3 when pressurised, find:

    a) the value of this pressure;

    b) the value of the hoop and longitudinal stresses.

4. A steel pipe of internal diameter 50 mm and external diameter 90 mm is subject to an internal pressure of 15 MPa. Determine the radial and hoop stresses at the inner surface.

5. A thick cylinder has an internal diameter of 200 mm and external diameter of 300 mm. If the cylinder is subject to an internal pressure of 70 MPa and an external pressure of 40 MPa, determine:

    a) the hoop and radial stress at the inner and outer surfaces of the cylinder;

    b) the longitudinal stress, if the cylinder is assumed to have closed ends.

Chapter 5

Concentrically loaded columns and struts

A beam turned on end and loaded directly through its centre line may be considered to be a *column*. Under these circumstances the column is subject to *concentric* or *axial compressive loads* (Figure 5.1).

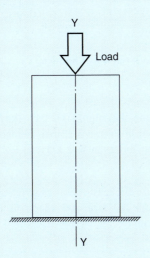

Y

Load

Y

Figure 5.1 Column subject to concentric compressive load

Structural members that are subject to concentric compressive loads are not only known as columns. Concrete *pillars* support axial compressive loads, using the fact that concrete is around 15 times stronger in compression than it is in tension. In steel girder construction, *struts* are used as the main compressive support members. In wooden structures, the axially loaded supports are often referred to as *posts*. There is often a distinction made between a column and a strut, whereby *columns* are those short thick members that generally fail by crushing when the compressive yield stress is exceeded, while the word *strut* is often used for those taller, more slender members that fail by *buckling* before the yield stress in compression is reached. We will use the word *column* or the word *strut* for structural members that are loaded axially and subject to compressive stresses, dependent on circumstances.

Key Point. *Columns* are generally considered short thick members that fail by crushing, while *struts* are taller, more slender members that fail by buckling

5.1 Slenderness ratio, radius of gyration and effective length

In a similar way to beams, the maximum concentric load that a column can support depends on the *material*

978-1-85617-775-7, Engineering Science, Mike Tooley & Lloyd Dingle

from which the column is manufactured and the *slenderness* of the column. The slenderness of a column is determined by considering column height, size and shape of cross-section. As you will see later, a measure of the shape of cross-section is given by the *radius of gyration*.

5.1.1 Slenderness ratio

As already mentioned, the slenderness of a column has a direct effect on the way in which the column may fail under compressive loads. Thus, if the column is *very long and thin*, it is likely to fail due to *buckling* well before it may have failed due to compressive stresses. Under these circumstances we tend to call columns *struts*. How do we define the slenderness of a column? Logic would suggest that we can use a ratio, the *slenderness ratio*. For a column, the slenderness ratio may be defined as:

$$\text{Slenderness ratio} = \frac{\text{Column length or height}}{\text{Minimum column width}}$$

Now this is fine for solid rectangular sections, such as posts used for wooden structures. However, we said earlier that the axial loads that can be sustained by columns are dependent on the shape of the column's cross-section. For rectangular columns, the above ratio provides an effective measure of whether or not a column will fail due to buckling. For other column cross-sections, such as I-sections, circular sections and L-sections, commonly found in steel columns, the above simple ratio has to be modified. This is where the radius of gyration is used. Then,

$$\text{Slenderness ratio} = \frac{\text{Effective length}}{\text{Least radius of gyration}} = \frac{L_e}{k} \tag{5.1}$$

Now, in order for us to be able to understand the use of this ratio, we need first to define the least radius of gyration and then the effective length.

Key Point. The slenderness ratio may be taken as a measure used to decide whether a member under compressive load is likely to fail due to compression or due to buckling

5.1.2 Radius of gyration

The *least radius of gyration takes into account both the size and shape of the column cross-section*, the size

being measured by the column cross-sectional area A and the shape effect being determined by considering the *second moment of area I* of the solid, i.e. *the moment of inertia* of the column. Thus:

$$\text{Radius of gyration } k = \sqrt{\frac{I}{A}} \tag{5.2}$$

However, the slenderness ratio is concerned with the *least* radius of gyration, which will be least when the moment of inertia is least. This occurs when I is taken about the axis where the least amount of material protrudes.

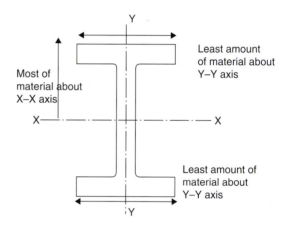

Figure 5.2 Cross-section of I-girder

So, for example, if a column is to have the *I-girder* cross-section shown in Figure 5.2, then the least moment of inertia will be about the Y–Y axis, so:

Key Point. The least radius of gyration takes into account both the size and shape of the column cross-section

Least radius of gyration k

$$= \sqrt{\frac{\text{Least moment of inertia}}{\text{Area of cross-section}}} = \sqrt{\frac{I_{\text{least}}}{A}} \tag{5.3}$$

In the case of our cross-section shown in Figure 5.2, the least radius of gyration will be given by $k = \sqrt{\frac{I_{YY}}{A}}$.

Example 5.1   A concrete column has a rectangular cross-section, with a breadth of 300 mm and a depth of 600 mm and an effective length of 3.4 m. For this column, calculate:

a)   the least radius of gyration;

b)   the slenderness ratio.

To find the least radius of gyration we must first find the least second moment of area (moment of inertia). In this case, this occurs about the Y–Y principal axis, that is the longitudinal axis, where:

$$I_{YY} = \frac{db^3}{12} = \frac{(600)(300)^3}{12}$$

$$= 1350 \times 106 \, \text{mm}^3. \text{ Then:}$$

a)   The least radius of gyration

$$k = \sqrt{\frac{I_{YY}}{A}} = \sqrt{\frac{1350 \times 10^6}{(300)(600)}} = 86.6 \, \text{mm}.$$

b)   The slenderness ratio:

$$\frac{L_e}{k} = \frac{3.4 \times 10^3}{86.6} = 39.26$$

So we can now find the slenderness ratio for a variety of column cross-sections, but what do we do with it?

Well, there is a direct comparison between the slenderness ratio of a column and its permissive *compressive strength* (σ_c). Thus, if we know the slenderness ratio, then we can determine the *compressive buckling strength of the column*. These comparisons for steel section columns are tabulated, for example, in BS 5950 Part 1:2000 or in BS EN 1993, which supersedes BS 5950.

By consulting these or similar tables, we can determine the compressive strength for a particular slenderness ratio, for example if the maximum compressive strength of a column about its least moment of inertia axis is 275 N/mm². Then, from tables in the BS EN standards, when the slenderness ratio is 80, the compressive strength is reduced to 181 N/mm². Codes of practice limit the slenderness ratio, for all practical purposes, to a maximum of 180. When the slenderness ratio is 180, then for the above example the compressive strength reduces to 54 N/mm², that is only one-fifth of the original compressive strength!

At slenderness ratios of 15 and below, the column will tend to fail at critical loads that are the same as those required for the column to fail in direct compression. Thus, as a 'rule of thumb', restricting the slenderness ratio to values of 15 or less means that failure due to buckling can generally be ignored. In order to find the slenderness ratio, you will have noted from Example 5.1 that it was necessary first to find the least radius of gyration. This may be achieved, as in the example, by first finding the least moment of inertia and then dividing by the area of the particular cross-section. For convenience, the formulae for finding k, for three of the most common steel cross-sections, are shown here in Table 5.1.

Key Point.   As a *rule of thumb*, failure due to buckling can generally be ignored in columns with a slenderness ratio of 15 or less

Table 5.1   Some formulae for finding k for various steel cross-sections

Steel column section	Least value of radius of gyration
Solid rectangular cross-section	$k = 0.289b$
Hollow square section	$k = \dfrac{1}{2}\sqrt{\dfrac{B^2 - b^2}{3}}$
Annulus	$k = \dfrac{\sqrt{D^2 + d^2}}{4}$

5.1.3   Effective length

So far we have not defined the *effective length*. This is directly related to the way in which the column is fixed at its ends.

Figure 5.3 shows the four basic methods of attachment for columns.

In Figure 5.3a, *both the ends are pinned*. This is considered to be the *basic case*, where for the purposes of analysis the effective length is equal to the actual length, i.e. $L_e = L$. Figure 5.3b shows *both ends are rigidly fixed*. Under these conditions, the middle half of the column behaves in the same manner as in the basic case, and for this reason the effective length is the actual length divided by two, i.e. $L_e = \dfrac{L}{2}$. Figure 5.3c shows the case where *one end is fixed and the other end is free*. Under these circumstances, the column behaves as though this length is only half of the effective length, so that the effective length becomes twice the actual length,

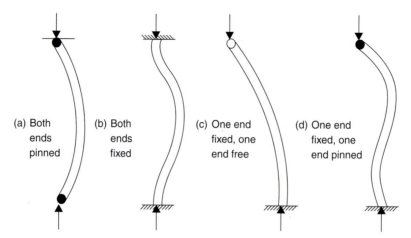

Figure 5.3 End fixtures for columns under compressive load

i.e. $L_e = 2L$. Finally, Figure 5.3d shows the case where *one end is fixed and the other is pin-jointed*. Analysis shows that under these circumstances the column has an effective length of approximately 0.7 times the actual length, i.e. $L_e = 0.7L$. These findings are tabulated in Table 5.2.

Table 5.2 Relationship between actual length L and effective length L_e, for various methods of attachment

Method of fixing	Both ends pinned	Both ends rigidly fixed	One end fixed, one end free	One end fixed, one end pinned
Effective length (L_e)	L	$\dfrac{L}{2}$	$2L$	$0.7L$

Note: BS 5950-1:2000 and BS EN 1993 provide a slight variant for conditions between total restraint and partial restraint. The latest and most appropriate BS and/or EN standards as appropriate must be consulted if current column/strut design information is required.

Key Point. Always consult the most appropriate BS and/or ISO/EN standards if current column/strut design information is required

In the next section we look at some of the theory that underpins the practical relationships given in the tables and also a little of the design information that may be found in the current BS and EN standards, mentioned earlier. In the meantime we conclude this section by considering an example that makes use of the data given in Table 5.1.

Example 5.2   A hollow-section steel column is shown in Figure 5.4, with the dimensions indicated. If the column is rigidly fixed at both ends and is subject to an axial load through its centroid, then, given that the compressive resistance of a column is the product of its compressive strength (200 N/mm²) and section area, determine for the column:

a)   the least radius of gyration;

b)   the slenderness ratio;

c)   the compressive resistance.

a)   From Table 5.1, the least radius of gyration,

$$k = \frac{\sqrt{D^2 + d^2}}{4} = \frac{\sqrt{200^2 + 150^2}}{4} = 62.5 \text{ mm}.$$

If tables are not available, then k may be found from $I = I_{XX} = I_{YY}$ for an annulus, that is $I_{least} = I = \dfrac{\pi\left(D^4 - d^4\right)}{64}$ and then, using the relationship $k = \sqrt{\dfrac{I_{least}}{A}}$, where the cross-sectional area of an annulus is given by $A = \dfrac{\pi\left(D^2 - d^2\right)}{4}$, you will find this yields the same result.

b)   To find the slenderness ratio, we just need the effective length, which is $L/2$ (from Table 5.2), that is $L_e = \dfrac{2600}{2} = 1300$ mm, so the

$$\text{slenderness ratio} = \frac{L_e}{k} = \frac{1300 \text{ mm}}{62.5 \text{ mm}} = 20.8$$

5.2.1   Euler theory

Euler theory is particularly useful because it enables us to derive expressions for buckling loads for struts of varying materials, cross-sections and lengths that have different types of end fixing (Figure 5.3).

Consider the axially loaded strut shown in Figure 5.5 with pinned ends, subject to the buckling load (P_e) that causes a deflection y in the strut at some distance x from one end of the strut.

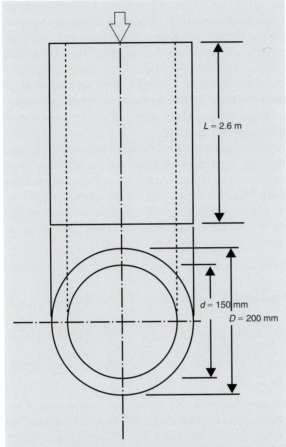

Figure 5.4 Hollow section steel column

c) Finally, the compressive resistance is equal to the compressive strength of the column (σ_c), multiplied by the section area (A). The section area for an annulus $A = \dfrac{\pi (D^2 - d^2)}{4} = \dfrac{\pi (200^2 - 150^2)}{4} = 4375\pi$, and so column compressive resistance $F_c = (200\,\text{N/mm}^2)(4375\pi\,\text{mm}^2) = 2.7489\,\text{MN}$.

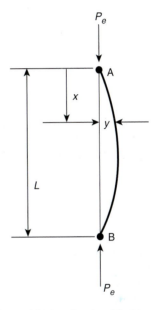

Figure 5.5 Strut with pinned ends subject to axial load

Using our bending theory (Chapter 2), the bending moment at the deflection y a distance x from one end is given by:

$$BM = EI\frac{d^2y}{dx^2} = -P_e y \qquad (5.4)$$

where, for the situation shown, the BM is considered positive, therefore the deflection will be negative, as given in equation 5.4.

From equation 5.4, $EI\dfrac{d^2y}{dx^2} + P_e y = 0$ and on division by EI we get that $\dfrac{d^2y}{dx^2} + \dfrac{P_e}{EI}y = 0$. Now, by making the substitution $m = \sqrt{\dfrac{P_e}{EI}}$, our equation may be written as:

$$\frac{d^2y}{dx^2} + m^2 y = 0 \qquad (5.5)$$

5.2   Euler's theory and the Rankine–Gordon relationship

In this section we look at Euler's fundamental strut theory and compare it with one of the empirical relationships that have been established through experimentation, in particular with the Rankine–Gordon relationship. We start with Euler theory, which is based on the assumption that a strut will fail due to buckling before it fails due to direct compression.

Now equation 5.5 is a second-order differential equation of a specific form known as a *harmonic DE*. It has a solution of the form $y = A \cos mx + B \sin mx$ and this is why we manipulated our BM equation in this way, in order to make use of this result.

Note: *Essential Mathematics 3*, on the companion website, explains how to find solutions for some of the simpler first- and second-order differential equations that will be of use to you during your study of engineering science.

So, using our substitution, a solution for our equation $\dfrac{d^2y}{dx^2} + \dfrac{P_e}{EI}y = 0$ is $y = A \cos \sqrt{\left(\dfrac{P_e}{EI}\right)}x + B \sin \sqrt{\left(\dfrac{P_e}{EI}\right)}x$. Now, to find a particular solution, we need to consider the *boundary conditions*, as we did previously when studying bending theory. Then, from Figure 5.5, at $x = 0$ then $y = 0$, so that, $0 = A \cos(0) + B \sin(0)$, therefore $A = 0$. Also at $x = L$, $y = 0$, where for a strut with pinned end joints, $L = L_e$, then $B \sin \sqrt{\left(\dfrac{P_e}{EI}\right)}L = 0$, so either $B = 0$ or $\sin \sqrt{\left(\dfrac{P_e}{EI}\right)}L = 0$. If $B = 0$ then, from Figure 5.5, $y = 0$ and so the strut would not have buckled. Therefore one solution is that $\sqrt{\left(\dfrac{P_e}{EI}\right)}L = \pi$. Under these circumstances, therefore, the *Euler buckling load* is found from the equation:

$$P_e = \frac{\pi^2 EI}{L^2} \tag{5.6}$$

We have chosen just *one solution* for $\sin \sqrt{\left(\dfrac{P_e}{EI}\right)}L = 0$ or $\sin mL = 0$ out of many, for example

$mL = 2\pi, 3\pi, 4\pi$ etc. Our chosen solution is known as the *fundamental* and all other solutions (in theory an infinite number) are known as the *harmonics*, each one creating a different mode of buckling. The lowest value corresponds with our fundamental mode of buckling illustrated in Figure 5.5, while if the strut is prevented from buckling in the fundamental mode, or subject to shock loads, then it may fail in the other modes illustrated in Figure 5.6.

In practice these higher modes of failure are rare, since the high stress associated with the fundamental failure mode ensures immediate buckle failure. The buckling loads for higher harmonic loads of failure may be found in a similar manner to the method we used to find the fundamental buckling load. These are shown, as appropriate, in Figure 5.6.

We also know that the failure stress $\sigma = \dfrac{P_e}{A}$, and from equation 5.3, where $k = \sqrt{\dfrac{I_{\text{least}}}{A}}$, we have that $I_{\text{least}} = Ak^2$, so that equation 5.6, $P = \dfrac{\pi^2 EI}{L^2}$, may be written as $\sigma = \dfrac{\pi^2 EAk^2}{AL^2}$ or as $\sigma = \dfrac{\pi^2 E}{\left(L/k\right)^2}$. Now, as the length of the strut decreases, the critical load increases and if the strut is short enough it will fail by direct compression before reaching the *Euler load* P_e, where here we are talking about failure due to the compressive collapsing stress (σ_c) or yield stress (σ_y) being attained before the Euler stress (σ_e) is reached. Once the Euler load has been reached, the resulting stress is given by

$$\sigma_e = \frac{\pi^2 E}{\left(L_{\text{eff}}/k\right)^2} \tag{5.7}$$

From equation 5.7 it can be seen that the Euler failure stress is inversely proportional to the square of

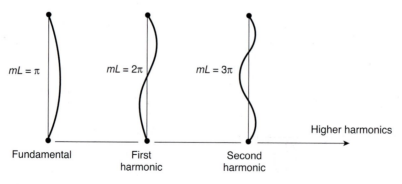

$mL = \pi$ Fundamental

$mL = 2\pi$ First harmonic

$mL = 3\pi$ Second harmonic

Higher harmonics

Figure 5.6 Fundamental and harmonic modes of strut failure

the slenderness ratio, $\dfrac{L_{\text{effective}}}{k}$. It has been shown by experiment that for slenderness ratio values between approximately 40 and 100, the results from Euler theory give failure values that are in excess of those found experimentally. This of course is totally unsatisfactory! Thus we need to place limits on the use of Euler theory when considering strut failure, where in practice it should not be used for struts with slenderness ratios less than approximately 120.

To overcome the problem of assessing failure in struts that have slenderness ratios in the approximate range of 40 to 120, we need to consider some other form of analysis. One such method, the *Rankine–Gordon formula*, we consider in the next section.

> **Key Point.** Always check the validity and limitations of Euler theory when using it to determine buckling loads

Buckling of a strut with pinned ends is considered to be the basic case, as explained above. However, we may also wish to find the Euler critical buckling load (P_e) for struts with other end fixings such as those illustrated in Table 5.2. These can be found from the differential equation of the deflection curve (equation 5.4), in a similar manner to the method we used for our basic pinned-end strut, except that we will need to consider the differing boundary conditions associated with each of the different attachment methods.

Another method for determining P_e for a variety of attachment methods is to consider the concept of effective length, which we looked at earlier. We know from Table 5.2 that for the basic case where both ends are pinned, the effective length is equal to the actual length of the strut, or $L_e = L$. So from equation 5.6, $P_e = \dfrac{\pi^2 EI}{L^2}$, we get that

$$P_e = \frac{\pi^2 EI}{L_e^2} \qquad (5.8)$$

where equation 5.8 now relates the Euler critical load to the *effective length* of the strut. Thus, for example, from Table 5.2, the critical load for a strut with *both ends rigidly fixed* is found by substituting $L_e = \dfrac{L}{2}$ into equation 5.8, so we get that $P_e = \dfrac{\pi^2 EI}{\left(\dfrac{L}{2}\right)^2}$ or $P_e = \dfrac{4\pi^2 EI}{L^2}$. The relationships between P_e and the end-fixing

conditions shown in Table 5.2 are given in Table 5.3. In addition, the corresponding expressions for the Euler critical stress and fixing method are given, based on the substitution $I_{\text{least}} = Ak^2$, used to find equation 5.7.

Table 5.3 Euler smallest critical loads and stresses for various methods of attachment

Method of fixing	Both ends pinned	Both ends rigidly fixed	One end fixed, one end free	One end fixed, one end pinned
Effective length (L_e)	L	$\dfrac{L}{2}$	$2L$	$\approx 0.7L$
Euler critical load (P_e)	$\dfrac{\pi^2 EI}{L^2}$	$\dfrac{4\pi^2 EI}{L^2}$	$\dfrac{\pi^2 EI}{4L^2}$	$\dfrac{2\pi^2 EI}{L^2}$
Euler critical stress (σ_e)	$\dfrac{\pi^2 E}{\left(L/k\right)^2}$	$\dfrac{4\pi^2 E}{\left(L/k\right)^2}$	$\dfrac{\pi^2 E}{4\left(L/k\right)^2}$	$\dfrac{2\pi^2 E}{\left(L/k\right)^2}$

In the above table $L =$ the actual length of the strut.

Example 5.3   A hollow-section steel column with an effective length $L_e = 12$ m has an external diameter of 30 cm and an internal diameter of 20 cm and is rigidly supported at its base. It is subject to a compressive load acting axially on its free end. If for the steel $E = 210$ GN/m², determine the:

a)   slenderness ratio for the column;

b)   least value of I;

c)   critical (safe) load that can be taken by the column; and

d)   corresponding critical stress.

Answers

a)   From Table 5.1, k for an annular cross-section is given by $k = \dfrac{\sqrt{D^2 + d^2}}{4} = \dfrac{\sqrt{0.3^2 + 0.2^2}}{4} = 0.09014$ and so with an effective length of 12 m the slenderness ratio $\dfrac{L}{k} = \dfrac{12}{0.09014} = 133.128$, so the *Euler method is valid*.

b) The least value of I for an annulus $= \dfrac{\pi\left(D^4 - d^4\right)}{64}$ (see Example 5.2a), so $I = \dfrac{\pi\left(0.3^4 - 0.2^4\right)}{64} = 3.19068 \times 10^{-4}\,\text{m}^4$. From Table 5.3, for a strut with one end fixed and one end free, we get $P_e = \dfrac{\pi^2 EI}{4L^2}$, so $P_e = \dfrac{(\pi^2)(210 \times 10^9)(3.19068 \times 10^{-4})}{(4)(12^2)}$ and $P_e = 1.1481\,\text{MN}$

c) Again, from Table 5.3, $\sigma_e = \dfrac{\pi^2 E}{4\left(L/k\right)^2} = \dfrac{(\pi^2)(210 \times 10^9)}{(4)(133.128)^2} = 29.236\,\text{MN/m}^2$.

5.2.2   The Rankine–Gordon relationship

In practice, many strut or column designs do not fall into the category of slenderness ratios applicable to Euler theory. As you have seen previously, short struts, with a slenderness ratio of less than 15, are likely to fail in compression before they fail in buckle, while the failure criteria for struts with slenderness ratios between 40 and 120 are not predicted at all well using Euler theory. Consequently a number of empirical design formulae have been established experimentally, to cater for this shortfall in the existing theory. These numerous design codes for struts and columns may be found in the current AISC, ISO, EN and BS standards. Space only permits us to consider one useful empirical design formula here, that is the *Rankine–Gordon relationship*.

Key Point. The Rankine–Gordon empirical relationship together with its appropriate design codes may be used to overcome some of the limitations concerning slenderness ratios associated with Euler theory

This relationship allows for the effect of direct compression and suggests that the strut will fail at a Rankine–Gordon load given by

$$\frac{1}{P_{RG}} = \frac{1}{P_e} + \frac{1}{P_c} \qquad (5.9)$$

where P_e is the critical Euler load and P_c is the critical failure load resulting from direct compression, as before. It can be seen from this relationship that if the strut is short, the Euler load would be very large and so $\dfrac{1}{P_e}$ would be negligibly small and can be ignored, so that $P_{RG} \approx P_c$; that is, according to Rankine–Gordon, the strut will fail in direct compression, as would be expected. Similarly, if the strut is very long then P_e would be very small and $\dfrac{1}{P_e}$ very large in comparison with $\dfrac{1}{P_c}$ so that in this case $P_{RG} \approx P_e$. Thus, this relationship is valid for struts with slenderness ratios at the extremes, as well as providing a reasonable estimate for failure loads of struts with intermediate values of slenderness ratio.

Now, from equation 5.9, writing the loads in terms of stresses and areas, we get that $\dfrac{1}{\sigma A} = \dfrac{1}{\sigma_e A} + \dfrac{1}{\sigma_c A}$. Since A is a common factor, then $\dfrac{1}{\sigma} = \dfrac{1}{\sigma_e} + \dfrac{1}{\sigma_c}$ or $\sigma = \dfrac{\sigma_e \sigma_c}{\sigma_e + \sigma_c}$ and, on division by σ_e, we may write that $\sigma = \dfrac{\sigma_c}{1 + \dfrac{\sigma_c}{\sigma_e}}$. Then from Table 5.3 for a strut pinned at both ends (our standard case), we have that $\sigma_e = \dfrac{\pi^2 E}{\left(L/k\right)^2}$ and on substitution we get that $\sigma = \dfrac{\sigma_c}{1 + \dfrac{\sigma_c}{\pi^2 E}\left(\dfrac{L}{k}\right)^2}$.

Or, where (a) is a constant for any given material and with other given end conditions, found experimentally, we may write that

$$\sigma = \frac{\sigma_c}{1 + a\left(\dfrac{L}{k}\right)^2} \qquad (5.10)$$

Also in terms of the Rankine–Gordon load we may write equation 5.10 as:

$$P_{RG} = \frac{\sigma_c A}{1 + a\left(\dfrac{L}{k}\right)^2} \qquad (5.11)$$

Typical values of 'a' for *pin-ended struts* are 0.00013 for mild steel, where $\sigma_c \approx 300\,\text{MN/m}^2$, and 0.006 for cast iron, where $\sigma_c \approx 550\,\text{MN/m}^2$.

Example 5.4 A hollow square-section cast iron column rigidly supported at both ends has external sides of 40 cm, a wall thickness of 5 cm and is 10 m long. Determine the slenderness ratio, the safe stress and safe load that can be taken by the column, using the Rankine–Gordon formula. Take the typical value for a for a cast iron strut with fixed ends as 0.00015 and σ_c for cast iron as given above, when using this formula.

We know that $k = \sqrt{\dfrac{I_{\text{least}}}{A}}$ and that $A = \dfrac{D^2 - d^2}{4} = \dfrac{0.4^2 - 0.3^2}{4} = 0.0549778 \text{ m}^2$ and, about the centroid, $I_{\text{least}} = \dfrac{D^4 - d^4}{12}$ for the square-section column. Then $I_{\text{least}} = \dfrac{0.4^4 - 0.3^4}{12} = 1.4667 \times 10^{-3} \text{ m}^4$ and so $k = \sqrt{\dfrac{1.4667 \times 10^{-3}}{0.0549778}} = 0.16333$. Since for a column rigidly supported at both ends the effective length is equal to $L/2 = 10/2 = 5$ m, the slenderness ratio in this case $\dfrac{L_{\text{eff}}}{k} = \dfrac{5.0}{0.16333} = 30.61$. Note that this slenderness ratio for the cast iron strut immediately precludes the use of Euler theory as a method of solution for this situation. Now, using equation 5.10 and the appropriate values for cast iron given in the question and earlier, we get that:

$$\sigma = \frac{\sigma_c}{1 + a\left(\dfrac{L}{k}\right)^2} = \left[\frac{(550 \times 10^6)(0.0549778)}{1 + (0.00016)(30.61)^2}\right]$$

$$= 201.699 \text{ MN/m}^2.$$

Also, the safe load $P_{RG} = \sigma A = (201.699 \times 10^6)(0.0549778) = 11.0889 \text{ MN}.$

5.3 Chapter summary

In this chapter we have taken a first look at *concentrically loaded* columns and struts. Eccentrically loaded struts have not been considered, although after studying this chapter it would be a relatively simple matter to determine the effects and consequences of eccentric loading by, for example, extending the use of specific design codes and empirical formulae to include those for eccentric loading of struts, such as the Perry–Robertson formula. However, in the interests of space and the introductory nature of this subject, a study of eccentrically loaded struts and columns has not been included.

We defined in general terms the difference between columns and struts, where the former were considered to be short thick members that take compressive loads, while struts were considered to be long thin members that were likely to fail due to buckle before failing due to crushing. However, no attempt was made to set hard and fast definitions, and the names column or strut have been used freely throughout this chapter, as appropriate.

The concepts of *slenderness ratio, radius of gyration* and *effective length* were defined and their uses with respect to column and strut design were considered. Apart from the simplest of solid cross-sections, equation 5.1 should be used to determine the slenderness ratio for the most used column cross-sections, that is: Slenderness ratio $= \dfrac{\text{Effective length}}{\text{Least radius of gyration}} = \dfrac{L_e}{k}$. This equation required us to be able to define and understand the significance of the radius of gyration (equation 5.2) and in particular to differentiate between the radius of gyration and the least radius of gyration $k = \sqrt{\dfrac{\text{Least moment of inertia}}{\text{Area of cross-section}}} = \sqrt{\dfrac{I_{\text{least}}}{A}}$ (equation 5.3), used when defining the slenderness ratio. Table 5.1 details a few formulae for finding k for some *steel* column cross-sections. This table should not be used for columns or struts that are manufactured from materials other than steel.

The effective length of struts was considered next and the relationships between the effective length and type of end fixings were detailed in Table 5.2. Note also that you may be required to find the least moment of inertia (*second moment of area*) when determining k. Do be careful to select the appropriate formula for the *least second moment of area* for the various cross-sections, which acts about the centroid of the cross-section being considered (see Table 2.1 for second moments of area about *principal* axes).

The second part of the chapter was given over to establishing and using Euler theory and the empirical Rankine–Gordon relationship, where we found, from the solution of the differential equation for the deflection of a strut subject to an axial load (equation 5.4), two equations $P_e = \dfrac{\pi^2 EI}{L^2}$ (equation 5.6) and $\sigma_e = \dfrac{\pi^2 E}{\left(L_{\text{eff}}/k\right)^2}$

(equation 5.7) for the *Euler load* and *Euler stress* (respectively), using a strut that was *pinned at both ends* as our basis for analysis. Note that it was the fundamental solution of the differential equation that gave us our most valid results for the least Euler load (P_e) and Euler stress (σ_e) that resulted in failure by buckling before failure due to exceeding the yield strength of the strut. Table 5.3 provides a number of formulae for finding Euler loads and stresses, for a number of methods of attachment of the strut ends. These were derived using the concept of *effective length* (L_e). We also considered the limitations in the use of Euler theory and noted that it was not valid, in particular for slenderness ratios between circa 40 and 120 where Euler theory gave values that were in excess of those found empirically through experiment this was considered most unsatisfactory!

To overcome the limitations imposed by the use of Euler theory, a number of empirical relationships and design codes have been developed. we looked in particular at one of the most useful of these, the Rankine–Gordon relationship, where we derived equations for

the Rankine–Gordon stress $\sigma = \dfrac{\sigma_c}{1 + a\left(\dfrac{L}{k}\right)^2}$ (equation

5.10) and load $P_{RG} = \dfrac{\sigma_c A}{1 + a\left(\dfrac{L}{k}\right)^2}$ (equation 5.11), for

a pin-ended strut subject to concentric axial loads.

Typical values of 'a' for *pin-ended struts* were given, these being 0.00013 for mild steel, where $\sigma_c \approx 300\,\text{MN/m}^2$, and 0.006 for cast iron, where $\sigma_c \approx 550\,\text{MN/m}^2$. Note also that the values of the constant 'a' for struts rigidly supported at both ends have been found by experiment to be approximately one-quarter those given here for pin-ended struts. More details on all the constants found empirically for a range of relationships associated with the mode of failure of struts and columns may be found in the associated national (BS) and international (AISC, ISO, EN) design standards.

5.4   Review questions

1. A solid rectangular post is 15 cm × 10 cm in cross-section and has an effective length of 2.4 m. Determine its least radius of gyration and slenderness ratio.

2. A hollow concrete column has a rectangular cross-section with an external breadth of 300 mm and external depth of 600 mm. If the internal breadth is 240 mm and the internal depth is 520 mm, then, for a column with an effective length of 2.8 m, calculate:

   a)   the least radius of gyration;

   b)   the slenderness ratio.

3. Explain the circumstances under which Euler theory provides a reasonably accurate estimate of the load and stress under which a strut subject to a concentric load will fail due to buckle.

4. The rolled steel universal I-section strut shown in Figure 5.7, is subject to an axial load through its centroid.

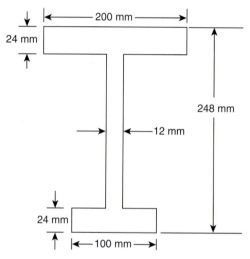

Figure 5.7 I-section strut subject to axial load through its centroid

The beam is 6 m long and is to be rigidly fixed at its base and free at the top end where the load is applied. If for the steel $E = 210\,\text{GN/m}^2$, determine, using Euler theory as appropriate:

   a)   the least value of I;

   b)   the slenderness ratio for the strut;

   c)   the safe load that can be taken by the strut; and

   d)   the corresponding safe stress.

5. A tubular stainless steel strut pinned at both ends is 3 m long and has an outer diameter of 40 mm and an internal diameter of 35 mm. Determine the

critical buckling loads for the strut, using both Euler theory and the Rankine–Gordon formula. Take the compressive yield stress for the strut as 315 MPa, the Rankine–Gordon constant $a = 0.00013$ and $E = 200$ GPa.

6. A steel I-section strut is to be manufactured with the cross-section shown in Figure 5.8. It is to be 8 m in length, fixed at its lower end and pin-jointed at its upper end. It is designed to take a compressive load of 120 kN and for the steel $E = 207$ GPa. Determine, using Euler theory, the minimum dimension for the breadth and depth of the strut cross-section in order for it to resist buckling loads.

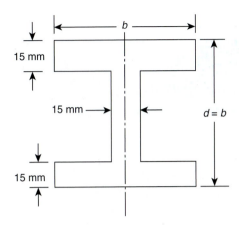

Figure 5.8 Design cross-section for strut

Chapter 6

Introduction to strain energy

In this chapter we consider the subject of the potential energy within a body that results from the application of external loads.

Strain energy or *resilience* (*the strain energy per unit volume*) is a form of potential energy possessed by a material when strained by an external force. The *work done* in producing the strain is stored as strain energy. If the material is strained by a force in tension, compression, bending or torsion, *within its elastic range*, it will return to its unstrained state after the removal of the external force by releasing the stored strain energy. The energy released is known as the *elastic strain energy* and is equal in magnitude to the work done in producing it.

6.1 Strain energy resulting from direct stress and pure shear stress

6.1.1 Strain energy from direct stress

A typical example of strain energy in an elastic material resulting from a *direct stress* occurs, for example, when a spring is compressed or extended. *The work done on the spring during the compression/extension is equal to the energy stored in the coils of the spring.* Since, in the case of a spring, the load required at any instant during the compression/extension is directly proportional to the amount of strain produced on the spring, this may be represented by a straight-line load–extension graph.

Consider the load–extension graph shown in Figure 6.1, where an elastic material (such as a steel bar with cross-sectional area A and length l) is subject to a gradually increasing tensile load (P).

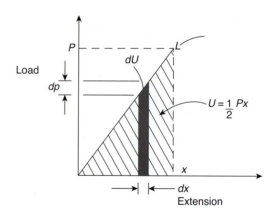

Figure 6.1 Load–extension diagram for an elastic material subject to a gradually applied tensile load

The external work done during a small increment of the extension (dx) is (from the graph) $dpdx$ and this external work is equal to the strain energy (dU) it produces in the material, i.e. $dU = dpdx$. This is represented by the shaded area in Figure 6.1. Therefore, the total work done (and so the total strain energy produced) as a result of the load being applied up to the limit of proportionality (L) of the material is given by

$$U = \frac{1}{2}Px \qquad (6.1)$$

that is, the total area under the graph (hatched area in Figure 6.1).

978-1-85617-775-7, Engineering Science, Mike Tooley & Lloyd Dingle

Now we know that the elastic modulus $E = \dfrac{\text{stress}}{\text{strain}}$ or

$E = \dfrac{P/A}{x/l} = \dfrac{Pl}{Ax}$, and so $x = \dfrac{Pl}{AE}$. Then $U = \dfrac{1}{2}P\dfrac{Pl}{AE}$ or

$$U = \frac{P^2 l}{2AE} \qquad (6.2)$$

Also, since stress $\sigma = \dfrac{P}{A}$ or $P = \sigma A$, then substituting in P in equation 6.1, we get that

$$U = \frac{\sigma^2 Al}{2E} \qquad (6.3)$$

We also know that Al is the volume of the bar, therefore:

$$U = \frac{\sigma^2}{2E} \times \text{volume} \qquad (6.4)$$

Also, the strain energy or *resilience* per unit volume of a material subject to direct compressive or tensile loads is given as:

$$U = \frac{\sigma^2}{2E} \qquad (6.5)$$

With specific reference to *springs*, we can see from equation 6.1 that if our elastic material is a spring then the strain energy within its coils is $U = \dfrac{1}{2}Px$. Since springs are rated according to their stiffness, where the *spring rating* (k) is defined as *the load required to produce unit compression or extension*, i.e. $k = \dfrac{P}{x}$ (N m), then $P = kx$. On substitution of this expression into equation 6.1, we get that the *strain energy of a spring when compressed or extended* is given by:

$$U = \frac{kx^2}{2} \qquad (6.6)$$

where this strain energy is measured in N m or Joules (J).

6.1.2   Strain energy from shear stress

If an elastic material such as the steel bar we considered earlier is subject to a load causing *pure shear*, then the strain energy stored per unit volume is represented by the area under the shear-stress (τ)/shear-strain (γ) curve shown in Figure 6.2.

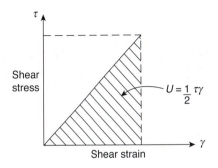

Figure 6.2 Graph of shear stress against shear strain for an elastic material subject to pure shear load

Then, from the graph, *the strain energy stored in the material per unit volume* is

$$U = \frac{1}{2}\tau\gamma \qquad (6.7)$$

Knowing, from your introductory work, that the *modulus of rigidity* or *shear modulus* $G = \dfrac{\tau}{\gamma}$ or $\gamma = \dfrac{\tau}{G}$, then, on substituting this expression into equation 6.7, we get that the strain energy stored in the material is

$U = \dfrac{\tau^2}{2G}$ per unit volume or that

$$U = \frac{\tau^2}{2G} \times \text{(volume of elastic material)} \qquad (6.8)$$

It is important that you are aware of the fact that equation 6.8 is only valid if the shear stress is uniform over the section of material concerned.

Now, in terms of the shear load (P_s) we know that the shear stress $\tau = \dfrac{P_s}{A}$ or that $P_s = \tau A$, and also that the volume of our elastic material, say of length (l), is equal to Al, so we have from equation 6.8 that

$U = \left[\left(\dfrac{P_s^2}{A^2}\right)\left(\dfrac{1}{2G}\right)\right](Al)$. Therefore, in terms of the load causing pure shear and the dimensions of the elastic material, we get that

$$U = \frac{P_s^2 l}{2AG} \qquad (6.9)$$

Key Point.   The equation $U = \dfrac{\tau^2}{2G} \times$ (volume of elastic material), relating the internal strain energy with the work done in shear, is only valid if the shear stress is uniform over the whole section of the material concerned

6.1.3   Strain energy considering the weight of the body

So far in our analysis of strain energy we have not considered the effect of strain due to the weight of the body itself. The following examples illustrate this effect.

Example 6.1   Determine the elastic strain energy stored in a bar suspended from one end, due to its own weight.

Figure 6.3 illustrates the situation.

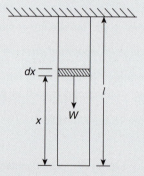

Figure 6.3   Bar suspended under its own weight

Consider the *element* of the bar of depth dx, shown in Figure 6.3. The weight force (W) of the bar below the element, and therefore acting on the element, is given by $W = \rho g A x$, where $\rho =$ the mass density (kg/m^3), $g =$ the acceleration due to gravity (m/s^2) and $A =$ the cross-sectional area of the bar. Now, substituting W for the load P in equation 6.2 and integrating will give the strain energy in the element due to the weight of the bar, i.e. substituting for W in $U = \dfrac{P^2 l}{2AE}$ where, for the element $l = dx$, we get $U = \dfrac{(\rho g A x)^2 dx}{2AE}$, so that the

strain energy in the element is $U = \dfrac{(\rho g)^2 A x^2 dx}{2E}$ and the total strain energy due to the weight of the bar is

$$U = \int_0^l \frac{(\rho g)^2 A x^2}{2E}\,dx \quad \text{or} \quad U = \frac{(\rho g)^2 A l^3}{6E}$$

$$(6.10)$$

In Example 6.2 we will find an expression for the strain energy due to both the weight of the bar and an external load suspended from one end of the bar.

Example 6.2   Determine the strain energy of a bar suspended at one end, if it is subject to a tensile load P in addition to the strain energy created by its own weight. Figure 6.4 illustrates the situation.

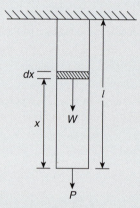

Figure 6.4   Suspended bar, subject to its own weight and an external load

In this case the axial load acting on the element is that due to the external tensile load P and the weight $W = \rho g A x$ (from Example 6.1), i.e. the tensile load acting on the element $= \rho g A x + P$. Substituting this expression for P in equation 6.2 as before, where $l = dx$, gives $U = \dfrac{(\rho g A x + P)^2 dx}{2AE}$ or, after multiplication of the brackets, gives $U = \dfrac{\rho^2 g^2 A^2 x^2}{2AE}dx + \dfrac{2P\rho g A x}{2AE}dx + \dfrac{P^2}{2AE}dx$, from which, after simplification, we get that $U = \dfrac{(\rho g)^2 A x^2}{2E}dx + \dfrac{P\rho g x}{E}dx + \dfrac{P^2}{2AE}dx$, and so the total strain energy in the bar, subject to its own weight and an external load, is found after integration of this expression between the limits $(0, l)$ as:

$$U = \frac{(\rho g)^2 A l^3}{6E} + \frac{P\rho g l^2}{2E} + \frac{P^2 l}{2AE} \qquad (6.11)$$

6.1.4 Strain energy due to impact/dynamic loading

So far we have considered the strain energy resulting from a gradually applied load. If, however, a body is subject to a dynamic, suddenly applied impact load, the initial stress, strain and strain energy set up within the body are much greater than if the load was gradually applied. However, these much greater loads are temporary and vary with time, gradually settling to the values obtained for the equivalent gradually applied loads.

Consider the situation illustrated in Figure 6.5, where a sliding weight $W = mg$, initially at rest, is dropped from a height h onto a flange and moves down with it (i.e. the weight and the flange display perfect plastic behaviour during impact), to produce an instantaneous extension x and a resulting instantaneous stress σ.

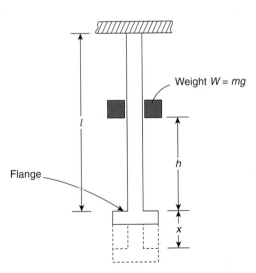

Figure 6.5 Sliding weight dropped from a height impacting with flange of suspended bar

Now, from the conservation of energy (assuming no external energy losses such as heat or sound, i.e. a perfectly plastic impact etc.) we may equate the potential energy $mg(h + x)$ or $W(h + x)$ when the weight is at rest to the gain in strain energy at impact, where the gain in

strain energy in terms of the stress is given by equation 6.4 as $U = \frac{\sigma^2}{2E} \times$ volume (Al), so that:

$$W(h + x) = \frac{\sigma^2 Al}{2E} \qquad (6.12)$$

If the extension (x) is very small compared with the length of the bar, we may ignore it to give the approximation for the resulting stress as $2EWh = \sigma^2 Al$ or $\sigma = \sqrt{\dfrac{2EWh}{Al}}$. But if the extension (x) is not considered small in terms of the length of the bar (l), then it needs to be taken into account when determining the resulting stress due to impact. Therefore, from the definition of Young's modulus where $E = \dfrac{\text{stress}}{\text{strain}}$ and where the strain $= \dfrac{x}{l}$ (from Figure 6.4), we get that $E = \dfrac{\sigma l}{x}$ or that $x = \dfrac{\sigma l}{E}$ and, on substituting this expression for x into equation 6.12, we get that

$$Wh + \frac{W\sigma l}{E} = \frac{\sigma^2 Al}{2E} \qquad (6.13)$$

$$\text{or} \quad \frac{\sigma^2 Al}{2E} - \frac{\sigma Wl}{E} - Wh = 0 \qquad (6.14)$$

which is a quadratic equation in terms of the variable (σ) and may be solved by applying the quadratic formula to give a precise equation for the instantaneous stress as:

$$\sigma = \frac{W}{A} \pm \sqrt{\left[\left(\frac{W}{A}\right)^2 - \frac{2EWh}{Al}\right]} \qquad (6.15)$$

Now, the first term after the equals sign in equation 6.15 is the value of the stress due to the static load acting on the bar. The positive expression under the square root sign is the additional stress at the point of maximum extension resulting from the impact load, while the minus expression under the square root sign is the stress that results from the first recoil of the bar.

Consider also what happens when $h = 0$, that is, when a dynamic load is suddenly applied to the bar. Then, from equation 6.15, the instantaneous stress is given as $\sigma = \dfrac{2W}{A}$, that is, the dynamic stress due to impact is *twice* the stress due to the static load.

Example 6.3   Assume that the suspended bar shown in Figure 6.5 has a diameter of 30 mm and is 3 m long and that a 20 kg sliding mass falls through a height $h = 250$ mm onto its flange without rebounding. If the modulus of the steel is $E = 207$ GN/m^2, determine the instantaneous stress of the bar at the moment of impact.

Now, from equation 6.14, where $\dfrac{\sigma^2 Al}{2E} - \dfrac{\sigma Wl}{E} - Wh = 0$, we can produce a quadratic equation that can be solved in order to determine the required stress.

The volume of the bar $Al = \dfrac{\pi d^2 l}{4} = \dfrac{\pi (30 \times 10^{-3})^2 (3)}{4} = 21.21 \times 10^{-4}$. Then we have that $\dfrac{\sigma^2 (21.21 \times 10^{-4})}{2 \times 207 \times 10^9} - \dfrac{\sigma (20 \times 9.81 \times 3)}{207 \times 10^9} - (20 \times 9.81 \times 0.3) = 0$, so that $\dfrac{\sigma^2}{19.52 \times 10^{13}} - \dfrac{2.84\sigma}{10^9} - 58.86 = 0$. On multiplication by (19.52×10^{13}) and after simplification, we get that $\sigma^2 - 55.44 \times 10^4 \sigma - 11489.47 \times 10^{12} = 0$. Then applying the quadratic formula gives $\sigma = \dfrac{55.44 \times 10^4 \pm \sqrt{307 \times 10^9 + 45957.9 \times 10^{12}}}{2}$ or $\sigma = \dfrac{55.44 \times 10^4 \pm 214.38 \times 10^6}{2}$, and so the instantaneous stress $\sigma = 214.93$ MN/m^2.

6.2   Strain energy in bending and torsion

6.2.1   Strain energy due to bending

Consider the element of length dx of a uniform beam, shown in Figure 6.6, that is subject to a constant bending moment, causing it to bend through an angle $d\theta$, with radius of bend R.

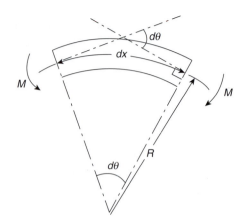

Figure 6.6   Element subject to a constant bending moment

The strain energy is equal to the work done by the bending moment, which from the geometry of the element shown $= \dfrac{1}{2} M d\theta$.

Also, because θ is in radians, we may write that $dx = R d\theta$ or $d\theta = \dfrac{dx}{R}$. From engineers' theory of bending, which you studied earlier, we know that $\dfrac{M}{I} = \dfrac{E}{R}$ or $\dfrac{1}{R} = \dfrac{M}{EI}$, so that $d\theta = \dfrac{dx}{R} = \dfrac{M}{EI} dx$.

Therefore the resulting strain energy within the element $= \left(\dfrac{M}{2}\right)\left(\dfrac{M}{EI} dx\right) = \dfrac{M^2}{2EI} dx$, and so the total strain energy resulting from the bending moment (that may vary) within the *uniform beam* of length (l) is given by

$$U = \int_0^l \frac{M^2}{2EI} dx \qquad (6.16)$$

If the bending moment (M) causing the bending is *constant* throughout the whole length of the beam, then equation 6.16, after integration of the only variable dx (between the limits), becomes

$$U = \frac{M^2 l}{2EI} \qquad (6.17)$$

Note that it is far more usual for the *bending moment* to *vary* along the length of the beam, so that equation 6.16 will be more useful for analysis, provided that an expression for the varying bending moment M in terms of dx is found prior to integration.

Key Point. When using equation 6.16, i.e. $U = \int_0^l \frac{M^2}{2EI}dx$, always remember to write the expression for the varying bending moment (M) in terms of the variable dx prior to integration

Note: For *simply supported beams* subject to a single load P, the work done by the load $= \frac{1}{2}Py$, where $y =$ the linear deflection of the beam at the point of application of the load. Under these circumstances this work is equal to the strain energy it produces, i.e.

$$U = \int_0^l \frac{M^2}{2EI}dx = \frac{1}{2}Py.$$

Example 6.4 A cantilever beam has a concentrated load P at its free end (Figure 6.7). Find, using equation 6.16 and ignoring shear, a formula for the strain energy in the beam due to the load.

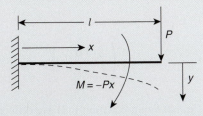

Figure 6.7 Cantilever beam subject to a concentrated load at its free end

Also, given that the beam is 1.8 m long, is made from steel with $E = 200$ GN/m², $I = 150 \times 10^{-6}$ m⁴, and that the applied load $P = 20$ kN, find the value in Joules of the strain energy.

By ignoring shear we may use equation 6.16, i.e. $U = \int_0^l \frac{M^2}{2EI}dx$. All we need do to apply this formula is to find a value for M in terms of the variable x. We know from our study of bending (see section 2.2) that at any point x along the beam the bending moment is simply given by $M = -Px$ (the minus sign indicating that the load causes the beam to 'hog'), so substituting this value into equation 6.16 gives $U = \int_0^l \frac{(-Px)^2}{2EI}dx = \int_0^l \frac{P^2x^2}{2EI}dx$. After integration, we get that $U = \frac{P^2l^3}{6EI}$, as required.

Now, to find the numerical value for this strain energy we simply substitute the values into this result, i.e. $U = \frac{(20 \times 10^3)^2(5.832)}{(6)(200 \times 10^9)(150 \times 10^{-6})}$

$$= 12.96 \, \text{J}$$

6.2.2 Strain energy due to a torque

Let us now consider strain energy due to torsion. Figure 6.8 shows an element of length dx of a uniform shaft subject to a uniform torque T, causing the element to twist through an angle (in radians) $d\theta$.

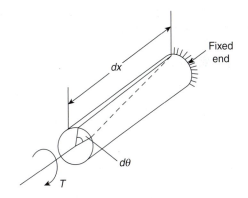

Figure 6.8 Element subject to a torque

In a similar way to bending, the strain energy due to the applied torque is equal to the work done by the torque, therefore strain energy $= \frac{1}{2}Td\theta$. We also know from the engineers' theory of torsion that $\frac{T}{J} = \frac{G\theta}{L}$, so in our case for the element $\frac{T}{J} = \frac{Gd\theta}{dx}$ or $d\theta = \frac{Tdx}{GJ}$, so $U = \left(\frac{T}{2}\right)\left(\frac{T}{GJ}dx\right)$. For the shaft length ($l$), where in practice the torque is usually *constant*, the total strain energy due to the applied torque is given by $U = \int_0^l \frac{T^2}{2GJ}dx$, and so, on integration of the only variable dx between the limits, we get that the:

Strain energy $U = \dfrac{T^2l}{2GJ}$ (6.18)

Key Point. In practice the torque causing the strain is usually constant, therefore the relationship $U = \dfrac{T^2 l}{2GJ}$ may normally be used

6.3   Castigliano's theorem

A very powerful and yet simple-to-use energy method for analysing deflections of externally loaded components is provided by Castigliano's theorem for deflection, which states that: *If the strain energy of a linearly elastic structure is expressed in terms of the external loads and is partially differentiated with respect to one of these loads, the result is a deflection at the point of application of the load in the direction of this load.* Now, in this expression for Castigliano's theorem, the terms *load* and *deflection* may generally be replaced equally well by *moment* and *angular deflection* when considering bending and torsional loads.

So, mathematically, if U is the *total* strain energy, then the deflection (say y_1) at the point of application of the load (P_1) in the direction of (P_1) is found to be

$$y_1 = \frac{\partial U}{\partial P_1} \qquad (6.19)$$

Similarly, $y_2 = \dfrac{\partial U}{\partial P_2}, \dots y_3 = \dfrac{\partial U}{\partial P_3}, \dots\dots y_n = \dfrac{\partial U}{\partial P_n}$.

Note that the relationship found in equation 6.19 is given here without proof.

Key Point. When using Castigliano's first theorem for deflection, the total strain energy within the material must be used when finding the deflection of the material at a particular point of application of a single load

6.3.1   Examples of the power of Castigliano's theorem

When varying loads cause different bending moments or varying torques it is necessary to use Castigliano's theorem. The following examples show the power of this theorem when we apply it to equations 6.2, 6.16 and 6.18 to obtain expressions for *axial*, *bending* and

torsional deflections, respectively, under varying load conditions.

First, using equation 6.2 for *axial* deflections, where $U = \dfrac{P^2 l}{2AE}$, we get that

$$y = \frac{\partial U}{\partial P} = \frac{\partial}{\partial P}\left(\frac{P^2 l}{2AE}\right) = \frac{Pl}{AE} \qquad (6.20)$$

Now, for bending deflections in *uniform cross-section* beams and curved bars, where equation 6.16 $U = \int\limits_0^l \dfrac{M^2}{2EI} dx$ may be written as $U = \dfrac{1}{2EI}\int\limits_0^l M^2 dx$, we get that

$$y = \frac{\partial U}{\partial P} = \frac{\partial}{\partial P}\left(\frac{1}{2EI}\int\limits_0^l M^2 dx\right) \text{ or}$$

$$y = \frac{\partial U}{\partial P} = \frac{1}{EI}\left(\int\limits_0^l \frac{\partial}{\partial P}(M^2) dx\right),$$

so that

$$y = \frac{1}{EI}\left(\int\limits_0^l \frac{\partial M}{\partial P} dx\right) \qquad (6.21)$$

Note that in obtaining equation 6.21 we differentiated (M^2) before integrating. This is valid since the differentiation was with respect to the variable (P) while the integration is with respect to a different variable (x). Do remember, however, that when finding the integral in equation 6.21, you must put the bending moment (M) in terms of the variable (x).

Also, for *torsional deflections*, from equation 6.18, where $U = \dfrac{T^2 l}{2GJ}$, then the rotation θ at any point where the torque is applied is found to be

$$\theta = \frac{\partial U}{\partial T} = \frac{\partial}{\partial T}\left(\frac{T^2 l}{2GJ}\right) = \frac{Tl}{GJ} \qquad (6.22)$$

Key Point. When using equation 6.22 remember that the angle of rotation θ due to a torque is measured in radians

Castigliano's theorem can also be used to find the deflection at a point *other than where the load or moment acts*, by following a simple procedure that involves the introduction of an imaginary load or moment. The procedure is as follows:

- Set up the equation for the *total strain energy U* acting on the body, *including the energy due to an imaginary load Q* that acts at the point where the deflection is to be found.
- Determine the expression for the *required deflection y* by taking the partial derivative $\left(\text{i.e. by taking}\right.$ $y = \dfrac{\partial U}{\partial Q}\bigg)$ of the total strain energy with respect to *the imaginary load or moment Q*.
- Solve the expression found in step 2 by equating Q to zero. This is allowable since Q was set up as an *imaginary* load or moment in the first place.

> **Key Point.** Castigliano's theorem is very useful in that it can be used to find the deflection at a point other than where the load or moment acts

The following example, based on the cantilever beam shown in Figure 6.7, illustrates the above procedure.

> **Example 6.5**   If the cantilever beam, illustrated in Figure 6.7, is now subject to a point load P acting at its centre, use the procedure set out above to determine:
>
> a)   an expression for the maximum deflection that takes place at the free end of the beam, neglecting shear;
>
> b)   the numerical value of this deflection if the point load at the centre of the beam $P = 25$ kN and if the beam has the same length and E and I values as in Example 6.4.

The new loading situation for the beam is shown in Figure 6.9, where it can be seen that the imaginary point load Q has been placed at the point where the deflection is required.

We are told that we may neglect shear strain energy; this is quite feasible since the beam is 1.8 m in length and the cross-section dimension of the beam is a relatively small fraction of the length, so the value of the strain energy due to shear is small in comparison with the strain energy due to bending.

From Figure 6.9, remembering your previous work on bending moments, it can be seen that the bending moment in the section of the beam before

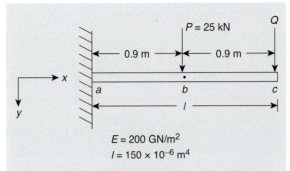

$E = 200$ GN/m^2
$I = 150 \times 10^{-6}$ m^4

Figure 6.9 Cantilever beam subject to a point load at its centre

the centre point load is $M_{ab} = P\left(x - \dfrac{l}{2}\right) + Q\left(x - l\right)$ and that the bending moment in the beam after the centre point load is $M_{bc} = Q\left(x - l\right)$. Then, from equation 6.16, i.e. $U = \int_0^l \dfrac{M^2}{2EI}\,dx$, we get the equation for the total strain energy as $U = \int_0^{l/2} \dfrac{M_{ab}^2}{2EI}\,dx + \int_{l/2}^{l} \dfrac{M_{bc}^2}{2EI}\,dx$ or

$$U = \frac{1}{2EI}\left(\int_0^{l/2} M_{ab}^2\,dx + \int_{l/2}^{l} M_{bc}^2\,dx\right).$$ Now, applying Castigliano's first theorem for deflection to this equation, i.e. $Y = \dfrac{\partial U}{\partial Q}$, we get that

$$y = \frac{\partial U}{\partial Q}$$

$$= \frac{1}{2EI}\left(\int_0^{l/2} \frac{\partial}{\partial Q}\left(M_{ab}^2\right)dx + \int_{l/2}^{l} \frac{\partial}{\partial Q}\left(M_{bc}^2\right)dx\right)$$

and, on partial differentiation with respect to Q, we get that

$$y = \frac{\partial U}{\partial Q} = \frac{1}{2EI}\left[\int_0^{l/2} 2M_{ab}\left(\frac{\partial M_{ab}}{\partial Q}\right)dx\right.$$

$$\left. + \int_{l/2}^{l} 2M_{bc}\left(\frac{\partial M_{bc}}{\partial Q}\right)dx\right].$$

On removal of the common factor 2, we get

$$y = \frac{\partial U}{\partial Q} = \frac{1}{EI}\left[\int_0^{l/2} M_{ab}\left(\frac{\partial M_{ab}}{\partial Q}\right)dx\right.$$

$$\left. + \int_{l/2}^{l} M_{bc}\left(\frac{\partial M_{bc}}{\partial Q}\right)dx\right] \qquad (1)$$

Now, from $M_{ab} = P\left(x - \frac{l}{2}\right) + Q(x - l)$, we get

that $\frac{\partial M_{ab}}{\partial Q} = x - l$ after partial differentiation with

respect to Q. Similarly, from $M_{bc} = Q(x - l)$, we

get that $\frac{\partial M_{bc}}{\partial Q} = x - l$. So, substituting these four

expressions into equation 1 gives

$$y = \frac{\partial U}{\partial Q}$$

$$= \frac{1}{EI}\left[\int_0^{l/2}\left[P\left(x - \frac{l}{2}\right) + Q(x - l)\right](x - l)dx\right.$$

$$\left. + \int_{l/2}^{l} Q(x - l)(x - l)dx\right]. \qquad (2)$$

Now, since Q is imaginary, we can at this stage legitimately equate it to zero, so with $Q = 0$, equation 2 becomes

$$y = \frac{1}{EI}\int_0^{l/2} P\left(x - \frac{l}{2}\right)(x - l)dx \quad \text{or}$$

$$y = \frac{P}{EI}\left[\int_0^{l/2} x^2 - \frac{3lx}{2} + \frac{l^2}{2}dx\right],$$

and on integration we get that

$$y = \frac{P}{EI}\left[\frac{x^3}{3} - \frac{3lx^2}{4} + \frac{l^2 x}{2}\right]_0^{l/2}$$

$$= \frac{P}{EI}\left(\frac{l^3}{24} - \frac{3l^3}{16} + \frac{l^3}{4}\right) = \frac{P}{EI}\left(\frac{5l^3}{48}\right),$$

therefore

$$y = \frac{5Pl^3}{48EI} \qquad (3)$$

I hope you have been able to follow this procedure and cope with the algebra!

Now for part (b). It is just a simple matter of evaluating equation 3, using the given values.

Therefore, from $y = \frac{5Pl^3}{48EI}$ and using consistent

units, we get that

$$y = \frac{5(25 \times 10^3)(1.8)^3}{48(200 \times 10^9)(150 \times 10^{-6})} = 0.0050625\,\text{m}$$

$$= 5.0625\,\text{mm}.$$

This is the deflection at the free end of the cantilever, at point (c).

6.4   Chapter summary

In this chapter you were introduced to the concept of strain energy, where engineering materials and structural components have been subjected to both gradually applied and suddenly applied external loads. The *work done* by these loads (within the elastic range of the material) was equated to the potential energy or *elastic strain energy* set up within the material as a result of this external loading.

In section 6.1.1, a mathematical expression equating the external work done with the internal strain energy was established using a load–extension graph and given

as equation 6.1, i.e. $U = \frac{1}{2}Px$. Using this equation

and the basic relationships between stress, strain and the elastic modulus E, several useful equations were established (equations 6.3 to 6.6), relating the internal strain energy to the stresses set up in materials when subject to direct compressive and tensile loads, including the definition for *resilience* which is simply *the internal strain energy per unit volume*. In section 6.1.2, a similar set of equations was established (based

on the fundamental equation 6.7, i.e. $U = \frac{1}{2}\tau\gamma$, relating

the strain energy stored in the material per unit volume to the work done in shear) for materials subjected to loads or forces that caused the pure shear stresses and strains to be set up within the material. In section 6.1.3, using examples, relationships were established for engineering materials (mainly in the form of bars and rods) that enabled *the internal strain energy to be found that resulted from their own weight, as well as from external loads*. In section 6.1.4 the stresses and elastic strains resulting from materials being subject to *sudden impact loads* were established, and from

equation 6.15, where $\sigma = \frac{W}{A} \pm \sqrt{\left[\left(\frac{W}{A}\right)^2 - \frac{2EWh}{Al}\right]}$,

we found that when materials were subject to impact (dynamic) loads, the stresses set up within the material were *twice* those due to static loads, i.e. $\sigma = \frac{2W}{A}$. This of course has significant implications for design engineers and all those involved with establishing safety standards for materials and structures subject to dynamic loading.

Section 6.2 was concerned with establishing and using relationships for internal strain energies that resulted from materials that were subject to bending and torsional loads. We used our knowledge of the engineers' theory of bending, i.e. that $\frac{M}{I} = \frac{E}{R}$, to help establish these relationships. We discovered that for normal engineering applications the *bending moments were not constant* throughout the length of the beam and, as a result, equation 6.16, that is $U = \int_0^l \frac{M^2}{2EI}dx$, was the most useful relationship for determining the total strain energy within the beam. When establishing the relationship between the total strain energy and torque, we found that *in practice, the torque causing the strain is usually constant*; thus equation 6.18, i.e. $U = \frac{T^2 l}{2GJ}$, can be used.

In section 6.3 you were introduced to *Castigliano's first theorem for deflections*, that is, for deflections due to linear loads and angular deflections due to moments. Finding the elastic deflections in beams and shafts and the loads and torques that cause them is of particular importance to engineers when they consider these beams and shafts for particular practical applications. The rather wordy definition of this theorem is more easily understood by considering the mathematical definition, where the *total strain energy U*, due to *all* the individual loads and/or moments that act on the material or structure, needs to be found, in order to find the *individual* deflection at a particular point and the

direction of application of an individual load or moment. Then this individual deflection (y_i) is found by taking the *partial derivative* of the strain energy with respect to the individual load or moment, as given in equation 6.19, i.e. $y_1 = \frac{\partial U}{\partial P_1}$, for the case of the deflection caused by a particular individual load.

If you are unfamiliar with partial differentiation please take a look at the website that accompanies this book at www.key2engineeringscience.com, where you will find some additional help in the *Essential Mathematics* sections. You may also have wondered about the fact that we only considered Castigliano's first theorem for deflections, given without proof, intimating that there are other theorems. This is in fact true but, due to space considerations and the fact that this section provides only an introduction to Castigliano's methods, his other theorem and its corollaries were not considered. However, for the sake of completeness Castigliano also devised a method for determining *forces/loads* in an elastic structure, as well as displacements, where the theorem for forces or loads states:

> The partial derivative of the strain energy (U) of elastic materials or structures with respect to any specific displacement (y_i) is equal to the corresponding load or force (P_i), provided that the strain energy is expressed as a function of the displacements.

So, mathematically, Castigliano's force theorem may be expressed as $P_i = \frac{\partial U}{\partial y_i}$. Castigliano's theorems related to moments, torques and angular displacements rather than just to loads and displacements, may also be treated separately, adding to the number of Castigliano's methods, or they may be treated in a *generalised* way and so be encompassed into the versions of the theorems we have discussed in this chapter.

As well as finding the deflections of a load or moment at its point of application, Castigliano's first theorem can also be used *to find deflections at any chosen points within the material or structure that are remote from the point of application* of any individual load or moment. To do this you were introduced to a three-step procedure that you should follow when using Castigliano's first theorem for displacement, in order to simplify the analysis. This procedure is repeated here for your convenience.

- Set up the equation for the total strain energy U acting on the body, including the energy due to an imaginary load Q that acts at the point where the load is to be found.

- Determine the expression for the required deflection (y) by taking the partial derivative of $y = \dfrac{\partial U}{\partial Q}$, that is the PD of the total strain energy with respect to the imaginary load Q.
- Solve the expression found in step 2 by equating Q to zero immediately prior to the integration.

Example 6.5 illustrates the use of this procedure in finding the maximum displacement for a cantilever beam subject to a point load remote from where the maximum deflection takes place.

6.5   Review questions

1. A steel bar of density of $7800\,\text{kg/m}^3$, suspended under its own weight, is 2.5 m long and has a diameter of 20 mm. If the bar has a 20 kN load suspended from its low end and the modulus for the steel is $E = 200\,\text{GN/m}^2$, determine the total strain energy in the bar due to the load and to its own weight.

2. A metal bar suspended at its upper end has a collar attached to its lower end. The bar has a diameter of 30 mm and a length of 2 m, and the modulus of the metal is $80\,\text{GN/m}^2$. A sliding mass of 10 kg falls through a height $h = 40$ cm onto the collar without rebounding. Determine the instantaneous stress in the bar at the moment of impact.

3. If, for the situation described in question 1, the maximum stress in the metal bar is to be limited to $120\,\text{MN/m}^2$, what is now the maximum height at which the 10 kg mass can be dropped?

4. A steel beam 1.2 m long, having an elastic modulus $E = 200\,\text{GN/m}^2$, is simply supported at its ends and

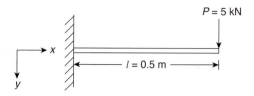

Figure 6.10 Cantilever beam for review question 6.5.5

carries a vertical point load of 15 kN at its centre. If the beam has a diameter of 60 mm, determine:

a) the second moment of area I of the beam;

b) the strain energy in the beam due to the load;

c) the deflection at the point of application of the load.

5. Figure 6.10 shows a short cantilever beam 0.5 m in length, 30 mm in diameter, subject to a load $P = 5$ kN at its free end. Given that the beam is made of a steel with an elastic modulus $E = 200\,\text{GN/m}^2$ and a shear modulus $G = 80\,\text{GN/m}^2$:

a) Find the strain energy in the beam that results from the load, ignoring shear.

b) Determine the maximum deflection using Castigliano's first theorem, ignoring shear.

c) Assuming that the shear stress is uniform over the whole beam and that for the beam the applied point load is equal to the shear load, i.e. $P = P_s$, determine the maximum deflection including shear.

d) If the 5 kN point load is now *applied at the centre of the beam* rather than at the free end, determine, again using Castigliano's theorem, the maximum deflection of the beam.

Chapter 7

Complex stress and strain

You will be familiar with simple stress and strain relationships from your previous study and from the material presented in section 1.5 of Chapter 1, where only stresses and strains in one dimension were considered and the stresses resulting from the loads were considered to be uniform across the whole cross-section of the material. Engineering materials and structures are generally subject to a *combination of loading situations* that set up a complex network of stresses and strains within the material. For example, a structure such as a roof truss for a building may be subject to a combination of tensile, compressive and torsional loads, resulting in a complex set of two- or three-dimensional stresses and resultant strains that may act in different directions on any particular *element* of the material within the truss.

In this chapter, we start by analysing complex stress systems using *stress elements*. These are obtained by isolating an infinitesimally small piece, at a point, of the material under load and then showing the stresses that act on all the surfaces of this element. The idea of stress elements and their use in determining the *direct* and *shear* stresses that result, within the material, is developed in the earlier sections of this chapter, where initially the stresses on planes oblique to direct one- and two-dimensional loads are analysed. *Mohr's stress circle* is then introduced, where a graphical representation of all possible states of normal and shear stress on any element of the material may be identified.

In the second part of the chapter we concentrate on complex strain, including volumetric strain and the relationship of strain to the *elastic constants* (*E*, *G* and *K*) and Poisson's ratio (*v*). We will also consider *Mohr's strain circle* and the *McClintock* method for determining the *principal strains* from *strain gauge rosette* arrangements.

7.1 Stresses on oblique planes

Now, we know from simple stress theory that stresses *normal* to the applied tensile or compressive force (*P*) may be determined using the simple relationship $\sigma = \dfrac{P}{A_0}$, where A_0 is the cross-sectional area normal to (at right angles to) the applied force. This simple relationship relies on the assumption that the stresses created by the applied force are equally distributed over the whole of the cross-sectional area. This assumption of course is not true in general, but by considering a *stress element* of the material, where the sides of the element are infinitesimally small, the stresses may be assumed to be uniformly distributed over the whole face. Therefore, when we are considering stresses within the material that may act on planes that are *other than* at right angles (normal) to the applied force, we will use the idea of stress elements to help us with the analysis.

> **Key Point.** A stress element is an infinitesimally small point in a material, where the stress acting on

978-1-85617-775-7, Engineering Science, Mike Tooley & Lloyd Dingle

any face of the element is deemed to be equal across the whole area of the face

Let us first consider the simplest of cases, where for a bar of rectangular cross-section that is subject to a single direct tensile load P (Figure 7.1) we will determine the *stresses* that act on the oblique plane (EF).

For the rectangular bar shown in Figure 7.1, we will assume that the *depth of the bar is unity* (i.e. the dimension in the *z-direction* of the Cartesian coordinate system shown on the figure is numerically equal to 1.0). Now, remembering Newton's laws, for static equilibrium within the bar, the *resultant* of the forces and thus the stresses acting on the part ABEF must be equal to the axial force P. This resultant tensile force P that acts on the plane EF may be resolved into two components, that is, *the component F_N normal to the plane* EF and the *component F_T tangential to this plane*. Thus, for the situation shown in Figure 7.1, the component of force *normal* to the plane is $F_N = P\cos\theta$ and the component of force *tangential* to the plane is $F_T = P\sin\theta$. Now, as a result of the normal and tangential forces acting on the oblique plane, corresponding *normal stresses* and *shear stresses* will be created.

Figure 7.2 shows a *two-dimensional* representation of a wedge from the corner of the *stress element* taken from a point on the oblique plane of our bar, with the direct stress (σ_x), the normal stress (σ_θ) and the shear stress (τ_θ) components marked on. It is this element that we will use to derive the relationships for the stresses acting at a point on this oblique plane, and because the faces on the stress element are infinitely small we can make the assumption that the stresses act evenly over the inclined face that we are considering.

Then, taking the depth of our wedge stress element to be unity, the normal stress and shear stress acting on the oblique plane can be determined by dividing the normal and shear forces that produce them by the area

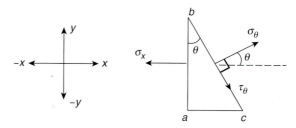

Figure 7.2 Stress element taken from oblique plane EF in Figure 7.1

of the oblique plane. Then, from Figure 7.2 and above, where the area A_0 is the cross-sectional area normal to the axial load, i.e. area ($ab \times$ unit depth), the area of the oblique plane (say A) is $A = \dfrac{A_0}{\cos\theta}$ (i.e. the area of plane bc in Figure 7.2). Therefore, since force normal to plane is $F_N = P\cos\theta$, the stress normal to the plane is $\sigma_\theta = \dfrac{F_N}{A} = \dfrac{P\cos\theta}{A_0/\cos\theta} = \dfrac{P}{A_0}\cos^2\theta$ and, again from above where $\sigma_x = \dfrac{P}{A_0}$ we have that

$$\sigma_\theta = \sigma_x \cos^2\theta \qquad (7.1)$$

Similarly, for the shear stress acting on the oblique plane, we get that $\tau_\theta = \dfrac{F_T}{A} = \dfrac{P\sin\theta}{A_0/\cos\theta} = \dfrac{P}{A_0}\sin\theta\cos\theta$ and so

$$\tau_\theta = -\sigma_x \sin\theta \cos\theta \qquad (7.2)$$

We can also write equation 7.2 in terms of a single trigonometric ratio that is more useful for analysis, by using the *trigonometric rule for products to sums*,

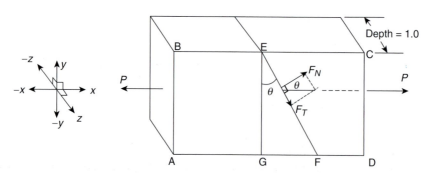

Figure 7.1 Direct tensile force P acting on a rectangular bar

i.e. $2 \sin A \cos A = \sin 2A$, so in our case, $\sin \theta \cos \theta = \frac{1}{2} \sin 2\theta$; therefore the shear stress acting on the plane

$$\tau_\theta = \frac{1}{2} \sigma_x \sin 2\theta \qquad (7.3)$$

From equation 7.1 it can be seen that the maximum value of the normal stress σ_θ occurs when the angle of inclination of the oblique plane upon which it acts is $\theta = 0$, where it is then equal to the direct stress σ_x. Also, from equation 7.3 it can be seen that the shear stress τ_θ has a maximum value when the angle of inclination of the oblique plane, upon which it acts, is $\theta = 45°$.

> **Key Point.** The shear stress has a maximum value when the angle of inclination of the plane on which it acts is 45° to the direct stresses

> **Example 7.1** A bar having a cross-sectional area of $1500\,mm^2$ is subject to an axial tensile load of $120\,kN$. Determine the stresses that act on a plane cut through the bar at an angle of $40°$.
>
> The direct tensile stress σ_x acting on the cross-section is $\sigma_x = \dfrac{P}{A_0} = \dfrac{(120 \times 10^3)}{(1500)} = 80\,N/mm^2$ or $80\,MN/m^2$. Then, from equation 7.1, we get that $\sigma_\theta = \sigma_x \cos^2 \theta = (80)(\cos 40°)^2 = 46.95\,MN/m^2$. Also, from equation 7.3, we get that $\tau_\theta = \dfrac{1}{2} \sigma_x \sin 2\theta = (0.5)(80)(\sin 80) = 39.4\,MN/m^2$.
>
> *Note*: The numerical values for the normal stress and shear stress acting on the inclined plane are fairly close together at this angle of inclination. This is because the angle of the plane is approaching 45°, where τ_θ reaches its maximum value of $40\,MN/m^2$, which is equal to $\dfrac{\sigma_x}{2}$. Furthermore, σ_θ, the stress normal to the oblique plane, also equals this value of $40\,MN/m^2$ at 45° and is also equal to $\dfrac{\sigma_x}{2}$. It is *always the case* that when the plane is inclined at 45°, then τ_θ max $= \dfrac{\sigma_x}{2}$ and $\sigma_\theta = \dfrac{\sigma_x}{2}$.

In the next section we will look at the situation where materials are separately subject to two mutually perpendicular direct stresses and then subject to pure shear stresses, after which we will consider a general two-dimensional stress system where direct stresses and shear stresses are considered in combination.

7.2 Two-dimensional direct stress, shear stress and combined stress systems

We first consider a *plane stress* system where a material is subject to two mutually perpendicular direct stresses, i.e. a system where the stresses are at right angles to one another on a two-dimensional plane.

7.2.1 Material subject to two mutually perpendicular direct stresses

Before we start the analysis that follows, we need to define a few terms that have a direct bearing on the approach we adopt. You have already met the concept of the *stress element* and will have noted that we always use a two-dimensional simplification of the element for the purposes of analysis of the stresses that act on the faces of the element, whether or not they act normal to or at some angle of inclination to the direct stresses. We will continue with this two-dimensional simplification of our stress element, which depicts a system in *plane stress*, where *only the x and y faces of the element are subjected to stresses and where all stresses act parallel to the x and y axes.*

> **Key Point.** Remember that we use a two-dimensional simplification of a three-dimensional stress element when considering materials under plane stress

A two-dimensional *plane stress* system is illustrated in Figure 7.3, where you can see that the mutually perpendicular *direct stresses* (σ_x, σ_y) that act at right angles to their corresponding faces on the element, have

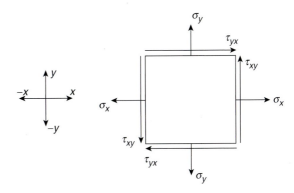

Figure 7.3 An element in plane stress with the direct and shear stresses marked on

been marked on along the x and y axes. Also included in the figure are the *shear stresses*, where each has a double letter subscript; the first letter indicates the face on which the shear stresses act and the second letter gives their direction on that face. So, for example, the shear stress marked as τ_{xy} can be seen to act on the x face of the element in the y direction.

> **Key Point.** Normal stresses are considered positive in tension, that is when they are directed away from the plane on which they act

Now we need to consider the convention for the *direction* of these stresses. In this convention, *normal stresses are positive in tension* that is, when they are *directed away from the plane* on which they act, whereas compressive stresses acting towards the plane on which they act are negative. With stresses in shear, the situation is slightly more difficult to envisage, where we find that *shear stresses are positive when they act in the positive direction of the relevant (x, y or z) axis in a plane on which the direct tensile stress is also in the positive direction of the axis*. Note also that if the *tensile* stress is in the *opposite direction* then the shear stress will be positive, if they also act in opposite directions to the positive directions of the appropriate axes. Figure 7.3 shows all the normal and shear stresses in their *positive* directions, in relation to the given x and y axes. I hope you can follow the argument given above and note the results shown in this figure. With respect to *shear couples*, we will, by convention, assign a *positive value to a couple that tends to turn the element clockwise* and vice-versa; for example, we use this convention when we consider Mohr's stress circle later on.

> **Key Point.** Shear stresses are positive when they act in the positive direction of the relevant (x, y or z) axis in a plane on which the direct tensile stress is also in the positive direction of the axis

Now, hoping that you are not too confused, and armed with all this information, we are in a position to start our analysis of these two-dimensional stress systems! Let us first consider the separate case where a material is subject to two *mutually perpendicular direct stresses*, as illustrated in Figure 7.4.

Figure 7.4 shows an *element* of *unit depth* in plane stress, i.e. where two mutually perpendicular direct stresses act on the faces of the element parallel to the x and y axes. A triangular wedge (abc) has been cut from the element at an angle θ to the y-axis, with the *normal*

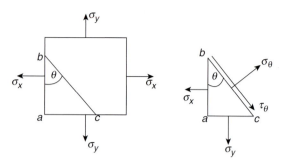

Figure 7.4 Element and wedge of element subject to direct stresses

and *shear* stresses that act on the plane (bc) shown in the figure.

For equilibrium of the wedge abc, *resolving forces perpendicular* (normal) to the plane bc then for our element with *unit depth* we have $\sigma_\theta \times bc \times 1 = (\sigma_x \times 1 \times ab \cos\theta) + (\sigma_y \times 1 \times ac \sin\theta)$ and on division by bc we get that $\sigma_\theta = \sigma_x \dfrac{ab}{bc}\cos\theta + \sigma_y \dfrac{ac}{bc}\sin\theta$ and, on substituting for the trigonometric ratios,

$$\sigma_\theta = \sigma_x \cos^2\theta + \sigma_y \sin^2\theta \qquad (7.4)$$

Now, by using the trigonometric identity *doubles to squares* and vice versa, i.e. $\cos 2\theta = 1 - 2\sin^2\theta = 2\cos^2\theta - 1$ we get that $\cos 2\theta - 1 = -2\sin^2\theta$ or $\dfrac{1}{2}(1 - \cos 2\theta) = \sin^2\theta$.

Also, from $\cos 2\theta = 2\cos^2\theta - 1$ we get that $\dfrac{1}{2}(\cos 2\theta + 1) = \cos^2\theta$.

Therefore, substituting these two expressions for $\cos^2\theta$ and $\sin^2\theta$ into equation 7.4, we get that $\sigma_\theta = \dfrac{\sigma_x}{2}(\cos 2\theta + 1) + \dfrac{\sigma_y}{2}(1 - \cos 2\theta)$ and, on expansion, $\sigma_\theta = \dfrac{\sigma_x}{2}\cos 2\theta + \dfrac{\sigma_x}{2} + \dfrac{\sigma_y}{2} - \dfrac{\sigma_y}{2}\cos 2\theta$, therefore,

$$\sigma_\theta = \frac{(\sigma_x + \sigma_y)}{2} + \frac{(\sigma_x - \sigma_y)}{2}\cos 2\theta \qquad (7.5)$$

Now resolving *forces parallel* to the plane bc, $\tau_\theta \times bc \times 1 = (\sigma_x \times 1 \times ab \sin\theta) - (\sigma_y \times 1 \times ac \cos\theta)$ and again, on division by bc, we get that $\tau_\theta = (\sigma_x \dfrac{ab}{bc}\sin\theta) - (\sigma_y \dfrac{ac}{bc}\cos\theta)$ and as before, substituting for the trigonometric ratios, we get that

$$\tau_\theta = \sigma_x \cos\theta \sin\theta - \sigma_y \sin\theta \cos\theta \qquad (7.6)$$

Also, on this occasion, if we use the *products* to *sums* trigonometric identities, we find after substitution and simplification that

$$\tau_\theta = \frac{(\sigma_x - \sigma_y)}{2} \sin 2\theta \qquad (7.7)$$

> **Key Point.** The *transformation equations* relating the normal stress and shear stress on a plane inclined to two mutually perpendicular direct stresses are,
>
> respectively, $\sigma_\theta = \dfrac{(\sigma_x + \sigma_y)}{2} + \dfrac{(\sigma_x - \sigma_y)}{2} \cos 2\theta$
>
> and $\tau_\theta = \dfrac{(\sigma_x - \sigma_y)}{2} \sin 2\theta$

In manipulating the equations in the above manner we now have relationships between the *normal stress* (equation 7.5), direct stresses and the angle of inclination of the oblique plane, as well as between the *shear stress* (equation 7.7), direct stresses and the angle of inclination of the oblique plane. Both these relationships will prove useful for further analysis.

Note: If you are having difficulties in following the algebra and trigonometry needed to verify any of the equations in this section, you should refer to the website that accompanies this book at www.key2engineeringscience.com where you will find several *Essential Mathematics* summaries that cover these fundamental mathematical topics.

7.2.2 Material subject to pure shear

We will now look at the situation where for our material (still under plane stress conditions) we consider the equilibrium of our element with only *shear stresses* acting on its faces instead of normal stresses. Figure 7.5 now shows our element and the wedge subject to shear stresses on its faces.

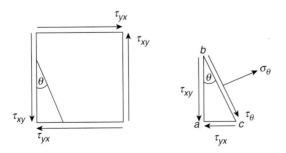

Figure 7.5 Element and wedge of element subject to shear stresses

Consider now the equilibrium of the wedge *abc* cut from our element, assuming as we did before that it has unit depth. Resolving *forces normal* to plane *bc*, we have $\sigma_\theta \times bc \times 1 = (\tau_{xy} \times 1 \times ab \sin \theta) + (\tau_{yx} \times 1 \times ac \cos \theta)$ and, on division by *bc* we get that $\sigma_\theta = \tau_{xy} \dfrac{ab}{bc} \sin \theta + \tau_{yx} \dfrac{ac}{bc} \cos \theta$. Substituting as before for the trigonometric ratios, we get that $\sigma_\theta = \tau_{xy} \cos \theta \sin \theta + \tau_{yx} \sin \theta \cos \theta$. Now, remembering the discussion we had at the end of section 7.1, concerning *complementary* shear stress, we know that the stresses on the *x* and *y* planes are of equal value; therefore we know that $\tau_{xy} = \tau_{yx}$, so

$$\sigma_\theta = 2\tau_{xy} \cos \theta \sin \theta \qquad (7.8)$$

Also, by applying the trigonometric identity for *products to sums*, as before, we get that

$$\sigma_\theta = \tau_{xy} \sin 2\theta \qquad (7.9)$$

Now, resolving *forces parallel* to the plane *bc*, we have $\tau_\theta \times bc \times 1 = -(\tau_{xy} \times 1 \times ab \cos \theta) + (\tau_{yx} \times 1 \times ac \sin \theta)$ and, again on division by *bc*, we get that $\tau_\theta = -\tau_{xy} \cos \theta \cos \theta + \tau_{yx} \sin \theta \sin \theta$. Remembering that for complementary shear stresses $\tau_{xy} = \tau_{yx}$, we get that

$$\tau_\theta = \tau_{xy} \sin^2 \theta - \tau_{xy} \cos^2 \theta \qquad (7.10)$$

Finally, again using the *squares to doubles* trigonometric identity, we may write equation 7.10 as

$$\tau_\theta = -\tau_{xy} \cos 2\theta \qquad (7.11)$$

> **Key Point.** The *transformation equations* relating the normal stress and shear stress on a plane inclined to a material subject to pure shear are $\sigma_\theta = \tau_{xy} \sin 2\theta$ and $\tau_\theta = -\tau_{xy} \cos 2\theta$, respectively

Note: The *negative sign* in equation 7.11 means that the *sense* (direction) of the shear stress on the oblique plane τ_θ is opposite to that shown in Figure 7.5, so that, irrespective of the way you show the direction of the stresses on the free body diagram, the *analytical* result will provide you with both the *magnitude* and *direction* of the stress under consideration.

7.2.3   Material subject to combined stresses

In the final part of this section we consider materials subject to combined, direct and shear stresses. Figure 7.6 shows the stress element for a complete two-dimensional plane stress system identical to that which we discussed earlier (see Figure 7.5).

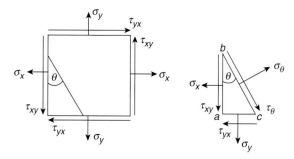

Figure 7.6 Element and wedge from element subject to combined direct and shear stresses

The stresses σ_x and σ_y are shown in Figure 7.6 as tensile stresses. However, these stresses may be compressive and may result from direct loads or loads due to bending. Similarly, the shear stresses may be as shown or completely reversed and result from pure shear load or from torsional loads.

Now logic would dictate that the analysis of this combined stress system may be achieved by summing the σ_θ conditions of stress illustrated in Figures 7.4 and 7.5, and indeed, this is the case! Thus, in the first instance, equations 7.5 and 7.9 may be combined to give

$$\sigma_\theta = \frac{(\sigma_x + \sigma_y)}{2} + \frac{(\sigma_x - \sigma_y)}{2}\cos 2\theta + \tau_{xy}\sin 2\theta \tag{7.12}$$

Similarly, we may combine equations 7.7 and 7.11 to give

$$\tau_\theta = \frac{(\sigma_x - \sigma_y)}{2}\sin 2\theta - \tau_{xy}\cos 2\theta \tag{7.13}$$

Equations 7.12 and 7.13 are often referred to as the *transformation equations* for *plane stress* because of the way they transform the stress components from one axis onto another. It is also important to note that because these equations were derived entirely by imposing equilibrium conditions on the stress system, these transformation equations may then be used for any *type of material* being considered. Thus their application is equally suitable for considering metals, composites etc.

Key Point.   The *general transformation equations* for a plane stress system are $\sigma_\theta = \dfrac{(\sigma_x + \sigma_y)}{2} + \dfrac{(\sigma_x - \sigma_y)}{2}\cos 2\theta + \tau_{xy}\sin 2\theta$ for normal stresses and $\tau_\theta = \dfrac{(\sigma_x - \sigma_y)}{2}\sin 2\theta - \tau_{xy}\cos 2\theta$ for shear stresses

---

Example 7.2   An element of a material in plane two-dimensional stress is subject to direct tensile stresses $\sigma_x = 120\,\text{MN/m}^2$ and $\sigma_y = 80\,\text{MN/m}^2$ that act at $90°$ to each other and also to shear stress $\tau_{xy} = 50\,\text{MN/m}^2$. Determine the stresses that act on this element of the material, when inclined at an angle of $30°$.

The stress acting normal to the face inclined at $30°$ is given from equation 7.12 as $\sigma_{30} = \dfrac{(120 + 80)}{2} + \dfrac{(120 - 80)}{2}\cos 60 + 50\sin 60 = 153.3\,\text{MN/m}^2$. Similarly, the stress that acts on the plane at right angles to σ_{30}, i.e. at $(30° + 90°) = 120°$, is again given from equation 7.12 as $\sigma_{120} = \dfrac{(120 + 80)}{2} + \dfrac{(120 - 80)}{2}\cos 240 + 50\sin 240 = 46.7\,\text{MN/m}^2$.

Now we also know from section 7.2.2 that the shear stresses on the x and y planes (at right angles to one another) are equal, i.e. that $\tau_{xy} = \tau_{yx} = 50\,\text{MN/m}^2$. Then the shear stress on the plane τ_{30} is, from equation 7.13, found to be $\tau_{30} = \dfrac{(120 - 80)}{2}\sin 60 - (50)(\cos 60) = -7.68\,\text{MN/m}^2$.

Then, from these results the stresses will act on the faces of the element as shown in Figure 7.7.

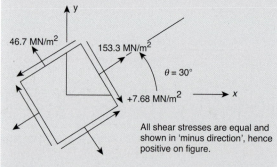

Figure 7.7 Resulting stresses acting on the face of the element

Note: The sum of the normal stresses is the same as the sum of the original direct stresses that acted on the *x* and *y* axes of the original element.

7.2.4   Principal stresses and maximum shear stress

We can find the *maximum and minimum stress* values that occur on *any plane* by applying the differential calculus, where σ_θ is a *maximum* or *minimum* when $\frac{d\sigma_\theta}{d\theta} = 0$. If you are unclear on finding maximum and minimum values you should refer to *Essential Mathematics 2* on the companion website.

So, from equation 7.12, where $\sigma_\theta = \frac{(\sigma_x + \sigma_y)}{2} + \frac{(\sigma_x - \sigma_y)}{2}\cos 2\theta + \tau_{xy}\sin 2\theta$, then differentiating with respect to θ and equating to zero, we get that $\frac{d\sigma_\theta}{d\theta} = -(\sigma_x - \sigma_y)\sin 2\theta + 2\tau_{xy}\cos 2\theta = 0$. Now, remembering the trigonometric identity $\frac{\sin\theta}{\cos\theta} = \tan\theta$ and dividing the equation $-(\sigma_x - \sigma_y)\sin 2\theta + 2\tau_{xy}\cos 2\theta = 0$ by $\cos 2\theta$, we get that $-(\sigma_x - \sigma_y)\tan 2\theta + 2\tau_{xy} = 0$ or $(\sigma_x - \sigma_y)\tan 2\theta = 2\tau_{xy}$ and so:

$$\tan 2\theta = \frac{2\tau_{xy}}{(\sigma_x - \sigma_y)} \qquad (7.14)$$

Figure 7.8 shows the sides of the element set up in relation to the stress values given in equation 7.14, and the use of Pythagoras to represent the hypotenuse in terms of these values, for the angle 2θ.

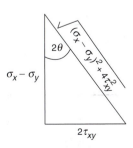

Figure 7.8 Representation of the stress values acting on the planes as given by equation 7.14

Also from Figure 7.8 we note that

$$\sin 2\theta = \frac{2\tau_{xy}}{\sqrt{(\sigma_x - \sigma_y)^2 + 4\tau_{xy}^2}}$$

$$\text{and } \cos 2\theta = \frac{\sigma_x - \sigma_y}{\sqrt{(\sigma_x - \sigma_y)^2 + 4\tau_{xy}^2}}.$$

So, substituting these two expressions for $\sin 2\theta$ and $\cos 2\theta$ into equation 7.12, the maximum and minimum values of σ_θ are given as

$$\sigma_\theta = \frac{\sigma_x + \sigma_y}{2} + \left[\left(\frac{\sigma_x - \sigma_y}{2}\right)\frac{\sigma_x - \sigma_y}{\sqrt{(\sigma_x - \sigma_y)^2 + 4\tau_{xy}^2}}\right]$$

$$+ \frac{(\tau_{xy})(2\tau_{xy})}{\sqrt{(\sigma_x - \sigma_y)^2 + 4\tau_{xy}^2}}.$$

After some algebraic manipulation, which involves multiplying every term by $\sqrt{(\sigma_x - \sigma_y)^2 + 4\tau_{xy}^2}$, squaring and simplifying, we find that

$$\sigma_1 \text{ or } \sigma_2 = \frac{1}{2}(\sigma_x + \sigma_y) \pm \frac{1}{2}\sqrt{(\sigma_x - \sigma_y)^2 + 4\tau_{xy}^2}$$

$$(7.15)$$

Now the *maximum* and *minimum* values of σ_θ are known as the *principal stresses* and are annotated as σ_1 and σ_2, as given in equation 7.5. Also, the solution of equation 7.14 produces two values for θ (the angle of the planes on which the principal stresses act), that are separated by 90°, i.e. $\theta = \frac{1}{2}\tan^{-1}\frac{2\tau_{xy}}{(\sigma_x - \sigma_y)}$ and $\theta = \frac{1}{2}\tan^{-1}\frac{2\tau_{xy}}{(\sigma_x - \sigma_y)} + 90$. Therefore, the planes on which the *principal stresses* act (the *principal planes*) are *mutually perpendicular*.

Key Point. The planes on which the principal stresses act are mutually perpendicular

Now, with respect to the *shear stresses* acting on these *principal planes*, if we substitute the above expressions we found for $\sin 2\theta$ and $\cos 2\theta$ from Figure 7.7 into equation 7.13, we find that

$$\tau_\theta = \left[\frac{(\sigma_x - \sigma_y)}{2} \frac{2\tau_{xy}}{\sqrt{(\sigma_x - \sigma_y)^2 + 4\tau_{xy}^2}} \right]$$

$$- \left[\frac{\tau_{xy}(\sigma_x - \sigma_y)}{\sqrt{(\sigma_x - \sigma_y)^2 + 4\tau_{xy}^2}} \right] = 0,$$

which shows that *the shear stresses acting on the principal planes are zero*, i.e. $\tau_\theta = 0$.

> **Key Point.**   Shear stresses on principal planes are zero

Figure 7.9 shows the original complex stress system alongside its *equivalent* principal stress system, where the shear stress acting on the principal planes is zero.

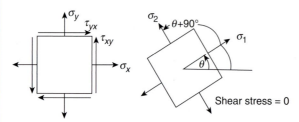

Figure 7.9 Original complex stress system together with its equivalent principal stress system

What about the *maximum shear stress* within the system? Well, if we look again at equation 7.7, where $\tau_\theta = \frac{(\sigma_x - \sigma_y)}{2} \sin 2\theta$, then it can be seen that $\tau_{max} = \frac{(\sigma_1 - \sigma_2)}{2}$ when $\theta = 45°$, that is when *the maximum shear stress acts at 45° to the principal planes*. So, substituting the relationship given by equation 7.15 for the principal stresses into the equation for τ_{max}, we get that

$$\tau_{max} = \frac{\frac{1}{2}\left[(\sigma_x + \sigma_y) + \left(\sqrt{(\sigma_x - \sigma_y)^2 + 4\tau_{xy}^2}\right)\right]}{2}$$

$$- \frac{\frac{1}{2}\left[(\sigma_x + \sigma_y) - \sqrt{(\sigma_x - \sigma_y)^2 + 4\tau_{xy}^2}\right]}{2}$$

from which, after simplification, we find that

$$\tau_{max} = \frac{1}{2}\sqrt{(\sigma_x - \sigma_y)^2 + 4\tau_{xy}^2} \qquad (7.16)$$

Example 7.3   An element in plane stress is subject to stresses $\sigma_x = 100 \, \text{MN/m}^2$, $\sigma_y = 40 \, \text{MN/m}^2$ and $\tau_{xy} = -30 \, \text{MN/m}^2$. Determine the magnitude and orientation of the principal stresses and maximum shear stress.

In order to find the magnitude of the principal stresses, with the information given, we can use equation 7.15, and to find the magnitude of the maximum shear stress we may use equation 7.16. The orientation of the principal stresses and so the orientation of the maximum shear stress can then be found by using equation 7.14.

So, from

$$\sigma_{1,2} = \frac{1}{2}(\sigma_x + \sigma_y) \pm \frac{1}{2}\sqrt{(\sigma_x - \sigma_y)^2 + 4\tau_{xy}^2}$$

we get that

$$\sigma_1 = \frac{1}{2}(100 + 40) + \frac{1}{2}\sqrt{(100 - 40)^2 + 4(-30)^2},$$

so $\sigma_1 = 70 + 42.43 = 112.43 \, \text{MN/m}^2$ and therefore $\sigma_2 = 70 - 42.43 = 27.57 \, \text{MN/m}^2$; also, since from equation 7.16 $\tau_{max} = \frac{1}{2}\sqrt{(\sigma_x - \sigma_y)^2 + 4\tau_{xy}^2}$, then $\tau_{max} = 42.43 \, \text{MN/m}^2$, as previously calculated in equation 7.15.

The orientation of the principal stress planes is found from, where $\theta_1 = \frac{1}{2}\tan^{-1}\frac{2\tau_{xy}}{(\sigma_x - \sigma_y)}$. Then $\theta_1 = \frac{1}{2}\tan^{-1}(-1) = -22.5°$. Also, the orientation of the plane on which σ_2 acts is $\theta_2 = -22.5 + 90 = 67.5°$.

The maximum shear stress $\tau_{max} = 42.43 \, \text{MN/m}^2$ acts on planes oriented at 45° to the principal planes.

Figure 7.10a shows the stress element with the magnitude and orientation of the principal stresses marked on, while Figure 7.10b shows the magnitude and orientation of the maximum shear stress. Note that the normal stresses that act on the planes of the shear stresses are *all equal* and may be found by taking the mathematical average of the direct stresses, i.e., in our case, $\sigma = \frac{\sigma_x + \sigma_y}{2} = \frac{100 + 40}{2} = 70 \, \text{MN/m}^2$.

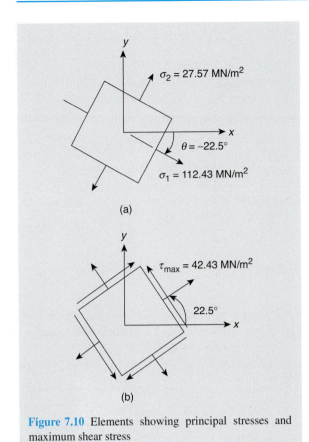

(a)

(b)

Figure 7.10 Elements showing principal stresses and maximum shear stress

7.3 Mohr's stress circle

Apart from the analytical determination of particular stresses acting on planes within a complete plane stress system, we may also use a graphical method to determine such stresses, known as *Mohr's stress circle*.

The method that follows for constructing Mohr's stress circle is based on the plane stress system shown in Figure 7.11.

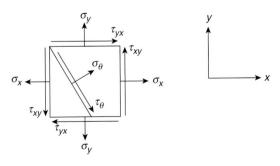

Figure 7.11 Plane stress system

1. We first set up orthogonal axes for the stresses, plotting the direct stress on the horizontal axis (abscissa) and the shear stress on the vertical axis (ordinate), with a datum or reference point (*P*) starting at their intersection, as shown in Figure 7.12.

2. Using our convention that tensile direct stresses are positive, compressive direct stresses are negative, and that shear stresses tending to turn the element clockwise are positive and shear stresses tending to turn the element anticlockwise are negative, we mark off from the datum the direct stresses σ_x and σ_y, where these tensile direct stresses are represented on the figure by $\sigma_x = PS$ and $\sigma_y = PQ$.

3. From Figure 7.11 it can be seen that the shear stress τ_{xy}, acting on the same plane as σ_x and its complement, tends to turn the element anticlockwise, so is *negative*, and as such is marked off on the figure *below* the σ axis, as line SA. The shear stress τ_{yx}, acting on the σ_y plane and its complement, is tending to turn the element clockwise so, from our convention, is *positive* and is marked off on the figure *above* the σ axis, as line QB.

4. The points BA are now joined to cut the σ axis at O.

5. The point O on the σ axis is the centre of a circle with radius OA, therefore this circle can now be drawn onto the figure.

6. *Note: Having constructed the circle in this manner, every point on its circumference represents the stress on any plane of the element oriented at an angle θ to the σ axis.* In the steps that follow, we verify the construction used to determine some of these stresses, i.e. $(\sigma_\theta, \tau_\theta, \tau_{max}, \sigma_1, \sigma_2)$. Draw on the radius OC so that angle AOC = 2θ and let angle AOS be ϕ.

7. Now, to find an *expression for σ_θ*, we note from Figure 7.12 that $PO = \frac{1}{2}(\sigma_x + \sigma_y)$ and $OS = \frac{1}{2}(\sigma_x - \sigma_y)$ or $(OS)^2 = \left(\frac{1}{2}(\sigma_x - \sigma_y)\right)^2$, also, from step 3 above, $\tau_{xy} = SA$ or $\tau_{xy}^2 = (SA)^2$. Then, from Pythagoras, $OA = \sqrt{(OS)^2 + (SA)^2}$; therefore

    $$radius \ OA = OC = \sqrt{\left(\frac{1}{2}(\sigma_x - \sigma_y)\right)^2 + \tau^2 xy}.$$

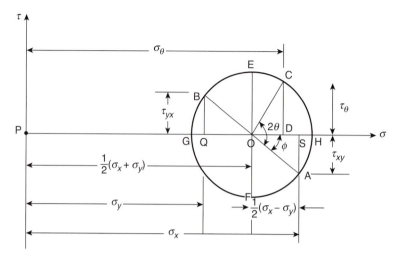

Figure 7.12 Mohr's stress circle

We also know from the figure that $PD = \sigma_\theta = PO + OD$, where $OD = OC\cos(2\theta - \beta)$ and, as $OA = OC$, then $\sigma_\theta = PO + OA\cos(2\theta - \phi)$. Now, once more using a trigonometric identity, i.e. $\cos(A - B) = \cos A\cos B + \sin A\sin B$, then in our case applying the identity, we get that $\sigma_\theta = PO + OA\cos\phi\cos 2\theta + OA\sin\phi\sin 2\theta$ and noting that

$$\tau_\theta PO = \frac{1}{2}(\sigma_x + \sigma_y), OA\cos\phi = OS = \frac{1}{2}(\sigma_x - \sigma_y)$$

and $OA\sin\phi = \tau_{xy}$, then finally the expression for the stress normal to the oblique plane is $\sigma_\theta = \frac{1}{2}(\sigma_x + \sigma_y) + \frac{1}{2}(\sigma_x - \sigma_y)\cos 2\theta + \tau_{xy}\sin 2\theta$. You will note that this equation for the stress normal to the oblique plane is identical to equation 7.12 and so verifies the values found from Mohr's circle for this stress.

8. Similarly, to find an *expression for* τ_θ from the stress circle, this time without all the algebraic steps, we note from Figure 7.12 that $\tau_\theta = CD = OC\sin(2\theta - \phi)$; so, applying the same trigonometric identity, as we did in step 7 above, we find that $\tau_\theta = PO + OA\cos\phi\sin 2\theta - OA\sin\phi\cos 2\theta$. Then substituting for $OA\cos\phi$ and $OA\sin\phi$ as before gives us the expression $\tau_\theta = \frac{1}{2}(\sigma_x + \sigma_y)\sin 2\theta - \tau_{xy}\cos 2\theta$ for the shear stress acting on the inclined plane, which of course is the same as that found (equation 7.13) when we considered the equilibrium of the stress element.

9. The *shear stress will be a maximum* at the top and bottom of the circle, i.e. at points E and F,

where lengths OE and OF are radii, which from Figure 7.12 we found in step 8 to be equal to

$$OA = OC = \sqrt{\left(\frac{1}{2}(\sigma_x - \sigma_y)\right)^2 + \tau^2 xy}, \text{ so that}$$

$$OE = OF = \tau_{max} = \pm\sqrt{\frac{1}{4}(\sigma_x - \sigma_y)^2 + \tau^2 xy} \text{ and}$$

so $OE = \tau_{max} = +\frac{1}{2}\sqrt{(\sigma_x - \sigma_y)^2 + 4\tau^2 xy}$ and $OF = \tau_{max} = -\frac{1}{2}\sqrt{(\sigma_x - \sigma_y)^2 + 4\tau^2 xy}$. Note that the direct stress corresponding to the maximum shear stress is given from Figure 7.12 as $PO = \frac{1}{2}(\sigma_x + \sigma_y)$.

10. We know that when the shear stress on the plane τ_θ is equal to zero, then line OE lies on the σ_x axis (plane) and is of the same magnitude as OH, where also $2\theta = \phi$ or $\theta = \frac{\phi}{2}$. So, from step 8, we find that $PO + OH = PH = \sigma_1 = \frac{1}{2}(\sigma_x + \sigma_y) + \frac{1}{2}\sqrt{(\sigma_x - \sigma_y)^2 + 4\tau^2 xy}$. Now, looking back and comparing this expression with equation 7.15, we note that *PH* represents the *maximum principal stress* σ_1, where the principal stress and principal plane are, from the figure, inclined at an angle of $\frac{\phi}{2}$ *to the* σ_x *plane.* Similarly, from $PO - OH = PG$ we find that the *minimum principal stress* $\sigma_2 = \frac{1}{2}(\sigma_x + \sigma_y) - \frac{1}{2}\sqrt{(\sigma_x - \sigma_y)^2 + 4\tau^2 xy}$, where the

principal stress and principal plane are inclined at an angle of $\dfrac{\phi}{2} + 90°$ to the σ_x plane. Remember also that $2\theta = \phi$ or $\theta = \dfrac{\phi}{2}$, and that these angles may be found from equation 7.14, that is from $\tan 2\theta = \dfrac{2\tau_{xy}}{(\sigma_x - \sigma_y)}$, provided that the direct stresses and shear stresses acting on the element, as appropriate, are known.

Please note that steps 7 to 10 given above validate the Mohr's circle construction method for finding $(\sigma_\theta, \tau_\theta, \tau_{max}, \sigma_1, \sigma_2)$. You will be pleased to know that using Mohr's circle to find all or some of these stresses is relatively straightforward, as the following example shows.

Key Point. Note that all angles drawn on Mohr's circle are double their actual value, i.e. they equal 2θ

Example 7.4 A material is subject to two orthogonal (perpendicular to each other) direct tensile stresses of 100 MN/m² and 40 MN/m², while a shear stress $\tau = 30$ MN/m² causes a positive couple (that is clockwise for the convention we are using for Mohr's circle) when it acts on the planes carrying the 40 MN/m² stress. Determine, using Mohr's circle, (a) the principal stresses, (b) the magnitude of the maximum shear stress, that act in the plane of this stress system, and (c) the inclination of the planes on which these stresses act.

Now, in order to construct Mohr's circle we first need to sketch the stress element on which we intend to base this construction. This is shown below in Figure 7.13.

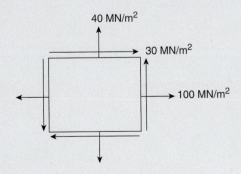

Figure 7.13 Stress element for Example 7.4

Now Mohr's circle is drawn, as shown in Figure 7.14, following the conventions and procedural steps 1–5, given earlier.

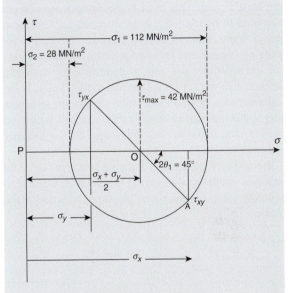

Figure 7.14 Mohr's circle for Example 7.4

Note:

1. All direct stresses are marked off to scale from the datum point P.

2. From the data, $\sigma_x = 100$ MN/m², $\sigma_y = 40$ MN/m², $\tau_{yx} = 30$ MN/m², $\tau_{xy} = -30$ MN/m² and $\dfrac{\sigma_x + \sigma_y}{2} = 70$ MN/m².

3. The principal stresses are measured as $\sigma_1 = 112$ MN/m², $\sigma_2 = 28$ MN/m², with σ_1 acting on the plane with angle of inclination $\theta_1 = -22.5°$, i.e. below the stress axis, while $\sigma_2 = 28$ MN/m² acts on the plane inclined at $\theta_2 = -22.5 + 90 = 67.5°$.

4. The maximum shear stress is measured as $\tau_{max} = 42$ MN/m² at an angle of 45° to the principal stresses axis.

5. Taking account of the convention used for the production of Mohr's circle, the measured values compare favourably with the exact values found in Example 7.3.

7.4   Strain

We now turn our attention to the strains that result from a material being stressed, in particular looking at elastic strain in engineering materials, since it is in the elastic range that most engineering materials are designed to be used. *Strain* in materials is a quantity, unlike stress, that can be measured, and through strain gauging systems and knowledge of the elastic constants for materials, we are able to determine the resulting stresses that are set up within materials. Thus it is the determination of strain that enables engineers to design and test a material or artefact that needs to withstand certain maximum stresses for a specific engineering purpose.

Before we look in detail at the various types of strain that may result from a material under stress, we need to reiterate what we understand by *plane stress* and *plane strain*.

Plane stress occurs, as is often the case in practice, when a material is stressed in two directions, say in the x and y directions with stress $z = 0$. However, under these circumstances plane strain may not result! For example, Figure 7.15 shows that even when a three-dimensional solid is subject to stress in, say, one direction, three-dimensional strain may result.

It can clearly be seen from Figure 7.15 that an extension from stress applied in the x-direction has also caused a contraction in the y and z directions.

Now we contrast this with the situation that results in *plane strain*, where the strain in one dimension is zero. Figure 7.16 shows a three-dimensional solid subject to plane strain in the x and y directions and where the strain in the z direction is zero.

Key Point.   Plane strain, for a three-dimensional system, will result when the strain in one dimension is zero

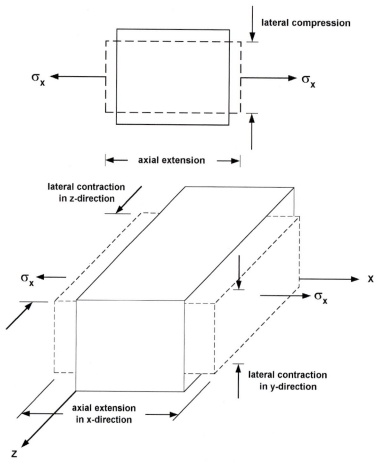

Figure 7.15 One-dimensional stress leading to three-dimensional strain

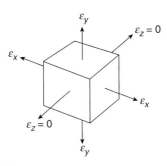

Figure 7.16 Plane strain in a three-dimensional solid

Note that zero strain in the z direction does not necessarily mean that there is zero stress in this direction. From the Poisson effect, there will also need to be a stress in the z direction that prevents the occurrence of this strain.

7.4.1 Principal strains

Let us start by reminding ourselves of the situation for a material subject to perpendicular stresses (σ_x, σ_y, σ_z), where no shear stresses are acting on the x, y and z planes. Under these circumstances we referred to them as principal stresses and labelled them (σ_1, σ_2, σ_3). The strains that result from these principal stresses are known as *principal strains*; they act in the corresponding x, y and z directions and are thus labelled as (ε_1, ε_2, ε_3). An illustration of the principal stresses and strains for the two-dimensional case is shown in Figure 7.17.

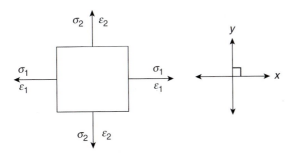

Figure 7.17 Two-dimensional principal stresses and strains

In section 1.7 of Chapter 1 you were introduced to the concept of Poisson's ratio and two-dimensional loading, where we found relationships (equations 1.20 and 1.21) for determining the total strains in the x and y directions from the combined perpendicular stresses, using Poisson's ratio. These relationships are repeated here as a starting point for our analysis of principal strains:

$$\varepsilon_x = \frac{\sigma_x}{E} - v\frac{\sigma_y}{E} \qquad (1.20)$$

$$\text{and } \varepsilon_y = \frac{\sigma_y}{E} - v\frac{\sigma_x}{E} \qquad (1.21)$$

If we extend the argument given in section 1.7 to three dimensions, you can see from Figure 7.15 that an axial extension of the bar (x direction) results in a lateral contraction in both the mutually perpendicular y and z directions. Then, by analogy, three-dimensional strain resulting from a tensile direct stress in the x direction is given by:

$$\varepsilon_x = \frac{\sigma_x}{E} - v\frac{\sigma_y}{E} - v\frac{\sigma_z}{E} \text{ or } \varepsilon_x = \frac{1}{E}(\sigma_x - v\sigma_y - v\sigma_z)$$
$$(7.17)$$

Similarly, the strains in the y and z directions in terms of the mutually perpendicular stresses are represented by:

$$\varepsilon_y = \frac{1}{E}(\sigma_y - v\sigma_x - v\sigma_z) \qquad (7.18)$$

$$\varepsilon_z = \frac{1}{E}(\sigma_z - v\sigma_x - v\sigma_y) \qquad (7.19)$$

In the absence of shear stresses on the face of the stress element shown in Figure 7.18, the stresses (σ_x, σ_y, σ_z) are in fact principal stresses.

Key Point. One-dimensional stress can result in three-dimensional strain

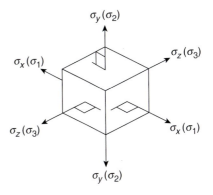

Figure 7.18 Three-dimensional principal stresses

So the *principal strains* in a given direction may be found from equations 7.17, 7.18 and 7.19, by writing them as follows:

$$\varepsilon_1 = \frac{1}{E}(\sigma_1 - v\sigma_2 - v\sigma_3) \qquad (7.20)$$

$$\varepsilon_2 = \frac{1}{E}(\sigma_2 - v\sigma_1 - v\sigma_2) \qquad (7.21)$$

$$\varepsilon_3 = \frac{1}{E}(\sigma_3 - v\sigma_1 - v\sigma_2) \qquad (7.22)$$

7.4.2 Bulk modulus and volumetric strain

You will already be familiar with the elastic constant E for direct loading and G for shear loading from the work we covered earlier on simple stress and strain in Chapter 1. However, there is a third member of the elastic constants known as the bulk modulus K, where $K = \dfrac{p}{\delta V/V}$, p is the volumetric pressure or *volumetric stress* (σ) and $\dfrac{\delta V}{V}$ is the change in volume over the original volume, i.e. the *volumetric strain* (ε_v). This equation may then be written as:

$$\text{Bulk modulus } (K) = \frac{\text{volumetric stress } (\sigma)}{\text{volumetric strain } (\varepsilon_v)} \qquad (7.23)$$

Key Point.

Bulk modulus $(K) = \dfrac{\text{volumetric stress } (\sigma)}{\text{volumetric strain } (\varepsilon_v)}$

We know from the definition that

$$\text{volumetric strain} = \frac{\text{change in volume } (\delta V)}{\text{original volume } (V)}.$$

Now it can be shown that the volumetric strain = the sum of the linear strains in the x, y and z directions, or

$$\varepsilon_v = \varepsilon_x + \varepsilon_y + \varepsilon_z \qquad (7.24)$$

Now, substituting equations 7.17 to 7.19 into equation 7.24, we produce an equation for the volumetric strain in terms of the mutually perpendicular stresses, that is:

$$\varepsilon_v = \frac{1}{E}(\sigma_x - v\sigma_y - v\sigma_z) + \frac{1}{E}(\sigma_y - v\sigma_x - v\sigma_z) + \frac{1}{E}(\sigma_z - v\sigma_x - v\sigma_y)$$ and on rearrangement we find that

$$\varepsilon_v = \frac{1}{E}(\sigma_x + \sigma_y + \sigma_z)(1 - 2v) \qquad (7.25)$$

Now let us consider the round bar shown in Figure 7.19, that is subject to a uniaxial stress (σ_x).

From the situation shown in Figure 7.19, we note that the stresses in the y and z planes are zero; therefore, from equation 7.25, we find that $\varepsilon_v = \dfrac{\sigma_x}{E}(1 - 2v)$. We also know that the volumetric strain is $\varepsilon_v = \dfrac{\delta V}{V}$ and, letting $\sigma_x = \sigma$, then the general equation for calculating the volumetric strain and/or the change in volume of a bar subject to an axial stress is given as

$$\varepsilon_v = \frac{\delta V}{V} = \frac{\sigma}{E}(1 - 2v) \qquad (7.26)$$

Example 7.5   A bar of circular cross-section is subject to a tensile load of 100 kN, which is within the elastic range. The bar is made from a steel with $E = 210$ GPa and Poisson's ratio = 0.3, it has a diameter of 25 mm and is 1.5 m long. Determine the extension of the bar, the decrease in diameter and the increase in volume of the bar.

Since we are told that the loaded bar is in the elastic range, we may find the axial strain from $\varepsilon = \dfrac{\sigma}{E}$, where $\sigma = \dfrac{load}{area} = \dfrac{10^5}{490.87} = 203.7$ N/mm^2, then the axial strain $\varepsilon = \dfrac{203.7\,\text{MPa}}{210\,\text{GPa}} = 0.00097$.

Now the extension (change in length) equals the product of the strain and the original length, that is, *extension* $= (0.00097 \times 1.5) = 1.455$ mm.

Circular bar subject to uniaxial tensile stress σ_x, causing lateral contraction and axial elongation with $\sigma_y = \sigma_z = 0$

Figure 7.19 Bar subject to uniaxial stress

To find the decrease in diameter, we first find the lateral strain, where from Poisson's ratio, which is lateral strain over longitudinal strain, $\varepsilon_{lat} = -v\varepsilon = -(0.3 \times 0.00097) = -0.000291$. That is a decrease, and so the *decrease in diameter* $= (0.000291 \times 25) = 0.007275$ mm.

From equation 7.26, the change in volume is given by $\delta V = V\varepsilon(1 - 2v)$, so that $\delta V = (490.87 \times 1500)(0.00097(1 - 0.6) = 286$ mm^3.

7.4.3   Relationship between the elastic constants and Poisson's ratio

In order to determine the relationships between constants E, G, K and v, we first need to establish the relationship between the *shear strain* (γ) and the maximum strain (ε_{max}) by considering an *element of a material in pure shear*. Consider Figure 7.20a, which shows a stress element subject to pure shear that produces the strained shape shown in Figure 7.20b.

The stress element ABCD (Figure 7.20a) is subject to pure shear by the stresses and complementary stresses (τ). The faces are distorted by these stresses into the rhombus shown in Figure 7.20b. The diagonal bd has lengthened, while the diagonal ac has shortened, as a result of the shear stresses (note that the corresponding strains would be very small in practice; they have been exaggerated in Figure 7.20b for the purpose of illustration).

Now, because the strains are very small, angles abd and bde are each approximately 45° and so the triangle abd has sides with the ratios $1 : 1 : \sqrt{2}$.

Now, by simple geometry, the length of the side $bd = \tau\sqrt{2}(1 + \varepsilon_{max})$, where ε_{max} is the strain along the diagonal bd. In Figure 7.20b the shear strain is given

by the angle of deformation in radians. You should now see that the angles bda and dba are both equal to $(\pi/4 - \gamma/2)$ and also, due to the extension of the diagonal bd, the angle dab is equal to $\left(\dfrac{\pi}{2} + \gamma\right)$. Now, using the cosine rule for triangle dab, we have in this case that $\left[\tau\sqrt{2}(1 + \varepsilon_{max})\right]^2 = \tau^2 + \tau^2 - 2\tau^2\cos\left(\dfrac{\pi}{2} + \gamma\right)$ or $2\tau^2(1 + \varepsilon_{max})^2 = 2\tau^2 - 2\tau^2\cos\left(\dfrac{\pi}{2} + \gamma\right)$, and on division by $2\tau^2$ we find that $(1 + \varepsilon_{max})^2 = 1 - \cos\left(\dfrac{\pi}{2} + \gamma\right)$. Now, using the trigonometric identity $\cos(\pi/2 + x) = -\sin x$ and expanding the LHS gives $1 + 2\varepsilon_{max} + (\varepsilon_{max})^2 = 1 + \gamma$. Also, since ε_{max} and γ are very small strains, we may use the approximation that $\sin\gamma = \gamma$ (remembering that the angle γ is in radians) and we may ignore $(\varepsilon_{max})^2$ when compared to $2\varepsilon_{max}$. So we now have that $1 + 2\varepsilon_{max} = 1 + \gamma$ or that the *strain on the diagonal* (the maximum strain) is given by

$$\varepsilon_{max} = \frac{\gamma}{2} \tag{7.27}$$

Now, remembering our definition for the shear modulus or modulus of rigidity G, given as equation 1.10 in Chapter 1, i.e. $\dfrac{\text{shear stress } (\tau)}{\text{shear strain } (\gamma)} = G$, then the shear strain $\gamma = \dfrac{\tau}{G}$ and therefore, from equation 7.27, we have that *the strain on the diagonal*

$$\varepsilon_{max} = \frac{\tau}{2G} \tag{7.28}$$

Now we know from sections 7.1 and 7.2.1 that an element subject to *pure shear* may be replaced by an

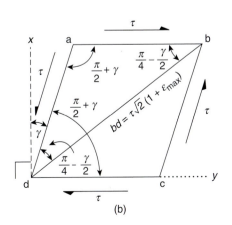

(a)   (b)

Figure 7.20 Stress element subject to pure shear

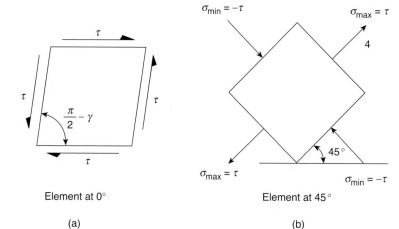

Figure 7.21 Stress resulting from pure shear

equivalent system of direct stresses acting at 45°, as shown in Figure 7.21, where one set of direct stresses are tensile and the other are compressive and where both sets are equal to the shear stress τ.

Then, using equation 7.20 for our *two-dimensional system* subject to *pure shear*, i.e. $\varepsilon_1 = \dfrac{1}{E}(\sigma_1 - v\sigma_2)$, where values with 'suffix 1', are maximum values, the direct stress system applied to the diagonal (i.e. at 45° to the shear stress) gives the *maximum strain* on the diagonal as $\varepsilon_{max} = \dfrac{1}{E}(\sigma_1 - v\sigma_2)$. From Figure 7.21b, we get, after substitution, that $\varepsilon_{max} = \dfrac{1}{E}(\sigma_1 - v(-\tau))$ or $\varepsilon_{max} = \dfrac{\tau}{E}(\sigma_1 + v)$, and substituting $= \dfrac{\tau}{2G}$ for ε_{max} from equation 7.28 gives $\dfrac{\tau}{2G} = \dfrac{\tau}{E}(1 + v)$, and so

$$G = \frac{E}{2(1 + v)} \qquad (7.29)$$

We will now, through an example, find a relationship that involves the *bulk modulus K*.

Example 7.6 Assume that the pressure vessel shown in Figure 7.22 contains a pressurised liquid, which subjects the vessel to volumetric (hydrostatic) stress. Show, stating all assumptions, that the elastic constants E and K and Poisson's ratio are related by the expression $K = \dfrac{E}{3(1 - 2v)}$.

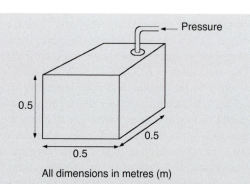

Figure 7.22 Pressure vessel for Example 7.6

In order to determine this relationship we need to remember one very important fact concerning the strain that results from volumetric stress. This is:

Volumetric strain = the sum of the linear strains in the x, y and z directions.

Now for hydrostatic stress, the stresses in the x, y and z directions are equal, so the resulting strains, subject to certain assumptions, are also equal. So: volumetric strain = 3 × linear strain.

The assumptions are that all surfaces of the pressure vessel are subject to equal pressure and have the same material properties. Then, sidewall stresses are equal and so volumetric strain is equal to three times the linear strain. So, from equation 7.17 where $\varepsilon_x = \dfrac{1}{E}(\sigma_x - v\sigma_y - v\sigma_z)$, the volumetric strain, on rearrangement, is given by

$$\varepsilon_v = \frac{3\sigma}{E}(1 - 2v) \qquad (7.30)$$

where $\sigma_x = \sigma_y = \sigma_z = \sigma$. Now we need an expression involving the bulk modulus K if we are to relate the elastic constants E and K with Poisson's ratio. We know that the bulk modulus

$$K = \frac{\text{volumetric stress}}{\text{volumetric strain}}, \text{ therefore}$$

$$\varepsilon_v = \frac{\sigma}{K} \qquad (7.31)$$

and from equations 7.30 and 7.31 we get that $\frac{\sigma}{K} = \frac{3\sigma}{E}(1 - 2v)$, and on rearrangement:

$$K = \frac{E}{3(1 - 2v)} \qquad (7.32)$$

equation 7.32 being the relationship we require.

In the next example we use our knowledge of thin-walled pressure vessels that we covered in Chapter 4.

Example 7.7 A cylindrical compressed air cylinder 1.5 m long and 0.75 m internal diameter has a wall thickness of 8 mm. Find the increase in volume when the cylinder is subject to an internal pressure of 3.5 MPa. Also find the values for the elastic constants G and K, given that $E = 207$ GPa and Poisson's ratio = 0.3.

For this problem the increase in volume can be found from Chapter 4 using equation 4.9. Then: change in volume $= \frac{pd}{4tE}(5 - 4v)V$, where you will remember that $p =$ internal pressure, $v =$ Poisson's ratio, $d =$ internal diameter, $V =$ internal volume and $t =$ wall thickness.

So, change in volume

$$= \frac{(3.5 \times 10^6)(0.75)(5 - 4 \times 0.3)(0.6627)}{(4)(0.008)(207 \times 10^9)}$$

$$= 0.000998 \, \text{m}^3.$$

Also, using equations 7.29 and 7.32,

$$G = \frac{E}{2(1 + v)} = \frac{(207 \times 10^9)}{3(1 + 0.3)} = 79.6 \, \text{GPa and}$$

$$K = \frac{E}{3(1 - 2v)} = \frac{(207 \times 10^9)}{3(1 - 0.6)} = 172.5 \, \text{GPa}.$$

There is one more useful relationship involving the constants G and K, that can be derived by combining equations 7.29 and 7.32 in the following manner.

We first transpose equations 7.29 and 7.32 to make Poisson's ratio the subject and then equate them, i.e.

$$v = \frac{E}{2G} - 1 \text{ and } v = \frac{1}{2} - \frac{E}{6K}, \text{ so that } \frac{E}{2G} - 1 = \frac{1}{2} - \frac{E}{6K},$$

from which, on rearrangement for E, we find that

$$E = \frac{9GK}{(G + 3K)} \qquad (7.33)$$

Key Point. The important relationships between the elastic constants and Poisson's ratio are

$$G = \frac{E}{2(1 + v)}, K = \frac{E}{3(1 - 2v)} \text{ and } E = \frac{9GK}{(G + 3K)}$$

7.4.4 Strain on an inclined plane and the transformation equations for strains

We may determine relationships for strain on an inclined plane in a similar manner to the way in which we found stresses on inclined planes.

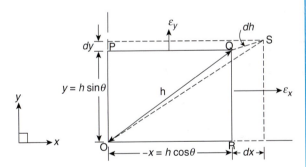

Figure 7.23 Element subject to plane strain

Consider the element of a material OPQR that is subject to plane strain, i.e. two mutually perpendicular strains (Figure 7.23), where for clarity the strain is grossly exaggerated. Now, if we wish to find the strain on the plane (h) inclined at angle θ, that is ε_θ, in terms of the principal strains ε_x, ε_y then this is quite easily done by considering the trigonometry of the situation.

Since the strain is very small and so the angle QSO is very small, then the change in length of OQ may be taken as QS, so that the strain on the plane $\varepsilon_\theta = \frac{\text{change in length}}{\text{original length}} = \frac{dh}{h}$ or $dh = \varepsilon_\theta h$.

Now the side $OR = x = h\cos\theta$ and the side $OP = y = h\sin\theta$. Knowing that the increase in length = the strain multiplied by the original length, then the increase in length of these sides is, respectively, $dx = \varepsilon_x x$ and $dy = \varepsilon_y y$. Also, from Pythagoras, $h^2 = x^2 + y^2$; therefore, we may write $\varepsilon_\theta h^2 = \varepsilon_x x^2 + \varepsilon_y y^2$, or that $\varepsilon_\theta h^2 = \varepsilon_x(h\cos\theta)^2 + \varepsilon_y(h\sin\theta)^2$, and after expanding and division by h^2 we have $\varepsilon_\theta = \varepsilon_x \cos^2\theta + \varepsilon_y \sin^2\theta$. Finally, using the products-to-sums trigonometric identity and after simplification, we find that

$$\varepsilon_\theta = \frac{1}{2}(\varepsilon_x + \varepsilon_y) + \frac{1}{2}(\varepsilon_x - \varepsilon_y)\cos 2\theta \quad (7.34)$$

You should compare equation 7.34 for two mutually perpendicular strains (plane strain) with equation 7.5 for two mutually perpendicular stresses!

For a material subject to plane strain, it can be shown that the complete *transformation equation* is given by

$$\varepsilon_\theta = \frac{1}{2}(\varepsilon_x + \varepsilon_y) + \frac{1}{2}(\varepsilon_x - \varepsilon_y)\cos 2\theta + \frac{1}{2}\gamma_{xy}\sin 2\theta \quad (7.35)$$

Once again you will note the similarities between the stress and strain *transformation* equations, by comparing equation 7.35 with equation 7.12.

It can also be shown that the *transformation* equation for *shear strain* is

$$\frac{1}{2}\gamma_\theta = -\frac{1}{2}(\varepsilon_x - \varepsilon_y)\sin 2\theta + \frac{1}{2}\gamma_{xy}\cos 2\theta \quad (7.36)$$

where, from Figure 7.23, you will note that the line OS moves *clockwise* with respect to the line OQ, so according to convention this is considered to be a *negative* shear strain, hence the minus sign in equation 7.36. This equation should be compared with equation 7.13, where apart from the sign change you will note that

$$\tau_\theta = \frac{\gamma_\theta}{2} \text{ and } -\tau_{xy} = \frac{\gamma_{xy}}{2}.$$

The maximum and minimum linear strains, or *principal strains*, are found in a similar way to their stress counterparts (equation 7.15) and can be calculated from the relationship

$$\varepsilon_1 \text{ or } \varepsilon_2 = \frac{1}{2}(\varepsilon_x + \varepsilon_y) \pm \frac{1}{2}\sqrt{(\varepsilon_x - \varepsilon_y)^2 + \gamma_{xy}^2} \quad (7.37)$$

These principal strains act in mutually perpendicular directions and the direction (orientation) of their axes

may be found from the relationship

$$\tan 2\theta = \frac{\gamma_{xy}}{(\varepsilon_x - \varepsilon_y)} \quad (7.38)$$

To complete our analysis of strain, we consider the *maximum and minimum shear strains* in the xy plane, which like their stress counterparts (equation 7.16) are found to lie on axes positioned at 45° to the principal strains and are given by the relationship

$$\frac{1}{2}\gamma_{max} = \pm\frac{1}{2}\sqrt{(\varepsilon_x - \varepsilon_y)^2 + \gamma_{xy}^2} \quad (7.39)$$

Key Point. Maximum shear strains (like their stress counterparts) occur on planes oriented at 45° to the planes on which the principal strain acts

Now, finally, you should compare equations 7.37, 7.38 and 7.39 with equations 7.15, 7.14 and 7.16, respectively, noting the similarities in form and the replacements for the variables, in particular noting that we replace σ with ε and τ with $\frac{\gamma}{2}$.

Example 7.8   An element of a material subject to plane strain has the following strains acting on it: $\varepsilon_x = 320 \times 10^{-6}$, $\varepsilon_y = 120 \times 10^{-6}$ and $\gamma_{xy} = 200 \times 10^{-6}$. Determine the:

a)   strains on the planes of the element when the element is oriented at an angle of 40° from the x-ordinate

b)   principal strains and their orientation

c)   maximum shear strains and their orientation

Answers

a)   To answer this part of the question we need to find both of the mutually perpendicular strains on the planes, which will be oriented at 40° and at $(40° + 90°)$ from the x-ordinate. Then, using equation 7.35, we get that:

$$\varepsilon_{40} = \frac{1}{2}(320 + 120) + \frac{1}{2}(320 - 120)\cos 80$$
$$+ \frac{1}{2}(200)\sin 80$$

$$\varepsilon_{40} = 220 + 100\cos 80 + 100\sin 80$$

$$\varepsilon_{40} = 220 + 100(0.1736) + 100(0.9848)$$

$$\varepsilon_{40} = 220 + 17.36 + 98.48 = 335.84.$$

Therefore, $\varepsilon_{40} = 335.84 \times 10^{-6}$ and similarly $\varepsilon_{130} = 220 + 100(-0.1736) + 100(-0.9848)$ or $\varepsilon_{130} = 104.16 \times 10^{-6}$.

b) The principal strains are found by applying equation 7.37, so that:

$$\varepsilon_{1,2} = \frac{1}{2}(320 + 120) \pm \frac{1}{2}\sqrt{(320 - 120)^2 + 200^2}$$

$$\varepsilon_{1,2} = 220 \pm \frac{1}{2}\sqrt{40000 + 40000}$$

$$\varepsilon_{1,2} = 220 \pm \frac{1}{2}(282.84) = 220 \pm 141.4$$

Therefore, $\varepsilon_1 = 361.4 \times 10^{-6}$ and $\varepsilon_2 = 78.6 \times 10^{-6}$

The orientation of the principal strains may be found by finding the *angles to the planes* on which they act, using equation 7.38. Then:

$$\tan 2\theta_1 = \frac{\gamma_{xy}}{(\varepsilon_x - \varepsilon_y)} = \frac{200}{320 - 120} = 1.000,$$

therefore $2\theta_1 = 45°$ so $\theta_1 = 22.5°$ and $\theta_2 = 112.5°$.

c) The maximum positive and negative shear strains may be calculated using equation 7.39, so that:

$$\frac{1}{2}\gamma_{max} = \pm\frac{1}{2}\sqrt{(\varepsilon_x - \varepsilon_y)^2 + \gamma_{xy}^2}$$

$$\frac{1}{2}\gamma_{max} = \pm\frac{1}{2}\sqrt{(320 - 120)^2 + 200^2}$$

$$\frac{1}{2}\gamma_{max} = \pm\frac{1}{2}\sqrt{80000}$$

$$\frac{1}{2}\gamma_{max} = +141.42 \text{ or } \frac{1}{2}\gamma_{max} = -141.42$$

So that $\gamma_{max\,1} = 282.84 \times 10^{-6}$ or $\gamma_{max\,2} = -282.84 \times 10^{-6}$.

Now these maximum shear strains will act along the planes oriented at 45° to the principal strains, that is at 67.5° or at 157.5°. To decide on which of these planes the shear strains will act, we will use equation 7.36, with $\theta_1 = 67.5°$. Then:

$$\frac{1}{2}\gamma_\theta = -\frac{1}{2}(\varepsilon_x - \varepsilon_y)\sin 2\theta_1 + \frac{1}{2}\gamma_{xy}\cos 2\theta_1$$

$$\frac{1}{2}\gamma_\theta = -\frac{1}{2}(200)\sin 135° + \frac{1}{2}(200)\cos 135°$$

$$\frac{1}{2}\gamma_\theta = -100(0.7071) + 100(-0.7071)$$

$$\frac{1}{2}\gamma_\theta = -70.71 - 70.71$$

$$\gamma_\theta = -282.84 \times 10^{-6}$$

Thus the plane of the element oriented at 67.5° has the maximum negative shear strain, $\gamma_{max\,2} = -282.84 \times 10^{-6}$, acting along it.

7.4.5 Mohr's strain circle

From our analysis of strain in section 7.4.4, you will have noted that the transformation equations for strains were found to be identical in form to their stress counterparts, with only the corresponding shear stress/strain variables changing in the following way,

$$\tau_\theta = \frac{\gamma_\theta}{2} \text{ and } -\tau_{xy} = \frac{\gamma_{xy}}{2}.$$

As a result of our analysis we are able to construct the Mohr's strain circle for a particular plane strain system, provided we plot *half of the shear strains* on an *inverted* shear axis and remember that the angles are doubled on Mohr's strain circle, as they were on his stress circle. So bearing in mind these changes and by substituting ε for σ, we can, following the same procedure as that given earlier in section 7.3, construct a Mohr's strain circle that enables the required strains to be found on its circumference in a similar manner to Mohr's stress circle.

Please note the reversal of the shear strain axis and the substitution of $\gamma/2$ for τ in Figure 7.24.

Key Point. For a Mohr's strain circle the variable on the shear axis changes in the following way,

$$-\tau_{xy} = \frac{\gamma_{xy}}{2}$$

Example 7.9 Construct a Mohr's strain circle for the data given in Example 7.8 and from it confirm the calculated values found for the strains in parts a), b) and c) of the task.

The required strain circle is shown in Figure 7.25. It is constructed taking into account the differences in parameters detailed in Example 7.8.

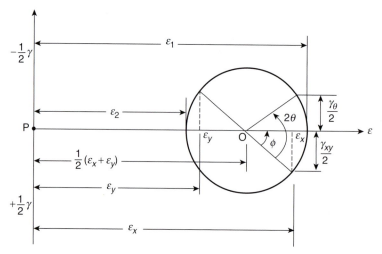

Figure 7.24 Mohr's strain circle

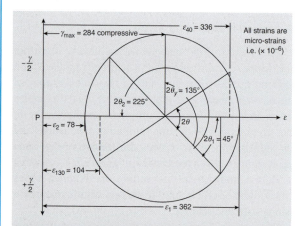

Figure 7.25 Mohr's strain circle for Example 7.9

From the construction, when drawn to scale, the following values (marked on the figure) may be measured: $\varepsilon_{40} = 336$, $\varepsilon_{130} = 104$, $\varepsilon_1 = 362$, $\varepsilon_2 = 79$, $\gamma_{max} = \pm284$; all strain values are micro (μ) strains. Please note the appropriate angles of the planes and to the planes, on which these strains act, are given on the construction.

7.5 Strain gauges

You will be aware, by what has been said before, that we cannot measure stresses directly but, in certain ways, we are able to measure strains and from these measurements we are able to estimate the stresses that act within engineering materials and structures.

To achieve the measurement of strain we employ the use of *strain gauges*. These come in various types but by far the most common are *electrical resistance* strain gauges. These consist of a fine looped wire, usually fixed to a concertinaed paper backing, as shown in Figure 7.26, that work on the principle that when strained (extended), the cross-section of the fine wire is altered and so, therefore, is the electrical resistance.

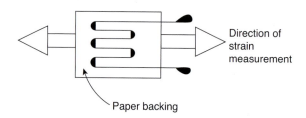

Figure 7.26 Typical electrical resistance strain gauge

They are designed to measure linear strain in *one direction*, that is in the direction in which they can expand, when the paper backing (which is rigidly attached to the material being strained) extends. They are not sensitive to lateral strain or shear strain oriented at large angles away from the strain measurement axis (see Figure 7.26). Thus in order to measure strain in small regions of the material's surface, we need to use more than one strain gauge; this we do in the form of a *strain gauge rosette*.

Key Point. Resistive type strain gauges are designed to measure strain in only one linear direction

7.5.1 Strain gauge rosettes

Strain gauge rosettes consist of two or more strain gauges located very close to one another and oriented at fixed angles with respect to each other. The gauges are mounted, again generally on paper, in tightly packed patterns and then the gauge mounting is fixed, normally by a strong adhesive, to the surface of the material or structure being measured. At least three independent strain readings are necessary in order to provide sufficient information for a two-dimensional plane strain system to be analysed. Figure 7.27 shows schematic drawings for two of the most popular rosette configurations currently in use.

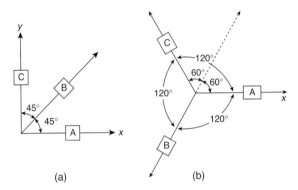

    (a)   (b)

Figure 7.27 Rectangular and delta strain gauge rosette configurations

Notice that the rectangular rosette has the three strain gauges oriented at 45° and 90° to each other (Figure 7.27a), while the delta (Δ) formation rosette is set up physically, in such a way that an angle of 120° exists between each gauge, but, as can be seen by the dotted gauge, B effectively bisects A and C, so that for the purposes of analysis, gauge B may be taken as being oriented at 60° from both A and C. Note that in both these arrangements, gauge A is aligned with the principal x-axis and also, in the case of the rectangular rosette, gauge C is aligned with the principal y-axis. This of course may not always be the case in practice.

> **Key Point.** At least three independent strain gauge readings are necessary to provide sufficient information for a two-dimensional plane strain system

7.5.2 Determining principal strains from rosettes

Principal strains may be obtained from rosette readings either by a graphical method or by analysis.

Graphical methods are recommended for their simplicity and ease with which the principal stresses may be found. Two graphical methods are commonly used, the McClintock method and Mohr's superimposition method, where a stress circle is superimposed onto a strain circle. Here, as you will see in the next section, we will concentrate on the *McClintock method*. No matter what method is used, the object of the exercise is to find, from strain gauge readings, the principal strains and so, in turn, determine the principal stresses that act within the material or structure under consideration.

One general *analytical method*, involves the use of equation 7.35, that is

$$\varepsilon_\theta = \frac{1}{2}(\varepsilon_x + \varepsilon_y) + \frac{1}{2}(\varepsilon_x - \varepsilon_y)\cos 2\theta + \frac{1}{2}\gamma_{xy}\sin 2\theta,$$

which will help us determine the principal strains from the three strain readings from a stain gauge rosette. This equation is applied to each of the three angles associated with the rosette gauge (see Figure 7.28 for a typical set-up), so that three equations are formed, which can be solved simultaneously for the three unknowns ε_x, ε_y, γ_{xy}. The principal strains can then be determined from equation 7.37, i.e. ε_1 or $\varepsilon_2 = \frac{1}{2}(\varepsilon_x + \varepsilon_y) \pm \frac{1}{2}\sqrt{(\varepsilon_x - \varepsilon_y)^2 + \gamma_{xy}^2}$. The algebra involved in finding the solutions to these simultaneous equations is not too difficult but is tedious and needs to be worked through carefully! This method can, however, be very much simplified if we *assume* that one of the strain gauges is aligned with the x-axis, where θ in equation 7.35 is zero, and under these circumstance we find, for example,

that $\varepsilon_1 = \frac{1}{2}(\varepsilon_x + \varepsilon_y) + \frac{1}{2}(\varepsilon_x - \varepsilon_y)\cos 0 + \frac{1}{2}\gamma_{xy}\sin 0$

or $\varepsilon_1 = \frac{1}{2}(\varepsilon_x + \varepsilon_y) + \frac{1}{2}(\varepsilon_x - \varepsilon_y)$, therefore, $\varepsilon_1 = \varepsilon_x$. Similarly we find that for a rectangular rosette, the strain ε_y is equal to the stress reading on the strain gauge oriented at 90° to the x-axis, as illustrated in Figure 7.27a, i.e. $\varepsilon_3 = \varepsilon_y$. The *direction* of the principal strain axes may be determined from equation 7.38, i.e.

from $\tan 2\theta = \dfrac{\gamma_{xy}}{(\varepsilon_x - \varepsilon_y)}$. We next apply a version of this

method to *a rectangular rosette configuration*.

7.5.3 Determining principal strains from a rectangular rosette

We know from above that the plane strain on the surface of the material being measured by a strain rosette is defined by finding ε_x, ε_y, γ_{xy} with respect to an arbitrary two-dimensional Cartesian axes system. We can also define the plane strain at the rosette by finding ε_1, ε_2 and θ, that is, by considering the principal strains and their directions. Either set of these parameters may be found provided we are able to obtain three independent strain readings; then, as mentioned in section 7.5.2, we can produce and solve the system of three equations for these unknowns.

The rectangular strain gauge rosette, shown in Figure 7.28, consists of three strain gauges that allow us to measure three independent strain readings at known angles from one another and so enable us to formulate the equations for the required unknowns.

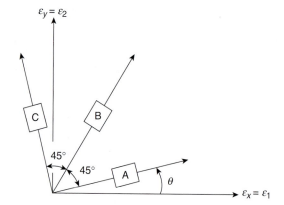

Figure 7.28 Rectangular rosette gauge arrangement

Now, by assuming that $\varepsilon_x = \varepsilon_1$ and $\varepsilon_y = \varepsilon_2$, equation 7.35, $\varepsilon_\theta = \frac{1}{2}(\varepsilon_x + \varepsilon_y) + \frac{1}{2}(\varepsilon_x - \varepsilon_y)\cos 2\theta + \frac{1}{2}\gamma_{xy}\sin 2\theta$ can be written in terms of the principal strains as $\varepsilon_\theta = \frac{1}{2}(\varepsilon_1 + \varepsilon_2) + \frac{1}{2}(\varepsilon_1 - \varepsilon_2)\cos 2\theta$, where we have omitted the term $\frac{1}{2}\gamma_{xy}\sin 2\theta$ because $\gamma_{xy} = 0$ *on the principal axes*. Note that this equation is simply a version of equation 7.34, after the exchange of variables.

Therefore, in our case, by considering the *arrangement* shown in Figure 7.28, where the gauges A, B and C are oriented at 45° to one another and gauge A is oriented at an angle θ, above ε_1, we can use $\varepsilon_\theta = \frac{1}{2}(\varepsilon_1 + \varepsilon_2) + \frac{1}{2}(\varepsilon_1 - \varepsilon_2)\cos 2\theta$ to determine the *principal strains* from the strain gauge readings.

So, for the angles of rotation along which the strains act (taken from the ε_1 axis) that is, for θ, $\theta + 45°$, and

$\theta + 90°$, we obtain the following equations:

$$\varepsilon_A = \frac{1}{2}(\varepsilon_1 + \varepsilon_2) + \frac{1}{2}(\varepsilon_1 - \varepsilon_2)\cos 2\theta \qquad (7.40a)$$

$$\varepsilon_B = \frac{1}{2}(\varepsilon_1 + \varepsilon_2) + \frac{1}{2}(\varepsilon_1 - \varepsilon_2)\cos(2\theta + 90°) \qquad (7.40b)$$

$$\varepsilon_C = \frac{1}{2}(\varepsilon_1 + \varepsilon_2) + \frac{1}{2}(\varepsilon_1 - \varepsilon_2)\cos(2\theta + 180°) \qquad (7.40c)$$

Now we have three equations with three unknowns, which we need to solve to find expressions for ε_1, ε_2 and θ in terms of the strain gauge readings (see Example 7.10). This may be achieved, as before, by using trigonometric identities, or by using matrix algebra or, more simply, by using a graphical method. Irrespective of the method chosen, the values for ε_1, ε_2 and θ under these circumstances may be obtained from the equations:

$$\varepsilon_1 = \frac{1}{2}(\varepsilon_a + \varepsilon_c) + \frac{1}{\sqrt{2}}\sqrt{(\varepsilon_a - \varepsilon_b)^2 + (\varepsilon_b - \varepsilon_c)^2},$$

$$\varepsilon_2 = \frac{1}{2}(\varepsilon_a + \varepsilon_c)$$

$$- \frac{1}{\sqrt{2}}\sqrt{(\varepsilon_a - \varepsilon_b)^2 + (\varepsilon_b - \varepsilon_c)^2} \text{ or}$$

$$\varepsilon_{1,2} = \frac{1}{2}(\varepsilon_a + \varepsilon_c) \pm \frac{1}{\sqrt{2}}\sqrt{(\varepsilon_a - \varepsilon_b)^2 + (\varepsilon_b - \varepsilon_c)^2}$$

$$(7.41)$$

$$\theta_{1,2} = \frac{1}{2}\tan^{-1}\left(\frac{\varepsilon_A - 2\varepsilon_B + \varepsilon_C}{\varepsilon_A - \varepsilon_C}\right) \qquad (7.42)$$

Note that the results given by equations 7.41 and 7.42 are *only valid for a rectangular rosette strain gauge system, with the gauges A, B and C, labelled in the order given in Figure 7.28.* If other labelling systems are used, then equations 7.40a to 7.40c need to be solved from scratch to produce the correct combinations of the variables.

We should not forget that one of the reasons for finding the principal strains in the first place was to determine the *principal stresses* that, in elastic materials, will act in the *same direction* as the strains. Provided we know the value of Poisson's ratio (v) for a material and its modulus of elasticity E (which can easily be determined from tensile tests), then we can easily calculate the principal stresses from the principal strains, using rearranged versions of equations 1.20 and 1.21, i.e. $\varepsilon_x = \dfrac{\sigma_x}{E} - v\dfrac{\sigma_y}{E}$ and $\varepsilon_y = \dfrac{\sigma_y}{E} - v\dfrac{\sigma_x}{E}$. Then, on rearrangement and after substituting ε_1, ε_2, σ_1 and σ_2 for ε_x, ε_y, σ_x and σ_y into these equations, we obtain:

$$\sigma_1 = \frac{E}{1 - v^2}(\varepsilon_1 + v\varepsilon_2) \qquad (7.43)$$

$$\sigma_2 = \frac{E}{1 - v^2}(\varepsilon_2 + v\varepsilon_1) \qquad (7.44)$$

Equations 7.43 and 7.44 enable us to find the principal stresses from the principal strains, resulting from any situation where the principal strains acting at a point in a particular material have been determined.

7.5.4 Determining principal strains from a delta rosette

From a similar analysis, using the strain transformation equations discussed above, we can determine the principal strains from the three strains that act upon on a *delta rosette*. Figure 7.29 shows a delta rosette strain gauge arrangement with the strain gauges *labelled in a specific order* and with strain gauge A at an angle θ above ε_1.

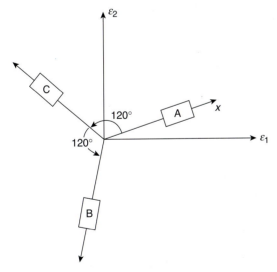

Figure 7.29 Delta rosette strain gauge arrangement

Note that although gauge B appears to be 240° from gauge A, in fact its line of action bisects gauges A and C, so that effectively gauge B reads the strain at 60° from gauge A. Then the *angles that need to be used* in equations 7.40a to 7.40c are, respectively, for gauge A, $\theta + 0°$, for gauge B, $\theta + 60°$ and for gauge C, $\theta + 120°$. Then, provided the labelling of the strain gauges follows

the order given here, it is found that the principal strains for this system are given by:

$$\varepsilon_{1,2} = \frac{1}{3}(\varepsilon_A + \varepsilon_B + \varepsilon_C)$$

$$\pm \frac{\sqrt{2}}{3}\sqrt{(\varepsilon_A - \varepsilon_B)^2 + (\varepsilon_B - \varepsilon_C)^2 + (\varepsilon_C - \varepsilon_A)^2}$$

$$(7.45)$$

$$\theta_{1,2} = \frac{1}{2}\tan^{-1}\left(\frac{\sqrt{3}(\varepsilon_C - \varepsilon_B)}{2\varepsilon_A - \varepsilon_B - \varepsilon_C}\right) \qquad (7.46)$$

Important note: It is very important to remember that when obtaining the principal strains from readings taken from strain gauge rosettes, the manufacturer's literature is followed precisely and, that due account is taken of the geometry of the arrangement of the strain gauges being used for the measurements. Thus equations 7.41, 7.42, 7.45 and 7.46 (which are in effect solutions) can only be used for the arrangement of the rectangular and delta rosettes illustrated in Figures 7.28 and 7.29. For any other arrangement, the principal strains would have to be determined by solving equations 7.40a to 7.40b on a case by case basis. In the example that follows we do just this.

> **Example 7.10** Three strain readings, taken from a *rectangular strain gauge* rosette attached to the surface of a material under stress, were as follows: gauge A, 400 μ, gauge B, 300 μ and gauge C, 100 μ, with gauges B and C being at angles of 45° and 90° measured anticlockwise from gauge A. Given that, for the material, $E = 200\ \text{GN/m}^2$ and Poisson's ratio $v = 0.3$, determine the:
>
> a) magnitude of the principal strains and the direction of their axes, principal stresses.
>
> Since we are dealing with the readings from a rectangular strain rosette, we can find the principal strains from the rosette strain reading using our modified version of equation 7.34, i.e. $\varepsilon_\theta = \frac{1}{2}(\varepsilon_1 + \varepsilon_2) + \frac{1}{2}(\varepsilon_1 - \varepsilon_2)\cos 2\theta$. Then we have, after inserting the micro-strains and the corresponding angles, the equations:
>
> $$400 = \frac{1}{2}(\varepsilon_1 + \varepsilon_2) + \frac{1}{2}(\varepsilon_1 - \varepsilon_2)\cos 2\theta$$

$$300 = \frac{1}{2}(\varepsilon_1 + \varepsilon_2) + \frac{1}{2}(\varepsilon_1 - \varepsilon_2)\cos(2\theta + 90°).$$

Also, $\cos(2\theta + 90°) = -\sin 2\theta$,

so we may write this equation as

$$300 = \frac{1}{2}(\varepsilon_1 + \varepsilon_2) - \frac{1}{2}(\varepsilon_1 - \varepsilon_2)\sin 2\theta$$

$$100 = \frac{1}{2}(\varepsilon_1 + \varepsilon_2) + \frac{1}{2}(\varepsilon_1 - \varepsilon_2)\cos(2\theta + 180°).$$

Also, $\cos(180° + 2\theta) = -\cos 2\theta$,

so we may write this equation as

$$100 = \frac{1}{2}(\varepsilon_1 + \varepsilon_2) - \frac{1}{2}(\varepsilon_1 - \varepsilon_2)\cos 2\theta$$

So we have that:

$$400 = \frac{1}{2}(\varepsilon_1 + \varepsilon_2) + \frac{1}{2}(\varepsilon_1 - \varepsilon_2)\cos 2\theta \quad (1)$$

$$300 = \frac{1}{2}(\varepsilon_1 + \varepsilon_2) - \frac{1}{2}(\varepsilon_1 - \varepsilon_2)\sin 2\theta \quad (2)$$

$$100 = \frac{1}{2}(\varepsilon_1 + \varepsilon_2) - \frac{1}{2}(\varepsilon_1 - \varepsilon_2)\cos 2\theta \quad (3)$$

We will now solve these equations simultaneously. Adding equation 1 to equation 3 gives

$$500 = (\varepsilon_1 + \varepsilon_2) \quad (4)$$

Then, substituting equation 4 into equation 1 gives

$$400 = 250 + \frac{1}{2}(\varepsilon_1 - \varepsilon_2)\cos 2\theta \quad \text{or}$$

$$150 = \frac{1}{2}(\varepsilon_1 - \varepsilon_2)\cos 2\theta \quad (5)$$

Also, substituting equation 4 into equation 2 gives

$$300 = 250 - \frac{1}{2}(\varepsilon_1 - \varepsilon_2)\sin 2\theta \quad \text{or}$$

$$50 = -\frac{1}{2}(\varepsilon_1 - \varepsilon_2)\sin 2\theta \quad (6)$$

Now you may wonder where all this is getting us, but remembering that $\tan\theta = \dfrac{\sin\theta}{\cos\theta}$, then, by dividing equation 6 by equation 5, we get

$$\frac{50 = -\frac{1}{2}(\varepsilon_1 - \varepsilon_2)\sin 2\theta}{150 = \frac{1}{2}(\varepsilon_1 - \varepsilon_2)\cos 2\theta} \quad \text{so that} \quad \frac{-\sin 2\theta}{\cos 2\theta} = \frac{1}{3}$$

or $\tan 2\theta = -\dfrac{1}{3}$ and $2\theta = -18.4°$. Now, because the sine is negative and the cosine is positive (then from your elementary trigonometry remembering CAST), the angle 2θ must lie in the fourth quadrant, i.e. $2\theta = 360 - 18.4° = 341.6°$ and $\theta = 170.8°$. Thus, gauge A is oriented at $170.8°$ anticlockwise from principal strain ε_1.

Then, substituting for 2θ into equation 5 gives

$$150 = \frac{1}{2}(\varepsilon_1 - \varepsilon_2)\cos 341.6° \text{ or } \frac{300}{0.9488} = \varepsilon_1 - \varepsilon_2 =$$

316.2. Now, by adding $316.2 = \varepsilon_1 - \varepsilon_2$ to equation 4, i.e. $[500 = (\varepsilon_1 + \varepsilon_2)] + [316.2 = \varepsilon_1 - \varepsilon_2]$, we find that $2\varepsilon_1 = 816.2$ and subtracting this equation from equation 4 gives $2\varepsilon_2 = 183.8$. Therefore, remembering that these are micro-strains, we find that

$$\varepsilon_1 = 408.1 \times 10^{-6},$$

$$\varepsilon_2 = 91.9 \times 10^{-6} \text{ and } \theta = 170.8°.$$

Now for part b). It is a relatively simple matter to find the principal stresses from the principal strains using equations 7.43 and 7.44, where

$$\sigma_1 = \frac{E}{1 - v^2}(\varepsilon_1 + v\varepsilon_2)$$

$$= \frac{200 \times 10^9}{1 - 0.09}\left[(408 \times 10^{-6}) + (0.3)(92 \times 10^{-6})\right]$$

$$= (219.8 \times 10^9)(435.6 \times 10^{-6}) = 95.74 \text{MN/m}^2.$$

Similarly,

$$\sigma_2 = \frac{E}{1 - v^2}(\varepsilon_2 + v\varepsilon_1) = (219.8 \times 10^9)(214.4 \times 10^{-6})$$

$$= 47.125 \text{MN/m}^2.$$

7.5.5 Graphical method for determining principal strains from a rosette

We now turn our attention to a graphical method for determining principal strains from three strain gauge rosette readings and, as mentioned earlier, we will concentrate on the McClintock method.

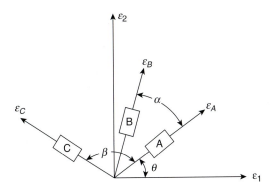

Figure 7.30 Three gauge rosette system with known strains

Consider the strain gauge arrangement shown in Figure 7.30. Using this method will enable us to produce Mohr's strain circle from the three known strain gauge readings, in known directions.

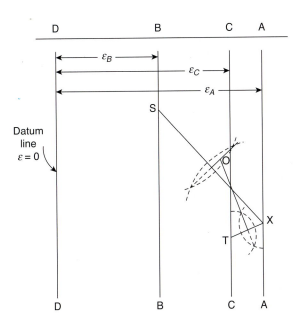

Figure 7.31 Illustration of the McClintock construction method showing construction lines to determine the centre of Mohr's strain circle

Method

1. Draw a vertical datum line DD, to represent zero direct strain. Then draw three vertical lines (AA, BB, CC) to scale at horizontal distances from this datum, to represent the known strains ε_A, ε_B and ε_C, as shown in Figure 7.31.

2. From any convenient point X on AA, draw a line at the known angle α anticlockwise to intersect

BB at S. Now draw another line through X at the known angle β anticlockwise to intersect CC at T.

3. Bisect XS and XT and draw perpendicular lines to intersect at O, which is the centre of Mohr's strain circle shown in Figure 7.31.

Now, from the construction, the principal strains ε_1, ε_2, and angle θ may be read off at the points indicated in Figure 7.32, once Mohr's strain circle has been drawn, with the centre at O and with radius OX. Note that with this construction, the angles α, β and θ have their *true* values, i.e. they *are not* drawn on at twice their value as with the complete Mohr's circle construction.

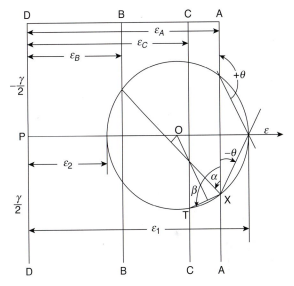

Figure 7.32 The complete strain circle, showing the principal strains and their direction using McClintock's method

Let us look now at an example, to illustrate this method.

Key Point. Provided the strain readings and the angle of orientation between the three strain gauges are known, the McClintock graphical method provides a simple way of determining reasonably accurate values for the principal strains

Example 7.11 Use McClintock's construction and the rectangular strain gauge rosette readings and positions, given in Example 7.10, to determine the principal strains and confirm the theoretical approach.

Thus we need to construct the Mohr's strain circle, using the McClintock method, where the strain readings are: gauge A, $400\,\mu$, gauge B, $300\,\mu$ and gauge C, $100\,\mu$, with gauges B and C being at angles of 45° and 90° measured *anticlockwise* from gauge A.

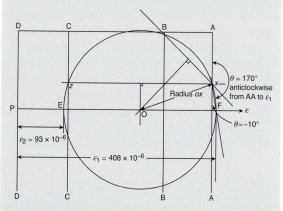

Figure 7.33 McClintock's construction for Example 7.11

Figure 7.33 shows the completed construction with some construction lines marked on. The radius for Mohr's strain circle $=$ OX and the principal strains, from the construction, are found to be $PF = \varepsilon_1 \simeq 408 \times 10^{-6}$ and $PE = \varepsilon_2 \simeq 93 \times 10^{-6}$, also $\theta \simeq 170°$ anticlockwise (positive) from gauge A, all these approximations being very close to the calculated values.

7.6 Chapter summary

In this chapter you have been introduced to the concepts, equations and methods used to solve simple engineering problems concerned with materials and structures subject to complex stress and strain. We started by considering the notion of the *stress element* and how, by considering this element to be infinitesimally small and so assuming the stress to act at a point within the material, we could ignore the fact that the stresses imposed on the material by the external loads would in practice vary considerably over the area of the material. We added the further provisos to our analysis that first we were only dealing with stresses in materials that obeyed *Hooke's law*, i.e. that were subject to stresses within their elastic range, and secondly that we would only be considering simplified *two-dimensional stress elements*, that is, materials subject to *plane stress*.

Sections 7.1 and 7.2 were primarily concerned (using the equilibrium of stress elements) with determining the stresses that occurred on planes inclined at some angle oblique to the direct stresses, mutually perpendicular stresses and resulting shear stresses that acted in various combinations on the material. The result of this analysis was to produce a series of *transformation equations*, to determine the magnitude and direction of stresses acting on any *inclined plane* of the material (at a point) that resulted from any two-dimensional *plane stress* situation. To help us formulate these transformation equations, we also followed the convention given in section 7.2.1 and Figure 7.3 that set the *positive and negative directions* for both the *normal (direct) stresses* and *shear stresses* that acted on the element. You should revise this section if you are still unclear about this convention, but also remember that, no matter what, *the analysis*, if correctly followed, will also *dictate the direction of these stresses*. The important transformation equations are repeated here, for your convenience and consolidation. The transformation equations are:

- For materials subject to two mutually perpendicular direct stresses,

$$\sigma_\theta = \frac{(\sigma_x + \sigma_y)}{2} + \frac{(\sigma_x - \sigma_y)}{2}\cos 2\theta \qquad (7.5)$$

$$\tau_\theta = \frac{(\sigma_x - \sigma_y)}{2}\sin 2\theta \qquad (7.7)$$

- For materials subject to pure shear conditions,

$$\tau_\theta = -\tau_{xy}\cos 2\theta \qquad (7.11)$$

For materials subject to combined direct and shear stresses,

$$\sigma_\theta = \frac{(\sigma_x + \sigma_y)}{2} + \frac{(\sigma_x - \sigma_y)}{2}\cos 2\theta + \tau_{xy}\sin 2\theta \qquad (7.12)$$

$$\tau_\theta = \frac{(\sigma_x - \sigma_y)}{2}\sin 2\theta - \tau_{xy}\cos 2\theta \qquad (7.13)$$

- Also, the equations relating the direct stresses to the principal stresses and maximum shear stress are:

$$\tan 2\theta = \frac{2\tau_{xy}}{(\sigma_x - \sigma_y)} \qquad (7.14)$$

$$\sigma_1 \text{ or } \sigma_2 = \frac{1}{2}(\sigma_x + \sigma_y) \pm \frac{1}{2}\sqrt{(\sigma_x - \sigma_y)^2 + 4\tau_{xy}^2} \qquad (7.15)$$

$$\tau_{max} = \frac{1}{2}\sqrt{\left(\sigma_x - \sigma_y\right)^2 + 4\tau_{xy}^2} \qquad (7.16)$$

In section 7.3 we considered *Mohr's stress circle*, where we looked at its *method of construction* (procedures 1 to 7) for a plane stress, and the information that could be obtained from it, by measuring the various stress components (σ_θ, τ_θ, τ_{max}, σ_1, σ_2) from the circle. Remember that for this circle the angles to the planes on which the stresses acted were all shown as being equal to 2θ. *Positive angles* are measured as is normal in trigonometry, *anticlockwise* from the positive horizontal (*x*-axis), unless stated otherwise. *Do not worry too much about steps 8–10* used to validate the construction, just make sure that you can construct, read and determine the required stress values and their orientation (see Example 7.4) from the construction.

In section 7.4, we started our brief study of strain by looking at the differences between plane stress and plane strain. Do remember that for *plane strain* (strain in only two directions) we may require three-dimensional stresses to act on the material, to achieve it, whereas *plane stress* occurs when the material (solid) is stressed in only two directions, say in the *xy* directions. However, three-dimensional strain can occur (due to the Poisson effect) as a result of a solid being stressed in one, two or three directions (see Figure 7.15).

In section 7.4.1 we formulated some relationships between the elastic constant *E*, *Poisson's ratio* and the principal stresses and strains, the most important of which are given by equations 7.20 to 7.22.

In section 7.4.2 we looked at *volumetric strain* and the definition of the *bulk modulus K* and formed one or two useful relationships such as equations 7.25 and 7.26 that we were able to apply in Example 7.5. In section 7.4.3 we derived yet more useful relationships between the *elastic constants E, G and K* and *Poisson's ratio* (*v*), in particular, those given by equations 7.29, 7.30, 7.32 and 7.33.

In section 7.4.4 we derived the strain transformation equations for strains of inclined planes, using the same type of analysis as we did when deriving their equivalents for stress. The two important equations relating the strain on the inclined plane with the combined strains acting on the material are given here for your convenience and to aid your revision.

- $$\varepsilon_\theta = \frac{1}{2}(\varepsilon_x + \varepsilon_y) + \frac{1}{2}(\varepsilon_x - \varepsilon_y)\cos 2\theta + \frac{1}{2}\gamma_{xy}\sin 2\theta \qquad (7.35)$$

- $$\frac{1}{2}\gamma_\theta = -\frac{1}{2}(\varepsilon_x - \varepsilon_y)\sin 2\theta + \frac{1}{2}\gamma_{xy}\cos 2\theta \qquad (7.36)$$

The principal strains and maximum/minimum shear strains may be found from equations 7.37 to 7.39. These are given here, again for your convenience and to aid your revision.

- $$\varepsilon_1 \text{ or } \varepsilon_2 = \frac{1}{2}(\varepsilon_x + \varepsilon_y) \pm \frac{1}{2}\sqrt{\left(\varepsilon_x - \varepsilon_y\right)^2 + \gamma_{xy}^2} \qquad (7.37)$$

- $$\tan 2\theta = \frac{\gamma_{xy}}{(\varepsilon_x - \varepsilon_y)} \qquad (7.38)$$

- $$\frac{1}{2}\gamma_{max} = \pm\frac{1}{2}\sqrt{\left(\varepsilon_x - \varepsilon_y\right)^2 + \gamma_{xy}^2} \qquad (7.39)$$

Now you should compare equations 7.37, 7.38 and 7.39 with equations 7.15, 7.14 and 7.16, respectively, noting the similarities in form and the replacements for the variables, in particular noting that we replace σ with ε and τ by $\frac{\gamma}{2}$. In section 7.4.5 we looked at the construction and interpretation of *Mohr's strain circle* and noted that its construction required us to follow the same procedure that we used when constructing Mohr's stress circle, the only difference being on the shear axes where $\tau_{xy} = -\frac{\gamma_{xy}}{2}$, that is, the shear axis was reversed and τ was replaced by $\frac{\gamma}{2}$. Example 7.9 shows the completed strain circle solution, for a particular problem.

The whole of section 7.5 was given over to the study of *strain gauges* and *strain gauge rosettes*. These gauges are used to measure strains in materials and engineering structures, during test or in-situ during service. The most common type of strain gauge is the *electrical resistance type* that measures strains in a particular *linear direction*, hence the need for *strain gauge rosettes*. Two types of strain gauge rosette were considered, these being the *rectangular* and *delta* strain gauge rosette configurations.

In section 7.5.2 we used equation 7.35 to find a set of equations that could be used to find the *principal strains* from *three* strain gauge readings, taken, in this case, from a *rectangular rosette*. These equations, 7.40a to 7.40c, were then solved and the *solution equations 7.41 and 7.42* could then be used to find the required *principal strains* and the direction of the axes on which they act. It was emphasised that these *solution equations could only be used* for the particular arrangement shown in Figure 7.28. Similarly, in section 7.5.3 for a delta rosette the solution equations 7.45 and 7.46 could only be used to find the principal strains for the particular delta rosette arrangement illustrated in

Figure 7.29. In Example 7.10 we showed how, for a rectangular rosette arrangement, we were able to solve equations 7.40a to 7.40c from scratch and so find the required principal strains and their orientation, *without* having to use the specific solution equations. Therefore, provided we have the *three* strain gauge readings and we know the *angles* between the gauges, we can always formulate equations similar to equations 7.40a to 7.40c for *any three-gauge rosette configuration* and solve the system simultaneously, in order to find the principal stresses.

Once the principal strains have been determined we may use equations 7.43 and 7.44:

$$\sigma_1 = \frac{E}{1 - v^2}(\varepsilon_1 + v\varepsilon_2) \qquad (7.43)$$

$$\sigma_2 = \frac{E}{1 - v^2}(\varepsilon_2 + v\varepsilon_1) \qquad (7.44)$$

to determine the *principal stresses* which act on the *surface of the material* at the position where the strain is being measured.

In section 7.5.5 we looked at the *McClintock method* for *graphically determining the principal strains* from a three-gauge rosette. The result of using this method is to produce a version of Mohr's strain circle, where reasonably accurate estimates for the principal strains and their orientation may be found from the construction. The full McClintock construction method is given in section 7.5.5 and illustrated in Figures 7.31 and 7.32. Example 7.11 shows how this method was applied to the strain gauge rosette problem solved analytically in Example 7.10.

7.7   Review questions

1. A circular rod having a diameter of 5 cm is subject to an axial load of 200 kN. Determine the stresses that act on a plane cut through the rod at an angle of 30°, stating any assumptions made when formulating your answers.

2. For question 1, state when the shear stress on the plane will be a maximum (τ_{max}), calculate its value and state the value of the corresponding tensile stress acting normal to the plane.

3. Explain the nature and use of *stress elements* when analysing complex stresses in engineering materials and structures.

4. An element of a material in plane two-dimensional stress is subject to two mutually perpendicular direct tensile stresses $\sigma_x = 150\,\mathrm{MN/m^2}$ and $\sigma_y = 100\,\mathrm{MN/m^2}$, and also to shear stress $\tau_{xy} = 50\,\mathrm{MN/m^2}$. Determine the stresses that act on this element of the material when inclined at an angle of 40°, and sketch the stress element, showing the resulting stresses on its faces.

5. An element in plane stress is subject to stresses $\sigma_x = 120\,\mathrm{MN/m^2}$, $\sigma_y = -40\,\mathrm{MN/m^2}$ (compressive) and $\tau_{xy} = -30\,\mathrm{MN/m^2}$. Determine the magnitude and orientation of the principal stresses and the maximum shear stress.

6. Construct a Mohr's stress circle and verify your answers for the magnitude and orientation of the principal stresses and maximum shear stress found for question 5.

7. The principal strains (ε_1 and ε_2) at a point on a loaded steel sheet are found to be 350×10^{-6} and 210×10^{-6}, respectively. If the modulus of elasticity $E = 210\,\mathrm{GN/m^2}$ and Poisson's ratio for the steel is 0.3, determine the corresponding principal stresses.

8. An aluminium alloy has a modulus of elasticity $E = 80\,\mathrm{MN/m^2}$ and a modulus of rigidity $G = 30\,\mathrm{MN/m^2}$. Determine the value of Poisson's ratio and the bulk modulus.

9. A rod 30 mm in diameter and 0.5 m in length is subject to an axial tensile load of 90 kN. If the rod extends in length by 1.1 mm and there is a decrease in diameter of 0.02 mm, determine the value of Poisson's ratio and the values of E, G and K.

10. A cylindrical aluminium container 75 cm long and 30 cm in diameter has a wall thickness of 5 mm. Find the increase in volume when the cylinder is subject to an internal pressure of 2.0 MPa. Also find the values for the elastic constants G and K, given that $E = 80\,\mathrm{GPa}$ and Poisson's ratio $v = 0.28$.

11. An element of a material subject to plane strain is strained in the following manner: $\varepsilon_x = 400 \times 10^{-6}$, $\varepsilon_y = 250 \times 10^{-6}$ and $\gamma_{xy} = 150 \times 10^{-6}$. Determine the:

    a) strains on the planes of the element when the element is oriented at an angle of 30° from the horizontal *x*-ordinate, principal strains and their orientation, maximum shear strains and their orientation, principal stresses, given that for the material $E = 210\,GN/m^2$ and Poisson's ratio $= 0.3$.

12. Construct a Mohr's strain circle for the data given in question 11 and from it confirm the calculated values found for the strains in parts a), b) and c) of the question.

13. Three strain readings, taken from a *rectangular strain gauge* rosette attached to the surface of a material under stress, were as follows: gauge A, 200×10^{-6}, gauge B, 250×10^{-6} and gauge C, 400×10^{-6}, with gauges B and C being at angles of $45°$ and $90°$ measured anticlockwise from gauge A. Find, by formulating and solving three simultaneous equations, the magnitudes of the principal strains and the direction of their axes.

14. Find the principal stresses from the principal strains you found for question 13, given that the material under stress has elastic modulus $E = 200\,GN/m^2$ and Poisson's ratio $= 0.3$.

15. Use McClintock's construction and the rectangular strain gauge rosette readings and positions, given in question 13, to determine the principal strains and confirm your theoretical approach.

Part II

Dynamics

In this part of the book we consider the dynamic behaviour of bodies, that is, the motion of bodies under the action of forces. The study of dynamics is normally split into two major fields, *kinematics*, the study of the motion of bodies without reference to the forces that cause the motion, and *kinetics*, where the motion of bodies is considered in relationship to the forces producing it. With the exception of the study of mechanisms in Chapter 9, we will not, in this short study of *dynamics*, differentiate a great deal between these two areas, rather concentrating on the macroscopic effects of forces and resulting motions in dynamic engineering systems and machinery.

We start our study of dynamics in Chapter 8 by looking at a number of fundamental concepts concerning linear and angular motion and the forces that create such motion, including a review of Newton's laws. We then consider momentum and inertia, together with the nature and effects of friction that acts on linear and angular motion machines and systems. Next, mechanical work and energy transfer are considered and their relationship to linear and angular motion systems is explored. Finally in Chapter 8, as a prelude to rotating machinery in Chapter 10, we take a brief look at circular motion and the forces created by such motion.

In Chapter 9, we first consider the motion of one or two single and multilink mechanisms. In particular,

we look at the nature of the relative velocities and accelerations of these mechanisms, together with the power and efficiencies of particular engineering machines that utilise such mechanisms. Then, very briefly, we consider the geometry and output motions of various types of cams. We will then look at a number of gyroscopic motion parameters, in particular gyroscopic rigidity, precession and reaction torque. Finally, one or two gyroscopic engineering applications are considered.

In Chapter 10 we will look at power transmission systems, where the application and dynamic parameters of belt drives, friction clutches, gears, screw drives, dynamic balancing, rotors and flywheels will be considered.

In Chapter 11, we conclude our short study of engineering dynamics with a brief look at the nature and effects of mechanical vibration. In particular, we consider oscillatory motion and the effects that external forces and system damping have on the resulting motion and outputs of oscillatory systems, in particular those displaying simple harmonic motion. Then we will look at natural vibration by considering a mass–damper system and the transverse vibration set up in cantilevers and other beams. Finally, damped vibration and forced vibration are looked at, again by modelling their effects using a mass–spring system.

Chapter 8

Fundamentals

8.1 Newton's laws

We start this chapter by considering Newton's laws of motion. In order to do this we need briefly to revise the concepts of speed, velocity, acceleration and momentum, which are fundamental to a proper understanding of these laws. You may already be familiar with these concepts, but they are presented here to assist those who are a little rusty or, for whatever reason, may have gaps in their basic knowledge.

8.1.1 Fundamental definitions

Here are a few fundamental definitions, which you will no doubt be familiar with but are worth emphasizing here, in order to understand their significance in Newton's laws of motion.

Speed may be defined as *distance per unit time*, or as *rate of change of position*. Speed takes no account of direction and is therefore a scalar quantity. The common SI units of speed are: metres per second (m/s) or kilometres per hour (km/h).

Velocity is defined as *rate of change of position in a specified direction*. Therefore, velocity is a *vector quantity* and the SI units for the magnitude of velocity are the SI units for speed, that is m/s. The direction of a velocity is not always quoted but it should be understood that the velocity is in some defined direction, even though that direction is not always stated. You will also remember, from *Essential Mathematics 2* on the companion website, that we may use the differential calculus to define instantaneous rates of change. Thus, mathematically, $v = \dfrac{ds}{dt}$, where v = velocity in metres

per second (m/s), s = change in position or displacement in metres (m) and t = time in seconds (s).

Acceleration is defined as *rate of change of velocity*. Acceleration is also a *vector quantity* and the SI unit of acceleration is m/s/s or m/s^2. Mathematically, acceleration $a = \dfrac{dv}{dt} = \dfrac{d^2s}{dt^2}$.

Equilibrium is a concept you have already met in your earlier study of statics, but for completeness, its relationship to dynamics is given here. *A body is said to be in equilibrium when it remains at rest or when it continues to move in a straight line with constant velocity.*

Momentum is the product of the mass of a body and its velocity (momentum = mv). *Any change in momentum requires a change in velocity, that is, acceleration.* This is why momentum is sometimes described as the *quantity of motion of a body.* It may be said that for a fixed quantity of matter to be in equilibrium, it must have constant momentum. The use of momentum will be seen next when we consider Newton's laws.

All matter resists change. The force resisting change in momentum is called *inertia*. The inertia of a body depends on its mass: the greater the mass, the greater the inertia. The inertia of a body is an innate (in-built) force that only becomes effective when acceleration (change in momentum) occurs. An applied force acts against inertia so as to accelerate (or tend to accelerate) a body.

You have already met the idea of *force* in your earlier study of statics, when we considered the forces acting on beams, columns, shells etc. We now look at the concept of force with respect to our study of dynamics, where, as in statics, the *applied force is called the action* and *the opposing force it produces is known as the reaction.*

Force is that which changes, or tends to change, the state of rest or uniform motion of a body. Forces that act on a body may be external (applied from outside the body), such as weight, or internal, such as the internal resistance of a material subject to a compression.

The effects of any force depend on its three characteristics – magnitude, direction, and point of application – as again you will remember from your study of statics. The difference between the forces tending to cause motion and those opposing motion is called the *resultant* or *out-of-balance force*. A body that has no out-of-balance external force acting on it is *in equilibrium and will not accelerate. A body that has such an out-of-balance force will accelerate at a rate dependent on the mass of the body and the magnitude of the out-of-balance force.* The necessary opposition that permits the existence of the out-of-balance force is provided by *the force of inertia.*

Key Point. A body is said to be in dynamic equilibrium when the external forces causing motion are equal to the external forces opposing motion, that is, when the acceleration is zero

8.1.2 Newton's laws of motion

Sir Isaac Newton (1642–1727) formulated, among other things, three laws of motion. These deal with the acceleration produced on a body by an external force. These long-standing laws assist engineers in many motion-related design problems.

Newton's first law of motion states that: *a body remains in a state of rest, or of uniform motion in a straight line, unless it is acted upon by some external resultant force.* Thus, if a body is moving it requires a force to cause it to accelerate or decelerate. The reason why a body behaves in accordance with Newton's first law is because of its *inertia*, which causes the body to resist the change of motion.

Newton's second law of motion states that: *the rate of change of momentum of a body is directly proportional to the force producing the change, and takes place in the direction in which the force acts.*

We also know, from our above definitions, that acceleration may be defined as change in velocity per unit time or rate of change in velocity.

If a force F (N) acts on a body of mass m (kg) for a time t (s), then the velocity changes uniformly from u (m/s) to v (m/s). Then:

The rate of change of momentum = $\dfrac{\text{change in momentum}}{\text{time taken}} = \dfrac{mv - mu}{t} = \dfrac{m(v - u)}{t}$,

but $a = \dfrac{v - u}{t}$, therefore rate of change of momentum $= ma$ and, according to Newton's second law, force is proportional to the rate of change of momentum, then $F \propto ma$ or $F = kma$, where k is a constant. The unit of force is chosen as that force which will give a mass of 1 kg an acceleration of 1 m/s². Therefore, by substitution into $F = kma$, we get $1 = k \times 1 \times 1$.

Therefore

$$F = ma \qquad (8.1)$$

It is also worthwhile at this point to remember that the weight force W of a body is the product of its mass and the acceleration due to gravity (g). Although this acceleration varies, dependent on geographical position, a good enough approximation to its value at sea level in temperate latitudes is 9.81 m/s², so we may define weight force based on Newton's second law as

$$W = mg \qquad (8.2)$$

Newton's third law of motion states that: *to every action there is an equal and opposite reaction.* So, for example, the compressive forces that result from the weight of a building, the *action*, are held in equilibrium by the *reaction* forces that occur inside the materials of the building's foundation. Another example is that of propulsion. An aircraft jet engine produces a stream of high-velocity gases at its exhaust, the *action*. These gases act on the airframe of the aircraft causing a *reaction*, which enables the aircraft to accelerate and increase speed for flight.

8.2 Linear equations of motion

You have already been introduced to the concepts of force, velocity, acceleration and now, Newton's laws. These concepts are further exploited through the use of the linear equations of motion. Look back now, and remind yourself of the relationship between mass, force, acceleration and Newton's laws.

The linear equations of motion rely for their derivation on the one very important fact that the *acceleration is assumed to be constant.*

Key Point. The linear equations of motion rely on the fact that the acceleration is constant

Even simple linear motion, motion along a straight line, can be difficult to deal with mathematically.

However, in the case where acceleration is constant it is possible to solve problems of motion by use of a velocity/time graph. We will now consider the derivation of the four standard equations of motion, using a graphical method.

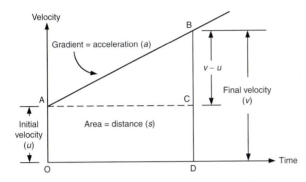

Figure 8.1 Velocity/time graph

The equations of motion, as you will already know, use standard symbols to represent the variables. These are:

s = distance in metres (m)
u = initial velocity (m/s)
v = final velocity (m/s)
a = acceleration (m/s^2)
t = time (s)

The velocity is plotted on the vertical axis and time on the horizontal axis. Constant velocity is represented by a horizontal straight line and acceleration by a sloping straight line. Deceleration or retardation is also represented by a sloping straight line, but with a negative slope.

Key Point. Acceleration and retardation are indicated by the slope of the velocity/time graph

By considering the velocity/time graph shown in Figure 8.1, we can establish the equation for distance.

The distance travelled in a given time is equal to the velocity (m/s) multiplied by the time (s). This is found from the graph by the area under the sloping line. In Figure 8.1, a body is accelerating from a velocity u to a velocity v in time t seconds. Now the distance travelled s = area under the graph, so that

$$s = ut + \left(\frac{v-u}{2}\right)t \quad \text{or} \quad s = \frac{(2u+v-u)t}{2},$$

and so $\quad s = \dfrac{(u+v)t}{2}$ \hfill (8.3)

In a similar manner to the above, one of the velocity equations can also be obtained from the velocity/time graph. Since the acceleration is the rate of change of velocity with respect to time, the value of the acceleration will be equal to the gradient of a velocity/time graph. Therefore, from Figure 8.1, we have:

$$\text{Gradient} = \text{acceleration} = \frac{\text{velocity}}{\text{time taken}};$$

therefore, from Figure 8.1, the acceleration is given by

$$a = \frac{v-u}{t} \tag{8.4}$$

or

$$v = u + at \tag{8.5}$$

Example 8.1 Derive the equations $s = ut + \dfrac{1}{2}at^2$ and $v^2 = u^2 + 2as$ using equations 8.3 to 8.5.

Substituting $v = u + at$ (equation 8.5) into equation 8.3 gives $s = \dfrac{(u+u+at)t}{2} = \dfrac{2ut + at^2}{2}$, and so

$$s = ut + \frac{1}{2}at^2 \tag{8.6}$$

Also, from equation 8.4, we get $t = \dfrac{v-u}{a}$, and substituting this expression into equation 8.3 gives $s = \left(\dfrac{u+v}{2}\right)\left(\dfrac{v-u}{a}\right)$, so $2as = (u+v)(v-u) = uv + v^2 - u^2 - uv$, and so

$$v^2 - u^2 = 2as \tag{8.7}$$

Here is an example of the use of the velocity/time graph.

Example 8.2 A body starts from rest and accelerates with constant acceleration of 2.0 m/s^2 up to a speed of 9 m/s. It then travels at 9 m/s for 15 s after which time it is retarded to a speed of 1 m/s. If the complete motion takes 24.5 s, find:

a)   the time taken to reach 9 m/s,
b)   the retardation,
c)   the total distance travelled.

The solution is made easier if we sketch a graph of the motion, as shown in Figure 8.2.

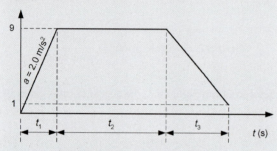

Figure 8.2 Velocity/time graph of the motion

a)   We have from the data that

$$u = 0 \text{ m/s}$$
$$v = 9 \text{ m/s}$$
$$a = 2 \text{ m/s}^2$$
$$t = ?$$

All we need to do is select an equation which contains all the variables listed above, that is, $v = u + at$, and on transposing for t and substituting the variables we get $t = \dfrac{9 - 0}{2}$ or $t = 4.5 \text{ s}$.

b)   The retardation is found in a similar manner. We have that $u = 9 \text{ m/s}$, $v = 2 \text{ m/s}$, $t = 5 \text{ s}$, $a = ?$

We again select an equation which contains the required variables, that is, $v = u + at$, and on transposing for a and substituting the variables we find that $a = \dfrac{1 - 9}{5} = -1.6 \text{ m/s}^2$ (the negative sign indicating a *retardation*).

c)   The total distance travelled requires us to sum the component distances travelled for the times t_1, t_2 and t_3. This problem is best solved by tabulating the variables for each stage, so:

$u_1 = 0 \text{ m/s}$	$u_2 = 9 \text{ m/s}$	$u_3 = 9 \text{ m/s}$
$v_1 = 9 \text{ m/s}$	$v_2 = 9 \text{ m/s}$	$v_3 = 1 \text{ m/s}$
$t_1 = 4.5 \text{ s}$	$t_2 = 15 \text{ s}$	$t_3 = 5 \text{ s}$
$s_1 = ?$	$s_2 = ?$	$s_3 = ?$

The appropriate equation is $s = \dfrac{(u + v)t}{2}$ and in each case we find that $s_1 = \dfrac{(0 + 9)4.5}{2}$, $s_2 = \dfrac{(9 + 9)15}{2}$ and $s_2 = \dfrac{(9 + 1)5}{2}$, so that $s_1 = 20.25$, $s_2 = 135$ and $s_3 = 25$.

Then total distance $s_T = 20.5 + 135 + 25 = 180.25 \text{ m}$.

The following example involves the use of Newton's second law as well as the equations of linear motion.

Example 8.3   A racing car of mass 1965 kg accelerates from 160 to 240 km/h in 3.5 s. If the air resistance is 2000 N/tonne, find the:

a)   average acceleration,
b)   force required to produce the acceleration,
c)   inertia force on the car.

(a)   We first need to convert the velocities to standard units, so $u = 160 \text{ km/h} = \dfrac{160 \times 1000}{60 \times 60} = 44.4 \text{ m/s}$ and $v = 160$ km/h $= \dfrac{240 \times 1000}{60 \times 60} = 66.6 \text{ m/s}$. We also know that $t = 3.5 \text{ s}$, so the acceleration is given from $a = \dfrac{v - u}{t} = \dfrac{66.6 - 44.4}{3.5} = 6.34 \text{ m/s}^2$.

(b)   The accelerating force is easily found using Newton's second law, where $F = ma = 1965 \text{ kg} \times 6.34 \text{ m/s}^2 = 12.46 \text{ kN}$.

(c)   From what has already been said you will be aware that the inertia force is equal to the accelerating force that overcomes it. Therefore the inertia force $= 12.46 \text{ kN}$.

We have used a graphical method for verifying the equations of motion. We could have equally well used the calculus, especially to find *instantaneous values* of velocity and acceleration. If we had, then by virtue of these being a snapshot at any instant in time, they would give correct values even if the acceleration was to vary. It is therefore worth remembering the relationships given earlier in our definitions. They were, using the differential calculus, that: $v = \dfrac{ds}{dt}$, $a = \dfrac{dv}{dt} = \dfrac{d^2s}{dt^2}$ or $a = v\dfrac{ds}{dt}$. Also, applying the integral calculus, i.e. integrating with respect to time, we get $\int a = \int \frac{d^2s}{dt^2}$ so $at +$ constant $= v$ and if we apply *boundary conditions* (see *Essential Mathematics 2* on the book's website) and let u be the value of the constant, then, at time $t = 0$, we find that $v = u + at$. Now, knowing that $\frac{ds}{dt} = v$, integrating again with respect to time, we get that $\int \frac{ds}{dt} = \int u + at$ or $s = ut + \frac{1}{2}at^2 +$ constant and so, by applying the *boundary condition* that the distance s is measured from the time when $t = 0$, so that the constant is zero, we therefore find that $s = ut + \frac{1}{2}at^2$.

Example 8.4   The motion of a body is modelled by the relationship $s = t^3 - 3.5t^2 + 2t + 4$, where s is the distance in metres and t is the time in seconds. Find:

a)   the velocity of the body after 5 seconds,

b)   the time when the body has a velocity of zero,

c)   its acceleration at time $t = 2$ s,

d)   when the acceleration of the body is zero.

Answers

a)   The velocity is given by $\frac{ds}{dt} = 3t^2 - 7t + 2$, therefore at $t = 5$ s the velocity $= 3(5)^2 - 7(5) + 2 = 42$ m/s.

b)   The body will have zero velocity when $\frac{ds}{dt} = 3t^2 - 7t + 2 = 0$, that is when $(3t - 1)(t - 2) = 0$ or when $t = \frac{1}{3}$ s or $t = 2$ s.

c)   The acceleration is found from $a = \frac{d^2s}{dt^2} = 6t - 7$ a When $t = 2$s, then $a = 5$ m/s^2.

d)   The acceleration is zero when $a = 6t - 7 = 0$, that is when $t = \frac{7}{6}$ s.

Key Point.   If $s =$ linear displacement, then $\frac{ds}{dt} =$ linear velocity and $\frac{d^2s}{dt^2} =$ linear acceleration

---

8.3   Angular motion

You have just met the equations for linear motion. There also exists a similar set of equations to solve engineering problems that involve angular motion as experienced, for example, in the rotation of a drive shaft. The linear equations of motion may be transformed to represent angular motion using a set of equations that we will refer to as the *transformation equations*. These are given below, followed by the equations of angular motion, which are tabulated alongside their linear equivalents.

Transformation equations

$$s = r\theta \qquad (8.8)$$

$$v = r\omega \qquad (8.9)$$

$$a = r\alpha \qquad (8.10)$$

where r is the radius of body from the centre of rotation and θ, ω and α are the angular displacement, angular velocity and angular acceleration, respectively. Thus these equations provide a relationship between the linear and angular variables, so that we are able, given the appropriate data, to find the linear or angular equivalents at will.

Key Point.   The linear (tangential) displacement, velocity and acceleration are related to their angular counterparts by the transformation equations

---

The *angular equations of motion*, with their linear equivalents, are shown in the table below.

Linear equations of motion		Angular equations of motion	
$s = \dfrac{(u + v)t}{2}$	(8.3)	$\theta = \dfrac{(\omega_1 + \omega_2)t}{2}$	(8.11)
$s = ut + \dfrac{1}{2}at^2$	(8.6)	$\theta = \omega_1 t + \dfrac{1}{2}\alpha t^2$	(8.12)
$v^2 - u^2 = 2as$	(8.7)	$\omega_2^2 = \omega_1^2 + 2\alpha\theta$	(8.13)
$a = \dfrac{v - u}{t}$	(8.4)	$\alpha = \dfrac{\omega_2 - \omega_1}{t}$	(8.14)

where

θ	$=$ the angular displacement in radians (rad)
t	$=$ time (s)
ω_1, ω_2	$=$ initial and final angular velocities (rad/s)
α	$=$ angular acceleration (rad/s^2)

We will now look at several simple examples that use the angular motion and transformation equations.

Key Point.   Angular displacement is measured in radians

Example 8.5 A 540 mm diameter wheel is rotating at $1500/\pi$ rpm. Determine the angular velocity of wheel in rad/s and the linear velocity of a point on the rim of the wheel.

Now, remembering that 2π rad $= 360°$, all we need do to find the angular velocity is to convert from rpm to rad/s, that is, the angular velocity (rad/s) $= 1500/\pi \times 2\pi/60 = 50$ rad/s.

Also, from the transformation equations, linear velocity $v = r\omega = (0.27)(50) = 13.5$ m/s.

Example 8.6 A pinion shown in Figure 8.3 is required to move with an initial angular velocity of 300 rpm and a final angular velocity of 600 rpm. If the increase takes place over 15 s, determine the linear acceleration of the rack. Assume a pinion radius of 180 mm.

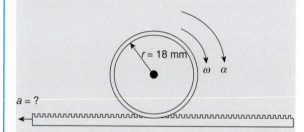

Figure 8.3 Rack and pinion

The velocities in radians per second are: 300 rpm $= 300 \times 2\pi/60 = 31.4$ rad/s and 600 rpm $= 600 \times 2\pi/60 = 62.8$ rad/s. Then, from $\alpha = \dfrac{\omega_2 - \omega_1}{t}$, the angular acceleration is $\alpha = \dfrac{62.8 - 31.4}{15} = 2.09$ rad/s². Now, from $a = r\alpha$, the linear acceleration is $a = (0.18)(2.09) = 0.377$ m/s².

Example 8.7 The armature of an electric motor rotating at 1500 rpm accelerates uniformly until it reaches a speed of 2500 rpm. During the accelerating period the armature makes 300 complete revolutions. Determine the angular acceleration and the time taken. Solving this type of problem is best achieved by first writing down the known values in the correct SI units.

So in this case we have: initial angular velocity of armature $\omega_1 = \dfrac{1500 \times 2\pi}{60} = 157$ rad/s. Similarly, $\omega_2 = \dfrac{2500 \times 2\pi}{60} = 262$ rad/s and $\theta = 300 \times 2\pi = 1885$ rad/s.

Then, using equation 8.13, that is $\omega_2^2 = \omega_1^2 + 2\alpha\theta$, we get $262^2 - 157^2 = 2\alpha(1885)$, or $\alpha = 11.7$ rad/s². Selecting an appropriate equation to find the time, that is using equation 8.14 where, after rearrangement, $\dfrac{\omega_2 - \omega_1}{\alpha} = t$, then $t = \dfrac{262 - 157}{11.7} = 9$ s.

Example 8.8 An aircraft sits on the runway ready for take-off. It has 1.4 m diameter wheels and accelerates uniformly from rest to 225 km/h (take-off speed) in 40 s.

Determine:

a) the angular acceleration of the undercarriage wheels;

b) the number of revolutions made by each wheel during the take-off run.

Apart from identifying all the known variables in the correct SI units, in this example it will also be necessary to consider a combination of linear and angular motion.

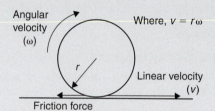

Figure 8.4 Relationship between linear and angular velocity

If we study Figure 8.4 we note that, in general, the angular rotation of the wheel causes linear motion along the ground, provided there are frictional forces sufficient to convert the rotating torque at the wheel into linear motion.

We have $v_1 = 0$ and $v_2 = \dfrac{225 \times 1000}{60 \times 60} = 62.5$ m/s. Using equation 8.9, that is $v = r\omega$, then $\omega_1 = 0$ and $\omega_2 = \dfrac{62.5}{0.7} = 89.29$ rad/s and the angular acceleration may be found from equation 8.14 as $\alpha = \dfrac{\omega_2 - \omega_1}{t} = \dfrac{89.29 - 0}{40} = 2.23$ rad/s².

Now the number of *radians* turned through by each wheel is found from equation 8.11, where $\theta = \dfrac{(\omega_1 + \omega_2)t}{2} = \dfrac{(0 + 89.29)40}{2} = 1785.8$ rad. So the number of revolutions turned through by each wheel is $\dfrac{1785.8}{2\pi} = 284.2$ revs.

8.3.1 Torque, moment of inertia and radius of gyration

Torque and moment of inertia

From Newton's third law, we know that to accelerate a mass we require a force such that $F = ma$. Now, in dealing with angular motion, we know that this force would be applied at a radius r from the centre of rotation and would thus create a turning moment or, more correctly, a torque T; thus, $T = Fr$ or $T = mar$. Since the linear acceleration $a = r\alpha$, then $T = m(r\alpha)r$ or:

$$T = m\alpha r^2 \qquad (8.15)$$

Now, from our equation for torque, the quantity mr^2 has a special significance. It is known as *the second moment of mass* or *moment of inertia* of a body about its axis of rotation. It is given the symbol I, thus $I = mr^2$, and the units of I are (kg/m²). The moment of inertia of a rotating body may be likened to the mass of a body subject to linear motion. Now, from equation 8.15, we may write the equation for torque as:

$$T = I\alpha \qquad (8.16)$$

The axis of rotation, if not stated, is normally obvious. For example, a flywheel or electric motor rotates about its centre, which we refer to as its polar axis. When giving values of I they should always be stated with respect to the reference axis. The mathematical derivation of the *moment of inertia I* is very similar to that we used to find the 'second moment of area' which also has the symbol I (look back at Chapter 2, section 2.4), except that *the moment of inertia of a body* about a given axis *is the sum of the product of each element of mass*, rather than element of area, *multiplied by the square of its distance from a particular axis*. The next example illustrates the mathematical procedure we adopt to find the mass moment of inertia.

> **Example 8.9** Find the moment of inertia of a rectangular lamina of length 10 cm and breadth 5 cm about an axis parallel to the 10 cm side and 10 cm from it (Figure 8.5). Take the *area density* of the lamina to be ρ kg/cm².

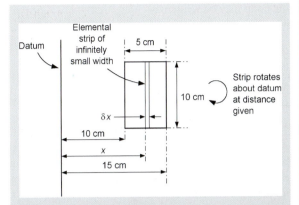

Figure 8.5 Figure for Example 8.9

Mass of element $= 10\rho\delta x$ kg and the second moment of mass for the element $I = 10\rho x^2 \delta x$ kg cm² and the second moment of mass (moment of inertia) for the rectangular lamina is

$$I = \int_{10}^{15} 10\rho x^2 dx = 10\rho \left[\frac{x^3}{3}\right]_{10}^{15} = 10\rho \left[\frac{15^2}{3} - \frac{10^2}{3}\right]$$

$$= 7916.67\rho \text{ kg cm}^2.$$

Radius of gyration (k)

In order to use the above definition for the moment of inertia I in a practical way, we need to be able to determine the radii at which the mass or masses are situated, as measured from the centre of rotation of the body (see Figure 8.6).

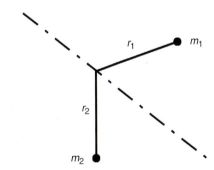

Figure 8.6 Rotating masses concentrated at a point

For most engineering components, the mass is distributed and not concentrated at any particular radius, so we need some way of finding an *equivalent radius* about which the whole mass of the rotating body is deemed to act. The radius of gyration k is the radius at which a mass M (equivalent to the whole mass of the body) would have to be situated so that its moment of inertia is equal to that of the body. So, the moment of the body may be written as

$$I = Mk^2. \tag{8.17}$$

> **Key Point.**   The radius of gyration (k) is the radius about which the whole of the rotating mass is deemed to act

All of this might, at first, appear a little confusing! In practice tables of values of k for common engineering shapes may be used. To enable you to tackle problems involving the inertia of rotating bodies, the values of k for some commonly occurring situations are given in Figure 8.7.

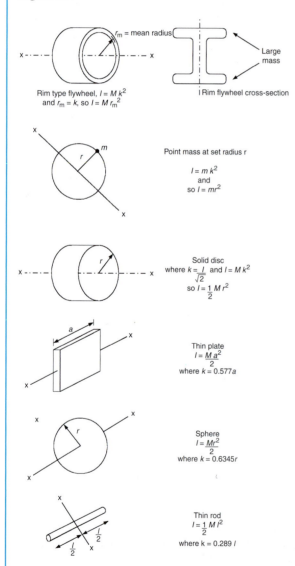

Figure 8.7 Definition of the moment of inertia for some engineering components

Example 8.10   A solid cylinder having a total mass of 140 kg and a diameter of 0.4 m is free to rotate about its polar axis. It accelerates from 750 to 1500 rpm in 15 s. There is a resistance to motion set up by a frictional torque of 1.1 N m as a result of worn bearings. Find the torque which must be applied to the cylinder to produce the motion.

We need to find the torque T from $T = I\alpha$ (equation 8.16), where for a solid disc $I = \frac{1}{2}Mr^2$ (Figure 8.7). The angular acceleration can be found using $\alpha = \frac{\omega_2 - \omega_1}{t}$ (equation 8.14), so $\omega_2 = \frac{1500 \times 2\pi}{60} = 157$ rad/s and $\omega_1 = \frac{750 \times 2\pi}{60} = 78.5$ rad/s, then $\alpha = \frac{157 - 78.5}{15} = 5.23$ rad/s^2. Now for the solid circular disc $I = \frac{1}{2}(140)(0.2)^2 = 2.8$ kg m^2 and from $T = I\alpha$ the net accelerating torque $T = (2.8)(5.23) = 14.64$ N m. Then, since net accelerating torque = applied torque − friction torque, we have 14.64 = applied torque − 1.1, giving the applied torque as 15.74 N m.

We have on occasions, including in the last example, mentioned *friction* and in order to fully understand the nature and effects of friction and be able to apply it to the dynamic engineering problems that appear later, this subject is covered in the next section.

8.4   Friction

When a surface is moved over another surface with which it is in contact, a resistance is set up opposing this motion. The value of the resistance will depend on the materials involved, the condition of the two surfaces, and the force holding the surfaces in contact; but the opposition to motion will always be present. This resistance to movement is said to be the result of *friction* between the surfaces.

We require a slightly greater force to start moving the surfaces (*static friction*) than we do to keep them moving (*sliding or dynamic friction*). As a result of numerous experiments, involving different surfaces in contact under different forces, a set of rules or laws has been established which, for all general purposes, materials in contact under the action of forces seem to obey. These rules are detailed below, together with one or two limitations on their use.

Laws of friction

1   The frictional forces always oppose the direction of motion, or the direction in which a body is tending to move.

2   The sliding friction force F opposing motion, once motion has started, is proportional to the normal force N that is pressing the two surfaces together, i.e. $F \propto N$.

3   The sliding frictional force is independent of the area of the surfaces in contact. Thus two pairs of surfaces in contact made of the same materials and in the same condition, with the same forces between them, but having different areas, will experience the same frictional forces opposing motion.

4   The frictional resistance is independent of the relative speed of the surfaces. This is not true for very low speeds or, in some cases, for fairly high speeds.

5   The frictional resistance at the start of sliding (static friction) is slightly greater than that encountered as motion continues (sliding friction).

6   The frictional resistance is dependent on the nature of the surfaces in contact, for example, the type of material, surface geometry, surface chemistry, etc.

> **Key Point.**   Friction always opposes the motion that produces it

From the above laws, we have established that the sliding frictional force F is proportional to the normal force N pressing the two surfaces together, that is $F \propto N$. You will remember from your mathematical study of proportion that in order to equate these forces we need to use a constant, the constant of proportionality i.e. $F = \mu N$. This constant μ is known as *the coefficient of friction* and in theory it has a maximum value of 1. Figure 8.8 shows the space diagram for the arrangement of forces on two horizontal surfaces in contact.

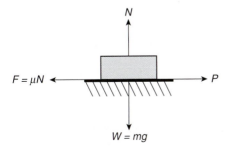

Figure 8.8 Space diagram for arrangement of forces

Note: The value of the force required to just start to move a body is greater than the force needed to keep the body moving. The difference in these two forces is due to the slightly higher value of the *coefficient of static friction* μ_s between the two surfaces when the body is stationary compared to the *coefficient of dynamic friction* μ_d when the body is rolling. It is the *coefficient of static friction* that we use in the examples that follow, since this is considered to be the *limiting friction coefficient*.

It is often difficult to visualise the nature and direction of all the forces that act on two bodies in contact, as well as resolving these forces into their component parts, and this can cause difficulties, which may be overcome by practice in producing space diagrams and resolving forces. Problems involving friction may be solved by *calculation* or by *drawing* and the following generalised example, involving the simple case of a block in contact with a horizontal surface, should help you to understand both methods of solution.

Consider again the arrangement of forces shown in Figure 8.8. If the block is in equilibrium, i.e. just on the point of moving, or moving with constant velocity, then we can find a solution by *calculation* by equating the horizontal and vertical forces as follows:

Resolving horizontally gives $P = F$ and resolving vertically gives $N = mg$, but from the laws of dry friction

$$F = \mu N \qquad (8.18)$$

So, from above, $F = \mu mg$ and so $P = \mu mg$.

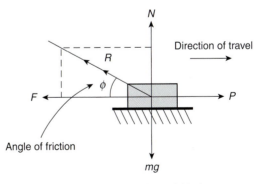

Figure 8.9 Space diagram for horizontal block

Now we know that a single *resultant* force, which can be found by *vector drawing*, can replace two or more forces. The space diagram for our horizontal block is shown in Figure 8.9, where F and N can be replaced by a resultant R at an angle ϕ to the normal force N. From Figure 8.9, it can be seen that,

$\dfrac{F}{R} = \sin \phi$ so $F = R \sin \phi$ and $\dfrac{N}{R} = \cos \phi$ so $N = R \cos \phi$, then $\dfrac{F}{N} = \dfrac{R \sin \phi}{R \cos \phi} = \tan \phi$. However, $\dfrac{F}{N} = \mu$, therefore

$$\mu = \tan \phi \qquad (8.19)$$

ϕ is known as the *angle of friction*. Once F and N have been replaced by R the problem becomes one of three coplanar forces, mg, P and R and can therefore be solved using the triangle of forces, which I am sure you are familiar with.

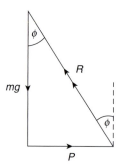

Figure 8.10   Vector diagram for horizontal block

Then, choosing a suitable scale, the vector drawing is constructed as shown in Figure 8.10.

Example 8.11   For the situation illustrated in Figure 8.11a, find the value of the force P to maintain equilibrium. We can solve this problem by calculation, resolving forces into their horizontal and vertical components, or we can solve it by vector drawing; both methods of solution are given in this example.

So, first, solution by calculation:

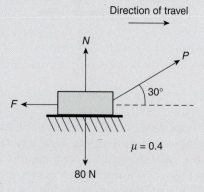

Figure 8.11a   Illustration of situation

Resolving forces horizontally and vertically, we get $F = P \cos 30°$ and $N + P \sin 30° = 80$ or $N = 80 - P \sin 30°$, respectively, but $F = \mu N$ and substituting for N from the above gives $F = \mu(80 - P \sin 30)$. We are given (from Figure 8.8a) that $\mu = 0.4$, and substituting $P \cos 30°$ for F in the equation $F = \mu(80 - P \sin 30)$ gives $P \cos 30 = \mu(80 - P \sin 30)$. After multiplying out the brackets and rearranging, we get $P \cos 30 + 0.4 P \sin 30 = 0.4(80)$, so $P(\cos 30 + 0.4 \sin 30) = 32$ and so $P = 30.02$ N.

Now the solution by drawing:

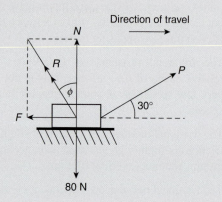

Figure 8.11b   Magnitude and direction of forces

The magnitude and direction of all known forces for our block are shown in Figure 8.11b. Remembering that $\mu = \tan \phi$, then $\tan \phi = \mu = 0.4$, therefore $\phi = \tan^{-1} 0.4$ and $\phi = 21.8°$.

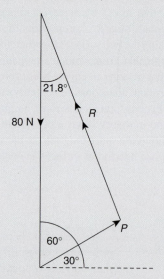

Figure 8.11c   Diagram showing force P

Then, from the resulting vector diagram (Figure 8.11c), we find that $P = 30$ N.

Key Point. The, coefficient of friction is given by the tangent of the friction angle

We finish this short introduction to friction by considering the forces acting on a body *at rest on an inclined plane* and then the forces that act on a body when *moving on an inclined plane*.

8.4.1 Forces on a body at rest on an inclined plan

Remember that the frictional resistance always acts in such a way as to oppose the direction in which the body is tending to move. So in Figure 8.12, where the body is in *limiting equilibrium* (i.e. on the point of slipping down the plane), the frictional resistance will act up the plane.

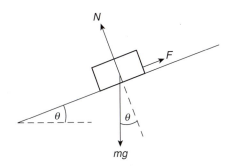

Figure 8.12 Force system for body in equilibrium on inclined plane

It can be seen that there are now three forces acting on this body, the weight *mg* acting vertically downwards, the normal force *N* acting perpendicular to the plane and the frictional resistance *F* acting parallel to the plane. These forces are in equilibrium and their values can be found by calculation or drawing.

Again, using simple trigonometry, we can resolve the forces parallel and perpendicular to the plane.

Resolving parallel to the plane, we get $F = mg \sin \theta$ and resolving perpendicular to the plane, we get $N = mg \cos \theta$. From $F = \mu N$ we see that $\mu = \tan \theta$.

Note that when, and only when, a body on an inclined plane is in *limiting equilibrium and no external forces act on the body* is the angle of slope θ equal to the angle of friction ϕ, i.e. $\theta = \phi$.

The drawing method would simply require us to produce a triangle of forces vector diagram, from which we could determine $\theta = \phi$ and μ.

8.4.2 Forces on a body moving up and down an inclined plane

Figure 8.13a shows the arrangement of forces acting on a body that is moving up an inclined plane and Figure 8.13b shows a similar arrangement when a body is moving down an inclined plane.

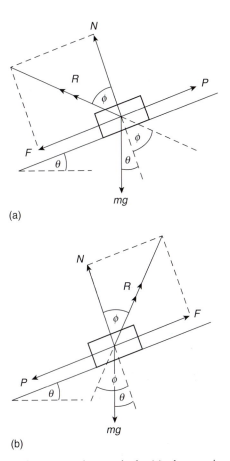

(a)

(b)

Figure 8.13 Forces acting on a body: (a) when moving up an inclined plane; (b) when moving down an inclined plane

Study both of these diagrams carefully, noting the arrangement of forces. Also note the clear distinction (in these cases) between the angle of friction ϕ and the angle of slope θ. The weight *mg* always acts vertically down and the frictional force *F* always opposes the force *P*, tending to cause motion either up or down the slope.

All problems involving bodies moving up or down an inclined plane can be solved by calculation or drawing. The resolution of forces and general vector diagrams for each case are detailed below.

a) *Forces on body moving up the plane* (Figure 8.13a):

Resolving forces horizontally, $P = F + mg \sin \theta$.

Resolving vertically, $N = mg\cos\theta$ so $F = \mu N = \mu mg\cos\theta$ and therefore $P = \mu mg\cos\theta + mg\sin\theta$, so

$$P = mg(\mu\cos\theta + \sin\theta) \qquad (8.20)$$

The solution by vector drawing will take the general form shown in Figure 8.14.

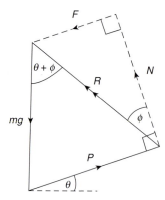

Figure 8.14 Solution by vector drawing when body is moving up the inclined plane

b) Forces on the body moving down the plane (Figure 8.13b)

Resolving forces horizontally, $P + mg\sin\theta = F$

Resolving vertically, $N = mg\cos\theta$ so $F = \mu N = \mu mg\cos\theta$ and therefore $P + mg\sin\theta = \mu mg\cos\theta$, so

$$P = mg(\mu\cos\theta - \sin\theta) \qquad (8.21)$$

Again, the solution by vector drawing will take the general form shown in Figure 8.15.

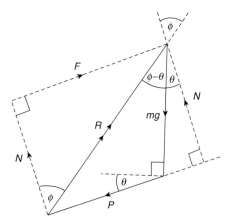

Figure 8.15 Solution by vector drawing when body is moving down the inclined plane

Example 8.12

a) A body of mass 400 kg is moved along a horizontal plane by a horizontal force of 850 N. Calculate the coefficient of friction.

b) The body then moves onto a plane made from the same material, inclined at 30° to the horizontal. A force P angled at 15° from the plane is used to pull the body up the plane with constant velocity. Determine the value of P.

a) The space diagram for the arrangement of forces is shown in Figure 8.16a.

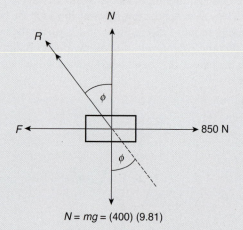

Figure 8.16a Space diagram for Example 8.12, part a

Then, by calculation, resolving forces horizontally from the figure, $F = 850$ N.

Resolving vertically, $N = (400)(9.81) = 3924$ N.

Then, from $F = \mu N$, $\mu = \dfrac{F}{N} = \dfrac{850}{3924}$ and $\mu = 0.217$.

Scale: 10 mm = 500 N

Figure 8.16b Vector drawing for Example 8.12, part a

Alternatively, from vector drawing (Figure 8.16b), we measure off the angle of friction as $\phi = 12.2°$ and $\tan^{-1} 12.2 = 0.217$, that is $\mu = 0.217$.

b) The space diagram for the arrangement of forces is shown in Figure 8.16c.

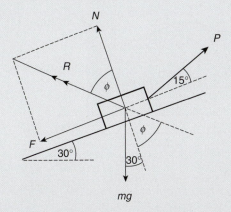

Figure 8.16c Space diagram for Example 8.12, part b

Then, by calculation, resolving forces horizontally from the figure,
$P \cos 15 = (400)(9.81) \sin 30 + F$.
Resolving vertically,
$N + P \sin 15 = (400)(9.81) \cos 30$
or $N = (400)(9.81) \cos 30 - P \sin 15$,
but $F = \mu N$,
so $F = 0.217((400)(9.81)(\cos 30) - (P \sin 15))$.
Substituting this expression for F into
$P \cos 15 = (400)(9.81) \sin 30 + F$
gives $P \cos 15 = (400)(9.81) \sin 30 + 0.217$
$((400)(9.81)(\cos 30) - (P \sin 15))$,
from which $P = 2641.2$ N.

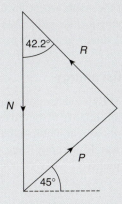

Vector drawing

Figure 8.16d Vector drawing for Example 8.12, part b

Also, if the vector drawing (Figure 8.16d) is drawn to scale, it will be found that $P = 2.6$ kN at 45° from the horizontal.

8.5   Energy

Energy may exist in many different forms, for example, mechanical, electrical, nuclear, chemical, heat, light and sound.

The principle of the conservation of energy states that: *energy may neither be created nor destroyed, only changed from one form to another.* There are many engineering examples of devices that transform energy; these include the:

- loudspeaker which transforms electrical to sound energy,
- petrol engine which transforms heat to mechanical energy,
- microphone which transforms sound to electrical energy,
- dynamo which transforms mechanical to electrical energy,
- battery which transforms chemical to electrical energy,
- filament bulb which transforms electrical to light energy.

Here, we are only concerned with the various forms of *mechanical energy* and its conservation. Provided no mechanical energy is transferred to or from a body, the total amount of mechanical energy possessed by a body remains constant, unless *mechanical work* is done. This concept is looked at next.

> **Key Point.** The total amount of mechanical energy possessed by a body remains constant, provided no mechanical energy is transferred and no work is done on or by the body

8.5.1   Work done

The *energy possessed by a body is its capacity to do work*. Mechanical work is done when a force overcomes a resistance and it moves through a distance.

Mechanical work may be defined as:

Mechanical work done (J)
   = force required to overcome the resistance (N)
   $\times$ *distance moved against the resistance* (m),

that is:

$$W = F \times d \qquad (8.22)$$

The SI unit of work is the Newton-metre (N m) or Joule, where $1\,J = 1\,N\,m$.

Note:

(a) No work is done unless there is both resistance and movement.

(b) The resistance and the force needed to overcome it are equal.

(c) The distance moved must be measured in exactly the opposite direction to that of the resistance being overcome.

The more common resistances to be overcome include: *friction*, *gravity* and *inertia* (the resistance to acceleration of the body), where:

the work done (WD) against friction = friction force × distance moved

WD against gravity = weight × gain in height

WD against inertia = inertia force × distance moved

Also note that the WD against inertia force is the out-of-balance force multiplied by the distance moved, or:

WD against the inertia force = mass × acceleration

× distance moved.

In any problem involving calculation of WD, the first task should be to identify the type of resistance to overcome. If, and only if, there is motion between surfaces in contact, is WD against friction. Similarly, only where there is a gain in height is there WD against gravity and only if a body is accelerated is WD against inertia (look back at our definition of inertia).

Example 8.13 A body of mass 30 kg is winched up from the ground at constant velocity through a vertical distance of 15 m. If a frictional resisting force of 100 N has to be overcome, calculate the total WD.

We are not told about air resistance so this may be ignored. Then the WD against gravity is equal to the weight multiplied by the gain in height, i.e. $WD = mgh = (30)(9.81)(15) = 4414.5\,J$ (assuming the standard average value for the acceleration due to gravity). Also, the WD against friction = $(100)(15) = 1500\,J$, so that the total work done $= 5914.5\,J$ or $5.9145\,kJ$.

Work done may be represented graphically and, for linear motion, this is shown in Figure 8.17a, where the force needed to overcome the resistance is plotted against the distance moved. The WD is then given by the area under the graph.

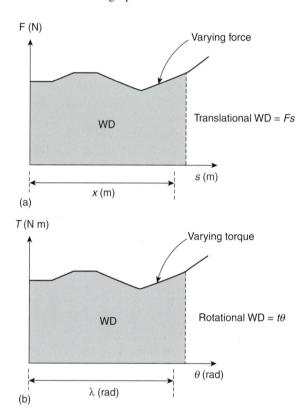

Figure 8.17 Work done

Figure 8.17b shows the situation for angular motion, where a varying torque T in N m is plotted against the angle turned through in *radian*. Again, the WD is given by the area under the graph, where the units are (N m × *radian*), then, noting that the radian has no dimensions, the units for work done remain as N m or Joules.

8.5.2 Potential energy

Potential energy (PE) is energy possessed by a body by virtue of its position, relative to some datum. The change in PE is equal to its weight multiplied by the change in height. Since the weight of a body $= mg$, then the change in PE may be written as:

Change in

$$PE = mg\Delta h \qquad (8.23)$$

Equation 8.23 is identical to the work done in overcoming gravity. So the work done in raising a mass to a height is equal to the PE it possesses at that height, assuming no external losses.

Example 8.14   A packing crate weighing 150 kg is moved 100 m by a conveyor system that has a frictional resisting force of 120 N. If the conveyor system is inclined at 30° to the horizontal, what is the total work done?

We need to find the work done against friction and the work done against gravity.

WD against friction = (100)(100) = 10000 N m

Now, gravitational acceleration acts vertically down, so we need to find the vertical height moved by the crate. Since the conveyor system is inclined at 30° to the horizontal, then the vertical component of height (that is the change in height as the crate travels 100 m up the conveyor) is given by $\Delta h = 100 \sin 30 = (100)(0.5) = 50$ m and so the $PE = mg\Delta h = (150)(9.81)(50) = 73575$ J.

So *total WD* = 10000 + 73575 = 83575 N m or 83.575 kN m (remembering that 1 N m = 1 J and WD, as opposed to energy, is often given in *Newton-metres* rather than *Joules*).

Key Point.   The SI unit for *work done (WD)* is the Newton-metre and the SI unit for *energy* is the Joule. These units are dimensionally identical, i.e. 1 N m = 1 J

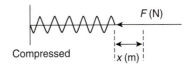

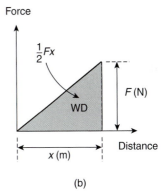

Figure 8.18 Spring system demonstrating strain energy

8.5.3 Strain energy

Strain energy is a particular form of PE possessed by a body that is deformed within its elastic range. You have studied the concept of strain previously in Chapter 7, in some detail, where we looked at materials under stress when subject to static loading, resulting in elastic strain. Here we consider briefly components that are strained under dynamic loading conditions.

Consider the spring arrangement shown in Figure 8.18. The force required to compress or extend the spring is $F = kx$, where k is the spring constant having units of N/m.

Figure 8.18a shows a helical coil spring in the unstrained, compressed and extended positions. The force required to move the spring varies in direct proportion to the distance moved (Figure 8.18b). Therefore the *strain energy of the spring (E_s) when compressed or extended = area under graph (force × distance moved)* $= \frac{1}{2}Fx$ and, since $F = kx$,

then the *strain energy of the spring in tension or compression*

$$E_s = \frac{1}{2}kx^2 \qquad (8.24)$$

A similar argument can be given for a spring that is subject to twisting or torsion about its centre (or polar axis). It can be shown that:

Strain energy of a spring when twisted

$$= \frac{1}{2}T\theta = \frac{1}{2}k_{tor}\theta^2 \qquad (8.25)$$

where θ = the angle of twist, T = the torque (N m) and k_{tor} is the spring stiffness or torque rate, that is the torque required to produce unit angular displacement. Note that the unit of strain energy, as for all other types of *energy*, is the Joule (J).

Key Point.   The spring stiffness k has units of N/m while the torque rate k_{tor} has units of N m/rad, that is, torque per unit angular displacement

Example 8.15   A drive shaft is subject to a torque rate of 30 MN/rad. What is the strain energy set up in the drive shaft when the angle of twist is $5°$?

All we need do here is use the relationship: strain energy in twist $= \frac{1}{2}k_{tor}\theta^2$. So, remembering that the angle of twist must be in radian, $\theta = \frac{(5)(2\pi)}{360} = 0.087$ rad and the *strain energy in twist* $= \frac{1}{2}(30 \times 10^6)(0.087)^2 = 113.54$ kJ.

8.5.4   Kinetic energy

Kinetic energy (KE) is energy possessed by a body by virtue of its motion. Translational KE, i.e. the KE of a body travelling in a linear direction (straight line), is given by the relationship: translational

$$KE_{tran} = \frac{mass \times (velocity)^2}{2} \quad or \quad KE_{tran} = \frac{1}{2}mv^2 \tag{8.26}$$

Flywheels are heavy wheel-shaped masses fitted to shafts in order to minimise sudden variations in the rotational speed of the shaft due to sudden changes in load. A flywheel is therefore a store of rotational KE. *Rotational KE* can be defined in a similar manner to translational (linear) KE, that is,

$$KE_{rot} = \frac{1}{2}I\omega^2 \tag{8.27}$$

where I is the moment of inertia (that you met earlier).

Example 8.16   Find the KE possessed by a railway train of mass 30 tonne, moving with a velocity of 120 km/h.

Note first that in many of these examples we need to convert km/h into m/s. This is easy, if you remember that there are 1000 metres in a kilometre, so that 120 km/h $= (120)(1000)$ m/h, and that there are 3600 seconds in an hour; therefore 120 kp $= \frac{(120)(1000)}{3600} = 33.33$ m/s. Then, remembering that 1 tonne $= 1000$ kg, the KE of the train is given by KE $= \frac{1}{2}mv^2 = (0.5)(30 \times 10^3)(33.33)^2 = 16.67$ MJ.

Example 8.17   Determine the total KE of a four wheel drive car which has a mass of 800 kg and is travelling at 50 km/h. Each wheel of the car has a mass of 15 kg, a diameter of 0.6 m and a radius of gyration of $k = 0.25$ m.

Then, $KE_{total} = KE_{tran} + KE_{rot}$, where 50 km/h $= 13.89$ m/s, and so $KE_{tran} = \frac{1}{2}mv^2 = (800)(13.89)^2 = 77.16$ kJ.

The moment of inertia for each wheel is $= Mk^2 = (15)(0.25)^2 = 0.9375$ kg m^2 and, from the transformation equations, $\omega = v/r = 13.89/0.3 = 46.3$ rad/s, therefore $KE_{rot} = \frac{1}{2}I\omega^2 = (0.5)(4 \times 0.9375)(46.3)^2 = 4.019$ kJ. Therefore, *total KE* of the car $= 77.16 + 4.019 = 81.18$ kJ.

8.5.5   Conservation of mechanical energy

From the definition of the conservation of energy we can deduce that the total amount of energy within certain defined boundaries will remain the same. When dealing with mechanical systems, the PE possessed by a body is frequently converted into KE and vice versa. If we ignore air frictional losses, we may write:

$$PE + KE = a\ constant$$

Thus, if a mass m falls freely from a height h above some datum, then, at any height above that datum:

$$Total\ energy = PE + KE$$

This important relationship is illustrated in Figure 8.19, where at the highest level above the datum, the PE is a maximum and is gradually converted into KE as the mass falls towards the datum. Immediately before impact, when height $h = 0$, the PE is zero and the KE is equal to the initial PE.

Since the total energy is constant, then: $mgh_1 = mgh_2 + \frac{1}{2}mv_2^2 = mgh_3 + \frac{1}{2}mv_3^2 = \frac{1}{2}mv_4^2$.

Immediately after impact with the datum surface, the mechanical KE is converted into other forms such as heat, strain and sound energy. If friction is present, then work is done overcoming the resistance due to friction and this is dissipated as heat. Then:

Initial energy = Final energy

+ Work done in overcoming frictional resistance

Note: KE is not always conserved in collisions. Where KE is conserved in a collision we refer to the collision

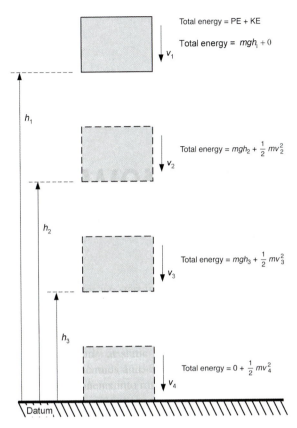

Figure 8.19 PE + KE a constant

The vertical height h is found using the sine ratio, that is $h = 10 \sin 10 = 1.736$ m, so increase in PE $= mgh = (2500)(9.81)(1.736) = 42.575$ kJ.

Now, using the relationship that total energy $=$ PE $+$ KE, immediately prior to the cargo breaking away KE $= 0$ and so PE $=$ total energy. Also, immediately prior to the cargo striking the base of the slope, PE $= 0$ and KE $=$ total energy (all other energy losses being ignored). So, at the base of the slope, KE $= 42575$ J, therefore $42575 = \frac{1}{2}mv^2$ and $v^2 = \dfrac{(2)(42575)}{2500} = 34.06$ so, $v = 5.84$ m/s.

8.6 Momentum

If a motor vehicle crashes into a stationary object such as a concrete post, the KE due to its motion is absorbed as work done in trying to alter the shape of the post! This, in turn, produces energy in the form of heat and sound which is absorbed by the environment. Therefore, no energy has been lost due to the collision but transferred from one form into another, in accordance with the conservation of energy law. In order, as engineers, to determine the loss of KE at impact, we first need to return to the concept of *momentum* you met earlier in this chapter.

as *elastic*; when KE is not conserved we refer to the collision as *inelastic*.

Key Point.   Kinetic energy of moving bodies is conserved if they collide elastically

Example 8.18   Cargo weighing 2500 kg breaks free from the top of the cargo ramp (Figure 8.20). Ignoring friction, determine the velocity of the cargo the instant it reaches the bottom of the ramp.

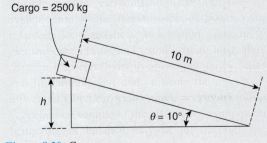

Figure 8.20 Cargo ramp

8.6.1   Impact and linear momentum

The law of conservation of linear momentum states that:

the total momentum of a system, in a particular direction, remains constant provided that the masses are unaltered and no external forces act on the system.

The above law may be applied to collisions and stated more simply as:

The total momentum before impact $=$

the total momentum after impact

Consider two bodies of different masses (m_1, m_2) that collide and then *move off together* after impact. Remembering that:

$$\text{Linear momentum } = mv \qquad (8.28)$$

then total momentum before impact $= m_1 v_1 + m_2 v_2$ and total momentum after impact $= m_1 v_a + m_2 v_a$ (where, in this case, $v_a =$ the velocity after impact, which because they move off together will be the same for both masses). So, these quantities may be equated, to give:

$$m_1v_1 + m_2v_2 = m_1v_a + m_2v_a \qquad (8.29)$$

Also, from the law of conservation of energy (assuming a perfect elastic collision), then:

Loss of KE at impact = *total KE before impact*

− total KE after impact

> **Key Point.** Linear momentum is the product of the mass of a body and its velocity

Example 8.19   A curling stone of mass 20.5 kg travelling at 4 m/s collides with another curling stone of mass 20.0 kg, travelling at 3 m/s in the opposite direction. Find the loss in KE if the two stones *remain together* directly after impact.

Using the momentum relationship, we found previously (equation 8.29), $m_1v_1 + m_2v_2 = m_1v_a + m_2v_a$, we get $(20.5)(4) + (20)(3) = 20.5v_a + 20v_a$ and so $v_a = 3.506$ m/s.

We now have velocities before and after impact, so using:

Loss of KE at impact = total KE before impact

− total KE after impact,

we get:

$$\text{Loss of KE} = \left[\frac{1}{2}(20.5)(4)^2 + \frac{1}{2}(20.0)(3)^2\right]$$

$$- \left[\frac{1}{2}(40.5)(3.506)^2\right]$$

and, on simplification, loss of KE = 164 + 90 − 249 = 5 J. This loss may be due to energy transfer on impact, such as heat and sound.

8.6.2   Rotational kinetic energy and angular momentum

From equation 8.27, where $\text{KE}_{\text{rot}} = \frac{1}{2}I\omega^2$, and the knowledge that bodies lose KE when they collide, we can write a relationship for the loss of rotational KE after impact, similar to the relationship for linear KE, that is:

The loss of rotational KE at impact = total rotational KE before impact − total rotational KE after impact. Then, using equation 8.27, this relationship may be expressed as:

Loss of rotational KE at impact

$$= \left[\frac{1}{2}I_1\omega_1^2 + \frac{1}{2}I_2\omega_2^2\right] - \left[\frac{1}{2}(I_1 + I_2)\omega_a^2\right].$$

The law of conservation of angular momentum is given as:

The total angular momentum of a mass system, rotating in a particular direction, remains constant; provided that the moments of inertia of the rotating masses remain unaltered and that no external torques act on the system.

So this law may be applied to collisions, where the moment of inertia of the individual masses remains constant after collision. Then, as with the linear case:

*Total angular momentum before impact =
total angular momentum after impact*

or

$$I_1\omega_1 + I_2\omega_2 = (I_1 + I_2)\omega_a \qquad (8.30)$$

where the angular momentum = $I\omega$ (which you should compare with the linear case).

Let us look at a couple of examples that should clarify the use of the above relationships.

> **Key Point.** Angular momentum is the product of the moment of inertia of a body and its angular velocity

Example 8.20   Determine the KE of a flywheel having a mass of 250 kg and a radius of gyration of 0.3 m, when rotating at 800 rpm.

Treating the flywheel as a solid disc, from Figure 8.7 we find that the moment of inertia $I = Mk^2$ or $I = (250)(0.3)^2 = 22.5$ kg m². Also, 800 rpm = $\frac{(800)2\pi}{60} = 83.78$ rad/s, so that $\text{KE}_{\text{rot}} = \frac{1}{2}I\omega^2 = (0.5)(22.5)(83.78)^2 = 78.96$ kJ.

Example 8.21   An electric motor drives a coaxial gearbox through a clutch. The motor armature has a mass of 50 kg and a radius of gyration of 18 cm, while the mass of the gearbox assembly is 500 kg with a radius of gyration of 24 cm. If the motor armature is revolving at 3000 rpm and the gearbox is rotating at 500 rpm in the same direction, what is the loss of KE when the clutch is engaged?

Let us first find the mass moment of inertia for the motor armature and for the gearbox, where (treating the armature and gearbox as solid discs) $I = Mk^2$. Then, for the motor armature, $I = (50)(0.18)^2 = 1.62 \, \text{kg m}^2$ and for the gearbox, $I = (500)(0.24)^2 = 28.8 \, \text{kg m}^2$. Also, the angular velocities in rad/s are, respectively, 314.16 rad/s and 52.34 rad/s. Then, to find the angular velocity after impact, we use the relationship for the conservation of angular momentum, that is:

Total angular momentum before impact =

total angular momentum after impact.

Then, $(1.62)(314.16) + (28.8)(52.34) = (1.62 + 28.8)\omega_a$ and, so $\omega_a = 66.28$ rad/s we are now in a position to find the loss in KE at impact. Then, from the loss of rotational KE at impact

$$= \left[\frac{1}{2}I_1\omega_1^2 + \frac{1}{2}I_2\omega_2^2\right] - \left[\frac{1}{2}(I_1 + I_2)\omega_a^2\right] \text{ we get}$$

Loss of rotational KE at impact =

$(0.5)(1.62)(314.16)^2 + (0.5)(28.8)(52.34)^2$

$- (0.5)(30.42)(66.28)^2 = 52.575 \, \text{kJ}.$

8.7 Power

Power is a measure of the rate at which work is done or the rate of change of energy. Power is therefore defined as: *the rate of doing work*. The SI unit of power is the Watt (W), i.e.:

$$\text{Power (W)} = \frac{\text{work done (J)}}{\text{time taken (s)}} = \frac{\text{energy change (J)}}{\text{time taken (s)}} \tag{8.31}$$

Or, if the body moves with constant velocity,

$$\text{Power (W)} = \text{force used (N)} \times \text{velocity (m/s)} \tag{8.32}$$

Note that the units are N m/s = J/s = Watt (W).

Example 8.22   A packing crate weighing 1000 N is loaded onto the back of a lorry by being dragged up an incline of 1 in 5 at a steady speed of 2 m/s. The frictional resistance to motion is 240 N. Calculate:

a)   the power needed to overcome friction,

b)   the power needed to overcome gravity,

c)   the total power needed.

a)   Power = frictional force × velocity along surface = 240 × 2 = 480 W.

b)   Power = weight × vertical component of velocity = 1000 × 2 × 1/5 = 400 W. Since there is no acceleration and therefore no work done against inertia,

c)   Total power = power for friction + power for gravity = 480 + 400 = 880 W.

Key Point.   Power may be defined as the rate of doing work *or* energy per unit time *or* the product of force and velocity

8.7.1   Power transmitted by a torque

Let us now consider power transmitted by a torque. You have already met the concept of torque in section 8.3.1 of this chapter, as well as in Chapter 7. Figure 8.21 illustrates the concept of power being transmitted by a torque in, for example, drive shafts.

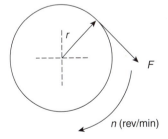

Figure 8.21   Power transmitted by a torque

Since the WD is equal to the force multiplied by the distance, then the WD in 1 rev = ($F \times 2\pi r$), but Fr is the torque T applied to the shaft, therefore the WD in 1 rev = $2\pi T$ Joules and in 1 minute WD = WD/rev × number of revs per minute (n) = $2\pi nT$. Therefore the WD in 1 s = $2\pi nT/60$ and, since work done per second is equal to power (1 J/s = 1 W), then

$$\text{Power (W) transmitted by a torque} = 2\pi nT/60 \tag{8.33}$$

You have already met the engineers' theory of twist in your study of torsion in section 3.2, where you will remember the relationship $\dfrac{\tau}{r} = \dfrac{T}{J}$, where:

- τ = the shear stress at distance r from the polar axis of the shaft,
- T = the twisting moment on the shaft,
- J = the polar second moment of area of the shaft.

We can use this relationship to calculate power transmitted by a torque in a shaft.

> **Key Point.**   Power transmitted by a torque may be found from the *work done per second* $= 2\pi nT/60$, where $n = the\ number\ of\ rpm$

Example 8.23   Calculate the power transmitted by a hollow circular shaft with an external diameter of 250 mm, when rotating at 100 rpm, if the maximum shear stress is 70 MN/m^2 and $J = 220 \times 10^6$ mm^4.

From engineers' theory of twist, $T = \dfrac{(70)(220 \times 10^6)}{125} = 123.2 \times 10^6$ N m $= 123.2 \times 10^3$ N mm. Note that we used the relationship 70 MN/m^2 = 70 N/mm^2, to keep the units consistent. Then, knowing that the power transmitted by a torque is $W = 2\pi nT/60$, we find that the power $= \dfrac{(2\pi)(100)(123.2 \times 10^3)}{60} = 1.2901$ MW.

8.8   Circular motion and forces of rotation

We considered angular motion in section 8.3. We now extend the notion of angular motion to include the accelerations and forces that act on bodies that move in a circular path, for example, motor vehicles that negotiate bends, aircraft that bank and turn or governors that have rotating bob weights to control the rate of opening of valves and the like. We start by considering the acceleration and forces that occur as a result of circular motion.

8.8.1   Centripetal acceleration and force

You will be aware of the relationship between linear and angular velocity through use of the transformation equations, where we know that $s = r\theta$, $v = r\omega$ and $a = r\alpha$. We can use these formulae to form relationships for the accelerations and forces that act on a body during circular motion. Consider Figure 8.22, where a point mass is subject to a rotational velocity.

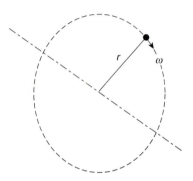

Figure 8.22 A point mass subject to a rotational velocity

From the figure it can be seen that the direction of motion of the mass must be continually changing, in order to produce the circular motion. Therefore, it is being subjected to an acceleration that is *acting towards the centre of rotation*; this is known as *centripetal acceleration*.

Now, from the transformation equations we know that the tangential velocity $v = r\omega$ and that the acceleration of the mass is equal to its change in velocity divided by the time taken for the mass to move through some angular distance θ. Thus the change in the tangential velocity of the mass divided by the time, that is its *acceleration*, is given as $\dfrac{v - u}{time} = \dfrac{r\omega\theta}{time}$, and since the angular velocity is equal to the angle turned through in radian divided by time, then $\dfrac{\theta}{time} = \omega$. Then, from above, the centripetal acceleration of the mass $= \omega^2 r$. Also, we know that $v = r\omega$ or $\omega = \dfrac{v}{r}$, so that centripetal acceleration $= \left(\dfrac{v}{r}\right)^2 r = \dfrac{v^2}{r}$; therefore we may write that:

$$\text{Centripetal acceleration} = \omega^2 r = \dfrac{v^2}{r} \tag{8.34}$$

Now when this acceleration acts on a mass, as in this case, it produces a force known as centripetal force, thus: *centripetal force* $(F_c) = mass \times$ *centripetal acceleration*. Then:

$$F_c = m\omega^2 r = \frac{mv^2}{r} \qquad (8.35)$$

Now, remembering Newton's third law, that to every action there is an equal and opposite reaction, the centripetal force acting on the mass pushing it towards the centre of rotation must be balanced by the radial force created by the mass maintaining the orbit. This force opposing the centripetal force is known as the *centrifugal force*. It is, for example, the centrifugal force created by your own body mass that keeps you pinned to the wall of a rotating fairground ride when the floor is lowered!

Key Point. The equal and opposite force to the centripetal force acting towards the centre of rotation is the inertia force or *centrifugal force* acting *away from* the centre of rotation

Example 8.24 An aircraft with a mass of 80000 kg is in a steady turn of radius 300 m, flying at 800 km/h. Determine the centripetal force required to hold the aircraft in the turn.

The tangential linear velocity of the aircraft $= \frac{800 \times 1000}{3600} = 222.2$ m/s and from $F_c = \frac{mv^2}{r}$ we find that: $F_c = \frac{(80000)(222.2)^2}{300} = 13.17$ MN.

Suppose a *conical pendulum* consists of a mass rotating in a horizontal circle at some angle θ from the vertical, supported from a pivot point by a non-extendable cord of negligible mass, as shown in Figure 8.23.

Since the acceleration towards the centre is balanced by an equal and opposite acceleration holding the mass in its circular path, then if we wish to find the force F in the cord or the angle θ for a given set of parameters, all we need do is resolve horizontally and/or vertically to find the force F or angle θ, or for that matter any other of the unknowns.

So, for the case illustrated in Figure 8.23, resolving horizontally we get $F\sin\theta = m\omega^2 r = m\frac{v^2}{r}$; also, resolving vertically, we get $F\cos\theta = mg$. From these two expressions, we find that $\dfrac{F\sin\theta = m\dfrac{v^2}{r}}{F\cos\theta = mg}$ or $\tan\theta = \dfrac{v^2}{gr}$.

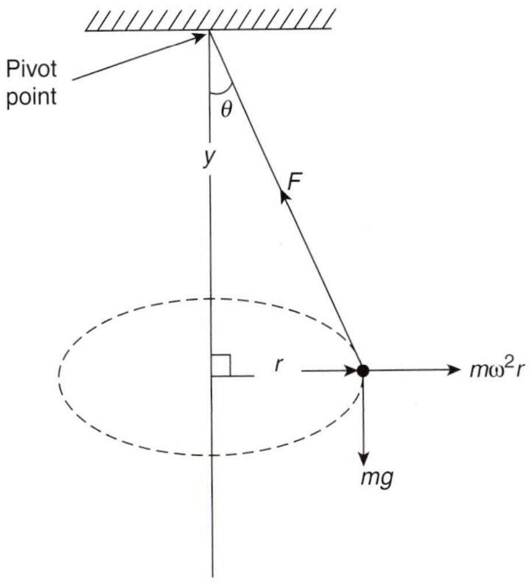

Figure 8.23 Conical pendulum arrangement

Then, for example, if we wish to find the tension force F in the strong cord of length 0.5 m and the radius of rotation r for a 10 kg mass rotating at 100 rpm, we proceed in the following manner.

We know from basic trigonometry that $r = 0.5\sin\theta$ and from the resolution of force horizontally we know that $F\sin\theta = 10(10.47)^2(0.5\sin\theta)$, and so $F = 548.1$ N. Also, from resolving vertically, $F\cos\theta = mg$ or $\cos\theta = \dfrac{mg}{F} = \dfrac{(10)(9.81)}{548.78} = 0.1788$, and so $\theta = 79.7°$. Now, since we know that $r = 0.5\sin\theta$, then $r = 0.5\sin 79.7°$ so that $r = 0.49$ m.

The above conical pendulum arrangement forms the basis of a particular type of engine governor, where bob weights move a central valve that meters the fuel and controls the engine speed.

8.8.2 Vehicle turning and banking

When cyclists round a bend, they bank (lean) inwards towards the bend. In a similar manner, when an aircraft enters a steady turn it banks so that the outer wing is higher than the inner wing in the turn. The reason for this banking is so that a component of force is acting towards the centre of the turn that counteracts the tangential forces trying to throw the vehicle out of the turn. We will use the case of the aircraft and that of the cyclist to analyse the forces that occur during a turn.

When an *aircraft turns*, centripetal acceleration holds the aircraft in the turn and this acceleration, together with the mass of the aircraft, results in a centripetal force (F_c) being created that acts towards the centre of the turn (Figure 8.24).

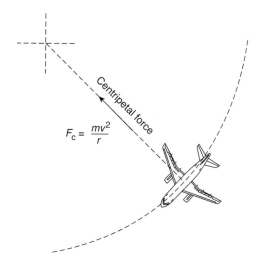

Figure 8.24 Centripetal force acting towards centre of turn

This centripetal force is opposed by an inertia force, i.e. by a *centrifugal force* that tends to throw the aircraft out of the turn. To counteract this inertia force, the horizontal component of the lift force is used. This component is created by banking the aircraft to produce a steady-state turn, as shown in Figure 8.25.

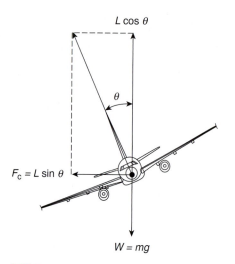

Figure 8.25 Forces acting in a correctly banked steady turn

The horizontal component of lift is equal to the centripetal force holding the aircraft in the turn. Then, resolving forces horizontally, we get that $L \sin \theta = \dfrac{mv^2}{r}$, where θ = angle of bank in radian, r = radius of turn in metres, L = lift force in Newtons, m = aircraft mass in kg, g = the acceleration due to gravity in m/s^2 and v = tangential linear velocity of aircraft in the turn in m/s.

Also from Figure 8.25, resolving vertically, we get that $L \cos \theta = W = mg$. Then, using the same trigonometric identity as we used before in section 8.8.1, we get $\dfrac{L \sin \theta}{L \cos \theta} = \dfrac{mv^2/r}{mg} = \dfrac{v^2}{gr}$, therefore:

$$\tan \theta = \frac{v^2}{gr} \qquad (8.36)$$

The relationship given in equation 8.36 is identical to that we found earlier for the conical pendulum!

Example 8.25 An aircraft enters a correctly banked turn of radius 2500 m at a velocity of 200 m/s. If the aircraft has a mass of 60000 kg, determine:

a) the centripetal force holding the aircraft in the turn,

b) the angle of bank.

a) This is given simply as $F_c = \dfrac{mv^2}{r} = \dfrac{(60000)(200)^2}{2500} = 960 \,\text{kN}$.

b) Since we do not have the lift force, we can only use the relationship $\tan \theta = \dfrac{v^2}{gr}$, so that:

$\tan \theta = \dfrac{(200)^2}{(2500)(9.81)} = 1.63099$, giving an angle of bank $\theta = 58.49°$.

Let us now return to the case of the cyclist, who is rounding a bend keeping to a circular path of constant radius. Figure 8.26 shows the arrangement of forces acting on and being created by the cyclist during a turn. These are: the *normal reaction force* (N) of the ground on the bicycle, *the friction force* (F) of the ground on the tyres and the weight ($W = mg$) acting vertically downwards from the centre of gravity (G) of the cycle/cyclist combination.

Note that the cyclist leans or banks into the turn to provide a component of weight force at a distance (a moment) necessary to counteract the overturning moment created by the centrifugal force trying to throw him/her out of the turn, and so is able to maintain equilibrium. Note also that the centripetal acceleration and the opposing inertia force due to the circular travel are deemed to act through the centre of gravity (G) of the cycle and cyclist combination.

Since the cycle/cyclist combination is in equilibrium during the turn, then the sum of the moments about

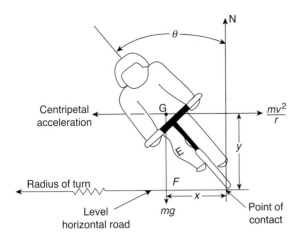

Figure 8.26 Cyclist rounding a bend in a circular path

any point must equate algebraically to zero. So, taking moments about the contact point between the bicycle tyre and the road, we get that $mgx = \left(\dfrac{mv^2}{r}\right)y$ or, on rearrangement, $\dfrac{x}{y} = \dfrac{v^2}{gr}$, and noting from Figure 8.26 that $\dfrac{x}{y} = \tan\theta$ we obtain the now familiar equation $\tan\theta = \dfrac{v^2}{gr}$!

Now resolving forces horizontally, as we have done before, gives $F = \dfrac{mv^2}{r}$. In this case it is the *frictional force* that counteracts the centrifugal inertia force. From your study of friction in section 8.4 you will remember that $F = \mu N$ *at the point of slip*. Therefore, in our case, $F = \dfrac{mv^2}{r} = \mu N$ and *slipping will occur* when $\dfrac{mv^2}{r} > \mu N$. From Figure 8.26 we also know that $N = mg$, so after substitution $\dfrac{mv^2}{r} > \mu mg$ or $\dfrac{v^2}{gr} > \mu$. It is also worth noting that for any particular radius of turn, the *limiting velocity immediately prior to slip is given by* $\dfrac{v^2}{gr} = \mu$.

Example 8.26 A racing cyclist with his/her bicycle has a combined mass of 100 kg and travels in a circular path around a curved bend of 20 m radius, while on a flat horizontal road. Determine:

a) the angle at which the combination needs to be tilted and the value of the force due to friction if the cyclist's velocity in the turn is 30 km/h,

b) the coefficient of friction between the tyres and the road if the maximum velocity that the cyclist can take the bend, before slip, is 40 km/h.

a) The angle of tilt (θ) that the cyclist needs to make in order to take the bend is given by the now familiar relationship $\tan\theta = \dfrac{v^2}{gr}$ where 30 km/h = 8.33 m/s, so $\tan\theta = \dfrac{(8.33)^2}{(9.81)(20)} = 0.3537$ and $\theta = 19.5°$. Now the value of the force due to friction may be found from the relationship $F = \dfrac{mv^2}{r}$, where $F = \dfrac{(100)(8.33)^2}{20} = 347$ N.

b) The coefficient of friction is easily found by remembering that the limiting velocity immediately before slip is given by the relationship $\dfrac{v^2}{gr} = \mu$, so that $\mu = \dfrac{(11.111)^2}{(9.81)(20)} = 0.63$.

8.9   Chapter summary

In this chapter we have reviewed or introduced a number of fundamental dynamic principles that will act as a foundation for further study of dynamics and will be useful, in their own right, during your study of engineering science. We started by reminding ourselves of some fundamental definitions that were necessary in order to fully understand the nature of Newton's laws, which are used as the foundation of classical mechanics. From Newton's laws, the linear equations of motion were derived, both graphically and analytically, and through use of the transformation equation the equations for angular motion were also found and then applied to simple engineering problems that also required knowledge of torque, moment of inertia and the radius of gyration.

In section 8.4 we considered *friction* and started by studying the 'laws' of friction, which have been found by experiment to model the behaviour of differing materials in contact subject to differing external conditions. We then used space diagrams to show the nature of all the forces acting on a body, together with analytical and vector-drawing resolution of forces, in order to solve simple engineering problems involving friction.

Section 8.5 was devoted to energy transfer, in particular to the various forms of mechanical energy and its conservation, where we covered work done, potential energy, strain energy (internal potential energy) and kinetic energy and their relationship to the dynamic behaviour of vehicles and bodies. The concepts and simple engineering applications associated with linear and angular momentum were introduced in section 8.6, including the relationship between rotational kinetic energy and angular momentum of rotating machines. Section 8.7 was given over to a very short study of power and power transmitted by a torque, again with reference to rotating machines.

Finally, in section 8.8, we looked at circular motion as a particular type of angular motion and considered the acceleration and forces that act on vehicles in circular motion, in particular, the conical pendulum, aircraft in turning flight and cyclists rounding bends in a circular path. This summary does not contain information related to the derivation or use of individual formulae, as it is felt that you will have met the majority of these formulae before and will be able to consolidate your understanding by attempting the practice review questions that follow.

8.10   Review questions

1. A thin cord can withstand a pulling force of 120 N. Calculate the maximum acceleration that can be given to a mass of 32 kg, assuming that the force to accelerate the mass is transmitted through the cord.

2. A Cessna 172 aircraft and a Boeing 747 aircraft are each given an acceleration of 5 m/s². To achieve this, the thrust force required by the Cessna's engines is 15 kN and the thrust force required by the Boeing 747 is 800 kN. Calculate the mass of each aircraft.

3. A car starts from rest and accelerates uniformly at 2 m/s² for 10 seconds. Calculate the distance travelled.

4. How long will it take a heavy vehicle to accelerate uniformly from 10 m/s to 15 m/s, if it covers a distance of 80 m in this time?

5. A train of mass 32 tonne accelerates uniformly from rest to 4 m/s². If the average resistance to motion is 1 kN per 1000 kg, determine: a) the force required to produce the acceleration, and b) the inertia force on the train.

6. A wheel has a diameter of 0.54 m and is rotating at $\frac{1500}{\pi}$ rpm. Calculate the angular velocity of the wheel in rad/s and the linear velocity of a point on the rim of the wheel.

7. A flywheel rotating at 20 rad/s increases its speed uniformly to 40 rad/s in 1.0 min. Sketch the angular velocity/time graph and determine: a) the angular acceleration of the flywheel, and b) the angle turned through by the flywheel in 1.0 minute, and so calculate the number of revolutions made by the flywheel in this time.

8. The flywheel of a cutting machine has a moment of inertia of 130 kg m². Its speed drops from 120 rpm to 90 rpm in 2 s. Determine: a) the deceleration of the flywheel, b) the braking torque.

9. A turbine and shaft assembly has a moment of inertia of 15 kg m². The assembly is accelerated from rest by the application of a torque of 40 N m. Determine the speed of the shaft after 20 s.

10. A rim-type flywheel gains 2.0 kJ of energy when its rotational speed is raised from 250 to 270 rpm. Find the required inertia (I) of the flywheel.

11. The armature of an electric motor has a mass of 50 kg and a radius of gyration of 150 mm. It is retarded uniformly by the application of a brake, from 2000 to 1350 rpm, during which time (t) the armature makes 850 complete revolutions. Find the retardation and braking torque.

12. On what variables does the value of frictional resistance depend?

13. 'The frictional resistance is independent of the relative speed of the surfaces, under all circumstances.' Is this statement true or false? You should give reasons for your decision.

14. Define a) the angle of friction, b) the coefficient of friction, and explain how they are related.

15. Sketch a space diagram that shows all the forces that act on a body moving with uniform velocity along a horizontal surface.

16. Explain the relationship between the angles θ and ϕ, a) when a body on a slope is in static equilibrium and b) when a body moves down a slope at constant velocity.

17. Sketch diagrams that show all the forces that act on a body when it is moving with constant velocity a) up a sloping surface, b) down a sloping surface.

18. For each case in question 17, resolve the horizontal and vertical components of these forces and show that for a body moving up the plane $P = \mu mg \cos \theta + mg \sin \theta$ and that for a body moving down the plane $P = \mu mg \cos \theta - mg \sin \theta$.

19. A load of mass 500 kg is positioned at the base of a sloping surface inclined at 30° to the horizontal. A force P, parallel to the plane, is then used to pull the body up the plane with constant velocity. If the coefficient of friction is 0.25, determine the value of the pulling force.

20. A crane raises a load of 1640 N to a height of 10 m in 8 s. Calculate the average power developed.

21. The scale of a spring balance which indicates weights up to 20 N extends over a length of 10 cm. Calculate the work done in pulling the balance out until it indicates 12 N.

22. A train having a mass of 15 tonne is brought to rest when striking the buffers at a terminus. The buffers consist of two springs in parallel, each having a spring constant of 120 kN/m that undergo a compression of 0.75 m when the train strikes the buffers. Find: a) the strain energy gained by the buffers, and b) the velocity of the train at the instant it strikes the buffers.

23. Find the KE of a mass of 2000 kg moving with a velocity of 40 kilometres per hour.

24. A motor vehicle starting from rest free-wheels down a slope whose gradient is 1 in 8. Neglecting all resistances to motion, find its velocity after travelling a distance of 200 m down the slope.

25. A railway carriage of mass 9000 kg is travelling at 10 kilometres per hour when it is shunted by a railway engine of mass 12000 kg travelling at 15 kilometres per hour in the same direction. After the shunt they move off together in the same direction. a) What is the velocity of engine and carriage immediately after the shunt, and b) what is the loss of KE resulting from the shunt?

26. A rotating disc has a mass of 20 kg and a radius of gyration of 45 mm. It is brought to rest from 1500 rpm in 150 revolutions by a braking torque. Determine: a) the angular retardation, b) the value of the braking torque.

27. An aircraft, at sea-level, enters a steady turn and is required to bank at an angle of 50°. If the radius of the turn is 2000 m, determine the velocity of the aircraft in the turn.

28. A racing cyclist rounds a bend of 45 m radius where the track, in the velodrome, is banked at an angle of 25°. If the coefficient of friction between the cycle tyres and the track is 0.25, find, by resolving forces parallel and perpendicular to the banked track, the maximum speed at which the cyclist can negotiate the bend.

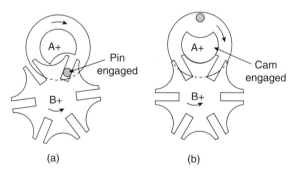

Chapter 9

Kinematics of mechanisms

This chapter is dedicated to the study of the kinematics of mechanisms. *Kinematics* is concerned with motion, without reference to forces, and *mechanisms* may be defined as those devices that transform some applied motion into some other desired motion. Thus a mechanism may change the direction, velocity or type of motion. A typical example of a mechanism is shown in Figure 9.1, where a Geneva wheel mechanism is illustrated.

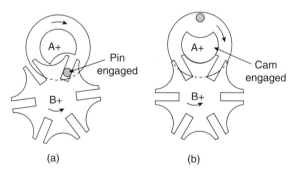

Figure 9.1 The Geneva mechanism

The Geneva mechanism consists of two wheels. The wheel centred at 'A' has a raised cam and a pin that protrudes from its surface and engages with the wheel centred at 'B' *once* every revolution, as shown in Figure 9.1a. During engagement, wheel B will rotate one-sixth of a revolution before the pin disengages, then, during the second half of the revolution of A, the cam on wheel A engages with wheel B (Figure 9.1b) but, as can be seen by the profile of the cam, no movement of

wheel B takes place until the pin on wheel A rotates to engage once more with wheel B. Wheel A is designed to continually rotate and, for each revolution of wheel A, wheel B will move one-sixth of a revolution in discrete steps. Examples of the use of the Geneva mechanism may be found in children's clockwork toys, watch/clock mechanisms and engineering stamping machines.

> **Key Point.** Kinematics is the study of motion without reference to forces and a mechanism may be defined as a device that transforms one form of applied motion into some other required motion

We begin our study of mechanism kinematics by considering a graphical method for finding the velocities and acceleration of mechanisms with rigid links when in planar motion. These parameters are of particular importance to engineers when considering the kinematics of a mechanism for a particular function. We then extend our knowledge by using *analytical* methods for determining the relative velocities and accelerations associated with the internal combustion engine slider-crank mechanism. Finally, we take a very brief look at cam mechanisms and their displacement diagrams.

9.1 Velocity and acceleration diagrams

9.1.1 Introduction to velocity diagrams

We start our study of *velocity diagrams* by first revising the idea of *vector motion* of a body in a plane (Figure 9.2), that is, planar motion. Figure 9.2a shows

978-1-85617-775-7, Engineering Science, Mike Tooley & Lloyd Dingle

the body rotating about the fixed point A, with angular velocity ω.

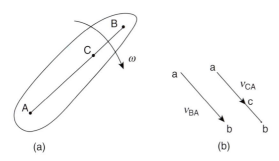

Figure 9.2 A rigid body subject to angular motion

A rigid body forming part of a mechanism (a link) will always be fixed in length, no matter what type of motion occurs. Thus the two arbitrary points A and B shown in Figure 9.2a will be a fixed distance apart along the line AB and therefore there can be no relative velocity between them. However, if the body is allowed to rotate, say about A, then there *will* now be a relative velocity between them. Therefore, under these circumstances, for a rigid link *the velocity of* B *relative to* A (v_{BA}) *can only occur in a direction perpendicular to the line* AB, and this must be true no matter what the motion of A. This fact is always used when constructing velocity diagrams to determine velocity parameters for rotating links pinned at one end, in planar mechanisms. Thus, in this case, as shown in Figure 9.2b, the relative velocity v_{BA} is represented by the vector **ab** of magnitude ωAB drawn *perpendicular* to AB, with the sense of the vector corresponding to the clockwise direction of the angular velocity of B relative to A.

Now if we consider any point on the link AB, say C, then, similarly, the velocity of C relative to A is given by $v_{CA} = \omega AC$. Therefore, by proportion,

$$\frac{v_{CA}}{v_{BA}} = \frac{\omega AC}{\omega AB} = \frac{\mathbf{ac}}{\mathbf{ab}}$$

and if point C is located somewhere along **ab**, then the velocity of C relative to A is given by **ac**. This relationship is quite easily understood if you consider the *tangential velocity* of, say, individuals on a rotating fairground ride, where their tangential velocity at any radius from the centre of rotation will be given by $v_{tan} = \omega r$ and their tangential velocity at any other radius may be found by considering the ratio of their radial distances from the centre of rotation, in a similar manner to the expression given above.

9.1.2 Velocity diagrams for the four-link mechanism

In order to clarify the nature of velocities and determine the velocities at certain points in a mechanism from velocity diagrams (using the relative velocity method), we will now consider the case of the *four-link mechanism* shown in Figure 9.3.

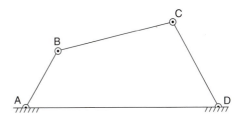

Figure 9.3 A four-link mechanism

From Figure 9.3 we can deduce that:

1. The absolute velocities of points A and D on the fixed link AD are both equal to zero, that is $v_A = v_D = 0$.

2. The absolute velocity of point B is its relative velocity to the fixed point A (v_{BA}), which from our argument, given previously, we know will be *perpendicular* to the direction of link AB. At the instant in time shown by its position on Figure 9.3, its magnitude will be $v_{BA} = \omega AB$.

3. The absolute velocity of point C, which is its relative velocity to the fixed point D (v_{CD}), has a direction perpendicular to link CD. Its magnitude is not known at this time.

4. The relative velocity of point C with respect to point B (v_{CB}) must be perpendicular to BC. Its magnitude and sense are not known at this time.

5. Note that the velocity of point B with respect to point C (v_{BC}) will also be perpendicular to the instantaneous position of link BC and will have the same magnitude, but it will be *opposite* in sense. This is also the case for the velocity of point A relative to point B (v_{AB}), where again its direction is perpendicular to the direction of link AB and its magnitude will be $v_{AB} = \omega AB$ but acting in the opposite sense.

Armed with the above information, it is possible to place all these vector quantities onto one diagram and determine all that we need to know about the velocities of the points within the mechanism, as desired. The following example illustrates the method, where we start by drawing a space diagram to scale.

Key Point. A space diagram drawn to scale may be used as the starting point to draw the vector diagram

Example 9.1 If, for the four-link mechanism illustrated in Figure 9.3, $\angle DAB = 60°$, link AD = 1.5 m, link BC = 1.0 m, link CD = 0.7 m and link AB = 0.5 m, and this link rotates at 25 rpm clockwise, determine:

i)   the velocity of a point P that sits halfway between point B and C on the link BC, and

ii) the angular velocity of the link CD.

Now, all of the above required velocities can be determined from the relative velocity diagram, once drawn. To assist you in understanding how we construct this diagram we start by drawing the space diagram of the situation described above, to scale, as shown in Figure 9.4.

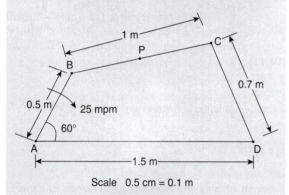

Scale   0.5 cm = 0.1 m

Figure 9.4 Space diagram

Now, following the steps 1 to 4 that we deduced earlier, we know that the velocities of points A and D are $v_A = v_D = 0$. We also know that the direction of the velocity of point B is perpendicular to the direction of the link AB and its magnitude is given by $v_{BA} = \omega AB$, where in this case the angular velocity $\omega = \dfrac{(25)(2\pi)}{60} = 2.62$ rad/s and so $v_{BA} = (2.62)(0.5) = 1.31$ m/s, acting in a clockwise sense. Now we are in a position to draw this vector as the starting point for our vector diagram, where we label the corresponding points in *lower case* letters, as shown in Figure 9.5.

Also, from point 3 above, we know that the direction of the absolute velocity of point C is its

velocity relative to the fixed point D which is perpendicular to the position of link DC, so we draw a line through 'd' and the vector point 'c' must lie on this line. Also, we know that the instantaneous velocity of point C relative to B must be perpendicular to the position of link CD, so we draw a line through 'b' perpendicular to the link CD. Now the position on the diagram where these two lines cross gives the position of point 'c', as shown in Figure 9.5.

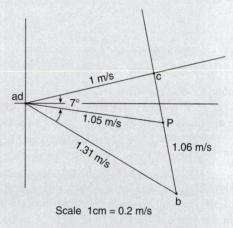

Scale   1 cm = 0.2 m/s

Figure 9.5 Velocity diagram

We set the scale for our velocity diagram once we knew the relative linear velocity of our rotating link. The velocity of the point 'p' is easily found by setting off a vector line from the fixed points 'a,d' that intersect the vector **bc** at its midpoint, as shown on Figure 9.5.

Now, using the scale 1 cm = 0.2 m, the magnitudes of the required velocities can be read off, as shown in Figure 9.5. Their direction and sense can also be determined from the diagram.

i)   The velocity of the point 'p' relative to the fixed link at 'a' is $v_{PA} = \mathbf{ap} = 1.05$ m/s in the direction shown on Figure 9.5.

ii) The linear velocity of the link CD, from the figure, $= cd = 1.0$ m/s, therefore the angular velocity $\omega_{CD} = \dfrac{\mathbf{cd}}{CD} = \dfrac{1.0}{0.7} = 1.43$ rad/s clockwise.

One final point: it should be remembered when constructing velocity diagrams that the displacements and velocities found for the mechanism are those at one instant in time, that is, they are *instantaneous* values.

9.1.3   Velocity diagram for an engine mechanism

Figure 9.6 shows a typical internal combustion engine mechanism that consists of a crank AB, a connecting rod BC and a piston C that is constrained to move linearly within cylinder walls.

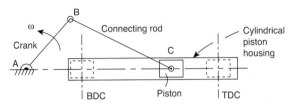

Figure 9.6 Internal combustion engine mechanism

The crank rotates so that when the crank and the connecting rod are aligned horizontally along AC, the piston is at its top position within the cylinder. This position is known as top dead centre (TDC), as shown in Figure 9.6. The bottom dead centre (BDC) position is achieved when the crank turns through an angle of 180° anticlockwise from its TDC position, at the point in time when the crank link and the connecting rod are folded to the left of the fixed point A.

Apart from the obvious engineering use of this mechanism, its study is important because it can easily be treated graphically (as you will see next) or analytically, which we cover later in section 9.2. The velocity diagram for this type of mechanism is constructed in an identical manner to the four-link mechanism that we considered in Example 9.1.

Example 9.2   Figure 9.7 shows the space diagram for an engine mechanism, where the crank AB rotates at 4000 rpm anticlockwise and, at the instant in time shown, $\angle$BAC $= 50°$. If the crank length AB $= 12.0$ cm and the connecting rod length BC $= 22.0$ cm, determine:

i)   the linear velocity of point B on the crank AB;

ii)   the velocity of the piston;

iii)   the angular velocity of the connecting rod BC;

iv)   the velocity of point P on the connecting rod, 7 cm from C.

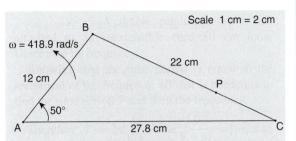

Figure 9.7 Space diagram for engine mechanism

i)   The linear velocity of the point B on the crank can be found from the information given, that is:

$$\omega = \frac{(4000)(2\pi)}{60} = 418.9 \text{ rad/s, so that}$$

$$v_{BA} = (418.9)(0.12) = 50.27 \text{ m/s.}$$

ii)   To determine the velocity of the piston C relative to A, we need to draw the velocity diagram (Figure 9.8). This is achieved by first drawing the vector for the crank, that is vector **ab**, which is known from the information given in the question and also from the answer found for part i). The *direction* of the piston velocity vector is also known because, in this arrangement, it is constrained to move in a horizontal straight line, so we mark a horizontal line of indeterminate length and then mark another vector to represent the direction of the velocity of point B, which is perpendicular to the direction given on the space diagram. Then where they intersect fixes point C, so that vector **ac** $= v_{CA} = 53$ m/s.

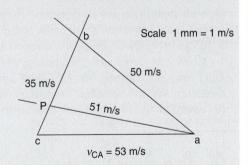

Figure 9.8 Velocity diagram for engine mechanism

iii)   The angular velocity of the connecting rod BC is found by reading off, from Figure 9.8, the linear velocity of B relative to C, that is vector **bc** $= 35$ m/s, and so

$$\omega_{BC} = \frac{v_{BC}}{\text{length of BC}} = \frac{35}{0.22} = 159 \text{ rad/s.}$$

iv)   This is found by proportion in a similar manner to the way in which we found the answer to part i) in Example 9.1, so that we mark off the vector line from point 'a' to cut the vector **bc** at point 'p' which is, in this case, 7/22 along **bc** from point 'c'. Then, from the diagram, we find by measurement that $\mathbf{ap} = v_{PA} = 51$ m/s.

There is one important aspect of velocity diagrams we have not covered, that is the situation where a link subject to *an angular velocity ω* has a free-sliding block attached to it (Figure 9.9), in which the link is free to slide towards the extremity of the link.

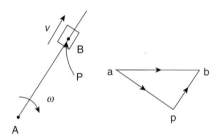

Figure 9.9 Rotating link with sliding block and associated velocity diagram

Figure 9.9 shows the rotating link where the *block B* is at some *point on the link* say P, at the instant in time we consider the relative velocities. Then, as before, the linear velocity of the point P on the link is ωAP, which acts perpendicular to AP and is represented by the vector **ap** (Figure 9.9). Now the velocity of the block B *relative* to the point P (as its passes through the point) is v, parallel to AP. Then the velocity of the block B relative to point A is represented by the vector **ab**, as shown in Figure 9.9.

We will leave velocity diagrams for the moment and turn our attention to the determination of acceleration within mechanisms, using vector drawing.

9.1.4   Acceleration diagrams

We now consider the *acceleration* of links and points on links of particular mechanisms in a similar manner to the way in which we found the velocities of these components earlier. It is in fact necessary to use the results of velocity diagrams to solve for accelerations, when using a vector drawing method.

Consider again a link AB that rotates with an angular velocity ω, and this time with an acceleration α. Then the *acceleration* of any point on the link relative to another will have two components; the first will be a *centripetal* component due to the angular velocity of the link, directed towards A, and the second will be a *tangential* component due to any acceleration of the link, perpendicular to the link AB, as shown in Figure 9.10.

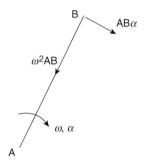

Figure 9.10 The centripetal and tangential acceleration of a link AB

The magnitude of the centripetal acceleration is found using equation 8.34, that is $a_c = \omega^2 r$, where in this case $r = $ AB, and the magnitude of the tangential acceleration is given by the transformation equation 8.10, that is $a_{\tan} = r\alpha$, where again $r = $ AB, as shown in Figure 9.10.

Key Point. Using the velocity diagram, the centripetal acceleration and tangential acceleration of a rotating link with velocity ω may be found from the relationships $a_c = \omega^2 r$ and $a_{\tan} = r\alpha$, respectively

Now, knowing the angular velocity of the link and the distance (length AB) of the link, the centripetal acceleration $\mathbf{a'b'_1}$ and tangential acceleration $\mathbf{b'_1 b'}$ vectors may be found and represented diagrammatically, as shown in Figure 9.11. Then, the acceleration of B relative to A (a_{BA}) is represented by the resultant vector $\mathbf{a'b'}$.

Please note: that the acceleration of, say, any point B is represented by $\mathbf{b'}$, where the superscript prime represents the differential coefficient of **b** with respect to time. This is consistent with the *rate of change function* that was covered in *Essential Mathematics 2* on the companion website, that is, the differential of, say, a point velocity **b** with respect to time $\frac{d\mathbf{b}}{dt} = \mathbf{b'}$. Thus, in our case, to distinguish between a *velocity* vector and an *acceleration* vector we use the *superscript prime* notation, while the *subscripts* 1,2 etc. are used for identification when constructing the acceleration diagram. Yet another alternative notation is the 'dot notation' where, for example, the rate of change of velocity with respect to time is $\dfrac{d\mathbf{b}}{dt} = \dot{\mathbf{b}}$, which again represents the acceleration. To avoid confusion, you should also remember that the rate of change of displacement with respect to time is $\dfrac{ds}{dt} = v$ and that $\dfrac{d^2s}{dt^2} = a$, so that in terms of prime notation we find that for displacement (s), $s' = v$ and $s'' = a$.

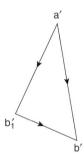

Figure 9.11 Acceleration diagram of link AB

> **Key Point.** Differentiating the velocity vector $\mathbf{v}$ once with respect to time yields the acceleration vector $\mathbf{v}'$

Example 9.3 Suppose that for the situation described in Example 9.1 for our four-link mechanism, the link AB is subject to an acceleration of $3.5\,\text{rad/s}^2$ for the point in time we used when constructing our velocity diagram (Figure 9.5) and we are required to find i) the *acceleration of the point* P on the link BC, and ii) the *angular acceleration of link* CD. Then, the required accelerations may be found from an *acceleration diagram* in a similar way to that in which we found the corresponding velocities from our velocity diagram. Collating all the information we have from the question and using the appropriate velocities we found from our velocity diagram (Figure 9.5), we deduce the following.

1) Starting with the acceleration of the link AB, that is the acceleration of A relative to B (a_{BA}), we know that this acceleration has two components: its centripetal acceleration given by $AB\omega^2$ and its tangential acceleration given by $AB\alpha$. Then, $a_{BA\,\text{centripetal}} = AB\omega^2 = (0.5)(2.62)^2 = 3.43\,\text{m/s}^2$, directed towards A, and $a_{BA\,\text{tan}} = AB\alpha = (0.5)(3.5) = 1.75\,\text{m/s}^2$,

acting in a direction perpendicular to AB and in a sense that increases the angular velocity, i.e. clockwise, since it is an acceleration.

2) For the *centripetal component* of the acceleration of the point C on the link BC with respect to B, that is $a_{CB\,\text{centripetal}}$, then $a_{CB\,\text{centripetal}} = BC\omega^2 = (1.0)(1.06)^2 = 1.12\,\text{m/s}^2$, acting towards B, where from the velocity diagram $\omega_{BC} = \dfrac{\mathbf{cd}}{CD} = \dfrac{1.06}{1.0} = 1.06\,\text{rad/s}$. The tangential acceleration of the point C on the link BC with respect to B, that is $a_{CB\,\text{tan}}$, has a direction that is perpendicular to CD but its magnitude is unknown.

3) Also, for the *centripetal component* of the acceleration of the point C on the link CD with respect to D, that is $a_{CD\,\text{centripetal}}$, then $a_{CD\,\text{centripetal}} = CD\omega^2 = (0.7)(1.43)^2 = 1.43\,\text{m/s}^2$, acting towards C, where from the velocity diagram $\omega_{CD} = \dfrac{\mathbf{cd}}{CD} = \dfrac{1.0}{0.7} = 1.43\,\text{rad/s}$. The tangential acceleration of the point C on the link CD with respect to D, that is $a_{CD\,\text{tan}}$, has a direction that is perpendicular to CD but its magnitude is unknown.

We now have sufficient information to construct our acceleration diagram, so with reference to Figure 9.12:

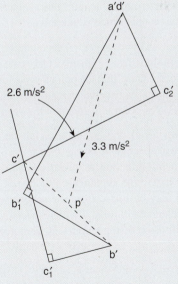

Figure 9.12 The acceleration diagram for Example 9.3

1)   The centripetal acceleration and tangential accelerations of B relative to A are drawn as $\mathbf{a'b'_1} = 3.43\,\text{m/s}^2$ and $\mathbf{b'_1 b'} = 1.75\,\text{m/s}^2$, the initial direction of the centripetal acceleration being taken from the space diagram (Figure 9.4) and the direction of the tangential acceleration acting perpendicular to it. Note also that at the points A and D on the mechanism, the acceleration, like the velocity, will be zero.

2)   The centripetal acceleration of C relative to B is drawn next and shown on the diagram as vector $\mathbf{b'c'_1} = 1.12\,\text{m/s}^2$. Again, as is always the case when known, the direction of the instantaneous centripetal acceleration is taken from the space diagram and the magnitude is determined from the velocity diagram. The tangential acceleration of the point C with respect to B is marked off perpendicular to $\mathbf{b'c'_1}$ passing through point $\mathbf{c'_1}$, with a vector line of indeterminate length, since at this stage the magnitude is unknown.

3)   The centripetal component of the acceleration of the point C, with respect to D, is drawn on as $\mathbf{d'c'_2} = 1.43\,\text{m/s}^2$. The tangential acceleration of the point C with respect to D is marked off perpendicular to $\mathbf{d'c'_2}$ passing through point $\mathbf{c'_2}$, with a vector line of indeterminate length, since again at this stage its magnitude is unknown.

4)   Where the two tangential acceleration vectors (from steps 2 and 3) intersect fixes the point $\mathbf{c'}$. Also, we know that the point P is positioned on the link halfway between B and C, so the acceleration of point P is given by the vector $\mathbf{a'p'}$.

Then, for i), the acceleration of the point P with respect to A is $a_{PA} = \mathbf{a'p'} = 3.3\,\text{m/s}^2$, with the direction as given on the diagram. For part ii) the angular acceleration of the link CD may be found from the *tangential* linear acceleration of point C with respect to point D, on division by the length CD. Therefore, from the diagram, $a_{CD} = \mathbf{c'_2 c'} = 2.6\,\text{m/s}^2$, so that

$$\alpha_{CD} = \frac{2.6}{CD} = \frac{2.6}{0.7} = 3.71\,\text{rad/s}^2, \text{ clockwise.}$$

We next consider the determination of velocities and accelerations in mechanisms, using an analytical method where we will concentrate on the simple engine slider-crank mechanism.

9.2   Displacement, velocity and acceleration analysis of an engine slider-crank mechanism

The velocities and accelerations of links in mechanisms can be analysed mathematically as well as being determined graphically. The analytical method has the advantage of allowing values to be calculated for all positions of a mechanism, as well as being able to find the desired accelerations by manipulating analytical expressions, whereas a velocity or acceleration diagram applies only to one position. Unfortunately, this analysis is not straightforward because the mathematics contains a series of expressions involving trigonometric terms that are cumbersome, complex and difficult to manipulate. The one exception that is relatively simple to analyse mathematically is the engine slider-crank mechanism; this is because it has only two inclined moving links, the crank and the connecting rod. It is for this reason, to simplify the mathematics, that we have chosen the engine slider-crank mechanism as our model.

9.2.1   Displacement analysis

We have said very little about displacement during our study of vector diagrams because in all graphical methods the action is frozen in time and the instantaneous displacement may be represented by our space diagram of the situation. In order to formulate some useful relationships for the mathematical analysis of displacement, we need again to start by freezing the action at some instant in time but, once formed, the relationships we develop may be used, for the slider-crank at least, to determine displacements and then velocities and accelerations, no matter what the geometry of the situation at the time.

Consider the simplified version of the engine slider-crank mechanism shown in Figure 9.13, where, at the instant in time shown, the crank AB of length r makes an angle θ with the centre line AC. The instantaneous displacement of the slider (piston) between the crank pin A and the slider C is shown as s and the connecting rod BC is shown with length l. Now, to simplify the analysis, we let $n = \dfrac{r}{l}$, i.e. we let n equal the ratio between the crank length and the connecting rod length, where in

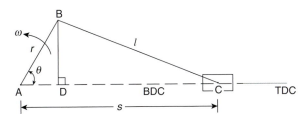

Figure 9.13 Simplified diagram of an engine slider-crank mechanism at some instant in time

practice in combustion engines the crank is normally between 1/3 and 1/5 the size of the connecting rod.

> **Key Point.**   In order to simplify our analysis, we let $n = \dfrac{r}{l}$, where $r = $ crank length and $l = $ connecting rod length

Now, when, say at time $t = 0$, $\theta = 0°$, then the displacement s will take a maximum value, that is the crank and connecting rod will both lie on the centre line AC and so position the piston (slider) at TDC, so at this point, $s_{max} = r + l$. If, at another moment in time, the crank rotates π rad($180°$), say anticlockwise, then the slider will be positioned at BDC and the minimum displacement will be $s_{min} = l - r$. We can manipulate these two expressions for maximum and minimum displacement in order to introduce the ratio n. As you will see in a moment, this is desirable in order to determine a *generalised expression* for the displacement of our slider-crank mechanism.

Then, remembering that $n = \dfrac{r}{l}$:

$$s_{max} = r + l = r\left(1 + \frac{l}{r}\right) = r\left(1 + \frac{1}{n}\right)$$

$$= \frac{r}{n}(n + 1) = l(n + 1)$$

$$s_{max} = r + l = l(n + 1) \qquad (9.1)$$

$$s_{min} = l - r = r\left(\frac{l}{r} - 1\right) = r\left(\frac{1}{n} - 1\right)$$

$$= \frac{r}{n}(1 - n) = l(1 - n)$$

$$s_{min} = l - r = l(1 - n) \qquad (9.2)$$

Now we will determine the *displacement s* for the instant in time shown in Figure 9.13, using the construction line BD. Then, from the right-angled triangle ABD, we find by elementary trigonometry that $BD = r \sin \theta$ and

$AD = r \cos \theta$. Also, from Pythagoras, for the triangle BCD we find that $(DC)^2 = (BC)^2 - (BD)^2$. Therefore,

$$DC = \sqrt{(BC)^2 - (BD)^2} = \sqrt{(l^2 - r^2 \sin^2 \theta)}.$$

Then, from equation 9.2, we may write that $DC = l(1 - n^2 \sin^2 \theta)^{1/2}$. Since $AD + DC = s$, then:

$$s = r \cos \theta + l(1 - n^2 \sin^2 \theta)^{1/2} \qquad (9.3)$$

If, for the instant in time shown in Figure 9.13, our crank is rotating with *constant* angular velocity, say in an anticlockwise direction, then from the differential calculus we know that $\dfrac{d\theta}{dt} = \omega$, that is, the rate of change of angular displacement with respect to any moment in time is equal to the angular velocity, where the angular displacement θ is measured in radian and the angular velocity ω is measured in rad/s. We also know that the angle the crank has turned through at time t, compared to its position at time $t = 0$, is given by $\theta = \omega t$.

Knowing this we can write a general form of equation 9.3 in terms of the parameters θ, ω and n by substituting $\theta = \omega t$ into equation 9.3 to give:

$$s = r \cos \omega t + l(1 - n^2 \sin^2 \omega t)^{1/2} \text{ or}$$

$$s = r\left[\cos \omega t + \frac{l}{r}(1 - n^2 \sin^2 \omega t)^{1/2}\right];$$

therefore

$$s = r\left[\cos \omega t + \frac{1}{n}(1 - n^2 \sin^2 \omega t)^{1/2}\right] \qquad (9.4)$$

> **Example 9.4**
>
> a) Confirm the distance of the piston pivot point from the crankshaft pivot (distance AC), as shown in the space diagram (Figure 9.7), using an analytical method.
>
> b) Given that the crankshaft for the slider-crank engine in Example 9.2 rotates at 900 rpm in an *anticlockwise* direction, at what time *after passing* the position of the crank shown in Figure 9.7 is the crank within $\pi/6$ radian of TDC?

a) We will use equation 9.3 for our solution, where $\theta = 50° = 0.873 \text{ rad} = \omega t$ and, from the dimensions given in Example 9.2, $n = \dfrac{r}{l} = \dfrac{12}{22} = 0.545$. Note that the ratio n, is very high and the dimensions chosen for Example 9.2 were selected on the basis of space constraints within the book, not good design!

Then $s = r\cos\theta + l(1 - n^2\sin^2\theta)^{1/2}$,

$$s = 0.12\cos 50 + 0.22\sqrt{1 - (0.545)^2(\sin 50)^2}$$

$$= 0.077 + 0.22\sqrt{1 - 0.17428} = 0.27691 \text{ m}$$

$$= 27.69 \text{ cm}.$$

This precise answer, obtained mathematically, is very much in accord with the value of 27.8 cm we obtained from our space diagram.

b) When $\omega = \dfrac{(900)(2\pi)}{60} = 30\pi \text{ rad/s}$ and $\theta = \dfrac{(50)(2\pi)}{360} = \dfrac{5\pi}{18} \text{ rad}$ then, from $\theta = \omega t$, the time $t = \left(\dfrac{5\pi}{18}\right)\left(\dfrac{1}{30\pi}\right) = 9.26 \text{ ms}$. Also, at $\dfrac{\pi}{6} \text{ rad} = 30°$ before TDC, the crank has travelled through an angle of $330 - 50 = 280° = \dfrac{560\pi}{360} \text{ rad}$, so the time period for this angular travel is $t = \dfrac{\theta}{\omega} = \dfrac{560\pi}{(360)(30\pi)} = 51.85 \text{ ms}$. Therefore, the time taken for the crank to travel from its initial position to within 30° of TDC is $t = 51.85 - 9.26 = 42.58 \text{ ms}$.

We now turn our attention to finding velocity and acceleration parameters for the engine slider-crank mechanism, using a mathematical method.

9.2.2 Velocity and acceleration analysis

Our starting point for the derivation of a relationship that will enable us to determine engine slider-crank velocities analytically is equation 9.4, which gives the relationship for the instantaneous displacement of the slider-crank, that is:

$$s = r\left[\cos\omega t + \frac{1}{n}(1 - n^2\sin^2\omega t)^{1/2}\right],$$

where from what we have said previously, $\theta = \omega t$ and $t =$ the time since the slider passed TDC. Now if we assume that the crank is subject to a constant angular velocity ω in an anticlockwise direction, then the velocity of the slider is given by $\dfrac{ds}{dt} = s'$, so differentiating equation 9.4 with respect to time will yield an equation for the velocity of the slider C, that is:

$$s'(t) = -r\omega\left[\sin\omega t + \frac{n\sin 2\omega t}{2\left(1 - n^2\sin^2\omega t\right)}\right] \quad (9.5)$$

From this equation we note that the slider velocity becomes zero when $\theta = \omega t = 0$ or π.

Similarly, we can obtain an equation for the acceleration of the slider by finding the second differential of equation 9.5, i.e.

$$\frac{d^2s}{dt^2} = s''(t) = -r\omega\left[\frac{d}{dt}(\sin\omega t) + \frac{n}{2}\frac{d}{dt}\left\{\frac{\sin 2\omega t}{\left(1 - n^2\sin^2\omega t\right)}\right\}\right]$$

so that, on differentiation, we get:

$$s''(t) = -r\omega^2\left[\cos\omega t + \frac{n}{4}\left\{\frac{4\cos 2\omega t(1 - n^2\sin^2\omega t) + n^2\sin^2 2\omega t}{\left(1 - n^2\sin^2\omega t\right)^{3/2}}\right\}\right] \quad (9.6)$$

One useful relationship that can be obtained from this complex equation is that found when we set the time $t = 0$, where we find that $s''(t) = -r\omega^2(1 + n)$, which is an expression that directly relates the acceleration of the slider to the centripetal acceleration of the crank.

Now, if you were able to differentiate equation 9.4 to obtain equations 9.5 and 9.6, then you have done extremely well! However, the complexity of equation 9.6 makes it particularly difficult to use, especially if *exact values* for the acceleration of the slider are required. This differential equation may be solved by applying numerical techniques; however, these methods of solution require careful analysis, even with the help of computer programs, so you will be pleased to know that such techniques will not be covered in this book!

However, we can *simplify* both the differentiation and indeed the equations themselves (9.5 and 9.6) by applying a slightly different approach, from which we will be able to determine *acceptable approximations* for the slider velocities and accelerations, that are far easier to manipulate!

Our starting point this time is equation 9.3, where $s = r \cos \theta + l(1 - n^2 \sin^2 \theta)^{1/2}$. Then, appreciating that

(i) $s' = \left(\dfrac{ds}{d\theta} \right) \left(\dfrac{d\theta}{dt} \right) = \left(\dfrac{ds}{d\theta} \right) \omega$ and

(ii)   that the differential of the bracketed expression raised to the power ($\frac{1}{2}$) can be made easier by expanding it to form a series, using the *binomial theorem*. The terms in this series may then be truncated to give an approximation, to any degree of accuracy.

It is hoped that you are familiar with the use of the binomial theorem to determine a series sum (expansion), when fractional powers are involved. However, for those of you who are a little rusty, we will now show the full mathematical method for simplifying the expression under the square root in equation 9.3, using the binomial expansion.

The form of the binomial expansion we will use is valid for integer 'n' provided $-1 < x < 1$ and is also valid if $x = 1$ when $n > -1$ and if $x = -1$ when $n > 0$. Therefore in our case $n = 1/2$ (the square root), then $n > 0$, so the expansion is valid when $-1 \leq -2x \leq 1$, i.e., when $-1/2 \leq x \leq 1/2$. The binomial expansion that fits these criteria takes the form:

$$(1 + x)^n = 1 + nx + \frac{n(n-1)x^2}{1 \times 2} + \frac{n(n-1)(n-2)x^3}{1 \times 2 \times 3}$$

and in our case, letting $x = n^2 \sin^2 \theta$ and $n = 1/2$, we find that:

$$\left(1 - n^2 \sin^2 \theta \right)^{1/2} = 1 + \left(\tfrac{1}{2} \right) \left(-n^2 \sin^2 \theta \right)$$
$$+ \frac{\left(\tfrac{1}{2} \right) \left(-\tfrac{1}{2} \right)}{1 \times 2} \left(-n^2 \sin^2 \theta \right)^2$$
$$+ \frac{\left(\tfrac{1}{2} \right) \left(-\tfrac{1}{2} \right) \left(-\tfrac{3}{2} \right)}{1 \times 2 \times 3} \left(-n^2 \sin^2 \theta \right)^3 + \ldots$$

$$\left(1 - n^2 \sin^2 \theta \right)^{1/2} = 1 - \frac{n^2}{2} \sin^2 \theta - \frac{n^4}{8} \sin^4 \theta$$
$$- \frac{n^6}{16} \sin^6 \theta \ldots$$

Now, because the ratio $n = r/l$ is unlikely to exceed $1/3$ and $\sin \theta \leq 1$, the terms in the above expression quickly converge towards zero. Therefore a good approximation may be obtained by considering only the first two terms

of the expansion, i.e.

$$\left(1 - n^2 \sin^2 \theta \right)^{1/2} \approx 1 - \frac{n^2}{2} \sin^2 \theta.$$

Then, from $s = r \cos \theta + l(1 - n^2 \sin^2 \theta)^{1/2}$, a *good approximation* for the displacement of the slider may be written as:

$$s = r \cos \theta + l \left(1 - \frac{n^2}{2} \sin^2 \theta \right) \tag{9.7}$$

Now, rewriting equation 9.7 as $s = r \cos \theta + l - \frac{r^2}{2l} \sin^2 \theta$, remembering that

$$n = \frac{r}{l}, s' = \left(\frac{ds}{d\theta} \right) \left(\frac{d\theta}{dt} \right) = \left(\frac{ds}{d\theta} \right) \omega \text{ and } \theta = \omega t,$$

then, after differentiation, we obtain our expression for the velocity of the slider as $s' = -r\omega \left(\sin \theta + \frac{r}{2l} \sin 2\theta \right)$ or

$$s' = -r\omega \left(\sin \theta + \frac{n}{2} \sin 2\theta \right) \tag{9.8}$$

This equation shows that the velocity of the slider is negative, i.e. it is reduced as the crank angle θ increases.

If you are still unclear about how we obtained the differential, just remember to use the appropriate trigonometric identities, so that

$$\frac{d}{d\theta} \left(\frac{r}{2l} \sin^2 \theta \right) = \left(\frac{r}{2l} \right) \frac{d}{d\theta} \left(\sin^2 \theta \right) = \left(\frac{r}{2l} \right) 2 \sin \theta \cos \theta.$$

Using the trigonometric identity $\sin 2\theta = 2 \sin \theta \cos \theta$, we obtain the required differential

$$\frac{d}{d\theta} \left(\frac{r}{2l} \sin^2 \theta \right) = \frac{r}{2l} (\sin 2\theta).$$

Finally, with constant angular velocity of the crank, the expression for the *acceleration* of the slider, after differentiating equation 9.8, is found to be:

$$a = s'' = -r\omega^2 (\cos \theta + \frac{r}{l} \cos 2\theta) \tag{9.9}$$

I hope you have followed the above mathematical methods. It has required a fair bit of algebraic manipulation, even for the engine slider-crank which is considered to be one of the simpler mechanisms!

Key Point. Equations 9.8 and 9.9 give good approximations for the velocity and acceleration of the slider, respectively

Example 9.5 Figure 9.14 shows a slider-crank mechanism at some instant in time. Determine expressions for the horizontal and vertical components of the velocity of the point P on the link BC.

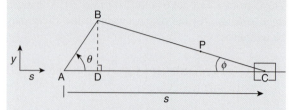

Figure 9.14 Slider-crank mechanism for Example 9.5

In order to establish the desired relationships we first need to identify the position in space of the fixed point P on the link BC. To help us we use the Cartesian coordinates labelled sy, the construction line BD and the two angles θ, ϕ, as shown in Figure 9.14.

Then the position of the point P in the horizontal direction is given as $s_p = s - p\cos\phi$ and the vertical position of the point P is given as $y_p = p\sin\phi$. You should make sure you understand how these two expressions were derived, by considering the geometry of the situation shown in Figure 9.14.

We also know from Figure 9.14 that $BD = r\sin\theta = l\sin\phi$, therefore $\dfrac{r}{l}\sin\theta = \sin\phi$ and, from the trigonometric identity $\sin^2\phi + \cos^2\phi = 1$, we know that $\cos\phi = \left(1 - \sin^2\phi\right)^{1/2}$. Therefore, after substitution, the horizontal position for the point P with respect to the crankpin A is

$$s_p = s - p\left(1 - \sin^2\phi\right)^{1/2} \text{ or}$$

$$s_p = s - p\left(1 - \frac{r^2}{l^2}\sin^2\theta\right)^{1/2}.$$

The vertical position of the point P with respect to the crankpin A is given by $y_p = \dfrac{pr}{l}\sin\theta$.

Then, remembering the use of the binomial theorem to determine a suitable estimate for the expression $\left(1 - \dfrac{r^2}{l^2}\sin^2\theta\right)^{1/2}$, we may write that

$$s_p = s - p\left(1 - \frac{r^2}{2l^2}\sin^2\theta\right).$$ Now we know from our earlier work that $s' = \left(\dfrac{ds}{d\theta}\right)\left(\dfrac{d\theta}{dt}\right) = \left(\dfrac{ds}{d\theta}\right)\omega$, so on differentiating s_p we obtain the required expression for the horizontal component of the velocity as

$$s' = -r\omega\left[\sin\theta + \frac{r}{2l}\left(1 - \frac{p}{l}\right)\sin 2\theta\right].$$

The vertical component of velocity is, after differentiating the above equation for position, found to be $y_p' = r\omega\left(\dfrac{p}{l}\cos\theta\right)$. It is also a simple process to find the expressions for the appropriate accelerations by differentiating these expressions once more. This is left as an exercise for you, later.

9.3 Cam mechanisms

9.3.1 Cams and followers

Another important component often used in engineering mechanisms is the cam, where a *cam and follower mechanism* is used to convert rotational motion into linear motion. A common example is the camshaft and valve gear within the reciprocating piston internal combustion engine used in road transport vehicles, whereby the cams on the shaft operate follower rods and rockers that are used to open and close the inlet and exhaust valves on top of the piston housing.

Figure 9.15 illustrates a typical arrangement for an internal combustion engine where, due to the 'pear

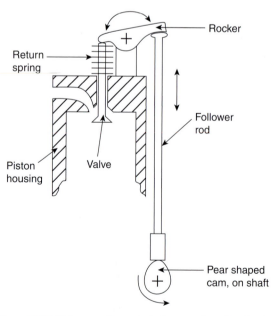

Figure 9.15 Valve gear for typical piston engine, showing just one value

shape' of the cam, its rotation causes the follower rod to move linearly up and down (reciprocate) and act on a rocker, which opens and closes the valve. The return movement of the follower rod may be directly due to gravity if the follower is positioned vertically above the cam, or, as illustrated, where the return action of the follower is assisted by a spring, which also is used to close the valve. An alternative to the particular type of mechanism illustrated in Figure 9.15 is that of the overhead cam, whereby the cam on the camshaft acts directly on top of the valve, avoiding the need for the follower rod and rocker arm.

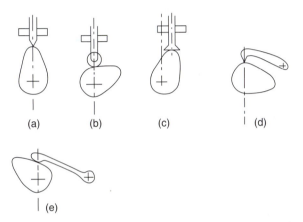

(a)   (b)   (c)   (d)

(e)

Figure 9.16 Typical cam and follower arrangements, a) shows a disc cam and in-line knife-point follower; b) shows a disc cam with an in-line roller follower; c) shows a disc cam with an offset flat-faced follower; d) shows a disc cam with an in-line knife-edge rocker follower; e) shows a disc cam with an in-line sliding and rocking follower

There are many different types of cam-follower arrangement and their selection will be dictated by the required geometrical constraints and particular application. Some typical arrangements are illustrated in Figure 9.16.

9.3.2 Cam profiles and displacement diagrams

In Figure 9.16 all disc cams have the same pear-shaped profile, this particular profile being typical of the cam design used for valve control in the internal combustion engine as illustrated in Figure 9.15. Many other cam profiles exist and are used in a variety of mechanisms. These include, among others, eccentric cams (Figure 9.18) and heart-shaped cams (which you will be asked to consider later).

Displacement diagrams are used to show the relationship between the angular motion (rotation) of a cam and the oscillatory linear motion of its follower. A displacement diagram, showing the path of the follower for one complete revolution of the cam, can easily be drawn by dividing the cam into suitable degree (or radian) subdivisions around its perimeter and then setting off the vertical displacement of the follower against the angular displacement of a *disc cam*, as illustrated in Figure 9.17.

Figure 9.18 provides one more illustration of a displacement diagram for an *eccentric* disc cam, again with a knife-edge follower, where in this case the angular displacement is shown in radian.

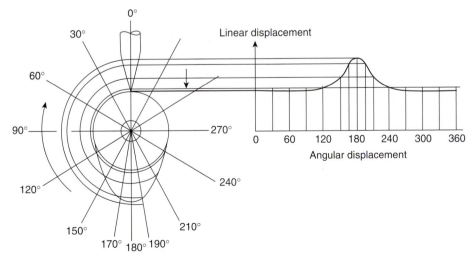

Figure 9.17 Disc cam and knife-edge follower displacement diagram

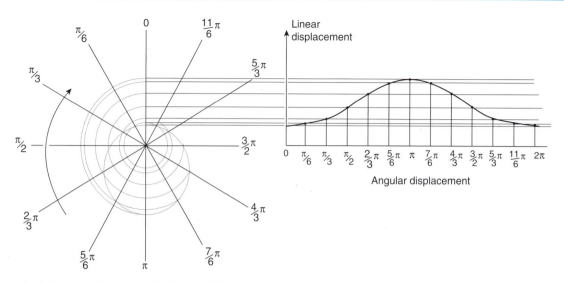

Figure 9.18 Eccentric disc cam and knife-edge follower displacement diagram

> **Key Point.** All types of cam and follower convert rotary to linear motion

9.4 Chapter summary

The aim of this chapter has been to introduce the kinematics of mechanisms, i.e. the motion of devices that convert one type of motion into another without the application of force. We started by considering, as an example, the Geneva mechanism which converts continuous angular motion into discrete step motion by the engagement and disengagement of a pin and angular cam.

We looked briefly at three aspects of mechanisms and their motion. These were:

- velocity and acceleration diagrams,
- the analysis of an engine slider-crank mechanism, and then,
- cam and follower mechanisms.

When constructing the velocity diagram, you should start from the space diagram of the situation drawn to scale, as discussed in Example 9.1, where a four-link (bar) mechanism was considered. Then, by deduction, careful labelling of relative velocities and by taking account of the information given with the problem, you should be able to draw the velocity diagram, again to scale (Figure 9.5), and determine from it the required velocities.

When determining accelerations, your starting point will normally involve using the velocities you obtained previously and considering the *components* of acceleration for any one of the links or points on the link. You may also need to use the transformation equations, $a_c = \omega^2 r$ and $a_{\tan} = r\alpha$, when finding the centripetal and tangential accelerations of a link. Note also that in this book we use the prime notation for the acceleration of a point on a link, thus the acceleration of a point P on a link is represented as $\mathbf{p}'$, with the normal emboldening to represent the original *vector quantity* (the velocity of the point P). Example 9.3 guides you through the procedure for finding the required accelerations for a four-link mechanism, while Figure 9.12 shows the acceleration diagram itself.

We next considered the analysis of an engine slider-crank mechanism, where we determined expressions for the displacement, velocity and acceleration of the links. The slider-crank was chosen for the *relative* simplicity of the analysis (because it has only two inclined moving links, the crank and the connecting rod) and because of its important application within the reciprocating piston internal combustion engine. Analytical methods really have only one fundamental advantage, that is, they allow any and every geometric position of the mechanism to be calculated from a general formula, unlike graphical methods that require each individual position to be redrawn. However, analytical methods are tedious, difficult and cumbersome and, for anything but the simplest case (the slider-crank), require the use of computers for their solution, hence why, in our study, we use only the engine slider-crank as our example for analysis.

Equations 9.1 and 9.2 enable us to find the maximum and minimum positions of the piston stroke, remembering that $n = \dfrac{r}{l}$, where $r =$ crank radius and $l =$ the length of the connecting rod. Equation 9.3 enables the displacement of the piston (slider) from the crankshaft pivot point to be determined for any angle of the crank, while equation 9.4 takes into account the angular velocity of the crankshaft and, for any instant in time t from some datum, we may use the substitution $\theta = \omega t$. Example 9.4 illustrates how we use these relationships to find the desired parameters. The velocity equation 9.5 is then found by differentiating equation 9.4 with respect to time. The acceleration equation 9.6 is found by differentiating equation 9.5 once more. The complexity of equation 9.6 prohibits a simple solution, so we then looked at a method for simplifying this equation, using the binomial theorem to produce a *good approximation* for the linear velocity and acceleration of the slider as given by equations 9.8 and 9.9, respectively. Example 9.5 shows how to find expressions for the horizontal and vertical components of the velocity of a point on the connecting link.

In the final part of this chapter we looked very briefly at *cam mechanisms*, where we started by illustrating and discussing a typical cam-follower arrangement for the valve gear of an internal combustion engine (Figure 9.15). You should note that many modern engines use the overhead cam principle to act on the valves directly, thus avoiding the need for the follower rods. Figure 9.16 illustrates some typical cam-follower arrangements.

9.5 Review questions

1. Describe what the study of 'kinetics' involves and explain what is meant by 'a mechanism', give two practical examples of the use of mechanisms.

2. Figure 9.19 shows a simplified sketch of an engine slider-crank mechanism, where the crank radius AB is 60 mm and the connecting rod BC is 240 mm in length. The crank rotates anticlockwise with uniform angular velocity of 300 rad/s. At a particular instant in time the crank angle CAB $= \dfrac{2\pi}{3}$ rad.

   a) Using a suitable scale, draw the space diagram for the slider-crank mechanism.

   b) By applying an appropriate procedure and using the information given, determine the

Figure 9.19 Slider-crank diagram for question 9.5.2

   linear velocity of the point B with respect to the fixed link at A.

   c) Draw the velocity diagram and determine from it the linear velocity of the slider C relative to A.

3. a) Using the information and results from question 2, construct the acceleration diagram for the slider-crank mechanism.

   b) Calculate the instantaneous angular acceleration of the link BC and the absolute linear acceleration of the slider (piston) C.

4. Explain and justify a simple approximation method for calculating the piston (slider) velocities and accelerations in a reciprocating piston engine.

5. Verify your answer to question 2c) and your answer for the linear acceleration of the slider in question 3b) by finding good approximations, using the analytical method explained in question 4.

6. Figure 9.20 shows a four-link mechanism ABDE, where at some instant in time link AB is 20° to AE and is rotating in a clockwise direction at 900 rev/min. The lengths of the links are as shown in the figure.

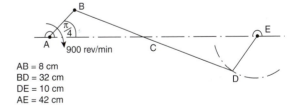

Figure 9.20 Four-link mechanism for question 9.5.6

   Find:

   a) The velocity and direction of motion of point C on the link BD, when C is at the halfway position along the link.

   b) The angular velocity of the link BD, that is the angular velocity of D relative to B.

7. Figure 9.21 shows a heart-shaped cam and concentric knife-edge follower mechanism.

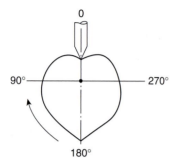

Figure 9.21 Heart-shaped cam and follower for question 9.5.7

Assuming the distance around the edge of this cam is 10 cm, sketch the displacement diagram for this cam and follower at intervals of 30° and comment on its shape.

Chapter 10

Power transmission systems

In this chapter we start with an investigation into power transmission between machines using belt drives, clutches and gear trains. We then turn our attention to the dynamics of rotating systems, looking particularly at static and dynamic balancing, flywheel dynamics and, finally, at the coupling of these components into rotational systems.

So, for example, the torque created by an electric motor may be converted into mechanical work by coupling the motor output to a gearbox, chain or belt drive. Motor vehicle transmission systems convert the torque from the engine into the torque at the wheels via a clutch, gearbox and drive shafts. The clutch is used to engage and disengage the drive from the engine and the gearbox assembly transmits the power to the wheel drive shafts, as dictated by the driving conditions.

Gears, shafts and bearings provide relatively rigid but positive power transmission and may be used in many diverse applications ranging, for example, from power transmission for the aircraft landing flap system on a jumbo jet to the drive mechanism for a child's toy.

Belts, ropes, chains and other similar elastic or flexible machine elements are used in conveying systems and in the transmission of power over relatively long distances. In addition, since these elements are elastic and usually quite long, they play an important part in absorbing shock loads and in damping out and isolating the effects of unwanted vibration.

A classic example of the effects of a component being out of balance may be felt through the steering wheel of a motor vehicle. A minor knock to a road wheel may be sufficient to put it out of balance and so cause rotational vibration which is felt back through the steering geometry. Any rotational machinery that is incorrectly balanced, or goes out of balance as a result of damage or a fault, will produce vibration, which is at best unacceptable and at worst may cause dangerous failure. In this section, you will be briefly introduced to the concept of balance and the techniques for determining out-of-balance moments.

We next consider flywheels. These components provide a valuable source of rotational energy for use in many engineering situations, such as machine presses, mills, cutting machines, and prime movers for transport and agricultural machinery. We look, finally, at the inertia and energies involved when transmission and rotational components are coupled together to form rotational systems. We begin our study of power transmission systems by looking at belt drives.

10.1 Belt drives

Power transmission from one drive shaft to another can be achieved by using drive pulleys attached to the shafts of machines, being connected by belts of varying cross-section and materials.

A wide variety of belts is available, including flat, round, V, toothed and notched. The simplest of these, and one of the most efficient, is the flat belt. Belts may be made from leather or a variety of reinforced elastomer materials. Flat belts are used with crowned pulleys, round and V-belts are used with grooved pulleys, while timing belts require toothed wheels or sprockets.

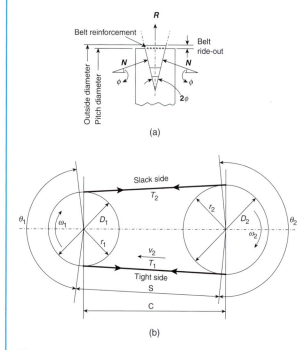

Figure 10.1 Basic belt drive geometry

To illustrate the set-up between a drive belt and pulleys, a V-belt system is shown in Figure 10.1.

Note that the pulley size is given by its pitch diameter and that in the case of a V-belt it rides slightly higher than the external diameter of the pulley (Figure 10.1a). The meaning of some of the more unfamiliar symbols in Figure 10.1, that you might not have encountered, are explained in the following paragraphs.

The speed ratio between the driving and driven pulleys is inversely proportional to the ratio of the pulley pitch diameters. This follows from the observation that there is no slipping (under normal loads). Thus the linear velocity of the pitch line of both pulleys is the same and equal to the belt velocity (v). So, from the transformation equations $v = r_1\omega_1 = r_2\omega_2$ or $v = \dfrac{D_1\omega_1}{2} = \dfrac{D_2\omega_2}{2}$, the *angular velocity ratio* (VR) is given by:

$$VR = \frac{\omega_1}{\omega_2} = \frac{D_2}{D_1} \qquad (10.1)$$

The angles θ_1 and θ_2 are known as the *angle of lap* or *angle of contact* (Figure 10.1b). This is measured between the contact length of the belt with the pulley and the angle subtended at the centre of the pulley.

The *torque* in belt drives and subsequent *transmitted power* is influenced by the differences in tension that exist within the belt during operation.

Key Point. The torque and thus the power transmitted by belt drives is directly related to the differences in tension within the belt during operation, and is given by $P = (T_1 - T_2)v$

The differences in tension result in a *tight side* and a *slack side* of the belt, as shown. These differences in tension influence the torque on the driving and driven pulley wheels, in the following way:

Torque on driving pulley $T_1 r_1 - T_2 r_1 = (T_1 - T_2)r_1$,

Torque on driven pulley $T_1 r_2 - T_2 r_2 = (T_1 - T_2)r_2$.

Now we know that power = torque × angular velocity, and since angular velocity is given by the linear speed divided by the radius of action, i.e. $\omega = v/r$, then we may write that:

$$\text{Transmitted power } P = (T_1 - T_2)v \qquad (10.2)$$

The *stresses* set up in the drive belt are due to:

(i) the tensile force in the belt, which is a maximum on the tight side of the belt;

(ii) the bending of the belt around the pulleys, which is a maximum as the tight side of the belt bends around the smaller diameter pulley;

(iii) the centrifugal forces created as the belt moves around the pulleys.

The maximum total stress occurs where the belt enters the smaller pulley and the bending stress is a major part. Thus there are recommended minimum pulley diameters for standard belts; using smaller diameters than recommended drastically reduces belt life.

10.1.1   Belt tension equations

Belts, when assembled on pulleys, are given an initial tension (T_0); you will be only too well aware of this fact if you have ever been involved in setting up the fan belt on a motor vehicle engine. If this initial tension is too slack the belt may slip below the maximum design power or, if too tight, undue stresses may be imposed on the belt, causing premature failure. The initial pre-tension may be determined by using the formula:

$$T_0 = \frac{T_1 + T_2}{2} \qquad (10.3)$$

This follows from the argument that since the belt is made from an elastic material, when travelling around the pulleys the belt length remains constant. Thus the increase in length on the tight side is equal to the reduction in length on the slack side, and so the corresponding tensions are given by $T_1 - T_0 = T_0 - T_2$, from which equation (10.3) results.

As already mentioned, when the belt passes over a pulley there is an increase in the stresses imposed within the belt. These are due to an increase in centrifugal force as the belt passes over the pulley, creating an increase in the tension of the belt.

Therefore, we have an increase in belt tension, additional to that of normal transmission power, which is due to the centrifugal forces set up in the belt as it passes over the pulleys. Therefore, our standard power equation (10.2) needs to be modified to take into account the effects of centrifugal tension. We can formulate this relationship leading to the required power equation by use of a little mathematical analysis, but in the interests of space, all we show here are the important results of such an analysis together with the relationships between the coefficient of friction (μ), the angle of lap (θ) and the belt tensions (T_1, T_2), which are given next.

The *additional centrifugal tension* (T_c) is given by

$$T_c = m_l v^2 \qquad (10.4)$$

where m_l is the *mass per unit length* of the belt and v = the belt linear velocity.

> **Key Point.** The centrifugal tension is directly related to the mass of the belt and its velocity squared and is given by $T_c = m_l v^2$

Also, from the analysis, it can be shown that

$$\frac{T_1 - T_c}{T_2 - T_c} = e^{\mu\theta} \qquad (10.5)$$

Under *low speed conditions* the centrifugal tension (which from equation 10.4 can be seen to be proportional to the square of the belt linear velocity) is relatively small and may be neglected to give the following relationship:

$$\frac{T_1}{T_2} = e^{\mu\theta} \qquad (10.6)$$

Also, as suggested above, we may modify our *power* equation (10.2) to take into account the centrifugal tension term (T_c) under high or low speed conditions.

So for *high speed conditions*, after rearranging and substituting equation 10.5 into equation 10.2, we get:

$$P = (T_1 - T_2)(1 - e^{-\mu\theta})v \qquad (10.7)$$

And for *low speed conditions*, after rearranging and substituting equation 10.6 into equation 10.2, we get:

$$P = T_1(1 - e^{-\mu\theta})v \qquad (10.8)$$

The above equations have been formulated on the basis of the geometry related to *flat belts* travelling over *flat belt pulleys*.

V-belt drives require us to replace μ by $\mu/\sin\phi$ where it appears in equations 10.5 to 10.8. Consider again the V-pulley drive shown in Figure 10.1a, where the groove angle is (2ϕ) and the reaction (N) acts at angle (ϕ), as shown. The normal reaction (R) between the belt and pulley is given by simple trigonometry as $R = 2N \sin\phi$ and so the friction force between the belt and the pulley is $2\mu N = \dfrac{\mu R}{\sin\phi}$. So in effect, going from flat to V-grooved belt drives, as mentioned previously, transforms μ to $\mu/\sin\phi$.

10.1.2 Belt drive design requirements

In order to select a drive belt and its associated pulley assemblies for a specified use, and to ensure proper installation of the drive, several important factors need to be considered. The basic data required for drive selection are listed below. They are:

- the rated power of the driving motor or other prime mover;
- the service factor, which is dependent on the time in use, operating environment and the type of duty performed by the drive system;
- belt power rating;
- belt length;
- size of driving and driven pulleys and centre distance between pulleys;
- correction factors for belt length and lap angle on smaller pulley;
- number of belts; and
- initial tension on belt

> **Key Point.** A primary consideration of belt drive design is that provision is made for belt adjustment and pre-tensioning

Many design decisions depend on the required usage and space constraints. Provision must be made for belt adjustment and pre-tensioning, so the distance between pulley centres must be adjustable. The lap angle on the smaller pulley should be greater than 120°. Shaft centres should be parallel and carefully aligned so that the belts track smoothly, particularly if grooved belts are used.

The operating environment should be taken into consideration where hostile chemicals, pollution or elevated temperatures may have an effect on the belt material. Consider other forms of drive, such as chains, if the operating environment presents difficulties.

From the above list, it can be seen that relationships between belt length, distance between centres, pulley diameters and lap angles are required if the correct belt drive is to be selected. Equation 10.9 relates the belt pitch length, L, the distance between centres, C, and the pulley pitch diameters, D, by:

$$L = 2C + 1.57(D_2 - D_1) + \frac{(D_2 - D_1)^2}{4C} \quad (10.9)$$

Also

$$C = B + \frac{\sqrt{B^2 - 32(D_2 - D_1)^2}}{16} \quad (10.10)$$

where $B = 4L - 6.28(D_2 + D_1)$ and $D_2 > D_1$.

The contact angle (*lap angle*) of the belt on each pulley is given by the relationships

$$\theta_1 = 180° - 2\sin^{-1}\left[\frac{D_2 - D_1}{2C}\right]$$

$$\theta_2 = 180° + 2\sin^{-1}\left[\frac{D_2 - D_1}{2C}\right] \quad (10.11)$$

One other useful formula (10.12) enables us to find the span length S (see Figure 10.1), given the distance between pulley centres and the pulley diameters.

$$S = \sqrt{C^2 - \left[\frac{(D_2 - D_1)}{2}\right]^2} \quad (10.12)$$

This relationship is important since it enables us to check for correct belt tension by measuring the force required to deflect the belt at mid-span. This length also has a direct bearing on the amount of vibration set up in the belt during operation, therefore we need to ensure that span length falls within required design limits.

Example 10.1   A flat belt drive system consists of two parallel pulleys of diameter 200 and 300 mm, which have a distance between centres of 500 mm. Given that the maximum belt tension is not to exceed 1.2 kN, the coefficient of friction between the belt and pulleys is 0.4 and the larger pulley rotates at 30 rad/s, find:

a)   the belt lap angle for the pulleys;

b)   the power transmitted by the system;

c)   the belt pitch length L;

d)   the pulley system span length between centres.

For part

a)   the *lap angles* are given by equation 10.11, so $\theta_1 = 180 - 2\sin^{-1}\left(\dfrac{300 - 200}{2 \times 500}\right) = 180 - 11.48 = 168.52°$ and $\theta_2 = 180 + 11.48 = 191.48°$.

b)   The maximum power transmitted will depend on the angle of lap of the smaller pulley. So, using equation 10.8 and assuming low speed conditions, we get:

$$P = T_1(1 - e^{-\mu\theta})v,$$

where linear velocity of belt $v_2 = r_2\omega_2 = (0.15)(30) = 4.5$ m/s (see Figure 10.1). Then lap angle for small pulley $\theta_1 = 2.94$ radian, so that

$$P = 1200\left(1 - e^{-(0.4)(2.94)}\right)(4.5)$$

or

$$P = 1200(1 - 0.3085)(4.5) = 3.734 \text{ kW}.$$

c)   The belt pitch length is found using equation 10.9, $L = 2C + 1.57(D_2 - D_1) + \dfrac{(D_2 - D_1)^2}{4C}$, where the distance between pulley centres C is given as 500 mm. Then: $L = 2(0.5) + 1.57(0.3 - 0.2) + \dfrac{(0.3 - 0.2)^2}{4(0.5)}$, so that $L = 1.162$ m.

d)   The span length S is given by equation 10.12. Then: $S = \sqrt{(0.5)^2 - \left[\dfrac{(0.3 - 0.2)}{2}\right]^2}$ and so $S = 0.497$ m.

10.2   Friction clutches

With the sliding surfaces encountered in most machine components, such as bearings, gears, levers and cams, it is desirable to minimise friction, so reducing energy loss, heat generation and wear. When considering friction in clutches and brakes, we try to *maximise* the friction between the driving and driven plates. This is achieved in dry clutches by choosing clutch plate materials with high coefficients of friction. Much research has been undertaken in materials designed to maximise the friction coefficient and minimise clutch disc wear, throughout a wide range of operating conditions, during its service life.

The primary function of a clutch is to permit smooth transition between connection and disconnection of the driving and driven components and, when engaged, to ensure that the maximum possible amount of available torque is transmitted.

Key Point.   The primary function of a clutch is to permit smooth transition between connection and disconnection of the drive and to transmit the maximum available amount of torque

Clutches are used where there is a constant rotational torque produced at the drive shaft by a prime mover such as an engine. The drive shaft can then be engaged through the clutch to drive any load at the output, without the need to alter the power produced by the prime mover. The classic application of this transmission principle is readily seen in automobiles, where the engine provides the power at the drive shaft which is converted to torque at the wheels. When we wish to stop the vehicle without switching off the engine, we disengage the drive using the clutch.

10.2.1   Disc clutches

A simple disc clutch, with one driving and one driven disc, is illustrated in Figure 10.2. The principle of operation is dependent on the driving friction created between the two clutch discs when they are forced together. The means of applying pressure to the clutch discs may be mechanical or hydraulic and, as with other components that rely on friction, they are designed to operate either dry or wet using oil.

The oil provides an effective cooling medium for the heat generated during clutch engagement and, instead of using a single disc (Figure 10.2), multiple discs may

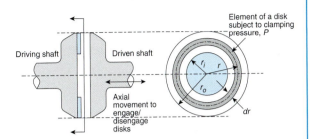

Figure 10.2 Simple disc clutch

be used to compensate for the inevitable drop in the coefficient of friction.

In a multi-disc clutch (Figure 10.3) the driving discs rotate with the input shaft; the clutch is engaged using hydraulic pressure which is generated from a master cylinder at the clutch pedal, or by some other means. The driving discs are constrained, normally by a key and keyway, to rotate with the input shaft. On engagement, the rotating drive discs are forced together with the driven disc to provide torque at the output. When the clutch is disengaged by releasing the hydraulic pressure, or through spring action, the discs are free to slide axially and so separate themselves.

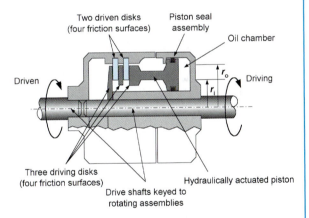

Figure 10.3 Hydraulically operated clutch assembly

From Figure 10.3 it can be seen that the driving discs include the backing plates, which only provide one friction surface, whereas the middle driving disc and the two driven discs each provide two contact surfaces. So in our example, the five discs provide a total of four driving and four driven contact surfaces.

10.2.2   Friction clutch equations

In order to estimate the torque and power from clutch-driven shafts, we need to use equations that relate factors such as: clutch size, friction coefficients, torque capacity, axial clamping force and pressure at the discs. When formulating such equations, we make two very important basic assumptions:

(i)   we assume uniform distribution of pressure at the disc interface;

(ii) we assume a uniform rate of wear at the disc interface.

The first assumption is valid for an unworn (new) accurately produced and installed clutch, with rigid outer discs. Consider a very small element of the disc surface (Figure 10.2), subject to an axial force normal to the friction surface. Then, if constant pressure p acts on the element at radius r and the element has width dr, the force on the element $= (2\pi\,dr)p$. So the total normal force acting on the area of contact is:

$$W = \int_{r_i}^{r_o} 2\pi prdr = \pi p(r_o^2 - r_i^2) \qquad (10.13)$$

where W is the *axial force clamping the driving and driven plates together*. Now we know that the friction torque that can be developed on the disc *element* is the product of the friction force and radius. The friction force is the product of the force on the element and the coefficient of friction. So we may write:

Torque on disc element $= (2\pi prdr)\mu r$ and the total torque that can be transmitted is:

$$T = \int_{r_i}^{r_o} 2\pi p\mu r^2 dr = \frac{2}{3}\pi p\mu(r_o^3 - r_i^3) \qquad (10.14)$$

Equation 10.14 represents the torque transmitted by one driving disc (plate) mating with one driven plate.

For clutches that have more than one set of discs, that is N discs, where N is an even number (see Figure 10.3), then:

$$T = \frac{2}{3}\pi p\mu(r_o^3 - r_i^3)N \qquad (10.15)$$

If we transpose equation 10.13 for p and substitute it into equation 10.15, we get:

$$T = \frac{2W\mu(r_o^3 - r_i^3)}{3(r_o^2 - r_i^2)}N \qquad (10.16)$$

This equation relates the torque capacity of the clutch with the axial clamping force.

We now *assume a uniform rate of wear* (assumption (ii)). We know that the rate at which wear takes place

between two rubbing surfaces is proportional to the rate at which it is rubbed (the *rubbing velocity*) multiplied by the *pressure applied*. So, as a pair of clutch plates slide over each other, their relative velocities are linear, our clutch plates rotate with angular velocity, so the rubbing velocity is dependent on the radius of rotation. From the above argument we may write that:

Wear $\propto$ pressure $\times$ velocity (where velocity is dependent on radius), so wear $\propto$ pressure $\times$ radius. Introducing the constant of proportionality, c, gives $pr = c$. Now, substituting $pr = c$ into equations 10.13 and 10.14 gives, respectively:

$$W = \int_{r_i}^{r_o} 2\pi c\,dr = 2\pi c(r_o - r_i) \qquad (10.17)$$

and

$$T = \int_{r_i}^{r_o} 2\pi c\mu r dr = \pi c\mu(r_o^2 - r_i^2)N \qquad (10.18)$$

So, on transposition of equation 10.17 for c and then substitution into equation 10.18 we get:

$$T = W\mu\left(\frac{r_o + r_i}{2}\right)N \qquad (10.19)$$

We have said that $pr = c$. Now it is known that the greatest pressure on the clutch plate friction surface occurs at the inside radius r_i. So, for a maximum allowable pressure (p_{max}) for the friction surface, we may write:

$$pr = c = p_{max}r_i \qquad (10.20)$$

Clutch plates are designed with a specified ratio between their inside and outside radii, to maximise the transmitted torque between surfaces. This ratio is given without proof as $r_i = 0.58r_o$; in practice clutches are designed within the approximate range $0.5r_o \le r_i \le 0.8r_o$.

Example 10.2   A multiple plate clutch, similar to that shown in Figure 10.3, needs to be able to transmit a torque of 160 N m. The external and internal diameters of the friction plates are 100 mm and 60 mm, respectively. If the friction coefficient between the rubbing surfaces is 0.25 and $p_{max} = 1.2$ MPa, determine the total number of discs required, *and* the axial clamping force. State all assumptions made.

Since this is a design problem, we make the justifiable assumption that the wear rate will be uniform at the clutch plate rubbing surfaces. We must also assume that the torque is shared equally by all the clutch discs and that the friction coefficient remains constant during operation. This last assumption is unlikely, and contingencies in the design would need to be made to ensure that the clutch met its service requirements. In practice, designers ensure that a new clutch is slightly over-engineered to allow for *bedding in*. After the initial bedding-in period wear tends to be uniform and friction coefficients are able to meet the clutch design specification under varying conditions, so the clutch assembly performs satisfactorily.

Using equation 10.18 after rearranging for N, where $c = p_{max}r_i$, then:

$$N = \frac{T}{\pi p_{max} r_i \mu (r_o^2 - r_i^2)},$$

so that

$$N = \frac{160}{\pi (1.2 \times 10^6)(30 \times 10^{-3})(0.25)(0.05^2 - 0.03^2)}$$

$$= 3.53.$$

Now since N must be an even integer, the nearest N above design requirements is $N = 4$. This is because the friction interfaces must transmit torque in parallel pairs. In Figure 10.3 there are four friction interfaces, which require *five* discs, the two outer discs having only one friction surface.

Now we also need to find the axial clamping force (W). This is given by equation 10.19 where:

$$T = W\mu \left(\frac{r_o + r_i}{2}\right)N,$$

so after transposition

$$W = \frac{T}{\mu N}\left(\frac{2}{r_o + r_i}\right) = \frac{160}{(0.25)(4)}\left(\frac{2}{0.05 + 0.03}\right)$$

$$= 4\,\text{kN}.$$

So we require a total of five discs and an axial clamping force of 4 kN.

Key Point. The friction clutch equations are based on the assumption that there is uniform distribution and uniform rate of wear at the disc interface

10.2.3   Cone clutches

A cone clutch has a *single pair* of conical friction surfaces (Figure 10.4) and is similar to a disc clutch, except that in the case of the disc clutch the coning angle takes a specific value of 90°. So the cone clutch may be thought of as the generalised case, of which the disc clutch is a special case. One of the advantages of a cone clutch, apart from its simplicity, is that due to the wedging action, the clamping force required is drastically reduced when compared to its disc clutch counterpart.

Key Point. A *cone clutch* has the advantage of simplicity and the significantly reduced clamping force when compared with its disc clutch counterpart

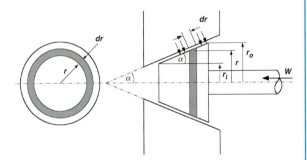

Figure 10.4 Cone clutch

The formulae for cone clutches are derived in a very similar way to those you met earlier concerning the plate clutch and V-belts. We therefore adopt a very similar mathematical approach.

Consider the element of width dr in Figure 10.4. Then, from the geometry, the area of this element is $= \dfrac{2\pi\,dr}{\sin\alpha}$. If p is again the normal pressure, then the normal force acting on the element is $= \dfrac{(2\pi\,dr)p}{\sin\alpha}$ and the corresponding axial clamping force $= (2\pi\,dr)p$. This expression is exactly the same as that shown earlier for the disc clutch element, so the total axial force acting on the mating cone surfaces is given as $W = \int_{r_i}^{r_o} 2\pi prdr$ (compare with equation 10.13). Then, assuming uniform pressure, $W = \pi p(r_o^2 - r_i^2)$, as before.

We also know that the friction force is equal to the product of the coefficient of friction (μ) and the normal reaction force. If we again consider the normal reaction force in terms of pressure, we may write:

Friction force $= \dfrac{(2\pi r \mu dr) p}{\sin \alpha}$ and the torque transmitted by the element is $= \dfrac{(2\pi r \mu dr) p r}{\sin \alpha}$. Then the total torque transmitted $T = \int_{r_i}^{r_o} 2\pi \mu p r^2 dr$ and, assuming uniform pressure:

$$T = \frac{2\pi \mu p}{3 \sin \alpha}(r_o^3 - r_i^3) \qquad (10.21)$$

and as before, eliminating p and using equation 10.13, we get that:

$$T = \frac{2\mu W}{3 \sin \alpha}\frac{(r_o^3 - r_i^3)}{(r_o^2 - r_i^2)} \qquad (10.22)$$

Finally, assuming uniform wear, we have that:

$$W = \pi p_{max} r_i (r_o - r_i) \qquad (10.23)$$

$$T = \frac{\pi \mu p_{max} r_i}{\sin \alpha}(r_o^2 - r_i^2) \qquad (10.24)$$

and in terms of W:

$$T = \frac{\mu W}{2 \sin \alpha}(r_o + r_i) \qquad (10.25)$$

Note that in equations 10.23 and 10.24 the constant 'c' has been replaced by $p_{max} r_i$, since $pr = c = p_{max} r_i$ (compare these equations with equations 10.18 and 10.19).

10.3   Gear trains

We continue our study of power transmission by looking at gears. The gearbox is designed to transfer and modify rotational motion and torque. For example, electric motor output shafts may rotate at relatively high rates, producing relatively low torque. A reduction gearbox interposed between the motor output and load, apart from reducing the speed to that required, would also increase the torque at the gearbox output by an amount equivalent to the gearbox ratio.

Key Point.   A gearbox is designed to transfer and modify rotational motion and torque

The gear wheel itself is a toothed device designed also to transmit rotary motion from one shaft to another. Gears are often used in transmission systems because they are rugged, durable and efficient transmission devices, providing rigid drives.

The simplest and most commonly used gear is the *spur gear*. These are designed to transmit motion between parallel shafts (Figure 10.5).

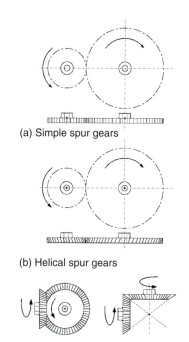

(a) Simple spur gears

(b) Helical spur gears

(c) Bevelled gears

Figure 10.5 Gear wheels

Spur gears can be helical, providing smooth and efficient meshing, which gives a quieter action and tends to reduce wear. The gears used in automobile transmission systems are often helical, for the above reasons (synchronised meshing or synchromesh). If the gear teeth are bevelled, they enable the gears to transmit motion at right angles. The common hand drill is a good example of the use of bevelled gears.

10.3.1   Spur gears

Figure 10.6 shows the basic geometry of two meshing spur gears. There are several important points to note about this geometry, which will help you when we consider gear trains.

The gear teeth profile is *involute*; this geometric profile is the curve generated by any point on a taut thread as it unwinds from a base circle. Once the gear teeth have been cut to this involute shape they will, if correctly spaced, mesh without jamming. It is interesting to note that this is the only shape where this can be achieved. The gear teeth involute profile is

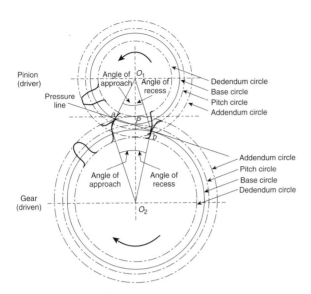

Figure 10.6 Geometry for two gears in mesh

extended outwards beyond the *pitch circle* (the point of contact of intermeshing teeth) by a distance called the *addendum*. Similarly, the tooth profiles are extended inwards from the pitch circle by an identical distance called the *dedendum*. When we refer to the diameter of gear wheels, we are always referring to the *pitch diameter*.

10.3.2 Simple gear train

Meshing spur gears which transmit rotational motion from an input shaft to an output shaft are referred to as *gear trains*. Figure 10.7 shows a simple gear train consisting of just two spur gears of differing size, with their corresponding direction of rotation being given.

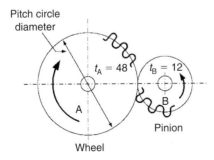

Figure 10.7 Simple gear train

If the larger gear *wheel* has 48 teeth (gear A) and the smaller *pinion* has 12 teeth (gear B), then the smaller gear will have to complete four revolutions for each

revolution of the larger gear. Now we know that the *VR* is defined as:

$$VR = \frac{\text{Distance moved by effort}}{\text{Distance moved by load}}$$

The *VR*, when related to our gear train, will depend on which of the gears is the driver (effort) and which the driven (load). The distance moved is dependent on the number of teeth on the gear wheel. In our example above, if the larger wheel is the driver, then we have:

$$VR = \frac{\text{Distance moved by driver}}{\text{Distance moved by driven}} = \frac{1}{4}$$

What this shows is the relationship between the number of gear teeth and the *VR*. When considering gears, the *VR* is identical to the *gear ratio*. Now, when the gear train is in motion, the ratio of the angular velocities of the gears is directly proportional to the angular distances travelled and, as can be seen from above, is the inverse ratio of the number of teeth on the gears. Thus the *gear ratio G* may be written formally as:

$$\text{Gear ratio } (G) = \frac{t_B}{t_A} = \frac{\omega_A}{\omega_B},$$

which for our gear pair gives $\frac{12}{48} = \frac{1}{4}$, so the gear ratio in this case is 1:4.

The gear teeth on any meshing pair must be the same size, therefore the number of teeth on a gear is directly proportional to its *pitch circle diameter*. So, taking into account this fact, the gear ratio formula may be written as:

$$G = \frac{\omega_A}{\omega_B} = \frac{t_B}{t_A} = \frac{d_B}{d_A} \qquad (10.26)$$

So, to summarise: *the gear ratio is inversely proportional to the teeth ratio, and so inversely proportional to the pitch circle diameter ratio.*

Example 10.3 A gearwheel (A) rotates with angular velocity of 32 rad/s, it has 24 teeth, it drives a gear B. If the gear ratio is 8:1, determine the velocity of B and the number of teeth on B.

From equation 10.26, $G = \frac{\omega_A}{\omega_B} = \frac{t_B}{t_A}$ or $\frac{8}{1} = \frac{32}{4} = \frac{t_B}{24}$ and so $t_B = 192$.

Therefore, the velocity of gear B = 4 rad/s and it has 192 teeth.

A simple gear train can, of course, carry more than two gears. When more than two gears are involved we still consider the gear train in terms of a combination of gear pairs. Figure 10.8 shows a simple gear train consisting of three gears.

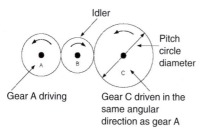

Figure 10.8 Gear train with idler

The overall gear ratio is the ratio of the input velocity over the output velocity, in our example this is ω_A/ω_C. If we consider the *VR*s of each gear pair sequentially from the input, we can determine the *VR* for the whole train. So, for the first pair of meshing gears, AB, the *VR* is given in the normal manner as ω_A/ω_B. Now, the second pair of meshing gears is ω_B/ω_C, so the overall gear ratio is:

$$G = \frac{\omega_A}{\omega_B} \times \frac{\omega_B}{\omega_C} = \frac{\omega_A}{\omega_C} \qquad (10.27)$$

You will note that the middle gear plays no part in the final gear ratio. This suggests that it is not really required. It is true that the size of the middle gear has no bearing on the final gear ratio, but you will note from Figure 10.8 that it does have the effect of *reversing* the direction of angular motion between the driving and driven gear. For this reason it is known as an *idler gear*.

10.3.3   Compound gear train

In a compound train, at least one shaft carries a compound gear; i.e. two wheels which rotate at the same angular velocity. The advantage of this type of arrangement is that compound gear trains can produce higher gear ratios without using the larger gear sizes that would be needed in a simple gear train. Examples of compound gear trains may be found in automobile gearboxes and in a variety of industrial machinery such as lathes and drilling machines.

Key Point. Compound gearboxes have the advantage of being more compact and having higher gear ratios than their simple gearbox counterpart

The following example illustrates how the gear ratio may be determined by considering the angular velocities of the gears and their relationship with the number of teeth.

Example 10.4 The compound gear train illustrated in Figure 10.9 consists of five gears, of which two, gears B and C, are on the same shaft. If gear A drives the system at 240 rpm (clockwise) and the number of teeth on each gear (N) are as given, determine the angular velocity of gear E at the output and the gear ratio for the system.

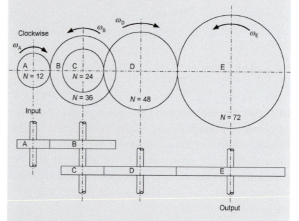

Figure 10.9 Compound gear train

Again, the system may be treated as gear pairs. Noting that gears B and C, being on the same shaft, rotate with the same angular velocity, then we proceed as follows.

From equation 10.27, for a simple gear train, the overall gear ratio (G) is given by:

$$G = \frac{\omega_A}{\omega_E} = \frac{\omega_A}{\omega_B} \times \frac{\omega_B}{\omega_C} \times \frac{\omega_C}{\omega_D} \times \frac{\omega_D}{\omega_E} \quad \text{where} \quad \omega_B = \omega_C,$$

so that the equation can be written as:

$$G = \frac{\omega_A}{\omega_E} = \frac{\omega_A}{\omega_B} \times 1 \times \frac{\omega_C}{\omega_D} \times \frac{\omega_D}{\omega_E}.$$

The angular velocity of gear E can then be found as follows:

$$\omega_E = \left(\frac{N_A}{N_B} \times 1 \times \frac{N_C}{N_D} \times \frac{N_D}{N_E} \right) \omega_A,$$

so

$$\omega_E = \left(\frac{12}{36} \times 1 \times \frac{24}{48} \times \frac{48}{72} \right) \omega_A = \frac{1}{9}\omega_A,$$

then

$$\omega_E = \frac{240}{9} = 26.67 \text{ rpm (anticlockwise)}.$$

The direction can be established as anticlockwise from the velocity arrows shown in Figure 10.9. Also note from above that $G = \dfrac{\omega_A}{\omega_E} \neq \dfrac{N_A}{N_E}$, i.e. the angular velocity ratio does not equal the ratio of the teeth on the gearwheels in this case. This is why considerable care is needed when solving compound gear problems.

10.3.4   Epicyclic gear trains

Unique gear ratios and movements can be obtained in a gear train by permitting some of the gear axes to rotate about others. Such gear combinations are called *planetary* or *epicyclic* gear trains. Epicyclic gear trains (Figure 10.10) always include a *sun* gear, a *planet* carrier or *link arm* and one or more *planet gears*.

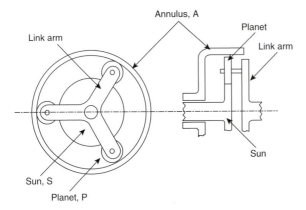

Figure 10.10 Epicyclic gear train

Often a ring is included which has internal gear teeth that mesh with the planet gear; this is known as the *annulus*. Epicyclic gear trains are unusual mechanisms in that they have two degrees of freedom. This means, for constrained motion, that any two of the elements in the train provide an input.

Epicyclic gear trains are used for most automatic gearboxes and some complex steering mechanisms. The major advantages of epicyclic mechanisms are the attainment of high gear ratios and the fact that they act as bearings as well as transmissions.

We will now attempt to analyse the motion of the gear train illustrated in Figure 10.10. The following procedure may be used.

(1)   First, imagine that the whole assembly is locked and rotated through 1 revolution, this has the effect of adding 1 revolution to all elements.

(2)   Next, apply a rotation of 1 to the gear which is to be fixed. This has the effect of cancelling the rotation given initially. Now the rotations of the other elements can be compared with the rotation of the fixed element by considering the number of teeth on each of the gears.

(3)   Construct a table for each of the elements and add the results obtained from the above actions. (When we sum the columns, using the above procedure, we are able to ensure that the fixed link has zero revolution.)

Example 10.5   An epicyclic gear train (Figure 10.10) has a fixed annulus (A) with 180 teeth, and a sun wheel (S) with 80 teeth.

i)   Determine the gear ratio between the sun and the link arm.

ii)   Find the number of teeth on the planet gears and so determine the gear ratio between the planet and the link arm.

So, following our procedure, we first lock the whole assembly and give it +1 revolution (so the annulus is zero when we sum). Next, we apply a −1 revolution to the gear specified as being at rest (so this element is zero when we sum), in this case the annulus. Then, to determine the arm in relation to the sun wheel, we fix the arm. Now, the number of revolutions of the sun wheel to that of the annulus equals the inverse ratio of their teeth. The results of these various actions are tabulated below.

Action	Effect			
	Arm	A	S	P
(1) Lock assembly and give all elements +1 revolution	+1	+1	+1	+1
(2) Fix the link arm and give −1 to A	0	−1	$+\dfrac{180}{80}$	$-\dfrac{180}{50}$
(3) Sum 1 and 2	(+1)	0	3.25	(−2.6)

Then a rotation of +1 for the link arm results in +3.25 for the sun wheel. *The gear ratio is 3.25:1.* Note that the number of teeth on the planet (P) was not required for part i), because it is in effect an idler gear between the sun and the annulus. The number of teeth on the planet gear given in the table above is determined as shown in part ii) below.

ii) Referring to Figure 10.10, we note from the geometry of the assembly that the diameter of the planet gear (P) is half the difference between the diameter of the annulus gear (A) and the sun gear (S), that is $d_P = \dfrac{d_A - d_S}{2}$, and since the gear pitch diameters are directly proportional to the number of teeth (see equation 10.26), we know that

$$t_P = \frac{t_A - t_S}{2} = \frac{180 - 80}{2} = 50.$$

Thus, as given in the table above, the required result for the gear ratio between the planet and the link arm is −2.6:1. The negative sign indicates that the angular velocity of the planet is in the opposite direction to that found for the sun. Since angular velocities determine the gear ratio, this would be expected.

10.3.5 Torque in gear trains

The torques and angular velocities at the input and output of a gear train are shown in Figure 10.11.

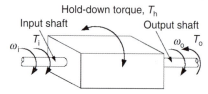

Figure 10.11 Gearbox torques and angular velocities

The gear train could take the form of a gearbox. The input torque will be in the same direction as the rotation of the input shaft, the torque at the output will act in the opposite direction to the rotation of the output shaft. This is because the torque created by the shaft driving the load at the output acts in opposition to the rotation of the drive shaft.

Under normal circumstances there will be a difference between the torque created at the input (T_i) and the torque created at the output (T_o). This difference produces a third torque, which creates a positive or negative turning moment that acts on the gear train casing. To counter this effect we need to ensure that

the gear train assembly is held down securely, hence we refer to this third torque as the *hold-down torque* (T_h). Under these circumstances, the gear train assembly is in equilibrium and so the sum of the torques is zero, or

$$T_i + T_h + T_o = 0 \qquad (10.28)$$

Also, if there are no losses through the gear train, then the input power equals the output power. We already know that power $P = T\omega$, and for our system the output torque acts in opposition to the input torque (in a negative sense), so we may write that $T_i\omega_i = -T_o\omega_o$.

The above assumes that no losses occur through the system; in practice there are always losses, due to bearing friction, viscous friction and gear drag. Now, since the efficiency of a machine is defined as:

$$\text{Efficiency} \, (\zeta) = \frac{\text{Power at output}}{\text{Power at input}}, \quad \text{then} \quad \zeta = \frac{T_o\omega_o}{T_i\omega_i} \qquad (10.29)$$

Key Point. Power from rotating shafts is the product of the shaft torque and angular velocity, that is $P = T\omega$

Example 10.6 A gearbox powered by a directly coupled electric motor is required to provide a minimum of 12 kW of power at the output, to drive a load at 125 rpm. The gearbox chosen consists of a compound train (Figure 10.12), which has an efficiency of 95%. Determine the:

i) gear ratio;

ii) electric motor output shaft velocity;

iii) power required from the electric motor;

iv) hold-down torque.

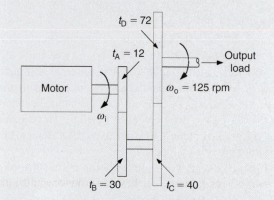

Figure 10.12 Motor gearbox assembly for Example 10.6

i) $G = \dfrac{t_B}{t_A} \times \dfrac{t_C}{t_B} \times \dfrac{t_D}{t_C} = \dfrac{t_D}{t_A}$ (inverse of teeth ratios). From the diagram, $G = \dfrac{30}{12} \times \dfrac{40}{30} \times \dfrac{72}{40} = \dfrac{72}{12} = \dfrac{6}{1} = 6 : 1$. ii) $\omega_i = 6\omega_o = (6)(125) = 750$ rpm. iii) Power required at input is given by the relationship: Efficiency $= \dfrac{\text{Power at output}}{\text{Power at input}}$, so $P_i = \dfrac{12}{0.95} = 12.63$ kW. iv) The hold-down torque may be found from equation 10.28, after finding the input and output torques. Then, since $T_i = \dfrac{P_i}{\omega_i}$ and $\omega_i = 78.54$ rad/s, we find that $T_i = \dfrac{12630}{78.54} = 160.8$ N m; similarly, at the output, $T_o = -\dfrac{12630}{13.09} = -916.7$ N m. Then, from equation 10.28, we get that $T_h = -T_o - T_i = 916.7 - 160.8 = 755.9$ N m.

From the results you will note that the torque required from the motor is six times smaller than that required at the output. This is one of the reasons for placing a gearbox between the motor and the load. We are then able to substantially reduce the size of the motor!

10.4 Balancing

As already mentioned in the introduction to this section, the effects of components being out of balance can be severe. You have already met the concept of *inertia* applied to rotational systems when you studied rotational kinetic energy in Chapter 8. Inertia forces and accelerations occur in all rotational machinery and such forces must be taken into account by the engineering designer. Here, we are concerned with the alleviation of inertia effects on rotating masses.

If a mass m is rigidly attached to a shaft which rotates with an angular velocity ω about a fixed axis and the centre of the mass is at a radius r from the axis, then we know from our previous work that the centrifugal force which acts on the shaft as a result of the rotating mass is given by the formula $F_c = m\omega^2 r$. When additional masses are added *which act in the same plane* (Figure 10.13), they also produce centrifugal forces on the shaft.

Then, for equilibrium *to balance* the shaft, we require that the sum of all the centrifugal forces $\sum m\omega^2 r = 0$.

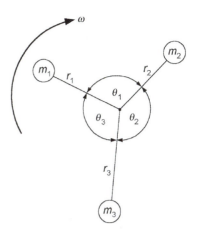

Figure 10.13 Fixed masses about centre of rotating shaft

Since all the masses are rigidly fixed, they all rotate with the same angular velocity, so ω may be omitted from the above expression to give $\sum mr = 0$. Now, to determine whether or not the shaft is in balance we can simply produce a vector polygon, where the vector product, mr, is drawn to represent each mass at its radius in the system.

Example 10.7 A rotating system consists of three masses rigidly fixed to a shaft, which all act in the same plane, as shown in Figure 10.14.

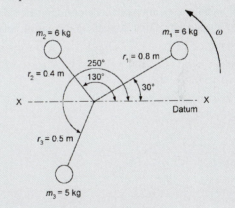

Figure 10.14 Rotating mass system

Determine the magnitude and direction of the mass required for balance, if it is to be set at a radius of 0.5 m from the centre of rotation of the shaft.

Using the information from Figure 10.14, we draw our vector polygon in which each component is mr, i.e. $m_1 r_1 = 4.8$ kg m, $m_2 r_2 = 2.4$ kg m and $m_3 r_3 = 2.5$ kg m. We can now read off the balancing couple from the vector diagram as 2.5 kg m, acting at an angle of 49°, as shown below.

So the required balancing mass acting at a radius of 0.5 m is $\dfrac{2.5 \text{ kg m}}{0.5 \text{ m}} = 5$ kg. Note that the common

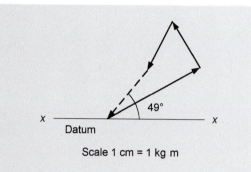

Scale 1 cm = 1 kg m

angular velocity, ω, was not needed to determine the out-of-balance mass.

If the masses rotate in *different planes*, then not only must the centrifugal forces be balanced but the moments of these forces about any plane of revolution must also be balanced. We are now considering rotating masses in three dimensions.

This system is illustrated in Figure 10.15, where masses now rotate about our shaft (AB), some distance (l) from a reference plane. The result of these masses rotating in *parallel* planes to the reference plane is that they create moment effects. Quite clearly, the magnitude of these moments will depend on the position chosen for the reference plane.

The technique we adopted for determining the balance state of the system is first, to transfer the out-of-balance

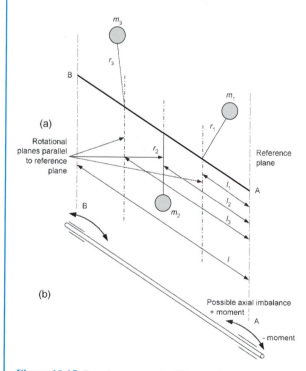

Figure 10.15 Rotating masses in different planes

couple to the reference plane by adding the individual moment couples, by vector addition, using a *moment polygon*. Each couple is given by the product $mr\omega^2 l$ (where l is the distance of each rotating mass from the reference plane). We then solve the *force polygon*, at the plane of reference, in the same way as in Example 10.7. Note that, as before, since all masses are rigidly attached to the shaft, they will all rotate with the *same angular velocity*, so this velocity may be ignored when constructing the moment polygon.

So we have a *two-stage* process which first involves the production of a *moment polygon* for the couples, from which we determine whether or not there is an out-of-balance moment. If there is, we will know the nature of this moment; i.e. its mass, radius of action, angular position and distance from the reference plane (mrl). The *force polygon* may then be constructed, which will include the mr, term from the moment polygon. We are then able to balance the system by placing the *balance* mass, derived from our polygons, at a position (l), angle (t) and radius (r) to achieve *dynamic balance* of the rotating system.

If the shaft is supported by bearings, radial balance across the bearing will be achieved by adopting the above method. However, axial forces may still exist at the bearing if this is not chosen as the reference plane for the unknown mass. This would result in longitudinal imbalance of the shaft (Figure 10.15b). There would then be a requirement for the addition of a *counter-balance mass*, to be placed in an appropriate position to one side of the shaft bearing. Thus there may be the need for two balance masses to ensure total dynamic balance throughout the system.

The above procedure should be readily understood by considering the following example.

Example 10.8   A rotating shaft is supported by bearings at A and B. It carries three disc cams which are represented by the equivalent concentrated masses shown in Figure 10.16a; their relative angular positions are also shown.

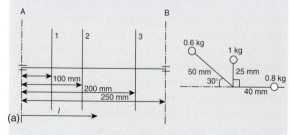

Figure 10.16a Rotating masses set-up

For the given situation, calculate the magnitude of the reaction at bearing A resulting from dynamic imbalance when the shaft rotates at 3000 rpm.

Using $F = mr\omega^2$ to determine the magnitude of the reaction force at A, we first need to establish the $m_A r_A$ value and then the $m_A r_A l_A$ moment value, to find force F_A.

To assist us we will set up a table, and by taking moments about 'B' we eliminate it from the calculation in the normal manner. Then, from the information given in Figure 10.16a, we find values as follows:

Plane	m(kg)	r(mm)	mr	l(m)	mrl
A	m_A	r_A	$m_A r_A$	0.25	$(m_A r_A)(0.25)$
1	1	25	25	0.2	5
2	0.8	40	32	0.1	3.2
3	0.6	50	30	0.05	1.5
B	m_B	r_B	$m_B r_B$	0	0

You should check that you arrive at the values shown in the above table, using the units indicated.

Now by constructing the *mr* diagram (using the values in the table), which we refer to as the *force vector* diagram (Figure 10.16b), we find that the magnitude $m_A r_A = 40$ and so the value for the moment $mrl = 10$. This may now be used to find the balancing moment, as shown in Figure 10.16c.

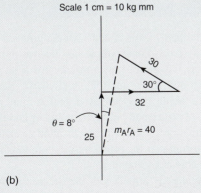

Scale 1 cm = 10 kg mm

(b)

Figure 10.16b Force vector diagram

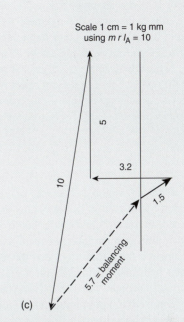

Scale 1 cm = 1 kg mm
using $m\,r\,l_A$ = 10

(c)

Figure 10.16c Moment vector diagram

Now the required reaction force at bearing A may be found since, from Figure 10.16c and the given length $l_A = 0.25$ m, we get that $m_A r_A = \dfrac{5.7 \,\text{kg(mm)(m)}}{0.25\,(\text{m})} = 22.8\,\text{kg mm}$, then

$$F_A = m_A r_A \omega^2 = (22.8 \times 10^{-3})\left(\frac{3000 \times 2\pi}{60}\right)2 = 2250\,\text{N}.$$

Note that reaction B may be found in a similar manner by taking moments about A.

10.5 Flywheels

A flywheel is an energy storage device. It absorbs mechanical energy by increasing its angular velocity and gives out energy by decreasing its angular velocity. The mechanical energy is given to the flywheel by a torque, which is often provided by a motor. It is accelerated to operating speed, thus increasing its inertial energy, which can be used to drive a load after the accelerating torque has been removed.

Key Point. A flywheel is an energy storage device that absorbs mechanical energy by increasing its angular velocity and gives out energy by decreasing its angular velocity

If a flywheel is given an input torque T_i, at angular velocity ω_i, this will cause the flywheel to increase speed. Then if a load or some other output acts on the flywheel, it will create an opposing torque T_o at angular velocity ω_o which will cause the flywheel to reduce speed. The difference between these torques will cause the flywheel to accelerate or retard, according to the relationship:

$$T_i(\omega_i) - T_o(\omega_o) = I\omega'$$

where, $\omega' = d\omega/dt$, which is the rate of change of angular velocity with respect to time or, as you know, the angular acceleration (α), and I, you should recognise as the second moment of inertia (see equation 8.16). We could have applied the torque from rest through some angle θ, then both angular displacement and angular velocity would become variables. Then the above relationship connecting input torque, output torque and acceleration would be written:

$$T_i(\theta_i, \omega_i) - T_o(\theta_o, \omega_o) = I\theta''$$

where $\theta'' = d^2\theta/dt$, which is the second derivative of angular displacement with respect to time, or again, acceleration (α). Now, assuming that the flywheel is mounted on a rigid shaft, then $\theta_i = \theta_o = \theta$ and so $\omega_i = \omega_o = \omega$. Then the above equation may now be written as:

$$T_i(\theta, \omega) - T_o(\theta, \omega) = I\theta'' = I\alpha \quad (10.30)$$

Key Point. Torque is equal to the product of the moment of inertia and the angular acceleration, that is $T = I\alpha$

Equation 10.30 can be solved by direct integration when the start values for angular displacement and angular velocity are known.

Now the *work input* to a flywheel $= T_i(\theta_1 - \theta_2)$, where the torque acts on the flywheel shaft through the angular displacement $= (\theta_1 - \theta_2)$, and similarly the *work output* from a flywheel $= T_o(\theta_3 - \theta_4)$, where again the torque acts on the shaft through some arbitrary angular displacement, say $(\theta_3 - \theta_4)$.

We can write these relationships in terms of *kinetic energy* (see equation 8.27). At $\theta = \theta_1$, the flywheel has a velocity ω_1 rad/s, and so its kinetic energy is: $KE_1 = \frac{1}{2}I\omega_1^2$. Similarly at θ_2, with velocity ω_2, $KE_2 = \frac{1}{2}I\omega_2^2$.

So the change in the kinetic energy of the flywheel is given by:

$$KE_2 - KE_1 = \frac{1}{2}I\left(\omega_2^2 - \omega_1^2\right) \quad (10.31)$$

Example 10.9 A steel cylindrical flywheel 500 mm in diameter has a width of 100 mm and is free to rotate about its polar axis. It is uniformly accelerated from rest and takes 15 s to reach an angular velocity of 800 rpm. Acting on the flywheel is a constant frictional torque of 1.5 N m. Taking the density of the steel to be 7800 kg/m³, determine the torque that must be applied to the flywheel to produce the motion.

To solve this problem, we will need not only to use equation 10.30, but also to refer back to Chapter 8 for the work we did on angular motion. Thus, to find the angular acceleration, we may use equation 8.14 written as $\omega_f = \omega_i + \alpha t$, so that $83.8 = 0 + 15\alpha$ (where the initial angular velocity is zero and 800 rpm = 83.8 rad/s), then $\alpha = 5.58$ rad/s². Now the mass moment of inertia I for a cylindrical disc is given by $I = \frac{1}{2}Mr^2$ (see Figure 8.7), so in order to find I, we first need to find the mass of the flywheel, that is $M = (0.25\pi)(0.5^2)(0.1)(7800) = 153.15$ kg. Then $I = (0.5)(153.15)(0.25^2) = 4.79$ kg m². Now, the *net* accelerating torque is found from equation 10.30, where $T_N = T_i - T_o = I\alpha$, so that $T_N = (T_i - 1.5) = (4.79)(5.58)$ and the accelerating torque $T_i = 28.23$ N m.

A further example of an integrated system involving the use of a flywheel is given in the next section, where we deal with coupled systems.

10.6 Coupled systems

In the final part of this chapter we are going to look at *coupled systems*. For example, an electric motor may be used as the initial power source for a lathe, with the rotary and linear motion required for machining operations being provided by a gearbox and selector mechanism positioned between the motor and the work piece. The motor vehicle provides another example of where the engine is the prime mover, coupled through a clutch to a gearbox and drive shafts.

We are already armed with the necessary underpinning principles from our previous work on belt drives, clutches, gear trains, rotating masses and flywheels. We may, however, also need to draw on our basic knowledge of dynamics from Chapter 8, particularly with respect to angular motion.

We will study this mainly through examples that are intended to draw together and consolidate our knowledge of power transmission systems. Let us first consider the inertia characteristics of a rotational system, in this case a gearbox, remembering our work on free-body diagrams from our study of shear force and bending moments!

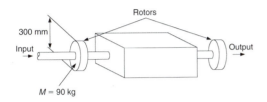

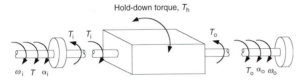

Figure 10.17 Gearbox assembly with rotors

Figure 10.17 shows a gearbox assembly with rotors attached to its input and output shaft. Let us assume that the gearbox ratio is a step-down; i.e. the input velocity is higher than the output velocity, and these are related to the overall gear ratio G by equation 10.27.

Now each of the rotors will have a mass moment of inertia which depends on their radius of gyration. For the purpose of our analysis we will assume the rotors are cylindrical. Then, if we draw the free-body diagram for the system *with respect to the input* (as shown in Figure 10.17), we are able to consider, with respect to any component in the system, the resulting torques, velocities and accelerations. We will use these ideas in the following example.

Example 10.10 A gearbox has rotors attached to its input and output as shown in Figure 10.17, where for the output rotor $I_o = 25\,\text{kg m}^2$. The input rotor is cylindrical, having a mass of 90 kg and a diameter of 300 mm. The angular velocity of the rotor on the input shaft is higher than that of the rotor at the output. If a torque of 10 N m is applied to the high-speed input shaft, calculate the value of the angular acceleration of the input shaft when $G = 5$ and the gearbox has an efficiency of 95%.

In order to solve this problem we first need to formulate the relationships, which enable us to determine the appropriate torques and accelerations for the system, from the point of view of the *input*.

The free-body diagram is shown as part of Figure 10.17 where the isolated system elements (when the torque is applied to the input rotor shaft) are shown for the input rotor and drive shaft, the gearbox assembly, and the output rotor and drive shaft. For the input rotor the torque, T, is opposed by the torque produced at the gearbox side of the shaft, so the equation of motion for I_i is given as:

$$T - T_i = I_i \alpha_i. \tag{a}$$

Compare equation (a) with equation 10.30. For the gearbox, the input and output torques are directly related to the gearbox ratio and gearbox efficiency by:

$$T_o = T_i G \zeta. \tag{b}$$

The relationship shown in equation (b) is true for a reduction gearbox, if you remember that one of the primary functions of a gearbox is to provide an increase in power at the output, so if the velocity at the output is reduced, as in this case, the torque is raised (see equation 10.29).

Now, at the output rotor, from the free-body diagram shown as part of Figure 10.17, we get the relationship:

$$T_o = I_o \alpha_o \tag{c}$$

and

$$G \alpha_o = \alpha_i. \tag{d}$$

Combining equations (b) and (c) gives $T_i G \zeta = I_o \alpha_o$, and substituting this combination for T_i into equation (a) gives $T - \dfrac{I_o \alpha_o}{G \zeta} = I_i \alpha_i$. Now, on substituting for α_o from equation (d) we get $T - \dfrac{I_o \alpha_i}{G^2 \zeta} = I_i \alpha_i$ which finally, after rearrangement, gives:

$$T = \alpha_i \left[I_i - \frac{I_o}{G^2 \zeta} \right] \tag{10.32}$$

So this equation tells us that the torque for the system (when for a reduction gearbox the torque is assumed to act at the high-velocity input side) is equal to the acceleration caused by this torque multiplied by the *equivalent inertia* (I_e) of the system, *which is the term in the square brackets in equation 10.32*.

As you can see, most of this example has been concerned with formulating the required relationships. The above technique can be used to develop system equations for many similar situations.

The required acceleration at the input can be found by direct substitution of the variables into equation 10.32, once we have determined (I_i). Then, mass moment of inertia for our cylindrical rotor is given by $I = \frac{1}{2}Mr^2$ where, in our case, $I_i = (0.5)(90)(0.15^2) = 1.01 \text{ kg m}^2$, then, from equation 10.32, the acceleration of the input shaft is $\alpha_i = \dfrac{10}{1.01 + 25/23.75} = 4.8 \text{ rad/s}^2$.

We will now consider another fairly simple system which consists of a shoe brake being used to retard a flywheel. The flywheel may be thought of as an annulus, where the internal thickness is considered, as a first approximation, to be negligible (Figure 10.18).

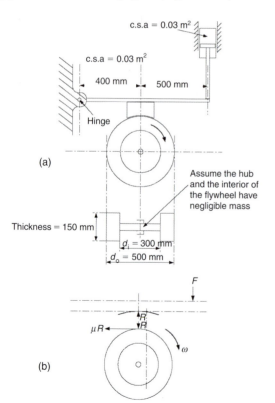

(a)

(b)

Figure 10.18 Flywheel and shoe brake system

Example 10.11   The steel flywheel shown in Figure 10.18 has an external diameter of 500 mm, an internal diameter of 300 mm, and the annulus formed by these diameters is 150 mm thick. The flywheel is retarded by means of a lever brake. Just prior to the application of the brake the flywheel is rotating with an angular velocity of 3500 rpm and then the brake is applied for 50 s. A piston assembly is used to activate the brake assembly, acting on the end of the lever. If the pressure applied by the piston is 3.5 kPa and the piston area is 0.03 m³, find:

i)   the angular velocity of the flywheel at the end of the braking period;

ii)   the power released from the flywheel during the braking period.

Take the flywheel density as 7800 kg/m³ and the coefficient of friction between the brake and the flywheel as 0.3.

Figure 10.18a shows the situation diagram with all relevant dimensions. Figure 10.18b shows the free-body diagram illustrating the braking reaction.

i)   The solution to this part of the problem relates to the equation $T = I\alpha$, which is a simplified form of equation 10.30, where, in our case, $T = T_B$ = torque due to braking. We need to find this and the mass moment of inertia of the flywheel. To find the braking torque we first determine the braking force; this is:

$$F = P \times A = (3500)(0.03) = 105 \text{ N}.$$

Now considering the equilibrium of the brake lever, the reaction force R is found by taking moments about the hinge (see free-body diagram), so $R \times 400 = (105)(500) = 131.25 \text{ N}$. Then, assuming the friction is limiting friction and using the fact that braking force $F = \mu R$, the braking torque $T_B = (\mu R)r$, where r is the radius of the flywheel. So $T_B = (0.3)(131.25)(0.25) = 9.84 \text{ N m}$.

Now the mass moment of inertia of a disc is given by $I = \frac{1}{2}Mr^2$, and logic suggests that I for an annulus will have the form $I = \frac{1}{2}M\left(R^2 - r^2\right)$, which it does. However, it is normally expressed as:

$$I = \frac{\pi \left(D_o^4 - d_i^4\right) tp}{32}$$

where t = the thickness of the flywheel and ρ = density of flywheel material, thus $tp = M$, and this form of the equation is often used as a first approximation for finding the moment of inertia, where the hub and inner disc of a flywheel are considered insignificant when compared with the bulk of the rim.

Now, $I_F = \dfrac{\pi\left(0.5^4 - 0.3^4\right)(0.15)(7800)}{32} =$ 6.25 kg m^2 and we find the angular retardation from $T_B = I\alpha$, where $\alpha = \dfrac{9.84\,\text{N m}}{6.25\,\text{kg m}^2} =$ 1.574 rad/s^2. Then the final angular velocity may be found using the equation $\omega_f = \omega_i + \alpha t$, where $\omega_f = 366.5 + (-1.574)(50) = 288$ rad/s or 2750 rpm.

ii) To find the power given out by the flywheel during braking, we first find the kinetic energy change. Then, using equation 10.31, where $KE_2 - KE_1 = \frac{1}{2}I\left(\omega_2^2 - \omega_1^2\right)$, the energy change $= (0.5)(6.25)\left(366.5^2 - 288^2\right) = 160.59$ J and so the power $P = J/s = 160.59/50 = 3.21$ kW.

10.6.1 Crank pressing machine

We now discuss a system for the operation of a *crank pressing machine*, which consists of a prime mover, gearbox and flywheel, to drive the press crank.

Let us first consider the selection of a power unit for a punch pressing machine. What information do we require to help with our selection? Well, at the very least, we would need information about the service load, the frequency of the pressing operation, the torque requirements and rotational velocities of the system.

Figure 10.19 shows the torque requirements for a typical pressing operation. Note that the torque requirement fluctuates, very high torque being required for the actual pressing operation, which is of course what we would expect.

We first determine the torque versus angular displacement graph (Figure 10.19). Let us assume that the press operates at 150 rpm and the pressing operation is completed in 0.3 s and takes place 30 times a minute. This is not an unreasonable assumption; many punch presses operate at much higher rates than this, though obviously this depends on the intricacy and force requirements of the product. Assume that the resisting torque during the pressing operation is 1800 N m.

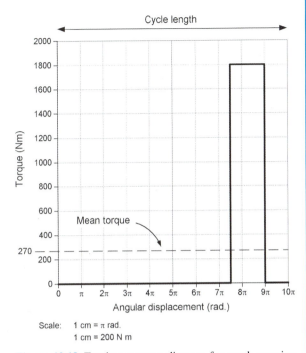

Figure 10.19 Turning moment diagram for crank pressing machine

On the basis of the above information we can deduce the angular displacement of the crank for one complete cycle of events, as shown in Figure 10.19. The punching operation takes place at 30 times a minute; therefore, the punch press crank will rotate 5 revolutions for each operation. So the angular displacement for one complete operation is $5 \times 2\pi = 10\pi$ rad.

The angular displacement of the crankshaft during the actual pressing operation is easily determined by comparing the time for one complete cycle of events (2 s) with the time for the actual pressing operation (0.3 s). So the angular displacement for the pressing operation $= (0.3 \times 10\pi/2) = 15\pi$ rad. We know that the high resisting torque of 1800 N m acts for 15% of the time required for one complete cycle of events. This torque would need to be provided by the prime mover if no flywheel was included in the system. So, what power would we require from our prime mover under these circumstances?

A suitable prime mover would be an electric motor and, for the task described above, this is likely to be three-phase AC. This type of motor is normally produced to run at synchronous speeds that depend on the number of pole pairs of the machine, 600, 900 and 1200 rpm, etc. We will choose a motor that runs at 1200 rpm, which will require *a reduction gearbox* of eight to one. What is the power required from our motor to provide the maximum torque, assuming that this peak torque needs to be *continuously* available from the motor?

Without losses we know that the power provided by the motor is equal to the power at the crankshaft to drive the press. We already know that the peak torque we require at the crank is 1800 N m. Then, knowing that:

$$T_i \omega_i = T_o \omega_o,$$

we find that

$$T_i = \frac{T_o \omega_o}{\omega_i} = \frac{(1800)(60)}{1200} = 90\,\text{N m}.$$

Note that in this case we may leave the angular velocities in rpm because here they are a ratio.

Now, if the motor has the capacity to deliver this torque *continuously*, then the *work* required from the motor in one complete cycle of events is $10\pi \times 90 = 900\pi$ N m. Power is the rate of doing work, so in 1 s the motor shaft rotates 20 revolutions, and the power required from our motor is $20 \times 900\pi = 56.5\,\text{kW}$.

It is obviously wasteful to provide such a large motor when the maximum torque requirements are only needed for a small fraction of the time. This is why our system *needs* the addition of a *flywheel*.

Let us assume that we require a flywheel to be fitted to our system, which keeps the press crank velocity between 140 and 160 rpm during pressing operations. In order to select such a flywheel we need to know the energy requirements of the system, and the total inertia that needs to be contributed by the flywheel during these pressing operations.

From our discussion so far I hope you recognise that the area under our torque–crank angle graph in Figure 10.19 represents the *energy requirements* for the press.

We assume as before that the energy in is equal to the energy out, and that there are no losses due to friction.

The total energy required during the pressing operation is given from our previous calculations as:

$$(1800\,\text{N m})(1.5\pi) = 8482$$

J (area under graph for pressing operation).

This energy is required to be spread over the complete operating cycle, i.e. spread over 10π rad, so the uniform torque required from the motor is given by $1800 \times 1.5\pi = T_m \times 10\pi$, where T_m = the uniform, or mean, motor torque, where, in this case, the mean torque $= 270$ N m (as shown on Figure 10.19).

Then the change in energy is $\Delta E = (T_{max} - T_{mean}) \times$ angular distance $= (1800 - 270)\,1.5\pi = 7210\,\text{J}$, therefore, the flywheel must supply 7210 J and therefore, the *motor* must supply $8482 - 7210 = 1272$ J. Note the drastic reduction in the energy required by the motor.

We are now in a position to determine the inertia required by the flywheel. Using equation 10.30, we have $7210 = \frac{1}{2}I\left(\omega_{max}^2 - \omega_{min}^2\right)$, where $\omega_{max} = 160\,\text{rpm} = 16.76\,\text{rad/s}$ and $\omega_{min} = 140\,\text{rpm} = 14.66\,\text{rad/s}$. Then the inertia of the flywheel is given by $I = \frac{2 \times 7210}{(280.9 - 214.9)} = 218.5\,\text{kg m}^2$.

This is where we end our discussion of this system. We could go on to make a specific motor selection, determine the requirements of a pulley and belt drive system and also determine the dimensional requirements of the flywheel, if we were so inclined.

Key Point. The turning moment diagram for the crank pressing machine clearly demonstrates how a flywheel may be used as an energy storage device, reducing the overall size and power requirements needed from the prime mover during pressing operations

We leave our discussion on the coupling of transmission and other systems, with one final example.

Example 10.12 An electric motor drives a cylindrical rotational load through a clutch (Figure 10.20). The armature of the motor has an external diameter of 400 mm and a mass of 60 kg, it may be treated as being cylindrical in shape. The rotor has a diameter of 500 mm and a mass of 480 kg. The motor armature revolves at 1200 rpm and the rotor at 450 rpm, both in the same direction. Then, assuming the moment of inertia of the driving clutch plate is 0.25 kg m² and that, after engagement, the moment of inertia of the whole clutch is 0.5 kg m², find the loss of kinetic energy when the clutch is engaged, ignoring all other inertia effects.

This problem requires us to find the angular momentum change as a result of the impact. Angular momentum is given by the product of moment of inertia and the angular velocity. Knowing this, we are able to find the angular velocity after impact, and then the loss of kinetic energy. The mass moment of inertia for a cylinder is determined simply from

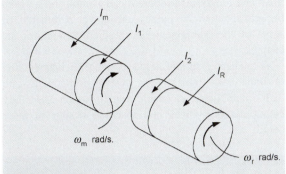

(a) Before impact

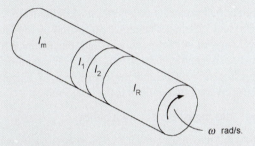

(b) After impact

Figure 10.20 Motor clutch assembly before and after impact

$I = \frac{1}{2}Mr^2$ in the normal way. Then, the moment of inertia of motor armature and driving clutch assembly, where I_1 = that of driving clutch assembly, is $= I_m + I_1 = (0.5)(60)(0.22) + 0.25 = 1.45 \text{ kg m}^2$. Moment of inertia of motor and clutch assembly, where I_2 = that of the driven clutch assembly, is

$$= I_R + I_2 = (0.5)(480)(0.25^2) = 15.5 \text{ kg m}^2.$$

Also, the angular velocity of the motor armature and driving clutch plate (same shaft) $= 125.66$ rad/s and the angular velocity of the rotor end driven clutch assembly (same shaft) $= 47.12$ rad/s.

Now, from the conservation of angular momentum we know that *the total angular moment before impact equals the total angular moment after impact*:

Therefore,

$$(I_m + I_1)\,\omega_m + (I_R + I_2)\,\omega_R = (I_m + I_1 + I_R + I_2)\,\omega_1,$$

where ω_f = final velocity after impact. So $(1.45)(125.66) + (15.5)(47.12) = (1.45 + 15.5)\omega_f$ and $\omega_f = 53.84$ rad/s.

Now to find the loss of kinetic energy at impact, we know that:

Loss of KE = Total KE before impact

$\qquad$ − total KE after impact

$\qquad = (KE \text{ of } I_m + I_1) + (KE \text{ of } I_R + I_2)$

$\qquad\quad - KE \text{ of } (I_m + I_1 + I_R + I_2)$

$\qquad = \frac{1}{2}(1.45)(125.66)^2 + \frac{1}{2}(15.5)(47.2)^2$

$\qquad\quad - \frac{1}{2}((16.5)(53.84)^2 = 4.799 \text{ kJ}.$

10.7 Chapter summary

This chapter has been concerned with power transmission systems, starting with direct belt drives that used both flat and V-belt methods for transmitting torque from a prime mover to that being driven. Next, the use and nature of clutches were introduced, in particular flat plate and cone-type friction clutches, where the prime mover is allowed to run continually at constant rpm and, upon engagement of the clutch, the prime mover drives the particular load as required. The use and nature of gears within drive systems was then introduced and the importance of gearboxes as power amplifiers was considered. Dynamic balance and balancing were then emphasised and one or two simple methods for calculating out-of-balance moments and restoring masses were introduced. We then considered flywheels, both separately and as used within coupled systems, in that they were used to store and release power for particular operations. Finally, coupled systems were studied, where a number of transmission components were considered collectively. In the final part of this section a crank pressing machine was considered in some detail.

Belt drives, as mentioned in the chapter, are used to transmit power from one drive shaft to another; the driven shaft may be some distance away from the driver (power source). Belts also have the advantage of absorbing shock loads and vibration felt within the system. Our analysis focused on flat-belt drives, where the transmitted power was found using equation 10.2, which was modified (equations 10.7 and 10.8) to take account of the additional centrifugal tension under high and low speed conditions. The belt drive design requirements were detailed in section 10.1.2 and, in particular, simple design equations were formulated for parameters such as lap angle, belt pitch and pulley system span, which are all important when considering a particular belt drive system.

In section 10.2, we first considered disc clutches. In particular, the simple single-disc mechanically operated clutch and the multidisc hydraulic clutch were looked at and a few useful design equations were established (equations 10.13 to 10.20), all of which were based on two important assumptions, which are repeated here for your convenience. These were:

(i) we assume uniform distribution of pressure at the disc interface;

(ii) we assume a uniform rate of wear at the disc interface.

We then considered cone clutches, where we established that, when compared to their disc clutch counterparts, they had the advantage of simplicity and the wedging action due to their geometry that enabled the clamping force to be reduced. Again, for this type of clutch some useful general design equations were established, in particular equations 10.22 to 10.25.

In section 10.3 we first considered gear wheels and then gear trains, where we discussed the geometry of spur gears and established some useful relationships for gear ratios of simple gear trains that took into consideration their angular velocity ratios, teeth ratios and diameter ratios as detailed in equations 10.26 and 10.27. Compound and epicyclic gears were then considered, where in particular we looked at one or two methods for determining gear ratios and inputs/outputs for these complex gear systems. In section 10.3.5 we looked at methods to establish the torque, power and efficiency of gear trains and saw how, in particular, the use of a gearbox enabled the power requirements from the prime mover (usually an electric motor) to be drastically reduced in order to power the output load.

In section 10.4 we looked at the need to ensure that transmission systems were kept 'in balance', using the concept of rotating masses to represent the inertia of the rotary components within the system. Examples 10.7 and 10.8 show the methods used to determine the necessary balance masses and reactions at important parts of the system, such as bearings.

In section 10.5 we introduced the concept of the flywheel as an energy-storing device. We looked at the torque, work and energy inputs and outputs from these devices and formulated the useful relationships given in equations 10.30 and 10.31. We then extended the use of flywheels and gearboxes into coupled systems, where we considered a motor/flywheel/brake combination, and

finally we looked at the operation of a crank press, which brought together the energy and power parameters of some of the components we studied throughout the chapter.

This chapter should serve as a useful introduction to some basic transmission system components and their interaction and enable you to go on and study engineering machines and machine design in more detail.

10.8 Review questions

1. A flat belt connects two pulleys with diameters of 200 and 120 mm. The larger pulley has an angular velocity of 400 rpm. Given that the distance C between centres is 400 mm, the belt maximum tension is 1200 N and the coefficient of friction between belt and pulleys is 0.3, Determine:

    a) the angle of lap of each pulley;

    b) the maximum power transmitted.

2. A single V-belt with $\phi = 18°$ and a unit weight of 0.25 kg/m is to be used to transmit 14 kW of power from a 180 mm diameter drive pulley rotating at 2000 rpm to a driving pulley rotating at 1500 rpm. The distance between the centres of the pulleys is 300 mm.

    a) If the coefficient of friction is 0.2 and the initial belt tension is sufficient to prevent slippage, determine the values of T_1 and T_2.

    b) Determine the maximum stress in the belt if its density is 1600 kg/m³. Note that the centripetal tension created by the centrifugal force acting on the belt is given by $T_c = m'v^2$, where $m' =$ the mass per unit length of the belt.

3. A multiplate clutch consists of five discs, each with an internal and external diameter of 120 and 200 mm, respectively. The axial clamping force exerted on the discs is 1.4 kN. If the coefficient of friction between the rubbing surfaces is 0.25, and uniform wear conditions may be assumed, determine the torque that can be transmitted by the clutch.

4. A motorised winch is shown in Figure 10.21. The motor supplies a torque of 10 N m at 1400 rpm. The winch drum rotates at 70 rpm and has a diameter of 220 mm. The overall efficiency of the winch is 80%.

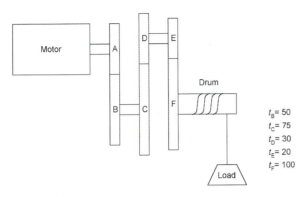

$t_B = 50$
$t_C = 75$
$t_D = 30$
$t_E = 20$
$t_F = 100$

Figure 10.21 Figure for question 10.7.4 – motorised winch

Determine the:

a)   number of teeth on gear A;

b)   power input to the winch;

c)   power output at the drum;

d)   load being raised.

5.   A flat belt 150 mm wide and 8 mm thick transmits 11.2 kW. The belt pulleys are parallel, and their axes are horizontally spaced 2 m apart. The driving pulley has a diameter of 150 mm and rotates at 1800 rev/min, turning so that the slack side of the belt is on top. The driven pulley has a diameter of 450 mm. If the density of the belt material is 1800 kg/m³, determine:

a)   the tension in the tight and slack sides of the belt if the coefficient of friction between the belt and the pulleys is 0.3;

b)   the length of the belt.

Chapter 11

Oscillatory motion and vibration

In this chapter we take a look at oscillatory motion, in particular harmonic motion, and see how this motion is applied to linear and transverse systems. We also consider the nature of vibration within these systems, both free and forced vibration, together with the effects of damping and resonance, which are all inextricably linked with oscillatory motion systems.

11.1 Simple harmonic motion

When a body oscillates backwards and forwards so that every part of its motion recurs regularly, we say that it has *periodic motion*. For example, a piston attached to a connecting rod and crankshaft, moving up and down inside the cylinder, has periodic motion, as you may have noticed when studying this system as a mechanism in section 9.2.

If we study the motion of the piston P carefully (Figure 11.1), we note that when the piston moves towards A its velocity *v* is from right to left. At A it comes instantaneously to rest and reverses direction. So before reaching A the piston must slow down, in other words the acceleration, *a*, must act in the opposite direction to the velocity. This is also true (in the opposite sense) when the piston reverses its direction and reaches B at the other end of its stroke. At the times in between these two extremities the piston is being accelerated in the same direction as its velocity.

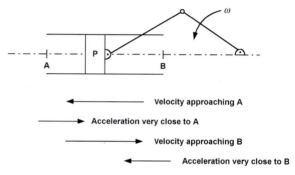

Figure 11.1 Reciprocal motion of slider-crank piston assembly

Neither the velocity nor the acceleration acting on the piston is uniform, and so they *cannot* be modelled using the methods for uniform acceleration. However, all is not lost, because although the periodic movement of the piston is complicated, its motion may be modelled using *simple harmonic motion* (SHM).

We may define SHM as a special periodic motion in which:

(i) the acceleration of the body is always directed towards a fixed point in its path;

(ii) The magnitude of the acceleration of the body is proportional to its distance from the fixed point.

Thus the motion is similar to the motion of the piston described, except that the *acceleration* has been defined in a particular way.

978-1-85617-775-7, Engineering Science, Mike Tooley & Lloyd Dingle

11.1.1 Analysis of SHM

Much of the following analysis of SHM should be familiar to you, having previously studied mechanisms in Chapter 9 and manipulated trigonometric identities in your previous study!

Consider the point P moving round in a circle of radius r with angular velocity ω (Figure 11.2a). We know from our study of the transformation equations of motion that its tangential speed $v = \omega r$. Then, as P rotates

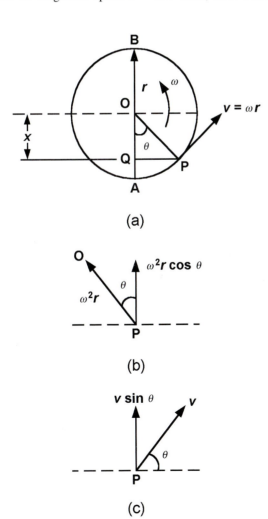

(a)

(b)

(c)

Figure 11.2 Analysis of simple harmonic motion

around the circumference of the circle, point Q will move vertically upwards as the point P rotates through one half-cycle from A to B. Then, immediately after P reaches position B, point Q will reverse direction for the other half cycle, travelling vertically down between A and B. Now, imagine that after leaving position A, the motion is frozen at a moment in time t, when the point is at position P as shown in Figure 11.2a. Then we can show that this point Q moves with SHM about the centre O.

The point Q moves as the point P moves, therefore the *acceleration of Q* is the component of the acceleration of P parallel to AB. Now the centripetal acceleration of P along PO is given by $\omega^2 r$. Then, using trigonometry and considering our arrangement (Figure 11.2b), the component of this acceleration parallel to AB is:

$$a = -\omega^2 r \cos\theta \qquad (11.1)$$

The negative sign results from our definition of SHM given above, where we stated that the acceleration is always directed *towards* a fixed point, in our case towards the point O. This is represented mathematically by use of the negative sign. Now we can also deduce (from Figure 11.2a) that $x = r\cos\theta$. Therefore, substituting this expression into equation 11.1 gives:

$$a = -\omega^2 x \qquad (11.2)$$

It is important to note how the acceleration of Q varies with different values of x. For example, the acceleration of Q will be zero when x is at the fixed point of rotation O. Also, the *acceleration* will be at a *maximum* when x is at the *limits of the oscillation*, i.e. at points A and B.

The time required for the point Q to make one complete oscillation from A to B and back is known as the *time period T*. The time period can be determined using $T = \dfrac{\text{Circumference of described circle}}{\text{Speed of Q}}$. Then:

$$T = \frac{2\pi r}{v} = \frac{2\pi}{\omega} \qquad (11.3)$$

(where, from our transformation equation, $v = \omega r$).

Also, note that the *frequency f* is the number of complete oscillations (cycles), back and forth, made in unit time.

The frequency is therefore the reciprocal of the time period T. The unit of frequency is the hertz (Hz), which is *one cycle per second*. Thus:

$$f = \frac{1}{T} \text{ cycles per second} = \frac{\omega}{2\pi} \text{ Hz} \qquad (11.4)$$

Key Point. The *frequency* (f) in Hertz (Hz), or cycles per second, is the *reciprocal* of the *time period*

Now the *velocity of Q* is the component of P's velocity parallel to AB, which from our arrangement (Figures 11.2a and 11.2c) is OQ = x. *Using the prime notation for the differential of the displacement x (see section 9.2.2)*, then the velocity of Q = $x' = -v \sin \theta$ or, since $v = \omega r$ then $x' = -\omega r \sin \theta$. Now, the variation of the velocity of Q, i.e. x', assuming we start at time zero and since $\theta = \omega t$, may be written as:

$$v = x' = -\omega r \sin \omega t \qquad (11.5)$$

Key Point. The notation for the differentiation process may take several forms, thus $x' = \dot{x} = \dfrac{dx}{dt}$ and $x'' = \ddot{x} = \dfrac{d^2 x}{dt^2}$ etc. are all equivalent

Note that $\sin \theta = \sin \omega t$ may be positive or negative, dependent on the value of $\theta = \omega t$. A negative sign is added to the above equations so that, by convention, when the velocity acts upwards it is negative and when acting downwards it is positive.

Equation 11.5 may be written in terms of angular displacement θ (as you have already seen) as $x' = -\omega r \sin \theta$ and, remembering that $\sin^2 \theta + \cos^2 \theta = 1$, then $x' = \pm \omega r \sqrt{1 - \cos^2 \theta}$, so that $x' = \pm \omega r \sqrt{1 - (x/r)^2}$. On multiplication by r, we get that:

$$v = x' = \pm \omega \sqrt{r^2 - x^2} \qquad (11.6)$$

In order to complete our analysis we need to consider the displacement of Q, i.e. the displacement of x, which can be seen from Figure 11.2a to be $x = AO \cos \theta$ or $x = r \cos \theta$ and, for the instant of time t being considered, may be written as:

$$x = r \cos \omega t \qquad (11.7)$$

The displacement, like the velocity and acceleration of Q or x, is sinusoidal (Figure 11.3).

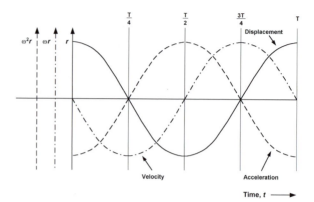

Figure 11.3 Displacement, velocity and acceleration of a point subject to SHM

Note from Figure 11.3 that when the velocity is zero, the acceleration is a maximum, and that the time t may be taken as zero anywhere in the cycle; *in this case* the time is zero when OP (the rotating point with radius r) is at A, i.e. when, by convention for the measurement of angles, $\theta = 270° = \dfrac{3\pi}{2}$ rad.

We could have approached the above analysis in reverse, i.e. having found the *displacement* $x = r \cos \omega t$ from the geometry of the situation illustrated in Figure 11.2a, we could then use the differential calculus to find the *velocity* and *acceleration*. Thus, after differentiating once, $v = x' = -\omega r \sin \omega t$ and after differentiating once more, $a = x'' = -\omega^2 r \cos \theta$. Equations 11.2 and 11.6 would then follow after we had carried out the same algebraic and trigonometric manipulation as before.

Example 11.1 A piston performs reciprocal motion in a straight line which can be modelled as SHM. Its velocity is 15 m/s when the displacement is 80 mm from the mid-position, and 3 m/s when the displacement is 160 mm from the mid-position. Determine:

i)   the frequency and amplitude of the motion;

ii) the acceleration when the displacement is 120 mm from the mid-position.

We first need to determine the distance the piston moves on either side of the fixed point, in our case, the mid-point. This is known as the *amplitude*. We use equation 11.6, i.e. $v = \omega \sqrt{r^2 - x^2}$, and substitute the two sets of values for the velocity and displacement given in the question, then solve for r, using simultaneous equations.

From given data, we have $x = 0.08$ m when $v = 15$ m/s and $x = 0.16$ m when $v = 3$ m/s, then we get that:

$$15 = \omega\sqrt{r^2 - 0.0064} \qquad (1)$$

and

$$3 = \omega\sqrt{r^2 - 0.0256} \qquad (2)$$

Dividing equation 1 by equation 2 eliminates ω and squaring both sides gives $25 = \dfrac{r^2 - 0.0064}{r^2 - 0.0256}$, and so $r = 162.5$ mm.

In order to determine the frequency, we first need to find the angular velocity from equation 1, then $15 = \omega\sqrt{0.0264 - 0.0064}$, so $\omega = 106$ rad/s and, using equation 11.4, $f = \dfrac{\omega}{2\pi} = \dfrac{106}{2\pi} = 16.9$ Hz.

To find the acceleration when $x = 0.12$ m we use equation 11.2, then $a = -\omega^2 x = -(106)^2(0.12)$, i.e. $a = 1348.3$ m/s^2 towards the centre of rotation.

11.2   Free vibration

All systems that possess mass and elasticity are capable of *free vibration*, i.e. vibration that takes place in the absence of the application of an external force. The primary interest in studying such systems is to establish their *natural frequency* (ω_n rad/s or $f = \dfrac{\omega_n}{2\pi}$ Hz), and in this section we use a simple spring–mass system as our model in order to establish equations that enable us to find the *natural frequency* of the system. In sections 11.3 and 11.4 we go on to consider what happens when systems are subject to *damped* and *forced vibration* and how natural frequencies of systems are affected by external forces, damping and resonance.

> **Key Point.** Free vibration takes place without the application of an applied force, where the body/system vibrates at its natural frequency

11.2.1   Modelling the motion of a spring–mass system

You will be aware that for a spring which obeys Hooke's law, the extension of the spring is directly proportional to the force applied to it. So it follows that the extension of the spring is directly proportional to the tension in the spring resulting from the applied force.

Consider the spring with the mass (m) attached to it hanging freely from a support as shown in Figure 11.4.

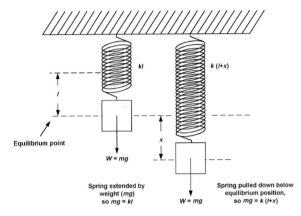

Figure 11.4 Reciprocal motion of spring–mass system

The mass exerts a downward tension mg on the spring, which stretches the spring by an amount l until the internal resistance force of the spring (kl) balances the tension force created by the mass. Then:

$$kl = mg \qquad (11.8)$$

where k is the *spring constant* or force required to stretch the spring by unit length and is measured in N/m. The spring constant is an inherent property of the spring and depends on its material properties and dimensions; $l =$ the length (in metres, m) the spring stretches in reaching equilibrium, $m =$ the mass in kg and $g =$ the acceleration due to gravity (m/s^2). Note that both sides of equation 11.8 have units of force.

> **Key Point.** $k =$ the spring constant; it has units of Newton per metre (N/m)

Suppose the mass is now pulled down a further distance x below its equilibrium position. The stretching tension acting downwards is now $k(l + x)$, which is also the tension in the spring acting upwards. Then we have that $mg = k(l + x)$, so the resultant restoring force acting upwards on the mass will be $F_R = k(l + x) - mg$ and, as from equation 11.8 $kl = mg$, then $F_R = kl + kx - kl$ or $F_R = kx$. When we release the mass it will oscillate up and down. If it has acceleration a at extension x then, by Newton's second law:

$$ma = -kx \qquad (11.9)$$

the negative sign resulting from the assumption that at the instant shown, the acceleration acts upwards or in the opposite direction to the displacement (x), which acts in the downward (positive) direction. Note also that, using our prime notation for differentiation, we may write equation 11.9 as $mx'' = -kx$.

Now, transposing equation 11.9 for a gives $a = -\dfrac{kx}{m}$. Also, from equation 11.2, $a = -\omega^2 x$ (where here we are dealing with the natural frequency ω_n), so by comparing these two equations we note that $k/m = \omega_n^2$ (where $\omega_n = $ *natural frequency of vibration* in rad/s) and, since k and m are fixed for any system, k/m is always positive.

> **Key Point.** The symbol used for the natural frequency in rad/s of free vibration is ω_n

Thus the motion of the spring is simple harmonic about the equilibrium point, provided the spring system obeys Hooke's law. Now, as the motion is simple harmonic, the time period for the spring–mass system is given by equation 11.3 as $T = \dfrac{2\pi}{\omega}$, and from $k/m = \omega_n^2$ we see that $\sqrt{\dfrac{k}{m}} = \omega_n$, so the time period T is:

$$T = 2\pi\sqrt{\frac{m}{k}} \tag{11.10}$$

We also know that the *frequency f* is equal to $\dfrac{1}{T}$, i.e.

$$f = \frac{1}{2\pi}\sqrt{\frac{k}{m}} \tag{11.11}$$

Using the above argument, we are able to derive a *differential equation* that models the SHM of the spring–mass system. So, rewriting equation 11.9 as $mx'' = -kx$ (using our prime notation) and transposing to give $x'' = -\dfrac{k}{m}x$ and then substituting $k/m = \omega_n^2$ into this equation, we find that:

$$x'' = -\omega_n^2 x \tag{11.12}$$

Then, by comparing equation 11.12 with equation 11.2, we are able to confirm that this motion is again *simple harmonic* and equation 11.12 may be written as:

$$x'' + \omega_n^2 x = 0 \tag{11.13}$$

This equation is a homogeneous second-order differential equation, often referred to as a *harmonic equation* (see *Essential Mathematics 3*, Examples EM 3.9 and

EM 3.10) that has the *general solution* $x = A\sin\omega_n t + B\cos\omega_n t$, where A and B are the constants of integration and $\omega_n = \sqrt{\dfrac{k}{m}}$ is the cyclic frequency. A particular solution may be found by applying boundary conditions (see *Essential Mathematics 2*, Example EM 2.6) to determine the values of A and B and so find the frequency of the motion. You will be asked to do this as an exercise at the end of the chapter.

The equivalent differential equation for *angular (rotational) natural vibration* is given as $J\dfrac{d^2\theta}{dt^2} = -S\theta$ or

$$\theta'' + \omega_n^2\theta = 0 \tag{11.14}$$

where $J = $ the rotational mass moment of inertia or second moment of area (see section 3.3), $S = $ the rotational stiffness, $\theta = $ the angle of rotation in radian and $\omega_n^2 = \dfrac{S}{J}$.

> **Key Point.** When considering the differential equation for natural *rotational* vibration, the rotational mass moment of inertia J replaces the mass m, the angle of rotation θ replaces the displacement x and the rotational stiffness S replaces the spring stiffness k

> **Example 11.2** A spiral spring is loaded with a mass of 5 kg which extends it by 100 mm. Calculate the time period of vertical oscillations and the natural frequency.
>
> The time period of the oscillations is given by equation 11.10, $T = 2\pi\sqrt{\dfrac{m}{k}}$, now $k = \dfrac{(5)(9.81)}{0.1m} = 490.5$ N/m. Then $T = 2\pi\sqrt{\dfrac{5}{490.5}} = 0.63$ s and the frequency $f = \dfrac{1}{T} = \dfrac{1}{0.63} = 1.59$ Hz.

11.2.2   Motion of a simple pendulum

The simple pendulum consists of a small bob of mass m (which is assumed to act as a particle). A light inextensible cord of length l to which the bob is attached, is suspended from a fixed point O. Figure 11.5 illustrates the situation when the pendulum is drawn aside and oscillates freely in the vertical plane along the arc of a circle, shown by the dotted line.

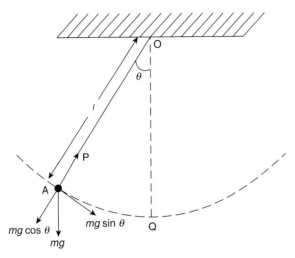

Figure 11.5 Motion of simple pendulum

It can be shown, by following a similar argument to that presented for the spring–mass system in section 11.2.1, that the pendulum for small angles θ (rad) describes SHM. Figure 11.5 shows the forces resulting from the weight of the bob, together with its radial and tangential components. Note that the tangential component $mg \sin \theta$ is the unbalanced restoring force acting towards the centre at Q of the oscillation, while the radial component $mg \cos \theta$ balances the force in the cord P. Then, if a is the acceleration of the bob along the arc at A due to the unbalanced restoring force, the equation of motion of the bob is given by:

$$ma = -mg \sin \theta \qquad (11.15)$$

Again, using our sign convention, the restoring force is towards Q and is therefore considered to be negative, while the displacement x is measured from point Q and is therefore positive. Now, when θ is small $\sin \theta = \theta$ when θ is in radian. Using this fact, equation 11.15 and, for SHM, the relationship $\omega_n^2 = g/l$, we can show that:

$$a = \frac{g}{l}x = -\omega_n^2 x \qquad (11.16)$$

The relationship given in equation 11.16 is SHM (compare with equation 11.2) and so the time period T for the motion is given by:

$$T = \frac{2\pi}{\omega_n} = 2\pi \sqrt{\frac{l}{g}} \qquad (11.17)$$

Compare this equation with equation 11.10!

Example 11.3 A simple pendulum has a bob attached to an inextensible cord 50 cm long. Determine its frequency.

Knowing that the time period is equal to the reciprocal of the frequency, all that is needed is to find the time period T and then the frequency.

Therefore: $T = 2\pi \sqrt{\dfrac{l}{g}} = 2\pi \sqrt{\dfrac{0.5}{9.81}} = 1.42 \text{ s}$ so

the frequency is $= \dfrac{1}{1.42} = 0.704 \text{ Hz}$

11.2.3 Natural vibration of a beam

Let us look at another example of natural vibration that involves a *beam* subject to a concentrated load (m) at its free end. Look back now at Table 2.3 in section 2.6, where the maximum deflection of a cantilever beam subject to a concentrated load is given by $y_{max} = -\dfrac{Wl^3}{3EI}$.

Example 11.4 Determine an expression for the natural frequency of the mass (m) positioned at the end of a cantilever beam, as shown in Figure 11.6.

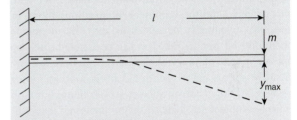

Figure 11.6 Cantilever beam subject to a point load

To simplify your analysis you may ignore the mass of the beam. Also, find a numerical value for this natural frequency when $m = 40 \times 10^3$ kg, $EI = 136.5$ MN m^2 and $l = 2.0$ m.

From Table 2.3 in section 2.6 we know that the maximum deflection of the cantilever beam illustrated in Figure 11.6 is given by the formula $y_{max} = -\dfrac{Wl^3}{3EI}$, where EI = the flexural rigidity of the beam, W = the load in Newton and, from our definition of the spring constant (i.e. spring stiffness) k, given above in section 11.2.1, we also know that the maximum deflection $y_{max} = -\dfrac{Wl^3}{3EI} = \dfrac{W}{k}$ where the units of k (in this case the beam stiffness) are N/m. Then, from this relationship, being aware that the minus sign is simply to indicate the direction of the deflection, which here we may ignore, we find that $k = \dfrac{3EI}{l^3}$ and, from equation 11.11, where $f = \dfrac{1}{2\pi}\sqrt{\dfrac{k}{m}}$, in the case of the cantilever beam the expression for the natural frequency is $f = \dfrac{1}{2\pi}\sqrt{\dfrac{3EI}{ml^3}}$.

Now the numerical value for the frequency is in this case $f = \dfrac{1}{2\pi}\sqrt{\dfrac{3(136.5 \times 10^6)}{(40000)(2^3)}} = 5.69\,\text{Hz}$.

The natural vibration of the beam considered here is known as *transverse vibration*, in that the oscillations resulting from the point load at the end of the cantilever beam act at right angles to (or transversely to) the length of the beam.

11.3   Damped natural vibration

If we again observe the motion of our spring–mass system (Figure 11.4), which is allowed to oscillate freely at its *natural frequency* after being extended, we note that after a time the amplitude of the oscillations of the mass eventually decreases to zero. Its motion is therefore not perfect SHM, it has been acted upon by the air which offers a resistance to its motion. We say that the spring–mass system has been *damped* by air resistance. All systems subject to oscillatory motion will be subject to some form of damping that will cause the vibration to die away unless subjected to some externally applied *harmonic force*, which may be applied directly or via some part of the system structure.

Key Point.   In *all* freely vibrating oscillatory systems the amplitude of the vibration will eventually die away, due to natural energy losses/transfer; thus their motion is not perfect SHM, they are subject to some form of *damping*

The *rate* at which our spring–mass system, or pendulum system, or any other type of body subject to oscillatory motion, is brought to rest, depends on the *degree of damping*.

For example, in the case of our spring–mass system the air provides *light damping*, because the number of oscillations that occur before the displacement of the motion is reduced to zero is large (Figure 11.7a).

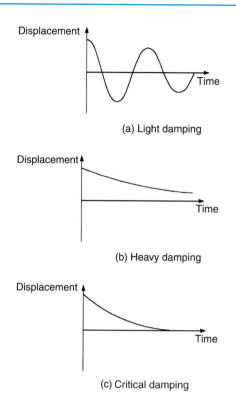

Figure 11.7 Effects of damping on a body subject to oscillatory motion

Similarly, a body subject to heavy damping has its displacement reduced quickly (Figure 11.7b). *Critical damping* occurs when the time taken for the displacement to become zero is a minimum (Figure 11.7c).

Engineering examples of damped oscillatory motion are numerous. For example, on traditional motor vehicle suspension systems, the motion of the springs is damped using oil shock absorbers. These prevent the onset of vibrations that are likely to make the control and handling of the vehicle difficult.

The design engineer needs to control the vibration when it is undesirable and to enhance the vibration when it is useful, so large machines may be placed on anti-vibration mountings to reduce unwanted vibration which, if left unchecked, would cause the loosening of parts and lead to possible malfunction or failure. Delicate instrument systems in aircraft are insulated from the vibrations set up by the aircraft engines and atmospheric conditions, again by placing anti-vibration mountings between the instrument assembly and the aircraft structure. Occasionally vibration is considered useful, for example, shakers in foundries and vibrators for testing machines.

You will remember from section 11.2.1, when we modelled the motion of the simple spring–mass system subject to free natural vibration, that we were able to represent this motion by a homogeneous second-order differential equation, i.e. equation 11.13, $x'' + \omega_n^2 x = 0$, where $k/m = \omega_n^2$, and as you will see this equation for *natural vibration* forms part of the equation we will develop for *damped natural vibration*.

To establish a relationship for damped natural vibration, let us assume that our spring–mass system vibrating freely is subject to damping. Diagrammatically, this situation is illustrated in Figure 11.8, where the damper is represented by a dashpot.

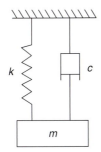

Figure 11.8 Model representing damped free vibration

Thus, Figure 11.8 shows our spring–mass system as before, but with the addition of a damper that exerts a damping force $F_d = cx'$, where c is the constant of proportionality, known as the *damping coefficient*, which has units of force per unit velocity, i.e. $N/_{m/s}$ or $Ns/_m$. Thus the accelerating force

$$mx'' = m\frac{d^2x}{dt^2} = -kx - c\frac{dx}{dt} \qquad (11.18)$$

> **Key Point.** The *damping coefficient* c has units $N/_{m/s}$ or $Ns/_m$

The negative signs result because the restoring force is in the opposite direction to the displacement (see equation 11.9) and the damping force is in the opposite direction to the velocity. Now, using our dash notation, dividing by m and rearranging, we may write equation 11.18 as:

$$x'' + \frac{c}{m}x' + \frac{k}{m}x = 0 \qquad (11.19)$$

Also, remembering $k/m = \omega_n^2$ and making the substitution $\beta = \dfrac{c}{2m}$, we may write that:

$$x'' + 2\beta x' + \omega_n^2 x = 0 \qquad (11.20)$$

Now, equations 11.19 and 11.20 show a common type of homogeneous differential equation (see *Essential Mathematics 3.1*) where, in the case of equation 11.20, the substitution $\beta = \dfrac{c}{2m}$ has been used, where β is referred to as the damping factor. Note also that if $c = 0$, then equation 11.20 is the same as equation 11.13, the harmonic differential equation we produced for free natural vibration (without damping). The full *general solution* of this type of equation is given in *Essential Mathematics 3.1*, where it takes the form $e^{px}(A\sin qx + B\cos qx)$ because the auxiliary equation formed from this type of equation has complex roots. In the particular case of equation 11.20, the solution is:

$$x = e^{-\beta t}\left[\left(A\sin\sqrt{\omega_n^2 - \beta^2}\right)t + B\cos\left(\sqrt{\omega_n^2 - \beta^2}\right)t\right] \qquad (11.21)$$

This solution tells us a great deal about the nature of damped free vibrational motion. The original equation has been derived by considering the three physical parameters k, m and c, i.e. by considering the *spring stiffness constant* k (N/m), the *mass* m (kg) and the *damping coefficient* c (N/m s^{-1}). The first term in equation 11.21 $(e^{-\beta t})$ is simply an exponential decay function with respect to time, where $\beta = \dfrac{c}{2m}$ is often referred to as the *damping factor* with units (s^{-1}). The expression under the *square root* signs in equation 11.21 provides a measure of the *damped natural frequency* ω_d in rad/s, i.e. $\omega_d = \sqrt{\omega_n^2 - \beta^2}$, or the frequency in hertz.

> **Key Point.** $\beta =$ the *damping factor* and has units (s^{-1})

$$f_d = \frac{\sqrt{\omega_n^2 - \beta^2}}{2\pi} \qquad (11.22)$$

Now, from equation 11.22 we can deduce that if the damping factor β is a lot less than ω_n, then the damped frequency ω_d will be close to the natural frequency and the mass will overshoot its equilibrium position. After a good many oscillations (of gradually decreasing amplitude) it will eventually come to rest; this *light damping* or *underdamped* situation is illustrated in Figure 11.7a. An increase in β but still *below* ω_n creates a *heavy damping* situation as shown in Figure 11.7b.

If β is equal to ω_n (*critical damping*) or is greater than ω_n, the mass, when disturbed, will return directly to its equilibrium position (Figure 11.7c).

Also, we know that the *periodic time* $T = \dfrac{1}{f_d}$, therefore from equation 11.22, we find that:

$$T = \frac{2\pi}{\sqrt{\omega_n^2 - \beta^2}} \qquad (11.23)$$

There is one more important parameter that we need to introduce in order to fully model free damped vibration. This is known as the *damping ratio* (ζ), which is the ratio of the damping factor β (the speed of the damping) to the undamped natural frequency ω_n, i.e.:

$$\zeta = \frac{\beta}{\omega_n} \qquad (11.24)$$

> **Key Point.** The damping ratio ζ is used to quantify the degree of damping of a system subject to oscillatory motion

We are able, using a range of numerical values, to *quantify the degree of damping* using this *damping ratio*. For example, when $\zeta = \dfrac{\beta}{\omega_n} = 1$ we have *critical damping* where the mass returns to the equilibrium position in the least time without oscillation, since from the above argument, when $\zeta = 1$, then $\beta = \omega_n$. Light damping is achieved with values of the damping ratio $\zeta \langle 0.1$, and under these circumstances there will be a significant number of oscillations before the mass settles at its equilibrium position. For moderate damping (ζ in the range 0.1 to less than 1) it is often more convenient to calculate the *amplitude ratio* over one cycle.

Then, knowing that the time period for one cycle is $T = \dfrac{2\pi}{\sqrt{\omega_n^2 - \beta^2}}$ and starting at $x = X_1$ when $t = 0$, then $x = X_2$. When $t = T = \dfrac{2\pi}{\sqrt{\omega_n^2 - \beta^2}}$ the amplitude ratio, from equation 11.21, is therefore:

$$\frac{X_1}{X_2} = \frac{e^0(A \sin 0 + B \cos 0)}{e^{-\beta T}(A \sin 2\pi + B \cos 2\pi)} = e^{\beta T},$$

so that the amplitude ratio

$$\frac{X_1}{X_2} = e^{\beta T} \qquad (11.25)$$

So

$$\ln \frac{X_1}{X_2} = \beta T = \delta = \frac{2\pi\beta}{\sqrt{\omega_n^2 - \beta^2}} \qquad (11.26)$$

where the term $\delta = \dfrac{2\pi\beta}{\sqrt{\omega_n^2 - \beta^2}}$ or, from equation 11.24,

$\delta = \dfrac{2\pi\zeta}{\sqrt{1 - \zeta^2}}$ is known as the *logarithmic decrement*.

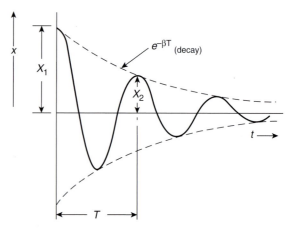

Figure 11.9 Moderately damped free vibration with amplitude ratio X_1/X_2

The resulting motion for moderately damped free vibration is illustrated in Figure 11.9.

> **Example 11.5**
>
> a) Derive an expression for the frequency of the damped oscillation in terms of the free natural frequency and the damping ratio.
>
> b) Determine the value of the damping ratio that will reduce the natural frequency to one-third of its value.
>
> c) Find the natural frequency in hertz and the damping ratio for a damped vibrating system, set up as shown in Figure 11.8, where $m = 15\,\text{kg}$, $k = 30\,\text{kN/m}$ and $c = 15\,\text{N s/m}$.
>
> a) From equation 11.24, $\beta = \zeta\omega_n$ and substituting for β in equation 11.22 gives:
>
> $$f_d = \frac{\sqrt{\omega_n^2 - (\zeta\omega_n)^2}}{2\pi}$$

On extracting the common factor we obtain the required expression,

$$f_d = \frac{\omega_n \sqrt{1 - (\zeta)^2}}{2\pi}.$$

Do take note of this useful relationship. It is often used when considering the frequency response of engineering systems!

b) Then, using $\omega_d = \omega_n \sqrt{1 - \zeta^2}$, we know that $\frac{\omega_d}{\omega_n} = \frac{1}{3} = \sqrt{1 - \zeta^2}$; therefore, $\zeta^2 = 1 - \frac{1}{9}$ and so the required damping ratio is $\zeta = 0.943$.

c) Then, using the relationship $k/m = \omega_n^2$, we find that $\omega_n = \sqrt{\frac{k}{m}} = \sqrt{\frac{30000}{15}} = 44.72$ rad/s, therefore $f_n = \frac{44.72}{2\pi} = 7.12$ Hz. Now the damping factor is given from the above analysis as $\beta = \frac{c}{2m} = \frac{15}{(2)(15)} = 0.5/\text{s}$; therefore, using equation 11.24, we find that $\zeta = \frac{\beta}{\omega_n} = \frac{0.5}{44.72} = 0.011$.

If we again consider *rotational damped natural vibration* as we did for rotational natural vibration in equation 11.14, all we need do is include the damping term. Then the equation may be written as:

$$\theta'' + 2P\theta' + \omega_n^2 \theta = 0 \qquad (11.27)$$

where J = the rotational mass moment of inertia or second moment of area, S = the rotational stiffness, θ = the angle of rotation in radian, $\omega_n^2 = \frac{S}{J}$ and $2P = \frac{c}{J}$.

11.4 Forced vibration

11.4.1 Forced natural vibration

If we again consider the spring–mass system as our model, then, if this system is subject to an external vibration (*forced vibration*), the *free natural vibration* of the system will eventually decay to zero and the system will settle down to a steady state and oscillate at the frequency of this forced vibration. An example of the use of external forced vibration is a foundry shaker that

subjects the melt to forced vibration to assist mixing and settlement.

Sinusoidal forces are often used to *model* forced vibration, since many types of oscillatory vibration may be formed from sinusoidal (harmonic) forces in combination. It is also true that many systems are indeed subject to this type of forced vibration, or a near variant, as a result of, for example, running machinery, seismic activity, weather conditions, etc.

Figure 11.10 shows our original spring–mass system, *without damping*, subject to sinusoidal *forced vibration*.

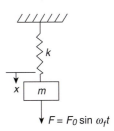

Figure 11.10 Sinusoidal forced vibration

From Figure 11.10 it can be seen that the applied sinusoidal force (sometimes referred to as the *harmonic excitation force*) is given by $F = F_0 \sin \omega t$, where $F_0 =$ the maximum *amplitude* of the excitation force and ω_f is the *frequency of the excitation force*. Now, the equation of motion will be $F_0 \sin \omega_f t - kx = m\frac{d^2 x}{dt^2}$ or, using our *dash* notation, it will be:

$$mx'' + kx = F_0 \sin \omega t \qquad (11.28)$$

Also, using our substitution $\omega_n^2 = \frac{k}{m}$ after dividing through by the mass m, we may write that:

$$x'' + \omega_n^2 x = F_0 \sin \omega_f t \qquad (11.29)$$

The solution to this particular type of differential equation consists of *two parts*: the *complementary function* (see *Essential Mathematics 3*, Examples EM 3.9 and EM 3.10), which is the solution to equation 11.13 that, in this particular case, we derived for the *free-vibration* (without damping) of the system; and the second part of the solution, the *particular integral* (*Essential Mathematics 3.2*) which represents, in this case, the steady-state *forced vibration* caused by the external harmonic excitation. Thus the general solution to equation 11.29 is:

$$x = A \sin \omega_n t + B \cos \omega_n t + \frac{F_0 \sin \omega_f t}{m(\omega_n^2 - \omega_f^2)} \quad (11.30)$$

As mentioned above, the first two terms of this solution $x = A \sin \omega_n t + B \cos \omega_n t$ (the complementary function) represent the *free vibration*, which decays to zero, to leave the final term $x = \frac{F_0 \sin \omega_f t}{m(\omega_n^2 - \omega_f^2)}$ (the particular integral) that represents the steady-state vibration and has an *amplitude* given by:

$$X = \frac{F_0}{m(\omega_n^2 - \omega_f^2)} \quad (11.31)$$

Key Point. Equation 11.31 enables us to find the *response amplitude* of the system to the forced vibration

Now, for this *freely* vibrating system, when the frequency of the external forced vibration (ω_f) equals the natural frequency of the body or system (ω_n), then we say *resonance* occurs. This is where the energy from the external driving force is most easily transferred to the body or system because no energy is required to maintain the vibrations of an *undamped system* at its natural frequency, so all the energy transferred from the driving force will be used to build up the *response amplitude* of vibration which, at *resonance*, will increase without limit! Hence the need to ensure such a system is adequately damped. If $\omega_f \langle \omega_n$, then the system vibrates *in phase* with the excitation force, and if $\omega_f \rangle \omega_n$, then the system vibrates 180° *out of phase* with the excitation force. Under these circumstances, when out of phase by 2π rad, the *amplitude* is given by $X = \frac{F_0}{m(\omega_f^2 - \omega_n^2)}$.

Now, with all the above argument concerning forced free vibration, we have only considered vibration that has been applied to the system by an externally applied excitation force. However, there is in fact another source of external vibration that can affect the system. This is through *oscillatory motion of the previously steady system mountings* via, for example, earth tremors to the solid base of a building, or from other machinery affecting the system mountings. Figure 11.11 shows our simple spring–mass model for this type of forced vibration.

Key Point. Forced vibration may be imparted to the system by the application of an external harmonic excitation force *or* through oscillatory motion of the previously steady system mountings

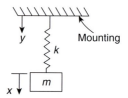

Figure 11.11 Forced vibration through system mountings

The motion of the mounting will have a displacement given by $y = Y_0 \sin \omega_f t$, with maximum amplitude Y_0. The change in the length of the spring is given by $(x - y)$ and so, since the force of the mass $\left(m \frac{d^2 x}{dt^2} = mx''\right)$ is equated to the spring restoring force $k(x - y)$ opposing the initial displacement, the equation of the motion will be $mx'' = -k(x - y) = -kx + ky$ so that $mx'' + kx = ky$ and, from $y = Y_0 \sin \omega_f t$, the equation of motion is:

$$mx'' + kx = Y_0 \sin \omega_f t \quad (11.32)$$

And again, on division by the mass m and then substituting for the undamped natural frequency $\omega_n^2 = \frac{k}{m}$, we obtain:

$$x'' + \omega_n^2 x = Y_0 \sin \omega_f t \quad (11.33)$$

The solution of this equation is almost identical to that for equation 11.30, i.e.:

$$x = A \sin \omega_n t + B \cos \omega_n t + \frac{kY_0 \sin \omega_f t}{m(\omega_n^2 - \omega_f^2)} \quad (11.34)$$

where, after the decay of the free vibration to zero, we get that:

$$x = \frac{kY_0 \sin \omega_f t}{m(\omega_n^2 - \omega_f^2)} \quad (11.35)$$

Note that the *amplitude* of the motion in this case is given as $X = \frac{kY_0}{m(\omega_n^2 - \omega_f^2)}$, provided $\omega_f < \omega_n$. We now turn our attention to forced damped vibration.

11.4.2 Forced damped vibration

We have spent some time looking at forced free vibration and have established a number of relationships to model the subsequent oscillatory motion of our spring–mass system. We now turn our attention briefly to the establishment of similar relationships for *damped systems* that are subject to *harmonic excitation*. Figure 11.12 shows our spring–mass model with damping, being subjected to forced vibration $F = F_0 \sin \omega_f t$.

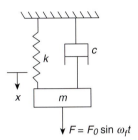

Figure 11.12 Forced vibration with damping

The differential equation modelling this motion is easily established if we start with the differential equation we found for the damped vibration case, i.e. equation 11.20, $x'' + 2\beta x' + \omega_n^2 x = 0$. All that is necessary is to add in the forcing component, so that we get:

$$x'' + 2\beta x' + \omega_n^2 x = F_0 \sin \omega_f t \qquad (11.36)$$

The solution to this equation consists of: the *complementary function*, i.e. equation 11.21, as before, where this solution models the *damped natural vibration* that eventually decays to zero; and the *particular integral* that models the steady-state forcing motion. Therefore the total solution to equation 11.36 is:

$$x = e^{-\beta t} \left[\left(A \sin \sqrt{\omega_n^2 - \beta^2} \right) t + B \cos \left(\sqrt{\omega_n^2 - \beta^2} \right) t \right]$$

$$+ \; \frac{F_0 \sin(\omega_f t - \phi)}{m\sqrt{(4\beta^2 \omega_f^2) + (\omega_n^2 - \omega_f^2)^2}}$$

where

$$x(t) = \frac{F_0 \sin(\omega_f t - \phi)}{m\sqrt{(4\beta^2 \omega_f^2) + (\omega_n^2 - \omega_f^2)^2}} \qquad (11.37)$$

The steady-state motion has a frequency of application ω_f rad/s and its *amplitude* is given by the expression

$$X = \frac{F_0}{m\sqrt{(4\beta^2 \omega_f^2) + (\omega_n^2 - \omega_f^2)^2}} \qquad (11.38)$$

with the *phase angle*

$$\phi = \tan^{-1} \frac{2\beta \omega_f}{\omega_n - \omega_f} \qquad (11.39)$$

The phase angle is *out of phase* with the *disturbing force* by the angle ϕ, as can be seen by the expression for the excitation (disturbing) force on the top line of equation 11.7. It can be seen that when $\omega_n = \omega_f$, the amplitude of the damped vibrations $= \dfrac{F_0}{2m\beta \omega_f}$, so that, unlike the forced undamped motion, the amplitude does not reach a maximum (resonance) when $\omega_n = \omega_f$; in fact, this now occurs when $X'(\omega_f) = 0$.

Now all this is very well, but in order to better visualise this motion in a concise graphical manner we can express equations 11.38 and 11.39 in a non-dimensional form that involves only the *damping ratio* and the *frequency ratio*. The non-dimensional equations for *the amplitude* $\left(\dfrac{Xk}{F_0} \right)$ and *phase angle* (ϕ) are given in equations 11.40 and 11.41, respectively.

$$\frac{Xk}{F_0} = \frac{1}{\sqrt{\left[1 - \left(\dfrac{\omega_f}{\omega_n} \right)^2 \right]^2 + \left[2\zeta \left(\dfrac{\omega_f}{\omega_n} \right) \right]^2}} \qquad (11.40)$$

$$\phi = \tan^{-1} \frac{2\zeta \left(\dfrac{\omega_f}{\omega_n} \right)}{1 - \left(\dfrac{\omega_f}{\omega_n} \right)^2} \qquad (11.41)$$

Equations 11.40 and 11.41 indicate that both the *amplitude* and *phase angle* are simply functions of the frequency ratio $\dfrac{\omega_f}{\omega_n}$ and damping ratio ζ. We can now establish what happens to the amplitude and phase of the motion with differing combinations of damping ratio and frequency ratio. We will attempt to do this using an example.

Example 11.6

a) Show, using base units, that the relationship for the amplitude $\dfrac{Xk}{F_0}$ is non-dimensional.

b) Draw graphs and describe the behaviour of the non-dimensional amplitude when $\zeta = 1.0, 0.5, 0.25, 0$ and when, for *each* of the damping ratios, the frequency ratio $\dfrac{\omega_f}{\omega_n} = 0, 0.5, 0.75, 1.0, 1.5, 2.0$.

c) Draw graphs of the phase angle against the values of the damping and frequency ratios given in part b) and explain the relationship between the two.

a) The derived units are $k = $ N/m, $X = $ metres (m), $F_0 = $ N, therefore, in base units:

$$\frac{Xk}{F_0} = \frac{(\text{m})\left(\dfrac{\text{kg m}}{\text{s}^2}\dfrac{1}{\text{m}}\right)}{\dfrac{\text{kg m}}{\text{s}^2}}$$

$$= (\text{m})\left(\frac{\text{kg m}}{\text{s}^2}\frac{1}{\text{m}}\right)\left(\frac{\text{s}^2}{\text{kg m}}\right) = 1$$

i.e. non-dimensional.

b) In order to produce reasonably accurate graphs of the non-dimensional amplitude $\dfrac{Xk}{F_0}$ against the required frequency ratios $\dfrac{\omega_f}{\omega_n}$, we produce the following table that also includes values for the phase angle (in degrees) for the damping and frequency ratios shown.

All of the above results have been calculated using equations 11.40 and 11.41. Figure 11.13 shows the plots of the non-dimensional amplitude against the frequency ratio, for varying damping.

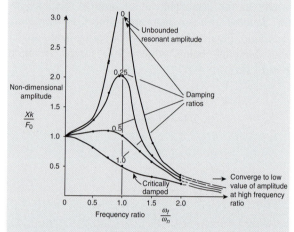

Figure 11.13 Graphs of amplitude against frequency ratio, for varying damping

Observations

i) When the damping ratio is zero, the amplitude is 1.0, irrespective of the degree of damping. In essence, this is the point where there is no forced vibration, and the system vibrates at its natural frequency.

ii) When the frequency ratio is close to 1.0, the degree of damping given by the damping ratio has a large influence on the amplitude. From the plots, where the damping ratio is any value *other than zero*,

ζ	1.0	1.0	1.0	1.0	1.0	1.0	0.5	0.5	0.5	0.5	0.5	0.5
ω_f/ω_n	0	0.5	0.75	1.0	1.5	2.0	0	0.5	0.75	1.0	1.5	2.0
Xk/F_0	1.0	0.8	0.64	0.5	0.308	0.2	1.0	1.109	1.15	1.0	0.512	0.277
ϕ	0	53	73.7	90	112	127	0	33.4	59.7	90	129.8	146.3
ζ	0.25	0.25	0.25	0.25	0.25	0.25	0	0	0	0	0	0
ω_f/ω_n	0	0.5	0.75	1.0	1.5	2.0	0	0.5	0.75	1.0	1.5	2.0
Xk/F_0	1.0	1.265	1.735	2.0	0.686	0.316	1.0	1.33	2.28	∞	0.809	0.33
ϕ	0	18.4	40.6	90	149	161	0	0	0	0	0	0

the maximum amplitude at *resonance* is reached in this region when $X'(\omega_f) = 0$, not at $\omega_f = \omega_n$ (refer back to the argument in section 11.4.2).

iii) When $\zeta = 0$ and $\dfrac{\omega_f}{\omega_n} = 1$, i.e. when the frequency of the forced harmonic oscillations equals the natural frequency of vibration (in an *undamped* system), the amplitude of the resulting oscillations is unbounded and increases to infinity.

iv) If the frequency ratio is very much greater than 1.0, the amplitude of the resulting vibrations decreases to a low value and all plots converge, irrespective of the degree of damping.

c) From the table, the plots of phase angle against frequency ratio are shown in Figure 11.14.

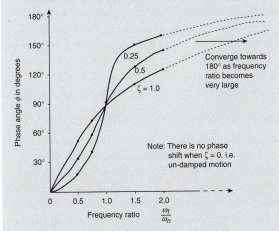

Figure 11.14 Graphs of phase angle against frequency ratio, for varying damping

More observations

i) When the frequency ratio is zero the phase angle is zero. This is to be expected because at this value of frequency ratio (where, by implication, $\omega_f = 0$) the displacement (X) is in phase with the spring force (kX).

ii) When the damping ratio is small, i.e. $\dfrac{\omega_f}{\omega_n} \ll 1$, then the resisting (inertia) force and damping force will be low, resulting in a small phase angle between the excitation force and the spring force, so that they are almost equal.

iii) When $\dfrac{\omega_f}{\omega_n} = 1$, then the excitation force lags the spring force by 90°.

iv) When the frequency ratio is very large, the phase angle 180° and, when infinitely large, the excitation force lags behind the spring force by 180°. These very high These very high values of the frequency ratio are not shown on our plots but if you put higher and higher values of the frequency ratio into equation 11.41, you will be able to see that no matter what the damping ratio is (apart from zero) the phase angle 180°, where we say that the frequencies of the spring the spring force and the excitation force are in 'anti-phase'. Note also the very small amplitudes that occur at these very high frequency ratios.

Before we leave the subject of forced damped vibration we need to consider one more system parameter of importance to design engineers; this is known as *transmissibility*. We have already discussed the way in which the excitation force can be transmitted from the system supports to the system itself and the need for damping to contain this unwanted vibration. It is also equally important to ensure that the vibration forces from heavy machinery and the like to their supports is kept to a minimum, using damping between the supports and the machinery. The ratio of the *transmitted force* (F_T) to the harmonic *excitation force* ($F_0 \sin \omega_f t$) is known as *transmissibility*.

Key Point. The transmissibility is the ratio of the transmitted force to the excitation force, i.e. F_T/F_0

It can be shown that this transmitted force from the machinery, which is to be isolated through springs and/or dampers positioned between the machinery and its supports, is given by:

$$F_T = kX\sqrt{1 + \left(\frac{2\zeta\omega_f}{\omega n}\right)^2} \qquad (11.42)$$

Then, also from equation 11.40, we find that the *response amplitude X* is:

$$X = \frac{F_0/k}{\sqrt{\left[1 - \left(\frac{\omega_f}{\omega_n}\right)^2\right]^2 + \left[2\zeta\left(\frac{\omega_f}{\omega_n}\right)\right]^2}} \quad (11.43)$$

So the *transmissibility*

$$\frac{F_T}{F_0} = \frac{1 + (2\zeta\omega_f/\omega_n)^2}{\left[1 - \left(\frac{\omega_f}{\omega_n}\right)^2\right]^2 + \left[2\zeta\left(\frac{\omega_f}{\omega_n}\right)\right]^2} \quad (11.44)$$

11.4.3 Resonance

From Example 11.6 and section 11.4.1, you will have noted that when, in a *freely* vibrating system, the frequency of the external forced vibration (ω_f) equals the natural frequency of the body or system (ω_n), *resonance* occurs. In order to mitigate the effects of *resonance* we needed to introduce a degree of damping into systems, and in order to decide on the necessary *amount of damping* we needed to understand the effect on the oscillatory output motion of systems when subjected to forced vibration. These effects can be seen by studying the graphs we plotted in Figures 11.13 and 11.14.

Resonance is generally considered troublesome, especially in mechanical systems. The classic example used to illustrate the effects of unwanted resonance is the Tacoma Bridge disaster in America in 1940.

The prevailing wind caused an oscillating force in resonance with the natural frequency of part of the bridge. Oscillations built up (as the above theory predicts) that were so large they destroyed the bridge.

The rudder of an aircraft is subject to forced vibrations that result from the aircraft engines, flight turbulence and aerodynamic loads imposed on the aircraft. In order to prevent damage from possible resonant frequencies, the rudder often has some form of hydraulic or viscous damping fitted. The loss of a flying control resulting from resonant vibration could result in an aircraft accident, with subsequent loss of life.

So, when might resonant frequencies be useful? One example is the electrical resonance that occurs when a radio circuit is tuned by making its natural frequency of oscillations equal to that of the incoming radio signals.

11.5 Chapter summary

In this chapter we first looked at periodic motion, in particular simple harmonic motion. We defined SHM as a special periodic motion in which:

(i) the acceleration of the body is always directed towards a fixed point in its path;

(ii) the magnitude of the acceleration of the body is proportional to its distance from the fixed point.

We concentrated on harmonic motion throughout the chapter because we may use it to model many situations in the real world, such as engineering structures and systems that are subject to vibration. During our analysis of SHM we derived several useful relationships concerned with the displacement, velocity and acceleration of the motion, in particular equations 11.1 to 11.7.

In section 11.2 we considered *free vibration* and modelled the motion resulting from free vibration using a spring–mass system, where the stiffness of the system was represented by the spring stiffness k, the spring was attached to the system mountings and supported the mass m. We noted from the model that the resulting steady free vibration was *simple harmonic* and had a natural frequency ω_n (rad/s) $= \sqrt{\frac{k}{m}}$ or $f = \frac{1}{2\pi}\sqrt{\frac{k}{m}}$. Also, from this relationship and from the differential equation we found that modelled the motion ($x'' + \omega_n^2 x = 0$, equation 11.13), we can see that the frequency does not depend on the amplitude (X) of the free vibration. Equation 11.14 is the equivalent differential equation for *rotational* free vibration. You should also note the general solution for harmonic differential equations like equation 11.13 is $x = A\sin\omega_n t + B\cos\omega_n t$. The motion of a simple pendulum was then considered, as yet another example of SHM.

In section 11.3 we initially considered *damped natural vibration,* where again using our spring–mass system, this time with the inclusion of a damper, we derived equation 11.19, $x'' + \frac{c}{m}x' + \frac{k}{m}x = 0$, which now includes the *damping coefficient c* in addition to the mass and stiffness parameters. Then, as with previous equations, we modified equation 11.19 by substituting for the *damping factor* $\beta = \frac{c}{2m}$ to give equation 11.20, $x'' + 2\beta x' + \omega_n^2 x = 0$, which has the general solution $x = e^{-\beta t}\left[\left(A\sin\sqrt{\omega_n^2 - \beta^2}\right)t + B\cos\left(\sqrt{\omega_n^2 - \beta^2}\right)t\right]$. You should be aware of the information this solution can give us about the nature of the motion of damped free vibration. We also introduced the damping ratio $\zeta = \dfrac{\beta}{\omega_n}$, which is a useful measure of the degree of damping. At the end of section 11.3.1 we detailed the differential equation for rotational damped natural vibration, equation 11.27, $\theta'' + 2P\theta' + \omega_n^2\theta = 0$, where the rotational parameters were the same as those detailed for equation 11.14.

In section 11.4 we looked at forced vibration, first considering forced natural vibration, then forced damped vibration.

In the first case (11.4.1) we looked at the two ways in which forced vibration can be applied to our spring–mass system (without damping), i.e. either by applying a harmonic forcing vibration to the system or where the initially steady system mountings are subject to vibrating forces from, for example, a building that is subject to seismic activity. We derived equation 11.29, $x'' + \omega_n^2 x = F_0 \sin \omega_f t$, which modelled the situation for forced harmonic excitation of the naturally vibrating system, with equations 11.30 and 11.31 providing the solutions that enabled the frequency and amplitude of the response to be determined, the frequency of the response being the same as the forcing vibration and the amplitude of the response being dependent on the forcing amplitude and frequency ratio. Equations 11.32 to 11.35 gave similar relationships and information for the forcing vibration that acted on the system *mountings*.

In section 11.4.2 we considered *forced damped vibration* where essentially, to estimate the nature of the response for this type of system, it is necessary to find the frequency ratio and damping ratio. Then, for example, either the equations 11.36 to 11.39 or the non-dimensional equations 11.40 and 11.41 can be used to find the *amplitude* and *phase angle* of the output vibration. Example 11.6 hopefully clarifies the nature of the response vibration for varying combinations of ζ and $\dfrac{\omega_f}{\omega_n}$. In reality, satisfactory system responses or indeed desired system responses are determined from tests where a range of forced harmonic vibrations and damping ratios are mathematically modelled in combination, in order to ensure the system is designed to meet the specified criteria. The level of mathematical rigour needed for this type of modelling, may be found in more advanced texts on vibration theory. We then introduced the idea of *transmissibility*, F_T/F_0, where equation 11.42 enables us to find the transmitted force, equation 11.43 the response amplitude and equation 11.44 enables us to find the transmissibility. Note that the transmitted force (F_T) is equal to the transmissibility

multiplied by the excitation (disturbing) force (F_0). Finally, in section 11.4.3, we briefly discussed the nature of *resonance* and you should now be aware of the circumstances under which resonance is likely to occur and the *degree of damping* that we need to use in order to control it.

11.6 Review questions

1. A particle moves with simple harmonic motion. Given that its acceleration is 12 m/s^2 when 8 cm from the mid-position, find the time period for the motion. Also, if the amplitude of the motion is 11 cm, find the linear velocity when the particle is 8 cm from the mid-position.

2. Write a short account of simple harmonic motion, explaining the terms time period, cycle, frequency and amplitude.

3. A simple pendulum is oscillating with an amplitude of 30 mm. If the pendulum is 80 cm long, determine the velocity of the bob as it passes through the mid-point of its oscillation.

4. A body of mass 2 kg hangs from a spiral spring. When the mass is pulled down 10 cm and released, it vibrates with SHM. If the time period for this motion is 0.4 s, find the linear velocity when it passes through the mid-position of its oscillation and the acceleration when it is 4 cm from the same mid-position.

5. A reciprocating piston operating in a straight line may be modelled by SHM. If its velocity is 10 m/s at a point when its displacement is 100 mm from its mid-position, determine the frequency and acceleration of the motion at this time if the *overall stroke* of the piston is 300 mm.

6. If, for the cantilever beam illustrated in Figure 11.6, the natural frequency of oscillation $f = 12$ Hz when it supports a mass of 50×10^3 kg at its free end, $E = 210$ GN and $I = 6.0 \times 10^{-4} \text{ m}^4$, determine the length ($l$) of the beam.

7. In the 'scotch yoke' mechanism illustrated in Figure 11.15, the crank has a radius of $OP = 12$ cm and rotates with a constant angular frequency of 5 Hz about its pivot point.

   If the subsequent horizontal motion of the slider is SHM, determine the velocity and acceleration of the slider, when it is 3 cm from the mid-point of its travel.

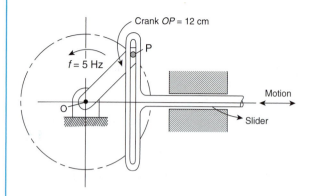

Figure 11.15 Scotch yoke mechanism for question 11.6.7

8. a) Show that $x = \sin \omega_n t$ is a solution to the equation $x'' + \omega_n^2 x = 0$.

    b) The oscillatory motion of the spring on a car suspension may be modelled by the equation $\dfrac{d^2 x}{dt^2} + 36x = 0$, where x is the displacement of the suspension spring from its equilibrium position after t seconds. Find i) a *general solution* for this equation of motion, and ii) a *particular solution for the displacement* when $t = 0$, $s = 0$ and $\dfrac{dx}{dt} = 24$.

9. A particular naturally vibrating system is subject to damping. The system consists of a small mass $m = 0.1$ kg, the system stiffness $k = 200$ N/m and the damping coefficient $c = 0.25$ kg/s. Determine the appropriate values of the natural and damped frequencies, the damping factor, damping ratio, periodic time and logarithmic decrement, and then sketch and describe the nature of the motion.

10. A generator has a mass of 150 kg and is supported by four springs *each* with stiffness $k = 200$ kN/m. If the rotation of the generator causes an out-of-balance excitation force of 5.0 kN when rotating at its operating frequency of 60 Hz and the system has a damping ratio $\zeta = 0.15$, determine the response amplitude and the transmissibility.

Part III

Thermodynamics

In this part of the book we study *thermodynamics*, with two additional chapters introducing *heat transfer* and *fluid mechanics*. These three topics are collectively known as the thermal sciences that are brought together here under the collective heading of thermodynamics.

Thermodynamics (literally translated as heat power) is concerned with the relationship between heat and mechanical energy and the conversion (*transfer*) of one to another. Thus, as engineers, we are concerned with designing, installing and maintaining plant and equipment based on thermodynamic (heat) principles and processes such as fuel-driven steam and other gas-driven plant, reciprocating-piston engines and gas turbine engines. Throughout our study of thermodynamics in this part of the book we will (due to space requirements) only be considering *gases* and *liquids* as the working fluid; a chapter on the website that accompanies this book at http://keytostudy.com/science/ will cover *vapours*, vapour processes and systems.

In Chapter 12 some *thermodynamic fundamentals* are reviewed that should be familiar to the reader. These include such topics as pressure, specific heat, latent heat and the gas laws that provide an essential foundation for the chapters that follow. Chapter 13 is concerned with closed and open *thermodynamic systems* and an introduction to the first and second laws of thermodynamics, in relation to these systems. In Chapter 14 we look at the *perfect gas processes* including constant pressure, constant volume, isothermal, isentropic and polytropic processes. The concepts of reversibility and reversible work are also covered. Chapter 15 introduces the concept of entropy and its use in determining end states for *thermodynamic cycles*. The Carnot cycle, Otto cycle, diesel cycle and constant pressure cycles are covered, together with their isentropic efficiencies. Chapter 16 is concerned with the practical cycles and efficiencies of *combustion engines*. In particular, we concentrate on the four-stroke piston engine and the closed and open gas turbine engine and their related practical cycles and efficiencies. An introduction to aircraft propulsion provides the platform for analysing the open system (turbojet) gas turbine engine cycle and related efficiencies. In Chapter 17, the separate subject of heat transfer is introduced. The subject matter is limited to *one-dimensional heat transfer* by conduction, convection and radiation. Fourier's conduction law is introduced and heat transfer by conduction and convection through composite wall and pipes is analysed. The concept of black and grey body radiation, together with the Stefan–Boltzmann law, are also covered. Finally, in Chapter 18, the separate subject of *fluid mechanics* is introduced, where first, topics in fluid statics such as thrust forces on immersed bodies and the buoyancy of immersed and floating bodies are briefly covered. In the second part of the chapter topics in fluid dynamics such as fluid momentum, Bernoulli theory and applications, fluid flow instruments, viscosity and energy losses in piped systems are briefly covered.

Chapter 12

Fundamentals

In this chapter we will consider some of the fundamental concepts, definitions and principles that underpin the study of thermodynamics and fluids, most of which you will probably have met before. We start by considering some very important thermo-fluid parameters, including density, pressure and its measurement, temperature and its measurement and linear expansion. We then discuss the nature of heat, in particular, specific heat and latent heat. Finally, we look at the expansion and contraction of gases and the use of the gas laws, including the two forms of the characteristic gas equation. It is hoped that after revising/studying these topics you will have the necessary foundation for the study of thermodynamics and fluids that follows.

12.1 Density and pressure

12.1.1 Density and relative density

The mass density of a body is defined as the mass per unit volume of the body. Thus density is a standard measure of the mass of a substance that is contained in a volume of one metre cubed (1 m^3). Combining the SI units for mass and volume gives the unit of density as kg/m^3 and, in symbols:

$$\rho = \frac{m}{V} \tag{12.1}$$

The amount of a fluid contained in a cubic metre will vary with temperature, although the expansion of liquids with rising temperature is insignificant when compared to gases. The standard densities quoted for liquids will be valid for all temperatures unless stated otherwise.

Key Point. The density of pure water at 4°C is taken as 1000 kg/m^3

Another parameter that is often used, particularly in fluid dynamics, is *specific volume* (v_s), where specific in this case means *per unit mass*, so that, in symbols, the specific volume is given as:

$$v_s = \frac{1}{\rho} \tag{12.2}$$

The units of specific volume will be m^3/kg (the inverse of the units of density).

Density may also be expressed in relative terms, i.e. in comparison to some datum value. The datum that forms the basis of relative density for *solid* and *liquid* substances is *water*. The *relative density* of a solid or liquid body is the ratio of the density of the body with that of the density of pure water measured at 4°C. That is:

$$\text{Relative density} = \frac{\text{density of the substance}}{\text{density of water}} \tag{12.3}$$

The relative density of water under these conditions is 1.0. Since relative density is a ratio it has no units. The old name for relative density was *specific gravity (SG)* and this is something you need to be aware of in case you meet this terminology in the future. Thus, to find the

978-1-85617-775-7, Engineering Science, Mike Tooley & Lloyd Dingle

relative density of any liquid or solid material, divide its density by 1000 kg/m^3.

Note that the base density used to find the relative density of a gas is *air*, so that the relative density of both water and air is 1.00. In practice the relative density of gases is rarely used. The standard sea-level value for the density of air is taken as 1.2256 kg/m^3 and this is the value we will use to find the relative densities of gases.

> **Key Point.** To find the relative density of any liquid or solid substance divide its density by 1000 kg/m^3, and to find the relative density of a gas divide its density by 1.2256 kg/m^3

One other useful parameter we often use when dealing with thermo-fluids is known as *specific weight* (γ). This is *weight per unit volume* and has SI units of N/m^3; it is the equivalent, symbolically, of:

$$\gamma = \rho g \qquad (12.4)$$

where, in equation 12.4, the exact value of the acceleration due to gravity $g = 9.80665$ m/s^2, but in this book we will use the approximation $g = 9.81$ m/s^2.

The *density* and *relative density* of some of the more common engineering substances are laid out in Table 12.1.

Table 12.1 Density and relative density of some common engineering substances

Substance	Density (kg/m^3)	Relative density
Gases		
Hydrogen	0.085	0.0695
Helium	0.169	0.138
Nitrogen	1.185	0.967
Air	1.2256	1.0
Oxygen	1.3543	1.105
Liquids		
Petroleum	720	0.72
Aviation fuel	780	0.78
Alcohol	790	0.79

Table 12.1 Continued

Substance	Density (kg/m^3)	Relative density
Kerosene	820	0.82
Hydraulic oil	850	0.85
Water	1000	1.00
Mercury	13600	13.6
Solids		
Ice	920	0.92
Polyethylene	1300–1500	1.3–1.5
Rubber	860–2000	0.86–2.0
Magnesium	1740	1.74
Concrete	2400	2.4
Glass	2400–2800	2.4–2.8
Aluminium	2700	2.7
Titanium	4507	4.507
Tin	7300	7.3
Steel	7800–7900	7.8–7.9
Iron	7870	7.87
Copper	8960	8.96
Lead	11340	11.34
Gold	19320	19.32

Example 12.1 A hydraulic cylindrical shock absorber has an internal diameter of 10 cm and is 1.2 m in length. It is half-filled with hydraulic oil and half-filled with nitrogen gas at standard atmospheric pressure. Determine (taking appropriate values from Table 12.1) the weight of the nitrogen and the weight of the hydraulic oil within the shock absorber.

All that is required is that we first find the internal volume of the shock absorber:

$$V = \left(\frac{\pi d^2}{4}\right)l \quad \text{then} \quad V = \left(\frac{\pi (0.1)^2}{4}\right)1.2 \quad \text{and}$$

$$V = 0.00785 \text{ m}^3.$$

Therefore the volume of the nitrogen and the volume of the oil are both $= 0.003925$ m³. Now the mass of each may be found using their densities. From equation 12.1, then $m = \rho V$ and for nitrogen $m = (1.185)(0.03925) = 0.0465$ kg. For the hydraulic oil the mass $m = (850)(0.03925) = 33.3625$ kg. Then the weight of the nitrogen $= 0.456$ N and the weight of the hydraulic oil $= 327.29$ N.

It can be seen from Example 12.1 that the weight of the nitrogen is virtually insignificant when compared with that of the oil. In a real shock absorber the nitrogen would be highly pressurised, increasing its density and mass many fold. Pressure/volume relationships are an important consideration when dealing with gases, as you will see when we discuss the gas laws.

Table 12.2 Some common SI units for pressure

Use	SI units
Pressure applied to solids	N m⁻², MN m⁻² or N/m², MN/m²
Fluid dynamic and hydrostatic pressure	Pascal (Pa) = 1 N/m²; kPa, MPa
Atmospheric pressure	1 bar = 10⁵ Pa = 10⁵ N/m²; mbar, millimetres of mercury (mm Hg)
Standard atmospheric pressure	101225 Pa or 101225 N/m² or 1.01225 bar or 760 mm Hg

Test your knowledge 12.1

1. What is the SI unit of density?

2. What is the essential difference between the values of relative density of solids and liquids compared with gases?

3. What is likely to happen to the density of pure water as its temperature increases?

4. Show that ρg has the equivalent units of specific weight.

5. A rectangular container has dimensions of 1.6 m × 1.0 m × 0.75 m. A certain liquid that fills the container has a mass of 864 kg. What are the density and relative density of this liquid?

12.1.2 Pressure and pressure law

You have met the concept of pressure before when considering pressure acting on solid materials in Chapter 1, where it was defined as force (N) per unit area (m²). There are in fact several types of pressure that you may not have met; these include *hydrostatic pressure*, i.e. the pressure created by stationary or slow-moving bulk liquid, *atmospheric pressure* due to the weight of the air above the earth acting on its surface, and *dynamic pressure* due to fluid movement, as well as the pressure applied to solids, mentioned above.

There are several common units for pressure, some of which we have been freely interchanging, as

appropriate, throughout Parts I and II of this book. To remind you, some of the more common SI units and their equivalents have been laid out in Table 12.2.

Now there are four basic factors or laws that govern the pressure within fluids. With reference to Figure 12.1, these *four laws* may be defined as follows:

(a) Pressure at a given depth in a fluid is equal in all directions (Figure 12.1a).

(b) Pressure at a given depth in a fluid is independent of the shape of the containing vessel in which it is held. In Figure 12.1b the pressure at points X, Y and Z is the same.

(c) Pressure acts at right angles to the surfaces of the containing vessel (Figure 12.1c).

(d) When a force is applied to a contained fluid, the pressure created is transmitted equally in all directions (Figure 12.1d).

The consequences of these laws or properties will become clear as you continue your study of thermofluids and, in particular, hydrostatic systems.

12.1.3 Hydrostatic and atmospheric pressure

Pressure at a point (or depth) in a liquid can be determined by considering the weight force of a fluid above the point. Consider Figure 12.2.

If the density of the liquid is known, then we may express the *mass* of the liquid in terms of its density and volume, where the volume $V = Ah$. Thus the mass

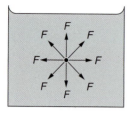

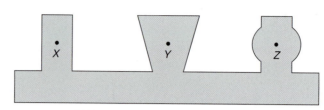

(a) Pressure at a given depth
is equal in all directions

(b) Pressure is independent of the shape
of the containing vessel at a given depth

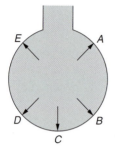

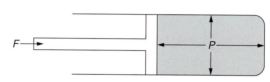

(c) Pressure acts at right angles
to the walls of the containing vessel

(d) Pressure transmitted through
a fluid is equal in all directions

Figure 12.1 Illustration of fluid pressure laws

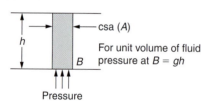

Figure 12.2 Pressure at a point

of liquid (m) = density × cross-sectional area (A) × height (h) or, in symbols:

$$m = \rho Ah \qquad (12.5)$$

Now, since the weight is equal to the mass multiplied by the acceleration due to gravity, then the weight is given by $W = \rho Agh$ and it follows that the pressure due to the weight of the liquid, the *hydrostatic pressure*, is equal to the weight divided by area A, i.e.

$$\text{Hydrostatic pressure} = \rho gh \qquad (12.6)$$

If standard SI units are used for *density* (ρ), *acceleration due to gravity* (g) and for the *height, head* or *elevation* (h), then the pressure (p) may be expressed in N/m² or pascal (Pa).

Note that the *atmospheric pressure* above the liquid was ignored; the above formula refers to *gauge pressure*.

This should always be remembered when using this formula.

> **Key Point.** ρgh = gauge pressure

> **Example 12.2** Find the head h of mercury corresponding to a pressure of 101.325 Pa. Take the density of mercury as 13600 kg/m³
>
> Pressure $p = \rho gh$ or
>
> $$h = \frac{p}{\rho g} = \frac{101325}{(13600)(9.81)}$$
>
> $$= 0.76 \text{ m or } 760 \text{ mm of Hg}$$

Therefore, this is the height of mercury needed to balance standard atmospheric pressure.

The air surrounding the earth has mass and is acted upon by the earth's gravity, thus it exerts a force over the earth's surface. This force per unit area is known as *atmospheric pressure*. At the earth's surface at sea level, under standard conditions (as laid down by the International Civil Aviation Authority), this *standard atmospheric pressure* is given the average value of 101325 N/m² = 101325 Pa or 1.0325 bar.

Outer space is a vacuum and is completely devoid of matter; consequently there is no pressure in this vacuum.

Therefore, pressure measurement relative to a vacuum is *absolute*. For most practical purposes it is only necessary to know how pressure varies from the earth's atmospheric pressure. A pressure gauge is designed to read zero when subject to atmospheric pressure; therefore, if a gauge is connected to a pressure vessel it will only read *gauge* pressure. So, to convert gauge pressure to absolute pressure, atmospheric pressure must be added to it, i.e. *Absolute pressure = gauge pressure + atmospheric pressure.*

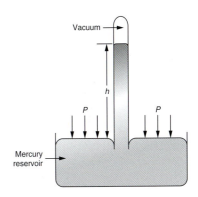

Figure 12.3 Simple mercury barometer

> **Example 12.3** Taking atmospheric pressure as $101325 \, \text{N/m}^2$, convert the following gauge pressures into absolute pressures. Give your answer in kPa.
>
> a) $400 \, \text{kN/m}^2$ b) $20 \, \text{MN/m}^2$ c) $5000 \, \text{Pa}$
>
> We know from the above that absolute pressure is equal to gauge pressure plus atmospheric pressure. Therefore, the only real problem here is to ensure the correct conversion of units.
>
>   Atmospheric pressure = $101.325 \, \text{kN/m}^2$, then:
>
> a) $400 + 101.325$ $= 501.325$
>
> b) $20000 + 101.325$ $= 20101.325$
>
> c) $5 + 101.325$ $= 106.325$ (remembering that $1 \text{Pa} = 1 \text{N/m}^2$ $= 1 \text{N m}^{-2}$)

12.1.4   Measurement of pressure

Devices used to measure pressure will depend on the magnitude (size) of the pressure, the accuracy of the desired readings and whether the pressure is static or dynamic. Here we are concerned with barometers to measure atmospheric pressure and the manometer to measure low pressure changes, such as might be encountered in a laboratory or from variations in flow through a wind tunnel. Further examples of dynamic pressure measurement due to fluid flow will be encountered later, when we study fluid dynamics in a little more detail.

The two most common types of barometer used to measure atmospheric pressure are the mercury and aneroid types. The simplest type of *mercury barometer* is illustrated in Figure 12.3. It consists of a mercury-filled tube, which is inverted and immersed in a reservoir of mercury.

The atmospheric pressure acting on the mercury reservoir is balanced by the pressure ρgh created by the mercury column. Thus the atmospheric pressure can

be calculated from the height of the column of mercury it can support.

The mechanism of an *aneroid barometer* is shown in Figure 12.4. It consists of an evacuated aneroid capsule, which is prevented from collapsing by a strong spring.

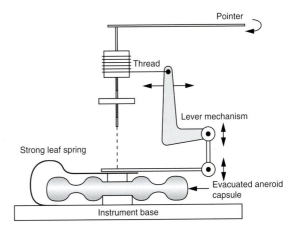

Figure 12.4 Aneroid barometer mechanism

Variations in pressure are felt on the capsule, which causes it to act on the spring. These spring movements are transmitted through gearing and amplified, causing a pointer to move over a calibrated scale.

A common laboratory device used for measuring low pressures is the U-tube manometer, shown in Figure 12.5. A fluid is placed in the tube to a certain level. When both ends of the tube are open to atmosphere, the level of the fluid in the two arms is equal. If one of the arms is connected to the source of pressure to be measured, it causes the fluid in the manometer to vary in height. This height variation is proportional to the pressure being measured.

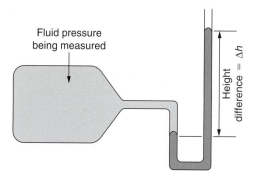

Figure 12.5 The U-tube manometer

The magnitude of the pressure being measured is the product of the difference in height between the two arms Δh, the density of the liquid in the manometer and the acceleration due to gravity, i.e. pressure being measured, *gauge pressure* $= \rho g \Delta h$.

Example 12.4 A mercury manometer is used to measure the pressure above atmospheric of a water pipe, the water being in contact with the mercury in the left-hand arm of the manometer. If the right-hand arm of the manometer is 0.4 m above the left-hand arm, determine the gauge pressure of the water. Take the density of mercury as 13600 kg/m^3.

We know that gauge pressure $= \rho g \Delta h = (13600)(9.81)(0.4) = 53366 \, \text{Pa}$.

12.2 Temperature, its measurement and thermal expansion

12.2.1 Temperature

You will already have met the idea of temperature in your previous studies, but you may not have formally defined it. A formal definition of temperature is as follows: *Temperature is a measure of the quantity of energy possessed by a body or substance. It measures the vibration of the molecules that form the substance.*

These molecular vibrations only cease when the temperature of the substance reaches *absolute zero*, i.e. $-273.15°$C. *Thermodynamic temperature (in the SI system) is always measured in Kelvin and is known as absolute temperature.*

You will also have met the Celsius (or centigrade) temperature scale and the way in which we convert degrees Celsius into Kelvin and vice-versa, but to remind you, see Example 12.5.

Key Point. Temperature measures the energy possessed by the vibration of the molecules that go to make up a substance

Example 12.5 Convert $60°$C into Kelvin and 650 K into degrees Celsius.

You need to remember that $1°$C $= 1$ K and that to convert degrees Celsius into Kelvin we simply add 273. Then $60°$ C $+ 273 = 333$ K.

Also, to convert Kelvin into degrees Celsius we subtract 273; then $650 - 273 = 377°$C.

Note that to be strictly accurate we should add 273.15, but for most practical purposes the approximate value of 273 is adequate.

12.2.2 Temperature measurement

The method used to measure temperature depends on the *degree of hotness* of the body or substance being measured. Measurement apparatus includes liquid-in-glass thermometers, resistance thermometers, thermistors and thermocouples.

All *thermometers* are based on some property of a material that changes when the material becomes colder or hotter. Liquid-in-glass thermometers use the fact that most liquids expand slightly when they are heated. Two common types of liquid-in-glass thermometer are mercury thermometers and alcohol thermometers. Both have relative advantages and disadvantages.

Alcohol thermometers are suitable for measuring temperatures down to $-115°$C and have a higher expansion rate than mercury, so a larger containing tube may be used. They have the disadvantage of requiring the addition of a colouring in order to be seen easily. Also, the alcohol tends to cling to the side of the glass tube and may separate.

Mercury thermometers conduct heat well and respond quickly to temperature change. They do not wet the sides of the tube and so flow well, in addition to being easily seen. Mercury has the disadvantage of freezing at $-39°$C and so is not suitable for measuring low temperatures. Mercury is also poisonous and special procedures must be followed in the event of spillage.

Resistance thermometers are based on the principle that electric current flow becomes increasingly more difficult with increase in temperature. They are used where a large temperature range is being measured, approximately $-200°$C to $1200°$C. *Thermistors* work along similar lines, except in this case they offer less and less resistance to the flow of electric current as temperature increases.

Thermocouple thermometers are based on the principle that when two different metal wires are joined at two junctions and each junction is subjected to a different temperature, a small current will flow. This current is amplified and used to power an analogue or digital temperature display. Thermocouple temperature sensors are often used to measure the temperatures inside engines. They can operate over a temperature range from about $-200°C$ to $1600°C$.

12.2.3   Thermal expansion

We have mentioned in our discussion on thermometers that certain liquids expand with increase in temperature. This is also the case with solids. *Thermal expansion* is dependent on the nature of the material and the magnitude of the temperature increase. We normally measure the linear expansion of solids, such as the increase in length of a bar of the material. With gases (as you have already seen) we measure volumetric or cubic expansion.

Every solid has a *linear expansivity value, i.e. the amount the material will expand in metres per Kelvin or per degree Celsius*. This expansivity value is often referred to as the *coefficient of linear expansion* (α); some typical values of (α) are given in Table 12.3.

Given the length of a material (l), its linear expansion coefficient (α) and the temperature rise (Δt), the *increase in its length* can be calculated using:

$$\text{Increase in length} = \alpha l(t_2 - t_1) \qquad (12.7)$$

Table 12.3 Some linear expansion coefficients for engineering materials

Material	Linear expansion coefficient $\alpha(/°C)$
Glass	9×10^{-6}
Cast iron	10×10^{-6}
Concrete	11×10^{-6}
Steel	12×10^{-6}
Copper	17×10^{-6}
Brass	19×10^{-6}
Aluminium	24×10^{-6}

Note that we are using lower case t to indicate temperature because when we find a temperature difference (Δt) we do not need to convert to Kelvin. However, if you are unsure about when to use Celsius, then for all thermodynamic problems always convert to Kelvin (absolute thermodynamic temperature).

For solids, an *estimate* of the cubic or volumetric expansion may be found using

$$\text{Change in volume} = 3\alpha V(t_2 - t_1) \qquad (12.8)$$

where V is the original volume.

A similar relationship exists for *superficial expansion*, where a body experiences a *change in area*. In this case the linear expansion coefficient is multiplied by 2, therefore:

$$\text{Change in area} = 2\alpha A(t_2 - t_1) \qquad (12.9)$$

where A is the original area.

Example 12.6   An aluminium rod is 0.9 m long at a temperature of $10°C$. What will be the length of the rod at $120°C$?

The increase in length of the rod $= \alpha l(t_2 - t_1) = (24 \times 10^{-6})(0.9)(110) = 2.376 \times 10^{-3}$ m, so length of rod at $120°C = 0.90238$ m.

Example 12.7   A steel bar has a length of 4.0 m at $10°C$.

a)   What will be the length of the bar when it is heated to $350°C$?

b)   If a sphere of diameter 15 cm is made from the same material, what will be the percentage increase in surface area if the sphere is subject to the same initial and final temperatures?

a)   Using $\alpha = 12 \times 10^{-6}$ from Table 12.3, the increase in length of the bar is given, from equation 12.7, by:

$$x = \alpha l(t_2 - t_1)$$
$$= (12 \times 10^{-6})(4.0)(350 - 10) = 0.0163 \text{ m}.$$

This can now be added to the original length, so final length $= 4.0 + 0.0163 = 4.0163$ m.

b) We first need to find the original surface area of the sphere, which is given by:

$$A = 4\pi r^2 = (4\pi)(0.075)^2 = 0.0707 \, \text{m}^2.$$

The increase in surface area of the sphere is given by equation 12.9; i.e. the increase $= 2\alpha A(t_2 - t_1) = (2)(12 \times 10^{-6})(0.0707)(340) = 5.769 \times 10^{-4} \, \text{m}^2.$

Therefore, percentage increase in area

$$= \left(\frac{\text{increase in area}}{\text{original area}} \right) \times 100$$

$$= \frac{5.769 \times 10^{-4}}{0.0707} = 0.82\%.$$

Test your knowledge 12.2

1) Define temperature.

2) Convert a) − 70°C into Kelvin, b) 288 K into °C.

3) You wish to record temperatures between −80°C and 20°C. Which type of thermometer would you choose, and why?

4) The coefficient of linear expansion for copper is 17×10^{-6}. What is the approximation for the superficial expansion coefficient for copper?

5) A metal rod is 30 cm long at a temperature of 18°C and 30.04 cm long when the temperature is increased to 130°C. What is the value of the coefficient of linear expansion of the metal?

12.3 Heat, specific heat and latent heat

12.3.1 Heat

The study of *heat energy* is a necessary foundation for our study of thermodynamics, where *engineering thermodynamics* is concerned with the relationship between heat, work and the properties of systems. As engineers we are concerned with the machines (engines) that convert heat energy from fuels into useful mechanical work. It is therefore appropriate that we

start our study of thermodynamics by considering the concept of *heat energy* itself.

Energy is the most important and fundamental physical property of the universe. Energy may be defined as *the capacity to do work*. More accurately, it may be defined as *the capacity to produce an effect*. These effects are apparent during the process of *energy transfer*.

Key Point. Heat is energy in transit and cannot be stored by matter

A modern idea of heat is that it is energy in transition and cannot be stored by matter. *Heat (Q)* may be defined as: *transient energy brought about by the interaction of bodies by virtue of their temperature difference when they communicate*. Matter possesses *stored energy* but not transient (moving) energy such as heat or work. Heat energy can only travel or transfer from a hot body to a cold body; it cannot travel uphill, Figure 12.6 illustrates this fact.

Key Point. Heat energy can only travel from a hot body to a cold body

---

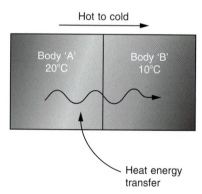

Figure 12.6 Heat energy transfer

When the temperatures of the two bodies illustrated in Figure 12.6 are equal and no change of state is taking place (for example, water to ice), then we say that they are in *thermal equilibrium*.

In fact, the *Zeroth Law of Thermodynamics* states that: *if two bodies are each in thermal equilibrium with a third body, they must also be in thermal equilibrium with each other.*

The unit of heat energy, as with all energy, is the *Joule*, which has identical units to that of the *unit of work*, the N m, as you know from your study of mechanical energy.

When heat flows, then a certain quantity of heat energy flows into or out of a material in a certain time.

> **Key Point.** Heat flow, when measured in Joules per second, is in fact thermal power in Watts

Therefore *heat flow* $\dot{Q}$ = energy/time or $\dot{Q}$ = J/s, is *power in Watts*. Thus, when heat flows into or out of a *thermal system*, power is being either absorbed or given out. *Heat energy flowing into a system is considered to be positive and heat energy flowing from a system is negative.*

12.3.2 Specific heat and thermal energy

From what has been said so far about heat and the expansion of heated materials, it will be apparent that different materials have different capacities for absorbing and transferring thermal (heat) energy.

The thermal energy needed to produce a temperature rise depends on the mass of the material; the type of material and the temperature rise to which the material is subjected. Therefore, the inherent ability of a material to absorb heat for a given mass and temperature rise is dependent on the material itself. This property of the material is known as its *specific heat capacity*.

In the SI system, *the specific heat capacity of a material is the same as the thermal energy required to produce a 1 K rise in temperature in a mass of 1 kg.*

From this definition it can be seen that the SI units of specific heat capacity are Joules per kilogram per Kelvin or, in symbols, J kg^{-1} K^{-1} or J/kg K. Some specific heat capacities for *engineering solids and liquids are* shown in Table 12.4. It should be noted that the value of specific heat capacities may vary slightly, dependent on the temperature at which the heat transfer takes place; therefore the table only gives average values.

> **Key Point.** The specific heat capacity of a substance is that required to raise the temperature of 1 kg of the substance by 1 K

Another way of defining the specific heat capacity of *any substance* is: *the amount of heat energy required to raise the temperature of unit mass of the substance through one degree, under specific conditions.*

In thermodynamics, two specified conditions are used, those of *constant volume* and *constant pressure*. With *gases the two specific heats do not have the same value* and it is essential that we distinguish between them.

If 1 kg of a gas is supplied with an amount of heat energy sufficient to raise its temperature by 1°C or 1 K while the volume of the gas remains constant, then the amount of heat energy supplied is known as the *specific heat capacity at constant volume* and is denoted by c_V.

Note that under these circumstances (Figure 12.7a) *no work is done*, but the gas has received an increase in *internal energy* (*U*) from the heat flowing into the system. *The specific heat at constant volume for air is c_V = 718 J/kg K*; this constant is well worth memorising, for your later work!

If 1 kg of a gas is supplied with an amount of heat energy sufficient to raise the temperature by 1°C or 1 K while the pressure is held constant, then the amount of heat energy supplied is known as the *specific heat capacity at constant pressure* and is denoted by c_p.

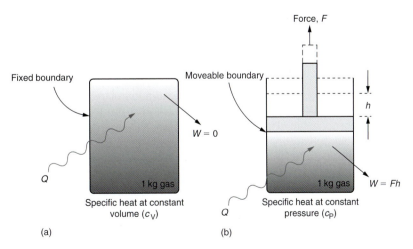

Figure 12.7 Comparison of constant volume and constant pressure specific heats

This implies that when the gas has been heated it will expand a distance h (Figure 12.7b), so work has been done. Thus, for the same amount of heat energy there has been an increase in internal energy (U) from the heat flowing into the system, *plus* work. The value of c_p is, therefore, greater than the corresponding value of c_V.

The *specific heat capacity at constant pressure for air* is $c_p = 1005$ J/kg K; again, this is a constant worth remembering.

> **Key Point.** For a gas, the specific heat at constant pressure will be greater than the specific heat at constant volume because work is done, as well as heat addition

Some common specific heat capacities for engineering solids, liquids and gases are shown in Tables 12.4 and 12.5. Note that in the table for gases both the specific heats at constant volume and constant pressure are given.

Table 12.4 Specific heat capacities for some solids and liquids

Substance	Specific heat capacity (J/kg K)
Aluminium	920
Cast iron	540
Steel	460
Copper	390
Alcohol	230
Ice	700
Mercury	140
Water	4200

Therefore, knowing the mass of any substance and its specific heat capacity, we are now in a position to be able to calculate the thermal energy required to produce any given temperature rise, from the equation:

$$\text{Thermal energy } Q = mc\Delta T \qquad (12.10)$$

where c = *specific heat capacity of the material*, with units J/kg K and ΔT (delta T) is the *temperature difference*. Note that in many books you will see ΔT written as $T_2 - T_1$, indicating the temperature

Table 12.5 Specific heat capacities for gases

Gas	Specific heat capacity (J/kg K)	
	(c_V)	(c_p)
Air	718	1005
Argon	520	710
Carbon dioxide	640	860
Nitrogen	743	1040
Oxygen	660	920
Propane	1490	1680

difference, so that the above formula for heat quantity may be written as $Q = mc(T_2 - T_1)$. Also, you will remember from our earlier discussion on heat that *heat flow* or *thermal power* $\dot{Q} = \dfrac{Q}{t} = \dfrac{\text{heat energy (Joules)}}{\text{time taken (seconds)}}$ in Watts, so that we may write in symbols:

$$\dot{Q} = \frac{mc\Delta T}{t} \qquad (12.11)$$

> **Example 12.8** How much thermal energy is required to raise the temperature of 5 kg of aluminium from 20°C to 40°C? Take the specific heat capacity for aluminium (from Table 12.4) as 920 J/kg K.
>
> All that is required is to substitute the appropriate values directly into equation 12.10:
>
> $$Q = mc\Delta T = (5)(920)(40 - 20)$$
> $$= 92000\,\text{J} = 92\,\text{kJ}.$$

12.3.3 Latent heat

The addition or rejection of thermal energy does not always give rise to a measurable temperature change. Instead, *a change in state, or phase change, may occur*, where the three states of matter are *solid, liquid* and *gas*. So, for example, when an electric kettle boils water, steam is given off. Water boils at 100°C (373 K) and it will remain at this temperature even though heat energy is being added to the water by the element of the kettle, until all the water has turned to steam. Since heat has been added and no measurable temperature rise has taken place, we refer to this heat energy as *latent* heat or hidden heat. When the addition of heat causes a measurable temperature rise in the substance, this is referred to as the addition of *sensible heat* or measurable heat. Figure 12.8 shows the effect of these state changes for water when it is heated from ice to steam. The same graph may be applied to all matter that can exist as a solid, liquid and gas or vapour.

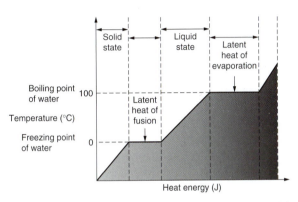

Figure 12.8 Change in temperature and state for water with continuous heat addition

> **Key Point.** Latent heat is heat added to a body without change in temperature

We refer to the thermal energy required to change a solid material into a liquid as *the latent heat of fusion*. For water, 334 kJ of thermal energy is required to change 1 kg of ice at 0°C into water at the same temperature. Thus the *specific latent heat of fusion for water* is 334 kJ. In the case of latent heat, *specific* refers to unit mass of the material, i.e. per kilogram.

So we define the *specific latent heat of fusion of a substance* as: *the thermal energy required to turn 1 kg of a substance from a liquid into a solid without change in temperature.*

Note: The specific latent heat of fusion is also known as the specific *enthalpy (h)* of fusion. As you will see in Chapter 13, where we study thermodynamic systems, *enthalpy* is used as a measure of the heat energy in a fluid when it is flowing and it is the sum of the internal energy *u* of the fluid and its pressure–volume energy (*pv*); in symbols, this specific enthalpy is given as $h = u + pv$.

If we wish to find the thermal energy required to change any amount of a substance from a *solid into a liquid*, then we use the relationship $Q = mL$, where *L* is the *specific latent heat of fusion* of the substance.

In a similar manner to the above argument: *the thermal energy required to change 1 kg of a substance from a liquid into a gas without change in temperature*

is known as the specific latent heat of vaporisation. If we wish to find the thermal energy required to change any amount of a substance from a *liquid into a gas*, we again use the relationship $Q = mL$, but in this case $L =$ the *specific latent heat of vaporisation* of the substance. The specific latent heat of vaporisation for *water is* $= 2.26$ MJ/kg K.

> **Example 12.9**
>
> a) How much heat energy is required to change 3 kg of ice at 0°C into water at 30°C?
>
> b) What thermal energy is required to condense 0.2 kg of steam into water at 100°C?
>
> a) The thermal energy required to convert ice at 0°C into water at 0°C is calculated using the equation $Q = mL$. Substituting values, we get $Q = (3)(334 \times 10^2) = 1.002$ MJ. Now the 3 kg of water formed has to be heated from 0°C to 30°C. The thermal energy required for this is calculated using equation 12.10. Then, taking the specific heat capacity for water from Table 12.4, we find in this case that $Q = mc\Delta T = (3)(4200)(30) = 378000$ J.
>   Then, the total thermal energy required $= 1.002 + 0.378 = 1.378$ MJ.
>
> b) In this case we simply use $Q = mL$ since we are converting steam to water at 100°C, which is the vaporisation temperature for turning water into steam. Then $Q = (0.2)(2.26 \times 10^6) = 452$ kJ.
>
> Note the large amounts of thermal energy required to change the state of a substance.

A liquid does not have to boil in order for it to change state. The nearer the temperature is to the boiling point of the liquid, the quicker the liquid will turn into a gas. At much lower temperatures the change may take place by a process of *evaporation*. The steam rising from a puddle, when the sun comes out after a rainstorm, is an example of evaporation, where water vapour forms as steam, well below the boiling point of the water.

> **Test your knowledge 12.3**
>
> 1. Define *specific heat* for a solid and explain the difference between the specific heat capacities for solids and those for gases.

2. Why, for a *gas*, is the specific heat at constant pressure greater than that at constant volume?

3. Write down the formula for thermal energy and state the meaning of each term.

4. Using the tables of specific heat capacities, determine the thermal energy required to raise the temperature of a copper container of mass 2 kg from 288 K to 348 K. If the time taken to achieve the temperature rise in the copper was 2 minutes, what is the amount of thermal energy being absorbed by the copper during this time?

5. If 3 kg of aluminium requires 54 kJ of energy to raise the temperature from 10°C to 30°C, find the specific heat capacity of the aluminium.

6. How much heat energy is required to change 2 kg of ice at 273 K into water at 313 K? Take the specific latent heat of fusion of water as 334 kJ/kg K and the specific heat of water as 4200 J/kg K.

12.4 Gases and the gas laws

12.4.1 Gases and gas pressure

In a gas the *forces of attraction are very small* and the *molecules can move freely*. Each molecule will, in accordance with Newton's first law of motion, travel in a straight line until it collides either with another molecule or with the walls of the container. Therefore a gas has no particular shape or volume but expands until it fills any vessel into which it has been introduced.

A model known as the *kinetic theory of gases* may be used, along with Newton's first law, to show that for a *perfect gas* (one that obeys the gas laws) the collisions of the gas molecules with the sides of the containing vessel produce *gas pressure*.

Thus the kinetic theory of gases is based on a perfect gas in which the following assumptions are made:

1. All collisions are perfectly elastic and take place in negligible time.

2. The volume occupied by the molecules is negligible compared with the volume of the gas.

3. Forces of attraction between molecules are negligible.

Then, using these assumptions, we know from our study of dynamics, and in particular Newton's second law, that:

$$\text{Force} \; = \; \text{rate of change of momentum} = \frac{m(v - u)}{t}$$

(see section 8.1.2) or

$$\text{Force} \; = \; \frac{\text{change in momentum}}{\text{time taken}}.$$

Then: Force × time = change in momentum or impulse = change in momentum = $m(v - u)$, so the *impulse force* created by molecular collisions with the walls of the containing vessel is: *Impulse = mass × velocity difference*, i.e. the difference in velocity $(v - u)$ of the molecule before and after impact.

In a practical situation there would be millions of such impacts occurring every second on each mm^2 of container surface and their combined effect is the creation of a *steady uniform pressure*.

It is from this kinetic theory that a *set of laws* has been established to model the behaviour of these *ideal* or *perfect gases*.

12.4.2 The gas laws

Now, from what has already been said, we know that the behaviour of gases may be modelled using *the ideal gas laws* where it is assumed that certain gas *processes* (expansion, compression etc.) take place without any external energy changes outside of the thermodynamic system being considered.

In the study of gases we have to consider the interactions between temperature, pressure and volume (remembering that density is mass per unit volume). A change in one of these characteristics always produces a corresponding change in at least one of the other two.

Unlike liquids and solids, gases have the characteristics of being easily compressible and of expanding or contracting readily in response to changes in temperature. Although the characteristics themselves vary in degree for different gases, certain basic laws can be applied to what we have already defined as a perfect gas. That is, a *perfect or ideal gas* is simply one which has been shown, through experiment, to follow or adhere very closely to these gas laws. In these experiments one factor, for example volume, is kept constant while the relationship between the other two is investigated. In this way it can be shown that:

1. *The pressure of a fixed mass of gas is directly proportional to its absolute temperature, provided the volume of the gas is kept constant.* In symbols, $p \propto T$ or, for a fixed mass of gas, provided V remains constant:

$$\frac{p}{T} = \text{constant} \qquad (12.12)$$

The above relationship is known as the *pressure law*.

You know, from the argument given previously, how the pressure of a gas is produced when the gas molecules, being in a state of perpetual motion, constantly bombard the sides of the gas-containing vessel. Therefore, if the temperature of the gas is increased, the velocity of the gas molecules will increase and so they will strike the walls of the container with increased force, which in turn creates an increase in the steady outward pressure from the gas. Figure 12.9 shows how the pressure of the gas varies with temperature.

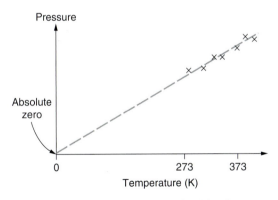

Figure 12.9 Pressure/temperature relationship of a gas

If the graph is 'extrapolated' downwards, in theory we will reach a temperature where the pressure is zero. This temperature is known as *absolute zero* and is approximately equal to -273 K. You met this value earlier, when you studied temperature. The relationship between the Kelvin scale and the Celsius scale is shown in Figure 12.10.

Key Point. When dealing with the gas equations or any thermodynamic relationship we always use absolute temperature (T) measured in Kelvin (K)

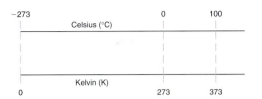

Figure 12.10 Kelvin–Celsius scales

Returning to the gas laws, it can also be shown experimentally that:

2. *The volume of a fixed mass of gas is directly proportional to its absolute temperature, provided the pressure of the gas remains constant.* So for a fixed mass of gas, provided m is fixed and p remains constant:

$$\frac{V}{T} = \text{constant} \qquad (12.13)$$

This relationship is known as *Charles's law*.

A further relationship occurs if we keep the *temperature* of the gas constant. This states that:

3. *The volume of a fixed mass of gas is inversely proportional to its pressure, provided the temperature of the gas is kept constant.*
In symbols, $p \propto \frac{1}{V}$ or, for a fixed mass of gas:

$$pV = \text{constant} \qquad (12.14)$$

This relationship is better known as *Boyle's law*. It is illustrated in Figure 12.11.

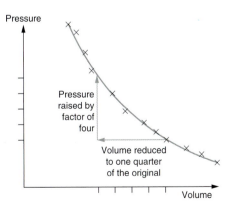

Figure 12.11 Boyle's pressure/volume relationship

In dealing with problems associated with the gas laws, remember that we assume that all gases are *ideal* or *perfect*. In reality no gas is ideal, but at low and medium pressures and temperatures many gases behave in an ideal way.

The pressure law, Charles's law and Boyle's law can all be expressed in terms of one single equation known as the *combined gas equation*. This is, for a fixed mass of gas:

$$\frac{pV}{T} = \text{constant} \qquad (12.15)$$

If we consider a fixed mass of gas before and after changes have taken place, then, from the combined gas equation, it follows that:

$$\frac{p_1 V_1}{T_1} = \frac{p_2 V_2}{T_2} \qquad (12.16)$$

where the subscript 1 is used for the *initial state* and subscript 2 for the *final state* of the gas. The above relationship is very useful for solving problems concerned with the gas laws.

Key Point.   A perfect gas is one that is assumed to obey the ideal gas laws

Example 12.10   A quantity of gas occupies a volume of 0.5 m^3. The pressure of the gas is 300 kPa when its temperature is $30°C$. What will be the pressure of the gas if it is compressed to half its volume and heated to a temperature of $140°C$?

When solving problems involving several variables, always tabulate the information given in appropriate units.

$$p_1 = 300 \text{ kPa}, \quad p_2 = ?$$

$$V_1 = 0.5 \text{ m}^3, \quad V_2 = 0.25 \text{ m}^3$$

$$T_1 = 303 \text{ K}, \quad T_2 = 413 \text{ K}$$

Remember to convert Celsius temperatures to Kelvin by adding 273 to the figure. Then, using equation 12.16 and after rearrangement, we find that:

$$p_2 = \frac{p_1 V_1 T_2}{T_1 V_1} = \frac{(300)(0.5)(413)}{(303)(0.25)} = 817 \text{ kPa}.$$

12.4.3   The characteristic gas equation

The combined gas law, equation 12.15, tells us that for a perfect gas with *unit* mass $\frac{pV}{T} = \text{constant}$. This relationship is of course true for *any fixed mass of gas*, so we can write that $\frac{pV}{T} = \text{mass} \times \text{constant } (R)$.

Now, for any perfect gas which obeys the ideal gas laws this constant R is *specific* to that particular gas, i.e. *R is the characteristic gas constant* or specific gas constant for the individual gas concerned. Therefore, *the characteristic gas equation* may be written as $\frac{pV}{T} = mR$ or as:

$$pV = mRT \qquad (12.17)$$

The units for the characteristic gas constant are the Joule per kilogram-Kelvin (J/kg K).

Note that when the above equation is used, both *absolute pressure and absolute temperature must be used*.

The characteristic gas constants for a number of gases are given in Table 12.6.

Table 12.6 Characteristic gas constants

Gas	Characteristic gas constant (J/kg K)
Hydrogen	4124
Helium	2077
Nitrogen	297
Air	287
Oxygen	260
Argon	208
Carbon dioxide	189

The characteristic gas constant for air, from the above table, is $R = 287$ J/kg K. This is related to the *specific heat capacities* (that you met earlier) for air in the following way, i.e.:

$$R = c_p - c_V \qquad (12.18)$$

You may wish to check this relationship by noting the values of R, c_p and c_V for air, as given in Table 12.5.

Note also that equation 12.18 is not only valid for air, it is also *valid for any perfect gas* that follows the ideal laws.

There is one more important relationship concerning the specific heat capacities that will be of use to you later, when you study thermodynamic processes in Chapter 14, and this is *the ratio of specific heats* (γ), thus:

$$\gamma = \frac{c_p}{c_V} \qquad (12.19)$$

Example 12.11 0.22 kg of gas at a temperature of 20°C and a pressure of 103 kPa occupies a volume of 0.18 m³.

If the c_V for the gas $= 720$ J/kg K, find:

a) the characteristic gas constant of the gas;

b) the specific heat capacity at constant pressure.

a) Using equation 12.17, then, after rearrangement, $R = \dfrac{pV}{mT} = \dfrac{(103 \times 10^3)(0.18)}{(0.22)(293)} = 288$ J/kg K.

b) From equation 12.18 we get that $c_p = R + c_V = 288 + 720 = 1008$ J/kg K.

Apart from the form of the characteristic gas equation given in equation 12.17, there is yet another way of representing this equation, this is in *molar* form. In order for us to derive this form of the equation, it is necessary to define the *mole*.

The mole is the amount of substance of a system which contains as many elementary entities as there are atoms in 12 grams of carbon-12.

When the mole is used, the elementary entities must be specified and may be atoms, molecules, ions, electrons or other particles. The unit symbol used for the mole is *mol* and in the SI system it is often convenient to use the *kmol*. Now, the mass per kilogram of any substance is known as the molar mass (M), so that:

$$m = nM \qquad (12.20)$$

where m is in kg, n is the number of *kilomoles* and M is in kg/kmol.

Relative masses of the various elements are commonly used, and physicists and chemists agreed in 1960 to give the value of 12 to the isotope 12 of carbon (this led to the definition of the mole, as given above). A scale is thus obtained of relative atomic mass or relative molecular mass. For example, the relative atomic mass of the oxygen element is 16, the *relative molecular mass* of oxygen gas (O_2) is approximately 32. The relative molecular mass is numerically equivalent to the molar mass, M, but is dimensionless.

Now, substituting for m from equation 12.20 into the characteristic gas equation 12.17 gives: $pV = nMRT$ or:

$$MR = \frac{pV}{nT} \qquad (12.21)$$

Now, *Avogadro's hypothesis* states that the volume of 1 mole of any gas is the same as the volume of 1 mole of any other gas, when the gases are at the same value of pressure (p) and temperature (T). This tells us that the quantity pV/nT is a constant for all gases. This constant is called the *molar gas constant* or *universal gas constant* and is given the symbol R_0, so $MR = R_0 = \dfrac{pV}{nT}$ or:

$$pV = nR_0 T \qquad (12.22)$$

Also, from $MR = R_0$,

$$R = \frac{R_0}{M} \qquad (12.23)$$

The value of R_0 has been shown to be 8214.4 J/kmol K.

Equation 12.22 enables us to find the specific (characteristic) gas constant for any gas when the molar mass is known; for example, for oxygen, which has a molar mass of 32 kg/kmol, the characteristic gas constant $R = \dfrac{R_0}{M} = \dfrac{8314.4}{32} = 259.8$ J/kg K.

Key Point. The *relative molecular mass* is numerically equal to the molar mass M but is *dimensionless*

Example 12.12 For the gas given in Example 12.11, find its molecular mass and its ratio of specific heats.

From Example 12.11 we know that this gas has a characteristic gas constant $R = \dfrac{pV}{mT} = 288$ J/kg K, so that it has a molecular mass $M = \dfrac{R_0}{R} = \dfrac{8314}{288} = 28.9$ kg/kmol. Also from Example 12.11 we know that $c_V = 720$ J/kg K and $c_p = 1008$ J/kg K so that, from equation 12.19, the ratio of specific heats is $\gamma = \dfrac{c_p}{c_V} = \dfrac{1008}{720} = 1.4$.

1. Define 'perfect gas'.

2. When using the *pressure law*, what parameter is assumed to remain constant?

3. According to Boyle's law, what happens to the volume of gas if the pressure is tripled?

4. Write down the SI units for the characteristic gas constant and explain their meaning in words.

5. The volume of a fixed mass of gas is 60 cm^3 at 27°C. If the pressure is kept constant, what is the volume at 127°C?

6. A sealed oxygen cylinder has a pressure of 10 bar at 0°C. Calculate:

   a) the new pressure when the temperature is 90°C;

   b) the temperature when the pressure increases by 50%.

7. A quantity of gas occupies a volume of 0.5 m^3. The pressure of the gas is 300 kPa when the temperature is 300 K, and the pressure is 900 kPa when the temperature is raised to 400 K. What is the new volume of the gas?

8. 0.25 kg of a gas at a temperature of 300 K and a pressure of 111.4 kPa occupies a volume of 0.2 m^3. If the c_V for the gas = 743 J/kg K, find:

   a) the characteristic gas constant;

   b) the molecular mass and relative molecular mass;

   c) the specific heat capacity at constant pressure;

   d) the ratio of specific heats.

12.5 Chapter summary

This chapter has been designed to provide you with some of the fundamental concepts, principles and definitions you will need for your study of thermodynamics and fluids within this part of the book.

We started by considering the key concepts of density and pressure. When considering *density* (equation 12.1) and *relative density* (equation 12.3) we introduced two further parameters, those of *specific volume* (equation 12.2, where v_s has units of m^3/kg) and *specific weight* (equation 12.4, where γ has units of N/m^3). You will see the usefulness of these two parameters when we study *fluid dynamics*. When discussing relative density we included the RD of gases, but in practice we rarely use this parameter because of the large changes in the density of gases with change in temperature. We went on to discuss pressure; Table 12.2 gives a range of SI units for pressure. *We will use these various units for pressure as appropriate* throughout this part of the book. You should therefore be *fluent in being able to convert from one set of units* to the other. We also defined *hydrostatic pressure* (equation 12.6) or *gauge pressure* and its relationship to *absolute pressure*, i.e. that *absolute pressure = gauge pressure + atmospheric pressure*. We then went on to look at one or two methods for measuring atmospheric and gauge pressure (Figures 12.3 to 12.5). In section 12.2 we also revised the idea of temperature and its measurement, using both the Celsius and Kelvin scales. This led us into the idea of temperature change and thermal expansion, where we found that changes in length, volume and surface area (equations 12.7–12.9) of a substance were dependent on the expansivity of the material and the temperature change to which the material had been subjected. Table 12.3 details the expansion coefficients for some of the more common engineering materials.

In section 12.3 we considered the nature of heat energy, where we discovered that *heat (thermal) energy (Q)* is *energy in transit* and so cannot be stored in matter but can only be brought about by the *interaction of bodies by virtue of their temperature difference*. The *thermal energy (Q)* required for any given temperature rise can be found using equation 12.10. When *heat flows*, *thermal power* is produced, since heat flow is

$$\dot{Q} = \frac{Q}{t} = \frac{\text{heat energy (Joules)}}{\text{time taken (seconds)}} \text{ (Watts); this may be}$$

found quantitatively using equation 12.11. *Specific heat capacities* for solids/liquids and gases were defined and some of the more common specific heat capacities are detailed in Tables 12.4 and 12.5. Do remember that for *gases* there are *two specific heat capacities* (c_V and c_p) that need to be taken into account, dependent on conditions, as you will see later when we consider thermodynamic processes in Chapter 14. In the final part of section 12.3 we looked at the idea of latent heat, where with heat addition there was no measurable rise in the temperature of the substance, rather the extra heat energy was used to change the state of this substance. We referred to this *specific latent heat* of a substance as the *latent heat of fusion* when the substance was converted from a *solid to a liquid* and as the *latent heat of vaporisation* when converted from a *liquid to a gas*, both without temperature change. To find the energy required to change the state of any amount of a substance we used the relationship $Q = mL$, where L = the *specific latent heat of fusion* (solid to liquid) or *vaporisation* (liquid to gas), dependent on the change of state being considered.

In section 12.4 we considered the nature of gases, gas pressure and the gas laws, including the derivation of the characteristic gas equation. We showed initially how pressure was created in contained gases on the basis of *Newton's laws* and the *kinetic theory of gases*, where this theory was based on certain assumptions about a *perfect gas*, i.e. a gas that obeys or very nearly obeys the ideal gas laws. You should remember that for *perfectly elastic collisions, momentum is conserved*. We looked in turn at the three basic gas laws, i.e. the *pressure law*, *Charles's law* and *Boyle's* law (see equations 12.12 to 12.14), and saw how they could be combined into the *combined gas law* (equations 12.15 and 12.16), which is particularly useful in determining parameters between an initial state and final state of the gas being considered. In the final part of this section we derived the *characteristic gas equation* and introduced the concept of the *characteristic gas constant R* which has units of J/kg K (see Table 12.6 for some common values). We also showed the relationship between R and the *specific heat capacities* (equation 12.18) which, together with the *ratio of specific heats* (equation 12.19) will prove useful in your later work on gases. Finally, we derived the characteristic gas equation in its molar form (equations 12.21 and 12.22), where we also defined the *mole, Avogadro's hypothesis*, the *molecular mass* (M in kg/kmol), the *relative molecular mass* and the *universal gas constant R_0*, where $R_0 = RM = 8214.4$ J/kmol K. Apart from these parameters being useful in the analysis of gas processes, they will also be useful later when we consider *combustion processes*.

12.6 Review questions

1. A rectangular container has dimensions of 2.0 m × 0.5 m × 0.8 m. A certain liquid that fills this container has a mass of 700 kg. What is the density and relative density of this liquid?

2. A metal rod is 25 cm long at a temperature of 20°C and 25.05 cm long when the temperature is increased to 140°C. What is the value of the coefficient of linear expansion of the metal?

3. How much heat energy is required to raise the temperature of 5 kg of aluminium by 30°C? Take the specific heat capacity of aluminium from Table 12.4.

4. Define: a) temperature; b) heat; c) heat flow.

5. Define the specific latent heat of fusion *and* of vaporisation.

6. How much thermal energy is required to turn 5 kg of ice at 0°C into water at 50°C? Take the specific latent heat of ice as 334 kJ/kg K and the specific heat of water as 4200 J/kg K.

7. A 200 W electric heater is embedded in a 1 kg block of ice. Taking the specific latent heat of ice as that given in question 6, determine:

   a) the mass of ice that has melted after 10 minutes;

   b) the length of time it takes for all the ice to melt.

8. Define: a) a perfect gas; b) Boyle's law; c) the combined gas law.

9. The air pressure of a motor vehicle tyre is 2 bar at 290 K. The temperature rises to 320 K after the vehicle has been travelling. Assuming there is no change in the volume of the tyre, determine the new tyre pressure in pascal.

10. Define: a) c_V b) c_p c) γ d) R.

11. 0.06 m³ of air is contained in a sealed cylinder at a pressure of 1.39 bar and a temperature of 17°C. A quantity of heat energy is supplied to increase the temperature of the air to 97°C. Given that the characteristic gas constant for air is 287 J/kg K and its coefficient $c_V = 718$ J/kg K, determine:

   a) the mass of the air;

   b) the density of the air at the initial conditions;

   c) the heat energy supplied during the process;

   d) the final pressure of the air.

12. 0.18 kg of a gas at a temperature of 288 K and a pressure of 103 kPa occupies a volume of 0.15 m³. If the c_V for the gas = 722 J/kg K, find:

   a) the characteristic gas constant;

   b) the molecular mass and relative molecular mass;

   c) the specific heat capacity at constant pressure;

   d) the ratio of specific heats.

Chapter 13

Thermodynamic systems

In this chapter we cover the ideas and concepts associated with thermodynamic systems. In particular, we start with some basic definitions and look at the properties that help to define the state of the system. We then look at closed and open systems and the application of the first law of thermodynamics to these systems. Next, we consider the transfer of work, energy and power in and out of the system, using the non-flow (NFEE) and the steady flow (SFEE) energy equations, and introduce the idea of enthalpy and specific enthalpy of the working fluid in open systems. Finally, we look briefly at the second law of thermodynamics in terms of the direct acting and reverse acting heat engine and its use as a measure of system thermal efficiency, leaving the more difficult concept of entropy until Chapter 15.

13.1 System definitions and properties

A thermodynamic system may be defined as: a particular quantity of a thermodynamic substance (normally a compressible fluid such as a vapour or gas) that is surrounded by an identifiable boundary.

We often talk about the system boundary as the control surface, where the control surface encloses a region in space, the control volume, as follows:

a) Inside the space there is a fixed mass of fluid whose behaviour is being investigated.

b) Across the space there may be a mass transfer of fluid at the inlet and outlet whose thermodynamic state is changing and is under investigation.

As you will see later, case a) describes the situation for *closed systems*, while b) describes the situation for *open systems*.

We are particularly interested in thermodynamic systems which involve *working fluids* because these fluids enable the system *to do work* or *have work done upon it*. *Transient energies* in the form of *heat* (Q) and *work* (W) can cross the system boundaries, and as a result there will be a change in the stored *internal energy* (U) of the contained substance (the working fluid).

Key Point. A thermodynamic system is essentially a thermodynamic substance surrounded by an identifiable boundary

Now, in order for a system to be classified as a *thermodynamic system* it must contain the following elements:

- A working fluid or substance, i.e. the matter which may or may not cross the system boundaries (control surface), such as water, steam, air etc.
- A heat source
- A cold body to promote heat flow and enable heat energy transfer
- System boundaries (control surface), that may or may not be fixed.

978-1-85617-775-7, Engineering Science, Mike Tooley & Lloyd Dingle

The *property* of a *working fluid* may be defined as: an observable quantity or characteristic of the system, such as pressure (p), temperature (T), volume (V), internal energy (U) etc.

Now, the *state of a working fluid* may be defined by *any two unique and independent properties*. Boyle's law, for example, defines the state of the fluid by specifying the independent thermodynamic properties of *volume* and *pressure*.

Some of the more common *properties* (with units) that are used to define the *state* of the working matter within a thermodynamic system are given in Table 13.1.

Table 13.1 Thermodynamic properties

Property	Units
1) Absolute pressure (p)	pascal (Pa), bar, N/m^2
2) Absolute temperature (T)	Kelvin (K)
3) Volume (V)	cubic metres (m^3)
4) Internal energy (U)	Joule (J)
5) Enthalpy (H)	Joule (J)
6) Specific volume	m^3/kg
7) Specific internal energy (u)	J/kg
8) Specific enthalpy (h)	J/kg

Note: the *first five* properties given in Table 13.1 are known as *intensive* properties because they are *independent of the mass* of the working substance. The *last three* are *extensive* or *specific* properties because they are *dependent on the mass* of the working substance, as can be seen by their units.

Key Point. Extensive or specific properties of a thermodynamic system are dependent on the mass of the working substance

When a system working fluid is subject to a *process*, then the fluid will have started with one set of properties and ended with another, irrespective of how the process took place or what happened between the start and end states. For example, if a fluid within a system has an initial pressure p_1 and temperature T_1 and is then compressed, producing an increase in pressure and temperature to p_2 and T_2, respectively, then we say that the *fluid has undergone a process from state 1 to state 2*.

A system that undergoes *a set of processes* whereby the working fluid returns to its original state is said to have undergone *a cycle* of processes. Much more will be said about thermodynamic processes and cycles in Chapters 14 and 15.

13.2 Closed and open systems

13.2.1 Closed systems

In the **closed system** there is a *closed or fixed boundary containing a fixed amount of the system working fluid*, while only an exchange of heat and work energy may take place across the system boundaries (control surface). An *energy* diagram of a typical closed system is shown in Figure 13.1.

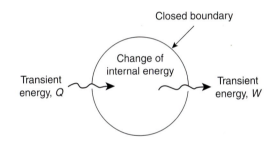

Figure 13.1 Closed system energy exchange

In Figure 13.1 the transient heat energy (Q) is seen entering the system, where it causes a change in internal energy ($\Delta U = U_2 - U_1$) and transient energy as work (W) is seen leaving the system.

Key Point. In a closed system there is no mass transfer of system fluid

The boundary of a closed system is not necessarily rigid. What makes the system closed is the fact that no mass transfer of the system working fluid takes place, although an interchange of heat and work (energy in transit) may take place.

A classic example of a *closed system* with *moving boundaries* that undergoes a *cycle of processes* is *the reciprocating piston internal combustion engine* (Figure 13.2).

Key Point. When a system undergoes a series of processes and finally returns to its initial state, we say that a *cycle* of processes has taken place

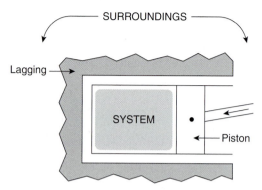

Figure 13.2 Internal combustion engine cylinder and piston assembly

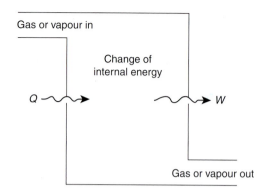

Figure 13.3 Energy exchange for a typical open system

The *closed boundary* (control surface) is formed by the crown of the piston, the cylinder walls and the cylinder head with the valves closed, the *transient energy* being in the form of combustible fuel that creates a sudden pressure wave which forces the piston down (out); therefore, as the piston moves, the boundaries of the system move. This movement causes the system to do *work* (force × distance) on its surroundings; in this case the piston connecting rod drives a crank to provide motive power.

Notice that in a *closed system, it requires movement of the system boundary for work to be done by the system or on the system.* Thus work (like heat) is a *transient energy* as mentioned previously; it is *not* contained within the system. Also, remember that in a closed system there is *no mass transfer of system fluid across the system boundary* while the interchange of the transient energies, heat (Q) and work (W), takes place.

13.2.2 Open systems

In the **open system** there is an open pathway across the system boundaries that allows the *mass transfer of fluid* to take place while the transient energies of heat (Q) and work (W) are being interchanged. This is the situation for case b) of our system definition where, *across the control volume, there may be a mass of fluid at the inlet and outlet whose thermodynamic state is changed as a result of its journey through the system.* An energy diagram for an open system is shown in Figure 13.3.

A practical example of an open system is an aircraft *gas turbine engine* (Figure 13.4). In this system there is a *transfer of mass across the system boundaries* in the form of a fluid flow that possesses its own *kinetic energy*, *pressure energy* and in some cases *potential energy*, as it enters the system.

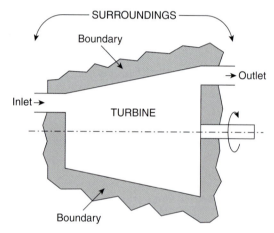

Figure 13.4 An open system gas turbine

This energetic fluid (air) as it passes through the system is subject to an interchange of transient energies in the form of heat and work, before exiting the system.

13.3 Closed systems and the first law of thermodynamics

The **first law of thermodynamics** is essentially a conservation law based on observations that heat and work are two mutually convertible forms of energy. When applied to open and closed systems, the law may be stated formally as follows:

When a system undergoes a thermodynamic cycle, then the net heat supplied to the system from its surroundings is equal to the net work done by the system on its surroundings.

In other words, *the total energy entering a system is equal to the total energy leaving the system*. Note that this statement takes no account of the energy losses that must occur in *real systems*. As you will see later, it is the *second law* that takes this fact into account.

13.3.1 First law of thermodynamics applied to a closed system

We now consider the application of the first law to a closed system, where it is represented diagrammatically in Figure 13.5.

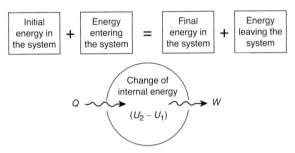

Figure 13.5 First law of thermodynamics applied to a closed system

From Figure 13.5 it can be seen that the initial internal energy is U_1 and the final internal energy is U_2, so *the change in internal energy* is shown as ($U_2 - U_1 = \Delta U$). So, in symbol form, the energy exchange may be represented as $U_1 + Q = U_2 + W$ (i.e. total energy in =total energy out). In its more normal form:

$$Q - W = (U_2 - U_1) = \Delta U \qquad (13.1)$$

Equation 13.1 is known as the *non-flow energy equation* (*NFEE*).

Heat and work energy transfer are given a sign convention, as shown in Figure 13.5. Heat entering a system is positive, work leaving a system is positive. Another way of expressing the same thing is to say that *heat supplied to the system, or done on the system, is positive* and that *work output or work done by the system is also positive*. Naturally, the inverse applies, i.e. heat done by the system or leaving the system is negative and work done on the system or entering the system is also negative.

Example 13.1   During a non-flow thermodynamic process the internal energy possessed by the working fluid within the system was increased from 10 kJ to 30 kJ, while 40 kJ of work was done by the system. What is the magnitude and direction of the heat energy transfer across the system during the process?

Using equation 13.1, $Q - W = (U_2 - U_1)$, where $U_1 = 10$ kJ, $U_2 = 30$ kJ and $W = 40$ kJ (positive work, i.e. work done by the system on the surroundings). Then, $Q - 40 = 30 - 10$ and so $Q = 60$ kJ and, since Q is *positive*, it must be *heat supplied to the system*, which may be represented by an arrow pointing into the system, as shown in Figure 13.5.

Equation 13.1 is one of the most important equations in thermodynamics; being concerned with the transfer of energy that takes place between the system and its surroundings. From this equation and *Joule's law*, we are able to deduce one or two more relationships that are fundamental to an understanding of *thermodynamic processes*, which you will study in Chapter 14, and will be of use here to aid your understanding of the behaviour of perfect gases in thermodynamic systems.

13.3.2 Joule's law and the NFEE

Joule carried out experiments from which he was able to deduce that the *change of internal energy* of a *perfect gas* within a system was *directly proportional to temperature*, i.e. $\Delta U \propto \Delta T$.

He also found that no matter what the type of process, or the way in which he applied the heat to the system, the constant of proportionality was always equal to mc_V (where c_V is the same *specific heat capacity at constant volume* you met earlier in section 12.3.2). Thus, from his experiments on a closed system, where both heat and work transfers may take place, Joule was able to establish the law that relates the changes in internal energy directly to changes in temperature as $\Delta U = mc_V \Delta T$, so, from Joule's law, we may write that:

$$\Delta U = mc_V(T_2 - T_1) \qquad (13.2)$$

Also, from the NFEE (equation 13.1), after substituting equation 13.2 for ΔU, we find that:

$$\Delta U = Q - W = mc_V(T_2 - T_1) \qquad (13.3)$$

> **Example 13.2** 0.4 kg of a perfect gas in an enclosed rigid metal cylinder is subject to 20 kJ of heat energy. If, for the gas, $c_V = 718$ J/kg K, determine the temperature change of the gas.
>
> This is a closed system with *rigid* boundaries (i.e. *constant volume*) where, for the perfect gas, no work takes place, and so all the heat addition is used to increase the internal energy. So, from equation 13.3, $Q = \Delta U$, so that
>
> $$\Delta U = mc_V(T_2 - T_1), \text{ so}$$
>
> $$(T_2 - T_1)$$
>
> $$= \frac{\Delta U}{mc_V} = \frac{20,000}{(0.4)(718)} = 69.6 \text{ K}.$$
>
> Therefore, the contained gas will rise in temperature by 69.6 K.

13.4 Open systems and the first law of thermodynamics

13.4.1 Open systems and the steady flow energy equation (SFEE)

In an **open system**, the fluid is continuously flowing in and out of the system while heat and work transfers take place. Therefore, we need to consider *all of the stored energies* possessed by the fluid as it passes through the system. These are:

1. Flow or pressure energy = pressure × volume = pV.

2. Potential energy = mgz (notice here we use z instead of h for height or elevation; the reason for this will become clear later).

3. Kinetic energy = $\frac{1}{2}mv^2$.

Now, applying the conservation of energy (the first law) to the open system shown in Figure 13.6, then:

Total energy in = total energy out.

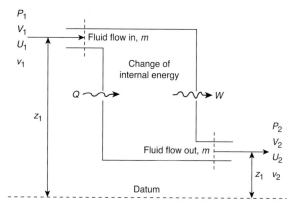

Figure 13.6 First law of thermodynamics applied to an open system

So, (transient energy in + stored energy in = transient energy out + stored energy out) or:

Heat energy $+ (\text{IE}_1 + \text{ press E}_1 + \text{ PE}_1 + \text{ KE}_1)$

$\qquad = \text{work energy} + (\text{IE}_2 + \text{ press E}_2 + \text{ PE}_2 + \text{KE}_1)$

So that in symbol form we have:

$$Q + U_1 + p_1V_1 + mgz_1 + \frac{1}{2}mv_1^2$$

$$= W + U_2 + p_2V_2 + mg_2 + \frac{1}{2}mv_1^2.$$

Rearranging gives

$$Q - W = (U_2 - U_1) + (p_2V_2 - p_1V_1)$$

$$+ (mgz_2 - mgz_1) + \left(\frac{1}{2}mv_2^2 - \frac{1}{2}mv_1^2\right)$$

$$(13.4)$$

Equation 13.4 is known as the *steady flow energy equation (SFEE)* and results from applying the first law, the conservation law, to the system. Like the NFEE, the SFEE is also a very important relationship in thermodynamics that will prove of great benefit later when you study flow processes and cycles.

13.4.2 Specific form of the SFEE

When dealing with *flow systems, where there is a mass transfer of fluid*, it is convenient to group the internal energy (U) and pressure energy (pV) of the fluid together. When this is done, another property of the fluid called *enthalpy* is used for this combination.

So, the enthalpy $(H) =$ internal energy $(U) +$ pressure energy (pV) and, in symbols:

$$H = U + pV \qquad (13.5)$$

Now it is also a feature of open systems that the *stored energy terms* are a function of *fluid mass flow rate* (kg/s). It is, therefore, convenient to work in *specific mass energies*, i.e. energy per kilogram of fluid (see Table 13.1, items 6, 7 and 8); i.e. in SI, the symbols and units for the individual specific energies are:

1. Specific internal energy (u) $\qquad = (\text{J/kg})$
2. Specific pressure energy (pV_s/kg) $\quad = (\text{J/kg})$
3. Specific enthalpy $(h = u + pV_s)$ $\quad = (\text{J/kg})$
4. Specific potential energy (gz) $\qquad = (\text{J/kg})$
5. Specific kinetic energy $(1/2\, v^2)$ $\qquad = (\text{J/kg})$

Then the steady-flow energy equation (SFEE) *in specific terms* may be written as:

$$Q - W = (u_1 - u_2) + (p_2 V_{s2} - p_1 V_{s1})$$
$$+ (gz_2 - gz_1) + \left(\frac{1}{2}v_2^2 - \frac{1}{2}v_1^2\right) \qquad (13.6)$$

Note:

1) You can see from the symbols given above that specific enthalpy is given the symbol h, which might have been confused with *height* in the potential energy term if we had used it. This is the reason for using z for height or elevation in thermodynamics.

2) Throughout our study of thermodynamics and fluids we will use the symbols (V) *for volume* and (V_s) for *specific volume*, in order to avoid confusion with our symbol (v) for velocity. In some textbooks, to avoid any possible confusion, they use the symbol C or c for velocity; however, in this book we are reserving this symbol for the *speed of sound*.

> **Key Point.** The enthalpy of a system fluid is its internal energy plus its pressure-volume energy

Note: Equation 13.6 infers that the *heat* and *work energy transfers* (in addition to all other energies in the equation) *are also in specific terms*, with units in J/kg. A little more needs to be said about the units we deal with when considering energy in specific terms, first that a *more common unit for the specific energies is to*

quote them in kJ/kg. When we do this, great care needs to be taken with the *specific kinetic energy terms*, that have units of $\dfrac{m^2}{s^2}$ *per unit mass*, from the manipulation of base units where

$$\frac{1}{2}v^2 = \frac{\text{kg m}^2}{\text{s}^2\text{kg}} = \frac{\text{m}^2}{\text{s}^2},$$

so that the units for any calculations involving the kinetic energy terms will always be in J/kg and can only be converted to kJ/kg *after* calculations have taken place (see Example 13.4 for what this means in practice).

Remember also that for open systems the *working fluid flows* through the system, so that *mass* and/or *volume flow rates* play an important part in determining the heat, work and power transfers that the fluid imparts to and from the system. Thus, the *mass flows* $(\dot{m})$ *rate* may be found from the relationship:

$$\dot{m} = \rho A v \qquad (13.7)$$

where $\dot{m}$, the *mass flow rate*, has units kg/s, the density ρ has units kg/m^3, the *cross-sectional area (A)* of the *control space* through which the fluid flows has units m^2 and the fluid velocity v has units m/s. From equation 13.7 it can easily be seen that the *volume flow rate* of the fluid $\dot{V}$, with units m^3/s, is given by:

$$\dot{V} = A v \qquad (13.8)$$

> **Key Point.** Do not confuse the symbols for volume (V), specific volume (V_s) or volume flow rate $(\dot{V})$ with velocity (v)

13.4.3 Specific enthalpy

There is yet one more important relationship that is used consistently in *steady flow* systems. This is the **change in specific enthalpy** of the working fluid between two set points in the system, specific enthalpy being a measure of the *total energy* in 1 kg of the *flowing fluid*. At this stage in our study of thermodynamic systems, we will only be considering *gases* as the fluid.

We know from our definition (equation 13.5) that enthalpy $H = U + pV$ and so specific enthalpy $h = u + pV$. Therefore, the *change in specific enthalpy* will be $h_2 - h_1 = (u_2 + p_2 V_2) - (u_1 + p_1 V_1) = (u_2 - u_1) + (p_2 V_2 - p_1 V_1)$ or:

$$h_2 - h_1 = (u_2 - u_1) + (p_2 V_{s2} - p_1 V_{s1}) \qquad (13.9)$$

Now, substituting equation 13.9 into equation 13.6 gives the SFEE in terms of *specific enthalpy change* as:

$$Q - W = (h_2 - h_1) + (mgz_2 - mgz_1)$$

$$+ \left(\frac{1}{2}v_2^2 - \frac{1}{2}v_1^2 \right) \qquad (13.10)$$

Also, for a perfect gas, we know from equation 13.2 that in specific terms $u_2 - u_1 = c_V(T_2 - T_1)$. Then, from equation 12.17, where $pV = mRT$, for *change* in the pV energy between two states, in specific terms (i.e. per unit mass) we may write that $p_2V_2 - p_1V_1 = R(T_2 - T_1)$. Therefore, substituting these relationships for $(u_2 - u_1)$ and $(p_2V_2 - p_1V_1)$ in equation 13.10 gives $h_2 - h_1 = (c_V + R)(T_2 - T_1)$, and remembering from equation 12.18 $(R = c_p - c_V)$ that $c_p = c_V + R$, we may then write that

$$h_2 - h_1 = c_p(T_2 - T_1) \qquad (13.11)$$

This relationship tells us that the total energy change of the gas is directly proportional to the change in temperature, as Joule predicted!

Example 13.3   At entry to a horizontal steady flow system, a gas has a specific enthalpy of 2000 kJ/kg and possesses 250 kJ/kg of kinetic energy. At the outlet from the system, the specific enthalpy is 1200 kJ/kg and there is negligible kinetic energy. If there is no heat energy transfer during the process, determine the magnitude and direction of the work done.

Using the specific enthalpy form of the SFEE equation, equation 13.10, we first note that the specific potential energy term $(gz_2 - gz_1) = 0$, since there is no change in height between fluid at entry and fluid at exit (horizontal). Also, there is negligible fluid kinetic energy at exit; in other words $1/2 \, v_2^2 = 0$ and during the process $Q = 0$.

Therefore, substituting appropriate values into the SFEE gives

$$0 - W = (1200 - 2000) + 0 + (0 - 250) \text{ so}$$

$$W = 1050 \text{ kJ/kg}$$

and, as the work is positive, then the work done by the system = 1050 kJ/kg.

In Example 13.4 we consider an *air compressor* as our *open system* where the heat and work transfers imparted on the working fluid as it enters the system, and the

resulting change of state as it exits the system, enable us to determine the required system parameters.

Example 13.4   Steady flow air enters the inlet of a horizontal compressor at a velocity of 10 m/s, a pressure of 1 bar and a temperature of 15°C, and exits the compressor flowing at 0.6 kg/s, with a velocity of 25 m/s, a pressure of 7 bar and a temperature of 65°C. If the input power to the compressor is 42 kW and for the air $c_p = 1.005$ kJ/kg K and $R = 0.287$ kJ/kg K, determine:

a)   the specific heat energy lost to the surrounding;

b)   the cross-sectional area of the compressor exit duct.

a)   To find the specific heat energy lost to the surroundings, we use equations 13.10 and 13.11. In our case, because the potential energy of the air = 0 (negligible difference in height between inlet and exit of the compressor), then the specific energy changes imparted to the air and the heat lost to the surrounding are given by the relationship:

$$Q - W = (h_2 - h_1) + \left(\frac{1}{2}v_2^2 - \frac{1}{2}v_1^2 \right).$$

Now, first, the *specific work input done on the air* (W) can be found from the input power to the compressor and the mass flow rate of the gas at the exit from the compressor.

Then, power in to the air = -42 kJ/s (power or work input per second done on the fluid, so negative) and mass flow rate of the air at exit $\dot{m} = 0.6$ kg/s; therefore, the specific work done on the gas at exit

$$W = -\frac{42 \text{ kJ/s}}{0.6 \text{ kg/s}} = -70 \text{ kJ/kg}.$$

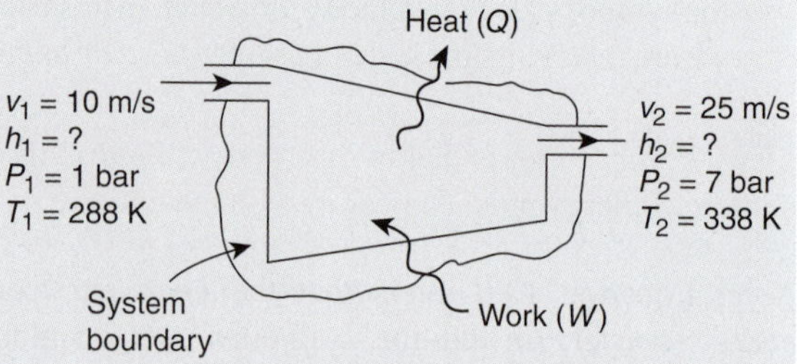

Heat (Q)

$v_1 = 10$ m/s
$h_1 = ?$
$P_1 = 1$ bar
$T_1 = 288$ K

$v_2 = 25$ m/s
$h_2 = ?$
$P_2 = 7$ bar
$T_2 = 338$ K

System boundary   Work (W)

Figure 13.7 Compressor inlet and outlet parameters for Example 13.4

Also, from equation 13.11, $h_2 - h_1 = c_p(T_2 - T_1)$. Then our equation, needed to determine the specific heat energy lost to the surroundings, will be:

$$Q - W = c_p(T_2 - T_1) + \left(\frac{1}{2}v_2^2 - \frac{1}{2}v_1^2\right)$$

and, on substitution of the values,

$$Q - (-70) = 1.005(338 - 288)$$
$$+ \left(\frac{25^2 - 10^2}{2 \times 10^3}\right), \text{ so } Q = -70 + 50.25$$
$$+ 0.2625 = -19.49 \text{ kJ/kg.}$$

That is, the specific heat energy lost to the surroundings (heat out) = 19.49 kJ/kg.

Note that for consistency of units, the *specific kinetic energy term* has been converted to kJ/kg on division by 1×10^3.

b) From the characteristic gas equation (12.17), then, at the exit, $p_2 V_{s2} = RT_2$, so that the specific volume at exit

$$V_{s2} = \frac{(287)(338)}{7 \times 10^5} = 0.1386 \text{ m}^3/\text{kg.}$$

Then, from equation 13.8, where $\dot{m} = \rho A v$ and the density

$$\rho = \frac{1}{V_s} = \frac{1}{0.1386} = 7.215 \text{ kg/m}^3,$$

we find that

$$A = \frac{\dot{m}}{\rho v} = \frac{0.6}{(7.215)(25)} = 3.326 \times 10^{-3} \text{m}^2.$$

Note again how, for consistency, we standardised the units when using the above equations.

13.5   Introduction to the second law of thermodynamics

Now, we know, according to the first law, that when a system undergoes *a complete cycle of processes* the net heat supplied is equal to the net work done. This is based on the principle of the conservation of energy and, for closed and open systems, it forms the basis upon which we derived the NFEE and the SFEE.

The second law of thermodynamics, which is also a natural law based on observation, makes it clear that although the net heat supplied is equal to the net work, ***the gross heat supplied must be greater than the net work done,*** because some heat must always be lost to the surroundings.

13.5.1   The second law and the heat engine

A heat engine (Figure 13.8) is a system operating in a complete cycle that produces work from a supply of heat energy.

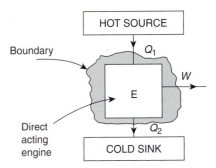

Figure 13.8 Representation of the heat engine

The second law implies that apart from a *source of heat*, there must also be a means for the rejection of heat in the form of a *heat sink* because the second law requires some heat to be rejected. In Figure 13.8 we have labelled the heat supplied from the source as Q_1 and the heat rejected to the sink as Q_2.

You might like to think of the heat engine in terms of the reciprocating piston internal combustion engine, where the *heat source* is the combustion of the injected fuel/air mix in the cylinder and the *heat sink* is the cooling fluid (water) to which the heat crossing the system boundary is rejected. From the *first law* we know that the *net heat supplied in a cycle = the net work done*. Symbolically, in the case of the heat engine shown in Figure 13.8, this is equal to:

$$Q_1 - Q_2 = W \tag{13.12}$$

Then, by the second law, the gross heat supplied must be greater than the net work done, i.e. $Q_1 > W$. Then the

thermal efficiency (η) of a heat engine

$$= \frac{\text{the net work done}}{\text{the gross heat supplied}} = \frac{W}{Q_1} \tag{13.13}$$

On substituting equation 13.12 into equation 13.13, we find that:

$$\eta = \frac{Q_1 - Q_2}{Q_1} = 1 - \frac{Q_2}{Q_1} \qquad (13.14)$$

It can be seen from the above argument that the second law implies that the thermal efficiency of a heat engine must be less than unity, or, as a percentage, *the thermal efficiency must be less than 100%*.

Also, from our model of the heat engine, a consequence of the second law is that *a temperature difference, no matter how small, is necessary before net work can be produced in a cycle*. This leads to a further statement (corollary) of the second law regarding the heat engine, which states that:

> *It is impossible for a heat engine to produce net work in a complete cycle if it exchanges heat only with bodies at a single fixed temperature.*

This may also be thought of as a restatement of the *zeroth law*, by which we know from earlier that no heat energy can be transferred between two bodies in contact at the same temperature, i.e. when in thermal equilibrium.

Before we leave the heat engine, we need to remember that the first and second laws work equally well for cycles that work in the *reverse* direction. The heat engine can be made to work in the reverse direction, provided there is a *work input* into the system (Figure 13.9) which is equal to the net heat rejected by the system.

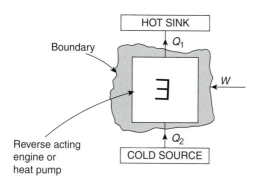

Figure 13.9 Representation of the reverse heat engine or heat pump

The reverse cycle heat engine is known as the *heat pump* (or refrigerator). In the heat pump cycle an amount of heat, Q_2, is supplied from the cold source and an amount of heat Q_1 is rejected to the hot sink. So, from the first law, $Q_1 = Q_2 + W$ and by the second law we know that work input is essential in order for heat to be transferred from a cold source to a hot sink (i.e. acting against the natural flow), so that the work input is greater than zero. The thermal efficiency of the heat pump is given in the same manner as that of the heat engine.

13.6 Chapter summary

In this chapter you have been introduced to thermodynamic systems. We have looked at the relationship between closed and open systems and the first law, and from this relationship we derived two very important thermodynamic equations and used them to determine heat, work and power parameters within the systems.

We started in section 13.1 by looking at some definitions and properties. You should, after studying this section, be able to define the requirements for a thermodynamic system; understand what is meant by the parameters (*property*, *state*, *process* and *cycle*) and be able to define and state the units for all the properties detailed in Table 13.1

In section 13.2 you were introduced to the idea of closed and open systems and you should be able to distinguish between the two, remembering essentially that *in a closed system there is no mass transfer of fluid across the system boundaries* and in addition, for a closed system with *rigid* boundaries, no work can take place.

Next, in section 13.3, we started by defining the *first law of thermodynamics*, which you should remember is essentially *a conservation law* based on repeatable observations. By applying the first law to a closed system, we derived the non-flow energy equation (NFEE). Equation 13.1 enables you to find heat and work transfers in and out of a closed system. You should also remember that *heat into the system is positive* and *work from (done by) the system is also positive*. Joule's experiments enabled him to form a law, as a result of his observations, that showed how the change in internal energy for a perfect gas was directly proportional to temperature (equation 13.2). Do remember that, no matter what the process, the change in internal energy is dependent upon temperature change only, and c_V must always be used to calculate ΔU no matter how the gas is heated.

In section 13.4 we looked at the relationship between the first law and open systems and derived the steady flow energy equation (SFEE), which takes into account not only the internal energy, work and heat transfers in the system but also the pressure energy, potential energy and kinetic energy of the working fluid and the exchanges of these energies to the fluid as it passes through the system. The *mass* and *volume flow rates* were also defined by equations 13.8 and 13.9, and you should note that in a *steady flow system the conservation of mass is preserved* so that, at any two points where the state of the working fluid is being considered, then, by the *conservation of mass law*, $\dot{m} = \rho_1 A_1 v_1 = \rho_2 A_2 v_2$. This relationship will prove particularly useful when we study fluid dynamics. We also considered the *specific form* of the SFEE and introduced the very useful parameters and relationships associated with *enthalpy* and *specific enthalpy* (equations 13.5, 13.9, 13.11), remembering that enthalpy provides a measure of the total energy of the working fluid, ignoring potential energy, which in gas and vapour systems is generally very low.

Finally, in section 13.5 we were briefly introduced to the *second law of thermodynamics* and its use in deriving a relationship for the *thermal efficiency* of a *heat engine* (equations 13.13, 13.14) and its inverse, the *heat pump*. The second law extends the ideas presented by the first law (equation 13.12), where we find that *the gross heat supplied must be greater than the net work done*; in other words we cannot have a 100% efficient heat engine because some of the heat energy must be lost and, in the case of the heat pump, the gross work supplied is essential for heat transfer from a cold source to a hot sink. The heat energy losses are measured by using the very abstract thermodynamic concept of *entropy*, which you will be introduced to in Chapter 15.

13.7   Review questions

1. Detail the criteria for classifying a thermodynamic system and give an example of a system that meets these criteria.

2. Define: a) control surface and control volume, b) state, c) extensive property, d) intensive property, e) process.

3. Explain the conditions needed for a closed system to transfer work.

4. Detail the sign convention for heat and work energy transfer.

5. If 35 kJ of heat energy is supplied to a closed system while the internal energy in the system is increased from 10 kJ to 50 kJ, determine the magnitude and direction of the work transfer.

6. If 37 kJ of heat energy is applied to a non-flow system while 54 kJ of work is done by the system, determine the magnitude and direction of the change in the internal energy of the working fluid.

7. a) Fully define specific heat capacity. b) Explain, for a perfect gas, why the specific heat capacity c_p always has a larger value than c_V.

8. Define: a) open system, b) specific enthalpy of a system fluid.

9. Explain the meaning of *each* of the terms in the *specific form* of the SFEE, giving the units for Q and W.

10. Calculate the mass *and* volume flow rate of air as it flows steadily through a duct of diameter 20 cm at 4 m/s. Take the density of the air as 1.225 kg/m^3.

11. In a steady flow process a compressor delivers 0.6 kg/s of air at a temperature of 9°C from inlet conditions of 1 bar and 2°C. The input power to the compressor is 54 kW and the velocity of the air at the inlet and exit is 14 m/s and 32 m/s, respectively. If for the air $c_p = 1005$ J/kg K, determine the specific heat energy loss to the surroundings.

12. With the aid of a diagram of a heat engine explain the second law of thermodynamics, in particular how the law modifies the first law with respect to heat and work transfers.

Chapter 14

Perfect gas processes

In this chapter we are going to *restrict* the working fluid involved in the processes to that of *gases* and *gas mixtures*, with a subsequent chapter also *only considering gas cycles and systems*. This restriction has been necessary in the interest of space within this book. However, you will find a separate chapter on the subject of *vapours*, their properties, processes, cycles and systems, including steam plant, on the website that accompanies this book, at www.key2engineeringscience.com

We start by considering the subject of reversibility and the stringent requirements needed for reversibility to be possible. We then look at the reversible displacement work transfer that may occur during the process. We then consider the heat, work and internal energy transfers for some of the most important perfect gas processes; these include the constant volume process, the constant pressure process, the isothermal process, the reversible adiabatic or isentropic process and finally the polytropic process. We end the chapter with a very quick look at Dalton's law and its application to gas mixtures and their analysis, as found, for example, in the combustion of air/fuel mixtures in combustion engines.

14.1 Reversibility and work

14.1.1 Reversibility

Before we consider any particular fluid processes you will need to understand the concept of *reversibility*.

You may remember from your study of oscillatory motion in section 11.3, that when a spring–mass system was freely oscillating, it was subject to a natural damping action caused by friction and air resistance, so that for each cycle the amplitude gradually reduced; thus each cycle was not repeatable and the motion eventually died away. We say that this motion is not repeatable, or is *irreversible*. Now, in a similar manner consider a pendulum undergoing oscillatory motion. You know that this is also subject to natural damping, but imagine this time that the pendulum swings from a frictionless pivot point and also that there is no air resistance or any other restriction to its motion. Then, under these fictional *ideal* circumstances, each part of the motion should *repeat* and *reverse* itself indefinitely, so we say that this motion is *reversible*. This analogy may be of help in understanding the circumstances required for a *process to be reversible*.

978-1-85617-775-7, Engineering Science, Mike Tooley & Lloyd Dingle

In its simplest sense:

A fluid system process is said to be reversible when it changes from one state to another and, at any instant during the process, an intermediate state point can be identified from any two properties that change as a result of the process.

With our pendulum analogy, for the motion being observed at any and all instants in time, we would find exactly the same amplitude and frequency (repeatability/reversibility) of motion at each of these *infinitely small time periods* during the motion.

Thus, for *reversibility*, the *fluid* undergoing the process must pass through an *infinitesimally large number of equilibrium states* (where the process is repeatable in the reverse direction). Figure 14.1a shows a representation of a reversible process where *unique equilibrium pressure and volume states* may be identified at *any time* during the process. Reversible processes are represented diagrammatically by *solid* lines (Figure 14.1a).

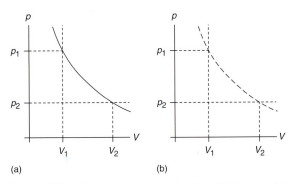

Figure 14.1 Diagrammatic representation of (a) reversible and (b) irreversible processes

In practical processes, because of energy transfers, the fluid undergoing a process *cannot be kept in equilibrium* in its intermediate states and a continuous path cannot be traced on a diagram of its properties. Such real processes are called *irreversible* and they are usually represented by a dashed line joining the end states (Figure 14.1b).

From what has been said above (including our analogy), in order to ensure reversibility and use the idea of reversibility, we need to prevent any energy transfers taking place anywhere during the process and ensure equilibrium is maintained (or very nearly). To achieve these stringent requirements for reversibility the system process would need to adhere to the following criteria:

- The process must be *frictionless*
- The difference in *pressure* between the fluid and its surroundings during the process must be infinitely small

- The difference in *temperature* between the fluid and its surroundings during the process must be infinitely small

> **Key Point.** For a process to be *reversible* it must be *frictionless* and the difference in *pressure and temperature* between the fluid and its surroundings must be *negligible*

Unfortunately, when the criteria for reversibility are examined, it can be seen that a reversible process is one that is negligibly removed from a state of equilibrium, and so a real heat engine or other machine which was to approach reversibility would be tortuously slow and very large! Therefore, in practice reversibility will never be attained; however, it does give us a standard of comparison for real processes and provides the target, as engineers, to which we should aim.

It is the usefulness of reversible processes as a ready means of comparison with real processes that suggests the need to study them, which of course we will do in this chapter. For example, when we study the *perfect gas processes* that follow, the fact that these processes are reversible is essential to the formulation of the relationships for work, heat and power transfers etc. that we will derive and use for comparison with real processes.

14.1.2   Reversible displacement work

We extend our definition and criteria for reversibility to that of reversible displacement work. Consider the *frictionless* piston and cylinder assembly, shown in Figure 14.2, that contains an ideal *frictionless fluid*.

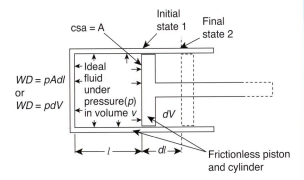

Figure 14.2 Frictionless piston and cylinder assembly

As we already know, the pressure exerted by the ideal fluid on the face of the frictionless piston creates a force, i.e. *Force* $= pA$, and if this force moves the piston initially some very small incremental distance, say dl, then the *incremental work done* by the system fluid on the piston (theoretically very close to equilibrium) will be $WD = pAdl$ and, because volume is equal to the change in incremental length times the area, the incremental *work done* $= pdV$.

Now, because we are talking about reversible work, the criteria given above must hold. Therefore, when *reversible work* takes place between any *two infinitesimally small states* during the process, the total amount of work done between, say, state 1 and state 2 can be found using the integral calculus. So, provided the pressure (p) can be expressed in terms of the volume (V), then:

$$\text{WD by fluid} = \int_1^2 pdV \qquad (14.1)$$

Equation 14.2 shows the work done per unit mass of fluid, i.e. equation 14.1 expressed in specific terms:

$$\text{Specific WD by fluid} = \int_1^2 pdV_s \qquad (14.2)$$

Please note the subtle difference between equations 14.1 and 14.2.

Processes involved with the *reversible expansion and contraction of ideal fluids* are modelled by the application of a rule or *law* relating pressure and volume that governs the process. Example 14.1 illustrates this idea.

> **Key Point.** The reversible displacement work done during a process may be found by integrating the law of the process between its start and end states

Example 14.1 In the piston and cylinder assembly illustrated in Figure 14.1, the fluid expands reversibly from an initial pressure of 4 bar and a specific volume of 0.2 m³/kg to a pressure of 1 bar according to the law $pV_s^2 = C$, where C is a constant. Sketch the expansion process on a p–V diagram and calculate the specific work done by the fluid.

From Figure 14.3, the specific work done is equal to the shaded area under the curve, where the equation of the curve is given by $p = \dfrac{C}{V_s^2}$ as shown.

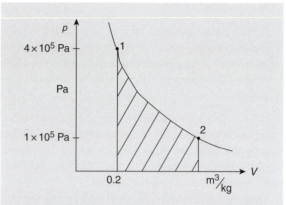

Figure 14.3 Pressure–volume curve for Example 14.1

Then $WD = \int_1^2 pdV_s$ and, in terms of the variable V_s the integral for the work done becomes:

$$WD = C\int_{V_{s1}}^{V_{s2}} \frac{dV_s}{V_s} = C\left(-\frac{1}{V_s}\right)_{V_{s1}}^{V_{s2}}.$$

Also, from the law of the process, $C = pV_s^2 = (4 \times 0.2^2) = 0.16$ bar and

$$V_{s2} = \sqrt{\frac{C}{p_2}} = \sqrt{\frac{0.16}{1.0}} = 0.4 \text{ m}^3/\text{kg}.$$

Therefore:

$$WD = C\left(-\frac{1}{V_s}\right)_{V_{s1}}^{V_{s2}}$$

$$= 16000 \text{ N/m}^2\left[-\left(\frac{1}{0.4} - \frac{1}{0.2}\right)\right] \text{ m}^3/\text{kg}$$

$$= 40 \text{ kN m/kg} = 40 \text{ kJ/kg}.$$

Note the careful manipulation of units during this calculation.

14.2 Perfect gas non-flow processes

14.2.1 Constant volume process

The *constant volume process for a perfect gas* is considered to be *a reversible non-flow* (closed system) *process* and, although you may not be aware of it, you have already met a constant volume process when we considered specific heat capacities in Chapter 12; look back at Figure 12.7a. This shows the working fluid being contained in a rigid vessel, so the system boundaries are immovable and *no work can be done* on or by the system.

> **Key Point.** In a perfect gas constant volume process no work is done

We know from the non-flow energy equation, NFEE (equation 13.1) that $Q - W = U_2 - U_1$ and, since for a constant volume process $W = 0$, then the heat transfer during the process is given by:

$$Q = U_2 - U_1 \qquad (14.3)$$

And, remembering from Joule's law and equation 13.4 that the change in internal energy for a perfect gas subject to a constant volume process is $U_2 - U_1 = mc_V(T_2 - T_1)$, then the *heat transfer (Q)* during the process, for a specific mass of gas, can also be represented by the relationship:

$$Q = mc_V(T_2 - T_1) \qquad (14.4)$$

This implies that for a constant volume process all the heat supplied is used to increase the internal energy of the working fluid. We can represent this reversible process for a perfect gas on a pressure (p–V) volume diagram, as shown in Figure 14.4a.

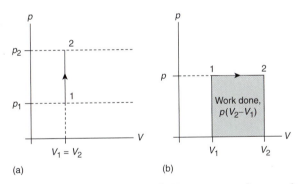

(a)   (b)

Figure 14.4 Representation of (a) constant volume and (b) constant pressure processes

Note that equations 14.3 and 14.4 can always be used with the variables quoted in specific terms, i.e. per unit mass, as we have done before. So, for example, the heat transferred during a constant volume process, per unit mass, would be given by $Q = c_{V_s}(T_2 - T_1)$ in J/kg or kJ/kg.

14.2.2 Constant pressure process

The *constant pressure (isobaric) non-flow process* for a perfect gas is also considered to be a *reversible* process. This process was also illustrated in section 12.3, Figure 12.7b, when we studied specific heat capacities. Consider again the pressure–volume diagrams shown in Figure 14.4a and 14.4b. It can be seen that when the boundary of the system is rigid, as in the constant volume process, then pressure rises when heat is supplied (vertical straight line). For a *constant pressure process* the boundary must move against an external resistance (maintaining a constant pressure, $p = p_1 = p_2$) as heat is supplied and work is done by the fluid on its surroundings.

Now, again from the definition of *specific heat capacity at constant pressure*, for a given mass of gas and change in temperature, the heat transferred (compare with equation 12.10) may be calculated from:

$$Q = mc_p(T_2 - T_1) \qquad (14.5)$$

The work done during the process is represented by the shaded rectangular area in Figure 14.3b and, as the pressure is constant at state 1 and state 2, then:

$$W = p(V_2 - V_1) \qquad (14.6)$$

Note from the NFEE that $Q - W = U_2 - U_1$, and we find that substituting equations 14.5 and 14.6 into the NFEE gives $Q - W = mc_p(T_2 - T_1) - p(V_2 - V_1) = U_2 - U_1$.

Now, we know that for a *constant volume process* all the heat addition is used to raise the internal energy, so, again from the NFEE, for the constant pressure process $mc_p(T_2 - T_1) - p(V_2 - V_1) = mc_V(T_2 - T_1) = (U_2 - U_1)$ or that *for the constant pressure process* $(U_2 - U_1) = mc_V(T_2 - T_1)$.

> **Key Point.** For a constant pressure process the work done is given by the product of the system pressure and change in volume

> **Example 14.2** 0.2 kg of air, initially at a temperature of 165°C, expands *reversibly* at a *constant pressure* of 7 bar until the volume is 0.05 m³. Using the standard values for R and c_p for air, determine the final temperature, heat transferred and work done.
>
> With the information presented in the question, we need first to use the characteristic gas equation to find the final temperature. So, using equation 12.17,

$pV_2 = mRT_2$, where for a constant pressure process $p = p_1 = p_2$ and, for air, $R = 287$ J/kg K and $c_p = 1005$ J/kg/K, we get that

$$T_2 = \frac{pV_2}{mR} = \frac{(7 \times 10^5)(.05)}{(0.2)(287)} = 609.75 \text{ K.}$$

The heat transferred is given by equation 14.5, so that:

$$Q = mc_p(T_2 - T_1) = (0.2)(1005)(609.75 - 438)$$

$$= 35.52 \text{ kJ,}$$

that is heat into the system.

To find the work done we use equation 14.6, $W = p(V_2 - V_1)$ where, again from the *characteristic gas equation* or, as it is sometimes known, *the equation of state*,

$$V_1 = \frac{mRT_1}{p} = \frac{(0.2)(287)(438)}{7 \times 10^5} = 0.035916 \text{ m}^3,$$

and so $W = 7 \times 10^5(0.05 - 0.035916) = 9.859$ kJ, and so, being positive, this is WD *by the air* on the surroundings.

14.2.3 Isothermal process

An isothermal non-flow process is one in which the temperature remains constant. This process you have met before, when we discussed Boyle's law! Look back at section 12.4.2 and in particular Figure 12.11 and equation 12.14, which gave the relationship that $pV = $ constant, remembering that this relationship only holds true if the *temperature remains constant*. Figure 14.5 shows an *isothermal process* that obeys Boyle's law,

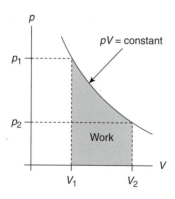

Figure 14.5 Isothermal process

where the area under the curve represents the work energy transfer between state 1 and state 2.

We also know from *the equation of state* that for a fixed mass of gas $pV = mRT$ and, if the temperature is constant, the mass (m) and R, are constants, so that $pV = $ constant, and so again it can be seen that it is Boyle's law that governs an *isothermal process*.

> **Key Point.** An isothermal process is one in which *temperature* remains constant

Following on from the ideas concerning *reversible displacement work* that we looked at in section 14.1.2, we can again apply the integral calculus to derive an expression for the displacement work of an isothermal process that obeys the law $pV = C$ (constant). In Figure 14.5 we note that *the work is equal to the shaded area under the graph*; then, as before, $W = \int_{V_1}^{V_2} p\, dV$, i.e. the total work is equal to the sum of all the infinitesimally small intermediate work states in the process. Then, substituting $p = \frac{C}{V}$, so that the integral is now in terms of one variable V, the *displacement work* is now given by

$$W = C \int_{V_1}^{V_2} \frac{1}{V} dV = C \left[\ln V\right]_{V_1}^{V_2}$$

so that $W = C\left[\ln V_2 - \ln V_1\right]$ or

$$W = p_1 V_1 \left[\ln \frac{V_2}{V_1}\right] \tag{14.7}$$

Equation 14.7 results from using the laws of logarithms and the law of the process $C = pV$. If we also use the substitution from the equation of state $pV = mRT$, then the *displacement work* for an isothermal process, where $T = T_1 = T_2$ is also given by:

$$W = mRT \left[\ln \frac{V_2}{V_1}\right] \tag{14.8}$$

We also know from the first law, applied to a closed system, that $Q - W = \Delta U$ or $Q = \Delta U + W$ and, since $T_2 = T_1$, $\Delta U = 0$. From equation 13.2, $\Delta U = mc_v(T_2 - T_1)$ and so $Q = W$, therefore

$$Q = W = p_1 V_1 \left[\ln \frac{V_2}{V_1}\right] \text{ and } Q = W = mRT \left[\ln \frac{V_2}{V_1}\right].$$

Do remember that an *isothermal process* is one that takes place *at constant temperature*, while *heat energy may still be transferred*.

Example 14.3   A mass of air (a perfect gas) occupying a volume of 0.15 m³ at 500 kPa pressure and 150°C expands isothermally and reversibly to a pressure of 100 kPa. Show the process on a p–V diagram and determine the work done and the heat transferred during the isothermal expansion. Use the standard values of c_V and R for air.

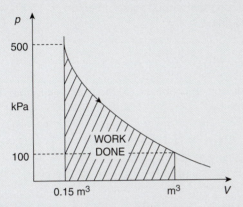

Figure 14.6 p–V diagram of the process for Example 14.3

The p–V (Figure 14.6) diagram shows the isothermal expansion, where the initial temperature $T_1 = T_2 = 423$ K. Now, because we are asked to find work, not specific work, we need to find the mass of the air and we can do this using our old friend the characteristic gas equation (or equation of state), where $p_1 V_1 = mRT_1$, so:

$$m = \frac{p_1 V_1}{RT_1} = \frac{(500 \times 10^3)(0.15)}{(287)(423)} = 0.618 \text{ kg.}$$

Now we know from equation 12.16 (the combined gas equation) that for a reversible isothermal process

$$p_1 V_1 = p_2 V_2 \text{ or } \frac{V_2}{V_1} = \frac{p_1}{p_2} = \frac{5}{1},$$

and using equation 14.8 we find that

$$W = mRT \left[\ln \frac{V_2}{V_1} \right] = (0.618)(287)(423)(\ln 5)$$

$$= 120748 \text{ J or } 120.748 \text{ kJ}$$

(work done by the air on the surroundings).
  Also, from the first law, $Q = \Delta U + W$ and from the argument given above, where for an isothermal process $\Delta U = 0$, we know that

$$Q = W = p_1 V_1 \left[\ln \frac{V_2}{V_1} \right],$$

so that the heat transferred to the system Q is also equal to the work, i.e. $Q = W = 120.748$ kJ.

14.2.4   Reversible adiabatic process

An adiabatic process is one in which no heat energy is transferred to or from the gas during the process. This process may be reversible or irreversible. We will concentrate here on reversible adiabatic processes involving a perfect gas as the working fluid. Thus, if a process takes place adiabatically, then $Q = 0$ and from the NFEE we find that $W = -\Delta U$.

Key Point.   An adiabatic process is one in which no heat is transferred to or from the working fluid

A *reversible adiabatic process* has further constraints placed upon it and must obey the following conditions:

i) $Q = 0$

ii)   It obeys the law $pV^\gamma = C$, where $\gamma = c_p/c_V$

iii)   There is no change in entropy during the process (see section 15.1).

Because of condition iii) the *reversible adiabatic process* is also referred to as an *isentropic process*, which literally means one-entropy.

Key Point.   In a reversible adiabatic or isentropic process there is no change in entropy during the process

Now, for reversible adiabatic processes $W = -\Delta U$ and from our previous work we know that $W = -\Delta U = -mc_V(T_2 - T_1)$, so $W = mc_V(T_1 - T_2)$. We also know from equation 12.18 that $c_p - c_V = R$ and from equation 12.19 that $\frac{c_p}{c_V} = \gamma$. Therefore, dividing equation 12.19 by c_V gives $\frac{c_p}{c_V} - 1 = \frac{R}{c_V}$, so on substituting $\frac{c_p}{c_V} = \gamma$ we get that $\gamma - 1 = \frac{R}{c_V}$ or:

$$c_V = \frac{R}{\gamma - 1} \qquad (14.9)$$

Also, from equations 12.19 and 14.9, $c_p = \gamma c_V = \dfrac{\gamma R}{\gamma - 1}$, so:

$$c_p = \frac{\gamma R}{\gamma - 1} \qquad (14.10)$$

Then, substituting equation 14.9 into the equation $W = mc_V(T_1 - T_2)$ gives:

$$W = \frac{mR(T_1 - T_2)}{(\gamma - 1)} \qquad (14.11)$$

Also, from the characteristic gas equation $pV = mRT$, after substitution, equation 14.11 may be written as:

$$W = \frac{p_1 V_1 - p_2 V_2}{\gamma - 1} \qquad (14.12)$$

To derive equation 14.12 we could just as easily have found the work done by this reversible adiabatic processes by integrating the law of the process between state 1 and state 2 (the area under the curve); that is, we find the definite integral

$$W = C \int_{V_1}^{V_2} \frac{1}{V^\gamma} dV.$$

We will use this method next in section 14.2.5, when we find the work done for a polytropic process.

There are also a number of useful equations relating p, V and T with the law for a reversible adiabatic process $pV^\gamma = C$ that are given here, without proof.

The first is a direct consequence of the law, i.e. $p_1 V_1^\gamma = p_2 V_2^\gamma$, so that:

$$\frac{p_1}{p_2} = \left(\frac{V_2}{V_1}\right)^\gamma \qquad (14.13)$$

It can also be shown that:

$$\frac{T_1}{T_2} = \left(\frac{V_2}{V_1}\right)^{\gamma - 1} \qquad (14.14)$$

$$\frac{T_1}{T_2} = \left(\frac{p_1}{p_2}\right)^{\gamma - 1/\gamma} \qquad (14.15)$$

Example 14.4 Air at a pressure of 1.1 bar and at 20°C, contained in a cylinder of volume $0.5\,\text{m}^3$, is compressed reversibly and adiabatically to a pressure of 5.5 bar. Taking for the air the normal values of c_p, c_V and R, determine the final temperature, final volume and the work done on the air in the cylinder during the process.

Using the normal values in equation 12.19, we find that

$$\gamma = \frac{c_p}{c_V} = \frac{1005}{718} = 1.4$$

and, from equation 14.15, the *final temperature* is given by

$$T_2 = T_1 \left(\frac{p_2}{p_1}\right)^{\gamma - 1/\gamma} = (293)\left(\frac{5.5}{1.1}\right)^{1.4 - 1/1.4}$$

$$= (293)(5)^{0.2857} = 464\,\text{K or }191°\text{C}.$$

The final volume may be found from equation 14.13, thus:

$$\frac{p_1}{p_2} = \left(\frac{V_2}{V_1}\right)^\gamma \quad \text{or} \quad \left(\frac{p_1}{p_2}\right)^{1/\gamma} = \frac{V_2}{V_1},$$

so final volume

$$V_2 = (V_1)\left(\frac{p_1}{p_2}\right)^{1/\gamma} = (0.5)\left(\frac{1.1}{5.5}\right)^{1/1.4} = 0.158\,\text{m}^3.$$

Finally, the work done on the air by the reversible adiabatic compression may be found from equation 14.12, so:

$$W = \frac{p_1 V_1 - p_2 V_2}{\gamma - 1}$$

$$= \frac{(1.1 \times 10^5)(0.50) - (5.5 \times 10^5)(0.1584)}{0.4}$$

$$= -80300\,\text{J} = -80.3\,\text{kJ}$$

(negative because the work is done on the fluid).

An alternative method to find the work done is to first find the *mass* of the air from the characteristic gas equation, so that

$$m = \frac{p_1 V_1}{RT_1} = \frac{(1.1 \times 10^5)(0.5)}{(287)(293)} = 0.654\,\text{kg}.$$

Then, for a reversal adiabatic process, we know that $W = mc_V(T_1 - T_2)$, so $W = (0.654)(718)(293 - 464) = -80296\,\text{J} = -80.3\,\text{kJ}$, as expected.

14.2.5 Polytropic process

The most general way of expressing a thermodynamic non-flow process is by means of the equation $pV^n = C$. This equation represents the general law for a *non-flow polytropic process where both heat and work energy may be transferred across the system boundary.*

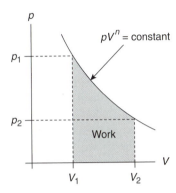

Figure 14.7 Curve for a polytropic process

> **Key Point.** In a polytropic process both heat and work energy may be transferred across the system boundary

In Figure 14.7 the area under the curve $pV^n = C$ represents the work energy transfer between state 1 and state 2 of the process. We can, in a similar manner to the non-flow isothermal process (Figure 14.4), find the total work (i.e. the sum of all the interim infinitesimal work states) by again considering the integral of the law between state 1 and state 2. For the polytropic process, the work done $W = \int_{V_1}^{V_2} p\,dV$ and from the law $p = \dfrac{C}{V^n}$ we find that

$$W = C \int_{V_1}^{V_2} \frac{1}{V^n} dV = C \int_{V_1}^{V_2} V^{-n} dV = \left[\frac{V^{1-n}}{1-n} \right]_{V_1}^{V_2}$$

when $n \neq 1$. Then, substituting for the constant C, where from the law $C = p_1 V_1^n = p_2 V_2^n$, we get that

$$W = \left(\frac{1}{1-n} \right) \left(p_2 V_2^n V_2^{1-n} - p_1 V_1^n V_1^{1-n} \right)$$

and, from *the laws of indices*,

$$W = \frac{\left(p_2 V_2^n V_2^{1-n} - p_1 V_1^n V_1^{1-n} \right)}{1-n} = \frac{p_2 V_2 - p_1 V_1}{1-n}.$$

Now, on multiplication of $W = \dfrac{p_2 V_2 - p_1 V_1}{1-n}$ by (-1) we find that the work done in a *non-flow polytropic process* is given by:

$$W = \frac{p_1 V_1 - p_2 V_2}{n-1} \qquad (14.16)$$

The work done may also be expressed in terms of temperature change, using once again the characteristic gas equation, so that on substituting for $p_1 V_1 = mRT_1$ and $p_2 V_2 = mRT_2$ in equation 14.16 we find that:

$$W = \frac{mR(T_1 - T_2)}{n-1} \qquad (14.17)$$

Now, from the NFEE, $Q = U_2 - U_1 + W$ and, after substituting equation 13.2, $\Delta U = mc_V(T_2 - T_1)$ for the change in internal energy and equation 14.17, given above, then:

$$Q = mc_V(T_2 - T_1) + \frac{mR(T_1 - T_2)}{n-1} \qquad (14.18)$$

Now equation 14.18 can be put into a more manageable form by using the substitution $c_V = \dfrac{R}{\gamma - 1}$ and simplifying. If this is done, then after a fair amount of algebraic manipulation we obtain the following expression for the heat transfer in a polytropic process:

$$Q = \left(\frac{\gamma - n}{\gamma - 1} \right) W \qquad (14.19)$$

Equation 14.19 expresses the heat transfer during a polytropic process as the product of *a ratio* and the *work done* on or by the gas during the process. The *direction of the heat transfer* will depend on the numerical values of γ and n, as well as on the direction of the work transfer.

There are also a number of useful equations for polytropic processes, given below, that relate p, V and T to the polytropic law $pV^n = C$. They take the same form as those we found earlier for the reversible adiabatic process.

Thus, from $pV^n = C$ we may write that $p_1 V_1^n = p_2 V_2^n$ or:

$$\frac{p_1}{p_2} = \left(\frac{V_2}{V_1} \right)^n \qquad (14.20)$$

Also:

$$\frac{T_1}{T_2} = \left(\frac{V_2}{V_1}\right)^{n-1} \qquad (14.21)$$

$$\frac{T_1}{T_2} = \left(\frac{p_1}{p_2}\right)^{n-1/n} \qquad (14.22)$$

Example 14.5 A volume of 0.7 m³ of a perfect gas at a pressure of 1.1 bar and a temperature of 290 K is compressed reversibly in a non-flow polytropic process according to the law $pV^{1.3} = C$ until the volume is 0.2 m³. Given that for the gas $c_p = 1.020$ kJ/kg K and $R = 0.295$ kJ/kg K, determine a) the mass of the gas, b) ΔU and c) the heat transfer during the process.

a) The mass can be found using the characteristic gas equation, so that

$$m = \frac{p_1 V_1}{RT_1} = \frac{(1.1 \times 10^5)(0.7)}{(295)(290)} = 0.9 \text{ kg}.$$

b) From equation 13.2,

$$\Delta U = mc_V(T_2 - T_1) \text{ and } \frac{T_1}{T_2} = \left(\frac{V_2}{V_1}\right)^{n-1} \text{ or }$$

$$T_2 = T_1\left(\frac{V_1}{V_2}\right)^{n-1} = (290)\left(\frac{0.7}{0.2}\right)^{0.3}$$

$$= 422.3 \text{ K}$$

and since, from equation 12.18, $c_V = 1.020 - 0.295 = 0.725$ kJ/kg K then:

$$\Delta U = (0.9)(0.725)(422.3 - 290) = 86.33 \text{ kJ}.$$

c) From equation 14.17,

$$W = \frac{mR(T_1 - T_2)}{n - 1}$$

$$= \frac{(0.9)(0.295)(290 - 422.3)}{1.3 - 1}$$

$$= -117.09 \text{ kJ}$$

so that, from the first law, $Q = \Delta U + W = 86.33 - 117.09 = -30.76$ kJ; therefore 30.76 kJ of heat energy is rejected by the gas to the surroundings.

14.3 Introduction to gas mixtures

We have spent some time looking at processes that involve only one perfect gas. Among the perfect gases we have considered is air, which as you know is a *mixture* of several gases, consisting primarily of oxygen (O_2) and nitrogen (N_2) with small amounts of argon (Ar) and carbon dioxide (CO_2), plus minute amounts of other trace gases that we will not concern ourselves with here. The individual components of a gas mixture do, however, exert their own pressure on the containing vessel and it can be shown that under set circumstances the sum of these individual (partial) pressures is equal to the total pressure exerted by the mixture of gases (see Dalton's law below). A study of gas mixtures and gas and vapour mixtures is very useful when considering the combustion of fuels in internal combustion engines.

14.3.1 Dalton's law

Dalton's law states that:

The pressure exerted by a mixture of gases (or gases and vapours), is the sum of the partial pressures that each would exert if it occupied alone the same space as the mixture and at the same temperature as the mixture.

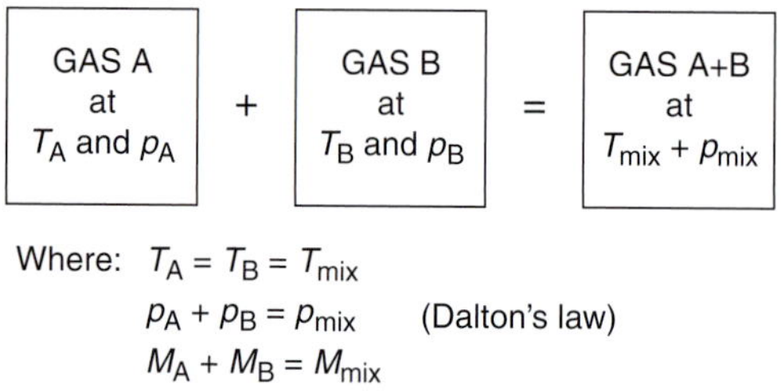

Where: $T_A = T_B = T_{mix}$
        $p_A + p_B = p_{mix}$ (Dalton's law)
        $M_A + M_B = M_{mix}$

Figure 14.8 Illustration of Dalton's law

This law is illustrated in Figure 14.8, where the individual gases (*A* and *B*) and a mixture of these two gases (*A* + *B*) are all enclosed in equal-volume containers at the same temperature, i.e. $V_A = V_B = V_{A+B}$ and $T_A = T_B = T_{A+B}$. Then under these circumstances, according to Dalton's law, the pressure exerted by the mixture is equal to the sum of the partial pressures of the constituents, i.e.

$$p_{mix} = p_A + p_B \qquad (14.23)$$

In fact, Dalton's law can be generalised for any number of component gases, provided the above criteria are observed, so:

$$p_{mix} = (p_A + p_B + p_C + \ldots\ldots\ldots p_n) \quad (14.24)$$

It is also true that because mass is conserved, $m_{mix} = (m_A + m_B + m_C + \ldots m_n)$. Now, by applying Dalton's law and knowing that *the partial pressure of each constituent gas is proportional to the number of moles*, then, as you will see next, we can carry out the analysis of these mixtures.

14.3.2 Analysis of mixtures

We can analyse gas or gas/vapour mixtures in two ways, by *volume* or by *weight* of the constituent gases. We refer to the former as *volumetric* analysis and to the latter as *gravimetric* analysis.

> **Key Point.** Mixtures may be analysed by volume (volumetric) or by weight (gravimetric) methods

Let us assume as a relatively good approximation (which we often do) that air is made up of just oxygen and nitrogen, where from your earlier work on the molar form of the characteristic gas equation ($pV = mR_0T$) in section 12.4.2, we know that the molar mass of oxygen is 32 kg/kmol and for nitrogen it is 28 kg/kmol. Then by volume there will be 21% oxygen and 79% nitrogen in air and by mass/weight there will be 23.3% oxygen and 76.7% nitrogen. These values are worth tabulating and memorising, since air is the most commonly used mixture in thermodynamic systems and this approximation is often used for analysis.

Table 14.1 Values for the approximate analysis of air

Constituent gas	Molecular mass/weight	Volumetric %	Gravimetric %
Oxygen (O_2)	32	21	23.3
Nitrogen	28	79	76.7

Now we have mentioned that the partial pressure of any constituent is proportional to the number of moles, let us see what this means in practice.

Consider equation 12.22, where $pV = nR_0T$. If the conditions of Dalton's law are met, V, R_0, T are constant; remembering that $R_0 = 8214.4$ J/kmol K, then

$p \propto n$. So if we assume that $pV = nR_0T, \ldots$ (1) gives the pressure and number of moles for the mixture (e.g. air), and $p_0V = n_0R_0T, \ldots$ (2) gives the pressure and number of moles for say, the oxygen, then, on division of (2) by (1), we get that:

$$\frac{p_0V = n_0R_0T}{pV = nR_0T},$$

so:

$$\frac{p_0 = n_0}{p = n} = \frac{\text{number of moles of oxygen}}{\text{number of moles of air}}.$$

Therefore the partial pressure of the oxygen (constituent) is given by:

$$p_0 = \frac{n_0 p}{n} \quad (14.25)$$

where we note that any partial pressure (in this case oxygen) is proportional to the number of moles, or the mole-fraction.

Also, from the characteristic gas equation in standard form, $pV = mRT$ (equation 12.17) and from $R = \frac{R_0}{M}$ (equation 12.23) we find that:

$$p = \frac{mR_0T}{MV} \quad (14.26)$$

so that in this form, equation 14.26 enables us to find any pressure or partial pressure, provided the appropriate values for m and M are known. Of course, if m and M are known, then the number of moles of any constituent gas or mixture is also known from the relationship $n = \frac{m}{M}$ (equation 12.20). The following very simple examples will help illustrate the way in which we go about the analysis.

> **Key Point.** The number of moles $(n) = \dfrac{\text{mass of the substance in kg (m)}}{\text{molecular weight } (M)}$

Example 14.6

a) Calculate the partial pressure of oxygen and nitrogen in air at a pressure of 1.1 bar.

b) A vessel of volume 0.6 m³ contains 2 kg of air at 288 K. Calculate the partial pressure of the oxygen and the nitrogen and the total pressure in the vessel.

a) Using volumetric analysis, we have from equation 14.25 and Table 14.1 that the partial pressure of oxygen $= \left(\dfrac{21}{100}\right)(1.1 \times 10^5) = 23100\,\text{Pa} = 0.231\,\text{bar}$ and the partial pressure of nitrogen $= \left(\dfrac{79}{100}\right)(1.1 \times 10^5) = 86900\,\text{Pa} = 0.869\,\text{bar}$.

b) Using gravimetric analysis, then, from Table 14.1, the mass of the oxygen $= \left(\dfrac{23.3}{100}\right)(2) = 0.466\,\text{kg}$ and the mass of the nitrogen $= \left(\dfrac{76.7}{100}\right)(2) = 1.534\,\text{kg}$. Then, from equation 14.25, $p = \dfrac{mR_0T}{MV}$ and the given volume of 0.6 m^3, the partial pressure of the oxygen is

$$p_0 = \frac{(0.466)(8314)(288)}{(32)(0.6)} = 58115\,\text{Pa} = 0.58115\,\text{bar}$$

and the partial pressure of the nitrogen is $p_0 = \dfrac{(1.534)(8314)(288)}{(28)(0.6)} = 218634\,\text{Pa} = 2.18634\,\text{bar}$.

Then, from Dalton's law and equation 14.23, the total pressure is:

$$p_{\text{air}} = p_0 + p_N = 0.58115 + 2.18634$$
$$= 2.76749\,\text{bar}.$$

Example 14.7 A mixture of 4 kg of oxygen, 8 kg of nitrogen and 10 kg of carbon monoxide is contained in a vessel with a volume of 10 m^3 at a temperature of 120°C. Determine the pressure exerted on the vessel by the mixture.

The molar mass/weight of carbon monoxide (CO) is found from the atomic weight of its constituents, where for the carbon C = 12 and for oxygen you will know that O = 16. Therefore the molar mass for carbon monoxide CO = 12 + 16 = 28.

Then, from the rearrangement of equation 14.20 we have $n = \dfrac{m}{M}$, so that for the mixture of gases

that occupies a volume of 10 m^3 at a temperature of 393 K, the number of kg-moles of each constituent will be:

$$O_2 = \frac{m}{M} = \frac{4}{32} = 0.125\,\text{kg-moles},$$

$$N_2 = \frac{m}{M} = \frac{8}{28} = 0.286\,\text{kg-moles},$$

$$CO_2 = \frac{m}{M} = \frac{10}{28} = 0.357\,\text{kg-moles}.$$

So the mixture consists of a total of $n = (0.125 + 0.286 + 0.357) = 0.768\,\text{kg-moles}$.

Then, from equation 12.12, where $p = \dfrac{nR_0T}{V} = \dfrac{(0.768)(8314)(293)}{10} = 187085\,\text{Pa}$, the total pressure exerted on the vessel by the mixture of gases is $p = 1.87085\,\text{bar}$.

14.4   Chapter summary

In this chapter we started by considering the concept of reversibility and reversible work, where we found that reversibility was possible when all the incremental stages in a process between an initial and final state remained infinitely close to the equilibrium position. To achieve this ideal, the process would need to adhere to the set of criteria identified in section 14.1.1. This of course meant that for real system processes, meeting these criteria would result in the need for the system to be very large and to operate infinitely slowly, so that in reality reversibility was something that thermodynamic engineering designers needed to aim towards, as a bench mark, but which could never be fully achieved in practice. We then extended our idea of reversibility to that of reversible displacement work, where we derived one or two relationships for reversible work using the integral calculus, noting that the work done on or by the ideal fluid in a reversible process, between the initial and final states, was given on a p–V diagram by the area under the curve.

We then went on to define some very important reversible processes, where perfect gases were used as the working fluid. You should be able to define each of these processes, i.e. constant volume, constant pressure, isothermal, adiabatic, reversible adiabatic/isentropic and polytropic processes. In addition you should understand the criteria and laws that govern these processes and be able to determine the heat, work and internal energy transfers for each, as appropriate. To act as an *aide-mémoire* and as a source of reference, Table 14.2 is given here to summarise the appropriate formulae and laws that will enable you to find the

Table 14.2 Reversible non-flow processes of perfect gases

1. Constant volume	
Law	$V = \text{constant}$
Heat transferred (Q)	$mc_V(T_2 - T_1)$
Change of internal energy (ΔU)	$mc_V(T_2 - T_1)$
Work done (W)	0

2. Constant pressure	
Law	$p = \text{constant}$
Heat transferred (Q)	$mc_p(T_2 - T_1)$
Change of internal energy (ΔU)	$mc_V(T_2 - T_1)$
Work done (W)	$p(V_2 - V_1)$

3. Isothermal	
Law	$pV = constant$
Heat transferred (Q)	$p_1 V_1 \left[\ln \dfrac{V_2}{V_1} \right]$ or $mRT \left[\ln \dfrac{V_2}{V_1} \right]$
Change of internal energy (ΔU)	0
Work done (W)	$p_1 V_1 \left[\ln \dfrac{V_2}{V_1} \right]$ or $mRT \left[\ln \dfrac{V_2}{V_1} \right]$

4. Reversible adiabatic or isentropic	
Law	$pV^\gamma = \text{constant}$
Heat transferred (Q)	0
Change of internal energy (ΔU)	$mc_V(T_2 - T_1)$
Work done (W)	$\dfrac{p_1 V_1 - p_2 V_2}{\gamma - 1}$

5. Polytropic	
Law	$pV^n = \text{constant}$
Heat transferred (Q)	$mc_V(T_2 - T_1) + \dfrac{mR(T_1 - T_2)}{n - 1}$ or $\left(\dfrac{\gamma - n}{\gamma - 1} \right) W$
Change of internal energy (ΔU)	$mc_V(T_2 - T_1)$
Work done (W)	$\dfrac{p_1 V_1 - p_2 V_2}{n - 1}$

appropriate heat, work and internal energy parameters associated with each of these processes.

In the final part of this chapter you saw how Dalton's law related the partial pressure of the gas/vapour constituents of a mixture to the pressure of the mixture when they all occupied the same volume at the same temperature. Also, you saw how we could analyse the properties of mixtures by using Dalton's law, together with the relationship between the partial pressure and the number of moles (the mole-fraction) and using the molar version of the characteristic gas equation. Remember also that there are two methods of analysis, volumetric and gravimetric.

Note: Throughout this chapter we have freely interchanged the SI methods for representing and measuring pressure and temperature, i.e. quoting them in bar, kPa, Pa, N/m^2, °C, K etc. You should be able to freely interchange these units when manipulating problems and know how to standardise them when using thermodynamic formulae.

14.5 Review questions

1. Explain the concept of reversibility, stating the criteria that the system process must adhere to in order to be deemed reversible.

2. A mass of 0.5 kg of a perfect gas at a pressure of 1.2 bar occupies a volume of 0.35 m^3 at 15°C. If $c_V = 720$ J/kg K, determine R, c_p and γ.

3. Calculate the work done by 0.5 kg of a perfect gas which expands reversibly from an initial pressure of 6 bar and a volume of 0.3 m^3 to a pressure of 1 bar according to the law $pV^{1.2} = C$.

4. 2.0 kg of air, initially at a temperature of 450 K, expands reversibly and at a constant pressure of 5 bar until the volume is 0.6 m^3. Using the standard values of R and c_p for air, determine the final temperature of the process and the heat and work transfers during the process.

5. A mass of air occupying a volume of 0.1 m^3 at 8 bar and 250°C expands isothermally to a pressure of 1 bar. Determine the mass of the air and the work and heat transferred during the process. Use the standard values of R and c_V for air.

6. Air at 1.1 bar and 20°C, initially occupying a volume of 0.15 m^3, is compressed isentropically to a pressure of 7.7 bar. Taking the normal values for the specific heats and characteristic gas constant for air, calculate the final temperature, final volume, the mass of air and the work done on the air.

7. A volume of 0.5 m^3 of a perfect gas at a pressure of 100 kPa and a temperature of 288 K is compressed reversibly in a process according to the law $pV^{1.5} = C$ until the volume is 0.15 m^3. Given that for the gas $c_p = 1.025$ kJ/kg K and $R = 0.298$ kJ/kg K, determine the mass of the gas, the change in internal energy and the work and heat transfers during the process.

8. 0.8 kg of gas at an initial temperature of 500°C expands reversibly and adiabatically until its temperature is 200°C and it has done 160 kJ of work. Taking $R_0 = 8214.4$ J/kmol K and $\gamma = 1.42$, determine c_p, c_V, R and the molecular weight of the gas.

Chapter 15

Thermal cycles

In this chapter we extend the idea of reversible processes to *reversible* and *irreversible* cycles, by first introducing the rather obtuse concepts associated with *entropy* and entropy change that takes place with irreversible processes. We take another look at the *heat engine* and *thermal efficiency* for ideal cycles that we touched upon towards the end of Chapter 13 when we introduced the second law of thermodynamics in section 13.5. The important cycles covered are the thermodynamically reversible *Carnot cycle*, the constant volume or *Otto cycle*, the *diesel cycle* and the *closed constant pressure cycle*. In Chapter 16 we will extend our knowledge of the Otto and constant pressure cycles to the practical cycles of the four-stroke spark ignition piston engine and to the closed and open system gas turbine engine. We start with the concept of entropy and its use (among other things) in being able to find the end-state of an irreversible process.

15.1 Entropy

15.1.1 Introduction

You have seen from section 14.1.2, when we were discussing *reversible* displacement *work*, how the *incremental work done* was given by $dW_{rev} = pdV$ and that *the total work done* during the process, between the two states, was given by $W_{rev} = \int_1^2 pdV$. Suppose now that a system undergoes the *same reversible process* from state 1 to state 2, and that during an infinitesimal (incremental) part of the process an amount of heat dQ_{rev} is transferred to or from the system, at temperature T; then a property, the *entropy*, can be defined such that the *change in entropy* during the *infinitesimal part of the process* is given by:

$$dQ_{rev} = TdS \qquad (15.1)$$

or

$$dS = \frac{dQ_{rev}}{T} \qquad (15.2)$$

Now the *total change in entropy* for the complete process is found by integrating $dS = \dfrac{dQ_{rev}}{T}$ between the two states. If this is done it can be shown (as you will see next) that for a change from T_1, p_1, V_1 to T_2, p_2, V_2 *the entropy change per unit mass* is given by:

$$s_2 - s_1 = c_p \ln\left(\frac{T_2}{T_1}\right) - R \ln\left(\frac{p_2}{p_1}\right) \qquad (15.3)$$

The units for change in entropy (in specific terms) are kJ/kg K.

So, what does the property of entropy or, more importantly to engineers, entropy change, signify? You will remember from our discussion about reversibility in section 14.1.1 that the ideal process was one in which friction losses were zero and the process was reversible. In real thermodynamic systems, such as combustion engines, turbines, heat pumps, compressors etc. we try and keep losses to a minimum and these will not only

978-1-85617-775-7, Engineering Science, Mike Tooley & Lloyd Dingle

include friction losses but also heat transfer losses; therefore, for real systems, we should try and make them *adiabatic* as well as *reversible*. Now we have already discussed *reversible adiabatic (isentropic) processes* in 14.2.4, where we found that the *additional criteria* for such reversible processes were that:

- No heat is transferred, i.e. $dQ_{rev} = 0$ and $Q = 0$
- There should be no change in entropy, i.e. $s_2 - s_1 = 0$, so that $c_p \ln\left(\dfrac{T_2}{T_1}\right) - R \ln\left(\dfrac{p_2}{p_1}\right) = 0$.

Thus, the ideal process with which comparisons can be made with real processes and cycles is the isentropic process. Hence the need to understand the concepts of entropy, entropy change and isentropic efficiencies, against which real cycles and systems may be measured.

> **Key Point.** For a process to be isentropic it must be adiabatic and must have constant entropy

Entropy may also be thought of in a slightly different way. The molecular movement of any gas or vapour (i.e. the distribution of heat energy between the internal energy and the pressure energy of the fluid) is considered to be an ordered process. However, *occasionally a degree of disorder is introduced into the process*, for example, skin friction of fluid flow in a duct, or turbulence within the fluid itself, and *this degree of disorder results in the creation of entropy*. This description of entropy is in line with our argument above, where *any deviation from the equilibrium position during the process results in an increase in entropy*. The greater the deviation, the greater the increase in entropy.

> **Key Point.** Entropy may be thought of as the degree of disorder in a process

15.1.2   Representing entropy

You have already seen in Chapter 14 that the conventional way of representing a process is through use of the *p–V* diagram. Look back at Figures 14.2, 14.3, 14.4, 14.6, etc., where for example, Figure 14.4 shows the *p–V* diagram for a reversible isothermal process. The development of these diagrams resulted from the fact that pressure and volume could be measured using engine indicators. However, temperature cannot be scaled onto these diagrams and, for open flow systems where the specific volume constantly changes, there is

no way of measuring it. Temperature and pressure can be measured with reasonable accuracy in flow systems and, since change in entropy (from equation 15.3) is related to temperature, we may represent this fact on a *temperature–entropy (T–S)* diagram, or a *(T–s)* diagram when we are considering entropy change in specific terms (see Figure 15.1).

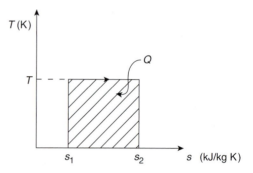

Figure 15.1 Representation of an isothermal process on a temperature–entropy (*T–s*) diagram

Figure 15.1 shows a reversible isothermal (constant temperature, *not* constant heat) process (an expansion) between state 1 and state 2 that results in a *change in entropy* (which of course will take place as a result of heat transfer across the system boundaries during the expansion), where the shaded area represents the heat transferred into the system as a result of the process.

> **Key Point.** Any datum may be chosen from which to measure entropy change

Thus, just as pressure is associated with work, *so temperature is associated with heat and it is a graph of temperature against entropy or specific entropy (T–s) that relates the two*, where, as mentioned, the area under the graph represents the heat energy transfer.

15.1.3   Change in entropy

We now turn to the derivation of equation 14.3 and the useful relationships we may obtain from it, for particular gas processes.

Please note at this stage that entropy changes for vapours can also be found. These are considered in the chapter on vapours and vapour systems you will find on the internet, at the companion website for this book.

Let us consider *unit mass of gas* that undergoes a change from state 1 conditions of T_1, p_1, V_1 to state 2 conditions T_2, p_2, V_2. From the first law (equation 13.1) we know that $Q = \Delta U - W$ so that for an incremental (infinitesimally small) heat transfer $dQ_{rev} = dU + W$ and, since infinitesimal work transfer $w = pdV$ (see argument in section 14.1.2 and equation 14.1), then in specific terms we may write the first law for a reversible process as $dQ_{rev} = du + pdV_s$ and, on division by temperature T, we obtain $\dfrac{dQ_{rev}}{T} = \dfrac{du}{T} + \dfrac{pdV_s}{T}$. Remembering from equation 13.2 that $dU = mc_v dT$ or, in *specific terms for unit mass, $du = c_{V_s} dT$*, then $\dfrac{dQ_{rev}}{T} = \dfrac{c_{V_s} dT}{T} + \dfrac{pdV_s}{T}$.

Now, from equation 15.2 $dS = \dfrac{dQ_{rev}}{T}$. Also, from equation 12.17, the characteristic gas equation (equation of state), again in specific terms for unit mass, we note that $pV_s = RT$ or $\dfrac{p}{T} = \dfrac{R}{V_s}$. So, after substitution in

$$\frac{dQ_{rev}}{T} = \frac{c_{V_s} dT}{T} + \frac{pdV_s}{T},$$

we find that

$$ds = \frac{c_{V_s} dT}{T} + \frac{RdV_s}{V_s},$$

then integrating:

$$\int_{s_1}^{s_2} ds = c_{V_s} \int_{T_1}^{T_2} \frac{dT}{T} + R \int_{V_{s1}}^{V_{s2}} \frac{dV_s}{V_s}.$$

Therefore:

$$s_2 - s_1 = c_{V_s} \ln \frac{T_2}{T_1} + R \ln \frac{V_{s2}}{V_{s1}} \qquad (15.4)$$

Note that for a reversible *constant volume* process, where $V_{s_1} = V_{s_2}$, the second term in equation 15.4 is eliminated since the Naperian log $\ln 1.0 = 0$.

Now we also know from the combined gas law that

$$\frac{p_1 V_{s_1}}{T_1} = \frac{p_2 V_{s_2}}{T_2} \text{ or } \left(\frac{T_2}{T_1} \times \frac{p_1}{p_2} \right) = \frac{V_{s_2}}{V_{s_1}},$$

so substituting into equation 15.4 gives

$$s_2 - s_1 = c_{V_s} \ln \frac{T_2}{T_1} + R \ln \left(\frac{T_2}{T_1} \times \frac{p_1}{p_2} \right)$$

and (remembering your laws of logarithms!) we get that

$$s_2 - s_1 = c_{V_s} \ln \frac{T_2}{T_1} + R \ln \frac{T_2}{T_1} + R \ln \frac{p_1}{p_2}$$

or

$$s_2 - s_1 = (c_{V_s} + R) \ln \frac{T_2}{T_1} + R \ln \frac{p_1}{p_2},$$

and remembering from equation 12.18 that $c_{p_s} = c_{V_s} + R$, then again on substitution we get that:

$$s_2 - s_1 = c_{p_s} \ln \left(\frac{T_2}{T_1} \right) + R \ln \left(\frac{p_1}{p_2} \right) \qquad (15.5)$$

which of course is our equation 15.3 with the pressure ratio inverted (hence the change in sign of the second term). Note that in the case of a *reversible constant pressure process* the second term in equation 15.3 or 15.5 is eliminated, in a similar manner to that for the constant volume case (equation 15.4). Do remember that equations 15.3, 15.4 and 15.5 give the *specific changes in entropy*, i.e. the entropy change per kilogram.

We also know that *for an isentropic process there is no change in entropy*, i.e. constant entropy, so that, as we have seen in section 15.1.1 above,

$$c_{p_s} \ln \left(\frac{T_2}{T_1} \right) - R \ln \left(\frac{p_2}{p_1} \right) = 0$$

or

$$c_{p_s} \ln \left(\frac{T_2}{T_1} \right) = R \ln \left(\frac{p_2}{p_1} \right),$$

from which, after anti-logging, we find that:

$$\frac{p_2}{p_1} = \left(\frac{T_2}{T_1} \right)^{\frac{c_{p_s}}{R}} \qquad (15.6)$$

and from equation 14.10 $\dfrac{c_p}{R} = \dfrac{\gamma}{\gamma - 1}$; therefore on substitution into $\dfrac{p_2}{p_1} = \left(\dfrac{T_2}{T_1} \right)^{\frac{c_{p_s}}{R}}$ we find that:

$$\frac{p_2}{p_1} = \left(\frac{T_2}{T_1} \right)^{\frac{\gamma}{\gamma - 1}} \qquad (15.7)$$

Compare equation 15.7 with equation 14.15, where the *inverse ratios* are shown! Equations 14.14, 14.15 and 15.5 show that for a *reversible adiabatic (isentropic) process* with fixed ratio γ of the *specific heats*, the temperature ratio depends solely on the volume and pressure ratios, as appropriate.

Example 15.1   1 kg of a perfect gas is heated at constant pressure from 38°C to 1100°C. Calculate the change of entropy. Take, for the gas, $c_p = 1.005$ kJ/kg K.

Using equation 15.3, after converting the temperature to Kelvin and remembering that $p_1 = p_2$ for a constant pressure process, we find that:

$$s_2 - s_1 = c_p \ln\left(\frac{T_2}{T_1}\right) - R \ln\left(\frac{p_2}{p_1}\right)$$

$$= (1005) \ln\left(\frac{1373}{311}\right) - R \ln(1)$$

$$= 1492.4 - 0.$$

Therefore, change in entropy $S_2 - S_1 = 1492.4$ J/K (since unit mass).

Example 15.2   A system consisting of 0.2 kg of gas is cooled at constant volume from 1370°C to 550°C. Calculate the decrease in entropy of the system, taking $c_V = 0.72$ kJ/kg K.

This is a constant volume process so it is appropriate to use equation 15.4. Then,

$$s_2 - s_1 = c_{V_s} \ln\frac{T_2}{T_1} + R \ln\frac{V_{s_2}}{V_{s_1}}$$

or

$$s_2 - s_1 = (7200) \ln\left(\frac{823}{1643}\right) + R \ln(1)$$

$$= -4977.5 \text{ J/kg K}.$$

This is the *specific* entropy decrease (negative), so the *total entropy decrease* of the system is $= (0.2)(4977.5) = 995.5$ J/K.

Example 15.3   0.5 kg of a perfect gas with volume 0.2 m³, at 2 bar and 293 K, expands adiabatically in a non-flow process until the final volume $V = 0.3$ m³ and the final pressure is 3 bar. During the process, 120 kJ of work is done on the gas. Find R, T_2, c_V, c_p of the gas and the change in entropy for the process.

Now, in order to use equation 15.3 or 15.5 to find the specific entropy change, we first need to find the required constants. We start by using the characteristic gas equation or equation of state, which is valid for both gases and vapours, no matter what the process.

Then $R = \dfrac{p_1 V_1}{m T_1} = \dfrac{(2 \times 10^5)(0.2)}{(0.5)(293)} = 273$ J/kg K

$$T_2 = \frac{p_2 V_2}{mR} = \frac{(3 \times 10^5)(0.3)}{(0.5)(273)} = 659.34 \text{ K}$$

Then, from the first law where $Q = \Delta U - W$ and where, for the adiabatic process, $Q = 0$, and where we know that $W = 120$ kJ, then

$$0 = mc_V(T_2 - T_1) - 120000 \quad \text{or}$$

$$c_V = \frac{120000}{(0.5)(659.34 - 293)} = 655.13 \text{ J/kg K},$$

then $c_p = c_V + R = 655.13 + 273 = 928.13$ J/kg K.

So, using equation 15.5,

$$s_2 - s_1 = c_{p_s} \ln\left(\frac{T_2}{T_1}\right) + R \ln\left(\frac{p_1}{p_2}\right),$$

we get that

$$s_2 - s_1 = (928.13) \ln\left(\frac{659.34}{293}\right) + (273) \ln\left(\frac{2}{3}\right)$$

$$= 642.02 \text{ J/kg K},$$

which is the specific entropy change, so the *entropy change* for the process is $S_2 - S_1 = (0.5)(642.02) = 321$ J/K.

15.2   The Carnot cycle

You have seen from section 13.5, when we considered the second law of thermodynamics, that the most efficient heat engine is one in which the complete cycle of system processes is reversible and the system works between the same two temperatures of the heat source and heat sink. The resulting *thermal efficiency* of such a system was given by equation 13.14, where $\eta = \dfrac{Q_1 - Q_2}{Q_1} = 1 - \dfrac{Q_2}{Q_1}$. One such system cycle that works on the same criteria as the heat engine and so provides the highest possible thermal efficiency is the *Carnot cycle*. This cycle may be used for both gases and vapours as the working fluid; however, in our version set out below, the working fluid will always be a *perfect gas*.

The Carnot cycle consists of four thermodynamically reversible processes as follows:

1–2   Heat supplied isothermally (at constant temperature T_1)

2–3   Isentropic expansion (T_1 to T_2)

3–4   Heat rejected isothermally (at constant temperature T_2)

4–1   Isentropic compression (T_2 to T_1)

This cycle of events is shown on a *T–s* diagram (Figure 15.2), which is applicable to *both gases and vapours*, and on a *p–V* diagram (Figure 15.3), which is only applicable to *perfect gases*.

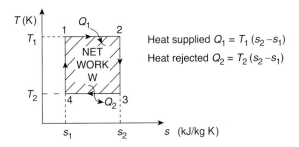

Figure 15.2 *T–s* diagram of the Carnot cycle

From Figure 15.2 and equation 15.1, it can be seen that the total heat supplied (Q_1) is given by the area under the graph below the process 1–2, i.e. $Q_1 = T_1(s_2 - s_1)$, and the heat rejected (Q_2) is given by the area under the graph below the process 3–4, i.e. $Q_2 = T_2(s_2 - s_1)$.

Now, from equations 13.12 and 13.14, the thermal efficiency $\eta = \dfrac{W}{Q_1} = \dfrac{Q_1 - Q_2}{Q_1} = 1 - \dfrac{Q_2}{Q_1}$, so that the thermal efficiency of the Carnot cycle

$$\eta_{\text{Carnot}} = \frac{T_1(s_2 - s_1) - T_2(s_2 - s_1)}{T_1(s_2 - s_1)} = 1 - \frac{T_2(s_2 - s_1)}{T_1(s_2 - s_1)}$$

or

$$\eta_{\text{Carnot}} = \frac{Q_1 - Q_2}{Q_1} = \frac{T_1 - T_2}{T_1} = 1 - \frac{T_2}{T_1} \quad (15.8)$$

Now it can be seen from equation 15.8 that, for any fixed lower temperature for heat rejection (T_2), the higher the temperature at which heat is supplied (T_1), the greater

will be the efficiency. These temperatures will obviously be dependent on the particular thermodynamic plant being considered. The Carnot efficiency is the optimum efficiency for which thermodynamic plant can be designed, and to attain it all the processes of the cycle must be thermodynamically reversible. In practice, combustion engines, compressors, turbines etc. cannot attain such conditions of operation.

> **Key Point.**   The Carnot efficiency is the optimum efficiency against which thermodynamic plant can be designed

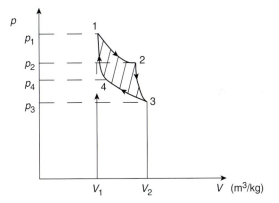

Figure 15.3 *p–V* diagram of the Carnot cycle

Figure 15.3 shows the Carnot cycle on a *p–V* diagram, where it can be seen that the gross work of the expansion processes (area $123V_2V_1$) and that of the work done on the gas by the compression processes (area $143V_2V_1$) leave very little net work done by the cycle (shaded area 1234). Now the ratio of the net work to the gross work is called the work ratio, i.e.

$$\text{Work ratio } r_W = \frac{\text{Net work}}{\text{Gross work}} \quad (15.9)$$

It can be seen from Figure 15.3 and the above argument that the work ratio for the Carnot cycle will be very low so, although it is very efficient, it is for this reason that the Carnot cycle is not aspired to because of its susceptibility to irreversibilities in practice. There is also another reason for not using the Carnot cycle in practice. Figure 15.3 shows that during isothermal heat addition and rejection the pressure is constantly changing. In practice it is much more efficient and easier to achieve if the gas is heated (as near as possible) at constant pressure (as in an open gas turbine engine) or constant volume (as in the reciprocating piston engine) conditions.

Example 15.4 In a Carnot cycle using air, the minimum and maximum pressures are 1 bar and 90 bar, respectively. If the cold temperature is 15°C and the maximum temperature reached in the cycle is 600°C, calculate the thermal efficiency and the work ratio.

The cycle is illustrated in Figure 15.4 on p–V and T–s diagrams.

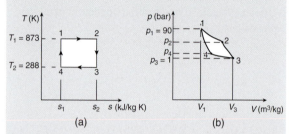

(a)   (b)

Figure 15.4 a) T–s diagram b) p–V diagram for Example 15.4

The Carnot efficiency is easily found from equation 15.8:

$$\eta_{\text{Carnot}} = \frac{T_1 - T_2}{T_1} = 1 - \frac{T_2}{T_1} = 1 - \frac{288}{873} = 0.6701,$$

so the $\eta_{\text{Carnot}} = 67.01\%$.

To find the net work, and the gross work due to the expansion, we need to find the entropy change $(s_2 - s_1)$ and to do this using

$$s_2 - s_1 = c_p \ln\left(\frac{T_2}{T_1}\right) - R \ln\left(\frac{p_2}{p_1}\right)$$

we first need to find p_2. This can be achieved by noting that the process 2–3 is an isentropic expansion, therefore we may use

$$\frac{p_2}{p_1} = \left(\frac{T_2}{T_1}\right)^{\frac{\gamma}{\gamma-1}}$$

(equation 15.7) and, by adopting our values for the process from figure 15.4, we find that

$$\frac{p_3}{p_2} = \left(\frac{T_3}{T_2}\right)^{\frac{\gamma}{\gamma-1}}, \text{ so } p_2 = p_3 \Big/ \left(\frac{T_3}{T_2}\right)^{\frac{\gamma}{\gamma-1}}$$

$$= 1 \Big/ \left(\frac{288}{873}\right)^{3.5} = \frac{1}{0.0206} = 48.49 \text{ bar}.$$

Then $s_2 - s_1 = c_p \ln\left(\frac{T_2}{T_1}\right) - R \ln\left(\frac{p_2}{p_1}\right)$ and, as process 1–2 is isothermal (constant temperature) and $p_1 = 90$ bar, $p_2 = 48.49$ bar, we get that

$$s_2 - s_1 = 0 - (0.287) \ln\left(\frac{48.49}{90}\right) = 0.1773 \text{ kJ/kg K}.$$

Therefore, the *net work* is given by $W_{\text{net}} = (T_1 - T_2)(s_2 - s_1) = (873 - 288)(0.1773) = 103.72 \text{ kJ/kg}$.

The *gross work due to expansion* is the sum of the work done in process (1–2) plus process (2–3).

Now, for process (1–2), from the first law, for an *isothermal process* $Q = W$ because for this process $\Delta U = 0$. Therefore the heat energy for process (1–2), which is equal to *the area* under the line (1–2) on the T–s diagram, is equal to the work, i.e. $W_{2,1} = (s_2 - s_1)T_1 = (0.1773)(873) = 154.8 \text{ kJ/kg}$.

For process (2–3), which is an isentropic process, $Q = 0$ and so, from the first law, for unit mass and the NFEE $W_{2,3} = c_V(T_2 - T_3)$ (see section 14.2.4), so $W_{2,3} = (0.718)(873 - 288) = 420.03 \text{ kJ/kg}$.

So *gross work* $= 154.8 + 420.03 = 574.83$ kJ/kg and then the *work ratio* $r_W = \dfrac{103.72}{574.83} = 0.1804$.

Note the very low work ratio for the Carnot cycle!

Key Point. The work ratio is considered the best measure of performance for the Carnot cycle

15.3 The Otto cycle

The *Otto cycle* is the ideal air standard cycle for the spark ignition piston engine. In this cycle it is assumed that the working fluid, air, behaves as a perfect gas and that there is no change in the composition of the air during the complete cycle. Heat transfer occurs at *constant volume* and there is *isentropic compression and expansion*.

This cycle differs from the practical engine cycle, in that the same quantity of working fluid is used repeatedly and so an induction and exhaust stroke are unnecessary.

Key Point. The Otto cycle is the ideal air standard cycle upon which the efficiency of the spark ignition piston engine is based

The thermodynamic processes making up a complete Otto cycle are illustrated in Figure 15.5 on the p–V

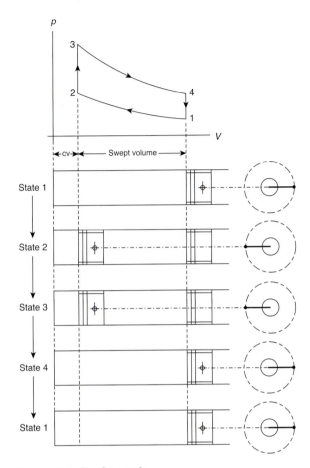

Figure 15.5 The Otto cycle

this can be used again with the first law to find the Otto cycle efficiency. Because the Otto cycle is the *air standard cycle* for the reciprocating piston internal combustion engine, its efficiency is known as the *air standard thermal efficiency*, which we will represent by the symbol $\eta_{\text{air standard}}$.

> **Key Point.** The air standard thermal efficiency
> $$\eta_{\text{air standard}} = \frac{\text{net work } (W_{\text{net}})}{\text{heat energy input } (Q_1)}$$

For the constant volume processes (2–3) and (4–1) the heat-in and the heat rejected during these processes, *per unit mass*, is given by $Q = c_V(T_2 - T_1)$, so that in our case we find that $Q_1 = c_V(T_3 - T_2)$ and $Q_2 = c_V(T_4 - T_1)$. Then $\eta = 1 - \dfrac{Q_2}{Q_1} = 1 - \dfrac{c_V(T_4 - T_1)}{c_V(T_3 - T_2)}$ and the *air standard thermal efficiency*, after cancelling the coefficients, will now be given by:

$$\eta_{\text{air standard}} = 1 - \frac{(T_4 - T_1)}{(T_3 - T_2)} \qquad (15.10)$$

Note that for the other two isentropic processes in the cycle $Q = 0$, so they play no part in the heat transferred to or from the system.

However, we do know that for an isentropic process $\dfrac{T_1}{T_2} = \left(\dfrac{V_2}{V_1}\right)^{\gamma - 1}$ (see equation 14.14) and so, for the two isentropic processes in the Otto cycle, we have that

$$\frac{T_4}{T_3} = \left(\frac{V_3}{V_4}\right)^{\gamma - 1} \qquad \text{and} \qquad \frac{T_1}{T_2} = \left(\frac{V_2}{V_1}\right)^{\gamma - 1}.$$

From Figure 15.4 we note that $\dfrac{V_3}{V_4} = \dfrac{V_2}{V_1}$, therefore $\dfrac{T_4}{T_3} = \dfrac{T_1}{T_2}$ and from this relationship, by inference,

$$\frac{T_4}{T_3} = \frac{T_1}{T_2} = \frac{T_4 - T_1}{T_3 - T_2}.$$

Also, by combining these results we note that

$$\frac{T_1}{T_2} = \left(\frac{V_2}{V_1}\right)^{\gamma - 1} = \frac{T_4 - T_1}{T_3 - T_2}$$

and (from the cycle) $\dfrac{V_1}{V_2} = $ the volume compression ratio r_V. Then

$$\left(\frac{V_2}{V_1}\right)^{\gamma - 1} = \left(\frac{1}{r_V}\right)^{\gamma - 1} = \frac{T_4 - T_1}{T_3 - T_2}$$

diagram, which is the most appropriate diagram for this cycle. A description of the processes for the Otto cycle is detailed below.

1–2 *Isentropic compression:* no heat transfer takes place, temperature and pressure increase and the volume decreases to the clearance volume.

2–3 *Reversible constant volume heating*: temperature and pressure increase.

3–4 *Isentropic expansion* (through swept volume): air expands and does work on the piston; pressure and temperature fall; no heat transfer takes place during the process.

4–1 *Reversible constant volume heat rejection* (cooling): pressure and temperature fall to original values.

We know from the second law that the efficiency is given as $\eta = \dfrac{W}{Q_1} = \dfrac{Q_1 - Q_2}{Q_1} = 1 - \dfrac{Q_2}{Q_1}$ and that

and so, from equation 15.10, the *air standard thermal efficiency* may be written as

$$\eta_{\text{air standard}} = 1 - \left(\frac{1}{r_V}\right)^{\gamma-1} \qquad (15.11)$$

Equations 15.10 and 15.11 offer two very convenient ways in which to find the air standard thermal efficiency, where it can be seen from equation 15.11 that this efficiency only depends on the volume compression ratio.

> **Example 15.5**   Calculate the Otto cycle air standard thermal efficiency for a reciprocating piston engine that has a swept volume of 210 cm³ and a clearance volume (CV) of 30 cm³.
>
> From Figure 15.5 it can be seen that the *total volume = swept volume + clearance volume*, and so the volume compression ratio
>
> $$r_V = \frac{\text{total volume}}{\text{clearance volume}} = \frac{210 + 30}{30} = 8 \text{ or } 8{:}1,$$
>
> as it is sometimes written. The air standard thermal efficiency is then given by equation 15.11 as $\eta_{\text{air standard}} = 1 - \left(\frac{1}{8}\right)^{1.4-1} = 1 - 0.435 = 0.5645$ or 56.45% efficient. Remembering that the ratio of specific heats for air $\gamma = 1.4$, this ratio does in fact vary with high temperature, as you will see later when we look at the real cycles of combustion engines.

15.3.1   Mean effective pressure

When considering reciprocating piston combustion engines (as opposed to fixed plant), whether petrol or diesel powered, the most effective measure for comparing the performance of these engines is provided by finding their *mean effective pressure*, where:

$$\text{Mean effective pressure (MEP)} = \frac{\text{net work output}}{\text{swept volume}} \qquad (15.12)$$

The MEP is effectively a measure of *engine work* and thus *power*, and is the average constant pressure that is assumed to act over the whole swept volume of the cycle that produces the same power output as the actual cycle. Figure 15.6 illustrates the concept of MEP.

> **Key Point.**   The MEP is the average constant pressure that acts over the whole swept volume of the cycle that produces the same work/power output as the actual cycle

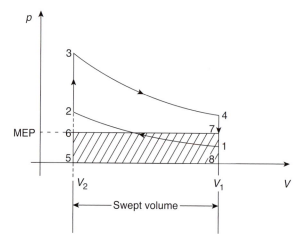

Figure 15.6 Illustration of MEP

In Figure 15.6 the shaded rectangle (56785) has the same area as that of the Otto cycle (12341), where the height of the rectangle is the MEP and its width is equal to the swept volume $(V_1 - V_2)$. Thus the equivalent work of the cycle W is given by the area of the shaded rectangle, i.e.

$$W = p_{\text{ME}}(V_1 - V_2) \qquad (15.13)$$

> **Example 15.6**   In an Otto cycle at the start of the compression the air is at a pressure of 1 bar and a temperature of 20°C. If the volume compression ratio is 9 and the maximum cycle temperature is 1750°C, determine the thermal efficiency, the net work and the MEP.
>
> This cycle is illustrated in Figure 15.7.
>
> In this case the thermal efficiency can easily be found from equation 15.11:
>
> $$\eta_{\text{air standard}} = 1 - \left(\frac{1}{r_V}\right)^{\gamma-1} = 1 - \left(\frac{1}{9}\right)^{0.4}$$
>
> $$= 1 - 0.415 = 0.585 = 58.5\%.$$

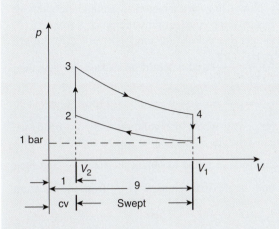

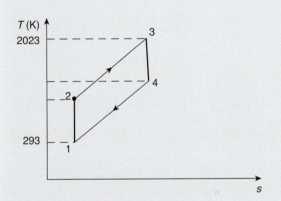

Figure 15.7 *p–V* and *T–s* diagrams for Example 15.6

Then, from $\eta = \dfrac{W}{Q_1} = \dfrac{Q_1 - Q_2}{Q_1} = 1 - \dfrac{Q_2}{Q_1}$ we note that $\eta = \dfrac{W}{Q_1}$ and also from above, where for this cycle the heat addition $Q_1 = c_V(T_3 - T_2)$, we find that $\eta_{\text{air standard}} = \dfrac{W_{\text{net}}}{c_V(T_3 - T_2)}$ or $W_{\text{net}} = (\eta_{\text{air standard}})\,c_V(T_3 - T_2)$. Then, from Figure 15.7 we can see that $T_3 = 2023$ K, and from equation 14.14 and the above analysis, we know that $\dfrac{T_1}{T_2} = \left(\dfrac{V_2}{V_1}\right)^{\gamma - 1}$, so that $T_2 = T_1\left(\dfrac{V_1}{V_2}\right)^{\gamma - 1}$ and again, from the *p–V* diagram in Figure 15.7, we see that $T_2 = T_1\,(r_V)^{\gamma - 1} = (293)(9)^{0.4} = 705.6$ K.

Therefore,

$$W_{\text{net}} = W_{\text{swept}} = (\eta_{\text{air standard}})\,c_V(T_3 - T_2)$$

$$= (0.585)(0.718)(2023 - 705.6) = 553.3 \text{ kJ/kg}$$

Now, in order to find the MEP from equation 15.12, we first need to find the swept volume $(V_1 - V_2)$ as in Figure 15.7. Since the working fluid is air, we can once again use the characteristic gas equation, which for unit mass is $p_1 V_1 = RT_1$ or $V_1 = \dfrac{RT_1}{p_1}$, knowing the volume compression ratio we note that $\dfrac{V_1}{V_2} = 9$ or $V_2 = \dfrac{V_1}{9}$, and so the swept volume

$$(V_1 - V_2) = \left(V_1 - \dfrac{V_1}{9}\right) = 0.8889 V_1,$$

where $V_1 = \dfrac{RT_1}{p_1} = \dfrac{(287)(293)}{(1 \times 10^5)} = 0.84\,\text{m}^3/\text{kg}$, so that the swept volume $(V_1 - V_2) = (0.8889)(0.84) = 0.747\,\text{m}^3/\text{kg}$.

So, knowing the swept volume and the net work or swept work, we can now find the MEP from equation 15.12 as:

$$\text{MEP} = \dfrac{553.3 \text{ kJ/kg}}{0.747 \text{ m}^3/\text{kg}} = 740.7 \text{ kJ/m}^3 \quad \text{or}$$

$$\text{MEP} = 740.7 \text{ kN/m}^2 = 7.407 \text{ bar}.$$

15.4 The diesel cycle

The original diesel engine had the fuel (powdered coal) forced in using compressed air. The ideal air standard diesel cycle described below and illustrated in Figure 15.8 is the cycle that most closely models this original diesel engine. As you will see later in section 15.4.1, modern oil engines or compression–ignition engines with direct fuel injection are based on the dual combustion cycle.

The *ideal air standard diesel cycle* (Figure 15.8) consists of four processes as follows:

1–2   isentropic compression

2–3   reversible constant pressure heat supply

3–4   isentropic expansion

4–1   reversible constant volume heat rejection

The air standard efficiency for this cycle is found in the same way as that for the Otto cycle, i.e. using $\eta = \dfrac{W}{Q_1} = \dfrac{Q_1 - Q_2}{Q_1} = 1 - \dfrac{Q_2}{Q_1}$, except in this case we

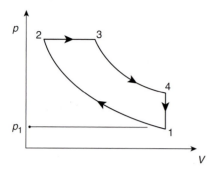

Figure 15.8 Illustration of the diesel cycle

get that

$$\eta = \frac{W}{Q_1} = \frac{Q_1 - Q_2}{Q_1} = \frac{c_p(T_3 - T_2) - c_V(T_4 - T_1)}{c_p(T_3 - T_2)},$$

so that:

$$\eta_{\text{air standard}} = 1 - \frac{(T_4 - T_1)}{\gamma(T_3 - T_2)} \qquad (15.14)$$

Note also that, as with the Otto cycle, the volume compression ratio $r_V = \dfrac{V_1}{V_2}$.

Example 15.7 A diesel engine with a compression ratio of 15:1 operates with an inlet pressure of 1 bar and an inlet temperature of 290 K. If the maximum cycle temperature is 1400 K, determine the air standard thermal efficiency based on the diesel cycle.

We have, from the question, temperatures $T_1 = 300\,\text{K}$ and $T_3 = 1400\,\text{K}$ and we know that for air $\gamma = 1.4$, so we need to find the other two temperatures and then use equation 15.14 to find the air standard efficiency.

From equation 14.14 we find that

$$\frac{T_1}{T_2} = \left(\frac{V_2}{V_1}\right)^{\gamma-1},$$

so

$$\frac{T_2}{T_1} = \left(\frac{V_1}{V_2}\right)^{\gamma-1}$$

and

$$T_2 = T_1 \left(\frac{V_1}{V_2}\right)^{\gamma-1} = 300 \left(\frac{15}{1}\right)^{0.4} = 886.25\,\text{K}.$$

Now, using the characteristic gas equation for unit mass for the constant pressure process from 2 to 3, we get that $p_2 V_2 = RT_2$ and $p_3 V_3 = RT_3$, therefore $\dfrac{p_2 V_2 = RT_2}{p_3 V_3 = RT_3}$ and, since the pressure is constant, then $\dfrac{V_2}{V_3} = \dfrac{T_2}{T_3} = \dfrac{886.25}{1400} = 0.633$. From Figure 15.8, observing where the volumes are positioned, we can now do a little bit of algebraic trickery with the volume ratios to obtain the ratio $\dfrac{V_4}{V_3}$ and then find the missing temperature T_4 from the equation $T_4 = T_3 \left(\dfrac{V_3}{V_4}\right)^{\gamma-1}$. Then

$$\frac{V_4}{V_3} = \frac{V_4 V_2}{V_2 V_3} = \frac{V_1 V_2}{V_2 V_3} = (r_V)\left(\frac{V_2}{V_3}\right)$$

$$= (15)(0.633) = 9.495,$$

so $T_4 = T_3 \left(\dfrac{V_3}{V_4}\right)^{\gamma-1} = 1400 \left(\dfrac{1}{9.495}\right)^{0.4} = 569\,\text{K}.$

Then, from equation 15.14, the air standard thermal efficiency is

$$\eta_{\text{air standard}} = 1 - \frac{(T_4 - T_1)}{\gamma(T_3 - T_2)}$$

$$= 1 - \frac{(569 - 300)}{(1.4)(1400 - 886.25)}$$

$$= 0.626 = 62.6\%.$$

15.4.1 Dual combustion cycle

As mentioned above, the dual combustion cycle, detailed here, is the cycle on which the modern compression–ignition oil engine operating cycle is based. In this cycle, the heat supplied to the system from the fuel is provided partly at constant volume and partly at constant pressure. Figure 15.9 illustrates the cycle.

The *ideal air standard dual combustion cycle* consists of five stages as follows:

1–2   isentropic compression

2–3   reversible constant volume heat addition

3–4   reversible constant pressure heat addition

4–5   isentropic expansion

4–1   reversible constant volume heat rejection

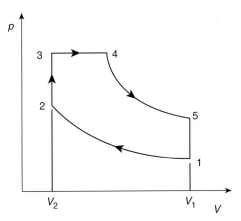

Figure 15.9 Illustration of the dual combustion cycle

> **Key Point.** The dual combustion cycle is the appropriate cycle upon which to model the modern oil diesel engine

As can be seen from the stages of this cycle and Figure 15.9, the heat from the fuel is supplied in two stages. In the first stage combustion takes place at constant volume and in the second stage combustion occurs at constant pressure. This two-stage combustion can lead to a rather complicated method of finding the air standard thermal efficiency because not only must the compression ratio (V_1/V_2) be taken into consideration but also the ratio of pressures and the ratio of the volumes (V_4/V_3), so in this book we will not be pursuing this method of finding the efficiency. Instead, we will stick with the standard definition, expressed in terms of temperatures.

Then the thermal air standard efficiency for the dual combustion cycle is found from our fundamental definition as

$$\eta = 1 - \frac{Q_2}{Q_1} = 1 - \frac{c_V(T_5 - T_1)}{c_V(T_3 - T_2) + c_p(T_4 - T_3)}$$

or, on division by c_V:

$$\eta = 1 - \frac{(T_5 - T_1)}{(T_3 - T_2) + \gamma(T_4 - T_3)} \qquad (15.15)$$

15.5 Constant pressure cycle

The *constant pressure cycle* often referred to as the *Joule cycle*, is the ideal cycle for the closed system

gas turbine plant. A typical gas turbine plant layout is shown in Figure 15.10. Such plant consists typically of a compressor, where the air is subject to isentropic compression, a combustor, where heat is added at constant pressure, a turbine, where the air is subject to isentropic expansion, and a heat exchanger, where the heat is rejected (extracted), again at constant pressure.

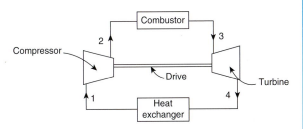

Figure 15.10 Typical closed gas turbine plant layout

The stages of the *ideal constant pressure cycle* are as follows:

1–2 isentropic compression

2–3 reversible constant pressure heat supply

3–4 isentropic expansion

4–1 reversible constant pressure heat rejection

This cycle is illustrated in Figure 15.11 on both a *p–V* and a *T–s* diagram. Note the lines of constant pressure superimposed on the *T–s* diagram for the heat addition and rejection. Note also from the *p–V* diagram that the pressure ratio for this cycle is $r_p = \dfrac{p_2}{p_1}$.

The air standard efficiency for this cycle can be determined in the usual way by starting with the basic definition. Then

$$\eta = \frac{W}{Q_1} = \frac{Q_1 - Q_2}{Q_1} = 1 - \frac{Q_2}{Q_1} = 1 - \frac{T_4 - T_1}{T_3 - T_2},$$

therefore:

$$\eta = 1 - \frac{T_4 - T_1}{T_3 - T_2} \qquad (15.16)$$

Now, from equation 14.15 and Figure 15.11 we note that

$$\frac{T_1}{T_2} = \left(\frac{p_1}{p_2}\right)^{\frac{\gamma-1}{\gamma}} \text{ and } \frac{T_4}{T_3} = \left(\frac{p_4}{p_3}\right)^{\frac{\gamma-1}{\gamma}},$$

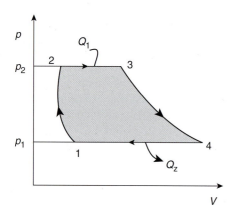

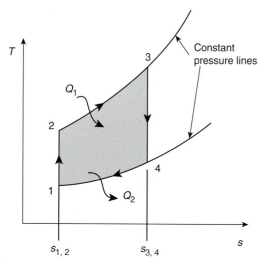

Figure 15.11 p–V and T–s diagrams for constant pressure cycle

also, from the above definition of the pressure ratio, $\frac{p_4}{p_3} = \frac{p_1}{p_2} = \frac{1}{r_p}$, and therefore by using inference, as we did for the Otto cycle, we find that

$$\frac{T_4}{T_3} = \frac{T_1}{T_2} = \frac{T_4 - T_1}{T_3 - T_2} = \left(\frac{1}{r_p}\right)^{\frac{\gamma-1}{\gamma}},$$

so the air standard thermal efficiency in terms of the pressure ratio is

$$\eta_{\text{air standard}} = 1 - \left(\frac{1}{r_p}\right)^{\frac{\gamma-1}{\gamma}} \qquad (15.17)$$

We can also formulate a relationship for the *work ratio* of the constant pressure cycle in a similar manner to what we found for the Carnot cycle.

The work ratio, from the situation given in Figure 15.11, may be found as follows:

$$r_W = \frac{\text{net work}}{\text{gross work}} = \frac{c_p(T_3 - T_4) - c_p(T_2 - T_1)}{c_p(T_3 - T_4)}$$

$$= 1 - \frac{T_2 - T_1}{T_3 - T_4}$$

and, again from the above explanation and equation 14.15, we know that

$$\frac{T_4}{T_3} = \frac{T_1}{T_2} = \left(\frac{1}{r_p}\right)^{\frac{\gamma-1}{\gamma}} \quad \text{or} \quad \frac{T_3}{T_4} = \frac{T_2}{T_1} = \left(r_p\right)^{\frac{\gamma-1}{\gamma}}.$$

Then $T_2 = T_1 \left(r_p\right)^{\frac{\gamma-1}{\gamma}}$ and $T_4 = T_3 / \left(r_p\right)^{\frac{\gamma-1}{\gamma}}$, then after substitution of these relationships for temperatures T_2 and T_4 into $r_W = 1 - \frac{T_2 - T_1}{T_3 - T_4}$ and a little bit of algebraic manipulation, we find that the work ratio for this cycle is given by:

$$\text{Work ratio} = 1 - \left(\frac{T_1}{T_3}\right)\left(r_p\right)^{\frac{\gamma-1}{\gamma}} \qquad (15.18)$$

Now this particular cycle, although closed, has to account for the *steady flow of air* around the cycle for its operation. Therefore, unlike all the other closed systems we have mentioned previously, we need to apply the steady flow energy equation (SFEE) to each process in the cycle, rather than as in previous cases where we used the non-flow energy equation (NFEE) to determine the non-flow parameters

You will remember that in the full version of the SFEE (equation 13.10) for open systems $Q - W = (h_2 - h_1) + (mgz_2 - mgz_1) + \left(\frac{1}{2}v_2^2 - \frac{1}{2}v_1^2\right)$, from section 13.4, we introduced the concept of *specific enthalpy change*. From the first law we were able to define the specific enthalpy change (ignoring changes in velocity for steady flow) by equation 13.9, $h_2 - h_1 = (u_2 - u_1) + (p_2 V_{s_2} - p_1 V_{s_1})$, and after application of the characteristic gas equation we were able to redefine specific enthalpy change, as in equation 13.11, i.e. $h_2 - h_1 = c_p(T_2 - T_1)$. This relationship is particularly useful in applying the work and heat transfers around the ideal *constant pressure cycle*.

Look again at Figure 15.10 and note that the SFEE may be applied to each of the following processes in the cycle resulting from the steady air flow, as follows: the work input from the air to the compressor $= (h_2 - h_1) = c_p(T_2 - T_1)$, the work output from the turbine $h_3 - h_4 = c_p(T_3 - T_4)$, the heat supplied to the air by the combustor $Q_1 = h_3 - h_2 = c_p(T_3 - T_2)$, the heat rejected (extracted) from the air by the heat exchanger $Q_2 = h_4 - h_1 = c_p(T_4 - T_1)$. It is upon these relationships for work and heat transfers for the perfect gas in a steady flow process that equation 15.16 was based, and these relationships will prove particularly useful when we consider open steady flow gas turbine cycles in Chapter 16.

Example 15.8 An air standard constant pressure cycle starts initially with air at 1 bar and 288 K that is compressed through a pressure ratio of 6, followed by constant pressure heat addition of 1400 kJ/kg. If these processes are followed by an isentropic expansion to the initial pressure and constant pressure cooling to the initial state, determine the thermal efficiency, mean effective pressure and work ratio for the cycle.

The *air standard thermal efficiency* is easily found using equation 15.17, i.e.

$$\eta_{\text{air standard}} = 1 - \left(\frac{1}{r_p}\right)^{\frac{\gamma-1}{\gamma}},$$

so that

$$\eta_{\text{air standard}} = 1 - \left(\frac{1}{6}\right)^{0.286} = 0.401 = 40.1\%.$$

In order to find the MEP we require the $\dfrac{\text{net work output}}{\text{swept volume}}$, where the net work $= c_p(T_3 - T_4) - c_p(T_2 - T_1)$, and in this case (see Figure 15.12) the swept volume is $V_{\text{swept}} = V_4 - V_2$. Then we proceed in the usual way to find the temperatures at the points in the cycle, and the appropriate volumes.

Then, from equation 14.15 on rearrangement, we have that

$$T_2 = T_1 \left(\frac{p_2}{p_1}\right)^{\frac{\gamma-1}{\gamma}} = 288(6)^{0.286} = 480.8 \text{ K}.$$

Now, we have no highest temperature but we do have the amount of constant pressure heat addition for process (2–3), i.e. for unit mass, $Q_1 = c_p(T_3 - T_2)$, so that $T_3 = T_2 + \dfrac{Q_1}{c_p}$, then

$$T_3 = 480.8 + \frac{1200}{1.005} = 1674.8 \text{ K}. \text{ Now we can find}$$

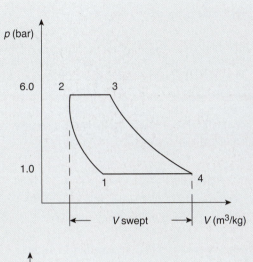

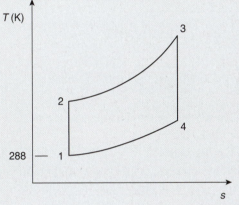

Figure 15.12 p–V and T–s diagrams for Example 15.8

temperature T_4 from the relationship for an isentropic process (expansion) given by $T_4 = T_3 / (r_p)^{\frac{\gamma-1}{\gamma}}$, then $T_4 = 1674.82 / (6)^{0.286} = 1003.3 \text{ K}$.

From the characteristic gas equation (the equation of state) for unit mass we have $V_1 = \dfrac{RT_1}{p_1} = \dfrac{(287)(288)}{(1 \times 10^5)} = 0.826 \text{ m}^3/\text{kg}$. Now, for the isentropic compression process (1–2) we have from equation 14.13 that $V_2 = V_1 \left(\dfrac{p_1}{p_2}\right)^{\frac{1}{\gamma}} = (0.826)\left(\dfrac{1}{6}\right)^{0.714} = 0.229 \text{ m}^3/\text{kg}$. Also, since $p_1 = p_4$ then, again from the equation of state, $V_4 = \dfrac{RT_4}{p_4} = \dfrac{(287)(1003.3)}{(1 \times 10^5)} = 2.88 \text{ m}^3/\text{kg}$.

Now the *net work*

$$= c_p(T_3 - T_4) - c_p(T_2 - T_1)$$

$$= (1.005)\,[(1672.8 - 1003.3) - (480.8 - 288)]$$

$$= 479.1\,\text{kJ/kg}$$

and the $V_{\text{swept}} = V_4 - V_2 = (2.88 - 0.229) = 2.65\,\text{m}^3/\text{kg}$, so that the

$$\text{MEP} = \frac{W_{\text{net}}}{V_{\text{swept}}} = \frac{479.1\,\text{kJ/kg}}{2.65\,\text{m}^3/\text{kg}}$$

$$= 180.79\,\text{kN/m}^2 = 1.81\,\text{bar}.$$

The *work ratio* is given by equation 15.18 as

$$r_w = 1 - \left(\frac{T_1}{T_3}\right)(r_p)^{\frac{\gamma-1}{\gamma}}$$

$$= 1 - \left(\frac{288}{1674.8}\right)(6)^{0.286} = 0.713.$$

Alternatively, we can find the *work ratio* from $r_W = \dfrac{\text{net work}}{\text{gross work}}$, where the *gross work* will be the turbine work $= c_p(T_3 - T_4) = (1.005)(1672.8 - 1003.3) = 672.85\,\text{kJ/kg}$, and the *net work* $= 479.1\,\text{kJ/kg}$ (as already found above), so the *work ratio* $= \dfrac{479.1}{672.85} = 0.712$, as before.

Note: In this chapter we have only considered perfect *gas* cycles and their air standard efficiencies and one or two other parameters that included the MEP and work ratio. In Chapter 16 we will look at both the practical constant volume cycle for reciprocating piston combustion engines and the constant pressure cycle for closed and open gas turbine engines and analyse other factors that we need to take into account when considering the efficiency and performance of these practical combustion engines. For example, we will need to consider, among other parameters, the *relative efficiency* of a particular piece of thermodynamic plant or equipment, such as the combustion engine, where the actual efficiency of the engine cycle is compared with the ideal efficiency of the cycle that most closely represents it.

15.6 Chapter summary

In this chapter we started by considering the concept of entropy as a measure of deviation from the ideal process, and found in particular that as engineers we considered *entropy change* as a way of measuring this deviation from the ideal. Note from equations 15.4 and 15.5, for constant volume and constant pressure processes, respectively, that the second term in each equation is eliminated and so the entropy change is given by the temperature ratios and the specific heat capacities, in each case.

In sections 15.2 to 15.5 we looked at the Carnot cycle, Otto cycle, diesel and dual combustion cycle and finally the constant pressure cycle or Joule cycle. You should understand that each of these ideal cycles may be used to model and compare real thermodynamic plant and engines and other prime movers, through use of air standard efficiencies. You should be able to identify and explain the processes within each cycle, sketch the cycles on (p–V) and (T–s) diagrams as appropriate and be able to calculate the air standard efficiencies of each cycle. Also, note that for all of these cycles we only considered air or other perfect gases as the working fluid, the study of vapour cycles, steam and steam plant being omitted in the interests of space but available for reference at the book website, www.key2engineeringscience.com. For your convenience, the cycle processes covered in this chapter, together with their air standard efficiency formulae, are tabulated in Table 15.1.

In addition to being able to find the efficiencies of these cycles you should also know how to find their work ratios and mean effective pressure (MEP), as appropriate. When determining these parameters you will have noticed the regular use made of the characteristic gas equation (equation of state) $pV = RT$ (for unit mass) and equations 14.13 to 14.15, i.e.

$$\frac{p_1}{p_2} = \left(\frac{V_2}{V_1}\right)^{\gamma}, \frac{T_1}{T_2} = \left(\frac{V_2}{V_1}\right)^{\gamma-1}$$

and

$$\frac{T_1}{T_2} = \left(\frac{p_1}{p_2}\right)^{\gamma-1/\gamma}, \text{ respectively.}$$

Remember to take care when using the laws of indices to transpose equations 14.13 to 14.15 or their equivalents. We will extend the use we have made of the Otto cycle and Joule (ideal constant pressure) cycle in Chapter 16, when we look at the real cycles for the spark ignition combustion engine and the gas turbine engine, for which they are the ideal.

operated on a *two-stroke cycle*, in which the processes and events in the four-stroke cycle described above are completed in two strokes of the piston (one revolution) rather than four (two revolutions). To achieve the two-stroke operation, the exhaust valve is opened before the end of the power stroke, the burnt gases being mostly cleared from the cylinder by the end of this stroke. To allow time for the burnt gases to be expelled and for the introduction of the charge in a spark ignition engine, the valves remain open in the early part of the next stroke, after which they are closed and compression begins. Like for like, two-stroke engines can produce more power than their four-stroke counterparts but not as efficiently. We will not pursue the two-stroke cycle any further in this chapter.

16.2 Internal combustion engine performance indicators

In practical internal combustion engines there are several well tried and tested methods for determining the equivalent performance of these engines. These include indicated power, brake power, specific fuel consumption and the related mechanical and thermal efficiencies, which we will take a look at in this section.

16.2.1 Indicated power

Engine indicators are used, as mentioned in section 15.3, to record the variation in cylinder pressure with volume; Figure 16.1 is a typical trace from an engine indicator. Now the area under this graph provides a measure of the work done by the gas on the piston, in the same way as the area under the ideal Otto cycle p–V diagram shown in Figures 15.5 and 15.6. The equivalent measure of this engine's net work (and thus power) was shown in Figure 15.6 to be the product of the MEP (p_{MEP}) and the swept volume.

> **Key Point.** The piston engine net work can be found from the product of MEP and the swept volume

Therefore, once the p_{MEP} is known, the work done on the piston by the charge can easily be found. The power resulting from this work is the indicated power = the net work done on the piston per unit time or per second. Thus if:

p_{MEP} = the mean effective pressure

A = cross-sectional area of the piston

L = stroke length

N = number of working strokes per second

then the average net force on the piston = $p_{MEP} \times A$.

The work done per stroke = $p_{MEP} \times A \times L$.

Power, i.e. the work done per working stroke per second, = $p_{MEP} \times A \times L \times N$ (when there are N/2 effective working strokes per cycle in a four-stroke engine).

Then *the indicated power* = $p_{MEP}LAN$ (16.1)

Note that if the correct SI units are used this relationship will give the indicated power in Watts.

> **Example 16.1** A four-cylinder spark ignition combustion engine has a cylinder bore of 15 cm and a stroke of 25 cm. During a test on the engine, when rotating at 3500 rpm, the results from an indicator diagram were as follows: area of indicator trace = 0.5 cm^2; length of trace = 0.4 cm. If the pressure scale on the trace is 25 kN/m^2 per cm height, determine the mean effective pressure and the indicated power.
>
> The mean height of trace = $\dfrac{0.5\,\text{cm}^2}{0.4\,\text{cm}}$ = 1.25 cm;
>
> therefore, using scale p_{MEP} = (1.25)(25) = 31.25 kN/m^2, the number of effective strokes per cycle, where for a four-stroke cycle $N = \dfrac{N}{2}$ is
>
> $= \dfrac{3500}{(2)(60)}$ = 29.17, and as there are four cylinders N = (29.17)(4) = 116.67.
>
> Now the indicated power = $p_{MEP}LAN$ =
>
> $(31.25)(0.25)\left(\dfrac{\pi(0.15)^2}{4}\right)$ (116.67) = 16.1 kW.

16.2.2 Brake power

The power delivered to the output shaft of an engine will always be less than the indicated power because of thermal losses due to friction and mechanical losses due to play in the bearings and/or in mechanical mechanisms. The way in which we measure the power delivered at the output shaft is to use some kind of *dynamometer* to measure *torque*. These instruments may vary from a simple rope *friction brake* (Figure 16.3) to a Froude hydraulic brake or modern electrical dynamometer; however, no matter what instrument is used, their purpose is to apply some kind of *braking torque* (T) at a specified rotational velocity often quoted in rpm.

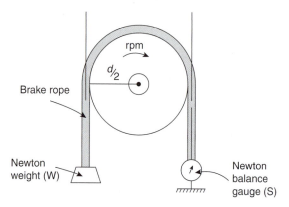

Figure 16.3 Simple rope friction brake

Figure 16.3 shows the set-up for a simple rope friction brake where the rope carries Newton weights (W) at one end and the other end is attached to a Newton balance (S) anchored to a rigid support. The rope is laid over the brake drum and since torque $T = Fr = F\dfrac{d}{2}$ then the difference in the torque created by the Newton weights and the Newton gauge will be the *braking torque*, i.e. $T = (W - S)\dfrac{d}{2}$ (N m). We know from our previous work that $power = T\omega$, where the units of ω are rad/s. Then, *if the rotational velocity (n) of the engine shaft is given in rpm*, as it often is, to convert it to rad/s we need to multiply it by the factor $\dfrac{2\pi}{60}$ so that our expression for the brake power will be:

$$Brake\ power = \frac{2\pi T n}{60} \qquad (16.2)$$

Key Point. The brake power is the power delivered at the output shaft of the engine and will always be less than the indicated power

16.2.3 Engine efficiencies and specific fuel consumption

We know, for example, that there are power losses, mainly due to friction in the pistons, bearings and mechanisms, that will always reduce the brake power measured at the output shaft when compared with the indicated power measure from within the engine. This leads us to the definition of the first of our efficiencies.

The engine **mechanical efficiency** (η_{mech}) compares the work done in unit time within the engine (the indicated power) with the useful work in unit time available at the output shaft (brake power). Therefore

$$\eta_{\text{mech}} = \frac{\text{brake power}}{\text{indicated power}} \qquad (16.3)$$

The brake power is sometimes still referred to in the old imperial units of brake horsepower (*bhp*), where 1 hp = 746 Watts.

Now, before we consider the thermal efficiencies we need to introduce a standard way of measuring the heat available from the fuel and then the rate at which the fuel is used. We determine the heat energy from the fuel using its **calorific value**. The calorie is a particularly small measure, so although we refer to the heat energy available from the fuel as its calorific value it has units of kJ/kg. There are in fact two values of the calorific value for each fuel. The *higher calorific value* (HCV) is determined experimentally and is achieved as a result of the steam formed in combustion condensing and thus giving up its latent heat. In practical cases of the combustion of fuels in engines, the steam formed as a result of the combustion does not condense and so does not give up its latent heat of evaporation. Under these circumstances the heat liberated by the fuel has a *lower calorific value* (LCV). Table 16.1 shows some typical calorific values of fuels, taken with *modifications* from Rodgers G and Mayhew Y., *Thermodynamic Properties of Fluids and Other Data* (Blackwell, 1988).

Table 16.1 Some typical calorific values for liquid fuels

Fuel	HCV kJ/kg	LCV kJ/kg
Gasoline (Petrol)	47300	44300
Kerosene	46300	43200
Paraffin	46000	41500
Diesel oil	46000	43250
Light fuel oil	44800	42100
Heavy fuel oil	44000	41400

We also need to introduce an *efficiency measure* for the fuel consumed in a specified time. This parameter is known as the **specific fuel consumption (SFC)** and is defined as: *the amount of fuel used per unit time* (usually hours) *divided by the brake power*, with normal units of kg/kWh, i.e.:

$$\text{Specific fuel consumption} = \frac{\text{Fuel used per unit time}}{\text{Brake power}}$$

$$(16.4)$$

> **Key Point.** The usual units for piston engine SFC are kg/kWh

The specific fuel consumption can be used with performance curves of power output and fuel consumption against speed, to determine the most economic speed, i.e. the optimum speed for minimum fuel consumption.

16.2.4 Engine thermal efficiencies

Having an understanding of indicated power, brake power and the calorific value of fuels, we can now define one or two related thermal efficiencies. We start with the **brake thermal efficiency**, defined as: *the ratio of the brake power to the power equivalent of the heat available from the fuel*, i.e.:

$$\eta_{\text{brake thermal}} = \frac{\text{Brake power}}{\text{Power available from the fuel}} \quad (16.5)$$

where *the power available from the fuel* (in Watts) is the product of the mass flow of the fuel $(\dot{m}_f)$ in kg/s and the calorific value of the fuel (CV) in J/kg.

The **indicated thermal efficiency** is very similar to the brake thermal efficiency except that indicated power is substituted for brake power, the units being the same, so that:

$$\eta_{\text{indicated thermal}} = \frac{\text{Indicated power}}{\text{Power available from the fuel}}$$

$$(16.6)$$

The actual thermal efficiency of a real engine is always less than the air standard efficiency we encountered in Chapter 15. Relating the two, the **relative efficiency** is defined as:

$$\eta_{\text{relative}} = \frac{\text{Indicated thermal efficiency}}{\text{Air standard efficiency}} \quad (16.7)$$

> **Key Point.** Always use the *indicated thermal efficiency* as the measure of the *actual thermal efficiency* of the engine, when finding the *relative efficiency*

Example 16.2   A single-cylinder oil engine using diesel oil, with a bore of 150 mm and a stroke of 250 mm, is subject to a test that gave the following results: engine speed 480 rpm; diesel oil fuel rate 42 g/min; MEP 8.1 bar; brake torque 250 N m. Using the calorific value for diesel oil given in Table 16.1, determine: a) the indicated power and brake power, b) the specific fuel consumption, c) the indicated and brake thermal efficiencies and d) the mechanical efficiency.

a)   For the four-stroke engine the number of power strokes per second is

$$= \frac{480}{(60 \times 2)} = 4,$$

therefore the indicated power, from equation 16.1,

$$= p_{\text{MEP}} LAN = (8.1 \times 10^5)(0.25)$$
$$\times \left(\frac{\pi (0.15)^2}{4} \right)(4) = 14313.9 \, \text{W}$$
$$= 14.314 \, \text{kW}.$$

From equation 16.2 the brake power

$$= \frac{2\pi T n}{60} = \frac{(2\pi)(250)(480)}{60} = 12566.4 \, \text{W}$$
$$= 12.566 \, \text{kW}.$$

b)   The fuel used $= 42 \, \text{g/min} = 0.7 \, \text{g/s}$ or $(0.7)(3600) \, \text{g/h} = 2.52 \, \text{kg/h}$. Then the *specific fuel consumption*, from equation 16.4,

$$= \frac{\text{Fuel used per unit time}}{\text{Brake power}} = \frac{2.52 \, \text{kg/h}}{12.566 \, \text{kW}}$$
$$= 0.2005 \, \text{kg/kWh}.$$

c)   The fuel power $= (\dot{m})(CV) = (0.7 \times 10^{-3} \, \text{kg/s})(43250 \, \text{kJ/kg}) = 30.275 \, \text{kW}$ so that the indicated thermal efficiency is, from equation 16.6,

$$\eta_{\text{indicated thermal}} = \frac{\text{Indicated power}}{\text{Power available from the fuel}}$$
$$= \frac{14.31 \, \text{kW}}{30.275 \, \text{kW}} = 47.3\%.$$

Similarly, from equation 16.5, the brake thermal efficiency is

$$\eta_{\text{brake thermal}} = \frac{\text{Brake power}}{\text{Power available from the fuel}}$$
$$= \frac{12.566\,\text{kW}}{30.275\,\text{kW}} = 41.5\%.$$

d) The mechanical efficiency is given by equation 16.3 as

$$\eta_{\text{mech}} = \frac{\text{Brake power}}{\text{Indicated power}}$$
$$= \frac{12.566}{14.314} = 0.878 = 87.8\%.$$

Example 16.3 A four-cylinder petrol engine operates on the four-stroke cycle. Each cylinder has a bore of 80 mm, a stroke of 120 mm and a clearance volume of 75.3 cm³. When the engine is operating at 4200 rpm, the fuel consumption is 12 kg/h, MEP is 7.5 bar and the torque developed is 120 N m.

Taking the calorific value of the fuel as 44300 kJ/kg, determine: a) the indicated power and brake power, b) the specific fuel consumption, c) the indicated and brake thermal efficiencies, d) the mechanical efficiency and e) the relative efficiency.

a) As before, the *number of power strokes per second for four cylinders*

$$= \frac{4200}{(60 \times 2)} \times 4 = 140$$

and so the *indicated power*

$$= p_{\text{MEP}}LAN = (7.5 \times 10^5)(0.12)$$
$$\times \left(\frac{\pi(0.08)^2}{4}\right)(140) = 63.334\,\text{kW}.$$

The *brake power*

$$= \frac{2\pi Tn}{60} = \frac{(2\pi)(120)(4200)}{60} = 52.78\,\text{kW}.$$

b) The *specific fuel consumption*

$$= \frac{\text{Fuel used per unit time}}{\text{brake power}} = \frac{12\,\text{kg/h}}{52.78\,\text{kW}}$$
$$= 0.227\,\text{kg/kWh}.$$

c) $$\eta_{\text{indicated thermal}} = \frac{\text{Indicated power}}{\text{Power available from the fuel}}$$
$$= \frac{\text{Indicated power}}{(\dot{m})(CV)}$$
$$= \frac{63.334\,\text{kW}}{(3.333 \times 10^{-3}\,\text{kg/s})(44300\,\text{kJ/kg})}$$
$$= \frac{63.334}{147.66} = 42.9\% \text{ and}$$

$$\eta_{\text{brake thermal}} = \frac{\text{Brake power}}{\text{Power available from the fuel}}$$
$$= \frac{52.78\,\text{kW}}{147.66\,\text{kW}} = 35.74\%$$

d) The mechanical efficiency is

$$\eta_{\text{mech}} = \frac{\text{Brake power}}{\text{Indicated power}} = \frac{52.78}{63.334} = 83.3\%$$

e) Before we can find the *relative efficiency* we first need to find the air standard efficiency, which we can do using equation 15.11, where for the Otto cycle

$$\eta_{\text{air standard}} = 1 - \left(\frac{1}{r_V}\right)^{\gamma-1}.$$

Now the compression ratio

$$r_V = \frac{\text{Total volume}}{\text{Clearance volume}}$$

and, from the information given, the swept volume

$$= \left(\frac{\pi(8)^2}{4}\right)(12) = 603.2\,\text{cm}^3$$

and so the total volume $= 603.19 + 75.3 = 678.49$, therefore $r_V = \frac{678.49}{75.3} = 9.01 \simeq 9.$

The $\eta_{\text{air standard}} = 1 - \left(\frac{1}{9}\right)^{0.4} = 1 - 0.415$
$$= 0.585 = 58.5\%,$$

therefore we find that the *relative efficiency* is

$$\eta_{\text{relative}} = \frac{\text{Indicated thermal efficiency}}{\text{Air standard efficiency}}$$
$$= \frac{0.429}{0.585} = 0.733 = 73.3\%.$$

16.3 The gas turbine engine

16.3.1 Introduction

The modern gas turbine engine has many applications. For example, it may be used as a power source for electrical generation or used to power ships and aircraft.

The working cycle of the gas turbine engine is similar to that of the four-stroke piston engine. In the gas turbine engine combustion occurs at a constant pressure, while in the piston engine it occurs at constant volume. In both engines there is an induction, compression, combustion and exhaust phase. You saw in section 15.3 that the ideal cycle upon which the closed gas turbine engine or plant is modelled is the *constant pressure or Joule cycle* (Figure 15.11).

As you already know, in the case of the piston engine we have a non-flow process whereas in the gas turbine engine we have a continuous flow process. In the gas turbine engine the lack of reciprocating parts gives smooth running and enables more energy to be released for a given engine size. Combustion occurs at constant pressure with an increase in volume, therefore the peak pressures that occur in the piston engine are avoided. This allows the use of lightweight, fabricated combustion chambers and lower octane fuels, although the higher flame temperatures require special materials to ensure a long life for combustion chamber components.

To ensure maximum thermal efficiency in the turbine, as you already know, we require the highest temperature of combustion (heat in) to give the greatest expansion of the gases. There has to be a limit on the temperature of the combusted gases as they enter the turbine, and this is dictated by the turbine materials. Additional cooling within the turbine helps to maximise the gas entry temperature to the turbine.

> **Key Point.** Within the limits of the materials, the higher the turbine entry temperature, the higher the turbine thermal efficiency

Although the closed gas turbine engine can be modelled reasonably accurately on the *constant pressure cycle* (Figure 15.12), in the practical cycle there will be thermodynamic and mechanical losses due to such things as:

- The air not being pure but containing other gases and water vapour.
- Heat being transferred to the materials of the compressor, turbine and exhaust units (open cycle),

so that they are not pure adiabatic or isentropic processes.

- Dynamic problems (in open gas turbines) such as turbulence and flame stability in the combustion chamber, whereby constant temperature and hence constant pressure cannot be maintained.
- Pressure losses as a result of the burnt air causing an increase in volume and hence a decrease in its density.
- Thermodynamic losses resulting from friction and play in mechanical mechanisms.

The losses in the compressor and turbine units of a *closed gas turbine engine* can be catered for by comparing the real cycle with the ideal constant pressure cycle through the use of isentropic efficiencies, which we discuss next. The open cycle gas turbine cannot be directly compared with the ideal constant pressure cycle because of the energy losses and complications that arise as a result of the combustion process. These are discussed briefly when we consider the jet engine as an open cycle gas turbine engine, later in this chapter.

16.3.2 Isentropic efficiency and the real cycle

Throughout Chapter 15 we looked at the air standard efficiencies of ideal cycles, where we assumed that, in the constant pressure cycle for example, the compression and expansion processes were *isentropic*. However, in real cycles, for example in the *closed gas-turbine plant*, with respect to the *compression* and *expansion processes*, the *real compression process* will require a *larger work input* to the compressor than in the ideal isentropic case, for the reasons highlighted above. The efficiency of the compressor, i.e. the *compressor isentropic efficiency* (η_C), is then defined as:

$$\eta_C = \frac{\text{Ideal specific enthalpy change}}{\text{Actual specific enthalpy change}} \qquad (16.8)$$

Hopefully, you understand why in *flow systems*, such as the gas turbine engine, we use *enthalpy* and *enthalpy change* as the measure of fluid energy/work. Do remember that enthalpy combines the *internal energy* and *pressure–volume energy* of the fluid flow.

> **Key Point.** A real compression process requires a larger work input to the compressor than that required for ideal isentropic compression

From the work we did on *steady flow processes* and *specific enthalpy change* in sections 13.4 and 15.5, you will also remember that the isentropic (ideal) work done on the compressor (W_C) is given by $h_2 - h_1 = c_p(T_2 - T_1)$. However, when dealing with isentropic efficiencies, *we represent ideal temperature by use of a prime*. Then the:

Ideal work done on the compressor

$$W_C = h_2 - h_1 = c_p(T_2' - T_1) \qquad (16.9)$$

So the isentropic compressor efficiency $\eta_C = \dfrac{c_p(T_2' - T_1)}{c_p(T_2 - T_1)}$. Now, to a close approximation the average *pressure ratio* (c_p) for the actual compression will be the same as the (c_p) in the ideal case, depending only on the mean temperature for the process. Therefore we may ignore any minor changes in (c_p) and cancel them so that the:

Isentropic compressor efficiency

$$\eta_C = \frac{(T_2' - T_1)}{(T_2 - T_1)} \qquad (16.10)$$

The difference between the *isentropic compression* and the *real compression* is illustrated on the *T–s* diagram of the *gas turbine constant pressure cycle* in Figure 16.4.

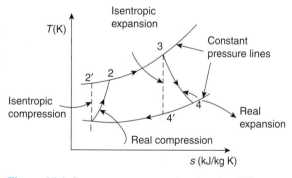

Figure 16.4 Constant pressure cycle, showing differences between isentropic and real compression and expansion processes

Now, if the *expansion process* of the gas in a real constant pressure cycle takes place in a *turbine*, the friction effects and other disturbances, as mentioned above, will cause the gas to leave the turbine *hotter* than it would do in an ideal expansion, which of course means the temperature *drop* through the turbine (T_3 to T_4) would *be less than* in the ideal case, resulting in *less* work being provided from the turbine. Then in a similar manner to that of the compressor, the

isentropic turbine efficiency η_T is the ratio of *the actual work output* (measured by the *enthalpy change*) to the *isentropic work output* for the same pressure ratio and inlet temperature, which from our definitions of specific enthalpy change is $\eta_T = \dfrac{c_p(T_3 - T_4)}{c_p(T_3 - T_4')}$. Also, making the same assumptions about the closeness of the values of c_p for the isentropic and real cases as we did for the compressor, then, after cancellation of these ratios, the:

Isentropic turbine efficiency

$$\eta_T = \frac{(T_3 - T_4)}{(T_3 - T_4')} \qquad (16.11)$$

Note that, as with the air standard efficiencies we considered in Chapter 15, the *isentropic efficiencies* given by equations 16.10 and 16.11 are in terms of temperature differences. Thus in order to find these efficiencies, we will need to make use of equations 14.13–14.15 and the characteristic gas equation, as we did in section 15.5.

Example 16.4 In a simple closed gas turbine engine consisting of a compressor, combustor, turbine and heat exchanger, the isentropic efficiency of the compressor is 85% and that of the turbine is 90%. The inlet air temperature to the compressor is 290 K, the compression ratio is 5:1 and the maximum temperature of the air in the engine is 1000 K. Assuming adiabatic compression and expansion, constant pressure heat addition and that the specific heat capacities for the air are also constant, determine: a) the specific work done on the compressor, b) the specific work output from the turbine, c) the specific net work from the engine and d) the cycle thermal efficiency.

a) We are told that the gas is air and the specific heat capacities are constant. Therefore, using standard values, $c_p = 1.005$ kJ/kg K, $c_V = 0.718$ kJ/kg K and so $\gamma = \dfrac{c_p}{c_V} = \dfrac{1.005}{0.718} = 1.4$, as we know. Now, from equation 14.15 $\dfrac{T_2'}{T_1} = \left(\dfrac{p_2}{p_1}\right)^{\frac{\gamma-1}{\gamma}}$, and after substitution of our values we find that $T_2' = (290)(5)^{0.286} = 459.5$ K and from equation 16.10 we get that $\eta_C = 0.85 = \dfrac{(459.5 - 290)}{(T_2 - 290)}$ and so $T_2 =$ 489.4 K. Then the *actual specific work* done on the compressor is given by $W_C = h_2 - h_1 = c_p(T_2 - T_1) = 1.005(489.4 - 290) = 200.4$ kJ/kg.

b) We follow a similar procedure to find the actual specific work done by the turbine. The isentropic exit temperature from the turbine is again found using equation 14.15 where, in this case, $\dfrac{T'_4}{T_3} = \left(\dfrac{p_4}{p_3}\right)^{\frac{\gamma-1}{\gamma}}$, so that $T'_4 = 1000\left(\dfrac{1}{5}\right)^{0.286} = 631$ K and, from equation 16.11, $0.9 = \dfrac{1000 - T_4}{1000 - 631}$. Therefore, $T_4 = 667.9$ K and the *actual specific work output from the turbine* $W_T = c_p(T_3 - T_4) = 1.005(1000 - 667.9) = 333.8$ kJ/kg.

c) The *specific net work output* is $W_{\text{net}} = W_T - W_C = 333.8 - 200.4 = 133.4$ kJ/kg.

d) The cycle thermal efficiency may be found from equation 13.12, where $\eta_{\text{thermal}} = \dfrac{W_{\text{net}}}{Q_1}$ and the heat transferred into the cycle $Q_1 = c_p(T_3 - T_2) = 1.005(1000 - 489.4) = 513.15$ kJ/kg, so that $\eta_{\text{thermal}} = \dfrac{W_{\text{net}}}{Q_1} = \dfrac{133.4}{513.15} = 0.2599 = 26\%$.

16.4   Aircraft propulsion

16.4.1   Introduction

Aircraft may be propelled using piston engines with a propeller, turbojet engines, turbofan engines or turboprop engines, the principles of aircraft propulsion all being the same. In this section we concentrate on propulsion by means of the *turbojet engine*. We take a brief look at the mechanics of aircraft propulsion and the overall efficiencies that are directly related to aircraft propulsion.

You know from Newton's third law that to every action there is an equal and opposite reaction. A jet propulsion engine takes in a working fluid (air), adds heat energy to it and accelerates this high-energy working fluid out of the engine exhaust in order to accelerate the aircraft in the opposite direction to the *thrust force* produced by the stream.

The thrust produced by the high-energy *gas* stream as it is ejected at the exhaust of the engine and the subsequent reaction force that accelerates the aircraft may be compared with the motion of an inflated balloon that is let free before the end is tied and accelerates forward as the airflow escapes through the neck of the balloon. The *thrust* force created by the airstream escaping from the neck of the balloon (the action) is counterbalanced by accelerating the mass of the balloon (the reaction) in the opposite direction.

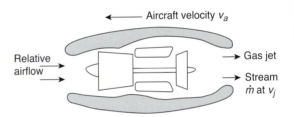

Figure 16.5 Aircraft turbojet propulsion engine

Consider the turbojet engine illustrated in Figure 16.5, which is fitted to an aircraft flying with airspeed (v_a). The engine working fluid flows at a rate of $\dot{m}_a$(kg/s) and the jet stream is expelled from the exhaust at a velocity of v_j (m/s). The product of these two quantities creates the *thrust force* exerted by the engine. This thrust force, that we refer to as the *total* or *gross thrust* (F_G), is given by:

$$F_G = \dot{m}_a v_j \qquad (16.12)$$

Now, as the aircraft flies through the air, the airflow entering the intake of the engine creates an opposing *drag force* that acts on the airframe and tries to retard the aircraft in flight. This *intake drag force* (F_D) that opposes the gross thrust force from the engine is given by the product of the air mass flow $\dot{m}$(kg/s) and the relative aircraft flight speed v_a (m/s), i.e.:

$$F_D = \dot{m}_a v_a \qquad (16.13)$$

The useful thrust force (the *net thrust* (F_N)) is given by the difference between the *gross thrust* and the *intake drag*, i.e. $F_N = \dot{m}v_j - \dot{m}v_a$, so:

$$F_N = \dot{m}_a \left(v_j - v_a\right) \qquad (16.14)$$

Note 1: There is one other component of thrust we have not yet considered, and that is *pressure thrust*. When the exhaust gas stream is not expelled completely to the pressure (p_6) outside the control volume of the engine, the remaining pressure of the gas jet inside the propelling nozzle of the engine (p_{05}, see section 16.5) exerts an *additional pressure thrust* over the gas jet exit area $\left(A_j\right)$ equal to $A_j(p_{05} - p_6)$. When this situation arises, i.e. when the nozzle is *chocking*, then the net thrust would be given by:

$$F_N = \dot{m}_a \left(v_j - v_a\right) + A_j(p_{05} - p_6). \qquad (16.15)$$

Note 2: We use the letter F rather than T to represent *thrust* in the above equations. This is because *thrust is a force*, which using the units given above will be measured in *Newton* (N), and also because we wish to avoid any confusion with the symbol we use for thermodynamic temperature.

We can find the *thrust power* (P_T) produced for an aircraft in flight by multiplying all the useful or *net thrust* (Newton) by the *flight speed* in (m/s). Then

$$P_T = \dot{m}_a \left(v_j - v_a\right) v_a + A_j(p_{05} - p_6)v_a \qquad (16.16)$$

16.4.2 Propulsion efficiencies

In aircraft propulsion units, including the turbojet engine, we are interested in the *thermal efficiency* as we were with the closed gas turbine engine and also the *propulsive efficiency* that shows how well the turbojet thrust is used to give useful power.

From the above discussion you now know that the turbojet engine produces its thrust by accelerating backwards a high-energy gas stream. The thermal efficiency of this process can be measured by the ratio of the increase in kinetic energy of air (the net work) as it passes through the engine over the energy content of the fuel input, so the thermal efficiency is, for *unit time*:

$$\eta_{\text{thermal}} = \frac{\text{Increase in KE of the gas stream}}{\text{Fuel energy input}} \qquad (16.17)$$

Now the increase in KE of the gas stream per second (i.e. *the rate of the energy input*) is given by $\frac{\dot{m}_a}{2} \left(v_j^2 - v_a^2\right)$ and the rate of energy input from the fuel is the mass flow rate of the fuel $\left(\dot{m}_f\right)$ kg/s multiplied by the calorific value

(CV) J/kg of the fuel, so the turbojet engine *thermal efficiency* may be expressed as:

$$\eta_{\text{thermal}} = \frac{\dfrac{\dot{m}_a}{2}\left(v_j^2 - v_a^2\right)}{\dot{m}_f CV} \qquad (16.18)$$

The *propulsive efficiency* $\left(\eta_{\text{prop}}\right)$ of a turbojet engine is defined as the ratio of the *thrust power* over *the rate of energy input* to the gas stream, i.e.:

$$\eta_{\text{prop}} = \frac{\text{Thrust power}}{\text{Rate of energy input}} \qquad (16.19)$$

Now, from equation 16.13, $P_T = \dot{m}\left(v_j - v_a\right)v_a$, and from the fact that *the rate of energy input is equal to the KE increase of the gas stream per unit time*, then:

$$\eta_{\text{prop}} = \frac{\dot{m}(v_j - v_a)v_a}{\dfrac{\dot{m}}{2}\left(v_j^2 - v_a^2\right)} = \frac{2(v_j - v_a)v_a}{\left(v_j^2 - v_a^2\right)}$$

$$= \frac{2(v_j - v_a)v_a}{(v_j - v_a)(v_j + v_a)} = \frac{2v_a}{v_j + v_a}$$

and on division by (v_a) we find that the *propulsive efficiency* is:

$$\eta_{\text{prop}} = \frac{2}{1 + v_j/v_a} \qquad (16.20)$$

Note: This very useful equation tells us immediately that the propulsive efficiency will be increased; the closer the jet velocity is to the velocity of the aircraft. Unfortunately, as we bring these velocities closer together, then from equations 16.13 and 16.18 both the thrust power and the thermal efficiency are reduced! There is a way out of this dilemma, and that is to increase the mass flow rate of the gas stream (*air flow*) ($\dot{m}_a$) through the engine. Then, as can be seen from $P_T = \dot{m}_a \left(v_j^2 - v_a^2\right)v_a$, the thrust power can be maintained. To achieve this increase in the mass flow of the air through the engine, *turbofan* and other high frontal area engines are used.

Finally, we can obtain an overall efficiency for the turbojet propulsion engine by combining the thermal efficiency with the propulsive efficiency. If this is done, the following expression can be derived for the *overall efficiency*:

$$\eta_{\text{overall}} = \frac{\dot{m}_a \left(v_j - v_a\right) v_a}{\dot{m}_f CV} \qquad (16.21)$$

We will now perform some sample calculations to see the effect of differing air mass flow rate and gas stream exit velocities on the performance of a turbojet engine. These simple calculations assume, among other things, that the flow rate of the air entering the engine remains constant as it enters the combustor and that the fuel flow rate is the same for the varying air mass flow rates

> **Key Point.** The overall thermal efficiency of the turbojet engine is the product of the thermal efficiency and the propulsive efficiency

Example 16.5   A turbojet engine aircraft is flying at 800 km/h when air flows into the engine at 50 kg/s while fuel with $CV = 43000$ kJ/kg enters the combustor at 0.05 kg/s. If the gas jet exit velocity $= 300$ m/s, determine the thermal, propulsive and overall efficiencies. If the air flow rate is now increased to 100 kg/s while the jet exit velocity is reduced to 280 m/s, again find the thermal, propulsive and overall efficiencies and comment on your results.

In the first case, the thermal efficiency is $\eta_{thermal} =$

$$\frac{\frac{\dot{m}}{2}\left(v_j^2 - v_a^2\right)}{\dot{m}_f CV} = \frac{25(300^2 - 222^2)}{(0.05)(43000 \times 10^3)} = 0.473 =$$

47.3%. The propulsive efficiency is $\eta_{prop} =$

$$\frac{2}{1 + v_j/v_a} = \frac{2}{1 + 300/222} = 0.85 = 85\%.$$ Now

the overall efficiency is found to be $\eta_{overall} =$

$$\frac{\dot{m}\left(v_j - v_a\right)v_a}{\dot{m}_f CV} = \frac{50(300 - 222)222}{(0.05)(43000 \times 10^3)} = 0.403 =$$

40.3%.

In the second case: $\eta_{thermal} = \dfrac{\frac{\dot{m}}{2}\left(v_j^2 - v_a^2\right)}{\dot{m}_f CV} =$

$$\frac{50(280^2 - 222^2)}{(0.05)(43000 \times 10^3)} = 0.677 = 67.7\%.$$ The

propulsive efficiency will be $\eta_{prop} = \dfrac{2}{1 + v_j/v_a} =$

$$\frac{2}{1 + 280/222} = 0.884 = 88.4\%.$$ Then in this

case the overall efficiency will be $\eta_{overall} =$

$$\frac{\dot{m}\left(v_j - v_a\right)v_a}{\dot{m}_f CV} = \frac{50(280 - 222)222}{(0.05)(43000 \times 10^3)} = 0.599 =$$

59.9%.

From the two sets of figures it can be seen that, for this over-simplified analysis, doubling the mass flow rate of the air leads to significant improvements in the thermal and overall efficiencies. Even with a reduction in the difference between the exhaust jet velocity and the aircraft velocity, the propulsive efficiency, in the second case, shows a slight improvement, this is why turbofan engines with large mass flows are the preferred choice for civil airliners that operate at cruise speeds around 900 to 1050 km/h.

16.5   The aircraft turbojet engine cycles and component efficiencies

Before we discuss the intricacies of the ideal and real cycles for the aircraft turbojet engine and the component efficiencies, we first need to look at the way in which we define *stagnation enthalpy, temperature and pressure*, which are directly applicable to the *fluid flow* through aircraft gas turbine engines that may themselves be either stationary or in motion.

16.5.1   Stagnation enthalpy, temperature and pressure

We start by considering once again the full version of the SFEE in specific terms, i.e. equation 13.6 where $Q - W = (u_1 - u_2) + (p_2 V_{s2} - p_1 V_{s1}) + (gz_2 - gz_1) + \left(\frac{1}{2}v_2^2 - \frac{1}{2}v_1^2\right)$. We defined *specific enthalpy* earlier in section 13.4.3 as $h = u + pV$ and *specific enthalpy change* as $h_2 - h_1 = (u_2 - u_1) + (p_1 V_1 - p_2 V_2)$, which we substituted into the SFEE to yield $Q - W = (h_2 - h_1) + (gz_2 - gz_1) + \left(\frac{1}{2}v_2^2 - \frac{1}{2}v_1^2\right)$. This equation shows that the energy change between any two positions is equal to the *enthalpy change* plus the changes in *potential* and *kinetic energy*. Now, for the air passing through an open gas turbine engine, the change in potential energy is generally negligible, so we will ignore it, unless told otherwise. So for energy changes through our turbojet engine this equation reduces to:

$$Q - W = (h_2 - h_1) + \left(\frac{1}{2}v_2^2 - \frac{1}{2}v_1^2\right) \qquad (16.22)$$

We also expressed the *specific enthalpy* of a *perfect gas* in terms of the specific heat capacity and temperature change to give the expression $h_1 = c_p T_1$, and for *specific enthalpy change* we obtained the expression:

$$Q - W = h_2 - h_1 = c_p(T_2 - T_1) \qquad (16.23)$$

Now, the internal energy plus the pressure–volume energy of the gas gives its enthalpy, but in flow systems there is also the kinetic energy of the gas to consider by virtue of its velocity at any point in the system (equation 16.22). We can add the kinetic energy of the gas to the existing *static enthalpy* to give the *total enthalpy* of the gas flow. Then, from what has been said, when the *static enthalpy* of the gas at any point is (h) and the *kinetic energy* of this gas at the same point is $\dfrac{v^2}{2}$ the *total specific enthalpy* or **stagnation specific enthalpy** (h_0), as it is more commonly known, is given as:

$$h_0 = h + \frac{v^2}{2} \qquad (16.24)$$

And when the fluid is a *perfect gas*, where $h = c_p T$ and the *total temperature* or *stagnation temperature* (T_0) is the corresponding temperature for the stagnation enthalpy, i.e. $h_0 = c_p T_0$, then, on division by c_p, we obtain:

$$T_0 = T + \frac{v^2}{2c_p} \qquad (16.25)$$

Equation 16.25 needs a little explanation, but first we must deal with the idea of stagnation enthalpy and stagnation temperature. Physically, *the stagnation enthalpy* (h_0) is the enthalpy that a gas stream of static enthalpy (h) and velocity (v) would possess when brought to rest adiabatically and without work transfer. Under these circumstances, equation 16.22 with no heat and work transfers becomes $0 = (h_0 - h) + \left(0 - \dfrac{1}{2}v^2\right)$ and from this we obtained equation 16.24.

When the gas stream described here is brought to rest adiabatically and without work transfer, then it will do so at a corresponding temperature known as the **total or stagnation temperature** T_0 (identified above in equation 16.25). The temperature T is then known as the *static temperature*. What about the expression $\dfrac{v^2}{2c_p}$? Well, to balance the equation this also needs to be a temperature and if we analyse the units of this

expression, it is found to be a temperature that we refer to as the *dynamic temperature*, to distinguish it from the static temperature. Thus equation 16.25 is telling us that the

Stagnation or total temperature

  = *static temperature + dynamic temperature*

so that $(T_0 - T)$ is the ram temperature rise. Now when this same gas flow is slowed down and the temperature rises, there is also simultaneously a pressure rise. This resulting pressure rise is the **stagnation or total pressure** (p_0) and is defined in a similar way to the stagnation temperature, except there is the added restriction that the gas is not only brought to rest adiabatically but also reversibly, i.e. *isentropically*. Then, using equation 15.7, we define the *stagnation pressure* by:

$$p_0 = p \left(\frac{T_0}{T}\right)^{\frac{\gamma}{(\gamma-1)}} \qquad (16.26)$$

Note: Stagnation enthalpy, temperature and *pressure* are identified by the suffix (0) preceding the point or stage in the cycle to which they apply. Thus, for example, the stagnation pressure and temperature at the exit to the compressor in a simple gas turbine engine (Figure 16.6) would be represented as p_{02} and T_{02}, respectively.

16.5.2   The ideal aircraft turbojet cycle

The ideal turbojet cycle for an aircraft gas turbine engine is based on the sketch of the engine (Figure 16.6) showing the component layout and stages through which the intake air passes

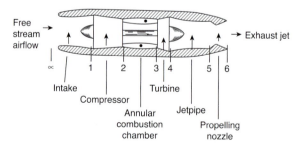

Figure 16.6 Component layout and stages in an aircraft turbojet gas turbine engine

At the heart of the aircraft turbojet engine is the gas generator shown as stages 1–4 in Figure 16.6.

This consists of the basic gas turbine engine as you know it, i.e. the compressor, combustor and turbine units, where instead of the fluid giving up its power to a heat exchanger (in the closed system) or powering a generator etc. in the open system plant, the high-velocity hot gases from the turbine pass through the jet pipe and out through the propelling nozzle, which creates a *thrust* to accelerate the aircraft in accordance with Newton's third law, as previously discussed. **The ideal turbojet cycle** is described below.

α–1 *The intake*: Air may enter the intake when either the engine is stationary or when the air is moving at some flight velocity (v_a) relative to the engine. With the engine stationary, the intake is considered adiabatic and without friction, so the air remains at constant stagnation temperature and pressure. When the aircraft is in flight, the free stream air is scooped up and accelerated to the airspeed of the aircraft. This action of work done on the air results in *an increase in the total or stagnation temperature and pressure* of the air (*the ram temperature and pressure rise*), which for our ideal cycle will be *isentropic*.

1–2 *Compression*: Here in the ideal situation *isentropic compression* takes place; the stagnation temperature and pressure increase.

2–3 *Heat input*: Heat is added to the air, ideally at constant stagnation pressure, in the combustor.

3–4 *Expansion*: In the ideal case, the gas is expanded isentropically on its journey through the turbine, while the turbine stagnation pressure and temperature fall.

4–5 *Jetpipe*: In the ideal case the airflow through the jetpipe will be adiabatic and without friction, so the stagnation temperature and pressure of the gas will remain constant.

5–6 *Propelling nozzle*: At entry to the nozzle the high-energy, high-temperature gas stream will be at a much higher stagnation pressure and temperature than the external ambient atmosphere and in the ideal case the gas flow will be adiabatic and reversible. The *stagnation* temperature and pressure will then remain constant but the *static* pressure and temperature will fall, resulting in an increase in dynamic pressure through an increase in velocity.

In Figure 16.7 the above ideal cycle for the turbojet is illustrated on both a *p–V* and a *T–s* diagram, in a similar manner to that for the closed gas turbine engine.

Note in Figure 16.7 that constant pressure heat addition occurs in the combustor, and that isentropic compressions and expansions take place in a similar manner to that of the closed gas turbine engine.

For the above ideal cycle the heat and work transfers involving the compressor, combustor and turbine may

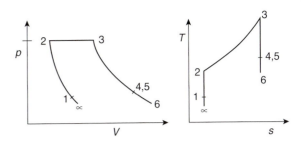

Figure 16.7 *p–V* and *T–s* diagrams for the ideal constant pressure turbojet gas turbine cycle

be determined in a similar manner to the way in which we found them in the closed-gas turbine cycle. The major difference is that we use stagnation enthalpy changes to determine them, where the enthalpies may be expressed in terms of temperature as in equation 16.23. The heat and work transfers for the other processes in the cycle may be determined as before, based on the nature of the process. All the heat and work transfers for the whole cycle are summarised next.

α–1 *The intake*: The steady flow of the air as it enters and travels through the intake is given from equation 16.24 as:

$$T_{01} = T_\alpha + \frac{v_a^2}{2c_p} \qquad (16.27)$$

where $(T_{01} - T_\alpha)$ is the *ram temperature rise* and (v_a) is the flight velocity relative to the engine. The stagnation pressure rise due to the ram effect occurs isentropically so, using equation 16.26, we get that

$$\frac{p_{01}}{p_\alpha} = \left(\frac{T_{01}}{T_\alpha}\right)^{\frac{\gamma}{(\gamma-1)}}.$$

1–2 *Compression*: The ideal compression is *adiabatic*, so $Q_{1-2} = 0$, and as the compressor is assumed to be horizontal (no potential energy change) then, from equation 16.22, for *mass air flow* ($\dot{m}_a$) kg/s the work per second (*power*) done on the air during compression is given by:

$$\dot{W}_{1-2} = \dot{m}_a(h_{01} - h_{02}) = \dot{m}_a c_p(T_{01} - T_{02}) \qquad (16.28)$$

Again, the ideal compression has the added restriction of also being *isentropic*, so the relationship $\dfrac{T_{02}}{T_{01}} = \left(\dfrac{p_{02}}{p_{01}}\right)^{\frac{(\gamma-1)}{\gamma}}$ may be used.

3–4 *Turbine*: The isentropic efficiency for the turbine is expressed in terms of stagnation temperatures and pressures, as was the case for the compressor. So:

Aircraft turbojet isentropic turbine compressor efficiency

$$\eta_T = \frac{(T_{03} - T_{04})}{(T_{03} - T'_{04})} \tag{16.40}$$

and:

Aircraft turbojet isentropic turbine efficiency

$$\eta_T = \frac{T_{03} - T_{04}}{T_{03}\left[1 - (p_{04}/p_{03})^{\frac{\gamma-1}{\gamma}}\right]} \tag{16.41}$$

4–5 *Jetpipe*: In the real jetpipe, there will be a *loss in stagnation pressure* within the gas stream as it passes through the jetpipe. This is found in a similar way to the pressure losses for the combustor, so the jetpipe pressure loss $= \frac{p_{05} - p_{04}}{p_{04}}$, which again is normally expressed as a percentage.

5–6 *Propelling nozzle*: The expansion of the gas through the propelling nozzle will be non-isentropic, so the nozzle efficiency may be expressed in a similar way to that of the turbine, i.e.:

$$\eta_{\text{Nozzle}} = \frac{T_{05} - T_6}{T_{03}\left[1 - (p_{05}/p_6)^{\frac{\gamma-1}{\gamma}}\right]} \tag{16.42}$$

Note: The above analysis of the aircraft turbojet ideal and real cycles and cycle efficiencies has been necessarily rather long and complicated, but the example that follows will hopefully show that, provided care is taken, analysing a real turbojet cycle is quite straightforward!

Example 16.6 A turbojet-powered aircraft is flying at 900 km/h at an altitude where the temperature is 255 K and the ambient atmospheric pressure is 54 kN/m², and where the following engine component efficiencies and other parameters apply:

Intake efficiency $\eta_i = 0.92$

Stagnation pressure ratio 8:1

Compressor efficiency $\eta_C = 0.88$

Combustion pressure loss of 4%

Turbine efficiency $\eta_T = 0.9$

Turbine inlet stagnation temperature $= 1400$ K

Jetpipe pressure loss of 4%

Assume that the expansion through the propelling nozzle is isentropic and that the nozzle is not *chocked*, so that there is *no* additional *pressure thrust*. Also take, for the air in the compressor, $c_p = 1.005$ kJ/kg K and $\gamma = 1.4$ and, for the gas from the turbine and throughout the remainder of the cycle, let the average values be $c_p = 1.2$ kJ/kg K and $\gamma = 1.33$.

Then, *assuming* that the power output from the turbine is equal to the power input from the compressor, and given that the engine air mass flow $m_a = 60$ kg/s and that for the fuel $CV = 43000$ kJ/kg, calculate the *net thrust* and the *specific fuel consumption* (SFC).

1) To determine the net thrust we need to work steadily and sequentially from when the air first enters the *intake*. Then the air stagnation temperature, from equation 16.27, is

$$T_{01} = T_\infty + \frac{v_a^2}{2c_p} = 255 + \frac{250^2}{(2)(1005)}$$

$$= 255 + 31.1 = 286.1 \text{ K},$$

where $v_a = 900$ km/h $= 250$ m/s. Also, from equation 16.35,

$$\frac{p_{01}}{p_\infty} = \left(1 + \frac{\eta_i v_a^2}{2c_p T_\infty}\right)^{\frac{\gamma}{(\gamma-1)}}$$

$$= \left(1 + \frac{0.92(250)^2}{(2)(1005)(255)}\right)^{3.5} = 1.45$$

and so $p_{01} = (54)(1.45) = 78.35$ kN/m².

2) Now, for the *compressor* we know from the isentropic relationship that

$$\frac{T'_{02}}{T_{01}} = \left(\frac{p_{02}}{p_{01}}\right)^{\frac{(\gamma-1)}{\gamma}} = 8^{0.286} = 1.813$$

and so $T'_{02} = (286.1)(1.813) = 518.7\,\text{K}$, then, from our definition of compressor efficiency

$$\eta_C = \frac{(T'_{02} - T_{01})}{(T_{02} - T_{01})}$$

and the information given, we find that

$$0.88 = \frac{(518.7 - 286.1)}{(T_{02} - 286.1)}$$

and so $T_{02} = 550.45\,\text{K}$. Then the power input to the compressor is, from equation 16.28, $\dot{W}_{1-2} = \dot{m}_a c_p (T_{01} - T_{02}) = (60)(1.005)(286.1 - 518.7) = -14026\,\text{kW}$ (negative because power is absorbed). Also note that at the compressor outlet $p_{02} = 8p_{01} = (8)(78.35) = 626.8\,\text{kN/m}^2$.

3) At the *exit* from the *combustion chamber* we know that $T_{03} = 1400\,\text{K}$ and so with a combustion pressure loss of 4% we find that $p_{03} = (0.96)(626.8) = 601.7\,\text{kN/m}^2$. Now, ignoring the enthalpy of the liquid fuel flow, to a good approximation the heat flow addition in the combustion chamber is given by equation 16.29 as $\dot{Q}_{2-3} = m_a (h_{03} - h_{02})$ or $\dot{Q}_{2-3} = m_a c_p (T_{03} - T_{02})$ and where, from the information given for the gas in the combustor, $c_p = 1.2\,\text{kJ/kg K}$ and $\gamma = 1.33$, then $\dot{Q}_{2-3} = (60)(1.2)(1400 - 550.45) = 61167.8\,\text{kW}$.

The fuel flow rate in kg/s is given by the ratio of the heat energy addition per second (kJ/s or power) divided by the calorific value of the fuel (kJ/s), so that the fuel flow rate

$$= \frac{61167.8\,\text{kJ/s}}{43000\,\text{kJ/kg}} = 1.4225\,\text{kg/s or } 5121\,\text{kg/h}$$

The fuel flow rate will be used later to find the SFC.

4) We know that for our analysis we must assume that the *turbine power must equal the compressor power* (see note in section 16.5.2; therefore:

$$\dot{m}_a (h_{03} - h_{04}) = \dot{m}_a (h_{02} - h_{01}) = 14026\,\text{kW}$$

or $\dot{Q} = \dot{m}_a c_p (T_{03} - T_{04}) = 14026\,\text{kW}$. Then

$(60)(1.2)(1400 - T_{04}) = 14026\,\text{kW}$, from which $T_{04} = 1205\,\text{K}$.

We can now, in a similar fashion to our approach with the compressor, use the turbine

efficiency equation to find T'_{04}. Then $\eta_T = \dfrac{(T_{03} - T_{04})}{(T_{03} - T'_{04})}$, so $0.9 = \dfrac{(1400 - 1205)}{(1400 - T'_{04})}$ and $T'_{04} = 1183.3\,\text{K}$.

Now, using the isentropic relationship $\dfrac{p_{03}}{p_{04}} = \left(\dfrac{T_{03}}{T'_{04}}\right)^{\frac{\gamma}{(\gamma-1)}}$ with $\gamma = 1.33$, we find that $\dfrac{p_{03}}{p_{04}} =$

$$\left(\frac{1400}{1183.3}\right)^4 = 1.959 \text{ and } p_{04} = \frac{p_{03}}{1.959} =$$

$$\frac{601.7}{1.959} = 307.15\,\text{kN/m}^2.$$

5) In the *jetpipe* we assume that because it will be well insulated the heat losses will be negligible, and of course no work is done, therefore $T_{05} = T_{04} = 1205\,\text{K}$. There is, however, a jetpipe stagnation pressure loss of 4%, so that $p_{05} = (0.96)p_{04} = (0.96)(307.15)$ so $p_{05} = 294.9\,\text{kN/m}^2$.

6) Flow through the propelling nozzle is assumed isentropic and the flow will expand to the *static pressure* outside the aircraft, i.e. $p_6 = 54\,\text{kN/m}^2$. Now, because we assume isentropic flow we can again use the relationship $\dfrac{T_{05}}{T_6} = \left(\dfrac{p_{05}}{p_6}\right)^{\frac{(\gamma-1)}{\gamma}}$ to find T_6, then with $\gamma = 1.33$ we find that $\dfrac{1205}{T_6} = \left(\dfrac{294.9}{54}\right)^{0.248} = 1.524$ and $T_6 = 790\,\text{K}$.

7) We are now in a position to find the velocity of the exhaust jet and then the net thrust, remembering that in this case there is no pressure thrust. Then, from equation 16.34: $v_6 = \sqrt{2c_p(T_{05} - T_6)} = \sqrt{2(1200)(1205 - 790)} = 998\,\text{m/s}$ and the net thrust is, from equation 16.14, $F_N = \dot{m}_a (v_j - v_a) = 60(998 - 250) = 44880\,\text{N}$. To find the specific fuel consumption (SFC) we use the relationship $SFC = \dfrac{\text{fuel flow rate}}{\text{net thrust}}$, then the $SFC = \dfrac{5121\,\text{kg/h}}{44880\,\text{N}} = 0.114\,\text{kg/N-h}$.

I hope you have followed the reasoning through all six stages of this process, noting how we found the required temperature and pressures and the fuel flow rate at each stage in the process, ending up with T_6 from which we found the jet exit velocity, then the net thrust, and finally the SFC!

The analysis of these cycles is quite a lengthy process and one for which many computer programs have now been written to solve for a differing range of the many variables. After the chapter summary that follows next, you will only be asked in the review questions to re-work this example using different values for the parameters, in order to acquaint yourself with the procedure.

16.6   Chapter summary

In this chapter we started in section 16.1 by considering the closed reciprocating piston internal combustion engine working cycle, having looked at the *ideal Otto cycle* in Chapter 15. You should be aware of the kinds of losses that occur in the practical cycle as a result of the design of the engine to make it functional. For example, we know in reality that the ignition process, the injection of the charge and the removal of the exhaust products cannot occur instantaneously, hence the need for *valve timing*. We went on in section 16.2 to discuss the engine performance indicators that are used to compare engines of a similar type and specification. You should be able to define *indicated power* and *brake power* and determine the required parameters for their associated *thermal and relative efficiencies*. Remember also that the *specific fuel consumption (SFC)* gives a measure of how fuel-efficient a particular engine is by comparing the fuel usage rate against the brake power. At the present time fuel economy is everything!

In section 16.3 we started with an introduction to the practical gas turbine engine, detailing its advantages over its reciprocating piston engine counterpart, where among other things you should note that the combustion process in gas turbine engines occurs ideally at constant pressure with a subsequent increase in volume which prevents the pressure peaks we get within the piston engine. Thus combustion components can be made lighter, and lower octane fuels may be used. Note also in this introduction the discussion on the losses that occur in the practical cycle when compared with the ideal constant pressure cycle. We then considered in section 16.3.2 the real cycle for the *closed gas turbine engine*. The *isentropic efficiencies* for the *compression and expansion processes* were derived and used to allow

for the losses in the real cycle, where these compressions and expansions are *non-isentropic*.

In section 16.4, as a prelude to the *open gas turbine cycles* for the turbojet engine, we introduced the concept of aircraft propulsion and how the output from aircraft propulsion units produce *thrust*, rather than brake power from a driveshaft. We found that the *gross thrust* was given by equation 16.12, before momentum drag (reverse thrust) was subtracted (equation 16.13). Then the *total net thrust*, which included the *pressure thrust* that acted on the propelling nozzle area (if the exit was chocked), was given by equation 16.15. We then, in section 16.4.2, considered the efficiencies that are applicable to all gas turbine aircraft propulsion units, where we defined the thermal and propulsive efficiencies and their combination that gave the overall efficiency of an aircraft gas turbine unit.

We moved on in section 16.5 to a discussion on aircraft turbojet cycles and efficiencies as an example of the open system gas turbine cycle. It was necessary in section 16.5.1 to define *stagnation enthalpy, temperature* and *pressure*, which as you will have seen are very necessary when analysing the efficiencies and performance of *moving* turbojet gas turbine engines fitted to aircraft. Think of the word *stagnation* as meaning *total* if you are having difficulties in understanding how stagnation values differ from the values you are familiar with. Thus, for example, the *stagnation enthalpy* is the sum total of the *static enthalpy* we are familiar with *and the dynamic enthalpy* created by the velocity of the airflow into and through the moving engine. We discussed the ideal cycle for the aircraft turbojet gas turbine engine in section 16.5.2, where the major differences between this cycle and the closed system ideal gas turbine constant pressure cycle were explained. These differences included the idea of doing work on the air as it entered the aircraft intake as a consequence of the aircraft's airspeed. This creates a rise in the *stagnation temperature and pressure* that results from the ram effect, which is known as the *ram temperature and pressure rise*. Other differences relate to the definition and units used for the *specific fuel consumption* (SFC) for the turbojet gas turbine engine (see the definition in section 16.5.3 and equation 16.39).

The last section in this chapter looked at the *real aircraft turbojet cycle* and the *component efficiencies* within this cycle. Although the procedure for analysing this cycle is slightly more complicated than was the case with the basic closed system gas turbine plant, many of the steps in the procedure are the same; for example, the use of the isentropic efficiencies for the compressor and the turbine was the same. In the ideal case the airflow through the intake was isentropic, and so the *intake isentropic efficiency* was introduced to cater for the losses in the real case. Stagnation pressure losses that occur in the practical combustion system were also catered for, using a pressure loss coefficient, and the efficiency of the combustion process itself was gauged by comparing the rise in enthalpy of combustion to the calorific value of the fuel used, as given by equation 16.38. Gas flow through the jetpipe was considered isentropic in the ideal case; however, in the real case pressure losses occur, which was again catered for by the introduction of a pressure loss coefficient that is often given as a percentage. Finally, in the real case the expansion of the gas stream through the propelling nozzle is not isentropic, so a nozzle efficiency was introduced to cater for this fact.

Example 16.6 set out the step-by-step approach to determining important performance criteria such as the *gross and net thrust*, the *thrust power* and the *SFC*. Provided you follow the stages as set out in this example, you should not have too many difficulties in determining the required design parameters from the analysis!

16.7 Review questions

1. Briefly discuss the losses that occur in the practical four-stroke cycle when compared with the ideal Otto cycle.

2. Explain the difference between indicated power and brake power.

3. A six-cylinder petrol engine operates on a four-stroke cycle. Each cylinder has a bore of 10 cm and a stroke of 15 cm, with a clearance volume of 147.3 cm³. When the engine is operating at 4000 rpm the fuel consumption is 16 kg/h, the MEP is 4 bar and the torque developed is 200 N m. Taking the CV for the fuel as 44300 kJ/kg, determine: a) the indicated power and the brake power, b) the SFC, c) the indicated and brake thermal efficiencies, d) the mechanical efficiency and e) the relative efficiency.

4. Explain why there is less work output from a real gas turbine than there would be from the equivalent ideal turbine.

5. In a simple closed gas turbine engine, as in Example 16.4, the isentropic efficiency of the compressor is 88% and that of the turbine is 90%. The inlet air temperature to the compressor is 300 K, the compression ratio is 7:1 and the maximum temperature of the air in the engine is 1400 K. Assuming adiabatic compression and expansion, constant pressure heat addition and that the specific heat capacities for the air are also constant, determine: a) the specific work done on the compressor, b) the specific work output from the turbine, c) the specific net work from the engine and d) the cycle thermal efficiency.

6. A turbojet-powered aircraft is flying at 1000 km/h at an altitude where the static pressure is 40 kN/m². Air enters the intake with a mass flow rate of 50 kg/s and exits the aircraft as a gas jet at a velocity of 700 m/s. The aircraft propelling nozzle has an exit diameter of 0.4 m and the nozzle is chocked with the gas flow at a pressure of 120 kN/m² at the nozzle. Determine: a) the gross thrust and b) the net thrust.

7. If, for the *same* turbojet-powered aircraft identified in question 6, fuel with a $CV = 43000$ kJ/kg enters the combustor at 0.4 kg/s with all other values being equal, determine: a) the thermal efficiency, b) the propulsive efficiency and c) the overall efficiency of the turbojet engine.

8. Define: a) stagnation specific enthalpy, b) stagnation temperature.

9. Why may the equation $\dot{Q}_{2-3} = m_a \left(h_{03} - h_{02} \right)$ be used as a good approximation for the heat flow addition to the gas in the combustor?

10. *Rework* Example 16.6, making the same assumptions, to *find the net thrust and SFC*, but using the following parameters: flight speed = 800 km/h, atmospheric pressure at altitude = 70 kN/m², temperature at flight altitude = 268 K, turbine inlet stagnation temperature = 1200 K, stagnation pressure ratio = 10:1. Use also the following related efficiencies: intake efficiency $\eta_i = 0.90$, compressor efficiency $\eta_C = 0.84$, combustion stagnation pressure loss of 6%, turbine efficiency $\eta_T = 0.9$, jetpipe pressure loss of 2%. As before, let $CV_{\text{fuel}} = 43000$ kJ/kg and, for the air in the compressor, $c_p = 1.005$ kJ/kg K and $\gamma = 1.4$ and, for the gas from the turbine and throughout the remainder of the cycle, $c_p = 1.2$ kJ/kg K and $\gamma = 1.33$.

Chapter 17

Introduction to heat transfer

17.1 Introduction

In the study of thermal science, *heat transfer* is considered to be a separate subject in its own right along with *thermodynamics* and *fluid mechanics*. In this chapter, therefore, we cannot hope to cover heat transfer in any depth. Rather, the aim of this chapter is to provide an introductory foundation to the subject. The in-depth study of heat transfer requires the use of some relatively sophisticated mathematics, involving partial differential equations that are considered beyond this introductory text. Therefore, for this reason and in the interests of space, we will be restricting our study to *one-dimensional* heat transfer. We start by considering heat transfer by conduction through solids and the use of Fourier's law. We then look at convection and heat transfer through solids, composites, cylinders and pipes. Finally we consider heat transfer by radiation, where we define black and grey body radiation; then, we derive and use the Stefan–Boltzmann law to solve simple problems involving heat transfer by radiation.

As you already know, from a thermodynamic point of view, the amount of heat transferred in a system as the result of a process is simply equal to the difference between the energy change of the system and the work done on or by the system.

When we consider engineering heat transfer we are primarily concerned with the determination of the *rate of heat transfer* for a specified difference in temperature. The dimensions of heaters, boilers, heat exchangers and refrigerators depend not only on the amount of heat to be transferred, but also on the rate of transfer under given external constraints. Engineering components such as transformers, electrical and mechanical machines, turbines and bearing assemblies require us to analyse heat transfer rates, in order to avoid conditions that will cause overheating and damage the equipment. These examples emphasise the importance of the solution of heat transfer problems in engineering, rather than just the use of thermodynamic reasoning alone.

Heat transfer may be defined as: the transmission of energy from one region to another as a result of a temperature difference between them.

In heat transfer, as in other branches of engineering science, the successful solution of problems requires us to make assumptions and simplifications. This enables us to formulate mathematical models whose solution is possible, either manually, or with the aid of computers and computer software. It is important to remember that all idealised models of real engineering problems should be treated with caution, since any simplified analysis may, in certain cases, severely limit the accuracy of the results.

Literature on heat transfer generally recognises three distinct modes of heat transmission, the names of which will be familiar to you, i.e. *conduction, convection* and *radiation*. Technically, only conduction and radiation are true heat transfer processes, because both of these transmission methods depend totally and utterly on a temperature difference being present. Convection also depends on the transportation of a mechanical mass. Nevertheless, since convection also accomplishes transmission of energy from high to low temperature regions, it is conventionally regarded as a heat transfer mechanism. We will now consider these heat transfer mechanisms separately, in a little more detail.

978-1-85617-775-7, Engineering Science, Mike Tooley & Lloyd Dingle

17.2 Conduction

17.2.1 Definition

Thermal conduction in solids and liquids seems to involve two processes. The first is concerned with atoms and molecules, the second with *free* electrons.

Atoms at a high temperature vibrate more vigorously about their equilibrium positions in the lattice than their cooler neighbours. Since atoms and molecules are bonded to one another, they pass on some of their vibrational energy. This energy transfer occurs from atoms of high vibrational energy to those of lower vibrational energy, without appreciable displacement. This energy transfer has a knock-on effect, since high vibrational energy atoms increase the energy in adjacent low vibrational energy atoms, which in turn causes them to vibrate more energetically, resulting in thermal conduction. Thermal conduction through solids and liquids by molecular transfer is illustrated in Figure 17.1a, while conduction in gases is illustrated in Figure 17.1b.

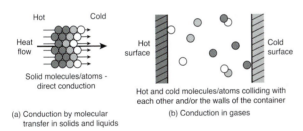

(a) Conduction by molecular transfer in solids and liquids

(b) Conduction in gases

Figure 17.1 Conduction by molecular transfer in *solids, liquids and gases*

The second process involves material with a ready supply of free electrons. Since electrons are considerably lighter than atoms, any gain in energy by electrons results in an increase in the electrons' velocity and they are able to pass this energy on quickly to cooler parts of the material.

> **Key Point.** Heat transfer by *conduction* involves two processes, one involving the energy transfer from high-energy vibrating atoms or molecules to those of lower energy and the second by the rapid transfer of energy between high-energy free electrons and their lower energy neighbours

This phenomenon is one of the reasons why electrical conductors, which have many free electrons, are also good thermal conductors. Do remember that metals are not the only good thermal conductors. The first mechanism described above, which does not rely on free electrons, is a very effective method of thermal conduction, especially at low temperatures. Examples of heat transfer by conduction are numerous. The exposed end of a metal spoon which is immersed in hot tea will eventually feel warm due to the conduction of energy through the spoon. On a cold winter's day there is likely to be a significant amount of energy lost through the walls of a heated room to the outside, primarily due to conduction. Yet another example of energy transfer by conduction is seen in the classic physics experiment where a copper rod is heated by a Bunsen burner and the rod becomes warm remote from the heat source.

The way in which heat may be transferred by conduction may be modelled using *Fourier's law*, as you will see next.

17.2.2 Fourier's law

As you already know from your study on thermodynamics, whenever a temperature gradient exists in a solid material, heat will flow from the high temperature to the low temperature region. The rate at which heat is transferred by conduction, $\dot{Q}$ is proportional to the temperature gradient dT/dx and the area A normal to the direction of the heat transfer (see Figure 17.2); i.e. $\dot{Q} \propto \dfrac{AdT}{dx}$. We now need a constant of proportionality for this relationship, which in this case is the *thermal conductivity* (k) of the material subject to the heat transfer. This constant k is an inherent property of the material that measures the ability of the material to conduct heat (Table 17.1) and has units of Watts per

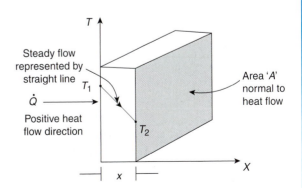

Figure 17.2 One-dimensional heat transfer by conduction

metre-Kelvin (W/m K). The formula then becomes:

$$\dot{Q} = -k\frac{AdT}{dx} \qquad (17.1)$$

The minus sign results from the second law of thermodynamics, where we know that heat flows from a higher to a lower temperature and in order for this heat flow to be positive (by convention) a minus sign is placed in front of equation 17.1 (see Figure 17.2). The mathematical relationship given by equation 17.1 was found by the French mathematician Fourier and, since this relationship has never been disproved, it is known as *Fourier's law*.

Provided the whole heat flow $\dot{Q}$(J/s) is *steady* and *perpendicular* to the surface through which it travels and the surface temperatures remain constant, then the temperature gradient dT/dx may be expressed in terms of the change in temperature $(T_2 - T_1)$ and the distance (x) across which the temperature change takes place (Figure 17.2). Then equation 17.1 may be written as $\dot{Q}_{cond} = -k\frac{A(T_2 - T_1)}{x}$ or as:

$$\dot{Q} = \frac{kA(T_1 - T_2)}{x} \qquad (17.2)$$

Note the switch of temperatures (for the temperature difference) in equation 17.2, made in order to ensure that the heat transfer is positive as it flows from hot to cold.

Another useful variant of equation 17.2 is to express conduction in terms of *the rate of heat transfer per unit area (q)*, i.e. $q = \dfrac{\dot{Q}}{A}$. The quantity q then has units of W/m^2, and is often referred to as the *heat flux*. It can be found from the relationship:

$$q = \frac{k(T_1 - T_2)}{x} \qquad (17.3)$$

For convenience, the thermal conductivities (k) of some common engineering materials are given in Table 17.1.

Example 17.1   The outer surface temperature of a pane of glass is 15°C and its inner-surface temperature is 25°C.

Calculate the rate of heat loss through the pane of glass, given that its thermal conductivity is 0.8 W/m K and it has dimensions of 1.0 m high, 0.5 m wide and 0.75 cm thick. Figure 17.3 illustrates the situation.

Table 17.1 Thermal conductivity of some common engineering materials

Material	Thermal conductivity (W/m K)
Aluminium	235
Brick	0.4–0.7
Cast iron	52
Concrete	0.8–1.3
Copper	400
Cork	0.05
Duralumin	166
Engine oil	0.15
Glass	0.8
Lead	35
Low carbon steel	43
Plastics	0.2–0.4
Rubber	0.15
Water	0.6

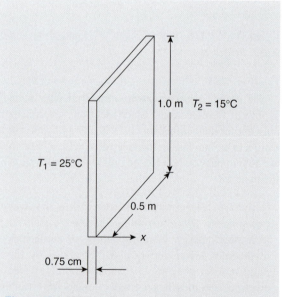

Figure 17.3 Heat conduction through a pane of glass

We make the assumptions that:

- Steady-state conditions prevail
- We are only considering conduction in the *x*-direction through the glass (one-dimensional conduction)
- The glass is homogeneous (i.e. it has constant properties throughout).

Under these circumstances we may now determine the heat transfer through the glass using equation 17.2. Then:

$$\dot{Q} = \frac{kA(T_1 - T_2)}{x}$$

$$= \frac{(0.8)(0.5)(25 - 15)}{0.0075}$$

$$= 533.33 \text{W}.$$

Note also that the *heat flux* in this case is $q = 1066.67 \text{ W/m}^2$, which can easily be found separately if desired from equation 17.3.

17.2.3 Heat transfer by conduction through solid composite walls

So far we have only considered plane walls, which consist of one homogeneous material. There are, however, many cases in practice when different materials are constructed in layers to form a composite wall. For example, furnace walls have an inner lining of refractory brick, followed by an insulating material and another layer of fire brick. There may also be an external layer of plaster or similar finishing material to complete the composite wall.

The flow of heat through a composite wall and the resistance to this heat flow by each of the materials is similar to the resistance to current flow set up in an electric circuit. You will find this *electrical analogy* very useful when solving problems concerned with heat transfer.

Consider the diagram of the composite wall shown in Figure 17.4. There are three layers of different materials of thickness (x_1, x_2, x_3) and with corresponding thermal conductivities (k_1, k_2, k_3). The internal wall temperature is T_0 and the material interface temperatures are T_1, T_2 and T_3, as shown.

If we think of the *flow of heat as being analogous to the flow of an electric current*, then we know that the heat flow is caused by a temperature difference while the current flow is caused by a difference in potential.

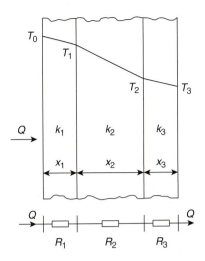

Figure 17.4 Heat transfer through a composite wall

So it is possible to think of *thermal resistance* in a similar manner to the way we think of *electrical resistance*. Then you will know from Ohm's law that $I = \frac{V}{R}$, so using this formula we can make comparisons with equation 17.2, $\dot{Q} = \frac{kA(T_1 - T_2)}{x}$, in the following manner. We let the heat flow $\dot{Q}$ be the equivalent of the current flow (I) and we let the temperature difference $(T_1 - T_2)$ be the equivalent of the potential difference V (volts). Then, substituting our electrical terms into equation 17.2 gives $I = \frac{kA(V)}{x}$. To complete the analogy, we see from this relation that $\frac{1}{R} = \frac{kA}{x}$ and so, using the equivalent electrical resistance, we find that the *thermal resistance* is:

$$R = \frac{x}{kA} \qquad (17.4)$$

Note that the *units of thermal resistance* will be in Kelvin per Watt (K/W). Now, our solid composite wall is the equivalent of three electrical resistors in series, as shown in Figure 17.4, so these resistors (and in fact any number of resistors) can be added to find the total *thermal resistance* (R_T) of the solid composite wall to the heat flow. Then:

$$R_T = R_1 + R_2 + R_3 + \ldots = \frac{x_1}{k_1 A} + \frac{x_2}{k_2 A} + \frac{x_3}{k_3 A} + \ldots \qquad (17.5)$$

It can be seen from equation 17.5 that in this case the area, A, *remains constant throughout the wall*, so it is usual to calculate the *total thermal resistance* for *unit surface area* in such problems. Again, if we continue with the electrical analogy then, again from $I = \dfrac{V}{R}$, we find that the expression for the *overall heat transfer* through the wall is given by $\dot{Q} = \dfrac{T_0 - T_3}{R_T}$ or, in *general terms*:

$$\dot{Q} = \frac{T_A - T_B}{R_T} \tag{17.6}$$

where the heat flow through the total width of the composite wall is from surface A to surface B.

Example 17.2 The composite wall of a building consists of an inner liner 10 mm thick ($k = 0.04$ W/m K), bricks 25 cm thick ($k = 0.5$ W/m K) and an outer protective concrete-based coating 30 mm thick ($k = 0.8$ W/m K). If the inner surface temperature of the wall is 25°C and the outer surface temperature is 10°C, find the *heat flux* through the composite wall.

We are required to find the *heat flux* through the composite wall, in other words we must find *the rate of heat transfer per unit area* (q). Then the total thermal resistance for *unit area* is given from equation 17.5 by $R_T = \dfrac{x_1}{k_1} + \dfrac{x_2}{k_2} + \dfrac{x_3}{k_3} + \ldots$ so that, in our case,

$$R_T = \frac{0.01}{0.04} + \frac{0.25}{0.5} + \frac{0.03}{0.8} = 0.7875.$$

Then, from equation 17.6,

$$q = \frac{t_A - t_B}{R_T} = \frac{25 - 10}{0.7875} = 19.05 \text{ W}.$$

Notice that we do not need to convert temperatures into Kelvin when finding temperature differences in the Celsius scale, which is why we use lower case (t) to indicate that temperatures are given in centigrade. Also note that the units of q are given in Watts because we are dividing the heat flux by unit area, i.e. A = 1.0.

17.3 Convection

In order to consider heat transfer from fluids to solids and vice versa, rather than just between solids, we need to understand how heat is transferred by convection.

17.3.1 Nature of convection

Heat transfer by *convection* consists of two mechanisms. In addition to energy transfer by random molecular motion (diffusion), there is also energy being transferred by the bulk motion of the fluid. So, in the presence of a temperature gradient, large numbers of molecules are moving together in bulk at the same time as the random motion of the individual molecules takes place. The cumulative effect of both of these energy transfer methods is referred to as *heat transfer by convection*. Figure 17.5 illustrates this motion.

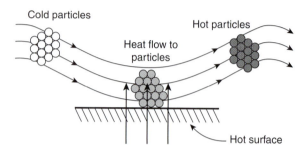

Figure 17.5 Heat transfer by convection

Key Point. Heat transfer by *convection* consists of two mechanisms: transfer by random molecular motion (diffusion) and transfer by the bulk motion of the fluid

---

We are especially interested as engineers in heat transfer by convection that occurs between a fluid in motion and a solid boundary surface. At the fluid–surface interface there is a region where the velocity of the fluid varies from zero at the solid surface to some finite value (u) associated with the flow velocity. This region of the fluid is known as the velocity boundary layer (shown as B_1, B_2 and the dotted lines in Figure 17.6).

Now, if the surface and fluid flow temperatures differ there is also a region of the fluid through which the temperature varies from T_s at the solid surface to T_f in the flow, as the heat energy is transferred. This region is called the thermal boundary layer and is shown in Figure 17.6 as the region identified by hatched lines.

The important point to note is that at the surface, where the bulk velocity is zero, all heat energy is transferred by random molecular motion, i.e. by *conduction*. As we move further away from the surface, the boundary layer grows and heat transfer is more and more dependent on the contribution made by the *bulk fluid motion*.

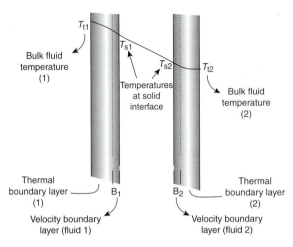

Figure 17.6 Thermal boundary layer at fluid/solid interface

Convection heat transfer may be classified according to the nature of the flow. When the fluid flow is caused by external means, we refer to heat transfer as forced convection. As an example, consider the use of a fan to provide forced convection air cooling of the electrical components within a personal computer. *Natural convection* is induced by forces that result from density differences caused by temperature variations within a fluid. Consider another example, say, electrical circuit boards. Air makes contact with the hot electrical components and experiences an increase in temperature, which causes a reduction in its density. In this lighter air, buoyancy forces cause vertical motion and the warm air rises. As it does so it is replaced by an inflow of cooler ambient air; these convection currents now act as the heat transfer mechanism.

Regardless of the heat transfer process, *the rate of heat transfer* that is of particular interest to engineers may be determined by using *Newton's law of cooling*. The appropriate rate equation is of the form:

$$\dot{Q} = hA\left(T_s - T_f\right) \qquad (17.7)$$

where (h) is called the *surface heat transfer coefficient*, with units of W/m^2 K.

17.3.2 Heat transfer with convection through composite walls

If the materials of the composite wall involve *heat transfer from fluids to solids and vice versa*, rather than just between solids, then there is a similar electrical analogy to that for heat transfer by conduction between solids. For fluids, as already mentioned, the *surface heat*

transfer coefficient (h) depends on both the type of fluid and the fluid velocity.

Then, in a similar way to the formulation of equation 17.4 using our electrical analogy, the *thermal resistance of a fluid at the interface with a solid* can be shown to be:

$$\text{Fluid thermal resistance} = \frac{1}{hA} \qquad (17.8)$$

Key Point. The relationship for thermal resistance $R = \dfrac{1}{hA}$, using the convection heat transfer coefficient, has *identical units* to its counterpart used for conduction, $R = \dfrac{x}{kA}$, which are K/W

The following example shows how we combine heat transfer by conduction through solids and heat transfer by convection at the solid–fluid interface.

Example 17.3 The wall of a house consists of an inner thermal block 125 mm wide and a 125 mm house brick separated by an air gap. The inner surface of the wall is at a temperature of 25°C and the outside air film temperature is 5°C. For simplicity, no additional insulating material has been included in the construction. Calculate the *heat flux* (the rate at which heat is lost per m^2) of the wall surface, when: the surface heat transfer coefficient from the outside wall to the outside air is 5 W/m^2 K, the resistance to heat flow of the air gap is 0.15 K/W and the thermal conductivities of thermal block and outer brick are 0.2 and 0.5 W/m K, respectively.

The situation for the wall is illustrated in Figure 17.7.

Using equation 17.4 for any resistance $R = \dfrac{x}{kA}$, the resistance of thermal block for *unit area* is given by

$$R_{\text{TB}} = \frac{125 \times 10^{-3}}{0.2} = 0.625 \text{ K/W.}$$

Similarly, the resistance of house brick

$$R_{\text{OB}} = \frac{125 \times 10^{-3}}{0.5} = 0.25 \text{ K/W.}$$

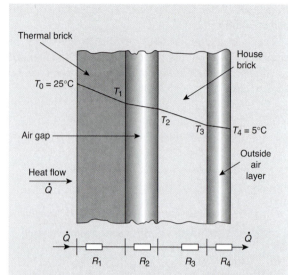

$R_1 = 0.625$ K/W $R_2 = 0.15$ K/W $R_3 = 0.25$ K/W $R_4 = 0.2$ K/W

R_1 = resistance of thermal brick
R_2 = resistance of air gap
R_3 = resistance of outer brick
R_4 = resistance of outside wall surface to the outside air

Figure 17.7 Heat transfer through a solid wall

Also, using equation 17.8 for the fluid, $R = \dfrac{1}{hA}$, the resistance of the outside wall surface to the outside air is $R_{\text{wall/air}} = \dfrac{1}{hA} = \dfrac{1}{5} = 0.2$ K/W. We are given the thermal resistance of the air gap as 0.15 K/W, therefore the total resistance $R_T = 0.625 + 0.25 + 0.2 + 0.15 = 1.225$ K/W.

Now, using equation 17.6, $\dot{Q} = \dfrac{t_0 - t_4}{R_T}$, we find the *heat flux* (rate of heat loss per square metre) to be $q = \dfrac{25 - 5}{1.225} = 16.3$ W.

Suppose we also wanted to find the interface temperature t_1 (Figure 17.7). We can apply the electrical analogy to each layer by again using equation 17.6. Then $16.3 = \dfrac{25 - t_1}{0.625}$, from which $t_1 = 14.8°$C. The interface temperatures t_2 and t_3 can be found in a similar way.

17.3.3 Heat transfer through cylindrical walls

Before we leave heat transfer by conduction and convection we will consider one more very common engineering situation, that of heat transfer through the wall of a cylinder. We again use our electrical analogy to find an expression for thermal resistance, but first it will be necessary to solve a simple differential equation in order to find an expression for the heat flow rate $\dot{Q}$.

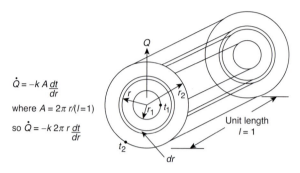

$\dot{Q} = -k A \dfrac{dt}{dr}$

where $A = 2\pi r/(l = 1)$

so $\dot{Q} = -k\, 2\pi r \dfrac{dt}{dr}$

Figure 17.8 Heat transfer through a cylindrical wall

> **Key Point.** For a plane wall the area perpendicular to the heat flow is constant, but this is not the case for cylindrical-walled objects

For a plane wall the area perpendicular to the heat flow is constant, but this is not the case for a cylinder wall. From Figure 17.8, at any radius within the cylinder, assuming *negligible axial and circumferential heat transfer*, we see that $\dot{Q} = -k2\pi rl\dfrac{dt}{dr}$ or $\dot{Q} = -k2\pi r\dfrac{dt}{dr}$ per unit length of cylinder. Separating the variables, $\dfrac{\dot{Q}dr}{r} = -k2\pi\, dt$ or $\dfrac{\dot{Q}}{2\pi k}\dfrac{dr}{r} = -dt$. Then, integrating between (r_2 and r_1), i.e. $\dfrac{\dot{Q}}{2\pi k}\displaystyle\int_{r_1}^{r_2}\dfrac{dr}{r} = -\int_{t_1}^{t_2} dt$, we have that $\dfrac{\dot{Q}}{2\pi k}\ln\dfrac{r_2}{r_1} = -(t_2 - t_1)$ and therefore, after transposition, *the heat transfer per unit length* is:

$$\dot{Q} = \frac{2\pi k(t_1 - t_2)}{\ln r_2/r_1} \qquad (17.9)$$

If we use our electrical analogy again ($I = V/R$ and $t_1 - t_2 = V$), as we did when finding the expression for the thermal resistance by conduction, equation 17.4, $R = \dfrac{x}{kA}$, then from equation 17.9 we can deduce that in this case, *the thermal resistance per unit length for a cylinder wall* is:

$$R = \frac{\ln (r_2/r_1)}{2\pi k} \qquad (17.10)$$

Example 17.4 A steel pipe carrying steam at 250°C has an internal diameter of 100 mm and an external diameter of 120 mm. The pipe is insulated using an inner layer of specially treated plastic-based composite 40 mm thick, with an outer layer of felt 50 mm thick. The ambient temperature surrounding the pipe is at 20°C.

Calculate the overall thermal resistance of the assembly and so find the heat loss per unit length of the pipe. Also, calculate the temperature on the outer surface of the pipe assembly. Assume that the heat transfer rates for the internal and external surfaces of the pipe assembly are 500 and 15 W/m² K and the thermal conductivities of the steel, plastic composite and felt are 45, 0.2 and 0.05 W/m K, respectively.

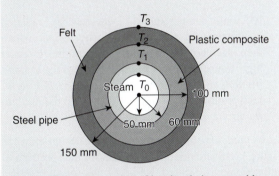

Figure 17.9 Cross-section of insulated pipe assembly

Figure 17.9 illustrates the situation with the appropriate dimensions shown. To find the total thermal resistance (R_T), we consider unit length of pipe. Then:

The thermal resistance of the *steam film*

$$R = \frac{1}{hA} = \frac{1}{(500)(2\pi \times 50 \times 10^{-3})} = 0.00637 \text{ K/W}$$

The thermal resistance of the *steel pipe*

$$R = \frac{\ln(r_2/r_1)}{2\pi k} = \frac{\ln(60/50)}{(2\pi)(45)} = 0.000645 \text{ K/W}$$

The thermal resistance of the *plastic composite*

$$R = \frac{\ln(r_2/r_1)}{2\pi k} = \frac{\ln(100/60)}{(2\pi)(0.2)} = 0.406 \text{ K/W}$$

The thermal resistance of the *felt*

$$R = \frac{\ln(r_2/r_1)}{2\pi k} = \frac{\ln(150/100)}{(2\pi)(0.05)} = 1.29 \text{ K/W}$$

The thermal resistance of the *air film*

$$R = \frac{1}{hA} = \frac{1}{(15)(2\pi \times 150 \times 10^{-3})}$$
$$= 0.0.0707 \text{ K/W}$$

and therefore, the *total thermal resistance of the pipe assembly* $R_T = 0.00637 + 0.000645 + 0.406 + 1.29 + 0.0707 = 1.774$ K/W.

To find the heat loss per unit length, we use equation 17.6 which in this case gives

$$\dot{Q} = \frac{t_0 - t_4}{1.774} = \frac{250 - 20}{1.774} = 129.7 \text{ W},$$

where (t_4) is the ambient temperature of the air surrounding the pipe assembly. The temperature on the *outer surface of the pipe* (t_3) is also found from equation 17.6 to be

$$129.7 = \frac{t_3 - t_4}{0.071} = \frac{t_3 - 20}{0.071} = 29.2°\text{C}.$$

Notice the insignificant amount of thermal resistance provided by the steel pipe. For this reason, to a first approximation, the thermal resistance of metal pipes is often ignored in these calculations. Imagine the energy losses if the pipe were not lagged!

We complete our very brief introduction to heat transfer by looking now at heat transfer by radiation.

17.4 Radiation

Thermal radiation is energy emitted by matter that is at some finite temperature. Although we are primarily interested in radiation from solid substances, radiation can also be emitted by liquids and gases. Thermal radiation is attributed to changes in electron energy within atoms or molecules. As electron energy levels change, energy is released which travels in the form of electromagnetic waves of varying wavelength.

When striking a body, the *emitted radiation* is either *absorbed by*, *reflected by* or *transmitted through the body*. If we assume that the quantity of radiation striking a body is unity, then we can state that:

$$\alpha + \gamma + \tau = 1 \tag{17.11}$$

where α is the fraction of incident radiation energy absorbed, called *the absorptivity*, γ is the fraction of energy reflected, called *the reflectivity*, and τ is the fraction of energy transmitted, called the *transmissivity*.

17.4.1 Black and grey body radiation

It is useful to define an ideal body which absorbs all the incident radiation which falls on it. This body is known as a *black body*. From Equation 17.11 it follows that a black body is one where $\alpha = 1$. So we say that a *black body is a perfect absorber*, because it absorbs the entire radiation incident upon it, irrespective of wavelength. We would also expect a black body to be the best possible *emitter* at any given temperature, otherwise its temperature would rise above that of its surroundings, which is not the case.

> **Key Point.** All incident radiation is absorbed by a black body

In practice, an almost perfect black body consists of an enclosure, such as a cylinder, with a dull black interior and a small hole (Figure 17.10).

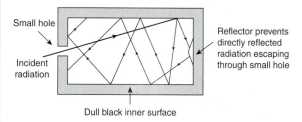

Figure 17.10 Black body design

Radiation entering the hole has little chance of escaping because any energy not absorbed when the wall of the cylinder is first struck will continue to strike other surfaces, gradually depleting its energy until none is left.

When the cylinder is heated by an external source, radiation appears from the hole (this radiation may be any colour according to its wavelength and temperature). Thus a *whole spectrum of energy* may be produced when a black body is heated, with the form of the curve shown in Figure 17.11.

Some important criteria can be deduced from a series of curves of this type, including:

- As the temperature rises the energy in each wave band increases, so the body will become brighter.

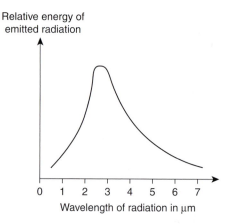

Figure 17.11 Radiated energy spectrum

- At some temperature, T, the energy radiated is a *maximum for a certain wavelength*, which then decreases with rising temperature. Even at high temperatures (around 1000 K) only *a small fraction* of the radiation appears as *visible light*; temperatures of around 4000 K are required to produce a maximum in the visible region.

So, by *modelling black bodies* using the design shown in Figure 17.10 and noting the criteria for the spectrum of radiated energy emitted, we know that for all practical purposes a *black body is both a perfect emitter and absorber*. Therefore, we can use this model to *compare* the *radiation emitted* by other bodies *without* these properties.

Then the *emissivity* (ε) of a body is the ratio of the energy emitted by the body to that emitted by a black body at the same temperature and wavelength. Such matter is known as a grey body.

A comparison of the *spectral emissive energy* of a tungsten filament lamp (non-black or *grey body*) when compared with that of the corresponding black body is shown in Figure 17.12.

> **Key Point.** The *emissivity* (ε) of a body is the ratio of the energy emitted by the body to that emitted by a black body at the same temperature and wavelength

17.4.2 The Stefan–Boltzmann law

It was found experimentally by Stefan, and subsequently proved theoretically by Boltzmann, that: *the total energy radiated by all wavelengths per unit area per unit time*

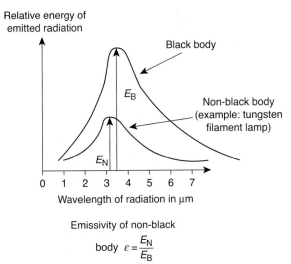

Emissivity of non-black
body $\varepsilon = \dfrac{E_N}{E_B}$

Figure 17.12 Grey body radiation

by a black body is directly proportional to the fourth power of the absolute temperature of the body.

$$E = \sigma T^4 \qquad (17.12)$$

where $\sigma = 5.672 \times 10^{-8}\,\text{W/m}^2\,\text{K}^4$ is the *Stefan–Boltzmann constant of proportionality*, found by experimentation.

Note also from our discussion on emissivity that *the energy emitted by a non-black body or grey body* is:

$$E = \varepsilon \sigma T^4 \qquad (17.13)$$

where ε is the *emissivity of the body*, as defined above.

When a body is very small when compared to its large surroundings, a negligible amount of radiation is reflected from the surroundings onto the body, so in effect the surroundings are black. Under these circumstances it can be shown that the rate of heat transfer from the body to its surroundings is

$$\dot{Q} = \varepsilon \sigma A \left(T_s^4 - T_{\text{sur}}^4\right) \qquad (17.14)$$

where T_s is the temperature of the solid body and T_{sur} is the temperature of the immediate surroundings, both in Kelvin (since powers are involved).

Now apart from radiation there are many instances where the surface of a body may transfer heat by convection from its surface to a surrounding gas (see Example 17.5). Then the total rate of heat transfer from the surface is the sum of the heat rates due to these two modes. The heat transfer rate for convection is given

by equation 17.7 and for radiation by equation 17.14. Then:

$$\dot{Q} = hA\left(T_s - T_f\right) + \varepsilon \sigma A \left(T_s^4 - T_{\text{sur}}^4\right) \qquad (17.15)$$

There are many occasions where it is necessary to express *the net radiation exchange* in the form

$$\dot{Q} = h_r A\left(T_s - T_{\text{sur}}\right) \qquad (17.16)$$

Also, from equation 17.14 where $Q = \varepsilon \sigma A \left(T_s^4 - T_{\text{sur}}^4\right)$ and equation 17.16 where $\dot{Q} = h_r A\left(T_s - T_{\text{sur}}\right)$, and on the assumption that $T_f = T_{\text{sur}}$, i.e. the temperature of the flow in the surroundings, then: $h_r A \left(T_s - T_{\text{sur}}\right) = \varepsilon \sigma A \left(T_s^4 - T_{\text{sur}}^4\right)$ and using the difference of two squares rule

$$h_r = \frac{\varepsilon \sigma A \left(T_s^2 - T_{\text{sur}}^2\right)\left(T_s^2 + T_{\text{sur}}^2\right)}{A\left(T_s - T_{\text{sur}}\right)}$$

$$= \frac{\varepsilon \sigma \left(T_s - T_{\text{sur}}\right)\left(T_s + T_{\text{sur}}\right)\left(T_s^2 + T_{\text{sur}}^2\right)}{\left(T_s - T_{\text{sur}}\right)},$$

so that:

$$h_r = \varepsilon \sigma \left(T_s + T_{\text{sur}}\right)\left(T_s^2 + T_{\text{sur}}^2\right) \qquad (17.17)$$

Equations 17.16 and 17.17 enable the *radiation heat transfer coefficients* to be determined in a similar way to equations 17.3 and 17.7 that we used for conduction and convection.

Example 17.5 A steam pipe without insulation passes through a large area of a factory in which the air and factory walls are at $20°\text{C}$. The external diameter of the pipe is 80 mm, its surface temperature is $220°\text{C}$, and it has an *emissivity* of 0.75. If the coefficient associated with free convection heat transfer from the surface to the air is $20\,\text{W/m}^2\,\text{K}$, what is the rate of heat loss from the surface per unit length of pipe?

From the above discussion, we make the *assumption* that radiation exchange between the pipe and the factory is between a small surface and a very much larger surface, so equation 17.14 derived from the Stefan–Boltzmann law applies. Also, heat loss from the pipe to the factory air is by *convection* and by *radiation exchange* with the factory walls.

Then, from equation 17.15, $\dot{Q} = hA\left(T_s - T_f\right) + \varepsilon\sigma A\left(T_s^4 - T_{\text{sur}}^4\right)$; therefore, in our case, the heat loss from the pipe is given by: $\dot{Q} = h\left(\pi dl\right)\left(T_s - T_{\text{air}}\right) + \varepsilon\sigma\left(\pi dl\right)\left(T_s^4 - T_{\text{wall}}^4\right)$, where $A = (\pi dl) =$ the surface area of the pipe and we are told that $T_{\text{air}} = T_{\text{wall}}$. Then, *the heat loss from the pipe per unit length* is:

$$\frac{\dot{Q}}{l} = \frac{\dot{Q}}{1} = (20)\left(\pi \times 80 \times 10^{-3} \times 1\right)(493 - 293)$$

$$+ (0.75)\left(5.672 \times 10^{-8}\right)$$

$$\left(\pi \times 80 \times 10^{-3} \times 1\right)\left(493^4 - 293^4\right)$$

$$= 1588\,\text{W}$$

Note, for the above calculation, that the *Stefan–Boltzmann* constant $\sigma = 5.672 \times 10^{-8}\,\text{W/m}^2\,\text{K}^4$ and $(l = 1)$.

17.5   Chapter summary

In this very brief introduction to heat transfer we have looked at one-dimensional heat flow by conduction, convection and radiation. You should now be able to explain the nature of these three methods of heat transfer.

We consider heat transfer by *conduction* in section 17.2, where we introduced what has now become known as Fourier's law, which states that: *the rate at which heat is transferred by conduction, $\dot{Q}$, is proportional to the temperature gradient dT/dx and the area A normal to the direction of the heat transfer.* The constant of proportionality was found to be the *thermal conductivity* (k), with units W/m K, which is an inherent property of the material and is a measure of its ability to conduct heat. We then considered heat flow in terms of temperature change and introduced equation 17.2, $\dot{Q} = \dfrac{kA(T_1 - T_2)}{x}$. You should be fully aware of the assumptions that govern the use of this equation. We then used an *electrical analogy* to derive several relationships that involved the thermal resistance and heat flux, and used these relationships in subsequent examples. This electrical analogy was used again in our subsequent study of convection and radiation.

In section 17.3 we dealt with *convection* and defined *the surface heat transfer coefficient* (h). We also showed how the formula for thermal resistance, equation 17.8, $R = \dfrac{1}{hA}$, could be used for problems through

composite walls that were subject to heat transfer by both *conduction* and *convection*. We then considered heat transfer through *cylindrical* walls and formulated the useful relationships for *heat transfer per unit length*, equation 17.9, $\dot{Q} = \dfrac{2\pi k(t_1 - t_2)}{\ln r_2/r_1}$, and *the thermal resistance per unit length*, equation 17.10, $R = \dfrac{\ln\left(r_2/r_1\right)}{2\pi k}$, both these relationships being used in Example 17.4.

Finally, in section 17.4 we looked briefly at radiation, where you saw that the total incident radiation can be absorbed, reflected and/or transmitted by a body. We then considered the concept of *black bodies* that have the capacity to absorb *all the radiation* that is incident upon them. You also saw that black bodies are the best possible emitters of radiation and that the *energy emitted* by all bodies was dependent on their *temperature*, the *wavelength* of the emitted radiation and the *emissivity* (ε) of the body concerned. Remember also that the *emissivity* (ε) of a body is the ratio of the energy emitted by the body to that emitted by a black body at the same temperature and wavelength, and that such matter is known as a *grey body*. We next considered the *Stefan–Boltzmann* law which gave us the mathematical relationship between the total energy radiated by a black body and the absolute temperature of the body, i.e. equation 17.13, where $E = \varepsilon\sigma T^4$. From this law we were able to formulate relationships for the heat transfer rate, equation 17.14, $\dot{Q} = \varepsilon\sigma A\left(T_s^4 - T_{\text{sur}}^4\right)$, by radiation, and the radiation heat transfer coefficient, equation 17.17, $h_r = \varepsilon\sigma\left(T_s + T_{\text{sur}}\right)\left(T_s^2 + T_{\text{sur}}^2\right)$, in a similar manner to the relationships we formed for their conduction and convection counterparts. We also found a relationship, equation 17.15, $\dot{Q} = hA\left(T_s - T_f\right) + \varepsilon A\left(T_s^4 - T_{\text{sur}}^4\right)$, that enables us to find the *heat transfer rate* for bodies subject to heat transfer by both convection and radiation.

17.6   Review questions

1. Briefly explain the nature of heat transfer by:
    a) conduction, b) convection, c) radiation.

2. The inner surface of a brick wall is at $30°\text{C}$ and the outer surface is at $5°\text{C}$. Calculate the rate of heat transfer per unit area of the wall surface, given that the wall is 30 cm thick and the thermal conductivity of the brick is 0.45 W/m K.

3. The temperature on the inside of a furnace wall 40 cm thick is $1800°\text{C}$ and the temperature on the outside of the 10 cm thick insulation is $40°\text{C}$. If the

thermal conductivity of the wall is 0.9 W/m K and that of the insulation is 0.2 W/m K; determine the heat transfer rate per unit area and the temperature at the interface of the wall and the insulation.

4. Calculate the rate of heat transfer by natural convection from a shed roof of area 10 m × 15 m if the roof surface temperature is 25°C, the air temperature is 8°C, and the average convection heat transfer coefficient is 10 W/m² K.

5. A steam pipe, 80 mm outside diameter, carries steam at 340°C. It is double-insulated by an inner layer of material 30 mm thick with a thermal conductivity of 0.2 W/m K and an outer layer of material 20 mm thick with thermal conductivity 0.15W/m K. If the surround temperature on the outside of the insulation is 40°C and the surface heat transfer coefficient is 1.2 W/m² K, calculate the rate of heat loss to the surroundings per unit length of the pipe.

6. A large pipe, 50 cm in diameter ($\varepsilon = 0.85$), carrying steam, has a surface temperature of 250°C. The pipe is located in a room at 15°C, and the convection heat transfer coefficient between the pipe surface and the air in the room is 20 W/m² K. Determine: a) the radiation heat transfer coefficient, b) the rate of heat loss per metre of pipe length.

Chapter 18

Introduction to fluid mechanics

Like heat transfer, *fluid mechanics* is also considered to be a separate branch of the thermal sciences. Fluid mechanics is split into two major areas: *fluid statics* and *fluid dynamics*. In this chapter we can do no more than provide a basic introduction to these two branches of fluid mechanics that will hopefully serve as a useful foundation for any further study you may wish to undertake.

We start by considering *fluid statics* which is the subject concerned with *fluids at rest*, where we cover topics that include sections on the pressures and thrust forces that act on immersed surfaces and the nature of these thrust forces on waterway structures, such as sluice gates, lock gates etc. We also look at the derivation and use of the buoyancy equation and its application to the buoyancy and equilibrium of bodies immersed in both gases and liquids.

We then move on to *fluid dynamics* (the study of fluids in motion), where we initially consider fluid momentum and the forces exerted by a fluid jet on stationary flat and curved surfaces. Next, the Bernoulli and continuity equations are introduced and applied to fluid flow measurement. We then turn our attention to viscosity and the viscous nature of fluids, including the use of Reynolds number for determining laminar and turbulent flow.

Friction losses in piped systems are then considered, where, from the general energy equation, Reynolds number and the Darcy equation, fluid friction losses in piped systems are determined. Finally, as an application of energy loss theory, we take a brief look at ways to estimate power losses from friction in plain bearings.

18.1 Thrust force on immersed surfaces

In this section we look at the thrust forces/pressures that are exerted by fluids at rest on immersed flat surfaces of regular shape and the nature of these forces on simple everyday engineering structures like sluice gates, lock gates and vertical separators. We then consider, using examples, the fluid thrust forces that act on immersed curved surfaces. In order to find resultant thrust forces and their point of action on these regular shaped surfaces, we need to remember how to define fluid pressure and know how to find the centroid of area of regular shaped laminae. Therefore, if you have forgotten about fluid pressure or how to find centroids and the second moment of area of regular laminae

978-1-85617-775-7, Engineering Science, Mike Tooley & Lloyd Dingle

(flat surfaces of regular thickness), you should look back at section 2.4 and section 12.1, to refresh your memory.

18.1.1   Thrust forces on flat surfaces immersed vertically

Consider a *vertical flat plate* that has one side of area A immersed in a fluid such as water, with its top edge in line with the free surface of the water, as shown in Figure 18.1.

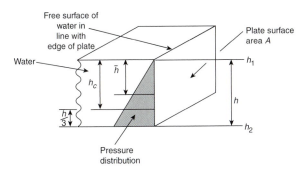

Figure 18.1 Plate with one surface immersed in water

From Figure 18.1 it can be seen that the *gauge pressure* of the water *at any point* on the flat plate is given by $p = \rho g h$, where the height of the water varies from zero at the air/water interface (h_1) to the depth of the plate (h_2). If we wish to find the *total thrust force* acting on the area of one side of the plate as a result of the pressure, then knowing that force $F_T = $ pressure $\times$ area or $F_T = \rho g h \times A$ we can find this force, but what value of the depth (h) do we choose? Well, apart from logic, it can be *proved* (as you will see in the next section) that this depth will be at the *centre of area* or *centroid* ($\bar{h}$) of the plate. For example, if the plate were circular this depth $\bar{h}$ would be at its centre, or point of balance of the plate.

In general, it is a fact that: *the total thrust force (or pressure force) (F_T) that acts on the area of an immersed surface is the product of the intensity of the pressure (P) at the centroid and the total area (A) of surface in contact.* Therefore, this *total thrust force* (F_T) acting on the surface of the vertical plate is:

$$F_T = \rho g \bar{h} A \qquad (18.1)$$

Note: The product $\bar{h}A$ in equation 18.1 is in fact the *first moment of area* of the submerged surface. It is also true that equation 18.1 is valid for all surfaces, whether they are flat or curved.

We now come to the question of the *pressure* or intensity of the pressure P that acts on the surface. This is not uniform but increases with depth, so that the *centre of this pressure* (the point at which the pressure across the whole immersed surface is deemed to act) will be towards the lower edge of the plate. Figure 18.1 shows the resulting *pressure distribution* for varying depth from the surface. This pressure distribution on the sides of the plate *is triangular*, and in such a case it can be shown by taking moments at the liquid surface, that is using *second moments of area*, that the *vertical distance to the centre of this pressure distribution* (h_c) on the plate will be two-thirds of the way down from the surface, or *one-third of the way up from its base*, as also shown in Figure 18.1 (see also Table 18.1).

> **Key Point.** For an immersed surface, $\bar{h} = $ the distance to its centroid, while $h_c = $ the distance to its centre of pressure; they are not the same

Do remember that the distance ($\bar{h}$) to the centre of area (centroid) of any immersed surface is not the same as the distance (h_c) to its centre of pressure, as will be shown in section 18.1.2, that follows.

> **Example 18.1** A *circular* flat plate 1.8 m in diameter is placed vertically in water so that the top of the plate is 0.6 m below the surface of the water. Find the total thrust force that acts on the plate. Assume the density of the water is, for this example and all other exercises, 1000 kg/m³, unless told otherwise.
>   The *vertical distance from the surface of the water to the centroid* of the circular flat plate is $\bar{h} = 0.6 + 0.9 = 1.5$ m, where the distance to the centre of the circle (its centroid) is $d/2$.
>   The area of the surface $A = \pi(0.9)^2 = 2.54$ m².
>   Then the total thrust force $F_T = \rho g \bar{h} A = (1000)(9.81)(1.5)(2.54) = 37.45$ kN.

> **Example 18.2** A vertical plate is used to separate two areas of a large industrial rectangular fuel storage tank as shown in Figure 18.2. The height of the fuel in the left-hand side (LHS) of the tank is $h_1 = 4.5$ m and in the right-hand side (RHS) $h_2 = 1.5$ m, and the tank is 6 m wide. Taking the density of the fuel as 7800 kg m⁻³, find a) the net horizontal thrust force on the separator plate, b) the line of action of the net thrust force.

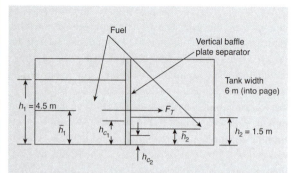

Figure 18.2 Side view of fuel storage tank with separator

From Figure 18.2 it can be seen that the area of fuel acting on the LHS of the separator plate $= 4.5 \times 6 = 27\,\text{m}^2$ and the area of fuel acting on the RHS $= 1.5 \times 6 = 9\,\text{m}^2$.

Then, LHS thrust force (from equation 18.1) $F_{T_1} = \rho g \bar{h}_1 A_1 = (780)(9.81)(2.25)(27) = 464.8\,\text{kN}$, where $\bar{h}_1 = 4.5/2 = 2.25\,\text{m}$.

Then, RHS thrust (from equation 18.1) $F_{T_2} = \rho g \bar{h}_2 A_2 = (780)(9.81)(0.75)(9) = 51.6\,\text{kN}$, where $\bar{h}_2 = 1.5/2 = 0.75\,\text{m}$.

Therefore the *net horizontal thrust force* acting on separator plate $F_T = 464.8 - 51.6 = 413.2\,\text{kN}$.

Now, in order to establish the *line of action* of the net thrust force, we need to *take moments* of the individual thrust forces and equate them to the net thrust force moment, using the depth to their *centres of pressure*. We know from above that the *distances to the centres of pressure* (from the base of the tank) acting on the LHS and RHS of the vertical separator plate will be $h_{c_1} = h_1/3$ and $h_{c_2} = h_2/3$, respectively. Then, taking moments about the base of the tank and equating them using the *principle of moments*, gives:

LHS thrust moment $h_{c_1} = (464.8)(1.5) = 697.2\,\text{kN m}$ (clockwise), where $h_{c_1} = 4.5/3 = 1.5\,\text{m}$.

RHS thrust moment $h_{c_2} = (51.6)(0.5) = 25.8\,\text{kN m}$ (anticlockwise), where $h_{c_2} = 1.5/3 = 0.5\,\text{m}$.

The net thrust force moment acting at a distance (h_c) from the base of the tank $F_M = (413.2)h_c$. Then, equating moments: $413.2 h_c = 697.2 - 25.8\,\text{kN m}$, therefore the *line of action of the net thrust* from the base $h_c = \dfrac{697.2 - 25.8}{413.2} = 1.62\,\text{m}$ (to the right).

18.1.2 Thrust forces on inclined flat surfaces

To help us find the *position of the centre of pressure* of a fluid acting on an inclined flat surface, we once again take moments *and* use the same techniques as we did in section 2.4, when we found the *centroid and second moment of area* of flat laminar surfaces. The following explanation illustrates the technique.

The vertical depth of the centre of pressure (h_c) for a surface inclined to the liquid surface may be found by taking moments about the point of intersection of the plane of the inclined surface and the liquid interface. Figure 18.3 shows the set-up for the inclined surface, which is shown in side view.

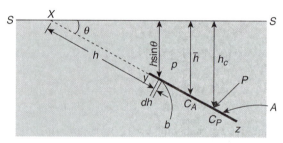

Key: A = area of surface
  S-S = free surface of liquid
  X = point of intersection of surface with liquid/air interface
  θ = angle of inclination to liquid/air interface
  p = pressure on elemental strip
  b = breadth of elemental strip (into paper)
  h = distance of elemental strip from X
  $\bar{h}$ = vertical depth to centroid (centre of area C_A)
  h_c = vertical depth to centre of pressure (C_P)
  P = total pressure acting on area A of surface at the centre of pressure (C_P)
  F_T = total thrust force, that is the product of total pressure (P) multiplied by total area of surface (A)

Figure 18.3 Set-up for the analysis of a flat surface inclined at angle θ to the surface of a liquid

Figure 18.3 shows a flat surface of length (yz) inclined at an angle θ at its *intersection* (X) with the liquid/air interface (SS). Consider the area of a horizontal elemental strip with breadth (b, into the paper), length (dh), at a distance (h) from (X). Then the area of incremental strip $= bdh$ and the pressure acting on the strip due to the depth of the liquid ($h \sin\theta$) is: $p = \rho g h \sin\theta$. Therefore the force on the elemental strip $F_s = (\rho g h \sin\theta)(bdh)$ or $F_s = (wh \sin\theta)(bdh)$, where ($w = \rho g$) is the *specific weight*, with units N/m^3, which is a convenient term to use in static pressure relationships and so will be used where appropriate from now on.

Then the *total thrust or pressure force* (F_T) resulting from the total pressure (P) *acting on the whole area* is the sum of all these elemental strips, i.e. $F_T = w \int bdhh \sin\theta$, and noting that $\int bdhh = the$ *first moment of area about X*, i.e. the total area A multiplied by the average distance of (h) from X, i.e. *the perpendicular distance to the centroid* $\bar{h}$, or $(A\bar{h})$, then *from F_T*:

we have that *the first moment of area*

$$= \int bdhh = \frac{A\bar{h}}{\sin\theta} \qquad (18.2)$$

and from $F_T = w\sin\theta \int bdhh = (w\sin\theta)\left(\dfrac{A\bar{h}}{\sin\theta}\right)$ we have that:

$$F_T = wA\bar{h} \qquad (18.3)$$

Compare equation 18.3 with equation 18.1. The above argument is proof of equation 18.1!

Now, taking moments about X:

Moment of force on strip from $F_s = (wh\sin\theta)(bdh)$ is $M_s = wh\sin\theta bdh \times h$ or $M_s = w\sin\theta bdh \times h^2$ and the sum of the moments of the forces on the incremental strips $M_T = w\sin\theta \int bdhh^2$. Now letting $I_X = the$ *second moment of area* of the surface about X, we get that:

$$I_X = \int bdhh^2 \qquad (18.4)$$

Then, the total moment $M_T = w\sin\theta I_X$. Also, from equations 18.2 and 18.3 and the above explanation for the first moment of area, where $F_T = wA\bar{h}$, the total moment may be expressed as $\dfrac{F_T(h_c)}{\sin\theta}$, i.e. the total force F_T acting at right angles *to the centre of pressure* of the submerged surface $\dfrac{h_c}{\sin\theta}$. Then the total moment $M_T = \dfrac{F_T(h_c)}{\sin\theta} = w\sin\theta I_X$, so that the *vertical distance to the centre of pressure* (h_c), the point at which the total force F_T is deemed to act, is $h_c = \dfrac{wI_X\sin^2\theta}{F_T}$ or, from equation 18.4:

$$h_c = \frac{I_X\sin^2\theta}{A\bar{h}} \qquad (18.5)$$

where, from the parallel axis theorem (section 2.4.3), i.e. $I_{BB} = I_{AA} + AS^2$, then:

$$I_X = I_{C_A} + \frac{A\bar{h}^2}{\sin^2\theta} \qquad (18.6)$$

Key Point. The parallel axis theorem enables us to find the second moment of area about any remote point, provided the second moment of area about the centroid of the body can be found

Equation 18.6 enables us to find I_X by first finding I_{C_A} (*second moment of area about centroid*), but without knowledge of the depth to the centre of pressure (h_c). Once I_X is found, if required, (h_c) can be determined using equation 18.5. The second moment of area about their centroids, for some common geometric shapes, is given in Table 18.1 (see also Table 2.1 in section 2.4.2).

Example 18.3 For the circular flat plate given in Example 18.1, find the depth to the centre of pressure from the surface.

Using the information given in Example 18.1 and Table 18.1, where $A = 2.54\,\text{m}^2$, $\bar{h} = 1.5$, the second moment of area about its centroid for the surface is

$$I_{C_A} = \frac{\pi d^4}{64} = \frac{\pi(1.8)^4}{64} = 0.515\,\text{m}^4.$$

Also, from

$$I_X = I_{C_A} + \frac{A\bar{h}^2}{\sin^2\theta},$$

where in this case $\sin^2\theta = 1$, then $I_X = I_{C_A} + A\bar{h}^2 = 0.515 + (2.54)(1.5)^2 = 6.23\,\text{m}^4$. Then, from equation 18.6, we have in this case that:

$$h_c = \frac{I_X}{A\bar{h}} = \frac{6.23}{(2.54)(1.5)} = 1.64\,\text{m}.$$

Example 18.4 A flat rectangular plate 2.5 m long and 1.5 m wide is immersed in water. The plane of the plate makes an angle of 60° with the surface of the water, with the 1.5 m edge of the plate parallel to and at a depth of 0.5 m below the surface of the water. Find the total thrust force that acts on the surface and the position of the centre of pressure with respect to one side of the plate.

Table 18.1 Centroids and second moments of area for some common regular geometric shapes

Shape	Geometry and position of centroid axis	Area A	Second moment of area I_{CC} about centroid axis
Rectangle		bd	$\dfrac{bd^3}{12}$
Triangle		$\dfrac{bh}{2}$	$\dfrac{bh^3}{36}$
Circle		πr^2 or $\dfrac{\pi d^2}{4}$	$\dfrac{\pi r^4}{4}$ or $\dfrac{\pi d^4}{64}$
Semicircle		$\dfrac{\pi r^2}{2}$ or $\dfrac{\pi d^2}{8}$	$0.1098r^4$ from $\dfrac{\left(9\pi^2 - 64\right) r^4}{72\pi}$
Trapezium	$\bar{h} = \dfrac{h(2a+b)}{3(a+b)}$	$A = \dfrac{h(a+b)}{2}$	$\dfrac{h^3 \left(a^2 + 4ab + b^2\right)}{36(a+b)}$

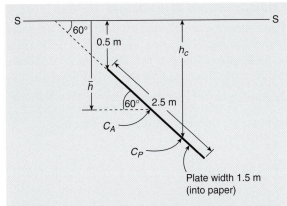

Figure 18.4 Situation for immersed plate (for Example 18.4)

From Figure 18.4 it can be seen that the *vertical depth to the centroid is* $\bar{h} = 0.5 + 0.5(2.5)\sin 60 = 1.583$ m and the area $A = (2.5)(1.5) = 3.75$ m^2.

Then (remembering $w = \rho g$) the total thrust force from equation 18.4 is $F_T = wA\bar{h} = (9810)(3.75)(1.583) = 58.23$ kN.

The second moment of area about the centroid of the plate is, from Table 18.1,

$$I_{C_A} = \frac{bd^3}{12} = \frac{(1.5)(2.5)^3}{12} = 1.953 \text{ m}^4,$$

then

$$I_X = I_{C_A} + \frac{A\bar{h}^2}{\sin^2\theta} = 1.953 + \frac{(3.75)(1.583)^2}{(0.866)^2}$$

$$= 14.483 \text{ m}^4.$$

so the *vertical depth of the centre of pressure*, from equation 18.6, is:

$$h_c = \frac{I_X \sin^2\theta}{A\bar{h}} = \frac{(14.483)(0.866)^2}{(3.75)(1.583)} = 1.83 \text{ m}.$$

Key Point.   The depth of the centre of pressure of an immersed body

$$= \frac{\text{Second moment of area of immersed body}}{\text{First moment of area of immersed body}}$$

18.1.3   Pressures and forces on lock gates

A practical problem concerned with fluid pressure is encountered when finding the forces that act on a lock gate. A pair of lock gates is shown in plan view in

Figure 18.5a and the side view of one gate in the direction A to B is shown in Figure 18.5b. The water pressure acting on the gates holds them both in contact at B, with the high water being on the left of the gates and the low water on their right.

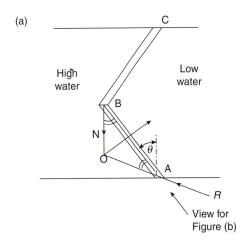

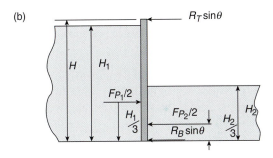

Figure 18.5 Lock gates

Consider first the single lock gate AB (Figure 18.5a)

The resultant of the water pressure force F_P acts at right angles to the centre of the gate. The gate BC acts on AB with a resultant force N acting at right angles (*normal*) to the point of contact of the two gates. The two gates have hinges fixed at their top and bottom, positioned at A and C. The *two hinges on side A will react with a total resultant force R*, the direction of which is not known at this time. However, assuming no leakage and still water, the gate AB is being held in equilibrium under the action of these three forces. If we assume that N and F_P intercept at O, then for equilibrium R must also pass through this point, so we can also fix the direction of R, as shown in Figure 18.5a. If we now let $\theta = $ *the angle of inclination from a line normal to side A of the lock* that will also be parallel to the direction of force N, then by geometry the angles $\angle OBA = \angle OAB = \theta$.

Then, resolving forces at point O *parallel* to lock gate AB, $N \cos \theta = R \cos \theta$, therefore $N = R$. Resolving forces at point O *normal* to lock gate AB, $F_P = N \sin \theta + R \sin \theta$ or $F_P = (N + R) \sin \theta$ and, as $N = R$, then $F_P = 2R \sin \theta$.

$$\text{Therefore } R = \frac{F_P}{2 \sin \theta} \qquad (18.7)$$

The inclination of the resultant force R on the hinge will be at angle θ to the centre line of the gate AB.

Consider the water pressure on the gate AB (Figure 18.5b)

The *components* of the reaction R at the top hinge R_T and bottom hinge R_B of gate AB parallel to side A, are $R_T \sin \theta$ and $R_B \sin \theta$, respectively and (H) is *the distance of the top hinge from the bottom of the gate*.

Also, if $P_1 =$ the *total high water pressure* and $P_2 =$ the *total low water pressure*, that act on *both gates*, then these pressures are split equally between the gates, so that the high and low water pressures acting only on gate AB will be $P_1 = \frac{1}{2}wH_1$ and $P_2 = \frac{1}{2}wH_2$, respectively, i.e. half the full value of the respective water pressures, where the heights of the high and low water are H_1 and H_2, as shown in Figure 18.5b.

Then the *resultant water forces* F_{P_1} and F_{P_2}, i.e. the forces that result from the water pressures on both sides of the single gate, will be:

$$F_{P_1} = \frac{wH_1}{2} \times \text{ wet area of gate } \quad \text{or}$$

$$F_{P_1} = \left(\frac{wH_1}{2} \right) (BH_1) = \frac{wBH_1^2}{2} \qquad (18.8)$$

$$F_{P_2} = \frac{wH_2}{2} \times \text{ wet area of gate } \quad \text{or}$$

$$F_{P_2} = \left(\frac{wH_2}{2} \right) (BH_2) = \frac{wBH_2^2}{2} \qquad (18.9)$$

where for F_{P_1} and F_{P_2}, $B =$ breadth of gate. Also, the *lines of action* of these resultant water forces will be at their centres of pressure, $H_1/3$ and $H_2/3$, respectively. Then $F_P = F_{P_1} - F_{P_2}$, so that:

$$F_P = \frac{wB}{2}(H_1^2 - H_2^2) \qquad (18.10)$$

Finding reaction at the hinges (R_B, R_T)

The hinges of gate AB will only take half of the water pressure acting on the gate, the other half being taken by the reaction of the gate BC. Therefore, remembering that $H =$ *the distance of the top hinge from the bottom of the gate* (where in this case the bottom hinge *is at* the bottom of the gate), then taking moments about X to eliminate R_B from our calculation, we get that:

$$R_T \sin \theta H = \left(\frac{F_{P_1}}{2} \times \frac{H_1}{3} \right) - \left(\frac{F_{P_2}}{2} \times \frac{H_2}{3} \right) \quad (18.11)$$

and, *resolving horizontally,*

$$\frac{F_{P_1}}{2} - \frac{F_{P_2}}{2} = R_B \sin \theta + R_T \sin \theta \qquad (18.12)$$

Key Point.   The hinges of a single closed lock gate take only half the water pressure acting on it, the other half of the pressure being taken by the reaction on the other lock gate

Equations 18.11 and 18.12 enable R_T and R_B to be found, while equation 18.7 may be used to find R, the resultant force on the hinges. *Take care in using equation 18.11 if moments are taken other than from the bottom of the gate, then the distances of the respective centres of pressure may change, dependent on where you take moments (see Example 18.5).*

Finding the point of action (y) of the resultant force F_P

We know from equations 18.10 that $F_P = \dfrac{wB}{2}(H_1^2 - H_2^2)$.

Then, to find the line of action (y), we need to take moments again in a similar fashion to the method we adopted above and in Example 18.2. Then, taking moments again from X,

$$F_P y = F_P y = \left(F_{P_1} \times \frac{H_1}{3} \right) - \left(F_{P_2} \times \frac{H_2}{3} \right)$$

so

$$F_P y = \left(\frac{wBH_1^2}{2} \times \frac{H_1}{3} \right) - \left(\frac{wBH_2^2}{2} \times \frac{H_2}{3} \right)$$

$$= \frac{wBH_1^3}{6} - \frac{wBH_2^3}{6},$$

therefore $F_P y = \dfrac{wB}{6}\left(H_1^3 - H_2^3\right)$ and, from equation 18.12 on division by F_P, we get that:

$$y = \frac{\dfrac{wB}{6}\left(H_1^3 - H_2^3\right)}{\dfrac{wB}{2}\left(H_1^2 - H_2^2\right)} \quad \text{and so} \quad y = \frac{1}{3}\frac{\left(H_1^3 - H_2^3\right)}{\left(H_1^2 - H_2^2\right)}$$

or

$$y = \frac{1}{3}\frac{\left(H_1^2 + H_1 H_2 + H_2^2\right)\left(H_1 - H_2\right)}{\left(H_1 + H_2\right)\left(H_1 - H_2\right)};$$

therefore,

$$y = \frac{\left(H_1^2 + H_1 H_2 + H_2^2\right)}{3\left(H_1 + H_2\right)} \tag{18.13}$$

Deriving equation 18.13 required a little bit of factorisation that I hope you were able to follow. A numerical example follows that makes use of the above equations.

Example 18.5   A set of lock gates, similar to those in Figure 18.6, are inclined so that each lock gate, of breadth 6 m, is at an angle $\theta = 30°$ to the normal of its adjacent side when the gates are closed. The water is at a depth of 7 m on one side of the closed gates and 3.5 m on the other. If the hinges on each gate are positioned 0.5 m and 7.5 m from the bottom of the lock, determine: a) the magnitude and position of the resultant water force on *each* gate, b) the magnitude and direction of the resultant force on the top and bottom hinges.

a)   For the water forces, from equations 18.8 and 18.9, where $B = 6$, $H_1 = 7$, $H_2 = 3.5$, we get that

$$F_{P_1} = \frac{wBH_1^2}{2} = \frac{(9810)(6)\left(7^2\right)}{2} = 1442\,\text{kN},$$

$$F_{P_2} = \frac{wBH_1^2}{2} = \frac{(9810)(6)\left(3.5^2\right)}{2} = 360.5\,\text{kN}.$$

Then $F_P = F_{P_1} - F_{P_2} = 1442 - 360.5 = 1081.5\,\text{kN}$, so that *the resultant water force*

$F_P = 1081.5\,\text{kN}$. The position of this force may be found using equation 18.13, then

$$y = \frac{\left(H_1^2 + H_1 H_2 + H_2^2\right)}{3\left(H_1 + H_2\right)}$$

$$= \frac{\left(7^2 + (7)(3.5) + 3.5^2\right)}{3\left(7 + 3.5\right)} = 2.72\,\text{m}$$

from bottom of lock.

b)   Figure 18.6 shows the forces and positions of the reactions on the top and bottom hinges, in a similar manner to those shown in Figure 18.5b.

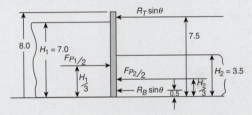

Figure 18.6 Water forces and hinge reactions (for Example 18.5)

Now, in this case we will take *moments about the bottom hinge* (which is positioned 0.5 m up from bottom of lock) to eliminate it from our calculation. So, from Figure 18.7,

$$\left(\frac{F_{P_1}}{2}\right)\left(\frac{7}{3} - 0.5\right)$$

$$= \left(\frac{F_{P_2}}{2}\right)\left(\frac{3.5}{3} - 0.5\right) + 7.0 R_T \sin\theta$$

so

$$\left(\frac{1442}{2}\right)(1.833)$$

$$= \left(\frac{360.5}{2}\right)(0.666) + 7.0 R_T \sin 30$$

and $3.5 R_T = 1321.59 - 120.17$, therefore the *reaction at the top hinge* is $R_T = 343.3\,\text{kN}$. Now, from equation 18.7 we find that

$$R = \frac{F_P}{2\sin\theta} = \frac{1081.5}{2\sin 30} = 1081.5\,\text{kN},$$

therefore $R_B = R - R_T = 1081.5 - 343.3 = 738.2\,\text{kN}$. The direction of the resultant force $R_T = 343.3\,\text{kN}$ on the hinge is $30°$ to the lock gate.

Note that in this case it was better to take moments for the individual situation rather than trying to manipulate equation 18.11.

18.1.4 Fluid pressure on a curved surface

The total fluid pressure force (F_P) on a *curved surface* and the position of the centre of pressure can be obtained by considering the force polygon for the forces causing equilibrium.

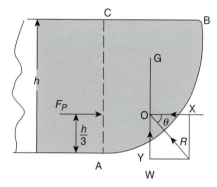

Figure 18.7 Pressure forces acting on a curved surface

Thus, for the curved surface AB shown in side view in Figure 18.7, the total force acting on the surface and its point of application can be obtained by considering the equilibrium of the *volume* of fluid ABC, where we assume *unit width* perpendicular to the surface ABC (i.e. into the paper). Then, from the figure, we see that:

F_P = *the total fluid pressure force* acting on the rectangular area (CB × unit width = 1);

W = *weight of the volume* of fluid (ABC × unit width = 1) acting at its centre of gravity G;

R = *resultant total reaction force* to the fluid pressure force acting on surface area (AB × unit width = 1).

Now, since the three forces F_p, W and R maintain the volume of fluid ABC in equilibrium, they will intersect at a common point O, where F_P and W intersect, as they did before in the case of the lock gate. Then, for the rectangular water face, as you have already found:

$F_P = wA\bar{h} = w \times AC \times \dfrac{AC}{2}$, so that for *any width*:

$$F_P = \frac{w\bar{h}^2}{2} \times \text{width} \qquad (18.14)$$

Now, the *resultant total reaction force R* can be found from the horizontal reaction XO to F_P and the vertical reaction YO to the weight W, using Pythagoras's theorem. Then $R = \sqrt{XO^2 + YO^2}$. The angle of inclination θ of R from the horizontal can also be found from these components, as shown in Example 18.5.

Example 18.6 A sluice gate of width 4 m is shaped as the quadrant of a circle with a radius of 2 m and is pivoted at its centre O, with the water level at the same height as the pivot (Figure 18.8).

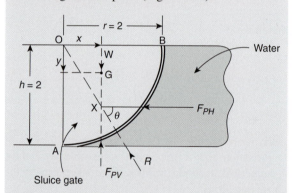

Figure 18.8 Sluice gate (for Example 18.6)

If the centroid of area for the quadrant of a circle is $\bar{x} = \bar{y} = \dfrac{4r}{3\pi}$ and the centre of gravity G of the water acts at its centroid, find the magnitude and direction of the resultant reaction force.

The position of G = $\dfrac{4r}{3\pi} = \dfrac{(4)(2)}{3\pi} = 0.85\,\text{m}$ from the pivot point in the x and y direction, so the horizontal and vertical components of the reaction force intersect at the point X and the line of action of the resultant R of the reaction force passes through this point as shown in Figure 18.8.

Then the horizontal component of the reaction force, i.e. the pressure that acts on the rectangular area (OA × width), is given by equation 18.14 as

$$F_{PH} = \frac{w\bar{h}^2}{2} \times \text{width} = \frac{(9810)(2^2)(4)}{2} = 78.48\,\text{kN}.$$

The vertical component of the reaction is given by the *weight of the fluid* that occupies the volume of the sluice gate, therefore

$$F_{PV} = w\left(\frac{\pi r^2}{4} \times \text{width}\right) = (9810)\left(\frac{\pi 2^2}{4} \times 4\right)$$

$$= 123.275\,\text{kN}.$$

The magnitude of the resultant reaction force is $F_P = \sqrt{(78.48)^2 + (123.275)^2} = 146.14$ kN. The direction of this resultant θ is given by the ratio $\tan\theta = \dfrac{123.275}{78.48} = 1.57$, so that $\theta = 57.5°$.

18.2 Buoyancy

If a body is floating in a fluid and is at rest, it will be in equilibrium in a vertical plane, so that the total upward force must equal the total downward force. This is true no matter whether the body is immersed in a liquid or a gas. The downward force on the body will be due to gravity, while the upward force will be due to the upward pressure of the fluid in which the body is suspended. This resultant upward pressure is known as the *buoyancy*.

Key Point. The resultant upward pressure acting on a floating body is its buoyancy

18.2.1 Buoyancy equation

Consider a large volume of fluid at rest and in equilibrium and imagine a small upright cylindrical element of this fluid, of cross-section δA and height δh, centred at some point within this large volume (Figure 18.9).

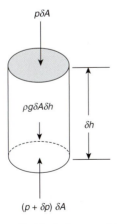

Figure 18.9 Right cylindrical element of fluid

Figure 18.9 shows the *forces* acting on the cylinder. There are also pressures acting on the vertical sides of the cylinder, but because the cylinder is in equilibrium they balance each other out and may be ignored.

It can be seen that there is a pressure (p) acting on the area of the upper surface and a larger pressure ($p + \delta p$) acting on the area of the lower surface due to the additional depth. The weight of the fluid contained in the cylinder is equal to the product of the density of the fluid, the volume of the cylinder and the acceleration due to gravity, i.e.

Weight of the fluid $= \rho\delta A\delta hg$ and so, for equilibrium, upward forces = downward forces or:

$$(p + \delta p)\,\delta A = p\delta A + \rho\delta A\delta hg.$$

Then, from this equation, we first note that after expanding and eliminating the expression $p\delta A$, we get that $\delta p\delta A = \rho\delta A\delta hg$, which tells us that the *resultant upward force = weight of the fluid displaced* in the cylinder, which of course is the *buoyancy*!

We now eliminate the common factor δA to give $\delta p = \rho g\delta h$ and, from the calculus, taking the limit as δh tends to zero, this becomes:

$$dp = \rho g dh$$

This is the *buoyancy equation*, which will be very familiar to you. It tells us that pressure varies with height!

Example 18.7 A rigid weather balloon of negligible mass is 1.2 m in diameter and filled with helium of density 0.165 kg/m³. Calculate the force required to prevent the balloon rising from the ground if the surrounding density of the air is 1.225 kg/m³. Find also the force required to completely submerge the balloon in water.

The volume of the helium in the spherical balloon $= \dfrac{4\pi r^3}{3} = \dfrac{4\pi(0.6)^3}{3} = 0.905$ m³. The weight of the helium in the balloon will therefore be $W_{he} = g\rho V = (9.81)(0.17)(0.905) = 1.51$ N. The weight of the air displaced at ground level (the buoyancy force) $W_a = g\rho_a V = (9.81)(1.225)(0.905) = 10.88$ N. Therefore the force required to prevent the balloon rising will be that force that maintains static equilibrium, i.e. $F = 10.88 - 1.51 = 9.37$ N.

Now the force required to submerge the balloon in water will be the weight of the water displaced: $W = g\rho V = (9810)(0.905) = 8.878$ kN, quite a lot!

18.2.2 Conditions for equilibrium of floating bodies

Bodies such as ships or boats that are at rest floating in still water are in *equilibrium* with their surroundings. In particular, the weight of such a vessel is equal to the weight of water they displace, that is the *buoyancy force* that keeps them afloat. Now if a boat is subject to a disturbance such as waves or winds it may *tilt* or pitch as a result of this disturbing force. If the boat (or any floating body) rights of its own accord, i.e. produces a *righting moment*, and returns to its original state of equilibrium after the disturbance, then we say that the boat is *stable*.

> **Key Point.** A vessel that is able to produce a righting moment that returns it to its original state of equilibrium is said to be *stable*

A body is considered *unstable* if a small displacement tends to cause a turning moment that displaces the body further from its equilibrium position. A body is considered to have *neutral stability* when it remains at rest in the position to which it was disturbed. With respect to boats, we are primarily concerned with the ability of the vessel to return to its equilibrium position after rolling (its lateral stability). When it rotates about its longitudinal axis, this type of built-in stability is important to avoid the boat capsizing! With aircraft, stability in pitch, that is stability about the lateral axis, and stability in yaw, that is stability about the vertical or normal axis, is also important. We will now consider the stability conditions for equilibrium of a boat.

The metacentre and metacentric height

Figure 18.10a shows the cross-section of a boat at rest floating on still water. The longitudinal axis of the boat will travel into the paper and the boat would tilt around the longitudinal axis if disturbed. The point G is the boat's *centre of gravity*, positioned on the waterline, and the point B is the boat's *centre of buoyancy*, i.e. the centre of gravity of the *upthrust* due to the displaced water. W is the weight of the boat and R is the total upthrust due to the displacement of the water that, for equilibrium, will equal the weight of the boat.

Now, if the ship tilts through some small angle θ (Figure 18.10b), then because the geometry of the displaced water has changed, the centre of buoyancy moves from B to B_1. The point M in the figure is known as the *metacentre*; it is the point at which the line of action of R, at its displaced position R_1, cuts the original vertical through the centre of gravity of the boat. If M is above the centre of gravity, as shown in Figure 18.10b,

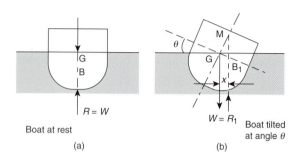

Figure 18.10 Vertical cross-section of a boat

then there is a restoring anticlockwise moment created by force R_1, taken about G, that tends to return the boat to its upright equilibrium position. If M is below G (think of a racing yacht keeled over), then the clockwise moment of R about G tends to increase the angle of tilt and the boat is unstable. The height GM is an important measure of the stability of the boat and is known as the *metacentric height*.

> **Key Point.** If the metacentre of a floating vessel is above the centre of gravity, then a restoring moment will be produced

We can find the metacentric height GM by first considering the restoring moment created by the line of action of $R = W$ moving to the position R_1. As the centre of buoyancy moves through the *distance x*, the restoring moment $= R_1x = Wx$. However, noting that for small angles $\tan\theta = \sin\theta = \theta$ rad, then the *length x* (B to B_1) is, from triangle MGB_1, equal to the length (GM $\times \theta$), which has a moment about G of $(W \times GM \times \theta)$. So, equating, we find that:

$$Wx = (W \times GM \times \theta) \qquad (18.15)$$

> **Key Point.** For small angles measured in radian we may assume that $\tan\theta = \sin\theta = \theta$ rad

> **Example 18.8** A stable ship has a displacement of 5000 tonnes in sea water. A cargo container weighing 400 kN moves 10.5 m across the deck, at right angles to the vertical plane of the longitudinal axis and towards the side rails. If the ship tilts by 5°, calculate the metacentric height.
>
> The overturning moment resulting from the moving cargo is $Wx = (400)(10.5) = 4.2$ MN m.

Since 1 tonne $= 1000\,\text{kg}$, the weight of the ship is given by $W = (9.81)(5 \times 10^6\,\text{kg}) = 49.05\,\text{MN}$ and the angle of tilt $5° = 0.087\,\text{rad}$. Therefore, the righting moment $= (49.05)(0.087)(\text{GM}) = 4.267\,\text{MN} \times \text{GM}$. We know that the overturning moment $=$ the righting moment. Therefore, from equation 18.15, $4.2 = 4.267 \times \text{GM}$ and the metacentric height $\text{GM} = \dfrac{4.2\,\text{MN m}}{4.267\,\text{MN}} = 0.984\,\text{m}$.

Deriving the relationship BM = I/V

Consider now Figure 18.11, where the *centre of gravity is below the water line*. Then, from a similar argument to above, we can again determine the *restoring moment* that results from the movement of the centre of buoyancy (B to B_1) but this time with respect to the length from *the centre of buoyancy* to *the metacentre* (BM) rather than GM.

Then, from Figure 18.11, the anticlockwise restoring moment is given by $(R \times BB_1)$ and again, for small angles, $\tan \theta = \sin \theta = \theta$ rad. Then the *length* BB_1 is (this time) from triangle MBB_1 equal to the length $(BM \times \theta)$. Therefore, equating moments $(R \times BB_1) = (R \times BM\theta)$ and remembering that the upthrust force $R = g\rho V$ (from Example 18.6), $(R \times BB_1) = (g\rho V \times BM\,\theta)$, so that:

The restoring moment $= (R \times BB_1) = (wV \times BM\,\theta)$
$$(18.16)$$

where $w = \rho g$, the specific weight in N/m^3, and $V = volume$ (m^3).

Now we can also express the distance from the centre of buoyancy to the metacentric height (BM) in another way, by using *second moments of area*.

Figure 18.11 shows a thin elemental strip with cross-sectional area δA in plan (see exploded view) at a distance y from the centre line of the boat. As the boat tilts, again through a small angle (θ rad), an arc of length ($y\theta$) is created. Therefore, the volume of this elemental strip is $\delta V = y\theta \delta A$. Then the weight of the water displaced by this volume will be $\delta W = wy\theta \delta A$ and the *moment of this weight force about* O will be $\delta M = wy\theta \delta A \times y$. Therefore, the restoring moment for the whole section (which runs through the length of the boat) can be found by integration, where *the total restoring moment* $= w\theta \int y^2 dA = w\theta I$. You should recognise that the integral $\int y^2 dA =$ the second moment of area (I). Therefore, from equation 18.16, where the restoring moment was given by $(R \times BB_1) = (wV \times BM\,\theta)$, then $(w\theta V)(BM) = w\theta I$, from which we get that the distance from the centre of buoyancy to the metacentre is:

$$BM = \frac{I}{V} \qquad (18.17)$$

Example 18.9 Assuming, for the same ship identified in Example 18.7, all existing conditions concerning the cargo and the angle of tilt prevail, the density of sea water is $1025\,\text{kg/m}^3$ and the second moment of area has a value $I = 8500\,\text{m}^4$, find the height of the ship's centre of gravity above its centre of buoyancy.

The volume of the water displaced $V = \dfrac{(5 \times 10^6)}{1025} = 4878\,\text{m}^3$. Then, from equation 18.17, the distance from the centre of buoyancy to the metacentre is $BM = \dfrac{I}{V} = \dfrac{8500}{4878} = 1.74\,\text{m}$. Now, from Example 18.7, since all existing conditions prevail, the metacentric height is $GM = 0.984\,\text{m}$. Therefore the distance from the centre of gravity to the centre of buoyancy $GB = BM - GM = 1.74 - 0.984 = 0.758\,\text{m}$.

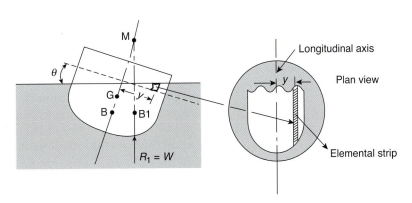

Figure 18.11 Vertical cross-section of a boat tilted at angle θ

18.3 Momentum of a fluid

In this section we consider the idea of the momentum of *fluids in motion* and in particular we look at the forces produced by the impact of a *jet* of fluid on *stationary surfaces*. You have already met the concept of momentum in section 8.1 where, from Newton's second law, we showed that force was the product of the mass of a moving body and the rate of change of velocity of the body, in other words $F = \dfrac{m(v-u)}{t}$. Then, knowing that $\dfrac{m}{t}$ is the mass flow rate of the body (kg/s) or, in this case, of the fluid, therefore *the force of a jet of fluid* is given by $F = \dot{m}(v_1 - v_2)$, where $\dot{m}$ is the mass flow rate, $v_1 =$ the initial velocity of the fluid jet before impact and $v_2 =$ final velocity of the fluid jet after impact.

Now, if a jet of fluid *impinges on a surface*, its velocity is changed. The *resultant force* needed to bring about this change in velocity is given by this formula:

The *resultant force* on the fluid $F = \dot{m}(v_2 - v_1)$

$$(18.18)$$

18.3.1 Impact of a fluid jet on stationary flat surfaces

Figure 18.12 shows a fluid jet with initial velocity (v) impacting on a stationary surface that is normal to the jet flow.

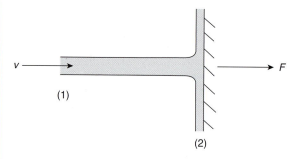

Figure 18.12 Fluid jet impacting on surface normal to jet

The force due to the motion of the jet in the *horizontal* direction is given by equation 18.18, where $v_1 = v$, i.e. $F = \dot{m}(v_2 - v)$. After impact at point (2) the velocity perpendicular to the plate $v_2 = 0$, so resultant force on the fluid perpendicular to the plate $= \dot{m}(0 - v)$ and, since $\dot{m} = \rho Av$ (the product of the fluid jet density, cross-sectional area and velocity), then the *resultant force on*

the jet $= -\rho Av^2$. Now you also know from Newton's third law that to every action there is an equal and opposite reaction, so that:

The resultant force exerted by the fluid on the plate

$= -$*the resultant force exerted by the plate on*

the fluid.

Hence, the *resultant force exerted by the fluid on the surface*:

$$F = \rho Av^2 = \rho \dot{Q}v = \dot{m}v \qquad (18.19)$$

where $\dot{Q} =$ the *volume flow rate* (m³/s) and $\dot{m} =$ mass flow rate in (kg/s), which you are familiar with. Take care not to mix up the symbol for volume flow rate with that for heat energy!

> **Key Point.** the mass flow rate $\dot{m}$ in (kg/s) = volume flow rate $\dot{Q}$ in (m³/s) × the mass density ρ in (kg/m³)

Figure 18.13 shows the situation for the impact of a jet on a fixed inclined flat surface.

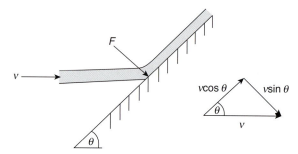

Figure 18.13 Fluid jet impacting on a fixed inclined flat surface

From the figure it can be seen that in this situation a component of the fluid jet flows along the plate after the impact.

Then, using the same argument as before and simple trigonometry, the resulting force on the fluid at impact is $= \dot{m}(v_2 - v_1) = m(0 - v\sin\theta) = -mv^2\sin\theta$. Then again, from Newton's third law and $\dot{m} = \rho Av$:

The resultant force exerted by the fluid on the surface:

$$F = \rho Av^2 \sin\theta \qquad (18.20)$$

Example 18.10   A jet of water 12.4 mm in diameter strikes a stationary flat plate. If the flow rate of the jet is 1.8 litres/s, what force is exerted on the plate when a) it is positioned normal to the jet, b) it makes an angle of 30° to the jet?

The cross-sectional area of the jet is

$$A = \frac{\pi d^2}{4} = \frac{\pi (0.0124)^2}{4} = 1.207 \times 10^{-4}.$$

The volume flow rate needs to be in m³/s, therefore all you need to know is that there are 1000 litres in a cubic metre, so that the volume flow rate $Q = 1.8/1000 = 1.8 \times 10^{-3} \text{m}^3/\text{s}$.

Then, from equation 18.19, $F = \rho A v^2 = \rho Q v = \dot{m} v$ or $\rho A v^2 = \rho Q v$ and, transposing to find the velocity, we get that

$$v = \frac{Q}{A} = \frac{1.8 \times 10^{-3}}{1.207 \times 10^{-4}} = 14.9 \text{ m/s}.$$

Therefore the force exerted by the jet on the plate is: $F = \rho Q v = (1000)(1.8 \times 10^{-3}) = 26.8 \text{ N}.$

18.3.2   Impact of a jet on a stationary curved surface

Consider the jet impacting on a curved surface as shown in Figure 18.14, where the jet is turned through an angle θ (with components F_x and F_y) between the fluid and the surface.

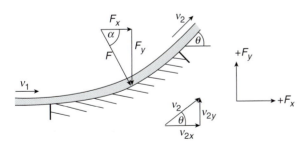

Figure 18.14 Fluid jet impacting on a curved surface

As you will see in the next section from Bernoulli's equation, the velocity of the jet before it is deflected equals the velocity of the jet after impact, i.e. $v_1 = v_2 = v$. Resolving forces in the x and y direction, we find that before deflection $v_{1x} = v$, $v_{1y} = 0$ and that after deflection $v_{2x} = v \cos \theta$, $v_{2y} = v \sin \theta$.

Then, from equation 18.8, $F = \dot{m}(v_2 - v_1)$. Therefore the forces *acting on the fluid* in the x and y directions are:

$$F_x = \dot{m}(v \cos \theta - v) = F_x = \dot{m}v(\cos \theta - 1) \text{ and}$$

$$F_y = \dot{m}(v \sin \theta - 0) = F_y = \dot{m}v \sin \theta.$$

Then again, using Newton's third law, we find that the forces F_x and F_y *that act on the surface* are:

$$\left[\begin{array}{l} F_x = -\dot{m}v(\cos \theta - 1) \\ F_y = -\dot{m}v \sin \theta \end{array} \right] \qquad (18.21)$$

Also, by Pythagoras:

$$\left[\begin{array}{l} F = \sqrt{F_x^2 + F_y^2} \\ \alpha = \tan^{-1}\left(\dfrac{F_y}{F_x}\right) \end{array} \right] \qquad (18.22)$$

where F_x and F_y are as given in equation 18.21.

Example 18.11   A water jet with velocity 30 m/s and diameter 20 mm is deflected through 60° by a smooth, fixed, curved vane. Find the force exerted on the vane by the jet.

The cross-sectional area of the jet is

$$A = \frac{\pi d^2}{4} = \frac{\pi (0.02)^2}{4} = \pi \times 10^{-4}.$$

The mass flow rate can now be found using equation 18.19, where $F = \rho A v^2 = \rho Q v = \dot{m} v$. Then $\dot{m} = \rho A v = (1000)(\pi \times 10^{-4})(30) = 9.42 \text{ kg/s}$. We can now find F_x and F_y using equation 18.21. Then $F_x = -\dot{m}v(\cos \theta - 1) = -(9.42)(30)(-0.5) = 141.3 \text{ N}$ and $F_y = -\dot{m}v(\sin \theta) = -(9.42)(30)(0.866) - 0.5) = -244.74 \text{ N}$. The force ($F$) exerted on the vane by the jet is given by equation 18.22 as

$$F = \sqrt{F_x^2 + F_y^2} = \sqrt{(141.3)^2 + (-244.74)^2}$$

$$= 282.6 \text{ N}.$$

18.3.3   Forces through a reducer

Consider fluid flowing through a reducer (or converging duct), as shown in Figure 18.15, which causes a resultant force F between the fluid and the walls of the reducer.

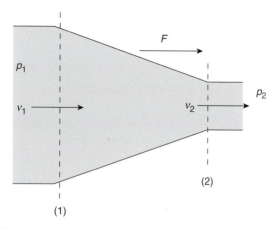

Figure 18.15 Fluid flow through a reducer

Bernoulli's equation (as you will see next) shows that there is a change in static pressure as the velocity increases through the reducer. So the forces *acting on the fluid* in the direction of the flow (Figure 18.15) are:

Force on fluid at inlet (1) $= p_1A_1$, force on fluid at outlet (2) $= -p_2A_2$, force on fluid due to walls of reducer $= -F$. Therefore the resultant force on the fluid $= p_1A_1 - p_2A_2 - F$.

Then again, from Newton's second law where the *resultant force = rate of change of momentum*, we find that $F = p_1A_1 - p_2A_2 - m(v_2 - v_1)$, or the *force acting on the walls of the reducer* is:

$$F = p_1A_1 - p_2A_2 - m(v_2 - v_1) \qquad (18.23)$$

The force (F) acting on the walls will be negative under some conditions. This helps explain the backward force felt when holding a hosepipe from which water is spouting.

Example 18.12 A tapered pipe is positioned horizontally. Water flowing at a mass flow rate of 16 kg/s enters the pipe where the cross-sectional area is 800 mm² at 20 m/s and leaves at 53 m/s where the cross-sectional area is 300 mm². Calculate the force exerted if the inlet pressure is 2.6 bar and the exit pressure is 1.2 bar.

In this very simple example all that you need to remember is to work in consistent units. So, using

equation 18.23 and working in standard units, we get that:

$$
\begin{aligned}
F &= p_1A_1 - p_2A_2 - m(v_2 - v_1) \\
&= (2.6 \times 10^5)(800 \times 10^{-6}) \\
&\quad - (1.2 \times 10^5)(300 \times 10^{-6}) \\
&\quad - 16(53.3 - 20) = -360.8\,\text{N}.
\end{aligned}
$$

Thus the force acting on the walls of the reducer is –361 N, pushing the pipe backwards.

We now move away from the momentum of a fluid flow and turn our attention to the energy contained in a fluid stream, as exemplified by Bernoulli's equation.

18.4 The Bernoulli equation

In this section we look at the conservation of *energy* in fluid systems, which in fact we have already covered when in section 13.4 we studied the steady flow energy equation (SFEE) for open systems. Here we consider *Bernoulli's equation* which, like the SFEE, is derived on the basis of the application of the *conservation of energy* to open systems in which fluid flows. We look first at the *continuity equation*, which again, although you might not be fully aware, you have used many times before, even in the last section where we considered the momentum of a liquid.

18.4.1 The equation of continuity

If a fluid is flowing through a tapered pipe or duct, then the velocity will change as the liquid flows from one section to another. Consider the circular pipe section shown in Figure 18.16.

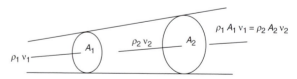

Figure 18.16 Continuity of flow along a tapered pipe

If the fluid flows at a steady rate in the pipe, i.e. the amount of fluid flowing past any point at any time is constant, and we assume that the pipe is flowing full, then, by the conservation of mass, *the mass of the fluid flowing at point 1 will be the same as the mass of the fluid flowing past point 2 at any moment in time.*

Then, for steady flow, since the *mass flow rates* ($\dot{m}$) are equal at both points in the pipe section, we have that $\dot{m}_1 = \dot{m}_2$ or $\rho_1 Q_1 = \rho_1 Q_2$, where ρ = mass density of the fluid in kg/m^3 and Q = the volume flow rate (m^3/s), which of course you know from your study of the last section.

Then, since Q = the *product* of the *area* of the section A(m^2) and the *velocity* at this section v(m/s), i.e. $Q = Av$, which again you already know, *the equation of continuity* becomes:

$$\rho_1 A_1 v_1 = \rho_2 A_2 v_2 \qquad (18.24)$$

If the fluid in our pipe or duct is a *liquid* it can be treated as virtually incompressible. Therefore, for steady *laminar* (see later) flow, *the density of the fluid will remain the same* as it passes through both points. If the fluid is a gas, such as air, this condition is not valid. When studying, for example, airflow over aircraft lift surfaces or air for Pitot-static instruments, compressibility effects have to be taken into account. In our present study of continuity we will only be concerned with incompressible flow. So assuming (for a liquid) equal densities at points 1 and 2, then:

$$A_1 v_1 = A_2 v_2 \qquad (18.25)$$

Note that equation 18.25 tells us that, at any point, the flow velocity is reduced if the area of section is increased.

> **Key Point.** For incompressible steady flow *the volume flow rate* remains the same, i.e. $A_1 v_1 = A_2 v_2$

Example 18.13 If the internal diameters of the pipe at points 1 and 2 (Figure 18.16) are 100 mm and 150 mm, respectively, and water is flowing steadily past point 1 with a constant velocity of 6 m/s, determine: a) the velocity at point 2; b) the volume flow rate; c) the mass flow rate; d) the weight flow rate.

a) The velocity of the water at point 2 is determined from equation 18.24, $A_1 v_1 = A_2 v_2$,

where in this case $v_1 = 6$ m/s, $A_1 = 7854$ mm^2 and $A_2 = 17672$ mm^2. Then,

$$v_2 = \frac{A_1 v_1}{A_2} = \frac{(7854)(6)}{17672} = 2.67 \text{ m/s}.$$

Now, because of the principle of continuity we may use the conditions at either section 1 or section 2. Choosing section 1 we have:

b) Volume flow rate $Q = A_1 v_1 = (7854 \times 10^{-6})$ m$^2 \times (6.0$ m/s$) = 0.047$ m^3/s.

c) The *mass flow rate* is, from equation 18.23 and part b), found to be: $\rho_1 A_1 v_1 = \rho_1 Q_1 = (1000)(0.047) = 47$ kg/s.

d) The weight flow rate ($\dot{W}$) is obtained by multiplying the mass flow rate by the acceleration due to gravity. Then $\dot{W} = g\dot{m} = (9.81)(47) = 461$ N/s.

18.4.2 Bernoulli equation for incompressible flow

We can apply the *conservation of energy* to fluid flow, provided we give due consideration to all of the appropriate forms of energy during the process. Now the *SFEE* (equation 13.4), which you met first in Chapter 13, details *all* the forms of energy we need to consider for a fluid flow process, so this equation is a good place to start in order to determine *Bernoulli's equation*.

The full version of the SFEE for energy changes between two states is given by equation 13.4 as:

$$Q - W = (U_2 - U_1) + (p_2 V_2 - p_1 V_1)$$
$$+ (mgh_2 - mgh_1) + \left(\frac{1}{2}mv_2^2 - \frac{1}{2}mv_1^2\right)$$

The subscripts 1 and 2 were used to indicate the energy conditions of the fluid at entry and at exit to the system. In the flow of a liquid in a pipe they can represent the two points of the liquid stream at the two cross-sections, as shown in Figure 18.16.

Now, when dealing with high energy fluid flows, as was the case when we considered the *gas* flow through a turbine engine in Chapter 16, the potential energy term in the SFEE was very small and could often be ignored. This is not the case for liquid flowing through an inclined pipe (see Figure 18.17) so it cannot be ignored. In the case of a liquid, the temperature change as it flows through the pipe is usually negligible, so that the *internal energy* change terms may be left out. If we assume that no heat is transferred and no work is done, i.e. $Q = W = 0$ as the liquid flows from point 1 to point 2, then the SFEE simplifies to

$$(p_2 V_2 - p_1 V_1) + (mgh_2 - mgh_1)$$

$$+ \left(\frac{1}{2} m v_2^2 - \frac{1}{2} m_{v_1}^2 = 0 \right)$$

or $\quad mgh_1 + \dfrac{1}{2} m v_1^2 + p_1 V_1 = mgh_2 + \dfrac{1}{2} m v_2^2 + p_2 V_2.$

Now we have already assumed, when we looked at the equation of continuity, that liquids were incompressible. Therefore, the density and the volume of the *steady* fluid flow will remain the same, so that $\rho_1 = \rho_2 = \rho$ and $V_1 = V_2 = V$. Then, on division by V, our equation becomes:

$$\frac{mgh_1}{V} + \frac{1}{2} \frac{m v_1^2}{V} + p_1 = \frac{mgh_2}{V} + \frac{1}{2} \frac{m h_2^2}{V} + p_2,$$

and noting that $\rho = \dfrac{m}{V}$, then

$$\rho g h_1 + \frac{1}{2} \rho v_1^2 + p_1 = \rho g h_2 + \frac{1}{2} \rho v_2^2 + p_2 \quad (18.26)$$

so that in general:

$$\rho g h + \frac{1}{2} \rho v^2 + p = \text{constant} \quad (18.27)$$

Equation 18.27 is known as **Bernoulli's equation**, where the three terms represent *potential energy*, *kinetic energy* and *flow work*, respectively. What about the units of the terms in equation 18.27? Well, written in this way, each term has the units of pressure N/m^2; you can check this for yourself! For some applications, for example when considering energy losses through pipes and components in a hydraulic system, it is convenient to express this equation in a different form. If we divide each term in equation 18.27 by ρg we obtain

$$h + \frac{v^2}{2g} + \frac{p}{\rho g} = \text{constant} \quad (18.28)$$

Each of the three terms now has units of length (m). In this form these lengths are referred to as *heads*, so equation 18.28 tells us that the *potential head + velocity head + pressure head = total head* (a constant) and, as you will see later, this form of the Bernoulli equation is very useful.

Key Point. In the head form of the Bernoulli equation the units of each term are in metres (m)

Example 18.14 In a pipe system (Figure 18.17) water is flowing from point 1 to point 2. At point 1 the pipe diameter is 50 mm, the water flow velocity is 3.5 m/s, with a pressure of 350 kPa (remember that 1 Pa = 1 N/m^2).

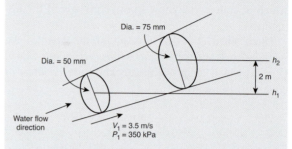

Figure 18.17 Pipe system

Point 2 is elevated 2.0 m above point 1 and has a diameter of 75 mm. Assuming there are no external energy losses or gains into or from the pipe system and that the density of the water is 1000 kg/m^3, determine: a) the velocity of the fluid at point 2, and b) the pressure of the fluid at point 2.

a) From the equation of continuity, where $A_1 v_1 = A_2 v_2$, we have in this case that: $v_1 = 6$ m/s, $A_1 = 1963$ mm^2 and $A_2 = 4418$ mm^2, and so

$$v_2 = \frac{A_1 v_1}{A_2} = \frac{(1963)(3.5)}{4418} = 1.55 \text{ m/s.}$$

b) Using equation 18.26 and rearranging, we get that $p_2 = p_1 + \rho g (h_1 - h_2) + \dfrac{1}{2} \rho \left(v_1^2 - v_2^2 \right)$, so that $p_2 = 350000 + 1000(9.81)(-2) + 500 \left(3.5^2 - 1.55^2 \right) = 335300$ N/m^2 or $p_2 = 335.3$ kPa.

Note that the term $\rho g = w$, the specific weight (N/m^3) we used earlier when studying buoyancy.

18.5 Application of Bernoulli to fluid flow measurement

In this section we look at two methods of measuring flow velocity and flow rate, using a venturi meter and an orifice meter that both rely on the application of Bernoulli theory.

18.5.1   The venturi meter

Figure 18.18 shows a venturi meter that provides a practical application of Bernoulli's equation and the equation of continuity that we have just been looking at.

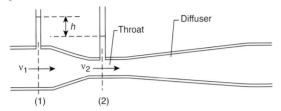

Figure 18.18 The venturi meter

The convergent section of the meter will reduce the area of the flow and hence, from continuity, increase the velocity of the flow. This increase in the velocity (from Bernoulli) will lead to a reduction of the pressure. Pressure tappings at (1) and (2) give a measure of the static pressure at the inlet and throat of the convergent section. The diffuser slows the velocity of the flow back to that of the unrestricted pipe. If the decrease in the velocity is too rapid, turbulence will occur and Bernoulli's equation will no longer be valid.

From Figure 18.18 and from the buoyancy equation we find that the difference in pressure between the tappings (the head) is

$$p_1 - p_2 = \rho g h \tag{1}$$

and, from Bernoulli, we may write this equation as

$$p_1 - p_2 = \frac{1}{2}\rho(v_2^2 - v_1^2) \tag{2}$$

We also know from the continuity equation 18.25 that

$$v_2 = \frac{A v_1}{A_2} \tag{3}$$

Substituting (3) into (2) and rearranging gives (4) as

$$p_1 - p_2 = \frac{1}{2}\rho v_1^2 \left[\left(\frac{A_1}{A_2} \right)^2 - 1 \right] \tag{4}$$

and then substituting (1) into (4) gives

$$\rho g h = \frac{1}{2}\rho v_1^2 \left[\left(\frac{A_1}{A_2} \right)^2 - 1 \right],$$

and on rearrangement we get that:

$$v_1 = \sqrt{ \frac{2gh}{ \left(\dfrac{A_1}{A_2} \right)^2 - 1 } } \tag{18.29}$$

Also, from continuity, $\dot{Q} = A_1 v_1$, therefore:

$$\dot{Q} = A_1 \sqrt{ \frac{2gh}{ \left(\dfrac{A_1}{A_2} \right)^2 - 1 } } \tag{18.30}$$

Note that in equation 18.30, h will be the only quantity that will vary as the volume flow rate varies, since the areas A_1 and A_2 are a fixed part of the design of the venturi meter.

Example 18.15   If, for the venturi meter shown in Figure 18.18, the *diameter* at point (1) is 30 mm and at point (2) it is 20 mm and *water* flows steadily into the meter with a velocity of 4 m/s at point (1), find: a) the mass flow rate, b) the velocity at the throat (point 2), and c) the loss in static pressure between (1) and (2) as an equivalent *head*.

a) From continuity, $\dot{m} = \rho A_1 v_1 = (1000)$ $(7.06 \times 10^{-4})(4) = 2.83$ kg/s, where

$$A_1 = \frac{\pi 0.03^2}{4} = 7.06 \times 10^{-4}\,\text{m}^2 .$$

b) $A_2 = \dfrac{\pi (0.02)^2}{4} = 3.14 \times 10^{-4}\,\text{m}^2$ and again from continuity

$$v_2 = \frac{A_1 v_1}{A_2} = \frac{(0.000706)(4)}{(0.000314)} = 9\,\text{m/s}.$$

c) From Bernoulli $p_1 - p_2 = \dfrac{1}{2}\rho(v_2^2 - v_1^2) =$ $500(9^2 - 4^2) = 10\,\text{kN/m}^2$. This is loss in static pressure, so from $p_1 - p_2 = \rho g h$, then $h = \dfrac{p_1 - p_2}{\rho g} = \dfrac{(10 \times 10^3)}{(1000)(9.81)} = 1.02$ m, i.e. the *head of water* reading on the manometer of the venturi meter $h = 1.02$ m.

18.5.2   The orifice meter

We first consider the flow through an orifice from a water tank or similar. Figure 18.19 illustrates the flow (from point 1 to point 2) that would result from water stored in a tank, through an orifice to atmosphere.

Bernoulli's equation can only be applied when the flow is truly axial, so it is necessary to measure the flow a short distance away from the circular orifice plate, i.e. at point (2), the vena contracta. The *vena contracta* is the

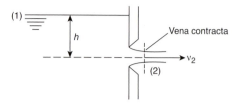

Figure 18.19 Flow through an orifice

minimum contracted area of the flow and occurs at about half a diameter away from the orifice. This is where we would take our pressure tapping for an orifice meter.

> **Key Point.** The *vena contracta* is the minimum contracted area of the flow that occurs a small distance downstream of the orifice

Then, from Bernoulli, between point (1) and (2):

$$\rho g h_1 + \frac{1}{2}\rho v_1^2 + p_1 = \rho g h_2 + \frac{1}{2}\rho v_2^2 + p_2,$$

where at the vena contracta (for axial flow) $h_2 = 0$, so that $\rho g h_1 + \frac{1}{2}\rho v_1^2 + p_1 = \frac{1}{2}\rho v_2^2 + p_2$. The velocity at point (1) in the tank $v_1 = 0$ and we assume that $p_1 = p_2$ since the tank and the orifice are open to atmosphere, so the Bernoulli equation reduces to $\rho g h = \frac{1}{2}\rho v_2^2$, so that:

$$v_2 = \sqrt{2gh} \qquad (18.31)$$

Then, assuming no losses, the velocity at the vena contracta is proportional to $\left(\sqrt{h}\right)$.

Figure 18.20 shows the orifice incorporated into an orifice meter, where friction losses are minimised by making the upstream side of the orifice sharp-edged.

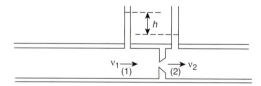

Figure 18.20 The orifice meter

The orifice meter, as shown in Figure 18.20, can be used to measure the flow rate in a pipeline, in a similar manner to the venturi meter. The equations developed to find the velocity and volume flow rate in the venturi meter (equations 18.29 and 18.30) are equally valid for the orifice meter.

> **Example 18.16**   A 100 mm diameter orifice plate is fitted into a 180 mm diameter pipe. If the pressure drop across the orifice is equivalent to a 90 mm head of fluid, find the volumetric flow rate of the fluid in the pipeline.
>   Using equation 18.30,
>
> $$\dot{Q} = A_1 \sqrt{\frac{2gh}{\left(\dfrac{A_1}{A_2}\right)^2 - 1}},$$
>
> where $A_1 = \dfrac{\pi(0.1)^2}{4} = 7.85 \times 10^{-3}\,\text{m}^2$ and $A_2 = \dfrac{\pi(0.18)^2}{4} = 0.0254\,\text{m}^2$, we find that the *volumetric flow rate*
>
> $$\dot{Q} = 0.0254 \sqrt{\frac{(2)(9.81)(90 \times 10^{-3})}{\left(\dfrac{0.0254}{7.85 \times 10^{-3}}\right)^2 - 1}} = 0.0226\,\text{m}^3/\text{s}.$$

18.6   Fluid viscosity

In this section we take a quick look at the nature of viscosity and its effect on fluid flow by considering the non-dimensional parameter known as *Reynolds number*.

18.6.1   The nature of viscosity

The ease with which a fluid flows is an indication of its viscosity. Cold heavy oils such as those used to lubricate large gearboxes have a high viscosity and flow very slowly, whereas petroleum spirit (an oil derivative) is extremely light and volatile and flows very easily and so has low viscosity. We thus define viscosity as the property of a fluid that offers resistance to the relative motion of the fluid molecules. The energy losses due to friction in a fluid are dependent on the fluid viscosity. We will look at friction losses in fluid systems later in this section.

> **Key Point.**   The viscosity of a fluid is a measure of its resistance to flow

As a fluid moves, there is developed in it a shear stress, the magnitude of which depends on the viscosity of the fluid. You have already met the concept of *shear stress* and should remember that it can be defined as the force required to slide one unit area of a substance over the other. It thus has units of N/m^2 and is denoted by the Greek letter tau (τ).

Figure 18.21 illustrates the concept of velocity change in a fluid by showing a thin layer of fluid (*boundary layer*) sandwiched between a fixed and a moving boundary.

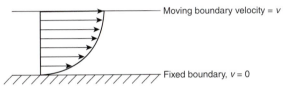

Figure 18.21 Velocity change at boundary layer

Now, a fundamental condition exists between a fluid and a boundary, where the velocity of the fluid (v) at the boundary surface is identical to that of the boundary. So in our case the velocity of the fluid *next to* the moving boundary also has velocity (v) and the fluid at the *fixed* or *stationary boundary* has *velocity zero*, as shown in Figure 18.21.

If the distance between the two surfaces is small, the rate of change of velocity (v) with respect to distance (y) may be taken as linear. The *velocity gradient* (or *shear rate*) is a measure of how the velocity changes and it is expressed as: the velocity gradient/shear rate $= \dfrac{dv}{dy}$. Now, from our definition of *shear stress* we know that *shear stress is directly proportional to the velocity gradient* $\left(\tau \propto \dfrac{dv}{dy}\right)$ and, using a *constant of proportionality* (μ), we find that:

$$\tau = \frac{\mu \, dv}{dy} \qquad (18.32)$$

The constant of proportionality μ is known as the *dynamic viscosity*. What are the SI units of μ? The units of dynamic viscosity are easily found by transposing equation 18.32 and looking at the units for the individual terms. Then $\mu = \dfrac{\tau \, dy}{dv} = \dfrac{\text{N}}{\text{m}^2} \dfrac{\text{m}}{\text{m/s}} = \dfrac{\text{N s}}{\text{m}^2}$, i.e. dynamic viscosity has units of $\left(\text{N s/m}^2\right)$ (*Newton-second per metre squared*). Alternative units expressing

μ in terms of *mass* can be obtained by substituting for the base units of force kg m/s^2, where we find that

$$\mu = \frac{\tau \, dy}{dv} = \left(\frac{\text{N}}{\text{m}}\right)\left(\frac{\text{s}}{\text{m}}\right) = \frac{\text{kg m s}}{\text{s}^2 \, \text{m}^2} = \frac{\text{kg}}{\text{m s}}, \text{ i.e. the}$$

dynamic viscosity may be expressed in units of (kg/m s), i.e. in *kilogrammes per metre-second*.

You will still find many textbooks and manufacturers' literature in which the obsolete *cgs* units for viscosity are quoted. In the interests of completeness they are also given here. The common units in this system are known as the *poise* and *centipoise*, where 1 poise = 0.1 N s/m^2 and 1 centipoise = 1×10^{-3} N s/m^2.

Many problems in fluid mechanics involve the use of the ratio of the *dynamic viscosity divided by the fluid mass density*. This ratio defines the *kinematic viscosity* (v_k), which is more often quoted in literature and data sheets. So:

$$\text{Kinematic viscosity } v_k = \frac{\mu}{\rho} \ (\text{m}^2/\text{s}) \qquad (18.33)$$

Note: The suffix k has been added to the usual symbol for kinematic viscosity (v) in order to avoid confusion with the standard symbol for velocity.

The *cgs* units for *kinematic viscosity* are the *stoke* and *centistoke*, where:

1 stoke = 1×10^{-4} m^2/s and 1 centistoke = 10^{-6} m^2/s.

The units for dynamic and kinematic viscosity can at first seem confusing, and you should take extreme care when using them. Remember that both the SI and possibly the older *cgs* system are used to accommodate the varying needs of manufacturers and users.

Example 18.17 The dynamic viscosity of pure water at 20°C is given as 1×10^{-3} N s/m^2. Determine the kinematic viscosity of the water in SI and *cgs* units, given that its density is 1000 kg/m^3.

From equation 18.33, $v_k = \dfrac{\mu}{\rho}$ and since the units for dynamic viscosity and density, given above, are in standard form, then $v_k = \dfrac{\mu}{\rho} = \dfrac{1 \times 10^{-3}}{1000} = 1 \times 10^{-6}$ m^2/s. Now, in *cgs* units we have we have that 1 centistoke = 10^{-6} m^2/s, so kinematic viscosity of water in centistokes is 1.0.

In Table 18.2 are listed some typical values for the dynamic and kinematic viscosity of a few common fluids, at given temperature, standard atmospheric pressure and selected densities.

Table 18.2 Typical values for the dynamic and kinematic viscosity of a few common fluids

Fluid	Temperature (°C)	Dynamic viscosity (N s/m²)	Kinematic viscosity (m²/s)
Air	0	1.72×10^{-5}	1.33×10^{-5}
Air	20	1.81×10^{-5}	1.51×10^{-5}
Water	20	1.0×10^{-3}	1.0×10^{-6}
Petroleum	25	2.87×10^{-4}	4.2×10^{-7}
Engine oil	20	3.57×10^{-1}	4.375×10^{-4}
Engine oil	80	1.9×10^{-1}	2.375×10^{-4}

18.6.2 Variation of viscosity with temperature

You will be aware from what was said earlier that cold gearbox oil has a high viscosity and is thus difficult to pour. In fact, in very cold climates motorists use some form of heater underneath the engine sump to keep the oil warm so that they are able to turn the engine over on the initial start of the day. From this example and many others, you are no doubt familiar with the fact that fluid viscosity varies with temperature. In my example, as the temperature of the oil (*liquid*) is *increased*, the viscosity *decreases* noticeably.

> **Key Point.** The viscosity of a *liquid* decreases with increase in temperature; it has less resistance to flow

It is important to realise that gases behave differently to liquids with respect to temperature change and viscosity. For gases the viscosity increases with increase in temperature, although the changes are generally smaller than those for liquids. Note the behaviour of air in Table 18.2, with change in temperature.

> **Key Point.** The viscosity of a *gas* increases slightly with increase in temperature

A measure of how greatly the viscosity of a fluid changes with temperature is given by its *viscosity index*, sometimes referred to as VI. Equipment, machinery and systems that are lubricated or operated by oils and hydraulic fluids may need to operate at extremes of temperature. In these circumstances the VI becomes especially important in determining fluid behaviour.

> **Key Point.** The viscosity index VI is a measure of how greatly the viscosity of the fluid changes with temperature

The viscosity index system uses a method whereby the fluid (particularly oils) is compared with two representative families of fluids at both 40°C and 100°C, one family showing a small change of viscosity between these temperatures and the other a large change. For oils, these families are chosen so that they have an *index number* of 100 and 0, respectively.

In general, the higher the VI the better the temperature viscosity behaviour; in other words, the higher the viscosity index, the smaller the change in viscosity with temperature. It is worth noting that changes in viscosity also occur with changes in *pressure*. In most liquids the viscosity increases with increase in pressure; this becomes significant at the pressures operating within hydraulic systems. There are thus certain similarities between the pressure sensitivity of a liquid and its temperature sensitivity, from which it might be deduced that liquids showing maximum change due to temperature would also be most affected by pressure. This is indeed the case; for example, oils of low VI increase in viscosity with pressure far more than those with a high VI.

18.6.3 Reynolds number, laminar and turbulent fluid flow

When analysing fluid flow it is often necessary to determine the characteristics of such flow. Consider the flow of air over the aircraft wing section shown in Figure 18.22a. It is represented by stream lines, which show the air flowing in layers in a smooth and regular manner. This type of flow is referred to as *laminar flow*.

Another example of laminar flow occurs when the flow through the nozzle of a water hose is adjusted

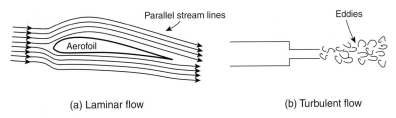

(a) Laminar flow　　　　　　　　(b) Turbulent flow

Figure 18.22 Illustration of laminar and turbulent flow

from the tap until a smooth flow of water is expelled from the nozzle. If you continue to open up the tap, eventually the water flows in a chaotic way, tumbling and turning over itself. This is an example of *turbulent flow* (Figure 18.22b).

Thus, more formally, in *laminar flow* the particles of fluid move smoothly in straight lines, although the velocity of the particles moving along one line may not necessarily be the same as that along another line. The *viscosity* of the fluid is of major importance in this type of flow.

In *turbulent flow* there is irregular and chaotic motion of the fluid particles such that a thorough mixing of the fluid takes place. In this type of flow, fluid inertia effects are dominant over viscous effects.

The behaviour of a fluid, particularly with regard to energy losses, is dependent on whether the flow is laminar or turbulent. For this reason it is important to have a way of predicting the type of flow, other than by direct observation which is often impossible and impractical in closed systems.

In the late nineteenth century Osborne Reynolds discovered that the flow condition in closed pipe systems depended on *fluid density, fluid viscosity, pipe diameter and the average velocity of flow*. His experiments showed that these parameters, in the form of a dimensionless number, could be used to decide whether the flow was laminar or turbulent.

This *dimensionless number* is known as **Reynolds Number** (Re) and is represented symbolically as:

$$(\text{Re}) = \frac{vd}{v_k} \qquad (18.34)$$

where: v = the *mean velocity* of the fluid, d = the *pipe diameter* and v_k = the *kinematic viscosity* of the fluid.

We can show that the Reynolds number is dimensionless by substituting standard SI units into equation 18.34.

Then, $\text{Re} = \left(\dfrac{\text{m(m)}}{\text{s}}\dfrac{\text{s}}{\text{m}^2}\right) = \textit{unity}$, a dimensionless number.

In *piped systems*, if a Reynolds number below about 2000 is obtained, the flow is assumed to be *laminar*,

and for a Reynolds number of about 2500 the flow is *turbulent*. Between the values of 2000 and 2500 the flow is defined as *critical* and may be laminar or turbulent, depending on external circumstances.

> **Example 18.18** Heavy oil flows in a pipe at an average velocity of 4 m/s. If the internal diameter of the pipe is 10 cm, determine whether the flow is laminar or turbulent. Take the density of the oil as 930 kg/m² and its dynamic viscosity as 1.08×10^{-1} N s/m².
>
> To answer this question we need to find the Reynolds number. Therefore we first find the kinematic viscosity of the oil; then, from equation 18.33,
> $$v_k = \frac{\mu}{\rho} = \frac{1.08 \times 10^{-1}}{930} = 1.16 \times 10^{-4}\ \text{m}^2/\text{s and}$$
> then, from equation 18.34, the Reynolds number
> $$(\text{Re}) = \frac{vd}{v_k} = \frac{(4)(0.1)}{1.16 \times 10^{-4}} = 3448.\ \text{Since the}$$
> Reynolds number > 2500, the flow is turbulent.

18.7　Friction losses in piped systems

When formulating the Bernoulli equation we made the assumption that there were no energy transfers to or from the fluid as it travelled along the pipeline between two particular points, whereas in reality, as you know from your previous study of the steady flow energy equation (SFEE), there are often energy losses resulting from pipe friction and losses through fixtures and fittings, as the fluid flows.

18.7.1　The general energy equation

Energy may be added and removed from a system as part of its function; for example, a pump within the system will add energy whereas a motor will remove energy. These further energy losses and gains may be represented as *heads* in a similar manner to those already given in Bernoulli's equation. Consider again the *head version* of the Bernoulli equation, where $h + \frac{v^2}{2g} + \frac{p}{\rho g} =$ constant or $h_1 + \frac{v_1^2}{2g} + \frac{p_1}{\rho g} = h_2 + \frac{v_2^2}{2g} + \frac{p_2}{\rho g}$. Now if we represent the energy transfers as the fluid travels from point 1 towards point 2 as *equivalent heads*, then: *energy added* to the system $= h_A$, *energy removed* from the system $= -h_R$ and *energy losses* $= -h_L$. Then Bernoulli's equation may be written as:

$$\frac{v_1^2}{2g} + \frac{p_1}{\rho g} + h_1 + h_A - h_R - h_L = \frac{v_2^2}{2g} + \frac{p_2}{\rho g} + h_2$$
(18.35)

Equation 18.35 is known as the *general energy equation for fluid flow*. The standard sign convention is used for the energy transfers, i.e. *energy added is positive* and *energy lost or removed from the system is negative*.

Note that the additional gains and losses of energy must occur in the direction of flow of the fluid. In the above form, equation 18.35 suggests flow from some arbitrary point 1 to point 2, read from left to right.

18.7.2 The Darcy equation and friction losses

In the general energy equation, the term (h_L) defines the *energy loss* from a fluid system. One component of this energy loss is due to *friction* in the flowing fluid. In particular, losses in straight pipes may be estimated from the Darcy equation, which can be written in terms of the *head loss* (h_L) as:

$$h_L = 4f\left(\frac{lv^2}{d2g}\right)$$
(18.36)

or, in terms of *the pressure* loss:

$$\Delta p = 4f\left(\frac{l\rho v^2}{d2g}\right)$$
(18.37)

where, $\Delta p =$ the pressure loss; $h_L =$ the head loss per unit weight; $f =$ the pipe friction factor; $l =$ the pipe length; $\rho =$ the fluid density; $v =$ the mean velocity of the fluid; and $g =$ acceleration due to gravity.

The **Darcy equation**, as shown in equations 18.36 and 18.37, may be determined from the general energy equation by considering the average shear stress at the pipe wall, created by the fluid flow.

The term (f) or **friction factor** in the Darcy equation depends on the *nature of the flow* in the pipe. For *laminar flow* only, the value of the friction factor depends on the

Reynolds number and can easily be calculated from the relationship $f = \frac{16}{\text{Re}}$.

If the flow in the pipe is *turbulent*, then the *friction factor can only be determined from experimental data*. The results of such data have been conveniently displayed on a Moody diagram, which enables us to estimate friction factors where turbulent flow is present.

It can also be seen from the Moody diagram (Figure 18.23) that the friction factor not only depends on the Reynolds number but also varies with surface roughness. In practice, when dealing with Moody diagrams, there is one important difference in terminology between the UK and USA versions of the diagram. In the US system, the quantity $4f$ quoted in the Darcy equation is simply termed f. This means that when using the US version of the Moody diagram the friction factor is given by $f = \frac{64}{N_R}$, where N_R is used instead of (Re) for Reynolds number.

Figure 18.23 shows the US version of the Moody diagram.

Example 18.19 Heavy oil with a kinematic viscosity of $1.2 \times 10^{-4}\,\text{m}^2/\text{s}$ flows 50 m through a 10 cm diameter pipe at a velocity of 1.5 m/s. Determine the energy loss resulting from the flow.

We first need to establish whether the flow is laminar or turbulent by calculating the Reynolds number. Then $(\text{Re}) = \frac{vd}{v_k} = \frac{(1.5)(0.1)}{1.2 \times 10^{-4}} = 1250$. Re < 2000 (*laminar*), therefore we may calculate (f) from the Reynolds number. Using the head loss form of the Darcy equation: $h_L = 4f\left(\frac{lv^2}{d2g}\right)$, where $f = \frac{16}{\text{Re}} = \frac{16}{1250} = 0.0128$, therefore

$$h_L = (4)(0.0128)\left(\frac{50 \times 1.5^2}{(0.1)(2)(9.81)}\right) = 2.94\,\text{m}.$$

This answer tells us that we lose 2.94 N m of energy for each Newton of oil as it flows along the 50 m pipe.

Example 18.20 Water at 20°C flows at 3 m/s in a 30 m plastic pipe with an internal diameter of 25 mm. Determine the friction factor (f) and the energy loss due to the flow.

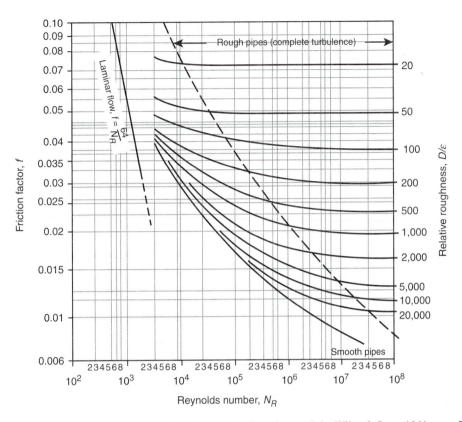

Figure 18.23 Moody diagram (reproduced from: Pao, R.H.F., *Fluid Mechanics*, John Wiley & Sons, 1961, page 284)

In order to determine whether or not the flow is laminar or turbulent, we must again find the Reynolds number.

We are not given the kinematic viscosity, but it can be found by consulting Table 18.2, where we find for water at 20°C that $v_k = 1.0 \times 10^{-6}\,\text{m}^2/\text{s}$. Then, $(Re) = \dfrac{vd}{v_k} = \dfrac{(3)(25 \times 10^{-3})}{1.0 \times 10^{-6}} = 75000$, which > 2000, hence the flow is *turbulent*.

Now, since the flow is turbulent we cannot use the simple formula given in the previous example to find the friction factor, we need to use the *Moody diagram*. So as well as the Reynolds number we need also to find the relative roughness (D/ε), where D is the pipe diameter and ε is the *average pipe wall roughness*.

Note that in the Moody diagram (Figure 18.23) the relative roughness is shown as a set of parametric curves. The wall roughness (ε) is often quoted in tables for different materials. Plastic pipes can be treated as smooth, so in this case, to read the value of the friction factor from the diagram, we move along the horizontal axis (note that the scales on

both axes are logarithmic) until we find 75000, which lies between 10^4 and 10^5 at the point three-quarters of the way between 6 and 8. Now follow this line up until you meet the roughness curve for plastic (smooth pipes, that starts from the bottom right-hand corner), and read across. You should find a friction factor which approximately equals 0.019.

Remember that, since we have the US version of the friction factor, this value 0.019 is the equivalent to $4f$ in our version of the Darcy equation. So, having found the friction factor, we can now use:

$$h_L = 4f\left(\frac{lv^2}{d2g}\right) \text{ where } 4f = 0.019 \text{ and so}$$

$$h_L = (0.019)\left(\frac{30 \times 3^2}{(25 \times 10^{-3})(2)(9.81)}\right) = 10.46\,\text{m}.$$

Head loss due to flow is 10.46 m.

Note: In this book we are only interested in energy losses in pipes of circular cross-section. *If the cross-section is non-circular* then, instead of using the diameter to find an appropriate Reynolds number, we use a characteristic dimension called the *hydraulic radius* where:

$$\text{Hydraulic radius } (R) = \frac{A}{L} = \frac{\text{cross-sectional area}}{\text{wetted perimeter}}.$$

> **Key Point.** In non-circular cross-section pipes and channels *the hydraulic radius* is used to calculate the Reynolds number

This dimension is particularly useful for open-section channels, sewerage systems and other mass fluid transport applications.

18.8 Energy loss in plain bearings

In this final section of the chapter we look at an application of the general energy equation that enables us to estimate power losses, primarily due to friction, within *plain bearing assemblies*. We first need to establish a little terminology and one or two basic concepts associated with lubrication and hydrodynamic bearings.

Plain bearings carry the load directly on supports, using a sliding motion, as opposed to roller bearings where balls or rollers are interposed between the sliding surfaces. Plain bearings are of two types: *journal or sleeve bearings*, which are cylindrical and support radial loads (those perpendicular to the shaft axis); and *thrust bearings*, which are generally flat and, in the case of a rotating shaft, support loads in the direction of the shaft axis (axial direction).

Figure 18.24 shows a crankshaft supported by two main bearings attached to the connecting rod by the connecting rod bearing. All three are journal bearings. Flanges on the main bearings act as thrust bearings that restrain the axial motion of the shaft. We now look a little more closely at these bearings, in our discussion on lubrication.

18.8.1 Lubrication

There are generally considered to be three different types of *bearing lubrication*:

- *Hydrodynamic lubrication* where the surfaces are completely separated by a film of lubricant and the loads generated between the surfaces are supported entirely by fluid pressure generated by the motion of the surfaces

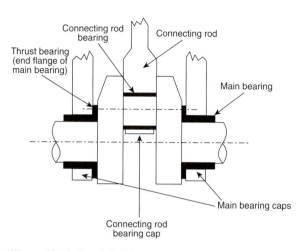

Figure 18.24 Crankshaft journal and thrust bearing

- *Mixed film lubrication* where the surface peaks of the rubbing surfaces are intermittently in contact and there is partial hydrodynamic support
- *Boundary lubrication* where surface contact is continuous, the lubricant being continually smeared over the surface, providing a renewable surface film which reduces friction and wear.

We will only be considering *hydrodynamic lubrication* as an example in this short section. In *hydrodynamic lubrication* the film thickness varies typically between about 0.008 mm at its thinnest point and 0.02 mm at its thickest point. Figure 18.25a shows a loaded bearing journal, subject to hydrodynamic lubrication, at rest.

The bearing clearance space is filled with oil, which has been squeezed out at the bottom by the load (W).

Slow clockwise rotation of the shaft to the right (18.25b) causes it to roll to the right. However, it stays in this position, unable to climb to the right, because of the lack of friction due to the lubricant; thus lubrication at the boundaries takes place. If the rotational speed of the shaft is progressively increased, more and more fluid is forced forward and adheres to the journal surface, causing sufficient pressure to build up to raise or float the shaft away from the boundary (Figure 18.25c). Hydrodynamic lubrication, therefore, depends on the rotational speed of the shaft (n), the dynamic viscosity of the oil μ (the higher the viscosity of the oil the lower the rotational speed required to float the shaft), and the bearing unit load (P) which is the load (W) divided by the bearing area (the diameter (D) multiplied by the length (L) of the bearing).

We can form a relationship between μ, n and P which optimises the efficiency of the hydrodynamic lubrication

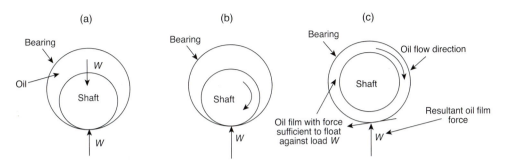

Figure 18.25 Hydrodynamic lubrication for journal bearing

of the journal bearing, where for hydrodynamic lubrication we optimise the relationship $\mu n/P$ for given conditions. So, for example, very smooth surfaces may be lubricated using low-viscosity oils; this reduces the effects of viscous friction, so under these circumstances a low value of $\mu n/P$ is appropriate.

In order to reduce power losses in bearings we should always endeavour to reduce the coefficient of friction between the surfaces. In *hydrodynamic lubrication* typical values for *the coefficient of friction* are 0.002 to 0.01, which are very low. This is why this form of lubrication is so effective!

Note: You will remember from your work in section 8.4 that the friction force between two rubbing surfaces is given by $F = \mu R$, where μ is the coefficient of friction between the surfaces and R is the normal reaction force created by the load applied to the surfaces. It is unfortunate that the symbol (μ) for the coefficient of friction is also the same as that used for dynamic viscosity. To avoid confusion, we will, *for this particular application*, use the symbol f for the coefficient of friction and the normal symbol μ for dynamic viscosity.

You should also be aware of the relationship between the torque created in a rotating shaft and the power required to drive this torque.

You will remember the formula for finding the power created by a torque from the work you did in section 3.4. Here we will be using the relationship $Power = 2\pi T_f n$, where in this case $T_f =$ the *friction torque* created by the bearings and the rotational velocity (n) is in *revolutions per second*.

18.8.2 Petroff method for estimating power loss

In 1833 the scientist Petroff analysed viscous friction drag in what we now know as hydrodynamic bearings. He devised the Petroff equation, which provides a

quick and simple method for obtaining reasonable estimates of coefficients of friction for lightly loaded bearings, enabling us to estimate power losses within such bearings. The derivation of the Petroff equation will not be given here; however, its use is simple, and should not present you with too many problems.

$$\text{Petroff equation } f = 2\pi^2 \left(\frac{\mu n}{P} \frac{R}{c} \right) \qquad (18.38)$$

where $f =$ coefficient of viscous friction; $\mu =$ dynamic viscosity (N/m^2 s); $n =$ rotational velocity (revs per second); $R =$ shaft radius (m); $c =$ radial clearance, i.e.: $c =$ (bearing diameter – shaft diameter)/2 ; and $P = \dfrac{W}{DL} = \dfrac{W}{A} = \dfrac{\text{load}}{\text{bearing area}}$.

Example 18.21 An 80 mm diameter shaft is supported by a bearing 60 mm in length with a *radial clearance* of 0.05 mm. It is lubricated by an oil having a dynamic viscosity of 60×10^{-3} N/m^{2}s (at the operating temperature). The shaft rotates at 480 rpm and carries a radial load of 4000 N. Determine an estimate for the bearing coefficient of friction and the power loss, using the Petroff method.

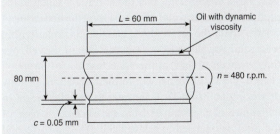

Figure 18.26 Journal bearing for Example 18.20

Figure 18.26 illustrates the journal bearing given in the question. In order to apply the Petroff method we must make the assumption that, for the loading conditions, there is no eccentricity between the bearing and the journal, and that no fluid flows in the axial direction. We must also assume that the simple friction condition $F_d = fR$ (where F_d = the friction drag force, and R = the radial load), applies.

Then, using Petroff's equation, the coefficient of viscous friction is given by $f = 2\pi^2 \left(\dfrac{\mu n}{P} \dfrac{R}{c} \right)$ as

$$f = 2\pi^2 \left(\frac{(60 \times 10 - 3)(8\,\text{rev/s})}{833.33} \frac{(40 \times 10^{-3})}{(0.05 \times 10^{-3})} \right)$$

$$= 0.0096,$$

where

$$P = \frac{L}{\text{bearing area}} = \frac{4000}{(0.08 \times 0.06)} = 833.33.$$

So the *frictional drag force* $F_d = fR = (0.0096)(4000) = 36.38$ N. Then the torque resulting from the drag force is $T_f = (36.38)(0.04) = 1.46$ N m, remembering that the torque due to the friction is equal to the frictional drag force multiplied by the shaft radius. Then, from power = $2\pi T_f n$, we find that the:

Power loss $= (2\pi)(1.46)(6\,\text{rev/s}) = 55$ Watts

18.9 Chapter summary

In this chapter you have been introduced to a few topics concerned with both fluid statics and fluid dynamics. We started by considering thrust forces on immersed surfaces, where you were introduced to equation 18.1, which essentially states that thrust force is equal to the product of specific weight ($w = \rho g$) and the first moment of area about the centroid $\bar{h}A$, and that this relationship was valid for all surfaces, whether flat or curved. We then went on to look at the relationships for thrust on curved surfaces, where we verified equation 18.1 by producing equation 18.3. In order to find the height of the centre of pressure h_c from the surface or base of the fluid, we needed to reintroduce the idea of second moments of area, which we applied in Examples 18.3 and 18.4 to find the distance to the centre of pressure

from the required datum. We finished this first section by considering the fluid forces that acted on lock gates and submerged curved surfaces. You should be able to calculate these forces, as in Examples 18.5 and 18.6.

We next considered the concept of buoyancy and the buoyancy equation $dp = \rho g dh$, where during the process of deriving this equation we were able to show that *the resultant upward force equalled the weight of the fluid displaced*, where this upward force was in fact the *buoyancy* that kept it in equilibrium. We also showed that pressure varied with height (which you knew already), but remember that for an immersed body increasing depth results in an increase in pressure, while increasing height results in a decrease in pressure acting on the body. Example 18.6 provided an application of the buoyancy equation to a balloon, while Examples 18.7 and 18.8 applied the theory to boats floating on water, where you should be able to determine the relative metacentric height and understand its significance with respect to the vessel's centre of gravity.

In section 18.3 we considered the momentum of a fluid jet, where equation 18.19, $F = \rho A v^2 = \rho \dot{Q} v = \dot{m} v$, showed us that the momentum force exerted by the jet on the stationary surface was not only equal to the mass flow rate times the fluid velocity but could also be found using the volume flow rate. You should make sure that you understand the units of the parameters in this equation. Forces and the reaction of forces caused by a jet of fluid can be applied to the assessment and design of pump impeller blades, compressors and turbines. Space prevented us from pursuing these applications, but the study of fluid momentum should nevertheless prove useful.

In section 18.4 we looked at the conservation of energy in piped fluid flow systems by first reviewing the idea of continuity and then introducing the Bernoulli equation for incompressible flow, where you will have noted the similarities between the Bernoulli equation and the SFEE that you met when studying the thermodynamics of open systems. Notice that although we derived the Bernoulli equation from an energy perspective in this section, it can also be derived on the basis of conservation of momentum. Do try and remember the limitations placed on the use of the Bernoulli equation, most importantly that it is valid only for steady *non-compressible* axial flow and that it does not take account of any friction losses in pipes or shock losses as the fluid passes through system components.

In section 18.5 we applied the continuity and Bernoulli equations to fluid flow measurement, where, using venturi and orifice meters, we were able to find changes in flow rates and fluid velocity. The theory we developed for these instruments did not take account of any energy losses that occur as the fluid flows through the instrument. These losses can be accounted for by introducing an error correction efficiency factor, known as the *coefficient of discharge* (C_d). Losses due to fluid turbulence and restriction to flow are very small within the venturi meter, where typically $C_d = 0.96$ to 0.98, while the cheaper and easily installed orifice meter is much less accurate, with typical coefficient of discharge values around $C_d = 0.6$ to 0.65.

In section 18.6 we considered the nature of *fluid viscosity*, i.e. the resistance to fluid flow of a viscous fluid. The viscosity of liquids decreases with increase in temperature, i.e. the fluid is able to flow more easily, but in the case of gases an increase in temperature causes a small increase in viscosity. Note that viscosity is directly proportional to the shear force created between the wall of the pipe and the fluid stream at the *boundary layer*. You should be able to differentiate between kinematic viscosity and dynamic viscosity and know the requirements for using equation 18.33, $v_k = \dfrac{\mu}{\rho}$, to find kinematic viscosity. In this same section we also introduced the Reynolds number, where you saw that this non-dimensional number that related fluid density, viscosity, pipe diameter and flow rate enabled us to determine whether the fluid flow in the pipe was laminar, turbulent or critical, without physically trying to assess the flow from observation.

We next considered, in section 18.7, friction losses in piped systems. Bernoulli's equation was extended to take account of the energy transfers (losses and gains) that will occur in piped fluid systems; this extended version of the Bernoulli equation we introduced as the *general energy equation*. Using this equation and the Darcy equation, we were able to find the losses in piped fluid flow due to friction. When using the Darcy equation to determine the head or pressure energy loss, it was necessary to find the *friction factor* and, dependent on the type of flow, this could be found in two ways. If the fluid flow was laminar (determined by the Reynolds number), then the friction factor could be found directly using the relationship $f = \dfrac{16}{(Re)}$; if the fluid flow was turbulent then we needed to find the friction factor from experiment, the results of which are conveniently displayed on a Moody diagram. You should be able to find the appropriate friction factor by reading off the results from Figure 18.23 or other versions of the diagram. Make sure you understand the circumstances under which you use the relationship

$f = \dfrac{16}{(Re)}$ or $f = \dfrac{64}{N_R}$ when finding the friction factor; Examples 18.18 and 18.19 illustrate their use.

Finally, in section 18.9, we looked briefly at determining energy losses due to friction in bearings. We introduced some common types of lubrication methods, concentrating on hydrodynamic lubrication, which is a particularly efficient method. *Hydrodynamic bearing lubrication* was found to depend directly on shaft speed, the dynamic viscosity of the oil and the load applied to the bearing. You then saw how, using these parameters, Petroff devised a method for quickly and simply obtaining estimates for *friction coefficients* (equation 18.38) that, once known, enable us to use the relationship Power $= 2\pi T_f n$ to find the power losses due to friction in hydrodynamic bearings.

18.10   Review questions

1. A rectangular flat plate 2 m in length and 1.5 m in depth is immersed vertically in water so that its top edge along its length is positioned parallel to and 0.75 m below the surface of the water. Given that the density of water is 1000 kg/m³, calculate the thrust force that acts on the plate.

2. A circular plate of diameter 2 m is immersed in fresh water with density 1000 kg/m³. The plane of the plate makes an angle of 30° with the surface of the water, with the *centre* of the circular plate being at a depth of 2.5 m vertically below the surface of the water. Using the appropriate value of *I* from Table 18.1, find the total thrust force that acts on the surface and the position of the centre of pressure with respect to one side of the plate.

3. Define metacentre and explain how the stability of a floating body depends on the position of the metacentre in relationship to the centre of gravity of the body.

4. A set of lock gates is inclined so that each lock gate, of breadth 6 m, is at an angle $\theta = 20°$ to the normal of its adjacent side when the gates are closed. The water is at a depth of 6 m on one side of the closed gates and 3 m on the other. If the hinges on each gate are positioned 0.5 m and 7 m from the bottom of the lock, determine: a) the magnitude and position of the resultant water force on *each* gate, b) the magnitude and direction of the resultant force on the top and bottom hinges.

5. A cylindrical buoy 2 m high, 1.5 m diameter, floating in sea water with density 1025 kg/m^3, has a mass of 900 kg. Find the metacentric height of the buoy and comment on your result.

6. A jet of water with a velocity of 56 m/s strikes a stationary flat plate at right angles and exerts a force of 480 N on it. If the relative density of the water is 1000 kg/m^3, what is the diameter of the jet?

7. Air enters an inclined venturi tube with a velocity of 50 m/s at a static pressure of 800 N/m^2. If the loss in height in the venturi is equal to 500 mm and the static pressure at the *throat* is −800 N/m^2, then, taking the density of air $\rho = 1.213$ kg/m^2, find the velocity at the throat of the venturi tube.

8. The reducer in a water pipe reduces the pipe diameter from 150 mm to 75 mm. The gauge pressure in the 150 mm diameter pipe is 350 kN/m^2 and the flow is 0.085 m^3/s. Find the force exerted on the reducer.

9. Figure 18.27 shows a wind tunnel set-up in a laboratory. The wind tunnel is circular in cross-section with a diameter upstream of the test section of 6 m and a test section diameter of 3 m. The test section is vented to atmosphere. Air flows through the wind tunnel with a velocity, at the test (working) section, of 250 km/h (kilometres per hour).

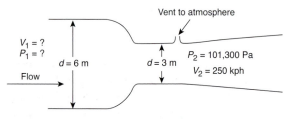

Figure 18.27 Wind tunnel set-up

Given that atmospheric pressure is 101300 Pa and the mass density of mercury is 13600 kg/m^3, find:

a) the upstream section velocity,

b) the upstream pressure and

c) the height of the mercury column being used to regulate the wind tunnel speed.

Note that the mercury column height results from the measured difference between the upstream and working section pressures.

10. Explain the nature of viscosity. Your answer should differentiate between dynamic and kinematic viscosity and include the effects on viscosity of temperature change, for both liquids and gases.

11. The dynamic viscosity of an engine oil is 0.25 N s/m^2. Determine its kinematic viscosity, given that its density is 750 kg/m^3.

12. Water at 20°C flows in a pipe at an average velocity of 3.5 m/s. If the internal diameter of the pipe is 120 mm, determine whether the flow is laminar or turbulent.

13. Find the *friction factor* if water is flowing at 10 m/s in a cast iron pipe having an inside diameter of 2.5 cm. The relative roughness of the pipe may be taken as approximately equal to 100.

14. Use the Darcy equation to determine the energy loss due to the flow of water in question 13.

15. A 100 mm diameter shaft is supported by a bearing 80 mm in length, with a radial clearance of 0.08 mm. The dynamic viscosity of the lubricating oil may be taken as 50×10^{-3} N s/m^2. If the shaft rotates at 900 rpm and carries a load of 6000 N, determine estimates for the bearing coefficient of friction and the power loss.

Part IV

Electrostatics and electromagnetism

Part IV provides an introduction to electrical science in the form of two chapters, one dealing with electrostatics and the other with electromagnetism. These two chapters are essential reading for anyone lacking a formal background in electrical science. The first of the two chapters begins by explaining the nature of electric charge and how it is quantified, then describes the ways that it can be concentrated and stored in capacitors. The chapter also introduces the important concept of the electric dipole that exists where two charges of equal magnitude but opposite polarity are placed in close proximity to one another. In the second of the two chapters we examine the comparable phenomenon that occurs when a magnetic dipole is created by a bar magnet having north and south poles at sits opposite extremities.

We explain the concepts of magnetic flux and how it is quantified, as well as describing ways in which it can be shaped and concentrated in inductors.

Electric and magnetic fields are supported by the medium in which they exist. In Chapter 19 we show that the ability of a particular medium to support electric charge is a key parameter in the specification of the dielectric material used in capacitors which act as repositories for electric charge, storing potential electrical energy in the presence of an electric field. In Chapter 20 we show the equivalent electromagnetic effect, where the ability of a particular medium to support magnetic flux is crucial in determining its ability to store potential electrical energy in the presence of a magnetic field.

Chapter 19

Electrostatics and capacitors

Electric charge is all around us. Indeed, many of the everyday items that we use in the home and at work rely for their operation on the existence of electric charge and the ability to make that charge do something useful. Electric charge is also present in the natural world and anyone who has experienced an electric storm cannot fail to have been awed by its effects. We begin this chapter by explaining what electric charge is and how it is quantified, before moving on to describe the ways in which electric charge can be concentrated and stored in capacitors.

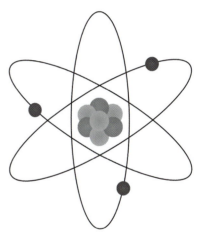

Figure 19.1 The Bohr model of the atom

19.1 The nature of electric charge

As you already know, all matter is made up of atoms or groups of atoms (*molecules*) bonded together in a particular way. In order to understand the nature of electrical charge we need to consider a simple model of the atom. This model, known as the Bohr model (see Figure 19.1), shows a single atom consisting of a central nucleus with orbiting electrons.

Within the nucleus there are *protons* which are *positively charged* and *neutrons* which, as their name implies, are electrically neutral and *have no charge*. Orbiting the nucleus are *electrons that have a negative charge, equal in magnitude (size) to the charge on the proton*. These electrons are approximately two thousand times lighter than the protons and neutrons in the nucleus.

In a stable atom the numbers of protons and electrons are equal, so that overall, the atom is neutral and has no charge. However, if we rub two particular materials together, electrons may be transferred from one to another. This alters the stability of the atom, leaving it with a net positive or negative charge. When an atom within a material *loses electrons* it becomes positively charged and is known as a *positive ion*; when an atom *gains an electron* it has a surplus negative charge and so is known as a *negative ion*. These differences in charge can cause what we know as *electrostatic* effects. For example, combing your hair with a nylon comb may result in a difference in charge between your hair and the rest of your body, resulting in your hair standing on end when your hand or some other differently charged body is brought close to it.

The number of electrons occupying a given orbit within an atom is predictable and is based on the position

978-1-85617-775-7, Engineering Science, Mike Tooley & Lloyd Dingle

of the element within the periodic table. The electrons in all atoms sit in a particular position (shell) dependent on their energy level. Each of these shells within the atom is filled by electrons from the nucleus outwards, as shown in Figure 19.2. The first, inner most, of these shells can have up to two electrons, the second shell can have up to eight and the third up to 18.

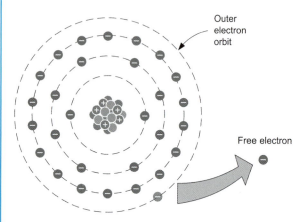

Figure 19.2 A material with a loosely bound electron in its outer shell

> **Key Point.** In a stable atom the number of positively charged protons present in the nucleus is equal to the total number of negatively charged electrons present in the shells that surround the nucleus. These electrons are in constant motion

If an electrical conductor has a deficit of electrons, it will exhibit a net positive charge. If, on the other hand, it has a surplus of electrons, it will exhibit a net negative charge. An imbalance in charge can be produced by friction (removing or depositing electrons using materials such as silk and fur, respectively) or induction (by attracting or repelling electrons using a second body which is respectively positively or negatively charged).

In order to conduct an electric current a material must contain charged particles. In solids (such as copper, lead, aluminium and carbon) it is the negatively charged electrons that are in motion. In liquids and gases, the current is carried by the part of a molecule that has acquired an electric charge. These are called *ions* and they can possess either a positive or a negative charge. Examples include *hydrogen ions* (H^+), *copper ions* (Cu^{++}) and *hydroxyl ions* (OH^-). It is worth noting that pure distilled water contains no ions and is thus a poor conductor of electricity, whereas salt water contains ions and is therefore a relatively good conductor of electricity.

> **Key Point.** An isolated charge will have positive polarity when there is a deficit of electrons and negative polarity when there is an excess of electrons

> **Key Point.** The flow of electric current in liquids and gases is made possible by means of positively or negatively charged molecules called ions. In a vacuum, current flow is made possible by means of a moving stream of negatively charged electrons, as in the cathode ray tube

19.2   Permittivity, electric flux density and field strength

Permittivity is the ability of a medium (such as air, glass or a ceramic material) to support the presence of an electric field. Since electric fields can be created in free space or vacuum, free space exhibits permittivity (i.e. the ability to support the existence of an electric field within it). The permittivity of free space is a fundamental physical constant which is used as the reference against which the permittivity of other materials is usually compared.

19.2.1   Electric flux density

The electric flux density (or *charge density*) is proportional to the magnitude of the electric charge (Q) and inversely proportional to the area (A) over which the charge is distributed (see Figure 19.3). The electric flux density is given by:

$$D = \frac{Q}{A}$$

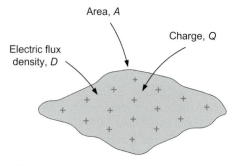

Figure 19.3 Electric flux density

where D is the electric flux density (in C/m^2), Q is the charge (in coulombs, C) and A is the area over which the charge is distributed (in m^2). Note that one coulomb of charge is equivalent to the charge carried by 6.25×10^{18} electrons.

Example 19.1 A charge of $40\,\mu$C appears on an aluminium disc having a diameter of 320 mm. Determine the charge density.

Solution

First we need to find the area of the disc using:

$$A = \pi r^2 = \pi \left(\frac{d}{2}\right)^2 = \pi \left(\frac{0.32}{2}\right)^2$$

$$= 3.142 \times 0.16^2 = 0.0804\,\text{m}^2$$

Now we can find the charge density using:

$$D = \frac{Q}{A}$$

where $Q = 40\,\mu$C and $A = 0.0804\,\text{m}^2$.
Thus:

$$E = \frac{40 \times 10^{-6}}{0.0804} = 497.5 \times 10^{-6}\,\text{C/m}^2$$

$$= 497.5\,\mu\text{C/m}^2.$$

Key Point. Electric flux density is the amount of electric charge present per unit area

Test your knowledge 19.1

A metal plate has dimensions 20 mm $\times$ 30 mm and receives a charge of $45\,\mu$C. What is the charge density on the plate?

19.2.2 Electric field strength

The electric field strength, E, is the ratio of applied electric potential, V, to distance, d, and is expressed in terms of volts per metre (V/m).

The strength of an electric field (E) is proportional to the applied potential difference and inversely proportional to the distance between the two conducting

surfaces (see Figure 19.4). The electric field strength is given by:

$$E = \frac{V}{d}$$

where E is the electric field strength (in V/m), V is the applied potential difference (in V) and d is the distance (in m).

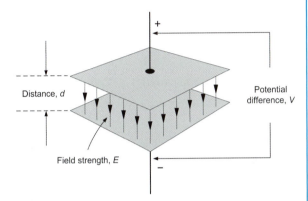

Figure 19.4 Electric field strength

Example 19.2 Two parallel copper plates are separated by a distance of 25 mm. Determine the electric field strength if the plates are connected to a 600 V DC supply.

Solution

The electric field strength will be given by:

$$E = \frac{V}{d}$$

where $V = 600$ V and $d = 25$ mm $= 0.025$ m.
Thus:

$$E = \frac{600}{0.025} = 24000\,\text{V/m} = 24\,\text{kV/m}$$

Example 19.3 The electric field strength between two parallel plates in a cathode ray tube is 18 kV/m. If the plates are separated by a distance of 21 mm, determine the potential difference that exists between the plates.

Solution

The electric field strength will be given by:

$$E = \frac{V}{d}$$

Rearranging this formula to make V the subject gives:

$$V = E \times d$$

Now $E = 18\,\text{kV/m} = 18000\,\text{V/m}$ and $d = 21\,\text{mm} = 0.021\,\text{m}$, thus:

$$V = 18000 \times 0.021 = 378\,\text{V}$$

Key Point. Electric field strength is the amount of applied electric potential per unit distance

Test your knowledge 19.2

Two parallel copper plates are separated by a distance of 100 mm. If the potential difference between the plates is 200 V, what will the electric field strength be?

Test your knowledge 19.3

The electric field between two parallel plates is 56 kV/m. If the plates are separated by a distance of 4 mm, determine the potential difference between the plates.

Test your knowledge 19.4

Two parallel copper plates are separated by an insulating material and connected to a 12 V battery. What must the thickness of the material be in order to produce an electric field strength of 4 kV/m?

19.2.3 Permittivity

Permittivity is defined as the ratio of electric flux density, D, to the applied electric field strength, E.

Hence:

$$\varepsilon = \frac{D}{E} = \frac{\left(\dfrac{Q}{A}\right)}{\left(\dfrac{V}{d}\right)}$$

where D is the electric flux density (in C/m^2), E is the electric field strength (in V/m), Q is the charge (in C), V is the applied potential difference (in V) and A is the area over which the charge is distributed (in m^2).

19.2.4 Permittivity of free space

The permittivity of free space, ε (also referred to as *vacuum permittivity* or *electric constant*) is normally given as 8.854×10^{-12}. In other words, a charge of 8.854×10^{-12} C would be produced when a potential difference of 1 V appears across the two opposite conducting surfaces of a one metre cube of free space, as shown in Figure 19.5.

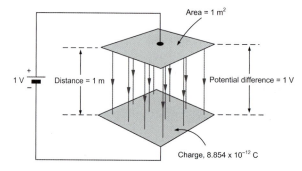

Figure 19.5 Permittivity of free space

The permittivity of free space, ε_0, permeability of free space, μ_0, and the speed of light, c, are related by the expression:

$$c = \frac{1}{\sqrt{\varepsilon_0 \mu_0}}$$

This implies that the speed of light is the reciprocal of the geometric mean of the electric and magnetic constants (respectively the permittivity and permeability of free space). This should not be an altogether surprising result since we know that electromagnetic waves (e.g. light and radio) travel through free space at the speed of light (measured as 3×10^8 m/s). We can use this value of c and the value of ε_0 in order to arrive at a value for the magnetic constant, μ_0, as follows:

$$c^2 = \frac{1}{\varepsilon_0 \mu_0}$$

Hence:

$$\mu_0 = \frac{1}{\varepsilon_0 c^2} = \frac{1}{8.854 \times 10^{-12} \times \left(3 \times 10^8\right)^2}$$

$$= \frac{10^{-4}}{8.854 \times 9} = 12.57 \times 10^{-7}$$

We shall be meeting μ_0 again in the next chapter.

> **Key Point.** Permittivity is the ability of a medium to support the presence of an electric field

19.2.5 Relative permittivity

It is often convenient to specify the permittivity of a particular dielectric material relative to that of air or free space, hence:

$$\varepsilon = \varepsilon_0 \times \varepsilon_r$$

where ε is the *absolute permittivity* of the material (in F/m), ε_0 is the permittivity of free space (8.854×10^{-12} F/m) and ε_r is the *relative permittivity* of the dielectric material.

Some typical dielectric materials and relative permittivities are given in the table below:

Dielectric material	Relative permittivity (free space = 1)
Air	1.0006 (i.e. 1!)
Aluminium oxide	7
Ceramic materials	15 to 500
Epoxy resin	4.0
Glass	5 to 10
Mica	3 to 7
Paper	2 to 2.5
Polystyrene	2.4 to 2.7
Polythene	2.2
Porcelain	6 to 7
Rubber	7
Teflon	2.1
Vacuum (or 'free space')	1

> **Key Point.** The permittivity of an insulating dielectric material is usually specified relative to the permittivity of free space

Test your knowledge 19.5

Two circular metal discs having a diameter of 80 mm are separated by a distance of 3.5 mm in a vacuum. If a potential of 250 V is applied to the discs, determine:

(a)   the electric field strength,

(b)   the electric flux density,

(c)   the charge present on the plates.

19.3   Force between charges

Consider two small charged bodies of negligible weight suspended in space as shown in Figure 19.6. If the two bodies have charges with the same polarity (i.e. either both positively or both negatively charged) the two bodies will move apart, indicating that a force of repulsion exists between them. If, on the other hand, the charges on the two bodies are unlike (i.e. one positively charged and one negatively charged) the two bodies will move together, indicating that a force of attraction exists between them. From this we can conclude that *like charges repel* and *unlike charges attract*.

> **Key Point.** Charges with the same polarity repel one another whilst charges with opposite polarity will attract one another

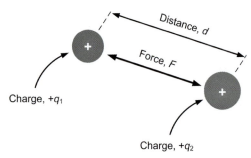

(a) Charges with the same polarity (repulsion)

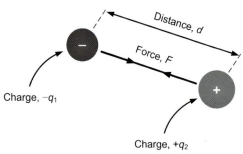

(b) Charges with the opposite polarity (attraction)

Figure 19.6 Force between charged bodies

19.3.1 Coulomb's law

Coulomb's law states that, if charged bodies exist at two points, the force of attraction (if the charges are of opposite charge) or repulsion (if of like charge) will be proportional to the product of the magnitude of the charges divided by the square of their distance apart. Thus:

$$F = \frac{kq_1q_2}{d^2}$$

where q_1 and q_2 are the charges present at the two points (in coulombs), d the distance separating the two points (in metres), F is the force (in Newtons), and k is a constant depending upon the medium in which the charges exist.

In vacuum or *free space*, $k = \dfrac{1}{4\pi\varepsilon_0}$

where ε_0 is the permittivity of free space (8.854×10^{-12} C/N m^2).

Combining the two previous equations gives:

$$F = \frac{q_1q_2}{4\pi\varepsilon_0 d^2}$$

or

$$F = \frac{q_1q_2}{4\pi \times 8.854 \times 10^{-12} \times d^2}\,\text{N}$$

If this formula looks complex, there are only a couple of things that you need to remember. The denominator simply consists of a constant ($4\pi \times 8.854 \times 10^{-12}$) multiplied by the square of the distance d. Thus we can rewrite the formula as:

$$F \propto \frac{q_1q_2}{d^2}$$

where the symbol $\propto$ denotes proportionality.

Example 19.4 Two charged particles are separated by a distance of 25 mm. Calculate the force between the two charges if one has a positive charge of $0.25\,\mu\text{C}$ and the other has a negative charge of $0.4\,\mu\text{C}$. What will the relative direction of the force be?

Solution

Now $F = \dfrac{q_1q_2}{4\pi \times 8.854 \times 10^{-12} \times d^2}$

where $q_1 = 0.25\,\mu\text{C} = 0.25 \times 10^{-6}\,\text{C}$, $q_2 = 0.4\,\mu\text{C} = 0.4 \times 10^{-6}\,\text{C}$ and $d = 2.5\,\text{mm} = 2.5 \times 10^{-3}$ m.

Thus,

$$F = \frac{0.25 \times 10^{-6} \times 0.4 \times 10^{-6}}{4\pi \times 8.854 \times 10^{-12} \times (2.5 \times 10^{-3})^2}$$

$$= \frac{0.1 \times 10^{-12}}{4\pi \times 8.854 \times 10^{-12} \times 6.25 \times 10^{-6}}$$

or

$$F = \frac{0.1}{4\pi \times 8.854 \times 6.25 \times 10^{-6}}$$

$$= \frac{0.1}{695.39 \times 10^{-6}} = 1.438 \times 10^2$$

Hence $F = 1.438 \times 10^2\,\text{N} = 143.8\,\text{N}$ (attraction).

Example 19.5 Two charged particles have the same positive charge and are separated by a distance of 10 mm. If the force between them is 0.1 N, determine the charge present.

Solution

Now $F = \dfrac{q_1q_2}{4\pi \times 8.854 \times 10^{-12} \times d^2}$

where $F = 0.1\,\text{N}$, $d = 0.01\,\text{m}$ and $q_1 = q_2 = q$.

Thus, $0.1 = \dfrac{q \times q}{4\pi \times 8.854 \times 10^{-12} \times (0.01)^2}$

Rearranging the formula to make q the subject gives:

$$q^2 = 0.1 \times 4\pi \times 8.854 \times 10^{-12} \times (0.01)^2$$

or

$$q = \sqrt{0.1 \times 4\pi \times 8.854 \times 10^{-12} \times (0.01)^2}$$
$$= \sqrt{4\pi \times 8.854 \times 10^{-17}}$$

hence

$$q = \sqrt{4\pi \times 8.8854 \times 10^{-17}}$$
$$= \sqrt{111.263 \times 10^{-17}} = \sqrt{11.1263 \times 10^{-16}}$$

thus

$$q = \sqrt{11.1263} \times \sqrt{10^{-16}}$$
$$= 3.336 \times 10^{-8}\,\text{C} = 3.336\,\mu\text{C}$$

Test your knowledge 19.6

List the factors that determine the force that exists between two isolated charges.

Test your knowledge 19.7

Two charges are separated in air by a distance of 1 mm. If the distance increases to 2 mm whilst the magnitude of the charges remains unchanged, by how much will the force between them change?

Test your knowledge 19.8

Two charged particles have identical positive charge and are separated in air by a distance of 2 mm. If the force between them is 0.4 N, determine the charge on each particle.

SCILAB

The force between two charged bodies can be easily calculated using a simple SCILAB script. The input dialog and output message boxes are shown in Figures 19.7 and 19.8, respectively.]

```
// Force between charges q1 and q2
// SCILAB source file: ex1901.sce
// Get data from dialog box
txt = ['q1 in uC';'q2 in uC';'d in mm'];
var = x_mdialog('Enter charge and distance',
txt,['1';'1';'1']);
q1 = 1E-6 * evstr(var(1));
q2 = 1E-6 * evstr(var(2));
d = 1E-3 * evstr(var(3));
// Permittivity of free space e0 in F/m
e0 = 8.854E-12;
// Calculate force between charges
F = (q1 * q2) / (4 * %pi * e0 * d^2);
// Display result
if F > 0 then
messagebox("Force (repulsion) = " + string(F)
+ " Newton");
else
messagebox("Force (attraction)= " + string(F)
+ " Newton");
end if
```

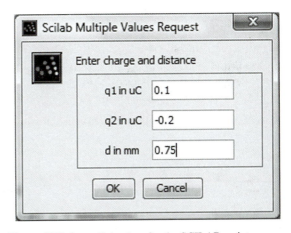

Figure 19.7 Input dialog box for the SCILAB script

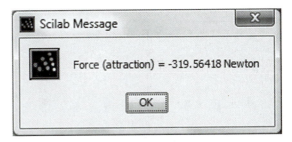

Figure 19.8 Output message box for the SCILAB script

19.3.2 Electric fields

The force exerted on a charged particle is a manifestation of the existence of an electric field. The electric field defines the direction and magnitude of a force on a charged object. The field itself is invisible to the human eye but can be drawn by constructing lines which indicate the motion of a free positive charge within the field; the number of field lines in a particular region being used to indicate the relative strength of the field at the point in question.

Figures 19.9 and 19.10 show the electric fields between isolated unlike and like charges, whilst Figure 19.11 shows the field which exists between two charged parallel metal plates (note the *fringing* that occurs at the edges of the plates).

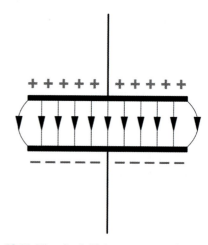

Figure 19.11 Electric field between two charged parallel metal plates

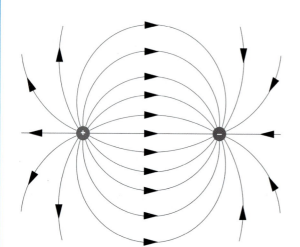

Figure 19.9 Electric field between isolated unlike charges

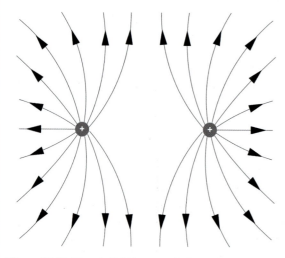

Figure 19.10 Electric field between isolated like charges

Test your knowledge 19.9

Sketch the electric field pattern that would exist between the arrangement of charges shown in Figure 19.12.

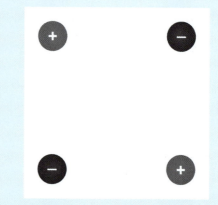

Figure 19.12 See Test your knowledge 19.9

19.4   Capacitors

Figure 19.11 shows the electric field that exists between two charged metal plates. This arrangement forms a simple *capacitor*. The quantity of electricity, Q, that can be stored in the electric field between the capacitor plates is proportional to the applied voltage, V, and the capacitance, C, of the capacitor. Thus:

$$Q = CV$$

where Q is the charge (in coulombs), C is the capacitance (in Farads) and V is the potential difference (in volts).

The relationship between Q and V is linear, as shown in Figure 19.13. The slope of this relationship (when Q is plotted against V) gives the capacitance. A steep slope corresponds to a high capacitance whilst a shallow slope corresponds to a low capacitance.

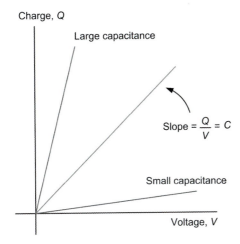

Figure 19.13 Charge plotted against voltage

The relationship between Q, C and V can be rearranged to make voltage or capacitance the subject, as shown below:

$$V = \frac{Q}{C} \text{ and } C = \frac{Q}{V}$$

Key Point. Capacitance is defined as the ratio of electric charge to electric potential for a capacitor

Example 19.6 A $10\,\mu F$ capacitor is charged to a potential of 250 V. Determine the charge stored.

The charge stored will be given by:

$$Q = CV = 10 \times 10^{-6} \times 250 = 2{,}500 \times 10^{-6}$$

$$= 2.5 \times 10^{-3} = 2.5\,\text{mC}$$

Example 19.7 A charge of $11\,\mu C$ is held in a 220 nF capacitor. What potential appears across the plates of the capacitor?

To find the voltage across the plates of the capacitor we need to rearrange the equation to make V the subject, as follows:

$$V = \frac{Q}{C} = \frac{11 \times 10^{-6}}{220 \times 10^{-9}} = 50\,\text{V}$$

Test your knowledge 19.10

A potential of 150 V is applied to a $220\,\mu F$ capacitor. What charge is present on the plates of the capacitor?

Test your knowledge 19.11

A charge of $350\,\mu C$ is to be placed on the plates of a capacitor of 470 nF. What voltage is needed to do this?

Test your knowledge 19.12

A charge of $6.5\,\mu C$ is produced when a potential of 1.2 kV is applied to two metal plates. What is the capacitance of this arrangement?

19.4.1   Capacitance

The amount of capacitance provided by a capacitor depends upon the physical dimensions of the capacitor (i.e., the size of the plates and the separation between them) and the dielectric material between the plates. The capacitance of a conventional parallel plate capacitor is given by:

$$C = \frac{\varepsilon_0 \varepsilon_r A}{d}$$

where C is the capacitance (in Farads), ε_0 is the *permittivity* of free space, ε_r is the *relative permittivity* (or *dielectric constant*) of the dielectric medium between the plates, A is the area of the plates (in square metres) and d is the separation between the plates (in metres). Recall that the *permittivity of free space*, ε, is 8.854×10^{-12} F/m.

Example 19.8 Two parallel metal plates each of area $0.2\,\text{m}^2$ are separated by an air gap of $1\,\text{mm}$. Determine the capacitance of this arrangement.

Here we must use the formula:

$$C = \frac{\varepsilon_0 \varepsilon_r A}{d}$$

where $A = 0.2\,\text{m}^2$, $d = 1 \times 10^{-3}\,\text{m}$, $\varepsilon_r = 1$, and $\varepsilon_0 = 8.854 \times 10^{-12}\,\text{F/m}$.

Hence:

$$C = \frac{8.854 \times 10^{-12} \times 1 \times 0.2}{1 \times 10^{-3}} = \frac{1.7708 \times 10^{-12}}{1 \times 10^{-3}}$$

$$= 1.7708 \times 10^{-9}\,\text{F} = 1.7708\,\text{nF}$$

Example 19.9 A capacitor of $1\,\text{nF}$ is required. If a dielectric material of thickness $0.5\,\text{mm}$ and relative permittivity 5.4 is available, determine the required plate area.

Rearranging the formula $C = \frac{\varepsilon_0 \varepsilon_r A}{d}$ to make A the subject gives:

$$A = \frac{Cd}{\varepsilon_0 \varepsilon_r} = \frac{1 \times 10^{-9} \times 0.5 \times 10^{-3}}{8.854 \times 10^{-12} \times 5.4}$$

$$= \frac{0.5 \times 10^{-12}}{47.811 \times 10^{-12}} = 0.0105\,\text{m}^2$$

thus $A = 0.0105\,\text{m}^2$ or $105\,\text{cm}^2$

Test your knowledge 19.13

A parallel plate capacitor has square plates of side $250\,\text{mm}$ separated by an air gap of $1.8\,\text{mm}$. Determine the capacitance of this arrangement.

Test your knowledge 19.14

A capacitor of $1.5\,\text{nF}$ is to be constructed from aluminium foil separated by a polystyrene dielectric having thickness of $0.1\,\text{mm}$ and relative permittivity 2.5. What area of foil is required?

SCILAB

The capacitance of a parallel plate capacitor can be easily calculated using a simple SCILAB script. The input dialog and output message boxes are shown in Figures 19.14 and 19.15, respectively.

```
// Capacitance of a parallel plate capacitor
// SCILAB source file: ex1902.sce
// Get data from dialogue box
txt = ['Relative permittivity of the
dielectric';'Area of plates in sq.
cm';'Plate separation in mm'];
var = x_mdialog('Enter relative
permittivity, plate area and
separation',txt,['15';'10';'1']);
er = evstr(var(1));
A = 1E-4 * evstr(var(2));
d = 1E-3 * evstr(var(3));
// Permeability of free space e0 in F/m
e = 8.854E-12;
// Calculate capacitance
C = (e * er * A) / d;
C = C * 1E+9
// Display result
messagebox("Capacitance = " + string(C)
+ " nF")
```

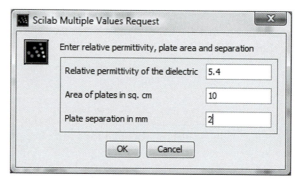

Figure 19.14 Input dialog box for the SCILAB script

Figure 19.15 Output message box for the SCILAB script

19.4.2   Multi-plate capacitors

In order to increase the capacitance of a capacitor, many practical components employ multiple plates (see Figure 19.16), in which case the capacitance is then given by:

$$C = \frac{\varepsilon_0 \varepsilon_r (n-1) A}{d}$$

where C is the capacitance (in Farads), ε_0 is the permittivity of free space, ε_r is the relative permittivity of the dielectric medium between the plates, n is the number of plates, A is the area of the plates (in square metres) and d is the separation between the plates (in metres).

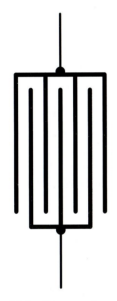

Figure 19.16 A multiple-plate capacitor

Example 19.10   A capacitor consists of six plates each of area $20\,\text{cm}^2$ separated by a dielectric of relative permittivity 4.5 and thickness 0.2 mm. Determine the capacitance of the capacitor.

Using $C = \dfrac{\varepsilon_0 \varepsilon_r (n-1) A}{d}$ gives:

$$C = \frac{8.854 \times 10^{-12} \times 4.5 \times (6-1) \times (20 \times 10^{-4})}{0.2 \times 10^{-3}}$$

$$= \frac{3984.3 \times 10^{-16}}{2 \times 10^{-4}} = 1992.15 \times 10^{-12}\,\text{F}$$

$$= 1.992\,\text{nF}$$

Test your knowledge 19.15

A parallel plate capacitor comprises 15 plates each having an area of $400\,\text{mm}^2$ and separated by polyester film of thickness 0.08 mm and relative permittivity 2.4. What capacitance does it have?

Test your knowledge 19.16

Write and test a SCILAB script that will calculate the capacitance of a multi-plate capacitor.

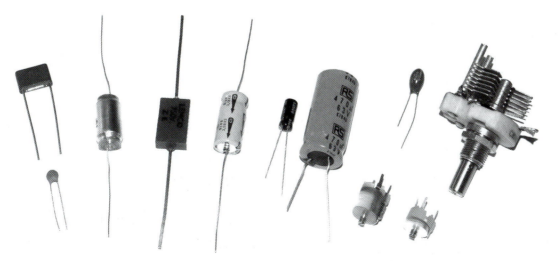

Figure 19.17 A selection of capacitors with values between 10 pF and 1000 μF and working voltage between 25 V and 400 V

19.5 Energy storage

The area under the linear relationship between Q and V that we met earlier in Figure 19.13 gives the energy stored in the capacitor. This area is shown shaded in Figure 19.18. By virtue of its triangular shape the area under the line is $\frac{1}{2}QV$ thus:

$$\text{Energy stored, } W = \frac{1}{2}QV$$

Combining this with the earlier relationship, $Q = CV$, gives:

$$W = \frac{1}{2}QV = \frac{1}{2} \times CV \times V = \frac{1}{2}CV^2$$

where W is the energy (in Joules), C is the capacitance (in Farads) and V is the potential difference (in volts).

This new relationship shows us that the energy stored in a capacitor is proportional to the product of the capacitance and the square of the potential difference between its plates.

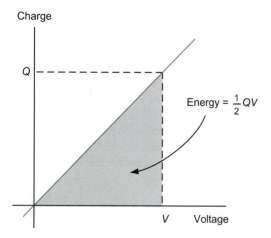

Figure 19.18 Energy stored in a capacitor

Example 19.11 A 100 μF capacitor is charged from a 20 V supply. How much energy is stored in the capacitor?

The energy stored in the capacitor will be given by:

$$W = \frac{1}{2}CV^2 = \frac{1}{2} \times 100 \times 10^{-6} \times (20)^2$$
$$= 50 \times 400 \times 10^{-4} = 20000 \times 10^{-4} = 2\,\text{J}$$

Example 19.12 A capacitor of 47 μF is required to store energy of 40 J. Determine the potential difference required to do this.

To find the potential difference (voltage) across the plates of the capacitor, we need to rearrange the equation to make V the subject, as follows:

$$V = \sqrt{\frac{2W}{C}} = \sqrt{\frac{2 \times 40}{47 \times 10^{-6}}} = \sqrt{\frac{80}{47} \times 10^6}$$
$$= \sqrt{1.702 \times 10^6} = 1.3 \times 10^3 = 1.3\,\text{kV}$$

Key Point. The energy stored in a capacitor is proportional to the square of the applied potential difference

Test your knowledge 19.17

A 220 μF capacitor is charged to a potential of 50 V. What energy is stored?

Test your knowledge 19.18

A 1.2 kV power supply is to be fitted with a capacitor that will store 0.5 J of energy. What value of capacitor is required?

19.6 Capacitors in series and parallel

In order to obtain a particular value of capacitance, fixed capacitors may be arranged in either series or parallel. Consider Figure 19.19, where C is the equivalent capacitance of the three capacitors (C_1, C_2 and C_3) connected in series.

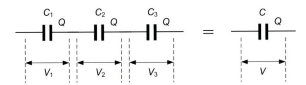

Figure 19.19 Three capacitors in series

The applied voltage, V, will be the sum of the voltages that appear across each capacitor. Thus:

$$V = V_1 + V_2 + V_3$$

Now, for each capacitor, the potential difference, V, across its plates will be given by the ratio of charge, Q, to capacitance, C. Hence:

$$V = \frac{Q}{C}, V_1 = \frac{Q_1}{C_1}, V_2 = \frac{Q_2}{C_2} \text{ and } V_3 = \frac{Q_3}{C_3}$$

Combining these equations gives:
In the series circuit the same charge, Q, appears across each capacitor, thus:

$$Q = Q_1 = Q_2 = Q_3$$

Hence:

$$\frac{Q}{C} = \frac{Q}{C_1} + \frac{Q}{C_2} + \frac{Q}{C_3}$$

from which:

$$\frac{1}{C} = \frac{1}{C_1} + \frac{1}{C_2} + \frac{1}{C_3}$$

When two capacitors are connected in series the equation becomes:

$$\frac{1}{C} = \frac{1}{C_1} + \frac{1}{C_2}$$

This can be arranged to give the slightly more convenient expression:

$$C = \frac{C_1 \times C_2}{C_1 + C_2}$$

Hence the equivalent capacitance of two capacitors connected in series can be found by taking the *product* of the two capacitance values and *dividing* it by the *sum* of the two capacitance values (in other words, *product over sum*).

Now consider Figure 19.20, where C is the equivalent capacitance of three capacitors (C_1, C_2 and C_3) connected in parallel.

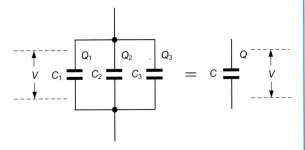

Figure 19.20 Three capacitors in parallel

The total charge present, Q, will be the sum of the charges that appear in each capacitor. Thus:

$$Q = Q_1 + Q_2 + Q_3$$

Now, for each capacitor, the charge present, Q, will be given by the product of the capacitance, C, and potential difference, V. Hence:

$$Q = CV, Q_1 = C_1V_1, Q_2 = C_2V_2 \text{ and } Q_3 = C_3V_3$$

Combining these equations gives:

$$CV = C_1V_1 + C_2V_2 + C_3V_3$$

In the parallel circuit the same voltage, V, appears across each capacitor, thus:

$$V = V_1 = V_2 = V_3$$

Hence:

$$CV = C_1V + C_2V + C_3V$$

From which:

$$C = C_1 + C_2 + C_3$$

When two capacitors are connected in parallel the equation becomes:

$$C = C_1 + C_2$$

Example 19.13 Capacitors of $2.2\,\mu F$ and $6.8\,\mu F$ are connected (a) in series and (b) in parallel. Determine the equivalent value of capacitance in each case.

(a)   Here we can use the simplified equation for *just two* capacitors connected in series:

$$C = \frac{C_1 \times C_2}{C_1 + C_2} = \frac{2.2 \times 6.8}{2.2 + 6.8}$$

$$= \frac{14.96}{9} = 1.66\,\mu F$$

(b)   Here we use the formula for two capacitors connected in parallel:

$$C = C_1 + C_2 = 2.2 + 6.8 = 9\,\mu F$$

Example 19.14 Capacitors of $2\,\mu F$ and $5\,\mu F$ are connected in series across a $100\,V$ DC supply. Determine: (a) the charge on each capacitor and (b) the voltage dropped across each capacitor.

(a)   First we need to find the equivalent value of capacitance, C, using the simplified equation for two capacitors in series:

$$C = \frac{C_1 \times C_2}{C_1 + C_2} = \frac{2 \times 5}{2 + 5} = \frac{10}{7} = 1.43\,\mu F$$

Next we can determine the charge (note that, since the capacitors are connected in series, the *same* charge will appear in each capacitor):

$$Q = CV = 1.43 \times 100 = 143\,\mu C$$

(b)   In order to determine the voltage dropped across each capacitor we can use:

$$V = \frac{Q}{C}$$

Hence, for the $2\,\mu F$ capacitor:

$$V_1 = \frac{Q}{C_1} = \frac{1.43 \times 10^{-6}}{2 \times 10^{-6}} = 71.5\,V$$

Similarly, for the $5\,\mu F$ capacitor:

$$V_2 = \frac{Q}{C_2} = \frac{1.43 \times 10^{-6}}{5 \times 10^{-6}} = 28.5\,V$$

We should now find that the total voltage ($100\,V$) applied to the series circuit is the sum of the two capacitor voltages, i.e.:

$$V = V_1 + V_2 = 71.5 + 28.5 = 100\,V$$

Test your knowledge 19.19

Three $22\,\mu F$ capacitors are connected in parallel and an additional $100\,\mu F$ capacitor is connected in series with the parallel combination. Determine the effective capacitance of this arrangement and the voltage drop that will appear across each capacitor when the series/parallel arrangement is connected to a $120\,V$ DC supply.

19.7   Chapter summary

In this chapter we explained that protons and electrons each carry a small electric charge; protons carry a single unit of positive charge whilst electrons carry a single unit of negative charge. The charge present on a single electron or a single proton is extremely small and it takes 6.25×10^{18} of these tiny charges to produce one coulomb of electric charge. If a body has exactly the same amount of positive and negative charge the net charge on the body will be zero.

Within the space surrounding an isolated charge or charged body a field will exist within which a force will be experienced by another isolated charge or charged body. Hence electrically charged bodies exert a force on one another. Like charges (i.e. charges with the same sign) repel one another while unlike charges (i.e. charges of different sign) will attract one another. The force that exists between charged bodies is proportional to the product of the charges and inversely proportional to the square of the distance between them.

Free space supports the existence of an electric field in the same way as other insulating dielectric materials such as ceramic and polystyrene. The ability of a particular material or medium to support electric charge is known as permittivity, a key parameter in the specification of the dielectric material used in capacitors.

Capacitors are invaluable electrical components that act as a repository for electric charge and store potential electrical energy in the form of an electric field between two conducting plates spaced by an insulating dielectric.

19.8   Review questions

Short answer questions

1. Show that the ratio of electric charge to applied potential (i.e. Q/V) for a 1 m³ volume of free space is equal to 8.854 pF.

2. What charge must be applied to a square metal plate having sides of 150 mm in order to produce a charge density of 2.5 C/m²?

3. If the plates of a capacitor are separated by a distance of 0.5 mm, what potential difference must be applied in order to produce an electric field strength of 25 kV/m?

4. Two charged particles have identical but opposite charges of 7.5 μC. If the charged particles are separated by a distance of 150 mm, determine the magnitude and direction of force acting on the particles.

5. Sketch the electric field pattern that would exist between the charges shown in Figure 19.21.

6. A charge of 960 μC appears on the plates of an 8 μF capacitor. What potential difference will appear across the capacitor's plates?

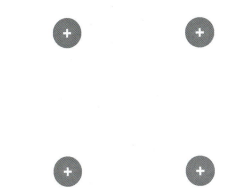

Figure 19.21 See Question 5

7. When 240 V is applied to the plates of a capacitor a charge of 60 μC appears on its plates. What is the value of the capacitor?

8. Two parallel metal plates measuring 10 cm by 10 cm are separated by a dielectric material having a thickness of 0.5 mm and a relative permittivity of 5.4. Determine the capacitance of this arrangement.

9. A 220 μF capacitor is required to store 11 mJ of energy. What voltage needs to appear across the plates of the capacitor?

10. Determine the equivalent capacitance of each of the networks of capacitors shown in Figure 19.22.

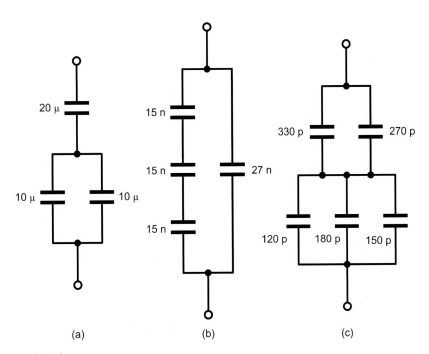

(a)   (b)   (c)

Figure 19.22 See Question 10

Long answer questions

11. A $5\,\mu F$ capacitor is charged to a potential of 100 V and then connected to an uncharged $2\,\mu F$ capacitor.

    (a) What will the final potential difference be?

    (b) What energy is stored in the system before and after the parallel connection of the two capacitors?

    (c) State any assumptions that you made in (b).

12. A capacitor consists of two square metal plates having sides of 200 mm separated by an air gap of 2 mm. This arrangement is charged to a potential of 100 V. After a short time the charging voltage is removed and a sheet of glass having a thickness of 2 mm and relative permittivity of 5.6 is inserted in the space between the plates. Determine:

    (a) the capacitance of the arrangement with and without the glass dielectric plate,

    (b) the potential difference that appears across the plates following the insertion of the glass dielectric plate. State any assumptions made.

13. Capacitors of $3\,\mu F$, $3\,\mu F$ and $5\,\mu F$ are available.

    (a) Determine the different values of capacitance that can be produced for all possible series/parallel arrangements of these components.

    (b) If all of the capacitors are rated at 100 V maximum, determine the maximum voltage that can be safely applied to each of the arrangements in (a).

Chapter 20

Electromagnetism and inductors

In the previous chapter we introduced the concept of charge and the electric fields that surround charges of similar or opposite polarity when placed in close proximity to one another. This latter arrangement of dissimilar charges is referred to as an *electric dipole*. A comparable phenomenon exists in magnetism, where a *magnetic dipole* is created by a bar magnet having north and south poles at its opposite extremities. Like electric charge, magnetic flux is present in the natural world and anyone who has used a compass for navigation will have benefited from its effects. We begin this chapter by explaining what magnetic flux is and how it is quantified, before moving on to describe the ways in which flux can be shaped and concentrated in inductors.

20.1 The nature of magnetic flux

If we could break up a bar magnet into tiny fragments, each of these fragments would behave as a *magnetic dipole*, not as a collection of isolated poles. In other words, a large magnet can be thought of as being made up from a large number of tiny magnetic dipoles where the north and south poles are inseparable, as shown in Figure 20.1. If we could further break up these tiny fragments into the electrons and nuclei that constitute its atoms, we would find that even these elementary particles behave as magnetic dipoles.

Magnetism is a phenomenon associated with the spinning motion of the fundamental charge-carrying

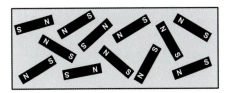

(a)   Magnetic dipoles randomly orientated. No overall field produced.

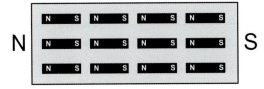

(b)   Magnetic dipoles aligned. Field lines produced in the surrounding space.

Figure 20.1 A bar magnet can be thought of as containing a very large number of tiny magnetic dipoles

978-1-85617-775-7, Engineering Science, Mike Tooley & Lloyd Dingle

particles. It is important to distinguish this from the normal orbital motion, which does not relate to magnetic properties.

Magnetic flux permeates the space that surrounds a permanent magnet. We cannot see this flux but we know that it exists because of its effects, notably the force that is experienced by a ferromagnetic material or that acts on a current-carrying conductor when suspended in a magnetic field.

To help us visualise magnetic flux we construct field lines, just as we did for the electrostatic charges that we met in the last chapter. These field lines follow the path that an imaginary free north pole would take, as shown in Figure 20.2.

> **Key Point.** Unlike electric charges (which can be isolated), magnetic poles can never be separated. In other words, magnetic poles always exist in pairs, one of which is a north pole and the other a south pole.

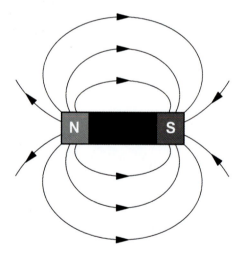

Figure 20.2 Magnetic field lines around a bar magnet

Figure 20.3 shows the magnetic field that exists around a conductor arranged to form a coil. The magnetic flux is concentrated in the centre of the coil and the magnetic field takes the same shape as that which would be produced by a bar magnet, as shown earlier in Figure 20.2. This arrangement forms a simple inductor.

With most coils and inductors we are not concerned so much with the total flux produced but more with how dense or concentrated the flux is. To increase

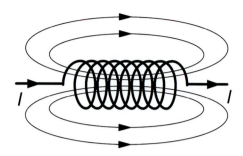

Figure 20.3 Magnetic field lines around a coil

the flux density we can introduce a ferromagnetic core material into the centre of a coil, as shown in Figure 20.4.

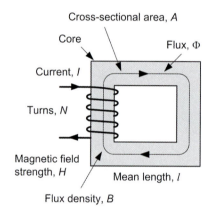

Figure 20.4 Use of a ferromagnetic core to concentrate flux

20.1.1 Magnetic field strength

The magnetic field strength, H, is the ratio of the product of the applied electric current, I, and the number of turns, N, to the perpendicular distance, d.

The magnetic field strength is given by:

$$H = \frac{NI}{l}$$

where H is the magnetic field strength (in A/m), A is the applied current (in A) and l is the mean length of the magnetic path (in m). Note that because 'turns' has no units, magnetic field strength should be expressed in terms of ampere per metre (A/m), but in some books and references you may find this given as '*ampere-turns per metre*'.

At this point it is well worth comparing magnetic field strength with electric field strength that we met in the previous chapter.

Quantity	Electric field strength	Magnetic field strength
Relationship	$E = \dfrac{V}{d}$	$H = \dfrac{NI}{l}$
Units	V/m	A/m

Example 20.1 A 400-turn coil is wound on a closed toroidal magnetic core having a mean diameter of 200 mm. What current should be applied to the coil in order to produce a magnetic field strength of 500 A/m?

Solution

First we need to find the mean length of the magnetic path using:

$$l = \pi d = 3.142 \times 200 \times 10^{-3} = 0.628\,\text{m}$$

Now we can find the magnetic field strength using:

$$H = \frac{NI}{l}$$

where $N = 400$, $l = 0.628$ m and $H = 500$ A/m

Thus:

$$I = \frac{H \times l}{N} = \frac{500 \times 0.628}{400} = 0.79\,\text{A}$$

Test your knowledge 20.1

A closed toroid is to be designed so that it produces a magnetic field strength of 2.5 kA/m when supplied with a current of 15 A. Determine the number of turns required if the mean length of the magnetic path is 300 mm.

20.2 Permeability and magnetic flux density

Ferromagnetic materials readily support the passage of magnetic flux. Such materials include iron, steel or ferrite (see page 365). Since magnetic fields can be created in free space or vacuum, free space exhibits permeability. The permeability of free space is a fundamental physical constant which is used as the reference against which the permeability of other materials is usually compared. Later on we shall explain this in more detail.

20.2.1 Permeability

Permeability is the ability of a medium (e.g. a material such as iron, steel or ferrite) to support the presence of a magnetic flux.

For any given magnetic medium, the ratio of flux density, B, to magnetic field strength, H, is a constant. This constant is known as the permeability of the medium and it is represented by the symbol μ. Hence:

$$\mu = \frac{B}{H}$$

Figure 20.5 shows the linear relationship that exists between B and H for some typical magnetic materials. It is important to note that there is a limit to the density of flux that these materials can support, and we refer to this as *saturation*. When this occurs, any further

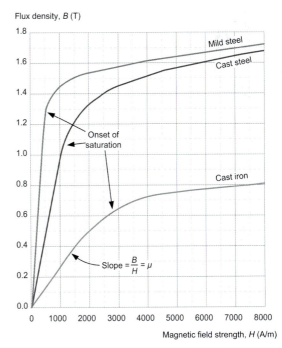

Figure 20.5 Relationship between B and H for some typical magnetic materials

increase in magnetic field strength (*magnetising force*) produces no significant further increase in flux density.

SCILAB

The relationship between magnetic flux density and magnetic field intensity can be easily illustrated using a simple SCILAB script. The graphical output produced by SCILAB for two typical sets of data is shown in Figure 20.6.

```
// Plot of B against H
// Filename: ex2001.sce
// Clear any existing plot
clf();
// Enter values for (in H/m) as a column
vector
h = [0;500;1000;1500;2000;2000;3000;3500;
4000];
// Enter corresponding values for B1 (in mT)
as a column vector
b1 = [0;450;890;1240;1405;1475;1505;1515;
1520;];
// Enter corresponding values for B2 (in mT)
as a column vector
b2 = [0;300;550;800;1005;1150;1205;1225;
1235];
// Plot data using crosses
plot2d(h,b1,style=-2);
// Plot data using circles
plot2d(h,b2,style=-3);
// Create a larger H-axis dataset
hi=0:25:4000;
// Generate interpolated data for B1
b1d=splin(h,b1);
b1i=interp(hi,h,b1,b1d); // hi and b1i are
the new data values
// Generate interpolated data for B2
b2d=splin(h,b2);
b2i=interp(hi,h,b2,b2d); // hi and b2i are
the new data values
// Plot interpolated data for B1
plot2d(hi,b1i,style=5);
// Plot interpolated data for B2
plot2d(hi,b2i,style=2);
// Print legend, grid and title
h = legend(['B1'; 'B2'],4, boxed=%f);
xgrid(); // add grid
xtitle('Flux density plotted against
magnetic field intensity');
```

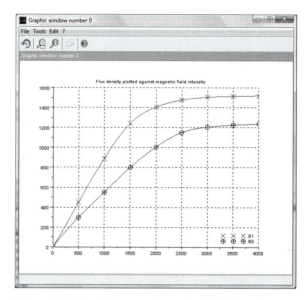

Figure 20.6 Graphical output from the SCILAB script

20.2.2 Magnetic flux density

The density of the magnetic flux (or *flux density*) is proportional to the magnitude of the magnetic flux (B) and inversely proportional to the area (A) over which the flux is distributed (see Figure 20.4). The magnetic flux density is given by:

$$B = \frac{\Phi}{A}$$

where B is the magnetic flux density (in T), Φ is the flux (in Wb) and A is the area (in m^2) over which the flux acts.

Example 20.2 Determine the flux density when a flux of 25 μT is present in a cylindrical ferrite rod which has a diameter of 15 mm.

Solution

First we need to find the cross-sectional area over which the flux acts:

$$A = \pi r^2 = \pi \left(\frac{d}{2}\right)^2 = \pi \left(\frac{15 \times 10^{-3}}{2}\right)^2$$

$$= 3.142 \times 56.25 \times 10^{-6} = 192.4 \times 10^{-6} \, \text{m}^2$$

Using $B = \dfrac{\Phi}{A}$, where $\Phi = 25 \times 10^{-6}$ and $A = 192.4 \times 10^{-6}$, gives:

$$B = \frac{\Phi}{A} = \frac{25 \times 10^{-6}}{192.4 \times 10^{-6}} = 0.13 \, \text{T}$$

Example 20.3 A 120-turn coil is tightly wound over a non-magnetic ring having a mean circumference of 300 mm and cross-sectional area 440 m². Calculate the total flux present when a current of 2.5 A is applied to the coil.

Solution

We can find the magnetic field strength using:

$$H = \frac{NI}{l}$$

where $N = 120$, $l = 0.3$ m and $I = 2.5$ A.

Thus:

$$H = \frac{N \times I}{l} = \frac{120 \times 2.5}{0.3} = 1000 \, \text{A/m}$$

Now, rearranging $\mu_0 = \dfrac{B}{H}$ gives $B = \mu_0 H$

Hence:

$$B = \mu_0 H = 4\pi \times 10^{-7} \times 1000$$

$$= 12.57 \times 10^{-4} = 1.257 \, \text{mT}$$

Finally, rearranging $B = \dfrac{\Phi}{A}$ gives $\Phi = BA$

Hence:

$$\Phi = 1.257 \times 10^{-3} \times 440 \times 10^{-6} = 553.1 \times 10^{-9}$$

$$= 0.553 \times 10^{-6} = 0.553 \, \mu\text{Wb}$$

20.2.3   Permeability of free space

The permeability of free space (also referred to as the *vacuum permeability* or *magnetic constant*), μ, is normally given as 12.57×10^{-7}.

Figure 20.7(a) shows a long straight conductor carrying a current of 1 A. The magnetic field lines around the conductor will take the form of a series of concentric circles but, to keep things as simple as possible, we've shown just one of these lines. The direction of current flow can be determined from the right-hand rule illustrated in Figures 20.8 and 20.9. Note that the '×' symbol on the two conductors conventionally indicates current flowing 'out of the paper' whereas a '•' symbol would indicate current flowing in the opposite direction, i.e. 'into the paper'.

Returning to Figure 20.7, the length of the magnetic path represented by this field line will be 2π m (i.e. π times the diameter of the field line shown in Figure 20.7a) and the magnetic field intensity will thus be $1/2\pi$ A/m.

Applying the relationship between flux density and magnetic field intensity gives:

$$\mu_0 = \frac{B}{H}$$

where μ_0 is the permeability of free space (also referred to as *vacuum permeability* or *magnetic constant*). μ is normally given as 12.57×10^{-7} H/m.

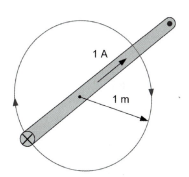

(a) A single current-carrying conductor

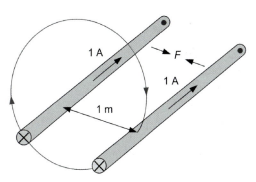

(b) Two current-carrying conductors

Figure 20.7 Field around a straight current-carrying conductor

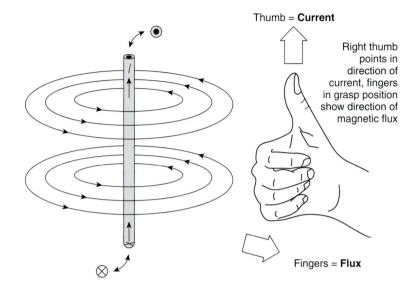

Figure 20.8 Right-hand rule for determining the direction of the field lines around a current-carrying conductor

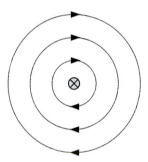

(a) Current flowing into the paper
(i.e. away from you)

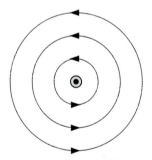

(b) Current flowing out of the paper
(i.e. towards you)

Figure 20.9 Labelling convention for the direction of current flow

The magnetic flux present in Figure 20.7(a) will be given by:

$$B = \mu_0 H = \mu_0 \times \frac{1}{2\pi} \text{ T}$$

In Figure 20.5(b) we have introduced a second conductor also carrying a current of 1 A but placed at exactly 1 m from the first conductor. The force on this conductor can be determined from:

$$F = BIl$$

where F is the force in Newton (N), B is the flux density in tesla (T), I is the current in ampere (A) and l is the length (m).

Hence the force per unit length acting on the conductor will be:

$$F = \mu_0 \times \frac{1}{2\pi} \text{ N}$$

from which:

$$\mu_0 = 2\pi F$$

From the definition of the ampere, the force acting between the two conductors in Figure 20.7(b) will be 2×10^{-7} N, hence:

$$\mu_0 = 2\pi \times 2 \times 10^{-7} = 4\pi \times 10^{-7} = 12.57 \times 10^{-7}$$

Once again, it is worth noting that the permeability of free space μ_0, the permittivity of free space, ε_0, and the speed of light c are related by the expression:

$$c = \frac{1}{\sqrt{\varepsilon_0 \mu_0}}$$

20.2.4   Relative permeability

It is often convenient to specify the permeability of a particular magnetic material, μ, relative to that of air or free space, hence:

$$\mu = \mu_0 \times \mu_r$$

where μ is the *absolute permeability* of the material (in H/m), μ_0 is the permeability of free space (12.57×10^{-7} H/m) and μ_r is the *relative permeability* of the core material.

Some typical core materials and their relative permeabilities are given in the table below:

Core material	Relative permeability (Free space = 1)
Vacuum or air	1
Ferrite	20 to 10000
Cast iron	100 to 250
Mild steel	200 to 800
Cast steel	300 to 900
Silicon iron	1000 to 5000
Mumetal	200 to 5000
Stalloy	500 to 6000

Note that *diamagnetic materials*, such as wood and water, have a relative permeability less than μ_r whilst *ferromagnetic materials*, such as iron and steel, have very high permeability. *Ferrite* materials are metallic oxides that have the characteristics of ceramics. They exhibit exceptional magnetic properties but, unlike materials such as iron and steel, they do not conduct electric current. A notable advantage of ferrite materials is that they can be pressed and extruded into complex shapes during manufacture.

Key Point. Permeability is the ability of a medium to support the presence of a magnetic field.

Key Point. The ampere is defined as the current which, if maintained in two straight parallel conductors of infinite length, of negligible circular cross-section, and placed 1 metre apart in vacuum, would produce between these conductors a force equal to 2×10^{-7} N per metre of length.

Test your knowledge 20.2

A toroidal electromagnet comprises 400 turns wound on a former that has a mean length of 550 mm. What current is required in order to produce a magnetic field strength of 750 A/m?

Test your knowledge 20.3

The data in the table below relates to a sample of a material to be used in the core of an inductor. Plot the *B–H* characteristic for the material and use this to estimate the onset of saturation and the maximum flux density that can be supported. Also determine the relative permeability of the core material when a magnetic field intensity of (a) 700 A/m and (b) 3500 A/m is applied. Explain why the permeability of the material falls rapidly between these two values.

H (A/m)	B (mT)
0	0
500	825
1000	1125
1500	1245
2000	1305
2500	1325
3000	1335
3500	1340
4000	1345

20.3   Force between conductors

By applying what we have just learned about magnetic field intensity, flux density and permeability, we can determine the force acting on the two current-carrying parallel conductors that we met in Figure 20.7(b). In order to do this we must first introduce Ampere's law.

20.3.1   Ampere's law

In the previous chapter we explained Coulomb's law which showed how the magnitude of the *electric field* in space is proportional to the *charge* which causes it. In a similar manner, Ampere's law shows how the *magnetic field* in the space surrounding a current-carrying conductor is directly proportional to the electric current that causes it.

Ampere's law states that for any closed loop path, the sum of the length elements times the magnetic field in the direction of the length element is equal to the *permeability* times the electric current enclosed in the loop.

The product of the flux density, B, and the length of the magnetic path, l, is equal to the product of the permeability, μ, of the magnetic medium, the number of turns and the current flowing, I. Thus:

$$Bl = \mu NI$$

Now $\mu = \dfrac{B}{H}$ and so $Bl = \dfrac{B}{H} NI$

Rearranging the expression in order to make H the subject gives:

$$H = \frac{NI}{l}$$

Applying this relationship for the case shown in Figure 20.7 (where $N = 1$ for the case of a single straight conductor), the force, H, acting on the conductor B due to the current in A will be given by:

$$H = \frac{NI}{l} = \frac{1 \times I_A}{2\pi d} = \frac{I_A}{2\pi d} \text{ A/m}$$

where I_A is the current in conductor A and d is the separation between the two conductors (note that $N = 1$ for a long straight wire).

The magnetic flux density at point B will be given by:

$$B = \frac{\mu_0 I_A}{2\pi d} \text{ T}$$

Thus the force acting on conductor B (per metre length) due to the current flowing in conductor A will be given by:

$$F = BIl = \frac{\mu_0 I_A}{2\pi d} \times I_B \times 1 = \frac{\mu_0 I_A I_B}{2\pi d}$$

$$= \frac{4\pi \times 10^{-7} \times I_A I_B}{2\pi d} = \frac{2 \times 10^{-7} \times I_A I_B}{d} \text{ N/m}$$

By the same reasoning, the force acting on conductor A (per metre length) due to the current flowing in conductor B will be given by:

$$F = BIl = \frac{\mu_0 I_B}{2\pi d} \times I_A \times 1 = \frac{\mu_0 I_B I_A}{2\pi d}$$

$$= \frac{4\pi \times 10^{-7} \times I_B I_A}{2\pi d} = \frac{2 \times 10^{-7} \times I_B I_A}{d} \text{ N/m}$$

Important note: In order to be able to apply Ampere's law in its simple form (see above) we need to know the shape and symmetry of the magnetic field. So, for example, we should only use the relationship to solve problems in which the magnetic field is predictable – for example, the field around a very long straight current-carrying conductor or within a coil tightly wound on a *toroidal* (i.e. doughnut-shaped) core.

Example 20.4 Calculate the flux density at a distance of 10 mm from the centre of a long straight conductor carrying a current of 500 A.

Solution

We can find the magnetic field strength using:

$$H = \frac{NI}{2\pi d}$$

where $N = 1$ (for a long straight wire), $I = 500$ A and $d = 0.01$ m.

Thus:

$$H = \frac{N \times I}{2\pi d} = \frac{1 \times 500}{2\pi \times 0.01} = 79.62 \times 10^2 \text{ A/m}$$

The flux density can be calculated from:

$$B = \mu_0 H = 12.57 \times 10^{-7} \times 79.62 \times 10^2$$

$$= 1000 \times 10^{-5} = 0.01 \text{ T}$$

Example 20.5 A flux of 0.05 Wb is developed in an electric motor over an area of 0.04 m². Calculate the force exerted on a conductor of effective length 0.1 m when a current of 50 A is present.

Solution

Now:

$$B = \frac{\Phi}{A} = \frac{0.05}{0.04} = 1.25 \text{ T}$$

Using $F = BIl$ gives:

$$F = 1.25 \times 50 \times 0.1 = 6.25 \text{ N}$$

List the factors that determine the force that exists between two parallel current-carrying conductors.

A flux density of $400\,\mu$T is produced at a distance of 50 mm from a long straight conductor suspended in air. What current is flowing in the conductor?

Two long air-spaced parallel conductors are separated by a distance of 40 mm between centres. Calculate the force acting on each conductor if they each carry a current of 600 A.

Key Point. Magnetic flux density is the amount of magnetic flux present per unit area.

Key Point. The force acting on a current-carrying conductor is proportional to the density of magnetic flux and the magnitude of the current flowing in the conductor.

SCILAB

The force between two current-carrying conductors in air can be easily calculated using a simple SCILAB script. The input dialog and result message boxes are shown in Figures 20.10 and 20.11, respectively.

```
// Force acting on parallel conductors in
air
// SCILAB source file: ex2502.sce
// Get data from dialog box
txt = ['Ia in A';'Ib in A';'d in mm'];
var = x_mdialog('Enter current in each
conductor and
separation',txt,['75';'75';'5.5']);
ia = evstr(var(1));
ib = evstr(var(2));
d = 1E-3 * evstr(var(3));
```

```
// Calculate force between conductors
F = (2E-7 * ia * ib) / d;
// Display result
messagebox("Force = " + string(F) +
" Newton");
```

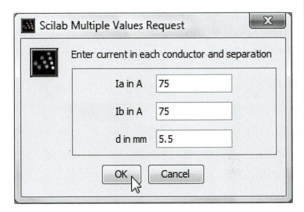

Figure 20.10 Input dialog box for the SCILAB script

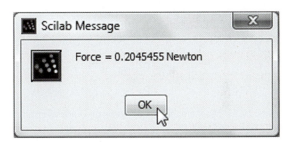

Figure 20.11 Output message box for the SCILAB script

20.3.2 Magnetic fields

The force exerted on a magnetic pole is a manifestation of the existence of a magnetic field. The magnetic field defines the direction and magnitude of a force on a free pole. As with electric fields, the field itself is invisible to the human eye but can be drawn by constructing lines that indicate the motion of a free north pole within the field; the number of field lines in a particular region being used to indicate the relative strength of the field at the point in question.

Figure 20.12 shows the magnetic fields that exist between various arrangements of bar magnets. Similar field patterns would be produced between the equivalent electromagnetic *solenoids* (i.e. coils carrying current). Figure 20.13 shows the field that exists in an air gap.

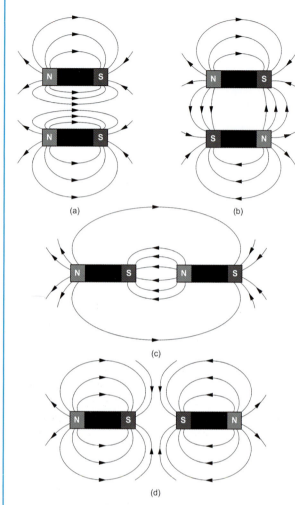

Figure 20.12 Magnetic field patterns around various arrangements of bar magnets

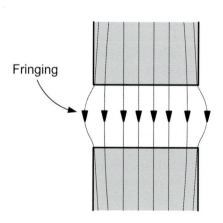

Figure 20.13 Magnetic field in an air gap

Fringing

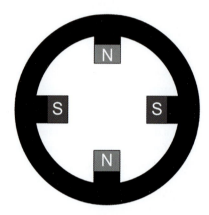

Figure 20.14 See Test your knowledge 20.7

20.4   Inductors

Figure 20.3 shows the magnetic field that exists in the space surrounding a current-carrying coil. This arrangement forms a simple *inductor*. The amount of flux, Φ, that will appear in the magnetic field around the coil is proportional to the applied current, I, and the *inductance*, L, of the inductor. Thus:

$$\Phi \propto LI$$

where Φ is the flux (in Tesla), L is the inductance and I is the applied current (in amps).

The relationship between Φ and I is linear for an ideal magnetic material (including air and 'free space'), as shown in Figure 20.15. The slope of this relationship (when $N\Phi$ is plotted against I) gives the inductance. Hence:

$$L = \frac{N\Phi}{I}$$

where N is the number of turns.

A steep slope (see Figure 20.15) corresponds to a high value of inductance whilst a shallow slope corresponds to a low value of inductance.

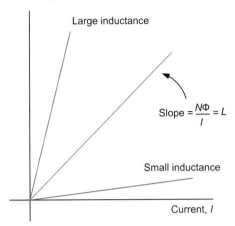

Figure 20.15 Flux plotted against current for an ideal core material

Key Point. Inductance is proportional to the ratio of magnetic flux to applied current.

Example 20.6 Determine the inductance of a 450-turn air-cored toroidal coil if it produces a flux density of 12 μT when a current of 6 A is applied.

The inductance of the coil will be given by:

$$L = \frac{N\Phi}{I} = \frac{450 \times 12 \times 10^{-6}}{6}$$

$$= 900 \times 10^{-6} = 900\,\mu\mathrm{H}$$

20.4.1 Inductance

The inductance of an inductor depends upon the physical dimensions of the inductor (e.g. the length and cross-sectional area over which the magnetic flux acts), the number of turns, and the permeability of the material of the core. For a close-wound inductor where there is no leakage of flux (and no disturbance from any external flux), the inductance of an inductor is given by:

$$L = \frac{N\Phi}{I}$$

Now $\Phi = BA$
Hence

$$L = \frac{NBA}{I}$$

Now $\dfrac{B}{H} = \mu_0\mu_r$, from which $B = \mu_0\mu_r H$, hence:

$$L = \frac{N(\mu_0\mu_r H)A}{I}$$

Now since $H = \dfrac{NI}{l}$ we have:

$$L = \frac{N\mu_0\mu_r\left(\frac{NI}{l}\right)A}{l}$$

from which:

$$L = \frac{\mu_0\mu_r N^2 A}{l}$$

where L is the inductance (in H), μ_0 is the permeability of free space (12.57 $\times$ 10^{-7} H/m), μ_r is the relative permeability of the magnetic core, l is the length of the core (in m), and A is the cross-sectional area of the core (in m^2). It is important to note that the foregoing relationship only holds true for simple close-wound coils in which all of the flux is uniformly concentrated within the coil. A selection of practical inductors with values ranging from 100 nH to 60 mH is shown in Figure 20.16.

Example 20.7 An inductor of 100 mH is required. If a closed magnetic core of length 20 cm, cross-sectional area 15 cm^2 and relative permeability 500 is available, determine the number of turns required.

Solution

Now $L = \dfrac{\mu_0\mu_r n^2 A}{l}$ and hence $n = \sqrt{\dfrac{Ll}{\mu_0\mu_r A}}$

Thus

$$n = \sqrt{\frac{Ll}{\mu_0\mu_r A}}$$

$$= \sqrt{\frac{100 \times 10^{-3} \times 20 \times 10^{-2}}{12.57 \times 10^{-7} \times 500 \times 15 \times 10^{-4}}}$$

$$= \sqrt{\frac{2 \times 10^{-2}}{94275 \times 10^{-11}}} = \sqrt{21215} = 146$$

Hence the inductor requires 146 turns of wire.

Figure 20.16 A selection of inductors with values between 100 nH and 60 mH and working current between 100 mA and 10 A

Test your knowledge 20.8

A 1.5 mH inductor consists of a coil of 400 close-wound turns on a toroidal ferrite ring. Determine the permeability of the ferrite if the ring has a mean diameter of 80 mm and a cross-sectional area of 400 mm².

SCILAB

The inductance of a solenoid can be easily calculated using the following SCILAB script:

```
// Inductance of a solenoid inductor
// SCILAB source file: ex2003.sce
// Get input data from dialogue box
txt = ['Relative permeability of the core';
'Number of turns';'Length of the
magnetic circuit (mm)';'Area of the
core (mm2)'];
var = x_mdialog('Enter relative
permeability, number of turns, length and
area',txt,['500';'1000';'100';'1600']);
ur = evstr(var(1));
n = evstr(var(2));
l = 1E-3 * evstr(var(3));
A = 1E-6 * evstr(var(4));
```

```
// Permeability of free space u0 in F/m
u0 = 12.57E-12;
// Calculate inductance
L = (u0 * ur * n^2 *l)/ A;
L = L * 1E+3;
// Display result
messagebox("Inductance = " + string(L) +
" mH")
```

20.5 Energy storage

The energy stored in an inductor is proportional to the product of the inductance, L, and the square of the current, I, flowing in it. Thus:

$$W = 0.5\,LI^2$$

where W is the energy (in Joules), L is the inductance (in Henry) and I is the current (in amps).

Example 20.8 A current of 1.5 A flows in an inductor of 5 H. Determine the energy stored.

Solution

Now $W = 0.5\,LI^2 = 0.5 \times 5 \times 1.5^2 = 6.620\,\text{J}$

Example 20.9 An inductor of 20 mH is required to store 2.5 J of energy. Determine the current that must be applied to the inductor.

Solution

Now $W = 0.5 LI^2$ and hence

$$I = \sqrt{\frac{W}{0.5L}} = \sqrt{\frac{2.5}{0.5 \times 20 \times 10^{-3}}}$$

$$= \sqrt{\frac{2.5}{0.5 \times 20 \times 10^{-3}}} = \sqrt{2.5 \times 10^2} = 15.8\,\text{A}$$

Key Point. The energy stored in an inductor is proportional to the square of the applied current.

Test your knowledge 20.9

Determine the amount of stored energy when a current of 1.5 A flows in an inductor of 10 H. By how much must the applied current be increased in order to double the amount of energy stored?

Test your knowledge 20.10

An inductor is required to store 550 mJ of energy when a current of 500 mA flows in it. What value of inductance is required?

20.6 Inductors in series

In order to obtain a particular value of inductance, fixed inductors may be arranged in either series or parallel. Consider Figure 20.17, where L is the equivalent inductance of the three inductors (L_1, L_2 and L_3) connected in series.

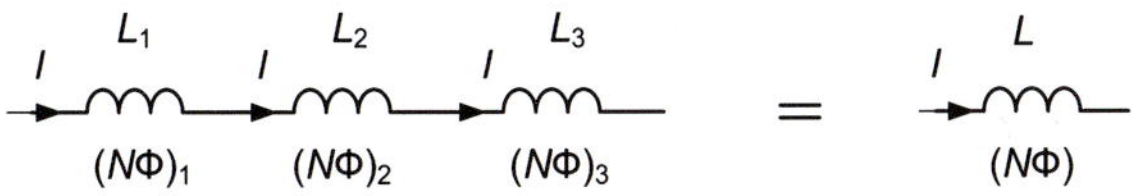

Figure 20.17 Three inductors in series

The applied current, I, will be common to all three components. Thus:

$$I = I_1 = I_2 = I_3$$

Now, for each inductor the product of the number of turns, N, and the flux, Φ (referred to as the *flux linkage*) for each inductor will be given by the expressions:

$$(N\Phi)_1 = L_1 I,\ (N\Phi)_2 = L_2 I \text{ and } (N\Phi)_3 = L_3 I$$

and for the series combination, the equivalent flux linkage:

$$(N\Phi) = LI$$

The total flux linked in the circuit will be the sum of the individual flux linkages:

$$(N\Phi) = (N\Phi)_1 + (N\Phi)_2 + (N\Phi)_3$$

from which:

$$LI = L_1 I + L_2 I + L_3 I$$

Hence:

$$L = L_1 + L_2 + L_3$$

Important note: The foregoing relationship only holds true if there is no coupling of magnetic flux between the inductors. Any coupling of flux would result in mutual inductance and the formula will then no longer be valid.

Example 20.10 Inductors of 2.2 H and 6.8 H are connected in series with a third unknown inductor. Determine the value of the unknown inductance if the inductance of the series combination is 20 H.

Solution

Here we use the formula for three inductors in series:

$$L = L_1 + L_2 + L_3$$

where $L = 20\,\text{H}$, $L_1 = 2.2\,\text{H}$ and $L_2 = 6.8\,\text{H}$.

Rearranging the equation gives:

$$L_3 = L - (L_1 + L_2) = 20 - (2.2 + 6.8) = 11\,\text{H}$$

20.7 Magnetic circuits and reluctance

When designing electromagnetic circuits we usually need to give careful consideration to the provision of a path through which the magnetic flux will flow. In some applications (such as the read/write head of a hard disk drive) we may need to concentrate on the magnetic flux in a gap, as shown in Figure 20.18(a). In others (such as a motor or loudspeaker) we may need to distribute the flux in a radial fashion, as shown in Figure 20.18(b). In either case we need to ensure that there is an effective path that will convey the flux from one point in the magnetic circuit to another. In order to do this we need to ensure that the *reluctance* of the circuit is as low as possible (you can think of reluctance as being analogous to electrical resistance).

The reluctance, S, of a magnetic circuit is directly proportional to its length, l, and inversely proportional to the product of its permeability, μ, and its cross-sectional area, A. Hence:

$$S = \frac{l}{\mu A} = \frac{l}{\mu_0 \mu_r A}$$

Now we know that $B = \dfrac{\Phi}{A}$, $\dfrac{B}{H} = \mu_0 \mu_r$ and $H = \dfrac{NI}{l}$. Combining these three relationships gives:

$$\Phi = BA = \mu_0 \mu_r HA = \mu_0 \mu_r A \frac{NI}{l}$$

$$= \frac{\mu_0 \mu_r A}{l} \times NI = \frac{1}{S} \times NI$$

Thus:

$$\Phi = \frac{NI}{S}$$

Making NI the subject of this yields the important result:

$$NI = S\Phi$$

Note that NI is often referred to as *magnetomotive force* (*m.m.f.*). This 'driving force' is analogous to the *electromotive force* that we will meet in the next chapter.

Example 20.11 A toroidal core of mean diameter 60 mm and cross-sectional area 96 mm² has a 0.5 mm radial air gap cut in it. If the relative permeability of the core material is 1200, determine the number of turns that need to be wound on the core in order to produce a flux of 0.05 mWb when a current of 500 mA is applied to the coil.

Solution

First we need to find the reluctance of each part of the magnetic circuit. We will do this separately for the toroidal core and for the air gap.

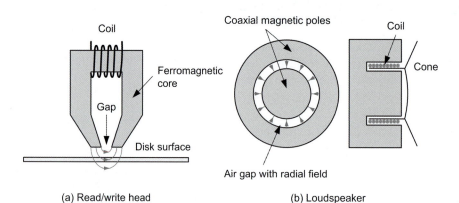

(a) Read/write head   (b) Loudspeaker

Figure 20.18 Some practical applications of electromagnetism

For the toroidal core we have:

$$S = \frac{l}{\mu_0 \mu_r A} = \frac{3.142 \times 60 \times 10^{-3}}{1200 \times 12.57 \times 10^{-7} \times 96 \times 10^{-6}}$$
$$= 130.2 \times 10^4$$

For the air gap we have:

$$S = \frac{l}{\mu_0 \mu_r A} = \frac{3.142 \times 0.5 \times 10^{-3}}{1 \times 12.57 \times 10^{-7} \times 96 \times 10^{-6}}$$
$$= 1302 \times 10^4$$

The total reluctance in the circuit will be the sum of these two individual reluctances, hence:

$$S_T = S_{core} + S_{air\ gap} = \left(130.2 \times 10^4\right)$$
$$+ \left(1302 \times 10^4\right) = 1432.2 \times 10^4$$

Next we can determine the number of turns required, using the relationship:

$$NI = S\Phi$$

Hence:

$$N = \frac{S\Phi}{I} = \frac{1432.2 \times 10^4 \times 0.05 \times 10^{-3}}{500 \times 10^{-3}}$$
$$= 0.1432 \times 10^4 = 1432 \text{ turns}$$

SCILAB

SCILAB can be used to perform calculations relating to m.m.f., flux and reluctance, as shown in the following sample script:

```
// Flux produced by an inductor
// SCILAB source file: ex2004.sce
// Get input data from dialogue box
txt = ['Mean length of the magnetic path
(mm)';'Cross-sectional area of the
core (mm2)';'Relative permeability of the
core';'Number of turns';'Current
(A)'];
var = x_mdialog('Enter parameters for the
inductor:',txt,['150';'100';'4800';'200';
'0.5']);
l = 1E-3*evstr(var(1));
```

```
a = 1E-6*evstr(var(2));
p = evstr(var(3));
n = evstr(var(4));
i = evstr(var(5));
// Calculate reluctance of the magnetic path
s = l/(12.57*1E-7*p*a);
// Calculate flux produced
phi =1E+3*(n * i)/s
// Display result
messagebox("Flux produced = " + string(phi)
+ " mWb")
```

Test your knowledge 20.12

Determine the reluctance of a closed magnetic circuit having mean length 240 mm, cross-sectional area 450 mm^2 and relative permeability 800.

Test your knowledge 20.13

A closed toroidal core has a mean diameter of 140 mm and a cross-sectional area of 250 mm^2. If the core has a relative permeability of 1100 and is wound with a 500-turn coil, determine the current that would need to be applied to the coil in order to produce a flux of 0.75 mWb.

20.8   Chapter summary

In this chapter we introduced the basic concepts of electromagnetism and magnetic circuits. We explained the nature of magnetic flux and showed how the magnetic field strength at a point depends on the applied electric current and the number of turns.

Within the space surrounding a current-carrying conductor, a field will exist within which a force will be experienced by an isolated magnetic pole. Hence magnetised bodies exert a force on one another. The force that acts on a current-carrying conductor is directly proportional to the product of the magnetic flux density and the current flowing in the conductor.

Free space supports the existence of a magnetic field in the same way as many other materials.

The ability of a particular material or medium to support magnetic flux is known as *permeability*, a key parameter in the specification of the core material used for inductors, transformers and other electromagnetic devices. *Ferromagnetic materials*, such as iron, steel and ferrite, can have very high permeability whilst *diamagnetic materials*, such as wood and water, have a relative permeability less than that of air or vacuum.

Inductors provide us with another means of storing energy. They act as a repository for magnetic flux, with energy stored in the form of a magnetic field permeating the space around a current-carrying conductor.

20.9   Review questions

Short answer questions

1. Show that the ratio of magnetic flux to applied current (i.e. Φ/I) for a 1 m³ volume of free space is equal to 1.257 μH.

2. A 500-turn coil is wound on a closed toroidal core having a mean length of 440 mm. What magnetic field strength will be produced when a current of 150 mA flows in the coil?

3. A closed toroidal inductor is to be designed so that a magnetic field strength of 500 A/m is generated when a current of 7.5 A is applied. If the mean circumference of the toroid is 440 mm, determine the number of turns required.

4. Two parallel bus bars, each 1 m in length are 150 mm apart. Determine the force between them when they each carry a current of 200 A.

5. Two overhead cables have to carry a maximum current of 200 A. What is the minimum distance that they must be separated if the maximum force between them is not to exceed 12 mN/m?

6. Which one of the field patterns shown in Figure 20.19 is correct?

7. A flux of 0.05 Wb is developed in a electric motor over an area of 0.04 m². Calculate the force exerted on a conductor of effective length 0.1 m if a current of 50 A is present.

8. An inductor has a closed magnetic core of length 400 mm, cross-sectional area 100 mm² and relative permeability 450. Determine the value of inductance if the coil has 250 turns of wire.

9. A current of 12 A flows in a 1.5 H inductor. What energy is stored in the inductor?

10. An inductor of 600 mH is required to store 400 mJ of energy. Determine the current that must be applied to the inductor.

Long answer questions

11. A 750-turn coil is wound on a toroidal core having a mean length of 0.25 m and a cross-sectional area of 0.075 m². If the core material has a permeability of 1600, determine the current required to produce a flux of 0.35 mWb.

12. The following data refer to a sample of a ferromagnetic material:

H (A/m)	B (mT)
0	0
500	400
1000	750
1500	950
2000	1075
2500	1150
3000	1200

Plot the B/H characteristic for the material and use it to determine the relative permeability of

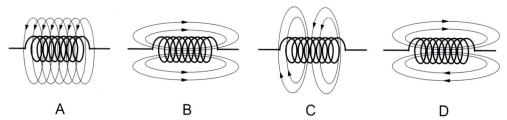

A   B   C   D

Figure 20.19 See Question 6

the material at (a) $H = 400\,\text{A/m}$ and (b) $H = 1200\,\text{A/m}$.

13. An air-cored coil has a length of $0.5\,\text{m}$ and a diameter of $50\,\text{mm}$.

If the coil is wound with 400 turns of wire, determine:

(a) the inductance of the coil;

(b) the energy stored in the field when a current of $10\,\text{A}$ flows in the coil.

Part V

Direct current

Part V provides an introduction to direct current and circuit theorems. In Chapter 21 we introduce direct current electricity and the fundamental nature of electric current, potential difference (voltage) and opposition to current flow (resistance). We also compare electric and magnetic circuits, showing that magnetic flux is analogous to electric current, magnetomotive force is analogous to electromotive force and reluctance of a magnetic circuit is analogous to resistance in an electric circuit.

We introduce the notion of specific resistance and look at how well different materials conduct electric current. We also examine the effect of temperature on resistance and the relationship that exists between electrical power, work and energy. Chapter 21 concludes with a brief mention of the internal 'loss' resistance that many practical components and circuits exhibit. We examine the effect of loss resistance on the voltage, current and power that can be delivered from a source that contains internal resistance.

Finally, Chapter 22 explains, with the aid of a number of worked examples, the use of circuit theorems as an aid to the solution of even the most complex of series/parallel direct current circuits containing multiple sources of e.m.f. and current.

Chapter 21

Current, voltage and resistance

This chapter has been included in order to provide a brief overview of electric current and *Ohm's law* for those who may not have had the benefit of any previous study of electrical principles. The chapter begins with an explanation of electric current as the organised movement of electric charge and shows how the resistance of a conductor depends on the physical dimensions of the conductor and the material of which it is made. If you already have a good understanding of this subject you should move on to the next chapter, which introduces circuit theorems and the solution of simple series and parallel networks of resistors.

21.1 The nature of electric current

Current is simply the rate of flow of electric charge. Thus, if more charge moves in a given time, more current will be flowing. If no charge moves then no current is flowing. The unit of electric current (defined in the previous chapter) is the ampere, A, and one ampere is equal to one coulomb of charge flowing past a point in one second.

Hence:

$$I = \frac{Q}{t}$$

where $t =$ time in seconds and $Q =$ charge in coulombs.

So, for example, if a steady current of 5 A flows for two minutes, then the amount of charge transferred will be:

$$Q = I \times t = 5 \times 120 = 600\,\text{C}$$

Alternatively, if 2400 coulombs of charge are transferred in one minute, the current flowing will be given by:

$$I = \frac{Q}{t} = \frac{2400}{60} = 40\,\text{A}$$

Electric current arises from the flow of electrons (or negative charge carriers) in a metallic conductor. The ability of an energy source (e.g. a battery) to produce a current within a conductor may be expressed in terms of *electromotive force* (e.m.f.). Whenever an e.m.f. is applied to a circuit, a potential difference (p.d.) exists. Both e.m.f. and p.d. are measured in volts (V). In many practical circuits there is *only one* e.m.f. present (the battery or supply), whereas a p.d. will be developed across *each* component present in the circuit.

The *conventional flow* of current in a circuit is from the point of more positive potential to the point of greatest negative potential (note that electrons move in the opposite direction!). *Direct current* results from the application of a direct e.m.f. (derived from batteries or a d.c. supply, such as a generator). An essential characteristic of such supplies is that the applied e.m.f. does not change its polarity (even though its value might be subject to some fluctuation).

978-1-85617-775-7, Engineering Science, Mike Tooley & Lloyd Dingle

For any conductor, the current flowing is directly proportional to the e.m.f. applied. The current flowing will also be dependent on the physical dimensions (length and cross-sectional area) and material of which the conductor is composed. The amount of current that will flow in a conductor when a given e.m.f. is applied is inversely proportional to its resistance. Resistance, therefore, may be thought of as an 'opposition to current flow'; the higher the resistance, the lower the current that will flow (assuming that the applied e.m.f. remains constant).

Example 21.1 A battery charger transfers a charge of 150 C to a battery in two hours. What charging current does this correspond to?

Solution

Here we must use $Q = I \times t$ (where $Q = 150\,\text{C}$ and $t = 2 \times 60 \times 60 = 7200\,\text{s}$). Rearranging the expression gives:

$$I = \frac{Q}{t} = \frac{1500}{7200} = 0.208\,\text{A} = 208\,\text{mA}$$

Test your knowledge 21.1

A power supply delivers a current of 2.5 A to a load. How much charge is transferred to the load in a time of 20 minutes?

Key Point. Electric current is the organised movement of charge carriers (electrons in metallic conductors)

Key Point. The conventional direction of current flow is opposite to that of the movement of charge carriers (electrons) in a metal conductor

Key Point. Electric current will flow from a point of greater electric potential to one with a lesser electric potential. The conventional direction of current flow is opposite to that of the movement of charge carriers (electrons) in a metal conductor

Key Point. Resistance can be thought of as an 'opposition to current flow'; the higher the resistance, the less the current that will flow, and vice versa

21.2 Ohm's law

Provided that temperature does not vary, the ratio of p.d. across the ends of a conductor to the current flowing in the conductor is a constant. This relationship is known as Ohm's law and it leads to the relationship:

$$\frac{V}{I} = \text{a constant} = R$$

where V is the *potential difference* (or voltage drop) in volts (V), I is the current in amps (A) and R is the resistance in ohms (Ω) (see Figure 21.1).

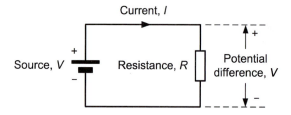

Figure 21.1 Voltage, current and resistance

The relationship between V and I is linear, as shown in Figure 21.2. The slope of this relationship (when V is plotted against I) gives the resistance. A steep slope corresponds to a high resistance whilst a shallow slope corresponds to a low resistance.

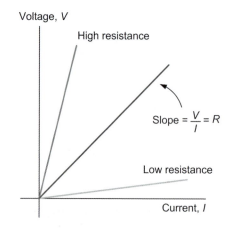

Figure 21.2 Voltage plotted against current

The relationship between V, I and R may be arranged to make V, I or R the subject, as follows:

$$V = I \times R \quad I = \frac{V}{R} \quad \text{and} \quad R = \frac{V}{I}$$

Example 21.2 A current of 0.1 A flows in a 56 Ω resistor. What voltage drop will be developed across the resistor?

Solution

Here we must use $V = I \times R$ (where $I = 0.1$ A and $R = 56\,\Omega$). Hence:

$$V = I \times R = 0.1 \times 56 = 5.6 \text{ V}$$

Example 21.3 An 18 Ω resistor is connected to a 9 V battery. What current will flow in the resistor?

Solution

Here we must use $I = \dfrac{V}{R}$ (where $V = 9$ V and $R = 18\,\Omega$). Hence:

$$I = \frac{V}{R} = \frac{9}{18} = 0.5 \text{ A}$$

Example 21.4 A voltage drop of 15 V appears across a resistor in which a current of 2 mA flows. What is the value of the resistance?

Solution

Here we must use $R = \dfrac{V}{I}$ (where $V = 15$ V and $I = 2 \times 10^{-3}$ A). Hence:

$$R = \frac{V}{I} = \frac{15}{2 \times 10^{-3}} = 7.5 \times 10^3 = 7.5 \text{ k}\Omega$$

Key Point. Ohm's law states that, provided that temperature does not vary, the ratio of p.d. across the ends of a conductor to the current flowing in the conductor is a constant

Key Point. When thinking about electric circuits, it can be useful to think of electromotive force (e.m.f.) as the 'cause' and potential difference (p.d.) as the 'effect'. The latter cannot exist without the former and both are measured in volts (V)

Test your knowledge 21.2

A current of 150 μA flows in a resistor of 22 kΩ. What voltage drop will appear across the resistor?

Test your knowledge 21.3

A 9 V power supply is to be tested at its rated output current of 450 mA. What value of load resistance should be used?

Test your knowledge 21.4

A resistance of 27 kΩ is connected to a 90 V power supply. What current will flow in the resistor?

21.3 Resistance and resistivity

The resistance of a metallic conductor is directly proportional to its length and inversely proportional to its area, as shown in Figure 21.3(a). The resistance is also directly proportional to its *specific resistance* or *resistivity*. The resistance, R, of a conductor is thus given by the formula:

$$R = \frac{\rho l}{A}$$

where R is the resistance (in Ω), ρ is the specific resistance (in Ω m), l is the length (in m) and A is the area in m².

Resistivity

Specific resistance or resistivity is a characteristic of a particular material and is defined as the resistance measured between the opposite faces of a cube having sides of unit length as shown in Figure 21.3(b). Some

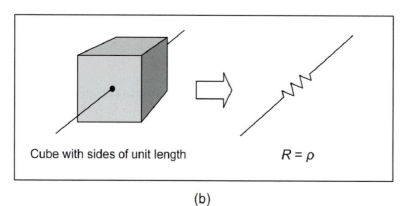

Length	Small
Cross-sectional area	Large
Resistance	Small
Length	Large
Cross-sectional area	Small
Resistance	Large

(a)

Cube with sides of unit length

$R = \rho$

(b)

Figure 21.3 Resistance and resistivity

typical materials used for electrical conductors and their specific resistivities are listed in the table below:

Material	Resistivity (Ω m at 20°C)
Silver	1.59×10^{-8}
Copper	1.68×10^{-8}
Copper (annealed)	1.72×10^{-8}
Gold	2.44×10^{-8}
Aluminium	2.82×10^{-8}
Tungsten	5.60×10^{-8}
Zinc	5.90×10^{-8}
Nickel	6.99×10^{-8}
Iron	1.0×10^{-7}
Platinum	1.06×10^{-7}
Tin	1.09×10^{-7}
Mild steel	1.38×10^{-7}

Material	Resistivity (Ω m at 20°C)
Lead	2.2×10^{-7}
Manganin	4.82×10^{-7}
Constantan	4.9×10^{-7}
Mercury	9.8×10^{-7}
Nichrome	1.10×10^{-6}

Key Point. Resistivity is a measure of how well a material opposes the movement of electric charge. Good conductors (such as silver and copper) have relatively low values of resistivity whilst poor conductors (such as plastics and ceramics) have very high values of resistivity

Example 21.5 A coil consists of an 8 m length of annealed copper wire having a cross-sectional area of 1 mm². Determine the resistance of the coil.

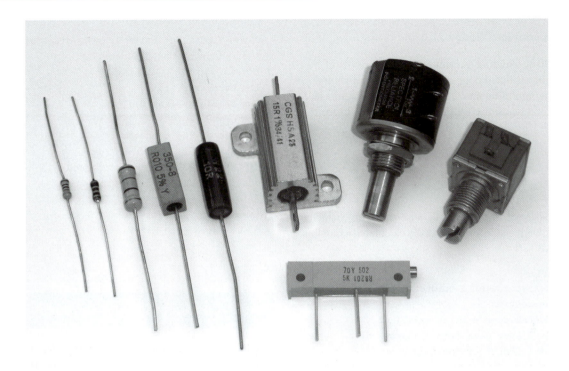

Figure 21.4 A selection of fixed and variable resistors with values between 1 Ω and 10 MΩ and power ratings between ¼ W and 10 W

Solution

Here we will use $R = \dfrac{\rho l}{A}$ (where $l = 8$ m and $A = 1$ mm$^2 = 1 \times 10^{-6}$ m^2)

From the table shown earlier, we find that the value of specific resistance, ρ, for annealed copper is 1.72×10^{-8} Ω m.

Hence:

$$R = \frac{\rho l}{A} = \frac{1.72 \times 10^{-8} \times 8}{1 \times 10^{-6}} = 13.76 \times 10^{-2}$$
$$= 0.1376 \; \Omega$$

Hence the resistance of the wire will be approximately 0.14 Ω.

Example 21.6   A wire having a specific resistance of 1.6×10^{-8} Ω m, length 20 m and cross-sectional area 1 mm^2 carries a current of 5 A. Determine the voltage drop between the ends of the wire.

Solution

First we must find the resistance of the wire and then we can find the voltage drop.

To find the resistance we use:

$$R = \frac{\rho l}{A} = \frac{1.6 \times 10^{-8} \times 20}{1 \times 10^{-6}} = 32 \times 10^{-2} = 0.32 \; \Omega$$

To find the voltage drop we can apply Ohm's law:

$$V = I \times R = 5 \times 0.32 = 1.6 \text{ V}$$

Hence a potential of 1.6 V will be dropped between the ends of the wire.

Test your knowledge 21.5

Determine the specific resistance of the material used for a 10 m length of wire having a diameter of 0.25 mm and a resistance of 5 Ω.

A 200 m length of transmission line consists of a pair of insulated copper conductors. What should be the minimum diameter of the two conductors if the loop resistance (out and back) is not to exceed 5 Ω?

SCILAB

The resistance of a length of wire of given diameter and known resistivity can be easily calculated using this SCILAB script:

```
// Resistance of a conductor having circular
cross-section
// SCILAB source file: ex2101.sce
// Get input data from dialogue box
txt = ['Length of the conductor (m)';
'Diameter of the conductor (mm)'; 'Specific
resistance of the material (x10-8)';];
var = x_mdialog('Enter length and diameter
of the conductor and specific resistance',
txt,['2.5';'0.2';'1.68';]);
l = evstr(var(1));
d = 1E-3*evstr(var(2));
p = 1E-8*evstr(var(3));
// Area of the conductor
a = %pi*((d/2)^2);
// Calculate resistance
r = p*l/a;
// Display result
messagebox("Resistance = " + string(r)
+ " ohm")
```

The resistivity of a material can be calculated from the next SCILAB script:

```
// Resistivity of the material of a
conductor having a circular cross-section
// SCILAB source file: ex2102.sce
// Get input data from dialogue box
txt = ['Length of the conductor (m)';
'Diameter of the conductor (mm)';
'Resistance of the conductor (ohm)';];
```

```
var = x_mdialog('Enter length, diameter and
resistance of the conductor',
txt,['25';'0.5';'2.5';]);
l = evstr(var(1));
d = 1E-3*evstr(var(2));
r = evstr(var(3));
// Area of the conductor
a = %pi*((d/2)^2);
// Calculate specific resistance
p = 1E8*r*a/l;
// Display result
messagebox("Resistivity = " + string(p) +
" ohm-metre (x10-8)")
```

21.4 Conductance and conductivity

Conductance is the inverse of resistance. The conductance, G, of a conductor is thus given by the relationship:

$$G = \frac{1}{R}$$

where G is the conductance (in siemens, S) and R is the resistance (in Ω).

Combining the foregoing relationship with the relationship that we met earlier between resistance and resistivity gives:

$$G = \frac{1}{R} = \frac{1}{\left(\dfrac{\rho l}{A}\right)} = \frac{A}{\rho l}$$

21.4.1 Conductivity

Conductivity is the ability of a material to convey an electric current. It is therefore the same as the reciprocal of the resistivity of the material. Hence:

$$\sigma = \frac{1}{\rho}$$

where σ is the conductivity (in $\Omega^{-1}\,\mathrm{m}^{-1}$) and ρ is the specific resistance (in Ω m). Combining this relationship with the one that we met earlier gives:

$$\sigma = \frac{l}{RA} = \frac{Gl}{A}$$

where σ is the conductivity (in $\Omega^{-1}m^{-1}$), R is the resistance (in Ω), l is the length (in m), G is the conductance (in S) and A is the area in m^2.

Some typical materials used for electrical conductors and their conductivities are listed in the table below:

Material	Conductivity (Sm^{-1} at 20°C)
Silver	6.3×10^7
Copper	6.0×10^7
Copper (annealed)	5.8×10^7
Gold	4.1×10^7
Aluminium	3.5×10^7
Nickel	1.43×10^7
Iron	1.0×10^7
Sea water	4.8
Deionised water	5.5×10^{-6}
Teflon	1×10^{-18}

Example 21.7 A marine electrical component comprises a brass bar having a length of 4.5 m and cross-sectional area 2.8×10^{-4} m^2. Determine the resistance of the bar given that the brass has a conductivity which is 28% that of copper. By how much can the cross-sectional area of the bar be reduced if it is to be replaced by copper?

Solution

The conductivity of brass, $\sigma = 0.28 \times 6.0 \times 10^7 = 1.68 \times 10^7\,\mathrm{S\,m^{-1}}$

The resistance of the bar can be calculated from:

$$R = \frac{l}{\sigma A} = \frac{4.5}{1.68 \times 10^7 \times 2.8 \times 10^{-4}}$$
$$= 0.957 \times 10^{-3} = 0.957\,\mathrm{m\Omega}$$

The reduction in cross-sectional area that could be achieved by replacing the brass bar with one made from copper will be:

$$\text{Reduction factor} = \frac{1}{0.28} = 3.57$$

21.5 Comparison of electric and magnetic circuits

It can be useful to make a comparison of electric circuits with the magnetic circuits that we met in the last chapter. We've summarised the main points in Figure 21.5 and the table below under the notions of 'force', 'opposition', 'flow' and 'ability to support flow':

Quantity	Electric circuit	Magnetic circuit
Force	Electromotive force (V)	Magnetomotive force (A)
	$V = IR$	$NI = S\Phi$
Opposition	Resistance (Ω)	Reluctance (A Wb^{-1})
	$R = \dfrac{\rho l}{A} = \dfrac{V}{I}$	$S = \dfrac{l}{\mu A} = \dfrac{NI}{\Phi}$
Flow	Current (A)	Flux (Wb)
	$I = \dfrac{V}{R}$	$\Phi = \dfrac{LI}{N}$
Ability to support flow	Conductivity ($\Omega^{-1}m^{-1}$)	Permeability (H m^{-1})
	$\sigma = \dfrac{l}{RA}$	$\mu = \dfrac{l}{SA}$

Note that, despite the similarities noted above, there are several important differences between electric and magnetic circuits. Firstly, because electric current requires a conductive path in order to flow, there are is no leakage or fringing path in an electric circuit that would be equivalent to the escape of flux from a magnetic circuit. Secondly, magnetic circuits are inherently non-linear due to the shape of the B–H characteristic of most ferromagnetic materials.

21.6 Temperature coefficient of resistance

You may have noticed from our definition of Ohm's law that the resistance of a resistor depends on the temperature. For most metallic conductors, resistance

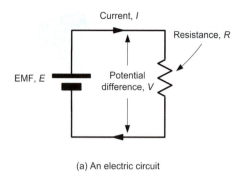

(a) An electric circuit

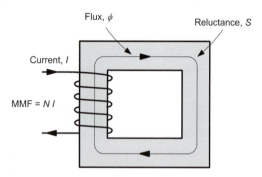

(b) A magnetic circuit

Figure 21.5 Comparison of electric and magnetic circuits

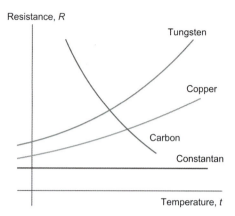

Figure 21.6 Variation of resistance with temperature for various materials

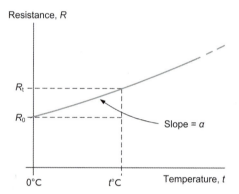

Figure 21.7 Straight-line approximation of Figure 21.6 over a limited temperature range

increases with temperature and we say that these materials have a *positive temperature coefficient*. For non-metallic conductors, such as carbon or semiconductor materials such as silicon or germanium, resistance falls with temperature and we say that these materials have a *negative temperature coefficient*. The variation of resistance with temperature for various materials is shown in Figure 21.6.

The resistance of a conductor, R_t, at a temperature, t, can be determined from the relationship:

$$R_t = R(1 + \alpha t + \beta t^2 + \gamma t^3 + ...)$$

where R is the resistance of the conductor at $0°C$ and α, β and γ are constants. In practice β and γ can usually be ignored and so we can approximate the relationship (see Figure 21.7) to:

$$R_t = R(1 + \alpha t)$$

where α is the temperature coefficient of resistance (in $°C^{-1}$).

The following table shows the temperature coefficient of resistance, α, of some common metals:

Metal	Temperature coefficient of resistance $°C^{-1}$
Nichrome	0.00017
Gold	0.0034
Silver	0.0038
Copper	0.0039
Aluminium	0.0040
Lead	0.0040
Tungsten	0.0045
Mild steel	0.0045
Nickel	0.0059

Example 21.8   A copper wire has a resistance of 12.5 Ω at 0°C. Determine the resistance of the wire at 125°C.

A wire having a specific resistance of 1.6 × 10^{-8} Ω m, length 20 m and cross-sectional area 1 mm^2 carries a current of 5 A. Determine the voltage drop between the ends of the wire.

Solution

To find the resistance at 125°C we use:

$$R_t = R_0(1 + \alpha t)$$

where $R_0 = 12.5$ Ω, $\alpha = 0.0039$°C^{-1} (from the table) and $t = 125$°C.
   Hence:

$$R_t = R_0(1 + \alpha t) = 12.5 \times (1 + (0.0039 \times 125))$$

$$= 12.5 \times (1 + 0.4875) = 18.6 \text{ Ω}$$

SCILAB

The resistance of a conductor at a given temperature can be calculated using the following SCILAB script (note that this calculation is based on a linear approximation and therefore will be less accurate for wide changes in temperature):

```
// Resistance at a given temperature
// SCILAB source file: ex2103.sce
// Get input data from dialogue box
txt = ['Resistance at 20 deg.C (ohm)';
'Temperature coefficient';
'Temperature (deg.C)';];
var = x_mdialog('Enter the resistance at
room temperature and temperature
coefficient', txt,['1000';'0.0039';'45';]);
r20 = evstr(var(1));
a = evstr(var(2));
t = evstr(var(3));
// Find resistance at 0 deg.C
r0 = r20/(1+(a*20));
// Find resistance at given temperature
rt = r0 * (1 + (a*t));
// Display result
messagebox("Resistance at " + string(t)
 + " deg.C = " + string(rt) + " Ohm")
```

21.7   Internal resistance

Whilst most resistors take the form of recognisable components (see Figure 21.4) not all of those present in a circuit are actually visible and tangible. This might sound strange to you at first but consider what happens when you connect a load to a battery. If the battery is perfect the voltage drop across the load will be identical to that of the e.m.f. of the battery when left unconnected. However, in practice the voltage on-load will be less than that of the unconnected battery. Why is this? The question can be easily answered if we take into account the fact that there is some resistance *inside* the battery. This resistance does not, of course, take the form of a recognisable component and is simply the aggregate resistance of the internal wiring of the battery and its electrolyte. In most cases this resistance is very small and can be neglected but in some cases (e.g. when the battery is exhausted) the resistance inside the battery (i.e. its *internal resistance*) may become appreciable. In practice, all components have some internal resistance. For example, an inductor will have some resistance (as well as inductance) due to the wire that is used in its winding, a generator will have some resistance due to its armature and brushes, and so on.

Since we now know how to solve problems involving voltage, current and resistance it's worth illustrating the effect of internal resistance with a simple example. Figure 21.8 shows what happens as the internal resistance of a battery increases. In Figure 21.8(a) a 'perfect' 10 V battery supplies a current of 1 A to a 10 Ω load. The output voltage of the battery (when *on-load*) is, as you would expect 10 V. In Figure 21.8(b) the

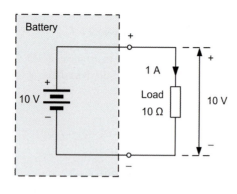

(a) Perfect battery

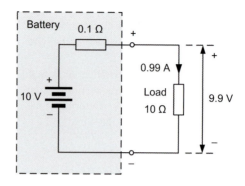

(b) Internal resistance = 0.1 Ω

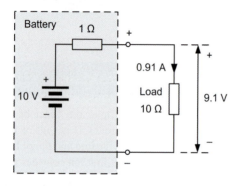

(c) Internal resistance = 1 Ω

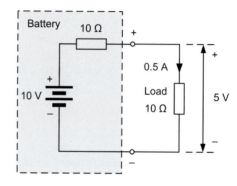

(d) Internal resistance = 10 Ω

Figure 21.8 Effect of internal resistance

battery has a relatively small value of internal resistance (0.1 Ω) and this causes the output current to fall to 0.99 A and the output voltage to be reduced (as a consequence) to 9.9 V. Figure 21.8(c) shows the effect of the internal resistance rising to 1 Ω. Here the output current has fallen to 0.91 A and the output voltage to 9.1 V. Finally, taking a more extreme case, Figure 21.8(d) shows the effect of the internal resistance rising to 10 Ω. In this situation, the output current is only 0.5 A and the voltage delivered to the load is a mere 5 V! We will be exploring this topic in much greater detail in the next chapter.

Key Point. Internal resistance places an additional series loss resistance into a circuit. The effect of this resistance is to limit output voltage and current and introduce an unwanted loss of power. As it is almost invariably unwanted, steps are usually taken to minimise this resistance in the design of electronic components and systems

Key Point. Internal resistance is important in a number of applications. When a battery goes 'flat' it's simply that its internal resistance has increased to a value that begins to limit the output voltage when current is drawn from the battery

Example 21.9 An inductor consists of a coil manufactured from 10.5 m of copper wire having a diameter of 0.5 mm. Determine the series loss resistance of the inductor and the voltage drop that will appear across its terminals when a current of 5 A flows in it.

Solution

To find the resistance of the inductor we use:

$$R = \frac{\rho l}{A} \quad \text{(where } l = 10.5\,\text{m} \quad \text{and}$$

$$A = 3.142 \times 0.25\,\text{mm}^2 = 0.196 \times 10^{-6}\,\text{m}^2)$$

From the table shown earlier we find that the value of specific resistance, ρ, for annealed copper is 1.72×10^{-8} Ω m.

Hence:

$$R = \frac{\rho l}{A} = \frac{1.72 \times 10^{-8} \times 10.5}{0.196 \times 10^{-6}}$$

$$= 92.14 \times 10^{-2} = 0.9214 \text{ Ω}$$

Next we can use Ohm's law to find the voltage drop:

$$V = IR = 5 \times 0.9214 = 4.607 \text{ V}$$

Note that if the inductor had been an ideal component, the winding resistance would have been zero and there would have been no voltage dropped across its terminals! The *equivalent circuit* of the inductor is shown in Figure 21.9.

Figure 21.9 Internal (loss) resistance of the inductor in Example 21.9

Example 21.10   A 90 V battery delivers 50 mA to a 1.5 kΩ load. What is the internal resistance of the battery and what current would it supply when connected to a load of 1 kΩ?

Solution

First we need to determine the on-load output voltage from the battery using Ohm's law:

$$V = IR = 50 \times 10^{-3} \times 1.5 \times 10^3 = 75 \text{ V}$$

The voltage dropped across the battery's internal resistance will be the difference between the off-load and on-load voltages, i.e.:

$$V_{loss} = V_{off\text{-}load} - V_{on\text{-}load} = 90 - 75 = 15 \text{ V}$$

We can now apply Ohm's law again to determine the internal loss resistance:

$$R_{internal} = \frac{V_{loss}}{I} = \frac{15}{50 \times 10^{-3}}$$

$$= 0.3 \times 10^3 = 300 \text{ Ω}$$

Finally, when connected to a 1 kΩ load the total resistance in the circuit will be:

$$R_{total} = R_{load} + R_{internal} = 1 \times 10^3 + 0.3 \times 10^3$$

$$= 1.3 \times 10^3 = 1.3 \text{ kΩ}$$

The current delivered by the battery will then be:

$$I = \frac{V_{off\text{-}load}}{R_{total}} = \frac{90}{1.3 \times 10^3} = 69.23 \times 10^{-3}$$

$$= 69.23 \text{ mA}$$

Finally, the voltage dropped across the 1 kΩ load will be:

$$V = IR = 69.23 \times 10^{-3} \times 1 \times 10^3 = 69.23 \text{ V}$$

Note that there are other (better) ways that we could have arrived at this solution as you will see in the next chapter!

Test your knowledge 21.9

A 24 V battery delivers 22.5 V when delivering 120 A to a starter motor. What is the internal resistance of the battery and what output voltage would it produce when supplying a load current of 200 A?

Test your knowledge 21.10

A coil is wound from a 4.8 m length of 0.75 mm diameter copper wire. Determine the loss resistance of the coil and the voltage that would be dropped across it when a current of 3.5 A is flowing in it.

21.8   Power, work and energy

From your study of dynamic engineering systems you will recall that energy can exist in many forms including kinetic energy, potential energy, heat energy, light energy etc. Kinetic energy is concerned with the movement of a body whilst potential energy is the energy that a body possesses due to its position. Energy can be defined as 'the ability to do work' whilst power can be defined as 'the rate at which work is done'.

In electrical circuits, energy is supplied by batteries or generators. It may also be stored in components such as capacitors and inductors. Electrical energy is converted into various other forms of energy by components such as resistors (producing heat), loudspeakers (producing sound energy) and light-emitting diodes (producing light).

The unit of energy is the Joule (J). Power is the rate of use of energy and is measured in Watts (W). A power of 1 W results from energy being used at the rate of 1 J per second. Thus:

$$P = \frac{E}{t}$$

where P is the power in Watts (W), E is the energy in Joules (J) and t is the time in seconds (s).

We can rearrange the previous formula to make E the subject, as follows:

$$E = P \times t$$

The power in a circuit is equivalent to the product of voltage and current. Hence:

$$P = I \times V$$

where P is the power in Watts (W), I is the current in amps (A) and V is the voltage in volts (V).

The formula may be rearranged to make P, I or V the subject, as follows:

$$P = I \times V \quad I = \frac{P}{V} \quad \text{and} \quad V = \frac{P}{I}$$

When a resistor gets hot it is dissipating power. In effect, a resistor is a device that converts electrical energy into heat energy. The amount of power dissipated in a resistor depends on the current flowing in the resistor. The more current that flows in the resistor the more power will be dissipated and the more electrical energy will be converted into heat.

It's important to note that the relationship between the current applied and the power dissipated is not linear – in fact it obeys a *square law*. In other words, the power dissipated in a resistor is proportional to the square of the applied current.

The relationship, $P = I \times V$, may be combined with that which results from Ohm's law (i.e. $V = I \times R$), to produce two further relationships. First, substituting for V gives:

$$P = I \times (I \times R) = I^2 R$$

Second, substituting for I gives:

$$P = \left(\frac{V}{R}\right) \times V = \frac{V^2}{R}$$

Example 21.11 A current of 1.5 A is drawn from a 3 V battery. What power is supplied?

Solution

Here we must use $P = I \times V$ (where $I = 1.5\,\text{A}$ and $V = 3\,\text{V}$):

$$P = I \times V = 1.5 \times 3 = 4.5\,\text{W}$$

Example 21.12 A voltage drop of 4 V appears across a resistor of 100 Ω. What power is dissipated in the resistor?

Solution

Here we must use $P = \dfrac{V^2}{R}$ (where $V = 4\,\text{V}$ and $R = 100\,\Omega$):

$$P = \frac{V^2}{R} = \frac{4^2}{100} = \frac{16}{100} = 0.16\,\text{W (or 160 mW)}.$$

Example 21.13 A current of 20 mA flows in a 1 kΩ resistor. What power is dissipated in the resistor and what energy is used if the current flows for 10 minutes?

Solution

Here we must use $P = I^2 R$ (where $I = 200\,\text{mA}$ and $R = 1000\,\Omega$):

$$P = I^2 R = 0.2^2 \times 1000 = 0.04 \times 1000 = 40\,\text{W}$$

To find the energy we need to use $E = P \times t$ (where $P = 40\,\text{W}$ and $t = 10$ minutes):

$$E = P \times t = 40 \times (10 \times 60) = 24000\,\text{J} = 24\,\text{kJ}$$

Key Point. Power is the rate at which energy is used and a power of one Watt results from energy being used at the rate of one Joule per second

Key Point. When a resistor gets hot it is dissipating a power proportional to the square of the current flowing in it or voltage dropped across it

Test your knowledge 21.11

An electric pump is rated at 24 V, 6 A. How much energy is required to operate the pump for 10 minutes?

Test your knowledge 21.12

A 15 Ω resistor is rated at 10 W. What is (a) the maximum current that should be allowed to flow in the resistor and (b) the maximum voltage drop that should be allowed to appear across the resistor?

Test your knowledge 21.13

Write and test a SCILAB script that will calculate the energy delivered to a load when given the current, voltage and time for which the load is connected to the supply.

21.9   Chapter summary

In this chapter we explained how electric current is the organised movement of electrons in metallic conductors from a point of greater electric potential to one with a lesser electric potential. We introduced resistance as an opposition to current flow and conductance as an ability to support current flow. We also noted the similarities and differences between electric and magnetic circuits.

Because of the losses associated with it, internal resistance is important in many electrical applications. We showed how this resistance has an effect on the output produced by a source of e.m.f. This is an

important topic that we shall return to later when we investigate the maximum power transfer theorem.

Finally, we introduced the concept of power as the rate at which energy is consumed and showed that, whilst energy can be stored in capacitors and inductors, it is dissipated as heat in resistors.

21.10   Review questions

Short answer questions

1. A charge of 60 mC is transferred between two points in a circuit in a time interval of 15 ms. What current will be flowing?

2. A current of 27 mA flows in a resistance of 3.3 kΩ. What voltage drop will appear across the resistor?

3. Determine the resistance of a 150 m length of wire having a resistivity of 2.82×10^{-8} Ω m and a diameter of 0.8 mm.

4. A copper telephone line has a diameter of 0.5 mm. If the resistance of the line is not to exceed 50 Ω, determine the maximum permissible line length. Take ρ for copper as 1.68×10^{-8} Ω m.

5. A 20 cm cube of metal alloy has a resistance of 5.5 μΩ measured across its opposite faces. Determine the conductivity of the alloy.

6. A resistor of 22 Ω carries a current of 0.5 A. What power is dissipated?

7. A 12 V power supply delivers a current of 70 mA for six hours. What energy is delivered?

8. A nominal 12 V battery delivers 10.5 V when supplying a current of 26 A. What is the internal resistance of the battery?

9. An aluminium bus bar has a resistance of 0.005 Ω at 20° C. Determine the power lost in the bus bar if it carries a current of 110 A and reaches a working temperature of 40°C.

10. A choke is manufactured from a 12 m length of copper wire having a diameter of 0.4 mm. Determine the loss resistance of the choke and the power that would be dissipated in it when a direct current of 1.2 A flows in it.

Long answer questions

11. A resistance tester comprises an accurate 5 V d.c. supply connected in series with a fixed resistor and

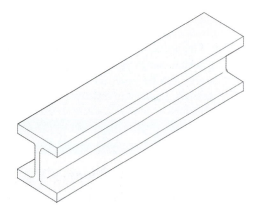

250

20

20

250

20

20

250

All dimensions in mm

Figure 21.10 See Question 13

a microammeter having a full-scale deflection of 50 μA and negligible internal resistance.

(a) Sketch a labelled circuit diagram of the resistance tester and explain briefly how the circuit works.

(b) Determine the value of the fixed resistor.

(c) Determine the range of resistance values that can be reliably measured if the minimum discernible indication on the meter is 2.5 μA.

12. The following data refers to a wirewound resistor of nominally 27 Ω:

Temperature, °C	0	10	20	30	40	50
Resistance, Ω		25.1	26.0	26.9	27.9	29.0	30.1

Plot a graph showing how the resistance varies with temperature and use it to determine the temperature coefficient of resistance. Also estimate the resistance at a temperature of 85°C.

13. The I-beam shown in Figure 21.10 is to be used as a bus bar having a length of 8 m. If the maximum voltage drop permissible between the ends of the bus bar is 20 mV when carrying a load current of 120 A, determine the required minimum conductivity of the material. Would aluminium be suitable?

Chapter 22

Circuit theorems

This section develops some of the basic electrical and electronic principles that we met in the previous chapter. It begins by explaining a number of useful circuit theorems including those developed by Thévenin and Norton. If you have previously studied Electrical and Electronic Principles at level 3 you should be quite comfortable with this material. If not, you should make sure that you fully understand this chapter before going further.

22.1 Kirchhoff's laws

Used on its own, Ohm's law is insufficient to determine the magnitude of the voltages and currents present in complex circuits. For these circuits we need to make use of two further laws; Kirchhoff's current law and Kirchhoff's voltage law.

22.1.1 Kirchhoff's current law

Kirchhoff's current law states that the algebraic sum of the currents present at a junction (or *node*) in a circuit is zero as shown in Figure 22.1.

Example 22.1 Determine the value of the missing current, I, shown in Figure 22.2.

Solution

By applying Kirchhoff's current law in Figure 22.2, and adopting the convention that currents flowing towards the junction are positive, we can say that:

$$+2 + 1.5 - 0.5 + I = 0$$

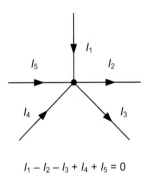

$$I_1 - I_2 - I_3 + I_4 + I_5 = 0$$

Convention:
Current flowing towards the junction is positive (+)
Current flowing away from the junction is negative (−)

Figure 22.1 Kirchhoff's current law

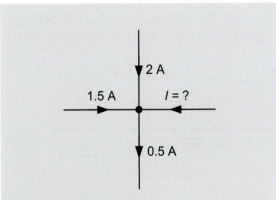

Figure 22.2 See Example 22.1

Note that we have shown I as positive. In other words we have assumed that it is flowing towards the junction.

978-1-85617-775-7, Engineering Science, Mike Tooley & Lloyd Dingle

Rearranging gives:

$$+3 + I = 0$$

Thus $I = -3\,\text{A}$

The negative answer tells us that I is actually flowing in the other direction, i.e. away from the junction.

Kirchhoff's voltage law

Kirchhoff's second law states that the algebraic sum of the potential drops present in a closed network (or *mesh*) is zero – see Figure 22.3.

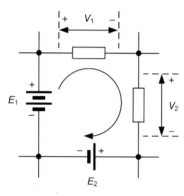

$$E_1 - V_1 - V_2 - E_2 = 0$$

Convention:
Move clockwise around the circuit starting with the positive terminal of the largest e.m.f.
Voltages acting in the same sense are positive (+)
Voltages acting in the opposite sense are negative (−)

Figure 22.3 Kirchhoff's voltage law

Example 22.2   Determine the value of the missing voltage, V, shown in Figure 22.4.

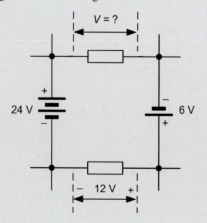

Figure 22.4 See Example 22.2

Solution

By applying Kirchhoff's voltage law in Figure 22.4, starting at the positive terminal of the largest e.m.f. and moving clockwise around the closed network, we can say that:

$$+24 - V + 6 - 12 = 0$$

Note that we have shown V as positive. In other words we have assumed that the more positive terminal of the resistor is the one on the left.
Rearranging gives:

$$+24 - V + 6 - 12 = 0$$

From which:

$$+18 - V = 0$$

Thus

$$V = +18\,\text{V}$$

The positive answer tells us that we have made a correct assumption concerning the polarity of the voltage drop, V, i.e. the more positive terminal is on the left.

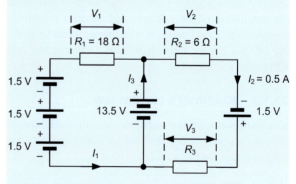
Key Point.   Kirchhoff's current law states that the algebraic sum of the currents present at a junction in a circuit is zero

Key Point. Kirchhoff's voltage law states that the algebraic sum of the potential drops present in a closed network is zero

22.2 Series and parallel circuit calculations

Ohm's law and Kirchhoff's laws can be used in combination to solve more complex series–parallel circuits. Before we show you how this is done, however, it's important to understand what we mean by the terms 'series' and 'parallel' circuit!

Figure 22.6 shows three circuits, each containing three resistors, R_1, R_2 and R_3. In Figure 22.6(a), the three resistors are connected one after another. We refer to this as a *series circuit*. In other words, the resistors are said to be connected *in series*. It's important to note that, in this series arrangement, *the same current flows through each resistor*.

In Figure 22.6(b), the three resistors are all connected across one another. We refer to this as a *parallel circuit*. In other words, the resistors are said to be connected *in parallel*. It's important to note that, in this parallel arrangement, *the same voltage appears across each resistor*.

In Figure 22.6(c), we have shown a mixture of these two types of connection. Here we can say that R_1 is connected in series with the parallel combination of R_2 and R_3. In other words, R_2 and R_3 are connected *in parallel* and R_1 is connected *in series* with the parallel combination.

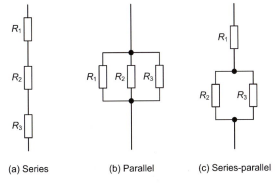

(a) Series   (b) Parallel   (c) Series-parallel

Figure 22.6 Series and parallel circuits

Example 22.3 For the circuit shown in Figure 22.7, determine:

(a) the voltage dropped across each resistor,

(b) the current drawn from the supply,

(c) the supply voltage.

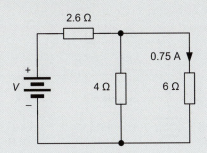

Figure 22.7 See Example 22.3

Solution

We need to solve this problem in several small stages. Since we know the current flowing in the 6 Ω resistor, we will start by finding the voltage dropped across it using Ohm's law:

$$V = I \times R = 0.75 \times 6 = 4.5\,\text{V}$$

Now the 4 Ω resistor is connected in parallel with the 6 Ω resistor. Hence the voltage drop across the 4 Ω resistor is also 4.5 V. We can now determine the current flowing in the 4 Ω resistor using Ohm's law:

$$I = \frac{V}{R} = \frac{4.5}{4} = 1.125\,\text{A}$$

Since we now know the current in both the 4 Ω and 6 Ω resistors, we can use Kirchhoff's law to find the current, I, in the 2.6 Ω resistor:

$$+I - 0.75\,\text{A} - 1.125 = 0$$

from which:

$$I = 1.875\,\text{A}$$

Since this current flows through the 2.6 Ω resistor it will also be equal to the current taken from the supply.

Next we can find the voltage drop across the 2.6 Ω resistor by applying Ohm's law:

$$V = I \times R = 1.875 \times 2.6 = 4.875\,\text{V}$$

Finally, we can apply Kirchhoff's voltage law in order to determine the supply voltage, V:

$$+V - 4.875 - 4.5 = 0$$

from which:

$$V = +9.375\,V$$

Hence the supply voltage is 9.375 V, as shown in Figure 22.8.

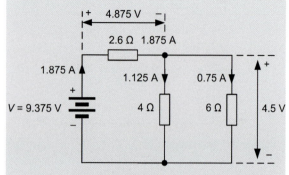

Figure 22.8 See Example 22.3

Test your knowledge 22.2

Determine the unknown voltages and currents shown in Figure 22.9. Also determine the value of R_3.

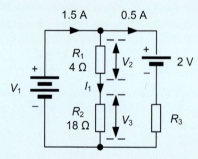

Figure 22.9 See Test your knowledge 22.2

22.2.1 Resistors in series

Consider the series arrangement of three resistors shown in Figure 22.10. By applying Kirchhoff's voltage law we know that:

$$V = V_1 + V_2 + V_3$$

We can also apply Ohm's law to each individual resistor and since the same current, I, flows in each:

$$V_1 = IR_1,\ V_2 = IR_2 \text{ and } V_3 = IR_3$$

In Figure 22.10, R is the equivalent value of the three resistors connected in series. For the equivalent circuit we can apply Ohm's law to arrive at:

$$V = IR$$

Combining the above relationships gives:

$$IR = IR_1 + IR_2 + IR_3$$

Hence:

$$R = R_1 + R_2 + R_3$$

From this we can conclude that, for a series circuit, the equivalent resistance of the circuit is simply the sum of the individual resistances.

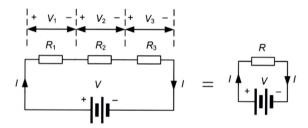

Figure 22.10 Series combination of resistors

22.2.2 Resistors in parallel

Turning to the parallel circuit arrangement in Figure 22.11, by applying Kirchhoff's current law we know that:

$$I = I_1 + I_2 + I_3$$

As before, we can apply Ohm's law to each individual resistor and since the same voltage appears across all three of them:

$$I_1 = \frac{V}{R_1}, I_2 = \frac{V}{R_2} \text{ and } I_3 = \frac{V}{R_3}$$

As before, R is the equivalent value of the three resistors connected in parallel. For the equivalent circuit we can apply Ohm's law to arrive at:

$$I = \frac{V}{R}$$

Combining the above relationships gives:

$$\frac{V}{R} = \frac{V}{R_1} + \frac{V}{R_2} + \frac{V}{R_3}$$

Hence:

$$\frac{1}{R} = \frac{1}{R_1} + \frac{1}{R_2} + \frac{1}{R_3}$$

From this we can conclude that, for a series circuit, the reciprocal of the equivalent resistance of the circuit is simply sum of the individual resistances. Note, also, that for the case of just two resistors in series we can simplify the relationship as follows:

$$\frac{1}{R} = \frac{1}{R_1} + \frac{1}{R_2}$$

Multiplying both sides by R_1R_2 gives:

$$\frac{R_1R_2}{R} = R_1R_2\left(\frac{1}{R_1} + \frac{1}{R_2}\right) = \frac{R_1R_2}{R_1} + \frac{R_1R_2}{R_2} = R_2 + R_1$$

from which:

$$R = \frac{R_1R_2}{R_1 + R_2}$$

It is important to note that this relationship only holds true for two resistors connected in parallel and can be more easily be remembered as 'product over sum'.

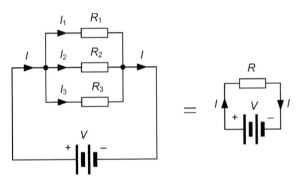

Figure 22.11 Parallel combination of resistors

Example 22.4   Resistors of 22 Ω, 47 Ω and 33 Ω are connected (a) in series and (b) in parallel. Determine the effective resistance in each case.

Solution

(a)   In the series circuit:

$$R = R_1 + R_2 + R_3$$

thus

$$R = 22 + 47 + 33 = 102 \ \Omega$$

(b)   In the parallel circuit:

$$\frac{1}{R} = \frac{1}{R_1} + \frac{1}{R_2} + \frac{1}{R_3}$$

thus

$$\frac{1}{R} = \frac{1}{22} + \frac{1}{47} + \frac{1}{33}$$

or

$$\frac{1}{R} = 0.045 + 0.021 + 0.03 = 0.096$$

thus

$$R = 10.42 \ \Omega$$

Example 22.5   Determine the effective resistance of the circuit shown in Figure 22.12.

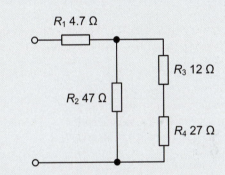

Figure 22.12   See Example 22.5

Solution

The circuit can be progressively simplified as shown in Figure 22.13.

The stages in this simplification are:

(a) R_3 and R_4 are in series and can be replaced by a single resistance (R_A) of $12 + 27 = 39 \ \Omega$

(b) R_A appears in parallel with R_2. These two resistors can be replaced by a single resistance (R_B) of:

$$\frac{39 \times 47}{39 + 47} = 21.3 \ \Omega$$

(c) R_B appears in series with R_1. These two resistors can be replaced by a single resistance (R) of $21.3 + 4.7 = 26 \ \Omega$

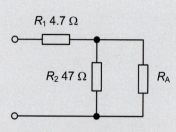

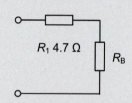

(a) R_A replaces the series combination of R_3 and R_4

(b) R_B replaces the parallel combination of R_a and R_B

(c) R replaces the series combination of R_1 and R_B

Figure 22.13 See Example 22.5

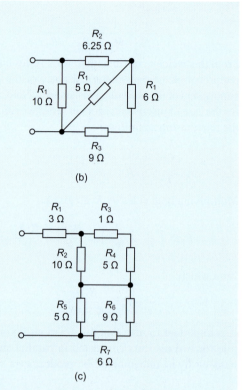

(b)

(c)

Figure 22.14 See Test your knowledge 22.3

Key Point. The resistance of a series combination of resistors is the sum of the individual resistance values

Key Point. The reciprocal of the resistance of a parallel combination of resistors is the sum of the reciprocals of the individual resistance values

22.3 The potential divider

The potential divider circuit (see Figure 22.15) is commonly used to reduce voltage levels in a circuit. The output voltage produced by the circuit is given by:

$$V_{out} = V_{in} \times \frac{R_2}{R_1 + R_2}$$

It is, however, important to note that the output voltage (V_{out}) will fall when current is drawn away from the arrangement.

Test your knowledge 22.3

Determine the equivalent resistance of each of the combinations of resistors shown in Figure 22.14.

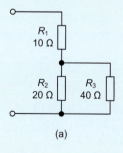

(a)

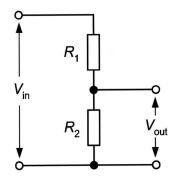

Figure 22.15 The potential divider

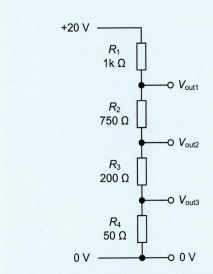

Figure 22.17 See Test your knowledge 22.4

Example 22.6 Determine the output voltage of the circuit shown in Figure 22.16

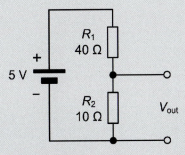

Figure 22.16 See Example 22.6

Solution

Here we can use the potential divider formula,

$$V_{out} = V_{in} \times \frac{R_2}{R_1 + R_2},$$

where $V_{in} = 5\,\text{V}$, $R_1 = 10\,\Omega$ and $R_2 = 40\,\Omega$, thus:

$$V_{out} = V_{in} \times \frac{R_2}{R_1 + R_2} = 5 \times \frac{10}{10 + 40}$$

$$= 5 \times \frac{1}{5} = 1\,\text{V}$$

Test your knowledge 22.4

Determine the output voltages produced by the voltmeter calibrator circuit shown in Figure 22.17.

22.4 The current divider

The current divider circuit (see Figure 22.18) is used to divert current from one branch of a circuit to another. The output current produced by the circuit is given by:

$$I_{out} = I_{in} \times \frac{R_1}{R_1 + R_2}$$

It is important to note that the output current (I_{out}) will fall when the load connected to the output terminals has any appreciable resistance.

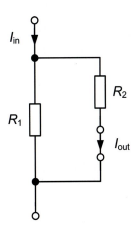

Figure 22.18 The current divider

Example 22.7 A moving coil meter requires a current of 1 mA to provide full-scale deflection. If the meter coil has a resistance of 100 Ω and is to be used as a milliammeter reading 5 mA full-scale, determine the value of the parallel 'shunt' resistor required.

Solution

This problem may sound a little complicated so it is worth taking a look at the equivalent circuit of the meter (Figure 22.19) and comparing it with the current divider shown in Figure 22.18.

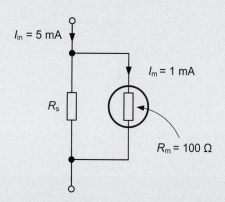

Figure 22.19 Meter circuit – see Example 22.7

We can apply the current divider formula, replacing I_{out} with I_m (the meter full-scale deflection current) and R_2 with R_m (the meter resistance). R_1 is the required value of shunt resistor, R_s.

From $I_{out} = I_{in} \times \dfrac{R_1}{R_1 + R_2}$ we can say that

$I_m = I_{in} \times \dfrac{R_s}{R_s + R_m}$ where $I_m = 1\,\text{mA}$, $I_{in} = 5\,\text{mA}$

and $R_2 = 100\,\Omega$.

Rearranging the formula gives:

$$I_m \times (R_s + R_m) = I_{in} \times R_s$$

thus

$$I_m R_s + I_m R_m = I_{in} \times R_s$$

or

$$I_{in} \times R_s - I_m R_s = I_m R_m$$

hence

$$R_s (I_{in} - I_m) = I_m R_m$$

and

$$R_s = \frac{I_m R_m}{I_{in} - I_m}$$

Now

$$I_m = 1\,\text{mA}, R_m = 100\,\Omega \text{ and } I_{in} = 5\,\text{mA}$$

thus

$$R_s = \frac{I_m R_m}{I_{in} - I_m} = \frac{1 \times 10^{-3} \times 100}{5 \times 10^{-3} - 1 \times 10^{-3}} = \frac{100}{4} = 25\,\Omega$$

Test your knowledge 22.5

A moving coil meter with a full-scale deflection of 100 μA and coil resistance 250 Ω is to be used as the basis of an ammeter reading 20 mA full-scale. Determine the value of shunt resistance required.

SCILAB

The resistance of a meter shunt can be easily found using the following SCILAB script:

```
// Resistance of a meter shunt
// SCILAB source file: ex2201.sce
// Get input data from dialogue box
txt = ['Meter full-scale deflection current
(mA)';'Resistance of meter coil (ohm)';
'Current to indicate at full-scale (A)';];
var = x_mdialog('Enter meter parameters and
required current indication',txt,
['1';'450';'2';]);
im = 1E-3*evstr(var(1));
rm = evstr(var(2));
iin = evstr(var(3));
// Calculate shunt resistance
rs = (im * rm)/(iin-im);
// Display result
messagebox("Required shunt resistance = "
+ string(rs) + "ohm")
```

22.5 The constant voltage source

Let's start by considering the properties of a 'perfect' battery. No matter how much current is drawn from it, this device – if it were to exist – would produce a constant voltage between its positive and negative terminals. In practice, and from work done in Chapter 29, you will be well aware that the terminal voltage of a battery falls progressively as we draw more current from it. We account for this fact by referring to its internal resistance, R_s (see Figure 22.20).

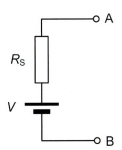

Figure 22.20 Internal resistance of a battery

The notion of internal resistance is fundamental to understanding the behaviour of electrical and electronic circuits and it is worth exploring this idea a little further before we take a look at circuit theorems in some detail. In Figure 22.20, the internal resistance is shown as a single discrete resistance, R_S connected in series with a voltage source. The source of e.m.f. has exactly the same voltage, V as that which would be measured between the battery's terminals when no current is being drawn from it. As more current is drawn from a battery, its terminal voltage will fall whereas the internal e.m.f. will remain constant. It might help to illustrate this with an example.

Example 22.8 A battery has an open-circuit (no-load) terminal voltage of 12 V and an internal resistance of 0.1 Ω. Determine the terminal voltage of the battery when it supplies currents of (a) 1 A, (b) 10 A and (c) 100 A.

Solution

The terminal voltage of the battery when 'on-load' will always be less than the terminal voltage when there is no load connected. The reduction in voltage (i.e. the voltage that is effectively 'lost' inside the battery) will be equal to the load current multiplied by the internal resistance. Applying Kirchhoff's voltage law to the circuit of Figure 22.21 gives:

$$V_L = V - (I_L R_S)$$

where V_L is the terminal voltage on-load, V is the terminal voltage with no load connected, I_L is the load current and R_S is the internal resistance of the battery.

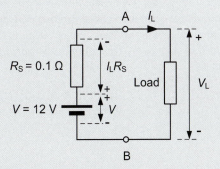

Figure 22.21 See Example 22.8

In case (a); $I = 1\,A$, $V = 12\,V$ and $R_S = 0.1\,\Omega$ and thus:

$$V_L = V - (I_L R_S) = 12 - (1 \times 0.1) = 12 - 0.1 = 11.9\,V$$

In case (b); $I = 10\,A$, $V = 12\,V$ and $R_S = 0.1\,\Omega$ and thus:

$$V_L = V - (I_L R_S) = 12 - (10 \times 0.1) = 12 - 1 = 11\,V$$

In case (c); $I = 1\,A$, $V = 12\,V$ and $R_S = 0.1\,\Omega$ and thus:

$$V_L = V - (I_L R_S) = 12 - (100 \times 0.1) = 12 - 10 = 2\,V$$

The previous example shows that, provided that the internal resistance of a battery is very much smaller than the resistance of any circuit connected to the battery's terminals, a battery will provide a reasonably constant source of voltage. In fact, when a battery becomes exhausted, its internal resistance begins to rise sharply. This, in turn, reduces the terminal voltage when we try to draw current from the battery as the following example shows.

Example 22.9 A battery has an open-circuit (no-load) terminal voltage of 9 V. If the battery is required to supply a load which has a resistance of 90 Ω, determine the terminal voltage of the battery when its internal resistance is (a) 1 Ω and (b) 10 Ω.

Solution

As we showed before, the terminal voltage of the battery when 'on-load' will always be less than the terminal voltage when there is no load connected. Applying Kirchhoff's voltage law gives:

$$V_L = V - (I_L R_S)$$

where V_L is the terminal voltage on-load, V is the terminal voltage with no load connected, I_L is the load current and R_S is the internal resistance of the battery.

Now

$$I_L = V_L / R_L$$

Thus

$$V_L = V - (V_L / R_L) \times R_S$$

or

$$V_L + (V_L / R_L) \times R_S = V$$

or

$$V_L (1 + (1/R_L) \times R_S) = V$$

or

$$V_L (1 + R_S / R_L) = V$$

Hence

$$V_L = V / (1 + R_S / R_L)$$

In case (a);

$$V = 12\,V \quad \text{and} \quad R_S = 1\,\Omega \quad \text{and} \quad R_L = 90\,\Omega$$

Thus

$$V_L = V / (1 + R_S / R_L) = 9/(1 + 1/90)$$
$$= 9/(1 + 0.011) = 9/1.011 = 8.902\,V$$

In case (b);

$$V = 12\,V \quad \text{and} \quad R_S = 10\,\Omega \quad \text{and} \quad R_L = 90\,\Omega$$

Thus

$$V_L = V / (1 + R_S / R_L) = 9/(1 + 10/90)$$
$$= 9/(1 + 0.111) = 9/1.111 = 8.101\,V$$

SCILAB

The internal resistance of a voltage source can be easily found using the following SCILAB script:

```
// Internal resistance of a voltage source
// SCILAB source file: ex222.sce
// Get input data from dialogue box
txt = ['No load output voltage (V)';
'On loaded output voltage (V)';
'Load current (A)';];
var = x_mdialog('Enter voltage on and
off-load and load current',
txt,['12';'10.5';'2.5';]);
v = evstr(var(1));
vl = evstr(var(2));
il = evstr(var(3));
// Calculate internal resistance
rs = (v - vl)/il;
// Display result
messagebox("Internal resistance of the
voltage source = " + string(rs) + "ohm")
```

A battery is the most obvious example of a voltage source. However, as we shall see later, any linear circuit with two terminals, no matter how complex, can be represented by an equivalent circuit based on a voltage source which uses just two components; a source of e.m.f., V, connected in series with an internal resistance, R_S. Furthermore, the voltage source can either be a source of direct current (i.e. a battery) or a source of alternating current (i.e. an a.c. generator, a signal source or the output of an oscillator).

An ideal voltage source – a *constant voltage source* – would have negligible internal resistance ($R_S = 0$) and its voltage/current characteristic would look like that shown in Figure 22.22.

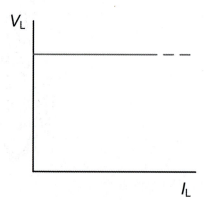

Figure 22.22 *V/I* characteristic for an ideal constant voltage source

In practice, $R_S \neq 0$ and the terminal voltage will fall as the load current, I_L, increases, as shown in Figure 22.23.

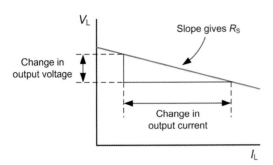

Figure 22.23 V/I characteristic for a real voltage source

The internal resistance, R_S, can be found from the slope of the graph in Figure 22.23 as follows:

$$R_S = \left(\frac{\text{change in output voltage, } V_L}{\text{corresponding change in output current, } I_L} \right)$$

The symbols commonly used for constant voltage d.c. and a.c. sources are shown in Figure 22.24.

Figure 22.24 Symbols used for constant voltage sources

Example 22.10   A battery has a no-load terminal voltage of 12 V. If the terminal voltage falls to 10.65 V when a load current of 1.5 A is supplied, determine the internal resistance of the battery.

Now:

$$R_S = \left(\frac{\text{change in output voltage, } V_L}{\text{corresponding change in output current, } I_L} \right)$$

$$= \left(\frac{12 - 10.65}{1.5 - 0} \right) = \frac{1.35}{1.5} = 0.9\ \Omega$$

Note that we could have approached this problem in a different way by considering the voltage 'lost' inside the battery (i.e. the voltage dropped across R_S). The lost voltage will be $(12 - 10.65) = 1.35$ V. In this condition, the current flowing is 1.5 A, hence applying Ohm's law gives:

$$R_S = V/I = 1.35/1.5 = 0.9\ \Omega$$

Test your knowledge 22.6

A nominal 28 V power supply has an internal resistance of 2.6 Ω. What current will it supply to a load of 11.4 Ω and what will the on-load output voltage be?

Thévenin's theorem

Thévenin's theorem states that:

Any two-terminal network can be replaced by an equivalent circuit consisting of a voltage source and a series resistance (or impedance if the source of voltage is an a.c. source) equal to the internal resistance (or internal impedance) seen looking into the two terminals.

In order to determine the Thévenin equivalent of a two-terminal network you need to:

(a) determine the voltage that will appear between the two terminals with no load connected to the terminals. This is the open-circuit voltage.

(b) replace any voltage sources with a short-circuit connection (or replace any current sources with an open circuit connection) and then determine the internal resistance of the circuit (i.e. the resistance that appears between the two terminals).

Example 22.11   Determine the Thévenin equivalent of the two-terminal network shown in Figure 22.25.

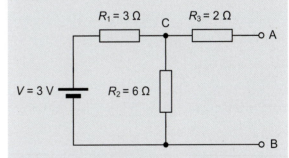

Figure 22.25 See Example 22.11

First we need to find the voltage that will appear between the two terminals, A and B, with no load connected. In this condition, no current will be drawn from the network and thus there will be no voltage dropped across R_3. The voltage that appears between A and B, V_{AB}, will then be identical to that which is dropped across R_2 (V_{CB}).

The voltage dropped across R_2 (with no load connected) can be found by applying the potential divider theorem, thus:

$$V_{CB} = V \times R_2/(R_1 + R_2) = 3 \times 6/(3+6) = 18/9 = 2\ V$$

Thus the open-circuit output voltage will be 2 V.

Next we need to find the resistance of the circuit with the battery replaced by a short-circuit as shown in Figure 22.26.

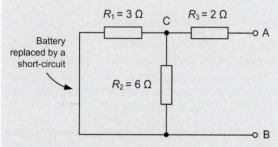

Figure 22.26 See Example 22.11

The circuit can be progressively reduced to a single resistor as shown in Figure 22.27.

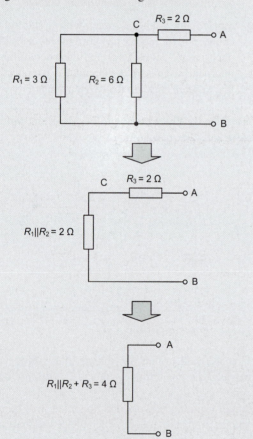

Figure 22.27 See Example 22.11

Thus the internal resistance is 4 Ω.

The Thévenin equivalent of the two-terminal network is shown in Figure 22.28.

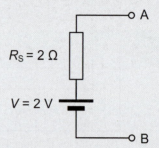

Figure 22.28 See Example 22.11

Example 22.12 Determine the current flowing in, and voltage dropped across, the 10 Ω resistor shown in Figure 22.29.

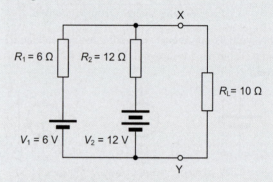

Figure 22.29 See Example 22.12

First we will determine the Thévenin equivalent circuit of the network.

With no load connected to the network (i.e. with R_L disconnected) the current supplied by the 12 V battery will flow in an anti-clockwise direction around the circuit as shown in Figure 22.30.

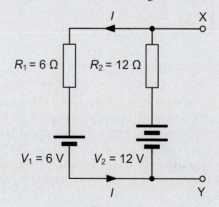

Figure 22.30 See Example 22.12

The total e.m.f. present in the circuit will be:

$$V = 12 - 6 = 6\,\text{V}$$

(note that the two batteries *oppose* one another)

The total resistance in the circuit is:

$$R = R_1 + R_2 = 6 + 12 = 18\,\Omega$$

(the two resistors are in series with one another)

Hence, the current flowing in the circuit when R_L is disconnected will be given by:

$$I = V/R = 6/18 = 0.333\,\text{A}$$

The voltage developed between the output terminals (X and Y) will be equal to 12 V *minus* the voltage dropped across the 12 Ω resistor (or 6 V *plus* the voltage dropped across 6 Ω resistor).

Hence

$$V_{XY} = 12 - (0.333 \times 12) = 12 - 4 = 8\,\text{V}$$

or

$$V_{XY} = 6 + (0.333 \times 6) = 6 + 2 = 8\,\text{V})$$

Thus the open-circuit output voltage will be 8 V.
The internal resistance with V_1 and V_2 replaced by short-circuit will be given by:

$$R = 6 \times 12/(6 + 12) = 72/18 = 4\,\Omega$$

Thus the internal resistance will be 4 Ω.
Figure 22.31 shows the Thévenin equivalent circuit of the network.

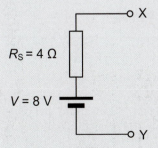

Figure 22.31 See Example 22.12

Now, with $R_L = 10\,\Omega$ connected between X and Y the circuit looks like that shown in Figure 22.32.

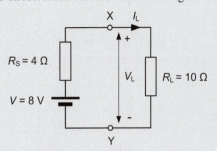

Figure 22.32 See Example 22.12

The total resistance present will be:

$$R = R_S + R_L = 4 + 10 = 14\,\Omega$$

The current drawn from the circuit in this condition will be given by:

$$I_L = V/R = 8/14 = 0.57\,\text{A}$$

Finally, the voltage dropped across R_L will be given by:

$$V_L = I \times R_L = 0.57 \times 10 = 5.7\,\text{V}$$

Example 22.13 Determine the current flowing in a 20 Ω resistor connected to the Wheatstone bridge shown in Figure 22.33.

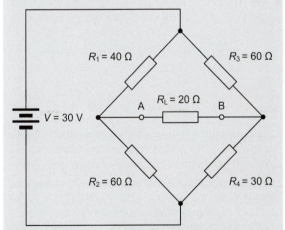

Figure 22.33 See Example 22.13

Once again, we will begin by determining the Thévenin equivalent circuit of the network. To make things a little easier, we will redraw the circuit so that it looks a little more familiar (see Figure 22.34).

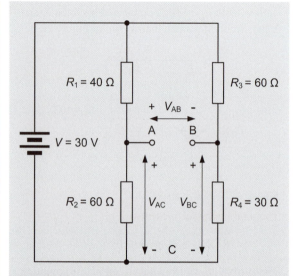

Figure 22.34 See Example 22.13

With A and B open-circuit, the circuit simply takes the form of two potential dividers (R_1 and R_2 forming one potential divider whilst R_3 and R_4 forms the other potential divider). It is thus a relatively simple matter to calculate the open-circuit voltage drop between A and B, since:

$$V_{AB} = V_{AC} - V_{BC} \text{ (from Kirchhoff's voltage law)}$$

where

$$V_{AC} = V \times R_2/(R_1 + R_2) = 30 \times 60/(40 + 60) = 18 \text{ V}$$

and

$$V_{BC} = V \times R_4/(R_3 + R_4) = 30 \times 30/(30 + 60) = 10 \text{ V}$$

Thus

$$V_{AB} = 18 - 10 = 8 \text{ V}$$

Next, to determine the internal resistance of the network with the battery replaced by a short-circuit, we can again redraw the circuit and progressively reduce it to one resistance. Figure 22.35 shows how this is done.

Thus the internal resistance is 44 Ω.

The Thévenin equivalent of the bridge network is shown in Figure 22.36.

Now, with $R_L = 20$ Ω connected between A and B, the circuit looks like that shown in Figure 22.37.

The total resistance present will be:

$$R = R_S + R_L = 44 + 20 = 64 \text{ Ω}$$

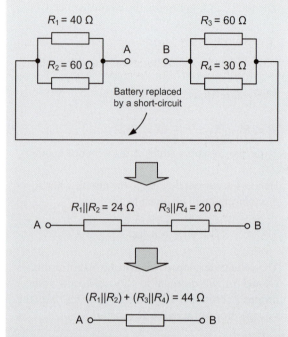

Figure 22.35 See Example 22.13

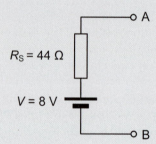

Figure 22.36 See Example 22.13

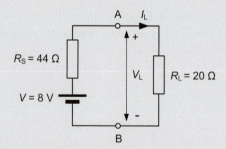

Figure 22.37 See Example 22.13

The current drawn from the circuit in this condition will be given by:

$$I_L = V/R = 8/64 = 0.125 \text{ A}$$

Test your knowledge 22.7

Determine the Thévenin equivalent circuit of the network shown in Figure 22.38 and use it to find the power that would be dissipated in an 8 Ω load connected to terminals P and Q.

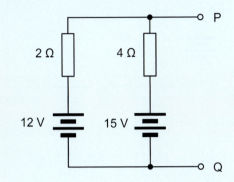

Figure 22.38 See Test your knowledge 22.7

Key Point. The Thévenin equivalent of a circuit consists of an ideal source of voltage connected in series with the internal resistance of the source

22.6   The constant current source

By now, you should be familiar with the techniques for analysing simple electric circuits by considering the voltage sources that may be present and their resulting effect on the circuit in terms of the current and voltage drops that they produce. However, when analysing some types of circuit it can often be more convenient to consider sources of current rather than voltage. In fact, any linear circuit with two terminals, no matter how complex, can be thought of as a current source; i.e. a supply of current, I, connected in parallel with an internal resistance, R_P. As with voltage sources, current sources can either be a source of direct or alternating current.

An ideal current source – *a constant current source* – would have infinite internal resistance ($R_p = \infty$) and its voltage/current characteristic would look like that shown in Figure 22.39.

In practice, $R_P \neq \infty$ and the current will fall as the load voltage, V_L, increases, as shown in Figure 22.40.

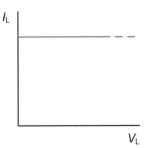

Figure 22.39 *V/I* characteristic for an ideal constant current source

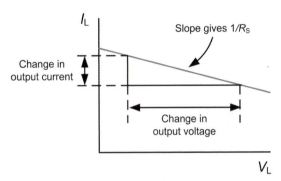

Figure 22.40 *V/I* characteristic for a real constant current source

The internal resistance, R_P, can be found from the reciprocal of the slope of the graph in Figure 22.40 as follows:

$$R_P = \cfrac{1}{\left(\cfrac{\text{change in output current, } I_L}{\text{corresponding change in output voltage, } V_L}\right)}$$

$$= \left(\cfrac{\text{change in output voltage, } V_L}{\text{corresponding change in output current, } I_L}\right)$$

The symbols commonly used for constant-current d.c. and a.c. sources are shown in Figure 22.41.

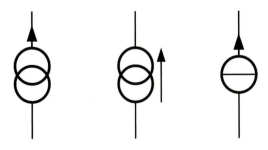

Figure 22.41 Symbols used for constant current sources

22.6.1 Norton's theorem

Norton's theorem states that:

Any two-terminal network can be replaced by an equivalent circuit consisting of a current source and a parallel resistance (or impedance if the source of current is an a.c. source) equal to the internal resistance (or internal impedance) seen looking into the two terminals.

In order to determine the Norton equivalent of a two-terminal network you need to:

(a) determine the current that will appear between the two terminals when they are short-circuited together; this is the short-circuit current;

(b) replace any current sources with an open-circuit connection (or replace any voltage sources with a short-circuit connection) and then determine the internal resistance of the circuit (i.e. the resistance that appears between the two terminals).

Example 22.14 Determine the Norton equivalent of the two-terminal network shown in Figure 22.42.

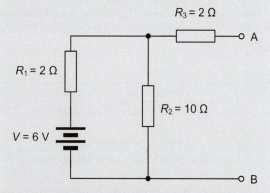

Figure 22.42 See Example 22.14

First we need to find the current that would flow between terminals A and B with the two terminals linked together by a short-circuit (see Figure 22.43).

The total resistance that appears across the 6 V battery with the output terminals linked together is given by:

$$R_1 + (R_2 \times R_3)/(R_2 + R_3)$$
$$= 2 + (2 \times 10)/(2 + 10)$$
$$= 2 + 20/12 = 3.67\,\Omega$$

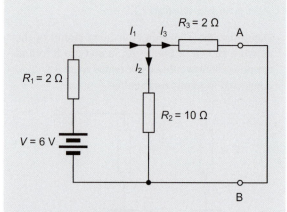

Figure 22.43 See Example 22.14

The current supplied by the battery, I_1, will then be given by:

$$I_1 = V/R = 6/3.67 = 1.63\,\text{A}$$

Using the current divider theorem, the current in R_3 (and that flowing between terminals A and B) will be given by

$$I_3 = I_1 \times R_2/(R_2 + R_3) = 1.63 \times 10/(10 + 2)$$
$$= 16.3/12 = 1.35\,\text{A}$$

Thus the short-circuit current will be 1.35 A.

Next we need to find the resistance of the circuit with the battery replaced by a short-circuit as shown in Figure 22.44.

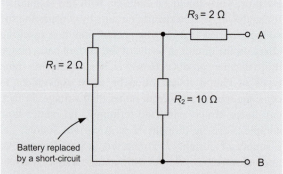

Figure 22.44 See Example 22.14

The circuit can be progressively reduced to a single resistor as shown in Figure 22.45.

Thus the internal resistance is 3.67 Ω.

The Norton equivalent of the two-terminal network is shown in Figure 22.46.

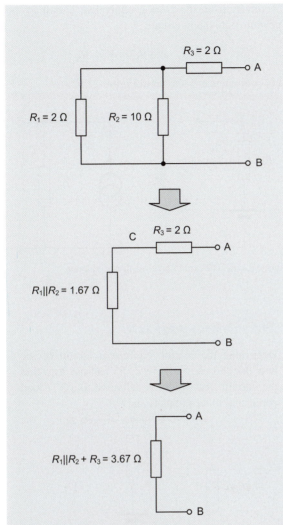

Figure 22.45 See Example 22.14

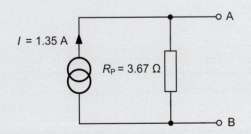

Figure 22.46 See Example 22.14

Example 22.15 Determine the voltage dropped across a 6 Ω resistor connected between X and Y in the circuit shown in Figure 22.47.

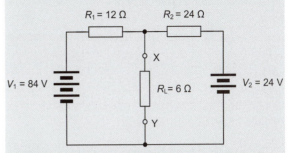

Figure 22.47 See Example 22.15

First we will determine the Norton equivalent circuit of the network. Since there are two voltage sources (V_1 and V_2) in this circuit, we can derive the Norton equivalent of each branch and then combine them together into one current source and one parallel resistance. This makes life a lot easier!

The left-hand branch of the circuit is equivalent to a current source of 7 A with an internal resistance of 12 Ω (see Figure 22.48).

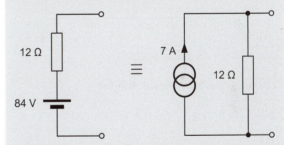

Figure 22.48 See Example 22.15

The right-hand branch of the circuit is equivalent to a current source of 1 A with an internal resistance of 24 Ω (see Figure 22.49).

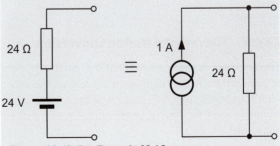

Figure 22.49 See Example 22.15

The combined effect of the two current sources is a single constant current generator of $(7 + 1) = 8\,\text{A}$ with a parallel resistance of $12 \times 24/(12 + 24) = 8\,\Omega$, as shown in Figure 22.50.

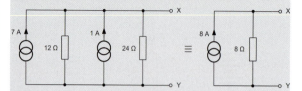

Figure 22.50 See Example 22.15

Now, with $R_\text{L} = 6\,\Omega$ connected between X and Y, the circuit looks like that shown in Figure 22.51.

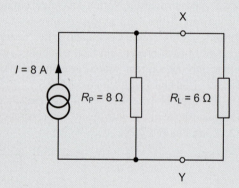

Figure 22.51 See Example 22.15

The total resistance present will be:

$$R = R_\text{P} \times R_\text{L}/(R_\text{P} \times R_\text{L}) = 8 \times 6/(8 + 6) = 3.43\,\Omega$$

When 8 A flows in this resistance, the voltage dropped across the parallel combination of resistors will be:

$$V_\text{L} = 8/3.43 = 2.33\,\text{A}$$

22.6.2 Thévenin to Norton conversion

The previous example shows that it is relatively easy to convert from one equivalent circuit to the other (don't overlook the fact that, whichever equivalent circuit is used to solve a problem, the result will be identical!).

To convert from the Thévenin equivalent circuit to the Norton equivalent circuit:

$$R_\text{P} = R_\text{S} \quad \text{and} \quad V_\text{TH} = I_\text{N} \times R_\text{P}$$

To convert from the Norton equivalent circuit to the Thévenin equivalent circuit:

$$R_\text{S} = R_\text{P} \quad \text{and} \quad I_\text{N} = V_\text{TH}/R_\text{S}$$

This equivalence is shown in Figure 22.52.

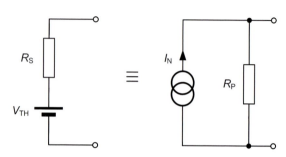

Figure 22.52 Thévenin and Norton equivalence

Test your knowledge 22.8

Determine the Norton equivalent circuit of the network shown in Figure 22.53 and use it to find the current that would be delivered to a 3 Ω load connected to terminals R and S.

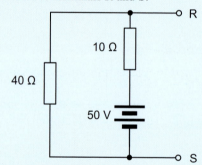

Figure 22.53 See Test your knowledge 22.8

Key Point. The Norton equivalent of a circuit consists of an ideal source of current connected in parallel with the internal resistance of the source

22.7 Superposition theorem

The superposition theorem is a simple yet powerful method that can be used to analyse networks containing a number of voltage sources and linear resistances.

The superposition theorem states that:

In any network containing more than one voltage source, the current in, or potential difference developed across, any branch can be found by considering the effects of each source separately and adding their effects. During this process, any temporarily omitted source must be replaced by its internal resistance (or a short-circuit if it is a perfect voltage source).

This might sound a little more complicated than it really is, so we shall explain the use of the superposition theorem with a simple example.

Example 22.16   Use the superposition theorem to determine the voltage dropped across the 6 Ω resistor in Figure 22.54.

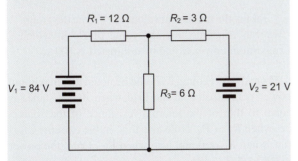

Figure 22.54 See Example 22.16

First we shall consider the effects of the left-hand voltage source (V_1) when taken on its own. Figure 22.55 shows the source voltage and currents (you can easily check these by calculation, if you wish!).

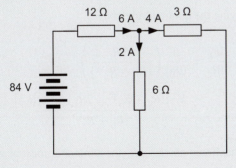

Figure 22.55 See Example 22.16

Next we shall consider the effects of the right-hand voltage source (V_2) when taken on its own. Figure 22.56 shows the source voltage and currents (again, you can easily check these by calculation).

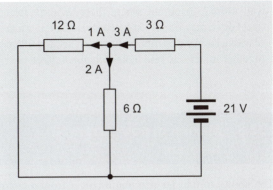

Figure 22.56 See Example 22.16

Finally, we can combine these two sets of results, taking into account the direction of current flow, to arrive at the values of current in the complete circuit. Figure 22.57 shows how this is done.

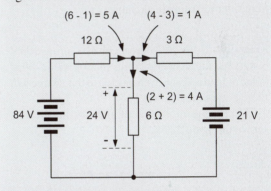

Figure 22.57 See Example 22.16

The current in the 6 Ω resistor is thus 4 A and the voltage dropped across it will be:

$$V = I \times R = 4 \times 6 = 24 \text{ V}$$

Test your knowledge 22.9

Use the superposition theorem to find the voltage dropped across the 6 Ω resistor in Figure 22.58.

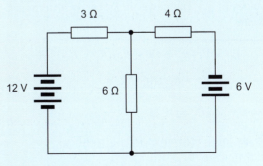

Figure 22.58 See Test your knowledge 22.9

22.8 Maximum power transfer theorem

The ability to transfer maximum power from a circuit to a load is crucial in a number of electrical and electronic applications. The relationship between the internal resistance of a source (which we met earlier), the resistance of the load to which it is connected and the power dissipated in that load can be easily illustrated by considering a simple example.

Example 22.17 A source consists of a 100 V battery having an internal resistance of 100 Ω. What value of load resistance connected to this source will receive maximum power?

The equivalent circuit of this arrangement is shown in Figure 22.59.

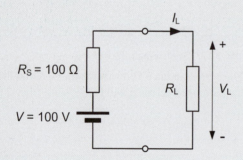

Figure 22.59 See Example 22.17

Now $P = V^2/R$, therefore we might consider what power would appear in resistor R_L for different values of R_L. The power in R_L can be determined from:

$$P_L = V_L^2/R_L = (V \times R_L/(R_L + R_S))^2/R_L$$
$$= V^2 \times R_L/(R_L + R_S)^2$$

Now, when $R_L = 10\,\Omega$:

$$P_L = (10/(10 + 100) \times 100)^2/10$$
$$= (10/110 \times 100)^2/10 = 8.26\,\text{W}$$

This process can be repeated for other values of R_L (in, say, the range 0 to 200 Ω) and a graph can be plotted from which the maximum value can be located, as shown in Figure 22.60.

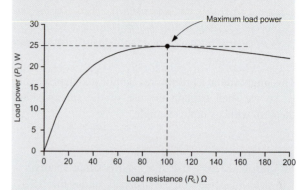

Figure 22.60 See Example 22.17

Taking the maximum value for P_L from the graph, we can conclude that the value of R_L in which maximum power would be dissipated is 100 Ω.

From the foregoing, it should be noted that maximum power is dissipated in the load when its resistance is equal to that of the internal resistance of the source, i.e. when $R_L = R_S$. Note also that, in the extreme cases (i.e. when $R_L = 0$ and also when $R_L = \infty$), the power in the load, $P_L = 0$.

A more powerful approach would be to apply differential calculus to prove that the maximum value for P_L occurs when $R_L = R_S$. The method is as follows:

$$P_L = V^2 \times R_L/(R_L + R_S)^2 = \frac{V^2 R_L}{(R_L + R_S)^2}$$

Applying the quotient rule gives:

$$\frac{dP_L}{dR_L} = \frac{d}{dR_L}\left(\frac{V^2 R_L}{(R_L + R_S)^2}\right)$$
$$= \frac{V^2\left[(R_L + R_S)^2 - 2R_L(R_L + R_S)\right]}{(R_L + R_S)^4}$$

For maximum power in R_L the numerator must be zero. Hence:

$$(R_L + R_S)^2 - 2R_L(R_L + R_S) = 0$$

thus

$$(R_L + R_S)^2 = 2R_L(R_L + R_S)$$

or

$$R_L + R_S = 2R_L$$

or

$$R_L = R_S$$

The maximum power transfer theorem states that, in the case of d.c. circuits:

Maximum power will be dissipated in a load when the resistance of the load is equal to the internal resistance of the source (i.e. when $R_L = R_S$).

Note that when the load resistance is matched to the source resistance we say that the system is *matched*. Later on we will be revisiting this topic when we look at how the maximum power transfer theorem relates to complex circuits that contain a combination of resistance, capacitance and inductance.

SCILAB

The maximum power transfer theorem can be neatly illustrated by the following SCILAB script that constructs a graph of power against load resistance (see Figure 22.61):

```
// Maximum power transfer theorem
// SCILAB source file: ex2203.sce
// Set up the plot window
rl=(0:0.1:10)';
vs = evstr(x_dialog('Voltage source (V)?',
'2')); // 2 V
rs = evstr(x_dialog('Internal resistance
(ohm)?','2')); // 2 ohm
//
re = rs + rl; // Effective resistance of
the circuit
il = vs ./ re; // Load current
ps = il^2 * rs; // Power loss in the
internal resistance
pi = il * vs; // Power produced by the
constant voltage source
pl = pi - ps; // Load power is the
difference between pi and ps
//
clf();
plot2d(rl,pl,style=[2]);
xgrid(color('lightgray')); // light
gray grid
xtitle('Maximum power transfer theorem',
'Load resistance','Load power');
```

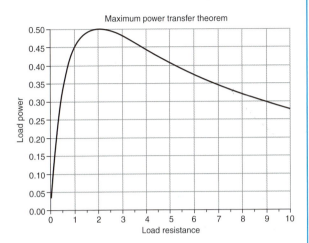

Figure 22.61 Graphical output from the SCILAB script. Note the maximum value on the graph corresponding to the matched load condition

Key Point. Maximum power will be transferred from a source to a load when the resistance of the load is identical to that of the load. We often refer to this as a *matched* condition

22.9 Chapter summary

This section developed some of the basic electrical and electronic principles that we met in the previous chapter. Starting with those of Thévenin and Norton, we introduced several theorems that can be used to solve network problems that cannot be solved using Ohm's law alone. We also showed how series, parallel and series/parallel combinations of resistors can be reduced to a single equivalent resistance.

Thévenin's theorem reduces a two-terminal network to an equivalent circuit consisting of a constant voltage source and a series resistance, whilst Norton's theorem reduces the same circuit to a constant current source and a parallel resistance. These techniques can be used interchangeably to solve circuits containing a number of resistances and sources.

The superposition theorem provides us with a further tool that can be used to solve complex circuits where several sources may be present. The theorem works by considering the effects of each source separately and then summing them to determine the current in, or potential difference developed across, any branch of the circuit. During this process, any temporarily omitted source must be replaced by its internal resistance.

The ability to transfer maximum power from a circuit to a load is crucial in a number of applications. The relationship between the internal resistance of a source and the resistance of the load is paramount in determining the amount of power that is actually transferred, and the maximum power transfer theorem shows us that maximum power is transferred between a source and a load when the two resistances are identical. This is often referred to as a *matched* condition.

22.10 Review questions

Short answer questions

1. Determine the current flowing in each resistor in Figure 22.62.

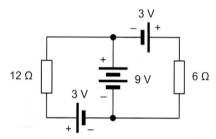

Figure 22.62 See Question 1

2. The terminal voltage of a battery falls from 24 V with no load connected to 21.6 V when supplying a current of 6 A. Determine the internal resistance of the battery. Also determine the terminal voltage of the battery when supplying a load current of 10 A.

3. Calculate the output voltage from the potential divider shown in Figure 22.63.

4. Determine the equivalent resistance of each of the networks shown in Figure 22.64.

5. Resistors of 10 Ω, 15 Ω and 22 Ω are available. Determine the resistance of each of the possible series, parallel and series/parallel combinations of these resistors.

6. A constant voltage source of 4.5 V and internal resistance 15 Ω is used to supply a circuit having a resistance of 75 Ω. What current will be supplied to the load?

7. A constant current source of 5 mA and internal resistance 15 kΩ is used to supply a circuit having a

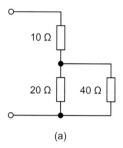

(a)

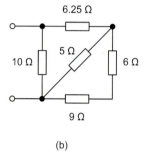

(b)

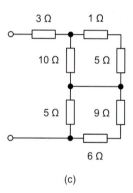

(c)

Figure 22.63 See Question 3

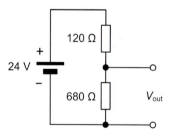

Figure 22.64 See Question 4

resistance of 35 kΩ. What voltage drop will appear across the load?

8. Determine the Thévenin equivalent of the circuit shown in Figure 22.65.

9. Determine the Norton equivalent of the circuit shown in Figure 22.65.

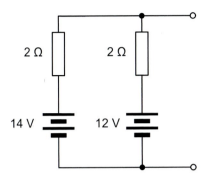

Figure 22.65 See Questions 8 and 9

10. Figure 22.66 shows a d.c. supply comprising three parallel-connected batteries. Determine the load power produced when this arrangement is used to supply a load having a resistance of 1.1 Ω.

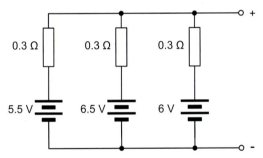

Figure 22.66 See Question 10

Long answer questions

11. A moving coil meter having a full-scale deflection of 1 mA and a coil resistance of 600 Ω is to be used as (a) a voltmeter reading 10 V full-scale and (b) an ammeter reading 10 mA full scale. Determine the value of series and shunt resistors required and sketch labelled circuit diagrams of each instrument.

12. An amplifier produces an output of 1.5 V into a load resistance of 600 Ω. If the output voltage increases to 2.5 V when the load resistance is disconnected, determine the internal resistance of the amplifier. Also determine the output voltage when a load resistance of 300 Ω is connected.

13. Use the superposition theorem to determine the power dissipated in the 8 Ω resistor shown in Figure 22.67.

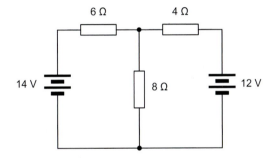

Figure 22.67 See Question 13

Part VI

Transients

In Parts IV and V we only consider what happens in an electric circuit when it is operating under steady-state conditions (in other words, where the voltage and current remains constant). However, in many electrical applications there can be significant changes in voltage and current and we need to be able to take these changes into account in our circuit analysis. Such transient conditions often exist for only a very brief time when there is a change from one steady-state condition to another, for example, when a switch is opened or closed, when a supply voltage is first applied to a circuit or when it is disconnected.

An analysis of circuits containing a mixture of resistance, capacitance and/or inductance requires the use of differential and integral calculus (see *Essential Mathematics 2*). By applying calculus theory we are able to develop an understanding of how rate of change impacts on the current and voltage in circuits that contain resistance together with capacitance or inductance.

In Chapter 23 we investigate the growth and decay of voltage and current in first-order circuits where resistance and capacitance or resistance and inductance are present. Chapter 24 further extends our analysis of transients to second-order circuits where resistance, capacitance and inductance are all present at the same time.

In order to develop an understanding of the behaviour of first- and second-order systems we show how Laplace transforms enable us to use straightforward algebraic equations to solve expressions in which differential and/or integral terms are present.

Chapter 23

Transients

Thus far we have only considered what happens in an electric circuit when it is operating under steady-state conditions (in other words, where there is no change in the applied voltage or current). However, in many electrical applications there can be significant changes in voltage and current and we need to be able to take these changes into account in our circuit analysis. These transient conditions often exist for only a very brief time when there is a change from one steady-state condition to another, for example, when a switch is opened or closed, when a supply voltage is first applied to a circuit or when it is disconnected. In order to fully understand the transient behaviour of a circuit it is necessary to be able to apply calculus, and so we shall start this chapter with an introduction to rate of change and how this impacts on the current and voltage in circuits that contain capacitance, inductance and resistance.

23.1 Rate of change

Change, or more specifically rate of change is important in many electrical and electronic applications and so it's important to explain just what we mean by this. In Chapter 21 we defined electric current as the rate of flow of electric charge. For the steady-state condition (where charge is being transferred at a uniform rate) we stated that:

$$I = \frac{Q}{t}$$

where I represents current, Q represents charge and t represents time.

A more meaningful definition covering all eventualities would be have been to say that:

$$i = \frac{dq}{dt}$$

where i represents the current at a particular instant of time and $\frac{dq}{dt}$ represents the rate at which charge is changing *at that time*.

If you find this difficult to understand you can simply think of $\frac{dq}{dt}$ as the slope of a graph of charge, q, plotted against time, t, at a given point in time, as shown in Figure 23.1. Note that if charge is transferred at a constant rate, the graph will simply be a straight line and the slope will be the same at all points along the line.

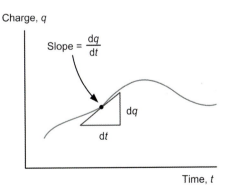

Figure 23.1 Instantaneous current, i, as a function of time, t

978-1-85617-775-7, Engineering Science, Mike Tooley & Lloyd Dingle

23.1.1 Capacitors and changing voltage

In Chapter 19 we introduced the following relationship that exists between charge, voltage and capacitance:

$$Q = CV$$

where Q is the charge (in C), C is the capacitance (in F) and V is the applied voltage (in V). In terms of changes in charge and voltage we can thus conclude that:

$$\frac{dq}{dt} = C\frac{dv}{dt}$$

We have also just seen that, for any electrical circuit:

$$i = \frac{dq}{dt}$$

Combining these two relationships leads us to the conclusion that:

$$i = C\frac{dv}{dt}$$

This important relationship tells us that the instantaneous current flowing in a capacitor is proportional to the rate of change of applied voltage.

Example 23.1 A 470 μF is connected to a supply voltage that changes at a uniform rate of 3 V/s. What current will flow in the capacitor?

Solution

Using $i = C\dfrac{dv}{dt}$ gives:

$$i = 470 \times 10^{-6} \times 3 = 1410 \times 10^{-6}$$

$$= 1.41 \times 10^{-3} = 1.41 \text{ mA}$$

Test your knowledge 23.1

A 22 μF capacitor is connected to a supply that changes from 20 V to 40 V in 250 ms. What current will flow in the capacitor?

Key Point. The current flowing in a capacitor at a particular instant of time is proportional to the rate of change of applied current

23.1.2 Inductors and changing current

Having looked at what happens when a changing voltage is applied to a capacitor, we will now look at what happens when a changing current is applied to an inductor. The e.m.f. induced within an inductor when the magnetic flux is changing is given by the relationship:

$$e = -N\frac{d\phi}{dt}$$

where N is the number of turns and $\dfrac{d\phi}{dt}$ is the rate of change of magnetic flux. The negative sign simply indicates that the induced e.m.f. acts in a direction that opposes the change that causes it.

From work that we did earlier in Chapter 20 we know that the following relationship exists between inductance, flux and current:

$$L = N\frac{\Phi}{I}$$

from which:

$$LI = N\Phi$$

where L is the inductance (in H), I is the current (in A), N is the number of turns, and Φ is the flux (in Wb). In terms of changes in charge and voltage we can thus conclude that:

$$L\frac{di}{dt} = N\frac{d\varphi}{dt}$$

Combining the previous relationships leads us to the conclusion that:

$$e = -L\frac{di}{dt}$$

This important relationship (known as *Lenz's law*) tells us that the instantaneous voltage dropped across an inductor is proportional to the rate of change of applied current.

Example 23.2 The current flowing in a 60 mH inductor changes at a uniform rate of 15 A/s. What voltage will be induced within the inductor?

Solution

Using $e = -L\dfrac{di}{dt}$ gives:

$$e = -60 \times 10^{-3} \times 15 = -900 \times 10^{-3} = -0.9 \text{ V}$$

Test your knowledge 23.2

A back e.m.f. of 180 V appears when a current changing at a rate of 36 A/s is applied to an inductor. What is the value of the inductance?

Key Point. The e.m.f. induced across the terminals of an inductor at a particular instant of time is proportional to the rate of change of applied current. This e.m.f. acts in a direction that opposes the change that causes it

23.1.3   Capacitance and resistance in series

Now consider the case in which current is flowing in a circuit comprising a series combination of resistance, R, and capacitance, C, as shown in Figure 23.2. Applying Kirchhoff's voltage law (see Chapter 22) to this circuit yields the following relationship:

$$V = iR + v_C = R\frac{dq}{dt} + v_C$$

Once again, we know, from work that we did earlier in Chapter 19, that $Q = CV$ and hence:

$$\frac{dq}{dt} = C\frac{dv_C}{dt}$$

Combining these two expressions gives:

$$V = CR\frac{dv_C}{dt} + v_C$$

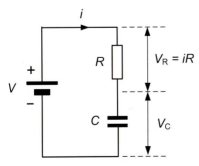

Figure 23.2 A C–R series circuit

Rearranging this expression gives:

$$V - v_C = CR\frac{dv_C}{dt}$$

Hence the rate of change of capacitor voltage with time will be given by:

$$\frac{dv_C}{dt} = \frac{1}{CR}(V - v_C)$$

from which:

$$dt = CR\frac{1}{(V - v_C)}dv_C$$

Integrating both sides of this expression gives:

$$t = CR\int\frac{1}{(V - v_C)}dv_C$$

$$t = -CR\ln(V - v_C) + k$$

The constant of integration, k, can be found by considering what happens when $t = 0$, as follows:
   When $t = 0$, $v_C = 0$ hence:

$$k = t + CR\ln(V - v_C) = 0 + CR\ln(V - 0) = CR\ln V$$

Hence:

$$t = -CR\ln(V - v_C) + CR\ln(V)$$

$$t = CR\ln(V) - CR\ln(V - v_C)$$

$$= CR(\ln(V) - \ln(V - v_C))$$

$$\frac{t}{CR} = \ln\left(\frac{V}{V - v_C}\right)$$

Using the laws of logarithms we arrive at:

$$e^{\left(\frac{t}{CR}\right)} = \frac{V}{V - v_C}$$

Inverting both sides of the expression gives:

$$e^{\left(-\frac{t}{CR}\right)} = \frac{V - v_C}{V} = 1 - \frac{v_C}{V}$$

Further rearrangement yields:

$$\frac{v_C}{V} = 1 - e^{\left(-\frac{t}{CR}\right)}$$

Hence:

$$v_C = V \left(1 - e^{-\frac{t}{CR}}\right)$$

You should note that this relationship contains an exponential term. We shall examine the implications of this in section 23.2.

23.1.4 Inductance and resistance in series

Next consider the case in which current is flowing in a circuit comprising of a series combination of resistance, R, and inductance, L, as shown in Figure 23.3. Once again we can use Kirchhoff's voltage law (see Chapter 22) to solve this circuit:

$$V = iR + v_L = iR + L\frac{di}{dt}$$

Rearranging this expression gives:

$$V - iR = L\frac{di}{dt}$$

We also know from work that we did earlier in Chapter 21 that $V = IR$, hence:

$$IR - iR = L\frac{di}{dt}$$

Hence that rate of change of inductor current with time will be given by:

$$\frac{di}{dt} = \frac{1}{L}(IR - iR) = \frac{R}{L}(I - i)$$

from which:

$$dt = \frac{L}{R}\frac{1}{(I - i)}di$$

Integrating both sides of this expression gives:

$$t = \frac{L}{R}\int \frac{1}{(I - i)}di$$

$$t = -\frac{L}{R}\ln(I - i) + k$$

Once again, the constant of integration, k, can be found by considering what happens when $t = 0$.

When $t = 0$, $i = 0$, hence:

$$k = t + \frac{L}{R}\ln(I - i) = 0 + \frac{L}{R}\ln(I - 0) = \frac{L}{R}\ln I$$

Hence:

$$t = -\frac{L}{R}\ln(I - i) + \frac{L}{R}\ln I$$

$$t = \frac{L}{R}\ln I - \frac{L}{R}\ln(I - i) = \frac{L}{R}(\ln I - \ln(I - i))$$

$$\frac{Rt}{L} = \ln\left(\frac{I}{I - i}\right)$$

Using the laws of logarithms we arrive at:

$$e^{\left(\frac{Rt}{L}\right)} = \frac{I}{I - i}$$

Inverting both sides of the expression gives:

$$e^{\left(-\frac{Rt}{L}\right)} = \frac{I - i}{I} = 1 - \frac{i}{I}$$

Further rearrangement yields:

$$\frac{i}{I} = 1 - e^{\left(-\frac{Rt}{L}\right)}$$

Hence:

$$i = I\left(1 - e^{-\frac{Rt}{L}}\right)$$

This is another important exponential relationship, the implications of which we will examine in much greater depth in section 23.3.

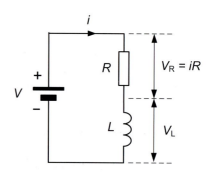

Figure 23.3 An *L–R* series circuit

23.2 C–R circuits

When a C–R network is connected to a constant voltage source (V), as shown in Figure 23.4 the voltage (v_C) across the (initially uncharged) capacitor will rise exponentially when the switch is closed, as shown in Figure 23.5. At the same time, the current in the circuit (i) will fall, as shown in Figure 23.6.

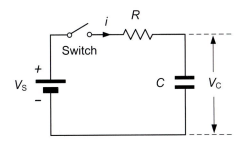

Figure 23.4 C–R circuit with energy supplied to C

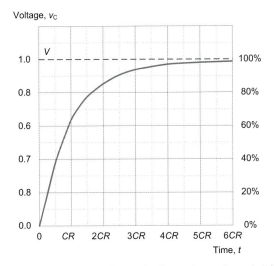

Figure 23.5 Exponential growth of capacitor voltage (v_c) in Figure 23.4

The rate of growth of voltage with time and decay of current with time will be dependent upon the product of capacitance and resistance. This value is known as the *time constant* of the circuit. Hence:

$$\text{Time constant}, t = C \times R$$

where C is the value of capacitance (in F), R is the resistance (in Ω) and t is the time constant (in s).

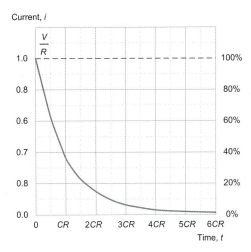

Figure 23.6 Exponential decay of current (i) in Figure 23.4

The exponential nature of growth and decay is vitally important in many applications so we shall now look at this in a little more detail.

23.2.1 Growth of voltage in a C–R circuit

In section 23.1 we showed how the voltage across the charging capacitor (v_C) grows exponentially with time (t) according to the relationship:

$$v_C = V\left(1 - e^{\frac{-t}{CR}}\right)$$

where v_C is the capacitor voltage (in V), V is the d.c. supply voltage (in V), t is the time (in s), and CR is the time constant of the circuit (equal to the product of capacitance, C, and resistance, R, in Ω).

The capacitor voltage will rise to approximately 63% of the supply voltage in a time interval equal to the time constant. At the end of the next interval of time equal to the time constant (i.e. after an elapsed time equal to 2CR) the voltage will have risen by 63% of the remainder, and so on. In theory, the capacitor will never quite become fully charged. However, after a period of time equal to 5CR, the capacitor voltage will to all intents and purposes be equal to the supply voltage. At this point the capacitor voltage will have risen to 99.3% of its final value and we can consider it to be fully charged.

Note that, if the initial rate of change of capacitor voltage had been sustained (i.e. $\dfrac{dv_C}{dt}$ at $t = 0$), the capacitor would have become fully charged (i.e. v_C would have become equal to V) in a time equal to the time constant (i.e. when $t = CR$).

The current in the charging circuit can be determined by using Ohm's law:

$$i = \frac{V - v_C}{R} = \frac{V - V\left(1 - e^{\frac{-t}{CR}}\right)}{R}$$

$$= \frac{V\left(1 - 1 + e^{\frac{-t}{CR}}\right)}{R} = \frac{V}{R}e^{\frac{-t}{CR}}$$

Hence, during charging, the current in the capacitor (i) decays exponentially with time (t) according to the relationship:

$$i = \frac{V}{R}e^{\frac{-t}{CR}}$$

where i is the current (in A), V is the d.c. supply voltage (in V), t is the time (in s), and CR is the time constant of the circuit (equal to the product of capacitance, C, and resistance, R, in s).

The current will fall to approximately 37% of the initial current in a time equal to the time constant. At the end of the next interval of time equal to the time constant (i.e. after a total time of $2CR$ has elapsed) the current will have fallen by a further 37% of the remainder, and so on.

Example 23.3   An initially uncharged capacitor of $1\,\mu\text{F}$ is charged from a 9 V d.c. supply via a 3.3 MΩ resistor. Determine the capacitor voltage 1 s after connecting the supply.

Solution

Using the relationship for exponential growth of voltage in the capacitor gives:

$$v_C = V\left(1 - e^{\frac{-t}{CR}}\right) = 9\left(1 - e^{\frac{-1}{1\times 10^{-6}\times 3.3\times 10^{6}}}\right)$$

$$= 9\left(1 - e^{\frac{-1}{3.3}}\right) = 9\left(1 - 0.738\right) = 2.358\ \text{V}$$

SCILAB

The following SCILAB script produces a plot of capacitor and resistor voltage for the charging case (see Figure 23.7):

```
// C-R circuit — growth of voltage on charge
// SCILAB source file: ex2301.sce
//
```

```
// Clear any previous graphic window
clf();
//
txt = ['Input resistance (R) in kohm';'Input
capacitance (C) in uF';'Supply voltage'];
var = x_mdialog('Enter resistance and
capacitance',txt,['100';'100';'100']);
r = evstr(var(1));
c = evstr(var(2));
vm = evstr(var(3));
c = 1E-6*c;
r = 1E+3*r;
//
T=c*r;
t = [0:1:10*T]'; // set time scale to 10CR
//
vc = vm*(1 - %e^(-t/(c*r)));
vr = vm-vc;
// Plot
plot2d(t,vc,style=5); // red line for
capacitor voltage
plot2d(t,vr,style=3); // green line for
resistor voltage
//
// Print grid and title
xgrid(); // add grid
xtitle('C-R charge - voltage plotted against
time');
hl=legend(['Capacitor voltage';'Resistor
voltage']);
```

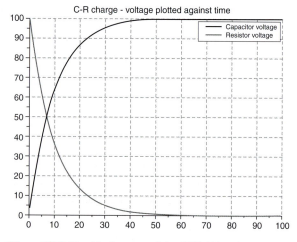

Figure 23.7 Graphical output of the SCILAB script

23.2.2   Decay of voltage in a *C–R* circuit

We have already explained that a charged capacitor contains a reservoir of energy stored in the form of an electric field. When the fully charged capacitor from Figure 23.4 is connected as shown in Figure 23.8, the capacitor will discharge through the resistor and the capacitor voltage (v_C) will fall exponentially with time, as shown in Figure 23.9. The current in the circuit (i) will also fall, as shown in Figure 23.10. The rate of discharge (i.e. the rate of decay of voltage with time) will once again be governed by the time constant of the circuit ($C \times R$).

The voltage developed across the discharging capacitor (v_C) varies with time (t) according to the relationship:

$$v_C = V_S e^{\frac{-t}{CR}}$$

where v_C is the capacitor voltage (in V), V_S, is the d.c. supply voltage (in V), t is the time (in s), and CR is the time constant of the circuit (equal to the product of capacitance, C, and resistance, R, in s).

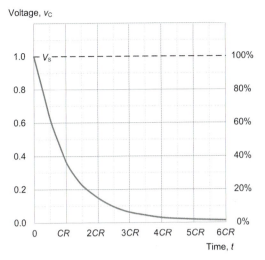

Figure 23.9 Exponential decay of capacitor voltage (v_C) in Figure 23.8

where i is the current (in A), V_S, is the d.c. supply voltage (in V), t is the time (in s), and CR is the time constant of the circuit (equal to the product of capacitance, C, and resistance, R, in s).

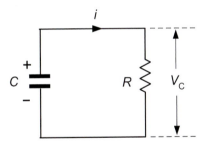

Figure 23.8 *C–R* circuit with energy released into *R*

The capacitor voltage will fall to approximately 37% of the initial voltage in a time equal to the time constant. At the end of the next interval of time equal to the time constant (i.e. after an elapsed time equal to $2CR$) the voltage will have fallen by 37% of the remainder, and so on. In theory, the capacitor will never quite become fully discharged. After a period of time equal to $5CR$, however, the capacitor voltage will to all intents and purposes be zero. At this point the capacitor voltage will have fallen below 1% of its initial value. At this point we can consider it to be fully discharged.

As with charging, the current in the capacitor (i) decays with time (t) according to the relationship:

$$i = \frac{V}{R} e^{\frac{-t}{CR}}$$

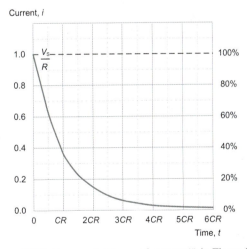

Figure 23.10 Exponential decay of current (i) in Figure 23.8

The current will fall to approximately 37% of the initial current in a time equal to the time constant. At the end of the next interval of time equal to the time constant (i.e. after a total time of $2CR$ has elapsed) the current will have fallen by a further 37% of the remainder, and so on.

Example 23.4 A 10 μF capacitor is charged to a potential of 20 V and then discharged through a 47 kΩ resistor. Determine the time taken for the capacitor voltage to fall below 10 V.

Solution

The formula for exponential decay of voltage in the capacitor is:

$$v_C = V e^{\frac{-t}{CR}}$$

Rearranging the relationship to make t the subject gives:

$$t = -CR \times \ln\left(\frac{v_C}{V}\right) = -0.47 \times \ln\left(\frac{10}{20}\right)$$

$$= -0.47 \times -0.693 = 0.325 \text{ s}$$

Key Point. The time constant of a C–R circuit is the product of the capacitance, C, and resistance, R

Key Point. The voltage across the plates of a charging capacitor grows exponentially at a rate determined by the time constant whereas the voltage across the plates of a discharging capacitor decays exponentially

Key Point. When a capacitor is charged, energy taken from the supply is stored in the electric field that appears between its plates. The capacitor can subsequently be discharged in order to release the stored energy

SCILAB

The following SCILAB script produces a plot of capacitor and resistor voltage for the discharging case:

```
// C-R circuit — decay of voltage on discharge
// SCILAB source file: ex2302.sce
//
// Clear any previous graphic window
clf();
//
txt = ['Input resistance (R) in kohm';'Input
capacitance (C) in uF';'Supply voltage'];
```

```
var = x_mdialog('Enter resistance and
capacitance',txt,['100';'100';'100']);
r = evstr(var(1));
c = evstr(var(2));
vm = evstr(var(3));
c = 1E-6*c;
r = 1E+3*r;
//
T=c*r;
t = [0:1:10*T]'; // set time scale to 10CR
//
vc = vm*(%e^(-t/(c*r)));
vr = vm-vc;
// Plot
plot2d(t,vc,style=5); // red line for
capacitor voltage
plot2d(t,vr,style=3); // green line for
resistor voltage
//
// Print grid and title
xgrid(); // add grid
xtitle('C-R discharge - voltage plotted
against time');
hl=legend(['Capacitor voltage';'Resistor
voltage']);
```

23.2.3 Tabular method

In order to simplify the mathematics of exponential growth and decay, the table below provides an alternative tabular method that may be used to determine the voltage and current in a C–R circuit

$\dfrac{t}{CR}$	k (ratio of instantaneous value to final value)	
	Exponential growth	Exponential decay
0.0	0.0000	1.0000
0.1	0.0951	0.9048
0.2	0.1812	0.8187 (see Example 23.5)
0.3	0.2591	0.7408
0.4	0.3296	0.6703
0.5	0.3935	0.6065

$\dfrac{t}{CR}$	k (ratio of instantaneous value to final value)	
	Exponential growth	Exponential decay
0.6	0.4511	0.5488
0.7	0.5034	0.4965
0.8	0.5506	0.4493
0.9	0.5934	0.4065
1.0	0.6321	0.3679
1.5	0.7769	0.2231
2.0	0.8647	0.1353
2.5	0.9179	0.0821
3.0	0.9502	0.0498
3.5	0.9698	0.0302
4.0	0.9817	0.0183
4.5	0.9889	0.0111
5.0	0.9933	0.0067

Example 23.5 A 150 μF capacitor is charged to a potential of 150 V. The capacitor is then removed from the charging source and connected to a 2 MΩ resistor. Assuming that there is no loss of energy, determine the capacitor voltage one minute later.

Solution

We will solve this problem using the tabular method rather than using the exponential formula. First we need to find the time constant:

$$C \times R = 150 \times 10^{-6} \times 2 \times 10^{6} = 300\,\text{s}$$

Next we find the ratio of t to CR. After 1 minute, $t = 60\,\text{s}$ therefore the ratio of t to CR is:

$$\frac{t}{CR} = \frac{60}{300} = 0.2$$

Referring to the table we find that when $t/CR = 0.2$, the ratio of instantaneous value to final value (k) for decay is 0.8187.

Thus:

$$\frac{v_\text{C}}{V} = 0.8187$$

Hence $v_\text{C} = 0.8187 \times 150 = 122.8\,\text{V}$

SCILAB

A more effective method of solving problems involving exponential growth or decay is the use of SCILAB scripts. The two examples that follow are for the charging (see Figures 23.11 and 23.12) and discharging cases, respectively:

```
// C-R circuit — voltage and current during
charge
// SCILAB source file: ex2303.sce
//
// Clear any previous graphic window
clf();
//
txt = ['Input resistance (R) in kohm';'Input
capacitance (C) in uF';'Supply
voltage in V';'Time in s'];
var = x_mdialog('Enter resistance and
capacitance',txt,['100';'100';'100';'1']);
r = evstr(var(1));
c = evstr(var(2));
vm = evstr(var(3));
t = evstr(var(4));
c = 1E-6*c;
r = 1E+3*r;
//
vc = vm*(1 - %e^(-t/(c*r)));
vr = vm-vc;
i = vr/r;
i = i*1E+3;
// Display results
messagebox(["Capacitor voltage = " +
string(vc) + " V" "Current =
" + string(i) + ",mA"]);
```

```
// C-R circuit — voltage and current during
discharge
// SCILAB source file: ex2304.sce
//
// Clear any previous graphic window
```

```
clf();
//
txt = ['Input resistance (R) in kohm';'Input
capacitance (C) in uF';'Supply
voltage in V';'Time in s'];
var = x_mdialog('Enter resistance and
capacitance',txt,['100';'100';'100';'1']);
r = evstr(var(1));
c = evstr(var(2));
vm = evstr(var(3));
t = evstr(var(4));
c = 1E-6*c;
r = 1E+3*r;
//
vc = vm*%e^(-t/(c*r));
vr = vc;
i = vr/r;
i = i*1E+3;
// Display results
messagebox(["Capacitor voltage = " +
string(vc) + " V" "Current =
" + string(i) + " mA"]);
```

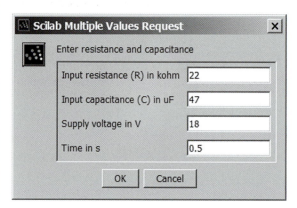

Figure 23.11 Input dialogue for the *C–R* circuit charge script

Figure 23.12 Output message box for the *C–R* circuit charge script

23.3 L–R circuits

As we saw earlier, inductors provide us with a means of storing electrical energy in the form of a magnetic field. It is also worth remembering that, in practice, every coil has both inductance and resistance and the circuit of Figure 23.13 shows these as two discrete components. In reality the inductance, L, and resistance, R, are both distributed throughout the component but it is convenient to treat the inductance and resistance as separate components in the analysis of the circuit.

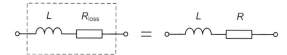

Figure 23.13 A real inductor has resistance as well as inductance

When the *L–R* network is connected to a constant voltage source (V), as shown in Figure 23.14, the current (i) flowing in the inductor will rise exponentially when the switch is closed, as shown in Figure 23.15. At the same time, the voltage dropped across the inductor (v_L) will fall, as shown in Figure 23.16

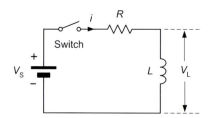

Figure 23.14 *L–R* circuit with energy supplied to *L*

The rate of growth of voltage with time and decay of current with time will be dependent upon the ratio of inductance to resistance. This value is known as the *time constant* of the circuit. Hence:

$$\text{Time constant, } t = \frac{L}{R}$$

where L is the value of inductance (in H), R is the resistance (in Ω), and t is the time constant (in s).

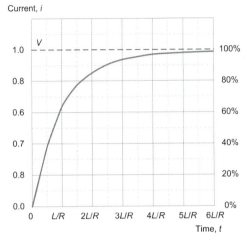

Current, i

Figure 23.15 Exponential growth of inductor current (i) in Figure 23.14

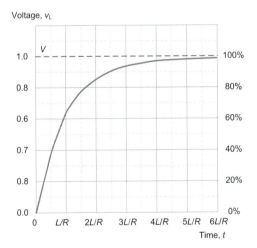

Voltage, v_L

Figure 23.16 Exponential decay of inductor voltage (v_L) in Figure 23.14

Key Point. All practical inductors possess resistance as well as inductance

23.3.1 Growth of current in an *L–R* circuit

In section 23.1 we showed how the current flowing in an inductor (i) grows exponentially with time (t) according to the relationship:

$$i = I\left(1 - e^{-\frac{Rt}{L}}\right) = \frac{V}{R}\left(1 - e^{-\frac{Rt}{L}}\right)$$

where i is the inductor current (in A), V is the d.c. supply voltage (in V), t is the time (in s), and L/R is the time constant of the circuit (equal to the ratio of inductance, L, to resistance, R).

The maximum value of current, I, flowing in the inductor will be equivalent to the ratio of supply voltage, V, to resistance, R. Hence:

$$i = \frac{V}{R}\left(1 - e^{-\frac{Rt}{L}}\right)$$

The inductor current will rise to approximately 63% of the maximum current (equivalent to the supply voltage, V divided by the resistance, R) in a time interval equal to the time constant. At the end of the next interval of time equal to the time constant (i.e. after an elapsed time equal to $2L/R$) the voltage will have risen by 63% of the remainder, and so on. In theory, the inductor will never quite become fully charged with flux. However, after a period of time equal to $5L/R$, the inductor current will to all intents and purposes be equal to the maximum value, V/R. At this point the inductor current will have risen to 99.3% of its final value and we can consider it to be fully charged with flux.

The voltage dropped across the inductor can be determined by using Ohm's law:

$$v_L = V - iR = V - I\left(1 - e^{\frac{-Rt}{L}}\right)R$$

$$= V - IR + IRe^{\frac{-Rt}{L}} = IR - IR + IRe^{\frac{-Rt}{L}} = Ve^{\frac{-Rt}{L}}$$

Hence, during the build-up of magnetic flux within the inductor, the voltage dropped across it (v_L) decays exponentially with time (t) according to the relationship:

$$v_L = Ve^{\frac{-Rt}{L}}$$

where v_L is the inductor voltage (in V), V is the d.c. supply voltage (in V), t is the time (in s), and L/R is the time constant of the circuit (equal to the ratio of inductance, L, to resistance, R).

The inductor voltage will fall to approximately 37% of the maximum current (equivalent to the supply voltage divided by the resistance) in a time equal to the time constant. At the end of the next interval of time equal to the time constant (i.e. after a total time of $2L/R$ has elapsed) the current will have fallen by a further 37% of the remainder, and so on.

Example 23.6   A coil having an inductance of 4 H and resistance of 16 Ω is connected to a 12 V d.c. supply. Determine the current supplied to the coil 0.1 s after it is connected to the supply.

Solution

Using the relationship for exponential growth of current in an inductor gives:

$$i = \frac{V}{R}\left(1 - e^{-\frac{Rt}{L}}\right) = \frac{12}{48}\left(1 - e^{-\frac{16\times0.1}{4}}\right)$$

$$= 0.25\left(1 - e^{-0.4}\right) = 0.25\left(1 - 0.67\right) = 0.0825 \text{ A}$$

Key Point.   The time constant of an L–R circuit is the ratio of inductance, L, to resistance, R

Key Point.   The current flowing in an inductor will grow exponentially when it is first connected to a supply. During this process the energy taken from the supply will be stored in the magnetic field that surrounds the inductor. When the supply current is subsequently interrupted the field will collapse and the energy will be released

SCILAB

The following SCILAB script produces a plot showing the growth of current in an L–R circuit:

```
// L-R circuit - growth of current
// SCILAB source file: ex2305.sce
//
// Clear any previous graphic window
clf();
//
txt = ['Input resistance (R) in ohm';'Input
inductance (L) in H';'Supply
voltage'];
var = x_mdialog('Enter resistance and
inductance',txt,['100';'100';'100']);
r = evstr(var(1));
l = evstr(var(2));
```

```
vm = evstr(var(3));
//
T=l/r;
t = [0:0.1:10*T]'; // set time scale to 10L/R
//
i = (vm/r)*(%e^(-r*t/l));
vl = vm-(i*r);
// Plot
plot2d(t,vl,style=5); // red line for
inductor voltage
//
// Print grid and title
xgrid(); // add grid
xtitle('L-R growth - current plotted
against time');
hl=legend(['Inductor current']);
```

23.3.2   Decay of current in an *L–R* circuit

When current is flowing in an inductor, energy is stored within it in the form of a magnetic field. When the inductor from Figure 23.14 is connected as shown in Figure 23.17, the energy stored in the inductor will be released and the current flowing in the inductor (i) will fall exponentially with time, as shown in Figure 23.18.

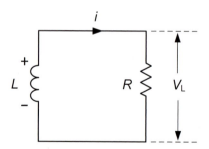

Figure 23.17 *L–R* circuit with energy released into R

The current flowing in the circuit (i) varies with time (t) according to the relationship:

$$i = \frac{V}{R}e^{-\frac{Rt}{L}}$$

where i is the inductor current (in A), V is the d.c. supply voltage (in V), t is the time (in s), and L/R is the time constant of the circuit (equal to the ratio of inductance, L, to resistance, R).

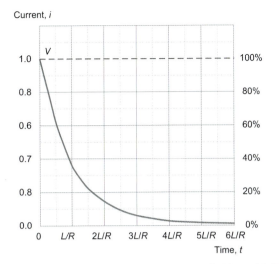

Figure 23.18 Exponential decay of current (i) in Figure 23.17

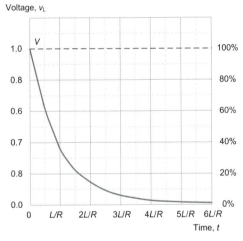

Figure 23.19 Exponential decay of inductor voltage (v_L) in Figure 23.17

The current will fall to approximately 37% of the initial current (equal to V/R) in a time equal to the time constant. At the end of the next interval of time equal to the time constant (i.e. after an elapsed time equal to $2L/R$) the current will have fallen by 37% of the remainder, and so on. In theory, the energy stored in the inductor will never quite become exhausted. After a period of time equal to $5L/R$, however, the inductor current voltage will to all intents and purposes be zero. At this point the current will have fallen below 1% of its initial value and we can consider the stored energy to be fully exhausted.

Similarly, the voltage that appears across both the inductor and the resistor will decay according to the relationship:

$$v_L = v_R = IRe^{\frac{-Rt}{L}}$$

where v_L is the inductor voltage and v_R is the resistor voltage (both in V), I is the initial current (in A), t is the time (in s), and L/R is the time constant of the circuit (equal to the ratio of inductance, L, to resistance, R).

The voltage will fall to approximately 37% of the initial voltage in a time equal to the time constant. At the end of the next interval of time equal to the time constant (i.e. after a total time of $2L/R$ has elapsed) the voltage will have fallen by a further 37% of the remainder, and so on.

> **Test your knowledge 23.4**
>
> Write and test a SCILAB script that will calculate the current in an L–R circuit for given values of inductance, resistance, supply voltage and time.

23.3.3 Tabular method

The table below provides an alternative tabular method that may be used to determine the voltage and current in an L–R circuit

$\dfrac{Rt}{L} = \dfrac{t}{\left(\dfrac{L}{R}\right)}$	k (ratio of instantaneous value to final value)	
	Exponential growth	**Exponential decay**
0.0	0.0000	1.0000
0.1	0.0951	0.9048
0.2	0.1812	0.8187
0.3	0.2591	0.7408
0.4	0.3296	0.6703
0.5	0.3935	0.6065
0.6	0.4511	0.5488
0.7	0.5034	0.4965

$\dfrac{Rt}{L} = \dfrac{t}{\left(\dfrac{L}{R}\right)}$	k (ratio of instantaneous value to final value)	
	Exponential growth	Exponential decay
0.8	0.5506	0.4493
0.9	0.5934	0.4065
1.0	0.6321	0.3679
1.5	0.7769	0.2231
2.0	0.8647 (see Example 23.7)	0.1353
2.5	0.9179	0.0821
3.0	0.9502	0.0498
3.5	0.9698	0.0302
4.0	0.9817	0.0183
4.5	0.9889	0.0111
5.0	0.9933	0.0067

Example 23.7 A 15 H inductor of negligible resistance is connected in series with a resistance of 10 Ω. The resulting combination is connected to a 100 V d.c. supply. Determine the current supplied to the circuit three seconds after connection.

Solution

We will solve this problem using the tabular method rather than using the exponential formula. First we need to find the time constant:

$$L/R = 15/10 = 1.5 \text{ s}$$

Next we find the ratio of t to L/R. After three seconds the ratio of t to L/R is:

$$\frac{t}{\left(\dfrac{L}{R}\right)} = \frac{3}{\left(\dfrac{15}{10}\right)} = \frac{3}{1.5} = 2$$

Referring to the table we find that when $Rt/L = 2$, the ratio of instantaneous value to final value (k) for decay is 0.8647.

Thus:

$$\frac{i}{I} = 0.8647$$

Hence: $i = 0.8647 \times I = 0.8647 \times \dfrac{V}{R} = 0.8647 \times \dfrac{50}{10} = 4.324 \text{ A}.$

23.4   Chapter summary

In electric circuits transients occur whenever there is a change of state. Even though they may last only a short time, they can involve significant changes in voltage and current. The following are the main points introduced in this chapter:

- When the voltage applied to a capacitor changes at a uniform rate a current will flow in the capacitor. The magnitude of this voltage will be directly proportional to the rate of change of voltage.
- When the current in an inductor changes at a uniform rate an induced e.m.f. will appear across the terminals of the inductor with a polarity that opposes the initial change. The magnitude of this voltage will be directly proportional to the rate of change of current.
- When a capacitor is charged from a constant voltage supply, the voltage across the capacitor's plates will increase exponentially at a rate determined by the time constant of the circuit. During this process, energy is taken from the supply to be stored in the electric field that appears between the capacitor's plates. The capacitor can subsequently be discharged in order to release the stored energy.
- When an inductor is connected to a constant voltage supply, the current flowing in the inductor will increase exponentially at a rate determined by the time constant of the circuit. During this process, energy is taken from the supply to be stored in the magnetic field surrounding the inductor. The stored energy will later be released when the flow of current is interrupted.

We shall further develop this important topic in the next chapter, when we will show how Laplace transforms enable us to use straightforward algebraic equations to solve expressions in which several differential or integral terms are present. These powerful techniques allow us to easily solve circuits in which all three types of component (inductance, capacitance and resistance) are present.

23.5   Review questions

Short answer questions

1. What voltage will appear across the terminals of a 10 H inductor if the current flowing in it increases uniformly at a rate of 40 A/s?

2. A 680 μF capacitor is connected to a supply voltage that falls at a uniform rate from 230 V to 120 V in 100 ms. What current will flow back into the supply?

3. A C–R circuit is to have a time constant of 150 ms. If a capacitor of 220 nF is to be used, determine the value of resistance required.

4. An L–R circuit comprises an inductance of 68 mH and a resistance of 470 Ω. Determine the time constant of the circuit.

5. A 22 μF capacitor is connected in series with a fixed resistor of 10 kΩ and a further variable resistor having a maximum value of 50 kΩ. Determine the range of time constant values that can be achieved with this arrangement.

6. A coil having an inductance of 12 H and a resistance of 180 Ω is connected to a 48 V d.c. supply. Calculate the current flowing in the circuit after a time of (a) 50 ms and (b) 100 ms from switching on.

7. A capacitor is charged from a 500 V d.c. supply via a resistance of 250 kΩ. If it takes 200 ms for the capacitor voltage to reach 300 V, determine the value of capacitance.

8. Prove that, if the initial rate of change could be sustained in a C–R circuit the final voltage would be reached in a time equal to the time constant.

9. Prove that, if the initial rate of change could be sustained in an L–R circuit the final current would be reached in a time equal to the time constant.

10. Explain why a very high voltage appears momentarily across the terminals of an inductor when the supply current is suddenly removed.

Long answer questions

11. A high-voltage power supply operates at 2.5 kV and its output is developed across a reservoir capacitor of 2 μF. Calculate the value of resistance required to discharge this capacitor to a voltage of 50 V within a time period of 200 ms after the supply has been switched off.

12. A relay coil operates from a 24 V supply and its contacts close whenever a current of 80 mA (or greater) flows in its coil. If the relay has an inductance of 20 H and a resistance of 150 Ω determine:

    (a) the steady-state current consumed by the relay;

    (b) the time taken for the relay to operate.

13. A solenoid has an inductance of 12 H and a resistance of 50 Ω and pulls in when a current of 0.25 A (or more) is applied to it. Determine the minimum supply voltage required to operate the solenoid within a time of 100 ms.

Chapter 24

Transients in R–L–C systems

In the previous chapter we described how circuits containing a combination of resistance and capacitance or resistance and inductance behave when subjected to a sudden change in voltage or current, known as a *transient*. In this chapter we will develop a better understanding of how these and more complex circuits behave by treating them as *systems* rather than networks of individual components.

24.1 First- and second-order systems

There are two types of system that we shall be concerned with: first-order and second-order systems:

*   A *first-order system* is one that can be modelled using a first-order differential equation. First-order electrical systems involve combinations of C and R or L and R.
*   A *second-order system* is one that can be modelled using a second-order differential equation. Second-order electrical systems involve combinations of all three types of component: L, C and R.

You met the time response of simple C–R and L–R systems in the last chapter. In the first case, the voltage developed across the capacitor in a C–R circuit will grow exponentially (and the current will decay exponentially) as the capacitor is charged from a constant voltage supply (see Figure 24.1). In the second case, the inductor voltage will decay exponentially (and the current will grow exponentially) as the inductor is charged from a constant voltage supply (see Figure 24.2).

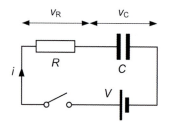

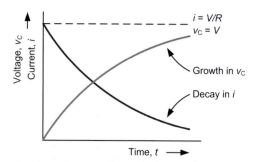

Figure 24.1 Time response of a series C–R network

978-1-85617-775-7, Engineering Science, Mike Tooley & Lloyd Dingle

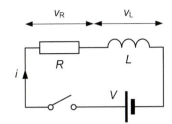

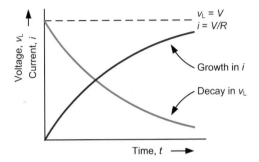

Figure 24.2 Time response of a series *L–R* network

The mathematical relationships between voltage, current and time for the simple series *C–R* and *L–R* circuits are as follows:

For Figure 24.1:

$$i = \frac{V}{R}e^{-\frac{t}{CR}} \text{ and } v_C = V\left(1 - e^{\frac{-t}{CR}}\right)$$

For Figure 24.2:

$$i = \frac{V}{R}\left(1 - e^{-\frac{Rt}{L}}\right) \text{ and } v_L = V\left(1 - e^{\frac{-Rt}{L}}\right)$$

The voltage impressed on a first- or second-order circuit is often referred to as the *forcing function* whilst the current is known as the *forced response*. In a simple case (as in Figures 24.1 and 24.2), once the switch has been closed, the forcing function is constant. However, this is not always the case as the forcing function may *itself* be a function of time. This added degree of complexity requires a different approach to the solution of these, apparently simple, circuits.
Consider a voltage of the form:

$$v = 100 + 50\sin\omega t \text{ V}$$

When used as a forcing function, this voltage can be thought of as constant step function (of amplitude 100 V) onto which is superimposed a sinusoidal variation (of amplitude 50 V). We will explore sinusoidal functions in the next chapter, but for the moment you just need to remember that forcing functions may not simply be constant values of voltage or current.

In general, we can describe a forcing function as a voltage, $v(t)$, where the brackets and the t remind us that the voltage is a *function of time*. The forced function, $i(t)$, would similarly be a function of time (again denoted by the brackets and the t).

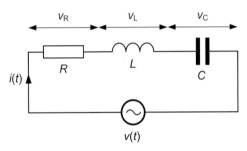

Figure 24.3 A series *L–C–R* circuit showing forcing function (voltage) and forced function (current)

Now take a look at Figure 24.3. Here a forcing function, $v(t)$, is applied to a simple *L–C–R* circuit. By applying Kirchhoff's voltage law we can deduce that:

$$v(t) = v_R + v_L + v_C$$

where $v_R = i(t)R$
and $v_L = L \times$ (rate of change of current with time)

$$= L\frac{di(t)}{dt}$$

and $v_C = \frac{1}{C} \times$ (area under the current–time graph)

$$= \frac{1}{C}\int i(t)dt$$

Thus $v(t) = i(t)R + L\dfrac{di(t)}{dt} + \dfrac{1}{C}\int i(t)dt$

This is an important relationship and you should have spotted that it is a second-order *differential equation*. Solving the equation can be problematic but is much simplified by using a technique called the Laplace transform.

24.2 Laplace transforms

The Laplace transform provides us with a means of transforming differential equations into straightforward algebraic equations, thus making the solution of expressions involving several differential or integral terms relatively simple.

The Laplace transform of a function of time, $f(t)$, is found by multiplying the function by e^{-st} and integrating the product between the limits of zero and infinity. The result (if it exists) is known as the Laplace transform of $f(t)$.

Thus:

$$F(s) = \mathcal{L}\{f(t)\} = \int_0^\infty e^{-st} f(t)\,dt$$

Frequently we need only to refer to a table of standard Laplace transforms. Some of the most useful of these are summarised in the table below:

$f(t)$	$F(s) = \mathcal{L}\{f(t)\}$	Comment
a	$\dfrac{1}{s}$	A constant
t	$\dfrac{1}{s^2}$	Ramp
e^{at}	$\dfrac{1}{s-a}$	Exponential growth
e^{-at}	$\dfrac{1}{s+a}$	Exponential decay
$e^{-at}\sin(\omega t)$	$\dfrac{\omega}{(s+a)^2 + \omega^2}$	Decaying sine
$e^{-at}\cos(\omega t)$	$\dfrac{s+a}{(s+a)^2 + \omega^2}$	Decaying cosine
$\sin(\omega t + \phi)$	$\dfrac{s\sin\phi + \omega\cos\phi}{s^2 + \omega^2}$	Sine plus phase angle
$\sin(\omega t)$	$\dfrac{\omega}{s^2 + \omega^2}$	Sine
$\cos(\omega t)$	$\dfrac{s}{s^2 + \omega^2}$	Cosine
$\dfrac{df(t)}{dt}$	$sF(s) - f(0)$	First differential
$\dfrac{d^2 f(t)}{dt^2}$	$s^2 F(s) - sf(0) - \dfrac{df(0)}{dt}$	Second differential
$\int f(t)dt$	$\dfrac{1}{s}F(s) + \dfrac{1}{s}F(0)$	Integral

The technique for solving problems using the Laplace transform is as follows:

1. Write down the basic expression for the circuit in terms of voltage, current and component values.

2. Transform the basic equation using the table of standard Laplace transforms (each term in the expression is transformed separately).

3. Simplify the transformed expression as far as possible. Insert initial values. Also insert component values where these are provided.

4. When you simplify the expression, try to arrange it into the same form as one of the standard forms in the table of standard Laplace transforms. Failure to do this will prevent you from performing the inverse transformation (i.e. converting the expression back into the time domain). Note that you may have to use partial fractions to produce an equation containing terms that conform to the standard form.

5. Use the table of standard Laplace transforms in reverse in order to obtain the inverse Laplace transform.

Don't panic if this is beginning to sound very complicated – the three examples that follow will take you through the process on a step-by-step basis!

Key Point. You can think of the Laplace transform as a device that translates a given function in the time domain, $f(t)$, into an equivalent function in the s-domain, $F(s)$. We can translate a differential equation into an equation in the s-domain, solve the equation, and then use the transform in reverse to convert the s-domain solution to an equivalent solution in the time domain. Figure 24.4 illustrates this process

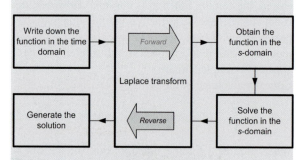

Figure 24.4 Laplace transformation process

Test your knowledge 24.1

A voltage is given by $v = 50\sin(100t + \pi/4)$ and is applied to a resistor of $100\,\Omega$. Use the table of standard Laplace transforms to write down an expression for the current flowing in the resistor in the s-domain.

Example 24.1 Use Laplace transforms to derive an expression for the current flowing in the circuit shown in Figure 24.5, given that $i = 0$ at $t = 0$.

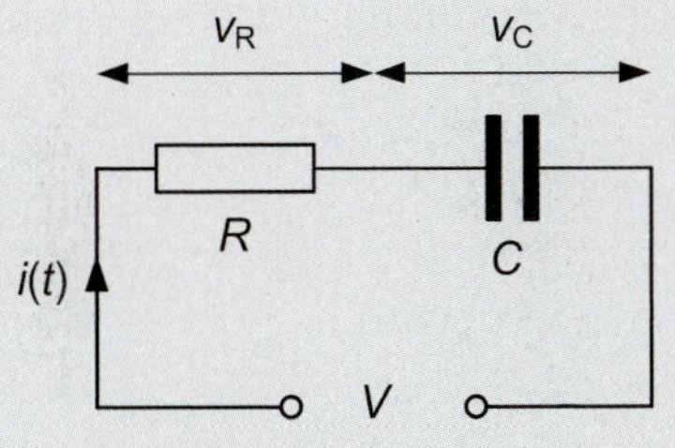

Figure 24.5 See Example 24.1

The voltage, V, will be the sum of the two voltages v_R and v_C, where $v_R = Ri(t)$ and $v_C = \frac{1}{C}\int i(t)\mathrm{d}t$. Note that, to remind us that the current is a function of time, we have used $i(t)$ to denote current rather than just i on its own.

Thus $V = v_R + v_C = Ri(t) + \frac{1}{C}\int i(t)\mathrm{d}t$

Applying the Laplace transform gives:

$$\mathcal{L}\{V\} = \mathcal{L}\{Ri(t) + \frac{1}{C}\int i(t)\mathrm{d}t\}$$

$$= \mathcal{L}\{Ri(t)\} + \mathcal{L}\left\{\frac{1}{C}\int i(t)\mathrm{d}t\right\}$$

$$\mathcal{L}\{V\} = R\mathcal{L}\{i(t)\} + \frac{1}{C}\mathcal{L}\left\{\int i(t)\mathrm{d}t\right\}$$

Using the table of standard Laplace transforms on page 436 gives:

$$\frac{V}{s} = RI(s) + \frac{1}{sC}[I(s) + I(0)]$$

Note that we are now in the s-domain (s replaces t) and the large I has replaced the small i (just as F replaces f in the general expression).

Now $i = 0$ at $t = 0$, thus $I(0) = 0$,

hence:

$$\frac{V}{s} = RI(s) + \frac{1}{sC}[I(s) - 0] = RI(s) + \frac{1}{sC}I(s)$$

$$= I(s) \times \left(R + \frac{1}{sC}\right)$$

or

$$I(s) = \frac{V}{s\left(R + \frac{1}{sC}\right)} = \frac{V}{Rs + \frac{1}{C}}$$

$$= \frac{V}{R}\left(\frac{1}{s + \frac{1}{CR}}\right).$$

Now $i(t) = \mathcal{L}^{-1}\{I(s)\}$, thus:

$$i(t) = \mathcal{L}^{-1}\{I(s)\} = \mathcal{L}^{-1}\left\{\frac{V}{R}\left(\frac{1}{s + \frac{1}{CR}}\right)\right\}$$

$$= \frac{V}{R}\mathcal{L}^{-1}\left\{\frac{1}{s + \frac{1}{CR}}\right\}$$

We now need to find the inverse transform of the foregoing equation. We can do this by examining the table of standard Laplace transforms on page 436, looking for $F(s)$ of the form $\frac{1}{s+a}$ (where $a = \frac{1}{CR}$).

Notice, from the table, that the inverse transform of $\frac{1}{s+a}$ is e^{-at}, i.e.:

$$\mathcal{L}^{-1}\left\{\frac{1}{s+a}\right\} = e^{-at}$$

Hence

$$i(t) = \mathcal{L}^{-1}\{I(s)\} = \frac{V}{R}\mathcal{L}^{-1}\left\{\frac{1}{s + \frac{1}{CR}}\right\} = \frac{V}{R}\left(\mathrm{e}^{-\frac{t}{CR}}\right)$$

Thus the current in the circuit is given by:

$$i(t) = \frac{V}{R}\left(\mathrm{e}^{-\frac{t}{CR}}\right) \text{ (exponential growth)}$$

Example 24.2 Use Laplace transforms to derive an expression for the current flowing in the circuit shown in Figure 24.6, given that $i = 0$ at $t = 0$.

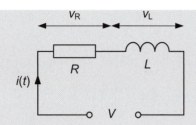

Figure 24.6 See Example 24.2

The voltage, V, will be the sum of the two voltages v_R and v_L, where $v_R = Ri(t)$ and $v_L = L\dfrac{di(t)}{dt}$ and $i(t)$ reminds us that the current, i, is a function of time.

Thus $V = v_R + v_L = Ri(t) + L\dfrac{di(t)}{dt}$

Applying the Laplace transform gives:

$$\mathcal{L}\{V\} = \mathcal{L}\left\{Ri(t) + L\frac{di(t)}{dt}\right\}$$

$$= \mathcal{L}\{Ri(t)\} + \mathcal{L}\left\{L\frac{di(t)}{dt}\right\}$$

$$\mathcal{L}\{V\} = R\mathcal{L}\{i(t)\} + \mathcal{L}L\left\{\frac{di(t)}{dt}\right\}$$

Using the table of standard Laplace transforms on page 436 gives:

$$\frac{V}{s} = RI(s) + sL[I(s) - i(0)]$$

Note that we are now in the s-domain (s replaces t) and the large I has replaced the small i (just as F replaces f in the general expression).

Now $i = 0$ at $t = 0$, thus $i(0) = 0$, hence:

$$\frac{V}{s} = RI(s) + sL[I(s) - 0]$$

$$= RI(s) + sLI(s) = I(s) \times (R + sL)$$

or

$$I(s) = \frac{V}{s(R + sL)}$$

We can simplify this expression by using partial fractions, thus:

$$\frac{V}{s(R + sL)} = \frac{A}{s} + \frac{B}{R + sL} = \frac{A(R + sL) + Bs}{s(R + sL)}$$

hence $V = A(R + sL) + Bs = AR + AsL + Bs$.

We now need to find the values of A, B and C.

When $s = 0$, $V = AR + 0$, thus $A = \dfrac{V}{R}$

When $s = \dfrac{-R}{L}$,

$$V = AR + A\left(\frac{-R}{L}\right) + B\left(\frac{-R}{L}\right)$$

$$= AR - AR - B\frac{R}{L}$$

thus $V = \dfrac{-BR}{L}$ or $B = \dfrac{-VL}{R}$

Replacing A and B with $\dfrac{V}{R}$ and $\dfrac{-VL}{R}$ respectively gives:

$$\frac{V}{s(R + sL)} = \frac{A}{s} + \frac{B}{R + sL} = \frac{V}{Rs} + \frac{\left(\frac{-VL}{R}\right)}{R + sL}$$

$$= \frac{V}{Rs} - \frac{VL}{R(R + sL)}$$

Thus

$$I(s) = \frac{V}{Rs} - \frac{VL}{R(R + sL)}$$

We need to find the inverse of the above equation, since:

$$i(t) = \mathcal{L}^{-1}\{I(s)\} = \mathcal{L}^{-1}\left\{\frac{V}{Rs} - \frac{VL}{R(R + sL)}\right\}$$

thus

$$i(t) = \mathcal{L}^{-1}\left\{\frac{V}{Rs} - \frac{VL}{R(R + sL)}\right\}$$

$$= \frac{V}{R}\mathcal{L}^{-1}\left\{\frac{1}{s} - \frac{L}{R + sL}\right\}$$

hence

$$i(t) = \frac{V}{R}\mathcal{L}^{-1}\left\{\frac{1}{s} - \frac{1}{s + \frac{R}{L}}\right\}$$

Referring once again to the table of standard Laplace transforms gives:

$$i(t) = \frac{V}{R}\left(1 - e^{\frac{-Rt}{L}}\right) \text{ (exponential decay)}$$

Example 24.3 Use Laplace transforms to derive an expression for the current flowing in the circuit shown in Figure 24.7, given that $i = 0$ at $t = 0$.

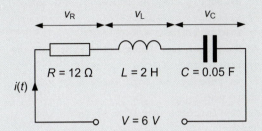

Figure 24.7 See Example 24.3

Here, the voltage, V, will be the sum of three voltages, v_R, v_L and v_C, where:

$$v_R = Ri(t), \ v_L = L\frac{di(t)}{dt} \text{ and } v_C = \frac{1}{C}\int i(t)\mathrm{d}t$$

Yet again, $i(t)$ reminds us that the current, i, is a function of time.

Thus

$$V = v_R + v_L + v_C$$

$$= Ri(t) + L\frac{di(t)}{dt} + \frac{1}{C}\int i(t)\mathrm{d}t$$

Applying the Laplace transform gives:

$$\mathcal{L}\{V\} = \mathcal{L}\left\{Ri(t) + L\frac{di(t)}{dt} + \frac{1}{C}\int i(t)\mathrm{d}t\right\}$$

$$\mathcal{L}\{V\} = R\mathcal{L}\{i(t)\} + L\mathcal{L}\left\{\frac{di(t)}{dt}\right\} + \mathcal{L}\left\{\frac{1}{C}\int i(t)\mathrm{d}t\right\}$$

Using the table of standard Laplace transforms on page 436 gives:

$$\frac{V}{s} = RI(s) + sL[I(s) - i(0)] + \frac{1}{sC}[I(s) + I(0)]$$

Now $i = 0$ at $t = 0$, thus $i(0) = 0$ and $I(0) = 0$, hence:

$$\frac{V}{s} = RI(s) + sL[I(s) - 0] + \frac{1}{sC}[I(s) + I(0)]$$

$$\frac{V}{s} = RI(s) + sLI(s) + \frac{1}{sC}I(s)$$

$$= I(s) \times \left(R + sL + \frac{1}{sC}\right)$$

or

$$I(s) = \frac{V}{s(R + sL + \frac{1}{sC})} = \frac{V}{sR + s^2L + \frac{1}{C}}$$

Now $V = 6\,\text{V}$, $L = 2\,\text{H}$, $R = 10\,\Omega$ and $C = 0.05\,\text{F}$, thus:

$$I(s) = \frac{V}{s^2L + sR + \frac{1}{C}} = \frac{6}{2s^2 + 12s + \frac{1}{0.05}}$$

$$= \frac{3}{s^2 + 6s + 10} \tag{i}$$

In order to perform the inverse Laplace transformation we need to express equation (i) in a form that resembles one of the standard forms listed earlier on page 436. The nearest form is as follows:

$$F(s) = \frac{\omega}{(s + a)^2 + \omega^2} \text{ which has}$$

$$f(t) = \mathrm{e}^{-at}\sin\omega t \text{ as its reverse transform.}$$

Fortunately, it's not too difficult to rearrange equation (i) into a more usable form:

$$I(s) = \frac{3}{s^2 + 6s + 10} = \frac{3}{s^2 + 6s + 9 + 1}$$

$$= \frac{3}{(s + 3)(s + 3) + 1}$$

or

$$I(s) = 3 \times \frac{1}{(s + 3)^2 + 1}$$

from which $a = 3$ and $\omega = 1$, thus the reverse transform is:

$$i(t) = \mathrm{e}^{-3t}\sin t \text{ (an exponentially}$$

$$\text{decaying sine wave)}$$

Test your knowledge 24.2

Write and test a SCILAB to obtain a graph of current plotted against time for a first-order *L–R* circuit comprising $L = 4\,\text{H}$ and $R = 6\,\Omega$ when connected to a 24 V constant voltage supply.

Test your knowledge 24.3

Use SCILAB to obtain a graph of current plotted against time for the second-order system in Example 24.3.

SCILAB

The following two scripts illustrate the use of the syslin and csim functions that are available in SCILAB, The first example shows how a first-order *C–R* system can be modelled whilst the second example illustrates the solution of a second-order *L–C–R* system. The graphical output of these two scripts is shown in Figures 24.8 and 24.9, respectively. Note how the current in the first-order system decays exponentially whilst that in the second-order system takes the form of an exponentially decaying sine wave.

```
// Current in a first order C-R system
// with 10V supply, 100 ohm in series with
40mF
// Filename: ex2401.sce
// V = 10, R = 100 and C = 40 x 10-3 = 0.04
// i(s) = V/(Rs + 1/C)
// Numerator and denominator must both be
expressed as
// polynomials of the form (as^2 + bs + c)
where the
// coefficients are [c b a] (note the order!)
// Numerator is thus (0 s^2) + (0 s) + V
// thus c = V = 10, b = 0 and a = 0
num = poly([10 0 0], 's', 'coeff');
// Denominator is (0 s^2+ Rs + 1/C) so
c = 1/0.04=25,
// b = 100 and a = 0
den = poly([25 100 0], 's', 'coeff');
// Range of time values to plot
t = 0.001: 0.1: 20; // Avoid zero
s = syslin('c', num/den);
gs = csim('impulse', t, s);
clf;
plot2d(t, gs, 5); // Red plot line
xgrid;
xtitle('Current in a first order (C-R)
system for a 10V supply','t (s)','i (A)');
```

```
// Current in a second order L-C-R
system
// 6V supply, L=2 H, C=0.05 F, R=1 ohm
// Filename: ex2402.sce
// i(s) = V/(Ls^2 + Rs + 1/C)
// Numerator and denominator must both be
expressed as
// polynomials of form (as^2 + bs + c)
where the
// coefficients are [c b a] (note the
order!)
// Numerator is thus (0 s^2) + (0 s) + V
// so c = V, b = 0 and a = 0
num = poly([6 0 0], 's', 'coeff');
// Denominator is (2s^2+ 1s + 1/0.05) so
c=1/0.05=20,
// b=1 and a=2
den = poly([20 1 2], 's', 'coeff');
// Range of time values to plot
t = 0.001: 0.1: 20; // Avoid zero
s = syslin('c', num/den);
gs = csim('impulse', t, s);
clf;
plot2d(t, gs, 5); // Red plot line
xgrid;
xtitle('Current in a second order (L-C-R)
system for a 6V supply','t (s)',
'i (A)');
```

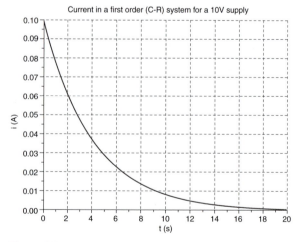

Figure 24.8 Graphical output of the SCILAB script showing how current decays exponentially with time in the first-order *C–R* system

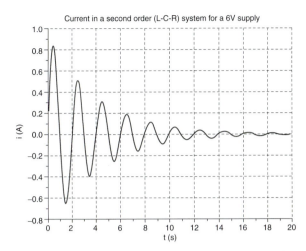

Current in a second order (L-C-R) system for a 6V supply

Figure 24.9 Graphical output of the SCILAB script showing how current takes the form of an exponentially decaying sine wave (i.e. a 'damped oscillation') in the second-order *L–C–R* system

24.3 Chapter summary

The Laplace transform provides us with a powerful means of solving first- and second-order systems in which a combination of resistance, capacitance and inductance is present. The Laplace transform allows us to translate a given function in the time domain, $f(t)$, into an equivalent function in the *s*-domain, $F(s)$. We can then solve the equation in the *s*-domain and use the transform in reverse to convert the *s*-domain solution to an equivalent solution in the time domain.

24.4 Review questions

1. A voltage given by $v = 20 \sin(314t + \pi/2)$ is applied to a resistor of $50\,\Omega$. Use the table of standard Laplace transforms to write down an expression for the current flowing in the resistor in the *s*-domain.

2. The current flowing in a circuit in the *s*-domain is given by:

$$I(s) = \frac{0.025}{0.025s + 1}\ \text{mA}$$

Use the inverse Laplace transform to write down an expression for the current in the time domain. Hence determine the current flowing in the circuit when $t = 12.5\,\text{ms}$.

3. Use Laplace transforms to derive an expression for the current flowing in the circuit shown in Figure 24.10, given that $i = 0$ at $t = 0$.

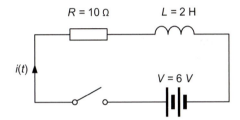

Figure 24.10 See Question 3

4. Use Laplace transforms to derive an expression for the current flowing in the circuit shown in Figure 24.11, given that $i = 0$ at $t = 0$.

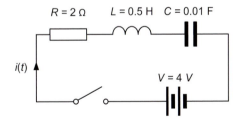

Figure 24.11 See Question 4

Part VII

Alternating current

Part VII is dedicated to circuits in which current flows alternately one way and then the other. These alternating current circuits are used widely for a.c. power distribution, amplifiers, filters and other analogue electronic applications. Chapter 25 begins by explaining fundamental terms and concepts such as waveforms, frequency and periodic time. It also explains how alternating voltage and current is measured and quantified. The chapter also explains the important concepts of reactance (both capacitive and inductive) and impedance.

Chapter 26 provides you with an introduction to the use of a simple yet powerful method of solving even the most complex of a.c. circuits. The chapter begins with an introduction to the system of complex notation, the j-operator and the Argand diagram and then explains how this can be used to solve circuits in which there is impedance (a series combination or resistance and reactance) or admittance (a parallel combination of resistance and reactance).

The phenomenon of resonance is explained in Chapter 27. We examine the effects of series and parallel resonance and introduce the relationship between Q-factor and bandwidth. We also explain the effects of

loading and damping on the performance of a resonant circuit and show how complex notation can be used to analyse resonant circuits.

Coupled magnetic circuits are explained in Chapter 28. We introduce mutual inductance and the coefficient of coupling before explaining the transformer principle. The chapter also provides a discussion of transformer losses, regulation, efficiency and the effect of loads that are not purely resistive. Chapter 29 takes this a stage further and provides an introduction to power and power factor in a.c. circuits.

In many practical applications the waveforms of voltage and current are not purely sinusoidal. Chapter 30 explains how a complex waveform comprises a fundamental component together with a number of harmonic components, each having a specific amplitude and phase relative to the fundamental. Chapter 30 also provides an introduction to Fourier analysis, a powerful technique based on the concept that all waveforms, whether continuous or discontinuous, can be expressed in terms of a convergent series. Finally, Chapter 31 concludes Part VII with an investigation of the power contained in a complex waveform.

Chapter 25

a.c. principles

If you have not studied a.c. theory before, this introductory section has been designed to quickly get you up to speed. If, on the other hand, you have previously studied Electrical and Electronic Principles at level 3 (or its equivalent) you should move on to Chapter 26 where you will be introduced to some powerful techniques for analysing a.c. circuits.

25.1   Alternating voltage and current

Unlike direct currents that have steady values and always flow in the same direction, alternating currents flow alternately one way and then the other. The voltage produced by an alternating current is thus partly positive and partly negative. An understanding of alternating currents and voltages is important in a number of applications including a.c. power distribution, amplifiers and filters.

25.1.1   Waveforms

A graph showing the variation of voltage or current present in a circuit is known as a *waveform*. Some common types of waveform are shown in Figure 25.1. Note that the waveforms of speech and music comprise many components at different frequencies and of different amplitudes, these waveforms are referred to as *complex*.

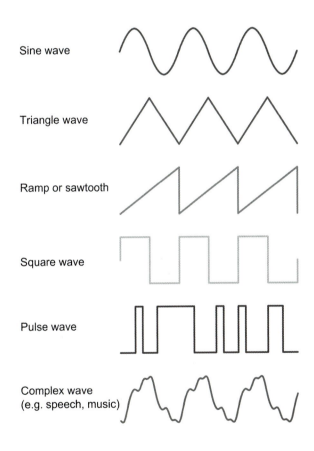

Sine wave

Triangle wave

Ramp or sawtooth

Square wave

Pulse wave

Complex wave (e.g. speech, music)

Figure 25.1 Various waveforms

The most fundamental waveform is a sine wave, as shown in Figure 25.2. The shape of this wave is defined by the circular motion that produces it, the instantaneous induced voltage being directly proportional to the sine of the angle between the electrical conductor (usually part

978-1-85617-775-7, Engineering Science, Mike Tooley & Lloyd Dingle

of a loop or coil) and the surrounding lines of magnetic flux. A sine wave comprises a single fundamental frequency with no harmonic components.

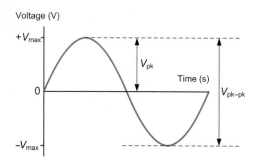

Figure 25.2 One cycle of a sine wave

The equation for the sinusoidal voltage shown in Figure 25.2, at a time, t, is:

$$v = V_m \sin(\omega t)$$

where v is the *instantaneous voltage*, V_m is the *maximum value* of voltage (also known as the *amplitude* or *peak value* of voltage) and ω is the rate of change of angle or *angular velocity* (in radians per second).

25.1.2 Frequency

The frequency of a repetitive waveform is the number of cycles of the waveform that occur in unit time. Frequency is expressed in hertz (Hz). A frequency of 1 Hz is equivalent to one cycle per second. Hence, if a voltage has a frequency of 400 Hz, four hundred cycles will occur in every second.

Since there are 2π radians in one complete revolution or cycle, a frequency of one cycle per second must be the same as an angular velocity of 2π radians per second. Hence, a frequency, f, is equivalent to:

$$f = \frac{\omega}{2\pi} \text{ Hz}$$

Alternatively, the angular velocity, ω, is given by:

$$\omega = 2\pi f \text{ rad/s}$$

We can thus express the instantaneous voltage in another way:

$$v = V_m \sin(2\pi f t)$$

Example 25.1   A sine-wave voltage has a maximum value of 100 V and a frequency of 50 Hz. Determine the instantaneous voltage present (a) 2.5 ms and (b) 15 ms from the start of the cycle.

We can determine the voltage at any instant of time using:

$$v = V_m \sin(2\pi f t)$$

where $V_m = 100$ V and $f = 50$ Hz
In (a), $t = 2.5$ ms, hence:

$$v = 100 \sin(2\pi \times 50 \times 0.0025) = 100 \sin(0.785)$$

$$= 100 \times 0.707 = 70.7 \text{ V}$$

In (b), $t = 15$ ms, hence:

$$v = 100 \sin(2\pi \times 50 \times 0.015) = 100 \sin(4.71)$$

$$= 100 \times -1 = -100 \text{ V}$$

Test your knowledge 25.1

Write down an expression for an alternating voltage with a maximum value of 155 V and a frequency of 400 Hz.

Test your knowledge 25.2

An alternating voltage is given by the expression $v = 340 \sin(200\pi t)$. Determine:

(a)   the maximum value of voltage

(b)   the voltage at 1 ms from the start of a cycle

(c)   the time, relative to the start of a cycle, at which the first negative peak of voltage will occur.

25.1.3 Periodic time

The periodic time of a waveform (Figure 25.3) is the time taken for one complete cycle of the wave. The relationship between periodic time and frequency is thus:

$$t = \frac{1}{f} \quad \text{or} \quad f = \frac{1}{t}$$

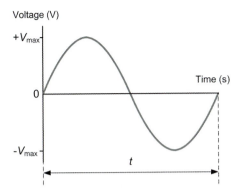

Figure 25.3 Periodic time

where t is the periodic time (in seconds) and f is the frequency (in Hz).

Example 25.2   What is the periodic time of a waveform with a frequency of 200 Hz?
   The periodic time can be found from:

$$t = \frac{1}{f} = \frac{1}{200} = 0.005 = 5\,\text{ms}$$

Example 25.3   What is the frequency of a waveform with a periodic time of 2.5 ms?
   The frequency can be found from:

$$f = \frac{1}{t} = \frac{1}{2.5 \times 10^{-3}} = 0.4 \times 10^3 = 400\,\text{Hz}$$

Key Point. Periodic time is the inverse of frequency. Thus as periodic time falls the frequency will increase, and vice versa

Key Point. Maximum value, peak value and amplitude are all synonymous terms. They all describe the maximum excursion of a waveform from its resting value

Test your knowledge 25.3

A broadcast frequency standard operates on a frequency of 5 MHz. What is the periodic time of the broadcast signal?

Test your knowledge 25.4

Write down an expression for the sinusoidal current shown in Figure 25.4. What will the instantaneous current be 3 ms from the start of a cycle?

Test your knowledge 25.5

A sine-wave voltage has an amplitude of 15 V and a period of 2.5 ms. Write down an expression for the voltage and use it to determine the time from the start of a cycle at which the voltage will reach 12 V.

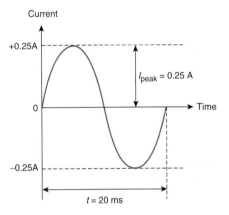

Figure 25.4 See Test your knowledge 25.4

SCILAB

The following SCILAB script produces a plot of a sine wave having given input parameters (see Figures 25.5 and 25.6):

```
// Sinusoidal waveform plotter
// SCILAB source file: ex2501.sce
//
// Clear any previous graphic window
clf();
//
txt = ['Input maximum value in V';'Input
frequency in Hz';'Input phase angle
in degrees';'Input time scale in seconds'];
var = x_mdialog('Enter waveform parameters',
txt,['100';'100';'45';'0.02']);
//
```

```
vm = evstr(var(1));
f = evstr(var(2));
phi = evstr(var(3));
tmax = evstr(var(4));
//
t = [0:0.0001:tmax]; // set time scale
v = vm*sin(2*%pi*f*t + phi/57.3); // convert
degrees to radians
// Plot
plot2d(t,v,style=5); // red line for voltage
// Print grid and title
xgrid(); // add grid
xtitle('Waveform plot');
xlabel('Time (s)');
ylabel('Voltage (V)');
```

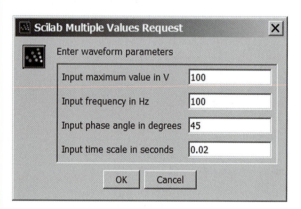

Figure 25.5 Input dialogue for the sinusoidal waveform plotter

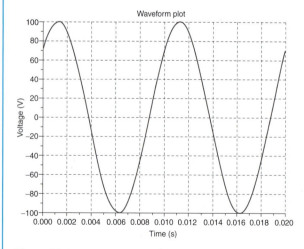

Figure 25.6 Graphical output of the SCILAB script

25.1.4 Average, peak, peak–to–peak and r.m.s. values

The *average* value of an alternating current which swings symmetrically above and below zero will obviously be zero when measured over a long period of time. Hence average values of currents and voltages are invariably taken over one complete half-cycle (either positive or negative) rather than over one complete full-cycle (which would result in an average value of zero).

The *peak* value (or *amplitude*) of a waveform is a measure of the extent of its voltage or current excursion from the resting value (usually zero). The *peak-to-peak* value for a wave that is symmetrical about its resting value is twice its peak value (i.e. $2V_m$).

The *r.m.s.* (or *effective*) value of an alternating voltage or current is the value which would produce the same heat energy in a resistor as a direct voltage or current of the same magnitude. Since the r.m.s. value of a waveform is very much dependent upon its shape, values are only meaningful when dealing with a waveform of known shape. Where the shape of a waveform is not specified, r.m.s. values are normally assumed to refer to sinusoidal conditions. Note also that where a variable is shown with no qualifying subscript (e.g. V or I) the quantity is usually assumed to be r.m.s. unless otherwise specified.

The following relationships apply to a sine wave:

$$V_{av} = 0.636 \times V_m$$

$$V_{pk-pk} = 2 \times V_m$$

$$V_{r.m.s.} = 0.707 \times V_m$$

$$V_m = 1.414 \times V_{r.m.s.}$$

Similar relationships apply to the corresponding alternating currents, thus:

$$I_{av} = 0.636 \times I_m$$

$$I_{pk-pk} = 2 \times I_m$$

$$I_{r.m.s.} = 0.707 \times I_m$$

$$I_m = 1.414 \times I_{r.m.s.}$$

25.1.5 Phase angle

When two waveforms of the same frequency are being compared it can be useful to compare them on the basis of the difference in angle that exists between their respective zero voltage crossing points (see Figure 25.7). This *phase angle* can take any value between $0°$ and $360°$ (or 0 and 2π radians).

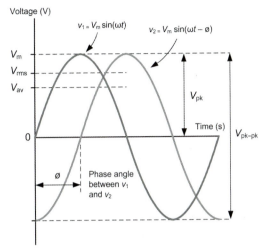

Figure 25.7 Average, peak, peak–peak and r.m.s. values and the phase angle between two waveforms of identical frequency

---

Example 25.4   A sinusoidal voltage has an r.m.s. value of 220 V. What is the peak value of the voltage?

Solution

Since $V_{\text{r.m.s.}} = 0.707V_{\text{m}}$,

$$V_{\text{m}} = V_{\text{r.m.s.}}/0.707 = 1.414 \times V_{\text{r.m.s.}}$$

Thus $V_{\text{m}} = 1.414 \times 220 = 311$ V

Example 25.5   A sinusoidal alternating current has a peak-to-peak value of 4 mA. What is its r.m.s. value?

First we must convert the peak-to-peak current into peak current:

Since $I_{\text{pk–pk}} = 2 \times I_{\text{m}}, I_{\text{m}} = 0.5 \times I_{\text{pk–pk}}$

Thus $I_{\text{m}} = 0.5 \times 4 = 2$ mA

Now we can convert the peak current into r.m.s. current using:

$$I_{\text{r.m.s.}} = 0.707 \times I_{\text{m}}$$

Thus $I_{\text{r.m.s.}} = 0.707 \times 2 = 1.414$ mA

Key Point.   Where a voltage or current is not specified with any qualifying subscript (e.g. V or I instead of V_{m} or $I_{\text{pk–pk}}$) the quantity is usually assumed to be an r.m.s. value unless otherwise specified

---

25.1.6   Alternating current in a resistor

Ohm's law is obeyed in an a.c. circuit just as it is in a d.c. circuit. Thus, when a sinusoidal voltage, V, is applied to a resistor, R (as shown in Figure 25.8), the current flowing in the resistor will be given by:

$$I = \frac{V}{R}$$

This relationship must also hold true for the instantaneous values of current, i, and voltage, v, thus:

$$i = \frac{v}{R}$$

and since $v = V_{\text{m}} \sin(\omega t)$

$$i = \frac{V_{\text{m}} \sin(\omega t)}{R}$$

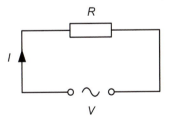

Figure 25.8 Alternating current in a resistor

The current and voltage will both have a sinusoidal shape and, since they rise and fall together, they are said to be *in-phase* with one another. We can represent this relationship by means of the *phasor diagram* shown in Figure 25.9. This diagram shows two rotating phasors (of magnitude I_{m} and V_{m}) rotating at an angular velocity, ω. The applied voltage (V_{m}) is referred to as the *reference phasor* and this is aligned with the horizontal axis (i.e. it has a phase angle of $0°$).

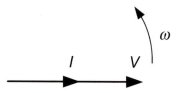

Figure 25.9 Phasor diagram for the circuit in Figure 25.8

Phasor diagrams provide us with a quick way of illustrating the relationships that exist between sinusoidal voltages and currents in a.c. circuits without having to draw lots of time-related waveforms. Figure 25.10 will help you to understand how the previous phasor diagram relates to time-related waveforms for the voltage and current in a resistor.

Example 25.6   A sinusoidal voltage with a peak-to-peak value of 20 V is applied to a resistor of 1 kΩ. What value of r.m.s. current will flow in the resistor?

This problem must be solved in several stages. First we will determine the peak-to-peak current in the resistor and then we shall convert this value into a corresponding r.m.s. quantity.

Since $I = \dfrac{V}{R}$ we can infer that $I_{pk-pk} = \dfrac{V_{pk-pk}}{R}$

Thus: $I_{pk-pk} = \dfrac{20}{1 \times 10^3} = 20 \times 10^{-3} = 20\,\text{mA}$

Next:

$$I_m = \frac{I_{pk-pk}}{2} = \frac{20}{2} = 10\,\text{mA}$$

Finally:

$$I_{r.m.s.} = 0.707 \times I_m = 0.707 \times 10 = 7.07\,\text{mA}$$

25.2   Reactance

We have already briefly mentioned reactance as a means of explaining the relationship that exists between the voltage and current in a capacitive or inductive circuit. Reactance is the equivalent of resistance (i.e. the ratio of voltage to current) in a circuit that contains pure capacitance or pure resistance. Reactance is given by the relationship:

$$X = \frac{V}{I}$$

where X is the reactance (in Ω), V is the r.m.s. voltage and I is the r.m.s. current.

25.2.1   Capacitive reactance

Earlier, in Chapter 23 we found that the instantaneous current flowing in a capacitor is proportional to the rate

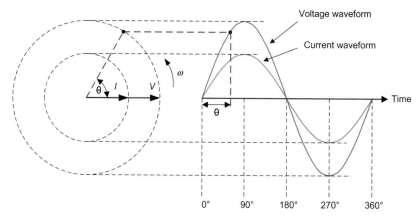

Figure 25.10 Phasor diagram and time-related waveforms

of change of applied voltage. Hence:

$$i = C\frac{dv}{dt}$$

The voltage applied to the capacitor will be given by:

$$v = V_m \sin(\omega t)$$

Combining this with the previous relationship yields:

$$i = C\frac{d(V_m \sin(\omega t))}{dt} = CV_m\omega\cos(\omega t)$$

This shows that the current (a cosine wave) will *lead* the voltage (a sine wave) by 90°.

The maximum value of current, i, will occur when $\cos(\omega t) = 1$, hence:

$$I_m = V_m C\omega$$

Dividing both sides by $\sqrt{2}$ to obtain the r.m.s. values of current and voltage yields:

$$I = VC\omega$$

From our earlier definition of reactance we know that the reactance of a capacitor, X_C, is simply the ratio of current flowing in the capacitor to the voltage appearing across it, hence:

$$X_C = \frac{V}{I} = \frac{1}{\omega C} = \frac{1}{2\pi fC}$$

25.2.2   Inductive reactance

Earlier in Chapter 23 we showed that the instantaneous voltage dropped across an inductor is proportional to the rate of change of applied current. In other words:

$$e = -L\frac{di}{dt}$$

The current flowing in the inductor will be given by:

$$i = I_m \sin(\omega t)$$

Combining this with the previous relationship yields:

$$e = -L\frac{d(I_m \sin(\omega t))}{dt} = -LI_m\omega\cos(\omega t)$$

The induced voltage, v, opposes the applied voltage (Lenz's law) and hence:

$$v = LI_m\omega\cos(\omega t)$$

This shows that the applied voltage (a cosine wave) will *lag* the current (a sine wave) by 90°.

The maximum value of voltage, v, will occur when $\cos(\omega t) = 1$, hence:

$$V_m = I_m L\omega$$

Dividing both sides by $\sqrt{2}$ to obtain the r.m.s. values of current and voltage yields:

$$V = LI\omega$$

From our earlier definition of reactance we know that the reactance of an inductor, X_L, is simply the ratio of current flowing in the inductor to the voltage appearing across it, hence:

$$X_L = \frac{V}{I} = \omega L = 2\pi fL$$

Example 25.7   A 1.5 μF capacitor is connected to a 110 V 50 Hz a.c. supply. What current will flow in the capacitor?

Rearranging the relationship $\dfrac{V}{I} = \dfrac{1}{2\pi fC}$ gives:

$$I = \frac{2\pi fC}{V} = \frac{6.28 \times 50 \times 15 \times 10^{-6}}{110} = 42.82\,\text{mA}$$

Example 25.8   A current of 15 mA flows in an inductor of negligible resistance when it is connected to a 24 V 400 Hz supply. What is the value of the inductance?

Rearranging the relationship $\dfrac{V}{I} = 2\pi fL$ gives:

$$L = \frac{V}{2\pi fI} = \frac{24}{6.28 \times 400 \times 15 \times 10^{-3}} = 0.637\,\text{mH}$$

Key Point.   Reactance is the equivalent of resistance in a circuit that contains pure capacitance or pure inductance

25.2.3 Variation of reactance with frequency

When an alternating voltage is applied to a capacitor or an inductor the amount of current flowing will depend upon the value of the capacitance or inductance and on the frequency of the voltage. In effect, capacitors and inductors oppose the flow of current in much the same way as a resistor. The important difference is that the effective resistance (or *reactance*) of the component varies with frequency (unlike the case of a pure resistance where the magnitude of the current *does not* change with frequency).

Capacitive reactance is inversely proportional to the frequency of the applied alternating current and can be determined from the formula that we met earlier:

$$X_C = \frac{1}{2\pi f C}$$

where X_C is the reactance (in Ω), f is the frequency (in Hz) and C is the capacitance (in F).

Since capacitive reactance is inversely proportional to frequency ($X_C \propto 1/f$), the graph of capacitive reactance plotted against frequency takes the form of a rectangular hyperbola (see Figure 25.11).

Inductive reactance is directly proportional to the frequency of the applied alternating current and can once again be determined from the formula that we met earlier:

$$X_L = 2\pi f L$$

where X_L is the reactance (in Ω), f is the frequency (in Hz) and L is the inductance (in H).

Since inductive reactance is directly proportional to frequency ($X_L \propto f$), the graph of inductive reactance

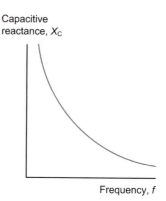

Figure 25.11 Variation of capacitive reactance with frequency

plotted against frequency takes the form of a straight line (see Figure 25.12).

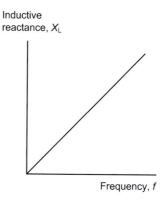

Figure 25.12 Variation of inductive reactance with frequency

Example 25.9 Determine the reactance of a 1 μF capacitor at (a) 100 Hz and (b) 10 kHz.

(a) At 100 Hz,

$$X_C = \frac{1}{2\pi \times 100 \times 1 \times 10^{-6}}$$

$$= 0.159 \times 10^4 = 1.59 \text{ k}\Omega$$

(b) At 10 kHz,

$$X_C = \frac{1}{2\pi \times 10 \times 10^3 \times 1 \times 10^{-6}}$$

$$= 0.159 \times 10^2 = 15.9 \text{ }\Omega$$

Example 25.10 Determine the reactance of a 10 mH inductor at (a) 100 Hz and (b) 10 kHz.

(a) At 100 Hz,

$$X_L = 2\pi \times 100 \times 10 \times 10^{-3} = 6.28\,\Omega$$

(b) At 10 kHz,

$$X_L = 2\pi \times 10000 \times 10 \times 10^{-3} = 628\,\Omega$$

Test your knowledge 25.10

Determine the reactance of a 60 mH inductor at (a) 20 Hz and (b) 4 kHz.

Test your knowledge 25.11

Determine the reactance of a 680 nF capacitor at (a) 400 Hz and (b) 20 kHz.

SCILAB

The following SCILAB script produces a plot of capacitive and inductive reactance against frequency (see Figure 25.13):

```
// Plot of xL and xC over the range 100 to
1000 Hz
// Filename: ex2502.sce
//
// Clear any existing plot
clf();
// Set up parameters
f = [100:10:1000]'; // Frequency range
100 Hz to 1 kHz
C = 1E-6; // C = 1 uF
L = 0.22; // L = 220 mH
// Plot the capacitive reactance in blue
plot2d(f,(2*%pi*f*C)^-1,2);
// Plot the inductive reactance in red
plot2d(f,2*%pi*f*L,5);
// Print legend, grid and title
h = legend(['Capacitive'; 'Inductive'],1,
boxed=%f);
xgrid(); // add grid
xtitle('Reactance plotted against
frequency');
```

25.2.4 Alternating current in a capacitor

When a sinusoidal voltage, V, is applied to a capacitor, C (as shown in Figure 25.14), the current flowing in the capacitor will be given by:

$$I = \frac{V}{X_C}$$

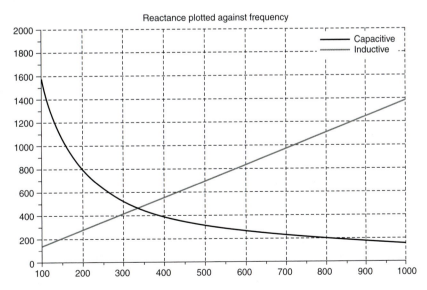

Figure 25.13 Graphical output of the SCILAB script

where X_C is the *reactance* of the capacitor. *Capacitive reactance*, like resistance, is measured in ohms (Ω).

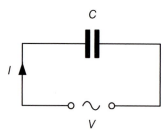

Figure 25.14 Alternating current in a capacitor

The current in a capacitor *leads* the applied voltage by a *phase angle* of 90° ($\pi/2$ radians) and, since $v = V\sin(\omega t)$:

$$i = \frac{V_{\mathrm{m}} \sin(\omega t + \frac{\pi}{2})}{X_C}$$

As before, the current and voltage both have a sinusoidal shape 90° apart and this relationship is illustrated in the phasor diagram shown in Figure 25.15. The applied voltage (V) is the *reference phasor* (with phase angle of 0°) whilst the current flowing (I) has a *leading phase angle* of 90°.

Figure 25.15 Phasor diagram for the circuit in Figure 25.14

25.2.5 Alternating current in an inductor

When a sinusoidal voltage, V, is applied to an inductor, L (as shown in Figure 25.16), the current flowing in the inductor will be given by:

$$I = \frac{V}{X_L}$$

where X_L is the *reactance* of the inductor. *Inductive reactance*, like resistance, is measured in ohms (Ω).

The current in an inductor *lags* the applied voltage by a *phase angle* of 90° ($\pi/2$ radians) and, since $v = V\sin(\omega t)$:

$$i = \frac{V_{\mathrm{m}} \sin(\omega t - \frac{\pi}{2})}{X_C}$$

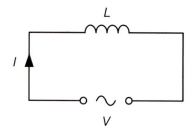

Figure 25.16 Alternating current in an inductor

As before, the current and voltage both have a sinusoidal shape 90° apart and this relationship is illustrated in the phasor diagram shown in Figure 25.17. The applied voltage, V, is the *reference phasor* (with phase angle of 0°) whilst the current flowing, I, has a *lagging phase angle* of 90°.

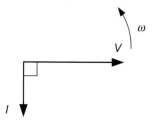

Figure 25.17 Phasor diagram for the circuit in Figure 25.16

Example 25.11 An r.m.s. current of 2.5 A flows when a capacitor is connected to a 50 Hz supply that delivers an r.m.s. voltage of 110 V. Derive an expression for the current flowing. What is the reactance of the capacitor?

This problem must be solved in several stages:

Next we can find the maximum voltage delivered by the supply:

$$V_{\mathrm{m}} = 1.414 \times V_{\mathrm{r.m.s}} = 1.414 \times 110 = 155.54\,\mathrm{A}$$

The supply voltage can then be described by the equation:

$$v = V_{\mathrm{m}} \sin(\omega t) = V_{\mathrm{m}} \sin(2\pi f t) = 155.54$$
$$\sin(2\pi \times 50 \times t) = 155.54 \sin(314 t)$$

Next we will determine the maximum current flowing in the capacitor:

$$I_{\mathrm{m}} = 1.414 \times I_{\mathrm{r.m.s}} = 1.414 \times 2.5 = 3.535\,\mathrm{A}$$

The supply current will be described by a similar equation to that for the supply voltage, but note that the current will lead the voltage by $\pi/2$:

$$i = I_\text{m} \sin(\omega t) = I_\text{m} \sin(2\pi f t) = 3.535 \sin(314t + \pi/2)$$

Finally, the reactance of the capacitor can be determined using:

$$X_\text{C} = \frac{V}{I} = \frac{V_\text{r.m.s.}}{I_\text{r.m.s.}} = \frac{V_\text{m}}{I_\text{m}} = \frac{155.54}{3.535} = 44\,\Omega$$

Test your knowledge 25.12

A voltage $v = 50\sin(314t)$ is first applied to a capacitor having a reactance of $25\,\Omega$ and then applied to an inductor having a reactance of $200\,\Omega$. Write down an expression for the current flowing in each case and use these expressions to determine the instantaneous value of current flowing 2.5 ms from the start of a cycle in both cases.

25.3   Impedance

When both resistance and reactance are present in a circuit we refer to the combined effect as impedance. Like resistance and reactance, impedance is simply the ratio of voltage to current and is given by the relationship:

$$Z = \frac{V}{I}$$

where Z is the impedance (in Ω), V is the r.m.s. voltage and I is the r.m.s. current.

Next we shall consider the impedance of various combinations of resistance and reactance.

25.3.1   Resistance and capacitance in series

When a sinusoidal voltage, V, is applied to a series circuit comprising resistance, R, and capacitance, C (as shown in Figure 25.18), the current flowing in the circuit will produce separate voltage drops across the resistor and capacitor (V_R and V_C, respectively). These two voltage drops will be 90° apart, with V_C lagging V_R. We can illustrate this relationship using the phasor diagram shown in Figure 25.19. Note that once again we have used current as the reference phasor in this series circuit.

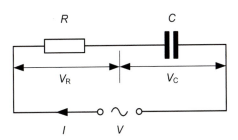

Figure 25.18 Resistance and capacitance in series

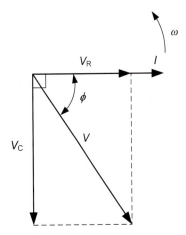

Figure 25.19 Phasor diagram for the circuit in Figure 25.18

From Figure 25.19 you should note that the supply voltage, V, is simply the result of adding the two voltage phasors, V_R and V_C. The angle between the supply voltage, V, and supply current, I, ϕ, is known as the *phase angle*.

Now $\sin\phi = \dfrac{V_\text{C}}{V}$, $\cos\phi = \dfrac{V_\text{R}}{V}$, and $\tan\phi = \dfrac{V_\text{C}}{V_\text{R}}$

Since $X_\text{C} = \dfrac{V_\text{C}}{I}$, $R = \dfrac{V_\text{R}}{I}$ and $Z = \dfrac{V}{I}$ (where Z is the *impedance* of the circuit), we can illustrate the relationship between X_C, R and Z using the *impedance triangle* shown in Figure 25.20.

Note that $Z = \sqrt{R^2 + X_\text{C}^2}$ and $\phi = \arctan\left(\dfrac{X_\text{C}}{R}\right)$

Example 25.12   A capacitor of 22 μF is connected in series with a 470 Ω resistor. If a sinusoidal current of 10 mA at 50 Hz flows in the circuit, determine:

(a) the voltage dropped across the capacitor

(b) the voltage dropped across the resistor

(c) the impedance of the circuit

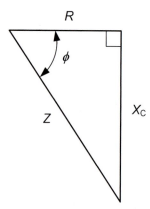

Figure 25.20 Impedance triangle for the circuit in Figure 25.18

(d)   the supply voltage

(e)   the phase angle.

(a) $V_C = IX_C = I \times \dfrac{1}{2\pi fC} = 10 \times 10^{-3} \times$
$\dfrac{1}{6.28 \times 50 \times 22 \times 10^{-6}} = 1.4\,V$

(b) $V_R = IR = 0.01 \times 470 = 4.7\,V$

(c) $Z = \sqrt{R^2 + X_C^2} = \sqrt{470^2 + 144.5^2} =$
$\sqrt{241780} = 491.7\,\Omega$

(d) $V = I \times Z = 10 \times 10^{-3} \times 491.7 = 4.91\,V$

(e) $\phi = \arctan\left(\dfrac{X_C}{R}\right) = \arctan\left(\dfrac{144.5}{470}\right)$
$= \arctan(0.3074) = 17.1°$

Test your knowledge 25.13

A resistor of $30\,\Omega$ is connected in series with a capacitive reactance of $40\,\Omega$. Determine the impedance of the circuit and the current flowing when the circuit is connected to a 115 V supply.

Test your knowledge 25.14

A capacitor of 220 nF is connected in series with a $220\,\Omega$ resistor. What current will flow in this circuit when it is connected to a 15 V 400 Hz supply, and what voltage will appear across each of the two components?

25.3.2   Resistance and inductance in series

When a sinusoidal voltage, V, is applied to a series circuit comprising resistance, R, and inductance, L (as shown in Figure 25.21) the current flowing in the circuit will produce separate voltage drops across the resistor and inductor (V_R and V_L, respectively). These two voltage drops will be 90° apart, with V_L leading V_R. We can illustrate this relationship using the phasor diagram shown in Figure 25.22. Note that we have used current as the reference phasor in this series circuit for the simple reason that the same current flows through each component (recall that earlier we used the applied voltage as the reference).

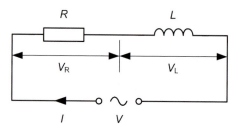

Figure 25.21   Resistance and inductance in series

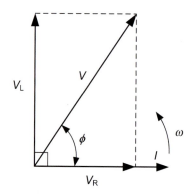

Figure 25.22   Phasor diagram for the circuit in Figure 25.21

From Figure 25.22 you should note that the supply voltage, V, is simply the result of adding the two phasors, V_R and V_L. The angle between the supply voltage, V, and supply current, I, ϕ, is known as the *phase angle*.

Now $\sin\phi = \dfrac{V_L}{V}$, $\cos\phi = \dfrac{V_R}{V}$, and $\tan\phi = \dfrac{V_L}{V_R}$

Since $X_L = \dfrac{V_L}{I}$, $R = \dfrac{V_R}{I}$ and $Z = \dfrac{V}{I}$

(where Z is the *impedance* of the circuit), we can illustrate the relationship between X_L, R and Z using the *impedance triangle* shown in Figure 25.23.

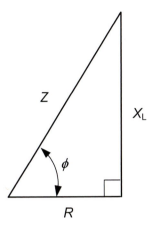

Figure 25.23 Impedance triangle for the circuit in Figure 25.21

Note that $Z = \sqrt{R^2 + X_L^2}$ and $\varphi = \arctan\left(\dfrac{X_L}{R}\right)$

Example 25.13 An inductor of 80 mH is connected in series with a 100 Ω resistor. If a sinusoidal current of 20 mA at 50 Hz flows in the circuit, determine:

(a) the voltage dropped across the inductor

(b) the voltage dropped across the resistor

(c) the impedance of the circuit

(d) the supply voltage

(e) the phase angle.

(a) $V_L = IX_L = I \times 2\pi f L = 0.02 \times 25.12 = 0.5\,\text{V}$

(b) $V_R = IR = 0.02 \times 100 = 2\,\text{V}$

(c) $Z = \sqrt{R^2 + X_L^2} = Z = \sqrt{R^2 + X_L^2}$
$= \sqrt{100^2 + 25.12^2} = \sqrt{10631} = 103.1\,\Omega$

(d) $V = I \times Z = 0.02 \times 103.1 = 2.06\,\text{V}$

(e) $\phi = \arctan\left(\dfrac{X_L}{R}\right) = \arctan\left(\dfrac{25.12}{100}\right)$
$= \arctan(0.2512) = 14.1°$

Test your knowledge 25.15

A coil is connected to a 50 V a.c. supply at 400 Hz. If the current supplied to the coil is 200 mA and the coil has a resistance of 60 Ω, determine the value of inductance.

Test your knowledge 25.16

An inductor of 2.2 mH and negligible resistance is connected in series with a 150 Ω resistor. What current will flow in this circuit when it is connected to a 48 V 10 kHz supply, and what voltage will appear across each of the two components?

25.3.3 Resistance, capacitance and inductance in series

When a sinusoidal voltage, V, is applied to a series circuit comprising resistance, R, capacitance, C, and inductance, L (as shown in Figure 25.24), the current flowing in the circuit will produce separate voltage drops across the resistor, inductor and capacitor (V_R, V_L and V_C, respectively). The voltage drop across the inductor will lead the applied current (and voltage dropped across V_R) by 90° whilst the voltage drop across the capacitor will lag the applied current (and voltage dropped across V_R) by 90°.

When the inductive reactance (X_L) is greater than the capacitive reactance (X_C), V_L will be greater than V_C and the resulting phasor diagram is shown in Figure 25.25. Conversely, when the capacitive reactance (X_C) is greater than the inductive reactance (X_L), V_C will be greater than V_L and the resulting phasor diagram will be shown in Figure 25.26. Note that once again we have used current as the reference phasor in this series circuit.

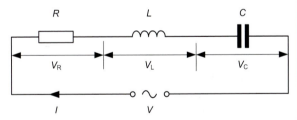

Figure 25.24 Resistance, inductance and capacitance in series

From Figures 25.25 and 25.26, you should note that the supply voltage, V, is simply the result of adding the three voltage phasors, V_L, V_C and V_R, and that the first stage in simplifying the diagram is that of resolving V_L and V_C into a single voltage ($V_L - V_C$ or $V_C - V_L$, depending upon whichever is the greater). Once again, the phase angle, ϕ, is the angle between the supply voltage and the current.

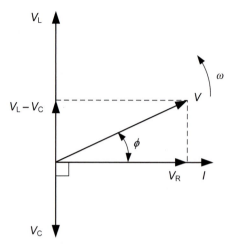

Figure 25.25 Phasor diagram for the circuit in Figure 25.24 when $X_L > X_C$

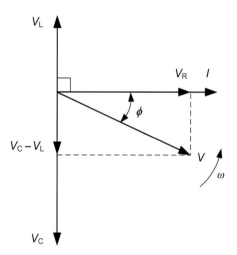

Figure 25.26 Phasor diagram for the circuit in Figure 25.24 when $X_C > X_L$

Figures 25.27 and 25.28, respectively, show the impedance triangle for the circuit for the cases when $X_L > X_C$ and $X_C > X_L$.

Note that, when $X_L > X_C$, $Z = \sqrt{R^2 + (X_L - X_C)^2}$ and $\phi = \arctan\left(\dfrac{X_L - X_C}{R}\right)$.

Similarly, when $X_C > X_L$, $Z = \sqrt{R^2 + (X_C - X_L)^2}$ and $\phi = \arctan\left(\dfrac{X_C - X_L}{R}\right)$.

Later, in Chapter 27, we will consider the special case that occurs when the capacitive reactance is equal to the inductive reactance (i.e. when $X_C = X_L$). When this occurs, the reactive components effectively cancel one another and the circuit behaves like a pure resistance!

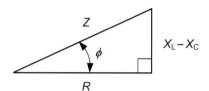

Figure 25.27 Impedance triangle for the circuit in Figure 25.24 when $X_L > X_C$

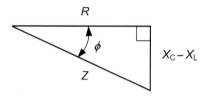

Figure 25.28 Impedance triangle for the circuit in Figure 25.24 when $X_C > X_L$

Example 25.14 A series circuit comprises an inductor of 80 mH, a resistor of 200 Ω and a capacitor of 22 μF. If a sinusoidal current of 40 mA at 50 Hz flows in this circuit, determine:

(a) the voltage developed across the inductor

(b) the voltage dropped across the capacitor

(c) the voltage dropped across the resistor

(d) the impedance of the circuit

(e) the supply voltage

(f) the phase angle.

(a) $V_L = IX_L = I \times 2\pi fL = 0.04 \times 25.12 = 1\,\text{V}$

(b) $V_C = IX_C = I \times \dfrac{1}{2\pi fC} = 40 \times 10^{-3} \times$

$\dfrac{1}{6.28 \times 50 \times 22 \times 10^{-6}} = 5.8\,\text{V}$

(c) $V_R = IR = 0.04 \times 200 = 8\,\text{V}$

(d) $Z = \sqrt{R^2 + (X_C - X_L)^2}$

$= \sqrt{200^2 + (144.5 - 25.12)^2}$

$= \sqrt{54252} = 232.9\,\Omega$

(e) $V = I \times Z = 0.04 \times 225.9 = 9.32\,\text{V}$

(f) $\phi = \arctan\left(\dfrac{X_C - X_L}{R}\right)$

$\quad = \arctan\left(\dfrac{144.5 - 25.12}{200}\right)$

$\quad = \arctan(0.597) = 30.8°$

Test your knowledge 25.17

A series circuit comprises an inductor of 80 mH, a resistor of 200 Ω and a capacitor of 22 μF. If a sinusoidal current of 40 mA at 50 Hz flows in this circuit, determine:

(a) the voltage developed across the inductor

(b) the voltage dropped across the capacitor

(c) the voltage dropped across the resistor

(d) the impedance of the circuit

(e) the supply voltage

(f) the phase angle.

Test your knowledge 25.18

Write and test a SCILAB script that will determine the impedance and phase angle of a series a.c. circuit containing given values of resistance, capacitance, inductance and frequency.

25.4 Chapter summary

This chapter has provided you with an introduction to the principles of a.c. circuits. It began with a description of some common types of waveform together with a detailed explanation of the parameters of sine waves. We also derived equations that will allow you to determine the value of voltage or current at a particular instant in time.

The chapter introduced you to the concept of reactance, impedance and phase angle in an a.c. circuit. We showed how reactance varies differently with frequency for capacitors and inductors and obtained relationships that will allow you to determine the reactance of a capacitor or an inductor at any given frequency. We also described two further techniques that can help you understand a.c. circuits; phasor diagrams and impedance triangles.

Finally, we investigated circuits in which a combination of resistance and reactance are present at the same time. We derived the relationships that allow us to determine the impedance of each type of circuit as well as the phase angle between the supply voltage and the current.

25.5 Review questions

Short answer questions

1. A resistance of 45 Ω is connected to a 150 V 400 Hz supply. Write down an expression for the instantaneous current flowing in the resistor and use it to determine the current that will be flowing in the resistor 3 ms from the start of a cycle.

2. A capacitor of 100 nF is used in a power line filter. Determine the frequency at which the capacitor will have a reactance of 10 Ω.

3. A resistor of 120 Ω is connected in series with a capacitive reactance of 200 Ω. Determine the impedance of the circuit and the current flowing when the circuit is connected to a 200 V a.c. supply.

4. A coil has an inductance of 200 mH and a resistance of 40 Ω. Determine:

    (a) the impedance of the coil at a frequency of 60 Hz

    (b) the current that will flow in the coil when it is connected to a 110 V 60 Hz supply.

5. A sinusoidal alternating voltage has a peak value of 160 V and a frequency of 60 Hz. Write down an expression for the instantaneous voltage and use it to determine the value of voltage at 3 ms from the start of a cycle.

6. An inductor is used in an aerial tuning unit. Determine the value of inductance required if the inductor is to have a reactance of 300 Ω at a frequency of 3.5 MHz.

7. A resistor of 120 Ω is connected in series with a capacitive reactance of 200 Ω. Determine the voltage developed across each component when the circuit is connected to a 50 V supply.

8. A capacitor of 2 μF is connected in series with a 100 Ω resistor across a 24 V 400 Hz a.c. supply. Determine the current that will be supplied to the circuit and the voltage that will be dropped across each component.

Long answer questions

9. A sinusoidal alternating current is specified by the equation: $i = 250 \sin(200 \pi t)$ mA. Determine:

   (a) the peak value of the current

   (b) the r.m.s. value of the current

   (c) the frequency of the current

   (d) the periodic time of the current

   (e) the instantaneous value of the current at $t = 2$ ms.

10. A voltage, $v = 17 \sin(100\pi t)$ V, is applied to a pure resistance of 68 Ω.

    (a) Sketch a fully labelled graph showing how the current flowing in the resistor varies over the period from 0 to 20 ms.

    (b) Mark the following on your graph and give values for each:

        (i) the peak value of current

        (ii) the periodic time.

    (c) Determine the instantaneous values of voltage and current at $t = 3.5$ ms.

11. A series circuit comprises an inductor of 60 mH, a resistor of 33 Ω and a capacitor of 47 μF. If a sinusoidal current of 50 mA at 50 Hz flows in this circuit, determine:

    (a) the voltage developed across the inductor

    (b) the voltage dropped across the capacitor

    (c) the voltage dropped across the resistor

    (d) the impedance of the circuit

    (e) the supply voltage

    (f) the phase angle.

Sketch a phasor diagram showing the voltages and current present in the circuit. Label your drawing clearly and indicate values.

Chapter 26

Complex impedance and admittance

This chapter provides you with an introduction to the use of complex notation and the j operator, which provides us with a convenient way of representing the effect of phase shift due to reactive components. Complex notation provides us with a simple yet powerful method of solving even the most complex of a.c. circuits. We will put these techniques to good use in later chapters, when we will be analysing complex networks.

26.1 Complex notation

Complex notation allows us to represent electrical quantities that have both *magnitude* and *direction* (you will already know that in other contexts we call these *vectors*). The *magnitude* is simply the amount of resistance, reactance, voltage or current, etc. In order to specify the *direction* of the quantity, we use an *operator* to denote the phase shift relative to the *reference* quantity (this is usually current for a series circuit and voltage for a parallel circuit). We call this operator 'j'.

Every complex number consists of a *real part* and an *imaginary part*. In an electrical context, the *real part* is that part of the complex quantity that is in phase with the reference quantity. The *imaginary part* (denoted by the j operator) is that part of the complex quantity that is at 90° to the reference.

26.1.1 The j operator

You can think of the j operator as a device that allows us to indicate a rotation or *phase shift* of 90°. A phase shift of +90° is represented by +j whilst a phase shift of −90° is represented by −j, as illustrated in Figure 26.1.

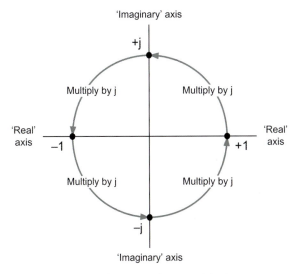

Figure 26.1 Successive phase change and the j operator

A *phasor* is simply an electrical vector. A *vector*, as you will doubtless recall, has magnitude (size) and direction (angle relative to some reference direction). The j operator can be used to rotate a phasor. Each successive multiplication by j has the effect of rotating the phasor through a further 90°.

The j operator has a value that is equal to $\sqrt{-1}$. Thus we can conclude that:

$$j = \sqrt{-1}$$

Multiplying j by j to give j^2 produces:

$$j^2 = \sqrt{-1} \times \sqrt{-1}$$

$$= -1$$

Multiplying again by j to give j^3 yields:

$$j^3 = \sqrt{-1} \times \sqrt{-1} \times \sqrt{-1}$$

$$= -1 \times \sqrt{-1}$$

$$= -j$$

A further multiplication by j gives:

$$j^4 = \sqrt{-1} \times \sqrt{-1} \times \sqrt{-1} \times \sqrt{-1}$$

$$= -1 \times -1$$

$$= +1$$

Note that because j is an operator and not a constant or variable, it should always precede the quantity to which it relates. Thus for example, a 10 Ω reactance would be referred to as j10, *not* 10j.

Test your knowledge 26.1

Show that:

(a) $j^5 - j^3 = 0$

(b) $j^4 - j^2 = 2$

(c) $(1 + j)(1 - j) = 2$

Key Point. Complex quantities have both real and imaginary parts. These two components are perpendicular to one another and thus the imaginary part has no effect along the line of action of the real part, and vice versa

---

Key Point. The j operator has the effect of rotating the direction of a voltage or current through an angle of 90°

The Argand diagram

The Argand diagram provides a useful method of visualising complex quantities and allowing us to solve problems graphically. In common with any ordinary '*x–y*' graph, the Argand diagram has two sets of axes as right angles, as shown in Figure 26.2. The horizontal axis is known as the *real axis* whilst the vertical axis is known as the imaginary axis (don't panic – the imaginary axis isn't really imaginary; we simply use the term to indicate that we are using this axis to plot values that are multiples of the j operator).

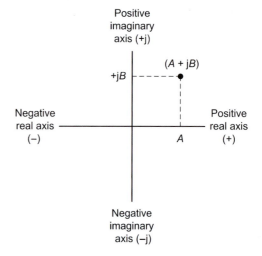

Figure 26.2 Argand diagram showing complex impedances

In Figure 26.2 we have plotted an impedance which has a real (resistive) part, A, and an imaginary (reactive) part, B. We can refer to this impedance as $(A + jB)$. The brackets help us to remember that the impedance is made up from two components; one imposing no phase shift whilst the other changes phase by 90°. Examples 26.1 and 26.2 show you how this works.

Example 26.1   A resistance of 4 Ω is connected in series with a capacitive reactance of 3 Ω.

(a)   Sketch a circuit and express this impedance in complex form

(b)   Plot the impedance on an Argand diagram

(c)   Sketch the impedance triangle for the circuit.

See Figure 26.3.

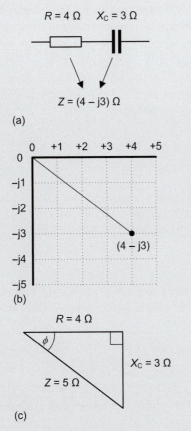

Figure 26.3 See Example 26.1

Example 26.2   A resistance of 4 Ω is connected in series with an inductive reactance of 3 Ω.

(a)   Sketch a circuit and express the impedance in complex form

(b)   Plot the impedance on an Argand diagram

(c)   Sketch the impedance triangle for the circuit.

See Figure 26.4.

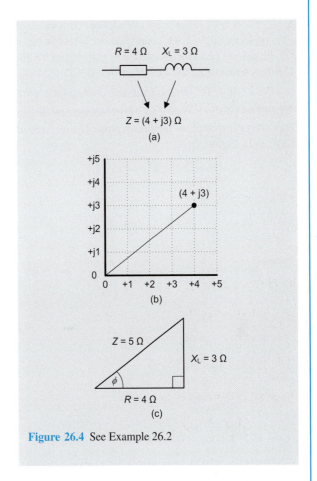

Figure 26.4 See Example 26.2

26.2   Series impedance

As you have just seen, the j operator and the Argand diagram provide us with a useful way of representing impedances. Any complex impedance can be represented by the relationship:

$$Z = R \pm jX$$

where Z represents impedance, R represents resistance and X represents reactance, all three quantities being measured in ohms.

The ± j term simply allows us to indicate whether the reactance is attributable to inductance, in which case the j term is positive (i.e. +j) or to capacitance (in which case the j term is negative (i.e. − j).

Consider, for example, the following impedances:

1. $Z_1 = 20 + j10$: this impedance comprises a resistance of 20 Ω connected in series with an inductive reactance (note the positive sign before the j term) of 10 Ω.

2. $Z_2 = 15 - j25$: this impedance comprises a resistance of 15 Ω connected in series with a capacitive reactance (note the negative sign before the j term) of 25 Ω.

3. $Z_3 = 30 + j0$: this impedance comprises a pure resistance of 30 Ω (there is no reactive component).

These three impedances are shown plotted on an Argand diagram in Figure 26.5.

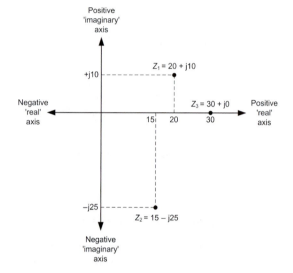

Figure 26.5 Impedances plotted on an Argand diagram

Voltages and currents can also take complex values. Consider the following:

1. $I_1 = 2 + j0.5$: this current is the result of an in-phase component of 2 A and a reactive component (at $+90°$) of 0.5 A.

2. $I_2 = 1 - j1.5$: this current is the result of an in-phase component of 1 A and a reactive component (at $-90°$) of 1.5 A.

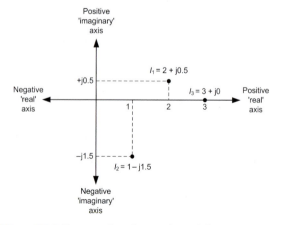

Figure 26.6 Currents plotted on an Argand diagram

3. $I_3 = 3 + j0$: this current is in phase and has a value of 3 A.

These three currents are shown plotted on an Argand diagram in Figure 26.6.

To determine the voltage dropped across a complex impedance we can apply the usual relationship:

$$V = IZ$$

where V is the voltage (V), I is the current (A) and Z is the impedance (Ω), all expressed in complex form.

Example 26.3 A current of 2 A flows in an impedance of $(100 + j120)$ Ω. Derive an expression, in complex form, for the voltage that will appear across the impedance.

Since $V = I \times Z$

$$v = 2 \times (100 + j120) = 200 + j240 \, \text{V}$$

Note that, in this example, we have assumed that the supply current is the *reference*. In other words, it could be expressed in complex form as $(2 + j0)$ A.

Example 26.4 An impedance of $(200 + j100)$ Ω is connected to a 100 V a.c. supply. Determine the current flowing and express your answer in complex form.

Since $I = \dfrac{V}{Z}$

$$I = \frac{100}{(200 + j100)} = \frac{100 \times (200 - j100)}{(200 + j100) \times (200 - j100)}$$

Note that we have multiplied the top and bottom by the *complex conjugate* in order to simplify the expression, reducing the denominator to a real number (i.e. no j term).

$$I = \frac{100 \times (200 - j100)}{(200^2 + 100^2)} = \frac{(2 \times 10^4 - j \times 10^4)}{(4 \times 10^4 + 1 \times 10^4)}$$

$$= \frac{(2 - j)}{5} = 0.4 - j0.2 \, \text{A}$$

Also note that we have assumed that the supply voltage is the *reference* quantity. In other words, the voltage could be expressed in complex form as $(100 + j0)$ V.

Test your knowledge 26.2

Identify the four impedances plotted on the Argand diagram shown in Figure 26.7.

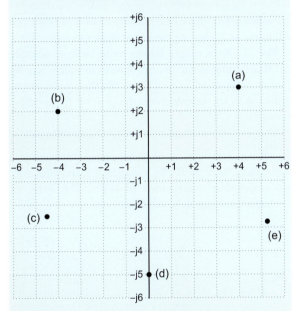

Figure 26.7 See Test your knowledge 26.2

Test your knowledge 26.3

Plot the following voltages on an Argand diagram:

(a) $30 + j40$ V

(b) $+ j20$ V

(c) $10 - j30$ V

(d) $-3 - j20$ V

Test your knowledge 26.4

(a) A capacitor having a reactance of 5 Ω is connected in series with a resistance of 5 Ω. Express this impedance in complex form.

(b) If an additional inductive reactance of 15 Ω is connected in series with the circuit in (a), write down the new impedance (once again in complex form).

Test your knowledge 26.5

Write down the impedance (in complex form) of each of the circuits shown in Figure 26.8.

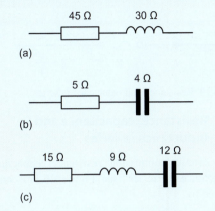

Figure 26.8 See Test your knowledge 26.5

26.2.1 Resistance and capacitance in series

A series circuit comprising resistance and capacitance (see Figure 26.9) can be represented by:

$$Z = R - jX_C \quad \text{or} \quad Z = R - \frac{j}{\omega C}$$

where $\omega = 2\pi f$.

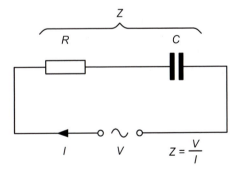

Figure 26.9 Resistance and capacitance in series

26.2.2 Resistance and inductance in series

A series circuit comprising resistance and inductance (see Figure 26.10) can be represented by:

$$Z = R + jX_L \quad \text{or} \quad Z = R + j\omega L$$

where $\omega = 2\pi f$.

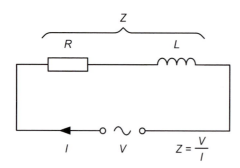

Figure 26.10 Resistance and inductance in series

26.2.3 Resistance, capacitance and inductance in series

A series circuit comprising resistance, capacitance and inductance in series (see Figure 26.11) can be represented by:

$$Z = R - jX_C + jX_L = R + j(X_L - X_C) = R + j\omega L - \frac{j}{\omega C}$$

where $\omega = 2\pi f$.

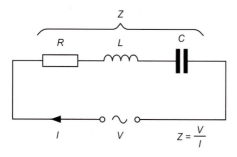

Figure 26.11 Resistance, capacitance and inductance in series

26.3 Parallel admittance

When dealing with parallel circuits it is much easier to work in terms of *admittance* (Y) rather than *impedance* (Z). Note that admittance is simply the reciprocal of impedance, hence:

$$Y = \frac{1}{Z}$$

The admittance of a circuit comprising resistance connected in parallel with reactance is given by:

$$Y = G \pm jB$$

where Y represents *admittance*, G represents *conductance* and B represents *susceptance*, and where:

$$G = \frac{1}{R} \quad \text{and} \quad B = \frac{1}{X}$$

All three quantities are measured in siemens (S).

The $\pm j$ term simply allows us to indicate whether the susceptance is due to capacitance (in which case the j term is positive (i.e. $+j$) or whether it is due to inductance (in which case the j term is negative (i.e. $-j$).

Consider, for example, the following admittances:

1. $Y_1 = 0.05 - j0.1$ this admittance comprises a conductance of 0.05 S connected in parallel with a negative susceptance of 0.1 S. The value of resistance can be found from:

$$R = \frac{1}{G} = \frac{1}{0.05} = 20\,\Omega$$

whilst the value of inductive reactance (note the minus sign before the j term) can be found from:

$$X = \frac{1}{B} = \frac{1}{0.1} = 10\,\Omega$$

2. $Y_2 = 0.2 + j0.05$ this admittance comprises a conductance of 0.2 S connected in parallel with a positive susceptance of 0.05 S. The value of resistance can be found from:

$$R = \frac{1}{G} = \frac{1}{0.2} = 5\,\Omega$$

whilst the value of capacitive reactance (note the plus sign before the j term) can be found from:

$$X = \frac{1}{B} = \frac{1}{0.01} = 20\,\Omega$$

To determine the current flowing in a complex admittance we can apply the usual relationship:

$$I = VY$$

where I is the current (A), V is the voltage (V) and Y is the admittance (S), all expressed in complex form.

Example 26.5   A voltage of 20 V appears across an admittance of $(0.1 + \text{j}0.25)\ \Omega$. Determine the current flowing and express your answer in complex form.

Since $I = V \times Y$

$$I = 20 \times (0.1 + \text{j}0.25) = 2 + \text{j}5\ \text{A}$$

Note that, in this example we have assumed that the supply voltage is the *reference*. In other words, it could be expressed in complex form as $(20 + \text{j}0)$ V.

26.3.1   Resistance and capacitance in parallel

A series circuit comprising resistance and capacitance (see Figure 26.12) can be represented by:

$$Y = G + \text{j}B = \frac{1}{R} + \frac{\text{j}}{X_\text{C}} = \frac{1}{R} + \text{j}\omega C$$

where $\omega = 2\pi f$, $G = \dfrac{1}{R}$ and $B = \dfrac{\text{j}}{X_\text{C}}$

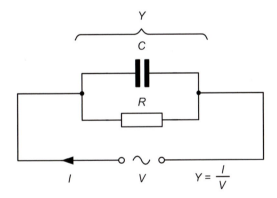

Figure 26.12 Resistance and capacitance in parallel

26.3.2   Resistance and inductance in parallel

A parallel circuit comprising resistance and inductance (see Figure 26.13) can be represented by:

$$Y = G + \text{j}B = \frac{1}{R} - \frac{\text{j}}{X_\text{L}} = \frac{1}{R} - \frac{\text{j}}{\omega L}$$

where $\omega = 2\pi f$, $G = \dfrac{1}{R}$ and $B = -\dfrac{\text{j}}{X_\text{L}}$

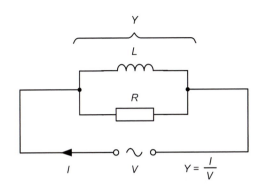

Figure 26.13 Resistance and inductance in parallel

26.3.3   Resistance, capacitance and inductance in parallel

Similarly, a series circuit comprising capacitance, inductance and resistance in parallel (see Figure 26.14) can be represented by:

$$Y = \frac{1}{R} + \frac{\text{j}}{X_\text{C}} - \frac{\text{j}}{X_\text{L}}$$

$$= \frac{1}{R} + \text{j}\omega C - \frac{\text{j}}{\omega L} = \frac{1}{R} + \text{j}\left(\omega C - \frac{1}{\omega L}\right)$$

where $\omega = 2\pi f$.

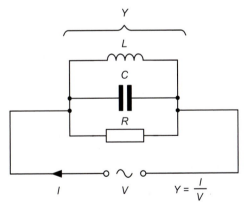

Figure 26.14 Resistance, capacitance and inductance in parallel

Test your knowledge 26.6

Write down the admittance (in complex form) of each of the circuits shown in Figure 26.15.

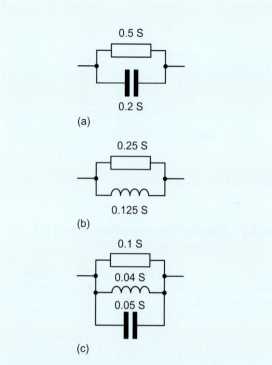

Figure 26.15 See Test your knowledge 26.6

Key Point. When solving parallel circuits it is often more convenient to work with admittance rather than impedance

26.4   Complex networks

Expressing impedances and admittances in complex form provides us with some useful ways of solving complex series/parallel networks. The basic rules of combination are as follows.

26.4.1   Series impedances

To determine the equivalent impedance of a series circuit we add together the individual impedances, thus:

$$Z = Z_1 + Z_2 + Z_3 \cdots + Z_n$$

26.4.2   Parallel impedances

To determine the reciprocal of the equivalent impedance of a parallel circuit we add together the reciprocals of the individual impedances, thus:

$$\frac{1}{Z} = \frac{1}{Z_1} + \frac{1}{Z_2} + \frac{1}{Z_3} \cdots + \frac{1}{Z_n}$$

When *only two* impedances are present we can simplify the expression to:

$$Z = \frac{Z_1 Z_2}{Z_1 + Z_2}$$

26.4.3   Parallel admittances

To determine the equivalent admittance of a parallel circuit we add together the individual admittances, thus:

$$Y = Y_1 + Y_2 + Y_3 \cdots + Y_n$$

26.4.4   Series admittances

To determine the reciprocal of the equivalent admittance of a series circuit we add together the reciprocals of the individual admittances, thus:

$$\frac{1}{Y} = \frac{1}{Y_1} + \frac{1}{Y_2} + \frac{1}{Y_3} \cdots + \frac{1}{Y_n}$$

When only two admittances are present we can simplify the expression to:

$$Y = \frac{Y_1 Y_2}{Y_1 + Y_2}$$

Note that it is usually easier to work with admittances when solving parallel circuits and with impedances in the case of series circuits. Converting from impedance to admittance and vice versa is usually fairly straightforward.

Example 26.6   Determine the impedance of the network shown in Figure 26.16 at a frequency of 400 Hz.

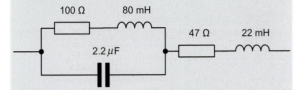

Figure 26.16 See Example 26.6

First we need to determine the inductive and capacitive reactance, as follows:

For the 80 mH inductor:

$$X_L = \omega L = 2\pi f L = 6.28 \times 400 \times 80 \times 10^{-3} = 201\,\Omega$$

For the 22 mH inductor:

$$X_L = \omega L = 2\pi f L = 6.28 \times 400 \times 22 \times 10^{-3} = 55.3\,\Omega$$

For the 2.2 μF capacitor:

$$X_C = \frac{1}{\omega C} = \frac{1}{2\pi f C} = \frac{0.159}{400 \times 2.2 \times 10^{-6}} = 181\,\Omega$$

We can now re-draw the circuit using complex notation as shown in Figure 26.17, where:

$$Z_1 = 100 + j201\,\Omega$$

$$Z_2 = -j181\,\Omega$$

$$Z_3 = 47 + j55.3\,\Omega$$

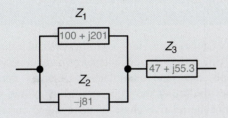

Figure 26.17 See Example 26.6

The network can be simplified in successive stages, as shown in Figure 26.18.

Z_4 is the parallel equivalent of Z_1 and Z_2, hence:

$$Z_4 = \frac{Z_1 Z_2}{Z_1 + Z_2} = \frac{(100 + j201) \times (-j181)}{(100 + j201) + (-j181)}$$

$$= \frac{(100 + j201) \times (-j181)}{(100 + j20)}$$

$$Z_4 = \frac{-j18100 + -j^2 36381}{100 + j20} = \frac{36381 - j18100}{100 + j20}$$

$$= \frac{(36381 - j18100)(100 - j20)}{(100 + j20)(100 - j20)}$$

$$Z_4 = \frac{3638100 - j727620 - j1810000 + j^2 362000}{100^2 + 20^2}$$

$$= \frac{3276100 - j2537620}{10400} = 315 - j244$$

Z_5 is the series equivalent of Z_4 and Z_3, hence:

$$Z_5 = Z_4 + Z_3 = (315 - j244) + (47 + j55.3)$$

$$= 362 - j188.7\,\Omega$$

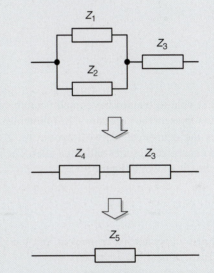

Figure 26.18 See Example 26.6

Test your knowledge 26.8

Find the equivalent impedance of the circuit shown in Figure 26.19 at a frequency of 400 Hz. Express your answer in complex form.

Test your knowledge 26.9

Find the equivalent admittance of the circuit shown in Figure 26.20 at a frequency of 400 Hz. Express your answer in complex form.

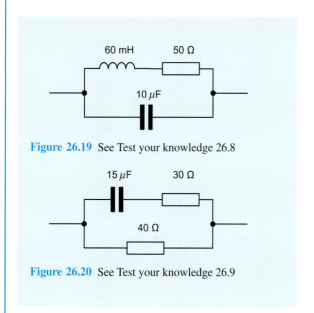

Figure 26.19 See Test your knowledge 26.8

Figure 26.20 See Test your knowledge 26.9

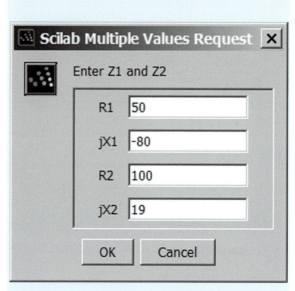

Figure 26.21 Input dialogue for the SCILAB script

SCILAB

SCILAB is able to manipulate complex numbers and uses %i to represent the j operator. The following script calculates the equivalent of two impedances, Z_1 and Z_2, when connected in series and in parallel. The input dialogue box is shown in Figure 26.21.

```
// Complex impedances in series and parallel
// SCILAB source file: ex2601.sce
// Get data from dialog box
txt = ['R1';'jX1';'R2';'jX2'];
var = x_mdialog('Enter Z1 and Z2',txt,
['50';'-80';'100';'19']);
R1 = evstr(var(1));
X1 = evstr(var(2));
R2 = evstr(var(3));
X2 = evstr(var(4));
// Expressions for Z1 and Z2
Z1 = R1 + (%i * X1);
Z2 = R2 + (%i * X2);
// Calculate series and parallel
equivalent impedances
Zs = Z1+Z2;
Zp = Z1*Z2/(Z1+Z2);
messagebox(["Series = " + string(Zs),
"Parallel = " + string(Zp)],"Results");
```

Test your knowledge 26.10

Write and test a SCILAB script that will calculate the equivalent of two admittances, Y_1 and Y_2, when connected (a) in parallel and (b) in series.

26.5   Chapter summary

Complex notation provides us with a powerful technique for solving series, parallel and series/parallel networks of impedances and admittances. Complex notation involves the use of the j operator which allows us to represent the relative phase of voltage and current in a circuit containing a combination of resistance and reactance.

Expressing impedances and admittances in complex form provides us with a set of useful and interchangeable techniques that we can use to solve even the most complex series/parallel networks. The basic rules of combination are:

- To determine the equivalent impedance of a series circuit we add together the individual impedances
- To determine the equivalent admittance of a parallel circuit we add together the individual admittances

- To determine the reciprocal of the equivalent impedance of a parallel circuit we add together the reciprocals of the individual impedances
- To determine the reciprocal of the equivalent admittance of a series circuit we add together the reciprocals of the individual admittances.

As a consequence, it is usually more convenient to work with impedance when solving series circuits and with admittance when solving parallel circuits. In the next chapter we shall be using complex notation to analyse circuits containing a mixture of resistance and reactance.

26.6   Review questions

Short answer questions

1. Determine the value of each of the following:

    (a) $1 + j^2$

    (b) $(2 + 3j)(2 - 3j)$

    (c) $\dfrac{(1 + j)}{(1 - j)}$

2. Plot the following impedances on an Argand diagram:

    (a) $(3 - j7)\,\Omega$

    (b) $(6 + j4)\,\Omega$

    (c) $(2 - j0)\,\Omega$

3. A resistance of $22\,\Omega$ is connected in series with a capacitive reactance of $15\,\Omega$. Write down an expression for the impedance of the circuit in complex form.

4. A coil has a resistance of $100\,\Omega$ and a reactance of $50\,\Omega$. Write down an expression for the impedance of the coil in complex form.

5. A resistance of $5\,\Omega$ is connected in parallel with a capacitive reactance of $4\,\Omega$. Write down an expression for the admittance of the circuit in complex form.

6. An admittance of $(0.4 + j0.1)\,S$ is connected in parallel with a conductance of $0.5\,S$. What will the combined admittance be and what current will flow in the circuit when a voltage of $20\,V$ is applied to it? Express your answer in complex form.

7. An impedance of $(8 + j12)\,\Omega$ is connected in series with a resistance of $10\,\Omega$. What will the combined impedance be and what voltage will appear across the circuit when a current of $3\,A$ is applied to it? Express your answer in complex form.

8. An impedance of $(30 + j40)\,\Omega$ is connected to a $50\,V$ a.c. supply. Determine the current flowing and express your answer in complex form.

9. Determine the voltage dropped across an impedance of $(5 + j4)\,\Omega$ when a current of $(15 - j12)\,A$ flows in it.

10. A current of $(9 - j3)\,A$ flows in a circuit when a voltage of $110\,V$ is applied. What is (a) the admittance and (b) the impedance of the circuit? Express your answers in complex form.

Long answer questions

11. Determine the voltage dropped across the circuit shown in Figure 26.22 when a current of $4.5\,A$ at $400\,Hz$ flows in it.

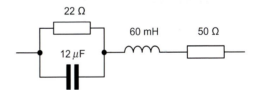

Figure 26.22 See Question 11

12. Determine the current supplied to the circuit shown in Figure 26.23 when a supply voltage of $110\,V$ at $400\,Hz$ is applied to it.

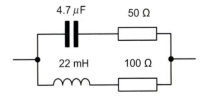

Figure 26.23 See Question 12

Chapter 27

Resonant circuits

A resonant circuit is one in which there is resistance and a combination of inductive and capacitive reactance. In such circuits there will be one particular frequency at which the current and voltage are exactly in phase and since the reactive components effectively cancel one another out, the circuit behaves like one that contains only pure resistance. This can be extremely useful in a number of applications that require circuits to be selective, e.g. accepting or rejecting current within a particular band of frequencies. This chapter provides you with an introduction to the behaviour and application of resonant circuits.

27.1   Series resonant circuits

Circuits that contain only a combination of resistance and capacitance or only a combination of resistance and inductance are 'non-resonant' because the voltage and current will not be in phase at any frequency (other than zero in the case of an inductance or infinity in the case of a capacitance!). More complex circuits, containing both types of reactance together with resistance, are described as 'resonant' since there will be one frequency at which the two reactive components will be equal but opposite. At this particular frequency (known as the *resonant frequency*) the effective reactance in the circuit will be zero and the voltage and current will be in phase. This can be a fairly difficult concept to grasp at first sight, so

we will explain it in terms of the way in which a resonant circuit behaves at different frequencies.

27.1.1   Series resonance

In Chapter 26 when we analysed the behaviour of series circuits comprising a combination of resistance, capacitance and inductance and noted the effect of inductive reactance, X_L, being greater than capacitive reactance, X_C, and vice versa. We did not, however, consider the notable effect that occurs when the two reactances are equal but of opposite sign. We will now do just that!

A series resonant circuit comprises inductance, resistance and capacitance connected in series, as shown in Figure 27.1.

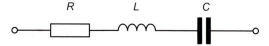

Figure 27.1 A series resonant circuit

At the resonant frequency, the inductive reactance, X_L, will be equal to the capacitive reactance, X_C. In this condition, the supply voltage will be in phase with the supply current. Furthermore, since $X_L = X_C$, the reactive components will cancel out (recall that they are 180° out of phase with one another) and the impedance of the circuit will take a minimum value, equal to the resistance, R. Putting this another way, the *net reactance* in the circuit will be zero and we will be left with resistance only.

At a frequency that is less than the resonant frequency (i.e. below resonance) the inductive reactance, X_L, will

978-1-85617-775-7, Engineering Science, Mike Tooley & Lloyd Dingle

be smaller than the capacitive reactance, X_C. In other words, $X_L < X_C$. In this condition, the supply voltage will lag the supply current by 90° (see page 455).

At a frequency that is greater than the resonant frequency (i.e. above resonance) the capacitive reactance, X_C, will be smaller than the inductive reactance, X_L. In other words, $X_C < X_L$. In this condition, the supply voltage will lead the supply current by 90° (see page 456).

The phasor diagram for a series resonant circuit at resonance is shown in Figure 27.2. Note how the two reactances, X_C and X_L, are equal but of opposite sign and the net reactance is therefore zero.

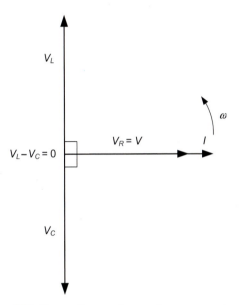

Figure 27.2 Phasor diagram for a series resonant circuit at resonance

SCILAB

The following SCILAB script produces a plot of net reactance against frequency when a capacitance of $1\,\mu\text{F}$ is connected in series with an inductance of 220 mH (see Figure 27.3). You might want to compare this with the variation of individual reactances shown in Figure 25.13 on page 453. Note that the frequency at which the two lines cross on Figure 25.13 is exactly the same as the frequency at which the net reactance curve passes through zero in Figure 27.3.

```
// Plot of net reactance over the range
100 to 1000 Hz
// Filename: ex2701.sce
```

```
// Clear any existing plot
clf();
// Set up parameters
f = [100:10:1000]';  // Frequency range
100 Hz to 1 kHz
C = 1E-6;  // C = 1 uF
L = 0.22;  // L = 220 mH
// Plot the net reactance in blue
plot2d(f,(2*%pi*f*L-(2*%pi*f*C)^-1),2);
// Print legend, grid and title
xgrid(); // add grid
xtitle('Net reactance plotted
against frequency');
```

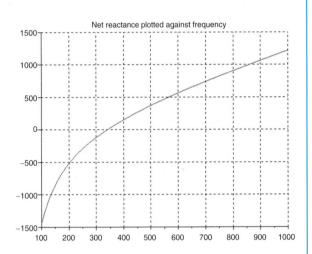

Figure 27.3 Graphical output of the SCILAB script

SCILAB's graph plotting functions can be put to good use to investigate the behaviour of a series resonant circuit above and below resonance. The script that follows shows how this is done. In this particular example we have used the capacitance and inductance values as in the previous example together with a series resistance of $100\,\Omega$ connected across a 2 V supply. It is worth running this script at frequencies over the range 200 Hz to 500 Hz and noting the effect on the voltages in the circuit. The resonant frequency (which can be determined from the previous script when the net reactance is zero) should produce inductor and capacitor voltages which are of the same size but 180°

out of phase (i.e. in *antiphase*). Figure 27.4 shows the voltage waveforms produced by the script at the default frequency (300 Hz).

```
/ Plot of voltage and current in a
series resonant circuit
// Filename: ex2702.sce
// First set up values for time
t=(0:0.00001:0.01)'; // Plot
range = 2 ms
clf(); // get rid of any earlier plot
// Set up circuit parameters
C = 1E-6; // C = 1 uF
L = 0.22; // L = 220 mH
R = 100; // R = 100 ohm
// Get frequency to use
var = x_mdialog('Enter frequency',
'Frequency
in Hz',['300']);
f = evstr(var(1));
//
// Calculate the inductive reactance
xL = 2 * %pi * f * L;
// Calculate the capacitive reactance
xC = (2 * %pi * f * C)^-1;
// Calculate the impedance
z = sqrt(R^2 + (xL - xC)^2);
// Calculate the voltage dropped across each
component when a
// constant voltage of 2 V is applied to the
circuit
i = 2 / z;
vl = i * xL * sin(2*%pi * f * t + %pi/2);
vc = i * xC * sin(2*%pi * f * t - %pi/2);
vr = i * R * sin(2*%pi * f * t);
// Plot the graphs of voltage against
frequency
plot2d(t,vl,5,rect=[0,-10,0.01,10]);
// plot in red
plot2d(t,vc,2,rect=[0,-10,0.01,10]);
// plot in blue
plot2d(t,vr,3,rect=[0,-10,0.01,10]);
// plot in green
xtitle('Voltages in a series L-C-R circuit
plotted against time','Time
(s)','Voltage (V)');
legend(['vL','vC','vR']);
```

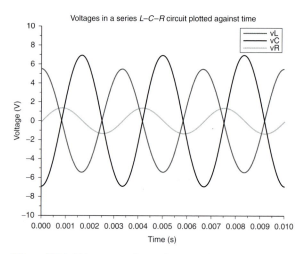

Figure 27.4 Voltage waveforms for the series resonant circuit at a frequency of 300 Hz

27.1.2 Resonant frequency

As we mentioned earlier, the reactive components in a series L–C–R circuit will effectively cancel each other out when circuit is resonant. We can thus determine the frequency of resonance, f_0, by simply equating the two reactive components, as follows:

$$X_L = X_C$$

thus $\dfrac{1}{2\pi f_0 C} = 2\pi f_0 L$

We need to make f_0 the subject of this equation:

$$f_0 = \frac{1}{4\pi^2 LC} = \frac{1}{2\pi \sqrt{LC}}$$

Example 27.1 A series circuit consists of $L = 60\,\text{mH}$, $R = 15\,\Omega$ and $C = 15\,\text{nF}$. Determine the frequency of resonance and the voltage dropped across each component at resonance if the circuit is connected to a 300 mV a.c. supply.

$$f_0 = \frac{1}{2\pi \sqrt{60 \times 10^{-3} \times 15 \times 10^{-9}}}$$

$$= \frac{1}{2\pi \times 30 \times 10^{-6}}$$

$$= 5.3 \times 10^3 = 5.3\,\text{kHz}$$

At resonance, the reactive components (X_L and X_C) will be equal but of opposite sign. The impedance at resonance will thus be R alone. We can therefore determine the supply current from:

$$I = \frac{V}{Z} = \frac{V}{R} = \frac{300 \times 10^{-3}}{15}$$
$$= 20 \times 10^{-3} = 20\,\text{mA}$$

At the resonant frequency the reactance of the inductor will be:

$$X_L = 2\pi f_o L$$
$$= 2\pi \times 5.3 \times 10^3 \times 60 \times 10^{-3} = 2\,\text{k}\Omega$$

At this frequency the voltage developed across the inductor will be given by:

$$V_L = I \times X_L = 20 \times 10^{-3} \times 2 \times 10^3 = 40\,\text{V}$$

Note that this voltage will lead the supply current by $90°$.

Since the reactance of the capacitor will be the same as that of the inductor, the voltage developed across the capacitor will be identical (but lagging the supply current by $90°$). Thus:

$$V_C = V_L = 40\,\text{V}$$

Test your knowledge 27.1

A series resonant circuit consists of a loss-free inductor $L = 1\,\text{mH}$, $C = 1.5\,\text{nF}$ and $R = 12\,\Omega$. Determine the resonant frequency of the circuit. Also determine the voltage that will be dropped across each component when a current of 100 mA flows at the circuit's resonant frequency.

Key Point. A resonant circuit contains a combination of inductance, capacitance and resistance. In such a circuit there will be one particular frequency at which the supply current and voltage are exactly in phase. At this frequency the circuit will behave like one of pure resistance

27.2 Parallel resonance

A parallel resonant circuit comprises inductance, resistance and capacitance connected in parallel, as shown in Figure 27.5. An important variation on this is when a capacitor is connected in parallel with a series combination of inductance and resistance, as shown in Figure 27.6.

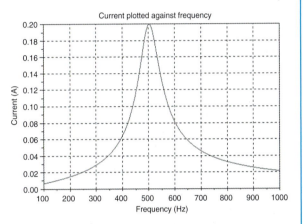

Figure 27.5 Parallel resonant circuit

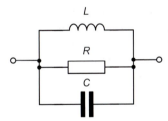

Figure 27.6 Alternative form of parallel resonant circuit

In the case of the circuit shown in Figure 27.6, the frequency of resonance can once again be determined by simply equating the two reactive components, as follows:

$$X_L = X_C$$

thus $\dfrac{1}{2\pi f_0 C} = 2\pi f_0 L$

We need to make f_0 the subject of this equation:

$$f_0 = \frac{1}{4\pi^2 LC} = \frac{1}{2\pi \sqrt{LC}}$$

Example 27.2   A parallel circuit consists of $L = 40\,\text{mH}$, $R = 1\,\text{k}\Omega$ and $C = 10\,\text{nF}$. Determine the frequency of resonance and the current in each component at resonance if the circuit is connected to a 2 V a.c. supply.

$$f_0 = \frac{1}{2\pi\sqrt{40 \times 10^{-3} \times 10 \times 10^{-9}}}$$

$$= \frac{1}{2\pi \times 20 \times 10^{-6}} = 7.96 \times 10^3 = 7.96\,\text{kHz}$$

At resonance, the reactive components (X_L and X_C) will be equal but of opposite sign. The impedance at resonance will thus be R alone. We can determine the supply current from:

$$I = \frac{V}{Z} = \frac{V}{R} = \frac{2}{1 \times 10^3} = 2 \times 10^{-3} = 2\,\text{mA}$$

The reactance of the inductor is equal to:

$$X_L = 2\pi f_0 L = 2\pi \times 7.96 \times 10^3 \times 40 \times 10^{-3}$$

$$= 2\pi \times 7.96 \times 40 = 2\,\text{k}\Omega$$

The current in the inductor will be given by:

$$I_L = \frac{V_L}{X_L} = \frac{2}{2 \times 10^3} = 1 \times 10^{-3} = 1\,\text{mA}$$

Note that this current lags the supply voltage by 90°.

Since the reactance of the capacitor will be the same as that of the inductor, the current in the capacitor will be identical (but leading the supply current by 90°). Thus:

$$I_C = I_L = 1\,\text{mA}$$

In the case of the circuit shown in Figure 27.6, the frequency of resonance is once again the frequency at which the supply voltage and current are in phase. Note that the current in the inductor, I_L, lags the supply voltage by an angle, ϕ (see Figure 27.7).

Now $I_L = \dfrac{V_L}{Z_L} = \dfrac{V}{\sqrt{R^2 + X_L^2}}$ and $I_C = \dfrac{V_L}{X_C}$

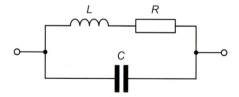

Figure 27.7 Phasor diagram for the circuit in Figure 27.6

At resonance, $I_C = I_L \sin\phi$ and $\sin\phi = \dfrac{X_L}{Z_L} = \dfrac{X_L}{\sqrt{R^2 + X_L^2}}$

Thus:

$$\frac{V}{X_C} = \frac{V}{\sqrt{R^2 + X_L^2}} \times \frac{X_L}{\sqrt{R^2 + X_L^2}} = \frac{V}{\left(\dfrac{1}{2\pi f_0 C}\right)}$$

$$= \frac{V}{\sqrt{R^2 + (2\pi f_0 L)^2}} \times \frac{2\pi f L}{\sqrt{R^2 + (2\pi f_0 L)^2}}$$

Rearranging, simplifying and dividing both sides by V gives:

$$2\pi f_0 C = \frac{1}{\sqrt{R^2 + (2\pi f_0 L)^2}} \times \frac{2\pi f_0 L}{\sqrt{R^2 + (2\pi f_0 L)^2}}$$

$$= \frac{2\pi f_0 L}{R^2 + (2\pi f_0 L)^2}$$

Further simplification gives:

$$C = \frac{L}{R^2 + (2\pi f_0 L)^2}$$

Thus:

$$C\left(R^2 + (2\pi f_0 L)^2\right) = L$$

$$R^2 + (2\pi f_0 L)^2 = \frac{L}{C}$$

$$(2\pi f_0 L)^2 = \frac{L}{C} - R^2$$

$$f_0^2 = \frac{1}{(2\pi L)^2}\left(\frac{L}{C} - R^2\right) = \frac{1}{(2\pi)^2}\left(\frac{L}{CL^2} - \frac{R^2}{L^2}\right)$$

Hence:

$$f_0 = \frac{1}{2\pi}\sqrt{\left(\frac{1}{LC} - \frac{R^2}{L^2}\right)}$$

At resonance, the impedance of the circuit shown in Figure 27.6 is called its *dynamic impedance*. This impedance is given by:

$$Z_d = \frac{V}{I}$$

From the phasor diagram of Figure 27.7:

$$I = \frac{I_C}{\tan\phi}$$

Thus, $Z_d = V\dfrac{\tan\phi}{I_C} = \dfrac{V}{I_C}\tan\phi = X_C\tan\phi$

Now, since $\tan\phi = \dfrac{X_L}{R}$

$$Z_d = X_C \times \frac{X_L}{R} = \frac{2\pi fL}{2\pi fC} \times \frac{1}{R} = \frac{L}{CR}$$

This value of impedance is known as the *dynamic impedance* of the circuit. You should note that that the dynamic impedance increases as the ratio of L to C increases.

> **Example 27.3** A coil having an inductance of $1\,\text{mH}$ and a resistance of $100\,\Omega$ is connected in parallel with a capacitor of $10\,\text{nF}$. Determine the frequency of resonance and the dynamic impedance of the circuit.
>
> Now
>
> $$f_0 = \frac{1}{2\pi}\sqrt{\left(\frac{1}{LC} - \frac{R^2}{L^2}\right)}$$
>
> $$= \frac{1}{2\pi}\sqrt{\left(\frac{1}{1\times 10^{-3}\times 10\times 10^{-9}} - \frac{100^2}{\left(1\times 10^{-3}\right)^2}\right)}$$
>
> $$f_0 = \frac{1}{2\pi}\sqrt{\left(10^{11} - \frac{100^4}{10^{-6}}\right)}$$
>
> $$= \frac{1}{2\pi}\sqrt{\left(10^{11} - 10^{10}\right)}$$
>
> $$= \frac{1}{2\pi}\sqrt{9\times 10^{10}}$$
>
> Thus:
>
> $$f_0 = \frac{3\times 10^5}{2\pi} = 47.7\times 10^3 = 47.7\,\text{kHz}$$
>
> The dynamic impedance is given by:
>
> $$Z_d = \frac{L}{CR} = \frac{1\times 10^{-3}}{10\times 10^{-9}\times 100} = 1\times 10^3 = 1\,\text{k}\Omega$$

Test your knowledge 27.2

A parallel resonant tuned circuit uses a loss-free inductor of $60\,\text{mH}$ connected in parallel with a resistance of $800\,\Omega$ and a capacitance of $100\,\text{nF}$. Determine the resonant frequency of the circuit. Also, determine the current that will flow in each component when a supply of $24\,\text{V}$ is connected to the circuit at its resonant frequency.

Test your knowledge 27.3

A coil having an inductance of $15\,\text{mH}$ and a resistance of $50\,\Omega$ is connected in parallel with a capacitor of $22\,\text{nF}$. Determine the frequency of resonance. Also determine the dynamic impedance of the circuit.

SCILAB

The following SCILAB script produces a plot of current against frequency for a series resonant circuit (see Figure 27.8). It can be instructive to experiment with different values of L, C and R – try doubling and halving each of the values in turn.

```
// Plot of current in a series resonant
circuit
// Filename: ex2703.sce
// First set up the table of values for
frequency
f = [100:1:1000]';
// Circuit parameters
C = 1E-6; // Capacitance = 1 uF
L = 100E-3 // Inductance = 100 mH
r = 50; // Resistance = 50 ohm
v = 10; // Supply = 10V
// Calculate the inductive reactance
xL = 2 * %pi * f * L;
// Calculate the capacitive reactance
xC = (2 * %pi * f * C)^-1;
// Calculate the impedance
z = sqrt(r^2 + (xL - xC)^2); // xL--xC is
the net reactance
// Calculate the current
```

```
i = v * z^-1;
// Get rid of any previous plot
clf();
plot(f,i); // and finally plot the curve
xgrid(); // add grid
xtitle('Current plotted against frequency',
'Frequency (Hz)','Current (A)');
// Add title and labels for the axes
```

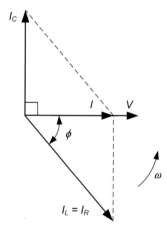

Figure 27.8 Graphical output of the SCILAB script

Key Point. The impedance of a series resonant circuit takes its lowest value (equivalent to the series resistance present in the circuit) at resonance. Conversely, the impedance of a parallel resonant circuit takes its highest value (equivalent to the parallel resistance present in the circuit) at resonance

Key Point. In terms of the current flowing, a series resonant circuit can be described as an *acceptor circuit* whilst a parallel resonant circuit can be described as a *rejector circuit*

27.3 Q-factor and bandwidth

The Q-factor (or *quality factor*) is a measure of the 'goodness' of a tuned circuit. Q-factor is sometimes also referred to as *voltage magnification factor*. Consider the circuit shown in Figure 27.9. The Q-factor of the circuit tells you how many times greater the inductor or capacitor voltage is than the supply voltage. The better

the circuit, the higher the voltage magnification and the greater the Q-factor.

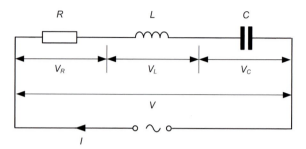

Figure 27.9 Voltages in a series resonant circuit

In Figure 27.9, $Q = \dfrac{V_L}{V_S} = \dfrac{V_C}{V_S}$

Since $V_L = I \times X_L = I \times 2\pi f L$ and $V_S = I \times Z = I \times R$ (at resonance):

$$Q = \frac{V_L}{V_S} = \frac{V_L}{V_R} = \frac{IX_L}{IR} = \frac{I \times 2\pi f L}{IR} = \frac{2\pi f L}{R} = \frac{\omega L}{R}$$

Similarly, since $V_C = I \times X_C = \dfrac{I}{2\pi f C}$ and $V_S = I \times Z = I \times R$ (at resonance):

$$Q = \frac{V_C}{V_S} = \frac{V_C}{V_R} = \frac{IX_C}{IR} = \frac{I}{2\pi f C \times IR} = \frac{1}{2\pi f C R} = \frac{1}{\omega C R}$$

For the equivalent parallel resonant circuit shown in Figure 27.10 the Q-factor is the *current magnification factor* and is defined as:

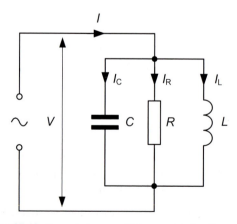

Figure 27.10 Currents in a parallel resonant circuit

$$Q = \frac{I_L}{I_R} = \frac{\left(\frac{V}{X_L}\right)}{\left(\frac{V}{R}\right)}$$

$$= \frac{VR}{VX_L} = \frac{R}{X_L} = \frac{R2\pi fL}{2\pi fL} = \frac{R}{\omega L}$$

Similarly,

$$Q = \frac{I_C}{I_R} = \frac{\left(\frac{V}{X_C}\right)}{\left(\frac{V}{R}\right)} = \frac{VR}{VX_C} = \frac{R}{X_C} = \frac{R}{\left(\frac{1}{2\pi fC}\right)} = R\omega C$$

Example 27.4 A parallel resonant circuit consists of $L = 10\,\mu\text{H}$, $C = 20\,\text{pF}$ and $R = 10\,\text{k}\Omega$. Determine the Q-factor of the circuit at resonance.

First, we need to find the resonant frequency from:

$$f_0 = \frac{1}{2\pi\sqrt{LC}} = \frac{1}{2\pi\sqrt{10 \times 10^{-6} \times 20 \times 10^{-12}}}$$

$$= \frac{1}{2\pi\sqrt{200 \times 10^{-18}}}$$

$$f_0 = \frac{0.159}{1.414 \times 10^{-8}} = 11.24 \times 10^6 = 11.24\,\text{MHz}$$

At resonance,

$$X_L = 2\pi f_0 L = 6.28 \times 11.24 \times 10^6 \times 10 \times 10^{-6}$$

$$= 705.9\,\Omega$$

Now $Q = R\omega L = 10000/705.9 = 14.2$

27.3.1 Bandwidth

As the Q-factor of a resonant circuit increases, the frequency response curve becomes sharper. As a consequence, the range of frequencies that lie within the scope of the tuned circuit's response (i.e. the *bandwidth* of the resonant circuit) is reduced. This relationship is illustrated in Figure 27.11.

We normally specify the range of frequencies that will be accepted by a resonant circuit by referring to the two *half-power frequencies*. These are the frequencies at which the power in a tuned circuit falls to 50% of its maximum value. At these frequencies:

1. The tuned circuit's resistive and reactive components (R and X) will be equal.

2. The phase angle (between current and voltage) will be $\pi/4$ (or $45°$).

3. Both current and voltage will have fallen to $\frac{1}{\sqrt{2}}$ or 0.707 of their maximum value.

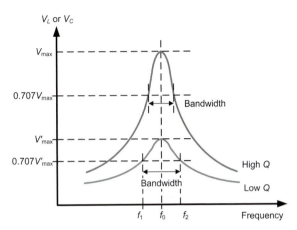

Figure 27.11 Relationship between Q-factor and bandwidth

At the upper of the two cut-off frequencies, f_2:

$$R = X_{\text{eff.}} \quad \text{(from 1 above)}$$

where $X_{\text{eff.}} = X_L - X_C$ (since $X_L > X_C$ above f_0) thus:

$$R = X_L - X_C = 2\pi f_2 L - \frac{1}{2\pi f_2 C}$$

Rearranging gives:

$$R - 2\pi f_2 L = -\frac{1}{2\pi f_2 C}$$

thus $2\pi f_2 \times (R - 2\pi f_2 L) = (2\pi f_2 \times R) - (2\pi f_2 \times 2\pi f_2 L) = -\frac{1}{C}$

or $\frac{1}{C} = -2\pi f_2 R + (2\pi f_2)^2 L$ ....(i)

whereas, at the lower of the two cut-off frequencies, f_1:

where $X_{\text{eff.}} = X_C - X_L$ (since $X_C > X_L$ below f_0)

thus:

$$R = X_C - X_L = \frac{1}{2\pi f_1 C} - 2\pi f_1 L$$

Rearranging gives:

$$R + 2\pi f_1 L = \frac{1}{2\pi f_1 C}$$

thus $2\pi f_1 \times (R + 2\pi f_1 L)$

$$= (2\pi f_1 \times R) + (2\pi f_1 \times 2\pi f_1 L) = \frac{1}{C}$$

or $\dfrac{1}{C} = 2\pi f_1 R + (2\pi f_1)^2 L$ ....(ii)

Equating (i) and (ii) gives:

$$-2\pi f_2 R + (2\pi f_2)^2 L = 2\pi f_1 R + (2\pi f_1)^2 L$$

$$(2\pi f_2)^2 L - (2\pi f_1)^2 L = 2\pi f_1 R + 2\pi f_2 R$$

$$(2\pi f_2)^2 - (2\pi f_1)^2 = \frac{R}{L} \times (2\pi f_1 + 2\pi f_2)$$

$$(2\pi f_2 + 2\pi f_1)(2\pi f_2 - 2\pi f_1) = \frac{R}{L} \times (2\pi f_1 + 2\pi f_2)$$

$$2\pi f_2 - 2\pi f_1 = \frac{R}{L}$$

$$f_2 - f_1 = \frac{R}{2\pi L}$$

But bandwidth, $f_w = f_2 - f_1$
thus:

$$f_w = \frac{R}{2\pi L}$$

Dividing both sides by f_0 gives:

$$\frac{f_w}{f_0} = \frac{R}{2\pi L f_0}$$

Earlier, we defined Q-factor as:

$$Q = \frac{X_L}{R} = \frac{2\pi f_0 L}{R}$$

Thus:

$$\frac{f_w}{f_0} = \frac{1}{Q}$$

or, $$Q = \frac{f_0}{f_w}$$

27.3.2 Loading and damping

Loading of a series tuned circuit occurs whenever another circuit is coupled to it. This causes the resonant circuit to be *damped* and the result is a reduction in Q-factor together with a corresponding increase in bandwidth (recall that bandwidth $= f_0/ - Q$).

In some cases, we might wish to deliberately introduce damping into a circuit in order to make it less selective, reducing the Q-factor and increasing the bandwidth. For a series resonant circuit, we must introduce the damping resistance in series with the existing components (L, C and R) whereas, for a parallel resonant circuit, the damping resistance must be connected in parallel with the existing components.

Example 27.5 A tuned circuit comprises a 400 μH inductor connected in series with a 100 pF capacitor and a resistor of 10 Ω. Determine the resonant frequency of the tuned circuit, its Q-factor and bandwidth.

$$f_0 = \frac{1}{2\pi \sqrt{400 \times 10^{-6} \times 100 \times 10^{-12}}}$$

$$= \frac{1}{2\pi \times 2 \times 10^{-7}} = \frac{0.159}{2 \times 10^{-7}}$$

$$= 0.795 \times 10^6 = 795 \text{ kHz}$$

The Q-factor is determined from:

$$Q = \frac{2\pi f_0 L}{R} = \frac{2\pi \times 795 \times 10^3 \times 400 \times 10^{-6}}{10}$$

$$= 199.7 \times 10^{-1} = 19.97$$

The bandwidth can be determined from:

$$f_w = \frac{f_0}{Q} = \frac{795 \times 10^3}{19.97} = 39.8 \times 10^3 = 39.8 \text{ kHz}$$

Example 27.6 A series resonant filter is to be centred on a frequency of 73 kHz. If the filter is to use a capacitor of 1 nF, determine the value of inductance required and the maximum value of series resistance that will produce a bandwidth of less than 1 kHz.

Rearranging $f_0 = \dfrac{1}{2\pi\sqrt{LC}}$ to make L the subject gives:

$$L = \frac{1}{4\pi^2 f_0^2 C} = \frac{1}{4\pi^2 \times 1 \times 10^{-9} \times (73 \times 10^3)^2}$$

$$= \frac{1}{4\pi^2 \times 5.329} = 4.75 \times 10^{-3} = 4.75\,\text{mH}$$

Now $f_{\text{w}} = \dfrac{f_0}{Q}$ and $Q = \dfrac{2\pi f_0 L}{R}$

Thus $f_{\text{w}} = f_0 \times \dfrac{R}{2\pi f_0 L}$, from which:

Hence $R = \dfrac{f_{\text{w}} \times 2\pi f_0 L}{f_0}$

$$= 1 \times 10^3 \times 2\pi \times 4.75 \times 10^{-3} = 29.8\,\Omega$$

Thus, to achieve a bandwidth of 1 kHz or less, the total series resistance must not exceed 29.8 Ω.

Test your knowledge 27.4

A parallel resonant circuit consists of a loss-free inductor $L = 220\,\mu\text{H}$, $C = 500\,\text{pF}$ and $R = 12\,\text{k}\Omega$. Determine the Q-factor and bandwidth of the resonant circuit and the value of additional resistance that must be connected in parallel with the circuit in order to reduce the Q-factor to 10.

Key Point. High values of Q-factor are associated with narrow bandwidth whilst low values of Q-factor are associated with wide bandwidth. Thus Q-factor provides us with a useful measure of the *selectivity* offered by a resonant circuit

---

27.4 Using complex notation to analyse resonant circuits

Earlier in this chapter we introduced series and parallel resonance and derived the expressions for resonant frequency, dynamic resistance and Q-factor. We can also make good use of complex notation when analysing resonant circuits. For example, the impedance of the

R–L–C series circuit shown earlier in Figures 27.1 and 27.9 is given by the expression:

$$Z = R + j\omega L - \frac{j}{\omega C} = R + j\left(\omega L - \frac{1}{\omega C}\right)$$

At resonance, $\omega L = \dfrac{1}{\omega C}$ and thus $Z = R + j(0) = R$

As you know, the circuit behaves like a pure resistor in this particular frequency and the supply current and voltage will be in phase.

At any frequency, f, we can determine the current flowing in a series resonant circuit and also the voltage dropped across each of the circuit elements present, as shown in Example 27.7.

Example 27.7 Determine the voltage dropped across each component in the circuit shown in Figure 27.12.

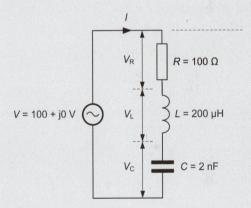

Figure 27.12 See Example 27.7

At 100 kHz, the inductive reactance is given by:

$$\begin{aligned}
j\omega L &= j \times 2\pi f \times L \\
&= j \times 6.28 \times 100 \times 10^3 \times 200 \times 10^{-6} \\
&= j125.6\,\Omega
\end{aligned}$$

Similarly, the capacitive reactance is given by:

$$\begin{aligned}
-\frac{j}{\omega C} &= -\frac{j}{2\pi f C} \\
&= -\frac{j}{6.28 \times 100 \times 10^3 \times 2 \times 10^{-9}} \\
&= -j\,796.2\,\Omega
\end{aligned}$$

The impedance of the circuit is given by:

$$Z = R + j\left(\omega C - \frac{1}{\omega C}\right)$$

$$= 100 + j(125.6 - 796.2) = 100 - j670.6\,\Omega$$

The current flowing in the circuit will be given by:

$$I = \frac{V}{Z} = \frac{100 + j0}{100 - j670.6} = 0.022 + j0.146\,\Omega$$

The voltage dropped across the capacitor will be given by:

$$V_C = I \times \frac{-j}{\omega C} = (0.022 + j0.146) \times -j796.2$$

$$= 116.25 - j17.52\,V$$

The voltage dropped across the inductor will be given by:

$$V_L = I \times j\omega L = (0.022 + j0.146) \times j125.6$$

$$= -18.33 + j2.76\,V$$

The voltage dropped across the resistor will be given by:

$$V_R = I \times R = (0.022 + j0.146) \times 100$$

$$= 2.2 + j14.6\,V$$

You might like to check these answers by adding the individual component voltages to see if they are equal to the supply voltage – they should be!

Test your knowledge 27.5

A coil having inductance of 10 mH and resistance of 15 Ω is connected in series with a capacitance of 2.2 μF. Determine the voltage dropped across each of the two components when the circuit is connected to a 10 V 400 Hz a.c. supply.

SCILAB

SCILAB is an extremely useful tool for plotting experimental data. The example that follows shows how SCILAB can generate interpolated data from a table of measured values of current in a series resonant circuit. The resulting data plot is shown in Figure 27.13. The script also calculates several important parameters of the series resonant circuit including resonant frequency, bandwidth and Q-factor (see Figure 27.14).

```
// Series resonant circuit experimental
data plot
// Filename: ex2704.sce
// Clear any previous graphic window
clf();
// Enter values for frequency as a column
vector
f=[1;2;3;4;5;6;7;8;9;10];
// Enter corresponding values for current as
a column vector
i=[3.6;7.5;11.7;11.4;8.6;6.7;5.5;4.6;3.9;
3.4];
// Plot raw data for v using crosses
plot2d("ln",f,i,style=-2);
// Create a larger frequency axis dataset
for interpolation
fi=1:0.01:10;
// Generate interpolated data for i
id=splin(f,i);
ii=interp(fi,f,i,id); // fi and ii are the
new data values for f and i
// Plot interpolated data for i
plot2d(fi,ii,style=5); // red line
// Find resonant frequency, fc
im=max(ii); // Find resonant frequency, fc
i=find(ii==im);
f0=(fi(1:1,i));
// Find lower and upper cut-off frequencies
i=find(ii>0.707*im); // i is the index
f1=min(fi(1:1,i));
f2=max(fi(1:1,i));
// Calculate bandwidth and Q-factor
bw=f2-f1;
q=f0/(f2-f1);
// Print grid and title
xgrid(); // add grid
xtitle('Current plotted against frequency');
// Print results
messagebox('f1 = '+string(f1)+' kHz;
f2 = '+string(f2)+' kHz; fc =
'+string(f0)+' kHz; Bandwidth = ' +
string(bw) + ' kHz; Q = '+string(q),
['Frequency response']);
```

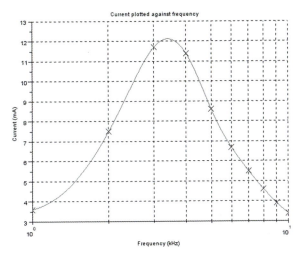

Figure 27.13 Graphical output of the SCILAB script

Figure 27.14 Parameters obtained from the measured data

Test your knowledge 27.6

If the value of capacitance, C, in the previous SCILAB script was 470 nF, find the value of inductance, L, and resistance, R.

27.5 Chapter summary

A resonant circuit is one in which there is a combination of inductance, capacitance and resistance. These components may be connected in series or parallel, but for both types of circuit there will be one particular frequency at which the current and voltage are exactly in phase. At this frequency the circuit will behave like one of pure resistance.

The goodness of a resonant circuit in terms of the sharpness of its frequency response and its ability to magnify voltage or current is expressed in terms of its Q-factor (or *quality factor*). The bandwidth of a resonant

circuit is the range of frequencies that fall within its *half-power* range.

High values of Q-factor are associated with narrow bandwidth whilst low values of Q-factor are commensurate with a wide bandwidth. In a series resonant circuit Q-factor is a measure of the *voltage magnification*, whilst in a parallel resonant circuit it is equivalent to the *current magnification* provided by the circuit.

27.6 Review questions

Short answer questions

1. A series tuned circuit consists of $L = 450\,\mu$H, $C = 1$ nF and $R = 30\,\Omega$. Determine:

    (a) the frequency at which maximum current will flow in the circuit

    (b) the Q-factor of the circuit at resonance

    (c) the bandwidth of the tuned circuit

2. If the tuned circuit in Question 1 is connected to a voltage source of $(2 + \text{j}0)$ V at a frequency of 500 kHz, determine the current that will flow and the voltage that will be developed across each component.

3. A parallel resonant circuit comprises $L = 200\,\mu$H, $C = 700$ pF and $R = 75$ kΩ.

    (a) Determine the resonant frequency, Q-factor and bandwidth of the tuned circuit

    (b) If the bandwidth is to be increased to 5 kHz, determine the value of additional parallel damping resistance required.

4. A series circuit consists of $L = 60$ mH, $R = 15\,\Omega$ and $C = 15$ nF. Determine the frequency of resonance and the voltage dropped across each component at resonance if the circuit is connected to a 300 mV a.c. supply at the resonant frequency.

5. The following data was obtained when a series L–C–R circuit was tested:

Frequency (Hz)	1	2	3	4	5	6	7	8	9	10
Current (mA)	1.4	2.5	4.7	9.4	9.9	6.5	4.3	3.2	2.5	2.2

Plot the frequency response of the L–C–R circuit and use it to determine the resonant frequency, bandwidth and Q-factor at resonance.

6. A coil having an inductance of 5 mH and a resistance of 150 Ω is connected in parallel with a capacitor of 20 nF. Determine the frequency of resonance and the dynamic impedance of the circuit.

7. The aerial tuned circuit of a long-wave receiver is to tune from 150 kHz to 300 kHz. Determine the required maximum and minimum values of tuning capacitor if the inductance of the aerial coil is 900 μH.

8. The parallel tuned circuit to be used in a variable-frequency oscillator comprises a coil of inductance 500 μH and resistance 45 Ω and a variable tuning capacitor having maximum and minimum values of 1000 pF and 100 pF, respectively. Determine the tuning range for the oscillator and the Q-factor of the oscillator tuned circuit at each end of the tuning range.

Chapter 28

Coupled magnetic circuits

Any circuit in which a change of magnetic flux is produced by a change of current is said to possess *self-inductance*. In turn, the change of flux will produce an e.m.f. across the terminals of the circuit. We refer to this as an *induced e.m.f.* When the flux produced by one coil links with another we say that the two coils exhibit *mutual inductance.* The more flux that links the two coils, the greater the amount of mutual inductance that exists between them. This important principle underpins the theory of the transformer.

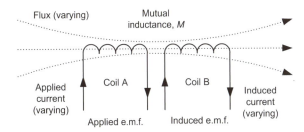

Figure 28.1 Mutual inductance

28.1 Mutual inductance

When two inductors are placed close to one another, the flux generated when a changing current flows in the first inductor will cut through the other inductor (see Figure 28.1). This changing flux will, in turn, induce a current in the second inductor. This effect is known as *mutual inductance* and it occurs whenever two inductors are *inductively coupled*. This is the principle of a very useful component, the *transformer*, which we shall meet later.

28.1.1 Coefficient of coupling

Let's assume that two coils, having inductances L_1 and L_2, are placed close together. The fraction of the total flux from one coil that links with the other coil is called the *coefficient of coupling*, k.

If all of the flux produced by L_1 links with, L_2, then the magnetic coupling between them is perfect and we can say that the *coefficient of coupling*, k, is unity (i.e. $k = 1$). Alternatively, if only half the flux produced by L_1 links with L_2, then $k = 0.5$.

Thus $k = \Phi_2 / \Phi_1$

where Φ_1 is the flux in the first coil (L_1) and Φ_2 is the flux in the second coil (L_2).

Example 28.1 A coil, L_1, produces a magnetic flux of 80 mWb. If 20 mWb appears in a second coil, L_2, determine the coefficient of coupling, k.

Now $k = \Phi_2 / \Phi_1$

where $\Phi_1 = 80$ mWb and $\Phi_2 = 20$ mWb

Thus $k = 20/80 = 0.25$

The unit of mutual inductance is the same as that for self-inductance. Two circuits are said to have a mutual

978-1-85617-775-7, Engineering Science, Mike Tooley & Lloyd Dingle

inductance of one Henry (H) if an e.m.f. of one volt is produced in one circuit when the current in the other circuit varies at a uniform rate of one ampere per second. Hence, if two circuits have a mutual inductance, M, and the current in one circuit (the *primary*) varies at a rate di/dt (A/s) the e.m.f. induced in the second circuit (the *secondary*) will be given by:

$$e = -M \times \text{(rate of change of current)}$$

$$\text{or:} \quad e = -M \times \frac{di}{dt} \text{ V} \qquad \text{(i)}$$

Note the minus sign which indicates that the e.m.f. in the second circuit opposes the increase in flux produced by the first circuit.

The e.m.f. induced in an inductor is given by:

$$e = -N \times \text{(rate of change of flux)}$$

$$\text{or:} \quad e = -N \times \frac{d\Phi}{dt} \text{ V}$$

where N is the number of turns and Φ is the flux (in Wb).

Hence the e.m.f. induced in the secondary coil of two magnetically coupled inductors will be given by:

$$e = -N_2 \times \frac{d\Phi}{dt} \text{ V} \qquad \text{(ii)}$$

where N_2 is the number of secondary turns.

Combining equations (i) and (ii) gives:

$$-M \times \frac{di}{dt} = -N_2 \times \frac{d\Phi}{dt}$$

Thus:
$$M = N_2 \times \frac{d\Phi}{dt} \times \frac{dt}{di} = N_2 \times \frac{d\Phi}{di}$$

If the relative permeability of the magnetic circuit remains constant, the rate of change of flux with current will also be a constant. Thus we can conclude that:

$$M = N_2 \times \frac{d\Phi}{di} = N_2 \times \frac{\Phi 2}{I1} \qquad \text{(iii)}$$

where Φ_2 is the flux linked with the secondary circuit and I_1 is the primary current.

Expression (iii) provides us with an alternative definition of mutual inductance. In this case, M is expressed in terms of the secondary flux and the primary current.

Example 28.2   Two coils have a mutual inductance of 500 mH. Determine the e.m.f. produced in one coil when the current in the other coil is increased at a uniform rate from 0.2 A to 0.8 A in a time of 10 ms.

Now $e = -M \times \dfrac{di}{dt}$

Since the current changes from 0.2 A to 0.8 A in 10 ms, the rate of change of current with time, $\dfrac{di}{dt}$, is $\dfrac{(0.8 - 0.2)}{0.01} = 600$ A/s.

Thus $e = -0.5 \times 600 = -300$ V

Example 28.3   An e.m.f. of 40 mV is induced in a coil when the current in a second coil is varied at a rate of 3.2 A/s. Determine the mutual inductance of the two coils.

Now $e = -M \times \dfrac{di}{dt}$

thus $M = -\dfrac{e}{\left(\dfrac{di}{dt}\right)} = \dfrac{0.04}{3.2} = 12.5$ mH

Test your knowledge 28.1

The coefficient of coupling between two coils is 0.35. If a flux of 1.5 mWb appears in the coil having the lesser inductance, what flux will be induced in the other coil?

Test your knowledge 28.2

Two coils have a mutual inductance of 60 mH. If a current falls at the rate of 15 A/s in one of the coils, determine the e.m.f. induced in the other coil.

Key Point.   When two inductors are placed close to one another, the flux generated when a changing current flows in the first inductor will cut through the other inductor. This changing flux will, in turn, induce a current in the second inductor

28.2   Coupled circuits

Consider the case of two coils that are coupled together so that all of the flux produced by one coil links with the other. These two coils are, in effect, perfectly coupled. Since the product of inductance (L) and current (I) is equal to the product of the number of turns (N) and the flux (Φ), we can deduce that:

$$L_1 I_1 = N_1 \Phi \quad \text{or} \quad L_1 = N_1 \Phi / I_1 \qquad \text{(i)}$$

Similarly:

$$L_2 I_2 = N_2 \Phi \quad \text{or} \quad L_2 = N_2 \Phi / I_2 \qquad \text{(ii)}$$

where L_1 and L_2 represent the inductance of the first and second coils, respectively, N_1 and N_2 represent the turns of the first and second coils, respectively and Φ is the flux shared by the two coils.

Since the product of reluctance (S) and flux (Φ) is equal to the product of the number of turns (N) and the current (I), we can deduce that:

$$S\Phi = N_1 I_1 = N_2 I_2 \quad \text{or} \quad \Phi = N_1 I_1 / S = N_2 I_2 / S$$

Hence, from (i):

$$L_1 = N_1 \Phi / I_1 = N_1 (N_1 \; I_1) / I_1 S = N_1^2 / S$$

and from (ii):

$$L_2 = N_2 \Phi / I_2 = N_2 (N_2 I_2) / I_2 = N_2^2 / S$$

Now $\quad L_1 \times L_2 = (N_1^2/S) \times (N_2^2/S) = N_1^2 N_2^2 / S^2 \qquad \text{(iii)}$

Now $\quad M = N_2 \times \dfrac{\Phi}{I_1}$ (see equation (i))

Multiplying top and bottom by N_1 gives:

$$M = N_1 N_2 \Phi / (N_1 I_1) = N_1 N_2 / S \qquad \text{(iv)}$$

Combining equations (iii) and (iv) gives:

$$M = \sqrt{L_1 L_2}$$

In deriving the foregoing expression we have assumed that the flux is perfectly coupled between the two coils. In practice, there will always be some leakage of flux. Furthermore, there may be cases when we do not wish to couple two inductive circuits tightly together. The general equation is:

$$M = k\sqrt{L_1 L_2}$$

where k is the coefficient of coupling (note that k is always less than 1).

Example 28.4   Two coils have self-inductances of 10 mH and 40 mH. If the two coils exhibit a mutual inductance of 5 mH, determine the coefficient of coupling.

Now $\quad M = k\sqrt{L_1 L_2}$

thus $k = \dfrac{M}{\sqrt{L_1 L_2}} = \dfrac{5}{\sqrt{400}} = 0.25$

Key Point.   By placing the coils close together or by winding them on the same closed magnetic core it is possible to obtain high values of mutual inductance. In this condition, we say that the two coils are *close coupled* or *tightly coupled*. Where the two coils are not in close proximity or are wound on separate magnetic cores, values of mutual inductance will be quite small. Under these circumstances we say that the two coils are *loose coupled*

28.2.1   Series connection of coupled coils

There are two ways of series connecting inductively coupled coils, either *series aiding* (so that the mutual inductance adds to the combined self-inductance of the two coils) or in *series opposition* (in which case the mutual inductance subtracts from the combined self-inductance of the two coils).

In the first, series aiding case, the effective inductance is given by:

$$L = L_1 + L_2 + 2M$$

whilst in the latter, series opposing case, the effective inductance is given by:

$$L = L_1 + L_2 - 2M$$

where L_1 and L_2 represent the inductance of the two coils and M is their mutual inductance.

Example 28.5 When two coils are connected in series, their effective inductance is found to be 1.2 H. When the connections to one of the two coils are reversed, the effective inductance is 0.8 H. If the coefficient of coupling is 0.5, find the inductance of each coil and also determine the mutual inductance.

In the first, series aiding case:

$$L = L_1 + L_2 + 2M \quad \text{or} \quad 1.2 = L_1 + L_2 + 2M$$

In the second, series opposing case:

$$L = L_1 + L_2 - 2M \quad \text{or} \quad 0.8 = L_1 + L_2 - 2M$$

Now $M = k\sqrt{L_1 L_2}$ thus $M = 0.5\sqrt{L_1 L_2}$
Thus:

$$1.2 = L_1 + L_2 + 2\left(k\sqrt{L_1 L_2}\right)$$

$$= L_1 + L_2 + 2\left(0.5\sqrt{L_1 L_2}\right)$$

$$= L_1 + L_2 + \sqrt{L_1 L_2} \tag{i}$$

and

$$0.8 = L_1 + L_2 - 2\left(k\sqrt{L_1 L_2}\right)$$

$$= L_1 + L_2 - 2\left(0.5\sqrt{L_1 L_2}\right)$$

$$= L_1 + L_2 - \sqrt{L_1 L_2} \tag{ii}$$

Adding (i) and (ii) gives:

$$1.2 + 0.8 = L_1 + L_2 + \sqrt{L_1 L_2} + L_1 + L_2 - \sqrt{L_1 L_2}$$

$$2 = 2(L_1 + L_2)$$

$$2 = 2(L_1 + L_2) \quad \text{or} \quad 1 = L_1 + L_2$$

$$\text{thus} \quad L_1 = 1 - L_2 \tag{iii}$$

Substituting for L_1 in equation (i) gives:

$$1.2 = (1 - L_2) + L_2 + \sqrt{(1 - L_2) \times L_2}$$

$$1.2 = 1 + \sqrt{L_2 - L_2^2}$$

$$0.2 = \sqrt{L_2 - L_2^2}$$

$$0.04 = L_2 - L_2^2$$

$$0 = -0.04 + L_2 - L_2^2$$

$$0 = 0.04 - L_2 + L_2^2$$

Solving this quadratic equation gives:

$$L_2 = 0.959\,\text{H} \quad \text{or} \quad 0.041\,\text{H}$$

Finally, from (iii):

$$L_1 = 0.041\,\text{H} \quad \text{or} \quad 0.959\,\text{H}$$

28.2.2 Dot notation

When two coupled circuits are drawn on a circuit diagram it is impossible to tell from the drawing the direction of the voltage induced in the second circuit (called the *secondary*) when a changing current is applied to the first circuit (known as the *primary*). When the direction of this induced voltage is important, we can mark the circuit with dots to indicate the direction of the currents and induced voltages, as shown in Figure 28.2.

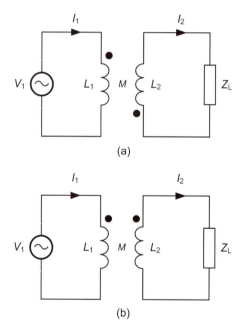

Figure 28.2 Dot notation

In Figure 28.2(a), the total magnetomotive force produced will be the sum of the magnetomotive force produced by each individual coil. In this condition, the mutual inductance, M, will be *positive*.

In Figure 28.2(b), the total magnetomotive force produced will be the difference of the magnetomotive force produced by each individual coil. In this condition, the mutual inductance, M, will be *negative*.

Assuming that the primary coil in Figure 28.2(a) has negligible resistance, the primary voltage, V_1, expressed using the complex notation that we met in Chapter 26 will be given by:

$$V_1 = (I_1 \times j\omega L_1) + (I_2 \times j\omega M)$$

Conversely, in Figure 28.2(b), the primary voltage, V_1, expressed using complex notation will be given by:

$$V_1 = (I_1 \times j\omega L_1) - (I_2 \times j\omega M)$$

where I_1 and I_2 are the primary and secondary currents, respectively, L_1 and L_2 are the primary and secondary inductances, respectively, ω is the angular velocity of the current and M is the mutual inductance of the two circuits.

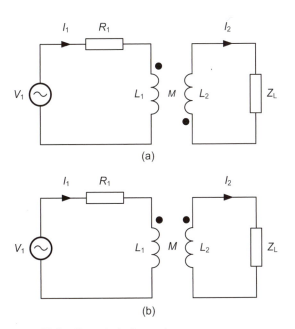

Figure 28.3 Effect of winding resistance

If the primary coil has resistance, as shown in Figure 28.3(a), the primary voltage, V_1, will be given by:

$$V_1 = I_1(R_1 + j\omega L_1) + (I_2 \times j\omega M)$$

Conversely, in Figure 28.3(b), the primary voltage, V_1, will be given by:

$$V_1 = I_1(R_1 + j\omega L_1) - (I_2 \times j\omega M)$$

28.3   Transformers

Transformers provide us with a means of coupling a.c. power from one circuit to another without a direct connection between the two. A further advantage of transformers is that voltage may be stepped-up (secondary voltage greater than primary voltage) or stepped-down (secondary voltage less than primary voltage). Since no increase in power is possible (like resistors capacitors and inductors, transformers are *passive* components) an increase in secondary voltage can only be achieved at the expense of a corresponding reduction in secondary current, and vice versa (in fact, the secondary power will be very slightly less than the primary power due to losses within the transformer).

Typical applications for transformers include stepping-up or stepping-down voltages in power supplies, coupling signals in audio and low-frequency amplifiers to achieve impedance matching and to isolate the d.c. potentials that may be present in certain types of circuit. The electrical characteristics of a transformer are determined by a number of factors including the core material and physical dimensions of the component. High-frequency transformers (i.e. those designed for operation above 10 kHz or so) normally use ferrite as the core material whereas low-frequency transformers (those designed for operation at frequencies below 10 kHz) use laminated steel cores.

The specifications for a transformer usually include the rated primary and secondary voltages and currents the required power rating (i.e. the rated power, usually expressed in volt-amperes, VA) which can be continuously delivered by the transformer under a given set of conditions), the frequency range for the component (usually stated as upper and lower working frequency limits), and the per-unit regulation of a transformer. As we shall see, this last specification is a measure of the ability of a transformer to maintain its rated output voltage under load.

Figure 28.4 Some typical small transformers for use at frequencies between 50 Hz and 20 kHz with ratings from 0.1 to 100 VA

28.3.1   The transformer principle

The principle of the transformer is illustrated in Figure 28.5. The primary and secondary windings are wound on a common low-reluctance magnetic core consisting of a number of steel laminations. All of the alternating flux generated by the primary winding is therefore coupled into the secondary winding (very little

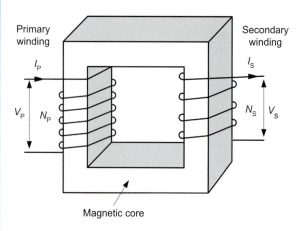

Figure 28.5 The principle of the transformer

flux escapes due to leakage). A sinusoidal current flowing in the primary winding produces a sinusoidal flux within the transformer core.

At any instant the flux, Φ, in the transformer core is given by the equation:

$$\Phi = \Phi_{max} \sin(\omega t)$$

where Φ_{max} is the maximum value of flux (in Wb), f is the frequency of the applied current and t is the time in seconds.

28.3.2   e.m.f. equation for a transformer

If a sine wave current is applied to the primary winding of a transformer, the flux will change from its negative maximum value, $-\Phi_M$, to its positive maximum value, $+\Phi_M$, in a time equal to half the periodic time of the current, $t/2$. The total change in flux over this time will of course be $2\Phi_M$. The average rate of change of flux in a transformer is thus given by:

$$\frac{\text{Total flux change}}{\text{Time}} = \frac{2\Phi_M}{\left(\dfrac{t}{2}\right)} = \frac{4\Phi_M}{t}$$

Now $f = 1/t$ thus:

$$\text{Average rate of change of flux} = \frac{4\Phi_M}{\left(\dfrac{1}{f}\right)} = 4f\Phi_M$$

The average e.m.f. induced per turn will be given by:

$$V/N = 4f\Phi_M$$

Hence for a primary winding with N_P turns, the average induced primary voltage will be given by:

$$V_P = 4N_P f \Phi_M$$

Similarly, for a secondary winding with N_2 turns, the average induced secondary voltage will be given by:

$$V_S = 4N_S f \Phi_M$$

Note that we have assumed that there is perfect flux linkage between the primary and secondary windings, i.e. $k = 1$.

For a sinusoidal voltage, the effective or r.m.s. value is 1.11 times the average value. Thus:

The r.m.s. value of the primary voltage (V_P) is given by:

$$V_P = 4.44 f N_P \Phi_{max}$$

Similarly, the r.m.s. value of the secondary voltage (V_P) is given by:

$$V_S = 4.44 f N_S \Phi_{max}$$

From these two relationships (and since the same magnetic flux appears in both the primary and secondary windings when the flux is perfectly linked) we can infer that:

$$\frac{V_P}{V_S} = \frac{N_P}{N_S}$$

where V_P and V_S are the primary and secondary voltages, respectively and N_P and N_S are the numbers of primary and secondary turns, respectively (see Figure 28.6).

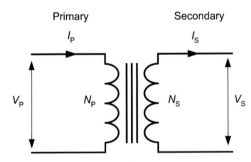

Primary | Secondary
I_P | I_S
V_P | N_P | N_S | V_S

Figure 28.6 Transformer turns, voltages and currents

Furthermore, assuming that no power is lost in the transformer (i.e. as long as the primary and secondary powers are the same) we can conclude that:

$$\frac{I_P}{I_S} = \frac{N_S}{N_P}$$

where I_S and I_P are the secondary and primary currents, respectively and N_P and N_S are the numbers of primary and secondary turns, respectively.

The ratio of primary turns to secondary turns (N_P/N_S) is known as the *turns ratio*.

Furthermore, since the ratio of primary turns to primary voltage is the same as the ratio of secondary

turns to secondary voltage, we can conclude that, for a particular transformer:

$$\text{Turns-per-volt (t.p.v.)} = \frac{V_P}{N_P} = \frac{V_S}{N_S}$$

The *turns-per-volt* rating can be useful when designing transformers with multiple secondary windings.

Once again, assuming that the transformer is 'loss free' and that no flux escapes, the power delivered to the secondary circuit will be the same as that in the primary circuit. Thus:

$$P_P = P_S \quad \text{or} \quad V_P \times I_P = E_S \times I_S$$

Thus:

$$\frac{V_P}{V_S} = \frac{I_S}{I_P} = \frac{N_P}{N_S}$$

Example 28.6   A transformer operates with a 220 V 50 Hz supply and has 800 primary turns. Determine the maximum value of flux present.

Now $V_P = 220$ V, $N_P = 800$ and $f = 50$ Hz

Since $V_P = 4.44 N_P f \Phi_M$ V

$$\Phi_M = V_P/4.44 N_P f = 220/(4.44 \times 800 \times 50)$$

$$= 220/177600 = 1.24 \text{ mWb.}$$

Example 28.7   A transformer has 2000 primary turns and 120 secondary turns. If the primary is connected to a 220 V a.c. mains supply, determine the secondary voltage.

Since $\dfrac{V_P}{V_S} = \dfrac{N_P}{N_S}$ we can conclude that:

$$V_S = \frac{V_P N_S}{N_P} = \frac{220 \times 120}{2000} = 13.2 \text{ V}$$

Example 28.8   A transformer has 1200 primary turns and is designed to operate with a 110 V a.c. supply. If the transformer is required to produce an output of 10 V, determine the number of secondary turns required.

Since $\dfrac{V_P}{V_S} = \dfrac{N_P}{N_S}$ we can conclude that:

$$N_S = \frac{N_P V_S}{V_P} = \frac{1200 \times 10}{110} = 109.1$$

Example 28.9 A transformer has a turns-per-volt rating of 1.2. How many turns are required to produce secondary outputs of (a) 50 V and (b) 350 V?

Here we will use $N_S = $ turns-per-volt $\times V_S$

(a) In the case of a 50 V secondary winding:

$$N_S = 1.5 \times 50 = 75 \text{ turns}$$

(b) In the case of a 350 V secondary winding:

$$N_S = 1.5 \times 350 = 525 \text{ turns}$$

Example 28.10 A transformer has 1200 primary turns and 60 secondary turns. Assuming that the transformer is loss-free, determine the primary current when a load current of 20 A is taken from the secondary.

Since $\dfrac{I_S}{I_P} = \dfrac{N_P}{N_S}$ we can conclude that:

$$I_P = \frac{I_S N_S}{N_P} = \frac{20 \times 60}{1200} = 1 \text{ A}$$

Example 28.11 A transformer has 455 primary turns and 66 secondary turns. If the primary winding is connected to a 110 V a.c. supply and the secondary is connected to an 8 Ω load, determine:

(a) the current that will flow in the secondary

(b) the supply current.

You may assume that the transformer is 'loss-free'.

Now $V_P = 110$ V, $N_P = 480$ and $N_S = 60$

$$V_P/V_S = I_S/I_P = N_P/N_S$$

thus $V_S = V_P \times (N_S/N_P) = 110 \times 66/455 = 16$ V

The current delivered to an 8 Ω load will thus be:

$$I_S = V_S/12 = 16/8 = 2 \text{ A}$$
$$I_P = I_S \times (N_S/N_P) = 2 \times 66/455 = 0.29 \text{ A}$$

Test your knowledge 28.5

A transformer is to be operated with a peak flux of 1.5 mWb. Calculate the number of primary turns required if the transformer is to be used with a 220 V 50 Hz supply.

Test your knowledge 28.6

A transformer has 480 primary turns and 120 secondary turns. If the primary is connected to a 110 V a.c. supply, determine the secondary voltage.

Test your knowledge 28.7

A step-down transformer has a 220 V primary and a 24 V secondary. If the secondary winding has 60 turns, how many turns are there on the primary?

Test your knowledge 28.8

A transformer has 440 primary turns and 1800 secondary turns. If the secondary supplies a current of 250 mA, determine the primary current (assume that the transformer is loss-free).

Test your knowledge 28.9

A 2:1 step-down transformer has 440 primary turns in which a current of 0.15 A flows when the transformer is connected to a 110 V a.c. supply. Assuming that the transformer is loss-free, determine the secondary current and the power delivered to the load.

Test your knowledge 28.10

A transformer provides 15 V from its secondary winding when the primary is fed from a 220 V supply. Determine the power that would be supplied to the primary when the secondary is connected to a 3.75 Ω load. Assume that the transformer is loss-free.

SCILAB

The following SCILAB script can be used to calculate the maximum flux produced in a transformer core:

```
// Maximum flux in a transformer core
// Filename: ex2801.sce
// Get transformer parameters
txt = ['Primary turns';'Primary
voltage (V)';'Frequency (Hz)'];
var = x_mdialog('Enter N1, E1 and f',txt,
['1200';'220';'50']);
n1 = evstr(var(1));
e1 = evstr(var(2));
f = evstr(var(3));
// Calculate the maximum flux in the core
p = e1 / (4.44 * n1 * f);
p = 1000 * p ; // display result in mWb
messagebox(['Maximum flux = ' + string(p) +
'mWb'],'Results');
```

Key Point. Transformers can be used to step up or step down alternating voltages. When used in a step-up arrangement the primary current will be correspondingly greater than the secondary current, and vice versa

Key Point. If a transformer is assumed to be loss-free, the power delivered by the secondary to a load (i.e. the product of secondary voltage and secondary current) will be identical to the power that is supplied to the primary (i.e. the product of primary voltage and primary current)

28.4 Equivalent circuit of a transformer

An *ideal transformer* would have coil windings that have negligible resistance and a magnetic core that is perfect. Furthermore, all the flux produced by the primary winding would be coupled into the secondary winding and no flux would be lost in the space surrounding the transformer. In practice, a real transformer suffers from a number of imperfections. These are summarised in the next few paragraphs.

28.4.1 Leakage flux

Not all of the magnetic flux produced by the primary winding of a transformer is coupled into its secondary winding. This is because some flux is lost into the space surrounding the transformer (even though this has a very much higher reluctance than that of the magnetic core).

The leakage flux increases with the primary and secondary current and its effect is the same as that produced by an inductive reactance connected in series with the primary and secondary windings. We can take this into account by including this additional inductance in the simplified equivalent circuit of a transformer shown in Figure 28.7.

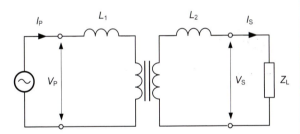

Figure 28.7 Transformer equivalent circuit showing the effect of leakage flux

28.4.2 Winding resistance

The windings of a real transformer exhibit both inductance *and* resistance. The primary resistance, R_1, effectively appears in series with the primary inductance, L_1, whilst the secondary resistance, R_2, effectively appears in series with the secondary inductance, L_2 (see Figure 28.8).

It is sometimes convenient to combine the primary and secondary resistances and inductances into a single

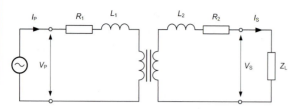

Figure 28.8 Transformer equivalent circuit showing the effect of leakage flux and winding resistance

pair of components connected in series with the primary circuit. We can do this by *referring* the secondary resistance and inductance to the primary circuit.

The amount of secondary resistance referred into the primary, R'_S, is given by:

$$R'_S = R_S \times \left(\frac{V_P}{V_S}\right)^2$$

thus the effective total primary resistance is given by:

$$R_e = R_P + R_S \left(\frac{V_P}{V_S}\right)^2$$

Similarly, the amount of secondary inductance referred into the primary, L'_2, is given by:

$$L'_S = R_L \times \left(\frac{V_P}{V_S}\right)^2$$

thus the effective total primary inductance is given by:

$$L_e = L_P + L_S \left(\frac{V_P}{V_S}\right)^2$$

The effective impedance, Z_e, appearing in series with the primary circuit (with both R_S and L_S referred into the primary) is thus given by:

$$Z_e = R_e + j\omega L_e$$

28.4.3   Magnetising current

In a real transformer, a small current is required in order to magnetise the transformer core. This current is present regardless of whether the transformer is connected to a load. This magnetising current is equivalent to the current that would flow in an inductance, L_m, connected in parallel with the primary winding.

28.4.4   Core losses

A real transformer also suffers from losses in its core. These losses are attributable to *eddy currents* (small currents induced in the laminated core material) and *hysteresis* (energy loss in the core due to an imperfect *B–H* characteristic). These losses can be combined together and represented by a resistance, R_{cl}, connected in parallel with the primary winding. The complete equivalent circuit for a transformer is shown in Figure 28.9.

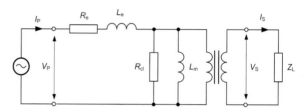

Figure 28.9 Transformer equivalent circuit showing the effect of leakage flux and core losses

Example 28.12   A transformer designed for operation at 50 Hz has 220 primary turns and 110 secondary turns. The primary and secondary resistances are 1.5 Ω and 0.5 Ω, respectively, whilst the primary and secondary leakage inductances are 20 mH and 10 mH, respectively. Assuming core losses are negligible, determine the equivalent impedance referred to the primary circuit.

Now $\dfrac{V_P}{V_S} = \dfrac{N_P}{N_S} = \dfrac{220}{110} = 2$

Hence $\left(\dfrac{V_P}{V_S}\right)^2 = 4$

The equivalent resistance referred to the primary is thus:

$$R_e = R_P + R_S \left(\frac{V_P}{V_S}\right)^2 = 1.5 + (0.5 \times 4) = 3.5 \ \Omega$$

Similarly, the equivalent inductance referred to the primary will be:

$$L_e = L_P + L_S \left(\frac{V_P}{V_S}\right)^2 = 0.02 + (0.01 \times 4) = 0.06 \ \text{H}$$

The impedance seen looking into the primary, expressed in complex form, will thus be:

$$Z_e = R_e + j\omega L_e = 3.5 + j(2\pi \times 50 \times 0.06)$$

$$= (3.5 + j18.84)\,\Omega$$

28.5 Transformer regulation and efficiency

The output voltage produced at the secondary of a real transformer falls progressively as the load imposed on the transformer increases (i.e. as the secondary current increases from its no-load value). The *voltage regulation* of a transformer is a measure of its ability to keep the secondary output voltage constant over the full range of output load currents (i.e. from *unloaded* to *full load*) at the same power factor. This change, when divided by the no-load output voltage, is referred to as the *per-unit regulation* for the transformer. This can be best illustrated by an example.

> **Example 28.13** A transformer produces an output voltage of 110 V under no-load conditions and an output voltage of 101 V when the full load is applied. Determine the per-unit regulation.
>
> The per-unit regulation can be determined for:
>
> $$\text{Per-unit regulation} = \frac{V_{S(\text{no-load})} - V_{S(\text{full-load})}}{V_{S(\text{no-load})}}$$
>
> $$= \frac{110 - 101}{110}$$
>
> $$= 0.081\,(\text{or } 8.1\%)$$

Most transformers operate with very high values of efficiency (typically between 85 and 95%). Despite this, in high-power applications the losses in a transformer cannot be completely neglected. Transformer losses can be divided into two types of loss:

- losses in the magnetic core (often referred to as *iron loss*);
- losses due to the resistance of the coil windings (often referred to as *copper loss*).

Iron loss can be further divided into *hysteresis loss* (energy lost in repeatedly cycling the magnetic flux in the core backwards and forwards) and *eddy current loss* (energy lost due to current circulating in the steel core).

Hysteresis loss can be reduced by using material for the magnetic core that is easily magnetised and has a very high permeability (see Figure 28.10 – note that energy loss is proportional to the area enclosed by the *B–H* curve). Eddy current loss can be reduced by laminating the core (using E and I laminations, for example) and also ensuring that a small gap is present. These laminations and gaps in the core help to ensure that there is no closed path for current to flow. Copper loss results from the resistance of the coil windings and it can be reduced by using wire of large diameter and low resistivity.

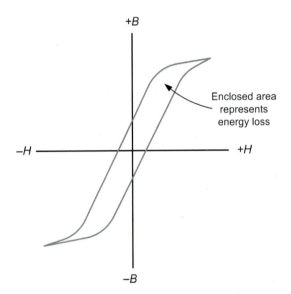

Figure 28.10 Transformer *B–H* curve and energy loss

It is important to note that, since the flux within a transformer varies only slightly between the no-load and full-load conditions, iron loss is substantially constant regardless of the load actually imposed on a transformer. Copper loss, on the other hand, is zero when a transformer is under no-load conditions and rises to a maximum at full load.

The efficiency of a transformer is given by:

$$\text{Efficiency} = \frac{\text{output power}}{\text{input power}} \times 100\%$$

from which

$$\text{Efficiency} = \frac{\text{input power} - \text{losses}}{\text{input power}} \times 100\%$$

and

$$\text{Efficiency} = 1 - \frac{\text{losses}}{\text{input power}} \times 100\%$$

As we have said, the losses present are attributable to iron loss and copper loss but the copper loss appears in both the primary and the secondary windings. Hence:

Efficiency

$$= 1 - \frac{\left(\begin{array}{c}\text{iron loss + primary copper loss} \\ \text{+secondary copper loss}\end{array}\right)}{\text{input power}} \times 100\%$$

28.5.1   Power factor

When the load on a transformer is purely resistive, the impedance, seen looking into its primary, also appears as resistance. However, when the load is partially reactive (i.e. when it is no longer purely resistive) the current and voltage in the primary and secondary will no longer be *in phase* and we need to take this into account when considering the power that is supplied to the transformer. We will be looking at this in much greater detail in Chapter 29 but, for now, you need to be aware that the true power supplied to a transformer may not be the same as the power that it apparently consumes and this needs to be taken into account whenever a transformer is operated at a significant level of power (i.e. more than a few hundred Watts) and the load is partially reactive.

The *true power* in an a.c. circuit is simply the average power that it consumes. Thus, for a resistive load:

$$\text{True power} = I^2 \times R$$

True power is measured in Watts (W).

The *apparent power* in an a.c. circuit is simply the product of the supply voltage and the supply current (these are the quantities that you would measure using a voltmeter and an ammeter). Thus:

$$\text{Apparent power} = V \times I$$

Apparent power is measured in volt-amperes (VA).

The *power factor* of an a.c. circuit is simply the ratio of the *true power* to the *apparent power*. Thus:

$$\text{Power factor} = \frac{\text{true power}}{\text{apparent power}}$$

In terms of units, power factor can be expressed as:

$$\text{Power factor} = \frac{\text{Watts}}{\text{volt-amperes}}$$

The power at the input of a transformer is given by the product of the apparent power and the power factor of the load. For example, the input power of a transformer operating at 1 kVA with a load power factor of 0.8 will be 800 W. We need to take this into account when calculating transformer losses and efficiency as illustrated in Example 28.14.

Example 28.14   A transformer rated at 500 VA has an iron loss of 3 W and a full-load copper loss (primary plus secondary) of 7 W. Calculate the efficiency of the transformer at 0.8 power factor.

The input power to the transformer will be given by the product of the apparent power (i.e. the transformer's VA rating) and the power factor.

Hence:

$$\text{Input power} = 0.8 \times 500 = 400\,\text{W}$$

Now

$$\text{Efficiency} = 1 - \frac{(7+3)}{400} \times 100\% = 97.5\%$$

Test your knowledge 28.11

A transformer produces an output voltage of 220 V under no-load conditions and an output voltage of 208 V when full load is applied. Determine the per-unit regulation.

Test your knowledge 28.12

A 1 kVA transformer has an iron loss of 15 W and a full-load copper loss (primary plus secondary) of 20 W. Determine the efficiency of the transformer at 0.9 power factor.

28.6   Transformer matching

When a load is connected to the secondary winding of a transformer, the impedance of the load is *reflected* by the transformer and appears as an impedance in the primary circuit, as shown in Figure 28.11. This provides us with a way of matching the impedance in one circuit to the impedance in another circuit without any significant power loss.

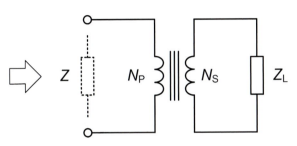

Figure 28.11 Impedance seen looking into a transformer

Assuming that the transformer is loss-free, the impedance seen 'looking into' the transformer's primary winding, Z, will be given by:

$$Z = V_P/I_P \qquad (i)$$

whereas the load impedance connected to the secondary winding of the transformer, Z_L, will be given by:

$$Z_L = V_S/I_S \qquad (ii)$$

Now the voltages, currents and 'turns ratio', N_P/N_S, are connected by the relationship:

$$\frac{V_P}{V_S} = \frac{I_S}{I_P} = \frac{N_P}{N_S} \qquad (iii)$$

thus $V_P = \dfrac{N_P}{N_S} \times V_S$

From (i) and (iii), $Z = \dfrac{N_P}{N_S} \times \dfrac{V_S}{I_P}$

But $V_S = I_S \times Z_L$ from (ii)

thus $Z = \dfrac{N_P}{N_S} \times \dfrac{I_S Z_L}{I_P}$

Now $\dfrac{I_S}{I_P} = \dfrac{N_P}{N_S}$ from (iii)

Thus $Z = \dfrac{N_P}{N_S} \times \dfrac{N_P}{N_S} \times Z_L = \left(\dfrac{N_P}{N_S}\right)^2 \times Z_L$

hence $Z = \left(\dfrac{N_P}{N_S}\right)^2 Z_L$ or $Z_L = \left(\dfrac{N_S}{N_P}\right)^2 Z$

Example 28.15   A loss-free transformer has a turns ratio of 20:1 and is connected to a 1.5 Ω pure resistive load. What impedance is seen at the input of the transformer?

Now $Z = \left(\dfrac{N_P}{N_S}\right)^2 \times Z_L = 20^2 \times 1.5 = 400 \times 1.5 = 600\,\Omega$

Example 28.16   A transformer having a turns ratio of 20:1 has its secondary connected to a load having an impedance of $(6 + j8)\ \Omega$. Determine the impedance seen looking into the primary winding.

Now $Z = \left(\dfrac{N_P}{N_S}\right)^2 Z_L$

where $N_P/N_S = 20$ and $Z_L = 6 + j8\,\Omega$

thus $Z = (20)^2(6 + j8) = 2400 + j3200\,\Omega$

Test your knowledge 28.13

The impedance seen looking into a transformer is 800 Ω. If the transformer has a turns ratio of 16:1, determine the impedance of the load.

Test your knowledge 28.14

Write and test a SCILAB script for determining the efficiency of a transformer when given the load power factor and VA rating.

Key Point.   The impedance seen looking into the primary of a transformer will be the same as the product of the impedance of the load and the *square* of the turns ratio. This provides us with a useful way of changing the impedance of a load

28.7 Chapter summary

In this chapter we saw that, when two inductors are placed close to one another, the current changing in the first inductor will produce a flux change that passes through the other inductor, and that this changing flux will, in turn, induce a current in the second inductor. This leads us to the transformer principle, in which a primary and secondary winding are tightly coupled together over a closed magnetic core. By varying the ratio of turns on the primary to those on the secondary we can use a transformer to step up or step down a.c. voltage without any significant loss of power.

Most transformers operate with very little loss of power and we frequently consider them to be 'loss free'. Where losses do occur, they may be due to imperfections in the magnetic core or due to energy wasted in the resistance of the coil. These are referred to as 'iron loss' and 'copper loss', respectively.

A load connected to the secondary winding of a transformer is seen as a corresponding impedance looking into the primary. This provides us with a way of matching the impedance in one circuit to the impedance in another circuit without any significant power loss by an appropriate choice of turns ratio.

28.8 Review questions

1. A flux of 60 μWb is produced in a coil. A second coil is wound over the first coil. Determine the flux that will be induced in this coil if the coefficient of coupling is 0.3.

2. Two coils exhibit a mutual inductance of 40 mH. Determine the e.m.f. produced in one coil when the current in the other coil is increased at a uniform rate from 100 mA to 600 mA in 0.50 ms.

3. An e.m.f. of 0.2 V is induced in a coil when the current in a second coil is varied at a rate of 1.5 A/s. Determine the mutual inductance that exists between the two coils.

4. Two coils have self-inductances of 60 mH and 100 mH. If the two coils exhibit a mutual inductance of 20 mH, determine the coefficient of coupling.

5. Two coupled inductors each have an inductance of 60 mH and negligible resistance. A supply of 1 V at 25 Hz is connected to one coil and a non-inductive load resistor of 300 Ω is connected to the other. If the coils have a mutual inductance of 20 mH, determine the supply current and the current in the load.

6. A transformer operates from a 110 V 60 Hz supply. If the transformer has 400 primary turns, determine the maximum flux present. By how much will the flux be increased if the transformer were to be used on a 50 Hz supply?

7. A 'loss-free' transformer has 900 primary turns and 225 secondary turns. If the primary winding is connected to a 220 V supply and the secondary is connected to an 11 Ω resistive load, determine:

   (a) the secondary voltage
   (b) the secondary current
   (c) the supply power.

8. Determine the regulation provided by a transformer if it produces an output voltage of 110 V under no-load conditions and an output voltage of 102 V when full load is applied.

9. A transformer rated at 50 VA has an iron loss of 1 W and full-load copper losses of 1.25 W in the primary and 0.75 W in the secondary. Determine the efficiency of the transformer assuming unit power factor.

10. A load resistance of 4 Ω is connected to the secondary winding of a loss-free transformer with a turns ratio of 16:1. What resistance will be seen looking into the primary winding and what power will be dissipated in the load when the primary is connected to a 28 V a.c. supply?

Chapter 29

Power, power factor and power factor correction

We have already shown how power in an a.c. circuit is dissipated only in resistors and not in purely reactive components such as inductors or capacitors. This merits further consideration and we need to be aware of the implications, particularly in applications where appreciable amounts of power may be present and where loads exhibit a significant amount of reactance.

29.1 Power in a.c. circuits

Earlier, in Chapter 21, we showed how the power in a d.c. circuit is simply the product of the applied voltage and current. However, there is an added complication in an a.c. circuit where the voltage and current do not rise and fall together (i.e. when they are not *in phase*). In this case we need to consider the instantaneous curve of power over a complete cycle of the supply in order to determine the relationship between voltage, current and power in the circuit. We will start by considering how the power varies over a complete cycle of the supply in the case of a pure resistive load and then move on to see what happens when a load is reactive rather than resistive.

29.1.1 Power in a pure resistance

The voltage dropped across a pure resistance rises and falls in sympathy with the current. The voltage and current are said to be *in phase*. Since the power at any instant is equal to the product of the *instantaneous voltage* and the *instantaneous current*, we can determine how the power supplied varies over a complete cycle of the supply current. This relationship is illustrated in Figure 29.1. From this you should note that:

(a) the power curve represents a cosine law at twice the frequency of the supply current, and

(b) all points on the power curve are positive throughout a complete cycle of the supply current.

From the foregoing, the power at any instant of time (known as the *instantaneous power*) is given by:

$$p = i \times v$$

where i is the *instantaneous current* and v is the *instantaneous voltage*.

since $i = I_m \sin(\omega t)$ and $v = V_m \sin(\omega t)$

$$p = i \times v = I_m \sin(\omega t) \times V_m \sin(\omega t) = I_m V_m \sin^2(\omega t)$$

From the double angle formulae, $\sin^2 \theta = \dfrac{1 - \cos(2\theta)}{2}$

978-1-85617-775-7, Engineering Science, Mike Tooley & Lloyd Dingle

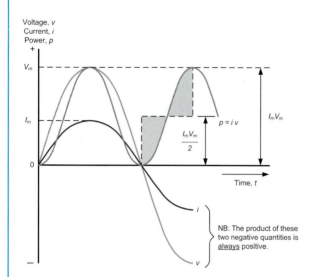

Figure 29.1 Voltage, current and power in a pure resistance

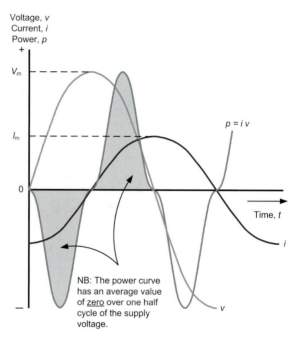

Figure 29.2 Voltage, current and power in a pure inductive reactance

Thus $p = \dfrac{I_m V_m (1 - \cos(2\omega t))}{2}$

There are a few important things to note from this result:

1. The power function is *always positive* (i.e. a graph of p plotted against t will always be above the x-axis, as shown in Figure 29.1)

2. The power function has *twice the frequency* of the current or voltage

3. The average value of power over a complete cycle is equal to $\dfrac{I_m V_m}{2}$

We normally express power in terms of the r.m.s. values of current and voltage. In this event, the *average power* over one complete cycle of current is given by:

$$P = I \times (IR) = I^2 R \text{ or } P = V \times \left(\frac{V}{R}\right) = \frac{V^2}{R}$$

where I and V are r.m.s. values of voltage and current.

29.1.2 Power in a pure reactance

We have already shown how the voltage dropped across a pure inductive reactance leads the current flowing in the inductor by an angle of 90°. Since the power at any instant is equal to the product of the instantaneous voltage and current, we can once again determine how the power supplied varies over a complete cycle of the supply current, as shown in Figure 29.2.

From this you should note that:

(a) the power curve is a sine function at twice the frequency of the supply current, and

(b) the power curve is partly positive and partly negative and the average value of power over a complete cycle of the supply current is zero.

Similarly, we have already shown how the voltage dropped across a pure capacitive reactance lags the current flowing in the capacitor by an angle of 90°. Since the power at any instant is equal to the product of the instantaneous voltage and current, we can once again determine how the power supplied varies over a complete cycle of the supply current, as shown in Figure 29.3.

From this you should note that, once again:

(a) the power curve represents a sine law at twice the frequency of the supply current, and

(b) the power curve is partly positive and partly negative and the average value of power over a complete cycle of the supply current is zero.

It's important to understand what's happening here! In circuits that contain only inductance or capacitance, power is taken from the supply on each cycle when the magnetic or electric fields are created and then returned

Voltage, v
Current, i
Power, p

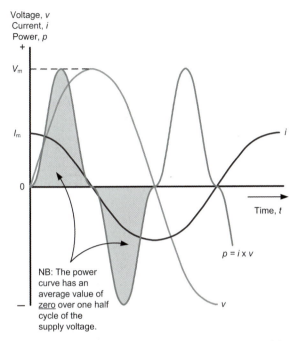

NB: The power curve has an average value of <u>zero</u> over one half cycle of the supply voltage.

Figure 29.3 Voltage, current and power in a pure capacitive reactance

to the supply when the fields later collapse. If the components are 'loss free' (i.e. if they have no internal resistance) the same amount of power is returned to the supply as is originally taken from it. The average power must therefore be zero.

Example 29.1 A sine-wave voltage has a maximum value of 100 V and a frequency of 50 Hz. Determine the instantaneous power in a $25\,\Omega$ resistive load 1.5 ms from the start of a cycle.

First we need to determine the maximum value of current flowing in the load from:

$$I_\text{m} = \frac{V_\text{m}}{R} = \frac{100}{25} = 4\,\text{A}$$

At $t = 1.5\,\text{ms}$, the instantaneous power, p, in the load can then be calculated from:

$$p = \frac{I_\text{m}V_\text{m}\,(1 - \cos(2\omega t))}{2}$$

$$= \frac{4 \times 100\,(1 - \cos(2 \times 2\pi f \times t))}{2}$$

From which:

$$p = 200\,(1 - \cos(2 \times 6.28 \times 50 \times 1.5 \times 10^{-3}))$$

$$= 200\,(1 - \cos(0.941))$$

$$p = 2\,(1 - 0.588) = 2 \times 0.412 = 82.4\,\text{W}$$

Test your knowledge 29.1

Show that the instantaneous power in a resistive load connected to a sinusoidal a.c. supply is given by:

$$p = \frac{I_\text{m}^2 R\,(1 - \cos(2\omega t))}{2}$$

Test your knowledge 29.2

Using a graph of the power function, show that the average power over a complete cycle of the supply is given by:

$$p = \frac{I_\text{m}^2 R}{2}$$

Key Point. In the case of a pure resistance, the power function (product of voltage and time) is always positive and has twice the frequency of the voltage and current waveforms

---

Key Point. In the case of a pure reactance, the power function is partly positive and partly negative and it has twice the frequency of the voltage and current waveforms. Over one complete cycle of voltage or current the average value of power in a pure reactance will be zero

---

29.2   Power factor

We met power factor briefly in Chapter 28 in relation to transformer ratings and efficiency but it is useful to distinguish between the true power and the apparent power in an a.c. circuit. It's important that you understand why they are different!

The *true power* in an a.c. circuit is simply the average power that it consumes. Thus:

$$\text{True power} = I^2 \times R$$

and is measured in Watts (W).

The *apparent power* in an a.c. circuit is simply the product of the supply voltage and the supply current (these are the quantities that you would measure using a voltmeter and ammeter). Thus:

$$\text{Apparent power} = V \times I$$

and is measured in volt-amperes (VA).

The *power factor* of an a.c. circuit is simply the ratio of the *true power* to the *apparent power*. Thus:

$$\text{Power factor} = \frac{\text{true power}}{\text{apparent power}}$$

Power factor provides us with an indication of how much of the power supplied to an a.c. circuit is converted into useful energy. A high power factor (i.e. a value close to 1 or *unity*) in an *L–R* or *C–R* circuit indicates that most of the energy taken from the supply is dissipated as heat produced by the resistor. A low power factor (i.e. a value close to zero) indicates that, despite the fact that current and voltage are supplied to the circuit, most of the energy is returned to the supply and very little of it is dissipated as heat.

In terms of units, power factor can be expressed as:

$$\text{Power factor} = \frac{\text{Watts}}{\text{volt-amperes}}$$

The current supplied to a reactive load can be considered to have two components acting at right angles. One of these components is in phase with the supply current and is known as the *active component of current*. The other component is 90° out of phase with the supply current (+90° in the case of a capacitive load and −90° in the case of an inductive load) and this component is known as the *reactive component of current*. The relationship between the active and reactive components of the supply current is illustrated in Figure 29.4.

From Figure 29.4, you should note that the active (i.e. *in-phase*) component of current is given by ($I \cos \phi$) whilst the reactive (i.e. 90° *out-of-phase*) component of current is given by ($I \sin \phi$).

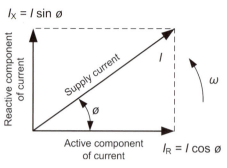

Figure 29.4 Relationship between active and reactive components of supply

The true power is given by the product of supply voltage and the active component of current. Thus:

$$\text{True power} = V \times (I \cos \phi) = VI \cos \phi \qquad \text{(i)}$$

Similarly, the reactive power is given by the product of supply voltage and the reactive component of current. Thus:

$$\text{Reactive power} = V \times (I \sin \phi) = VI \sin \phi \qquad \text{(ii)}$$

Also from Figure 29.4, the product of supply current and voltage (i.e. the apparent power) is given by:

$$\text{Apparent power} = V \times I \qquad \text{(iii)}$$

Combining formula (i) and formula (iii) allows us to express power factor in a different way:

$$\text{Power factor} = \frac{\text{true power}}{\text{apparent power}} = \frac{VI \cos \phi}{VI} = \cos \phi$$

Hence power factor $= \cos \phi$

Power factor can have a value that ranges between 0 (where there is no true power) and 1 (where the true power and the apparent power are the same).

We can also express power factor in terms of the ratio of resistance, R, and reactance, X, present in the load. This is done by referring to the impedance triangle (Figure 29.5), where:

$$\text{Power factor} = \cos \phi = \frac{R}{X}$$

The ratio of reactive power to true power can be determined by dividing equation (ii) by equation (i).

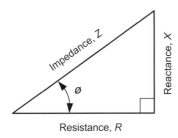

Figure 29.5 Impedance triangle

Thus:

$$\frac{\text{Reactive power}}{\text{True power}} = \frac{VI \sin \phi}{VI \cos \phi} = \tan \phi$$

Example 29.2 A power of 1 kW is supplied to an a.c. load that operates from a 220 V 50 Hz supply. If the supply current is 6 A, determine the power factor of the load.

$$\text{Power factor} = \frac{\text{true power}}{\text{apparent power}}$$

$$= \frac{1000}{220 \times 6} = \frac{1000}{1320} = 0.758$$

Example 29.3 The phase angle between the voltage and current supplied to an a.c. load is 25°. If the supply current is 3 A and the supply voltage is 220 V, determine the true power and reactive power in the load.

$$\text{Power factor} = \cos \phi = \cos 25° = 0.906$$

$$\text{True power} = VI \cos \phi = 220 \times 3 \times \cos 25°$$

$$= 220 \times 3 \times 0.906 = 598 \text{ W}$$

$$\text{Reactive power} = VI \sin \phi = 220 \times 3 \times \sin 25°$$

$$= 220 \times 3 \times 0.423 = 279 \text{ W}$$

Example 29.4 An a.c. load comprises an inductance of 0.5 H connected in series with a resistance of 85 Ω. If the load is to be used in conjunction with a 220 V 50 Hz a.c. supply, determine the power factor of the load and the supply current.

The reactance, X_L, of the inductor can be found from:

$$X_L = 2\pi f L = 2\pi \times 50 \times 0.5 = 157 \, \Omega$$

$$\text{Power factor} = \cos \varphi = \frac{R}{X_L} = \frac{85}{157} = 0.54$$

The supply current, I, can be calculated from:

$$I = \frac{V}{Z} = \frac{V}{\sqrt{(R^2 + X_L^2)}} = \frac{220}{\sqrt{(85^2 + 157^2)}}$$

$$= \frac{220}{\sqrt{(31874)}} = \frac{220}{178.5} = 1.23 \text{ A}$$

Test your knowledge 29.3

A current of 6.5 A flows in an a.c. load when it is connected to a 110 V a.c. supply. If the load has a power factor of 0.7, determine the true power in the load.

Test your knowledge 29.4

A load has an inductance of 450 mH and a resistance of 90 Ω. Determine the power factor of the load when operated (a) from a 50 Hz supply and (b) from a 60 Hz supply.

Test your knowledge 29.5

Determine the phase angle between supply voltage and supply current for a load having a power factor of 0.7.

Test your knowledge 29.6

A motor rated at 1.2 kW operates at an efficiency of 70%. Determine the power factor of the motor if it consumes 7.5 A when operated from a 230 V supply.

SCILAB

The following SCILAB script determines the impedance and power factor of an inductive load (see Figure 29.6):

```
// Power factor for an inductive load
// SCILAB source file: ex2901.sce
// Get data from dialog box
txt = ['Inductance (H)';'Resistance (ohm)';
'Frequency (Hz)'];
var = x_mdialog('Enter load parameters and
frequency',txt,['2';'350';'50']);
L = evstr(var(1));
r = evstr(var(2));
f = evstr(var(3));
// Calculate the inductive reactance
xL = 2 * %pi * f * L;
//
// Calculate the impedance
z = sqrt(r^2 + xL^2);
// Calculate the phase angle
pf =r/z;
// Display results
messagebox(["Impedance = " + string(z)
+ " ohm","Power factor = " +
string(pf)],"Results");
```

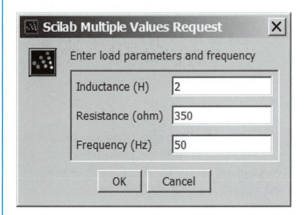

Figure 29.6 Input dialogue box for the SCILAB script

Key Point.   Power factor is the ratio of true power to apparent power

Key Point.   Even though voltage is present and current is flowing, no power is consumed in a purely reactive load

29.3   Power factor correction

When considering the utilisation of a.c. power, it is very important to understand the relationship between power factor and supply current. We have already shown that:

$$\text{Power factor} = \frac{\text{true power}}{\text{apparent power}}$$

Thus:

$$\text{Apparent power} = V \times I = \frac{\text{true power}}{\text{power factor}}$$

Recall that V and I are the quantities that would actually be measured by meters connected to an a.c. load. Furthermore, since V is normally a constant (i.e. the supply voltage), we can infer that:

$$I \propto \frac{\text{true power}}{\text{power factor}}$$

This relationship shows that the supply current is inversely proportional to the power factor. This has some important implications. For example, it explains why a large value of supply current will flow when a load has a low value of power factor. Conversely, the smallest value of supply current will occur when the power factor has its largest possible of 1.

It might help to explain the relationship between power factor and supply current by taking a few illustrative values. Let's assume that a motor produces a power of 1 kW and operates with an efficiency of 60%. The input power (in other words, the product of the voltage and current taken from the supply) will be (1 kW/0.6) or 1.667 kW. The supply current that would be required at various different power factors (assuming a 220 V a.c. supply) is given in the table below:

Power factor:	1.0	0.8	0.6	0.4	0.2	0.1
Supply current (A):	7.6	9.5	12.6	19	38	76

High values of power factor clearly result in the lowest values of supply current whilst low values of power factor can produce alarmingly high values of current which, in turn, can result in rupturing of fuses and overheating due to excessive I^2R losses.

Example 29.5 A power of 400 W is supplied to an a.c. load when it is connected to a 110 V 60 Hz supply. If the supply current is 6 A, determine the power factor of the load and the phase angle between the supply voltage and current.

$$\text{True power} = 400\,\text{W}$$

$$\text{Apparent power} = 110\,\text{V} \times 6\,\text{A} = 660\,\text{VA}$$

$$\text{Power factor} = \frac{\text{true power}}{\text{apparent power}}$$

$$= \frac{400}{660} = 0.606$$

$$\text{Power factor} = \cos\phi, \text{ thus}$$

$$\phi = \cos^{-1}(0.606) = 52.7°$$

Example 29.6 A motor produces an output of 750 W at an efficiency of 60% when operated from a 220 V a.c. mains supply. Determine the power factor of the motor if the supply current is 9.5 A. If the power factor is increased to 0.9 by means of a 'power factor correcting circuit', determine the new value of supply current.

$$\text{True power} = \frac{750}{0.6} = 1.25\,\text{kW}$$

$$\text{Power factor} = \frac{\text{true power}}{\text{apparent power}} = \frac{1.25 \times 10^3}{220 \times 9.5}$$

$$= \frac{1.25\,\text{kW}}{1.65\,\text{kVA}} = 0.76$$

Supply current (for a power factor of 0.9) =

$$\frac{1.25 \times 10^3}{\text{power factor}} = \frac{1.25 \times 10^3}{220 \times 0.9} = 6.3\,\text{A}$$

Steps must often be taken to improve the power factor of an a.c. load (see Figure 29.7). For an inductive load this can be achieved by connecting a capacitor in parallel with the load. Conversely, when the load is capacitive, the power factor can be improved by connecting an inductor in parallel with the load. In order to distinguish between the two types of load (inductive and capacitive) we often specify the power factor as either *lagging* (in the case of an inductor) or *leading* (in the case of a capacitor).

In either case, the object of introducing a component of *opposite* reactance is to reduce the overall phase

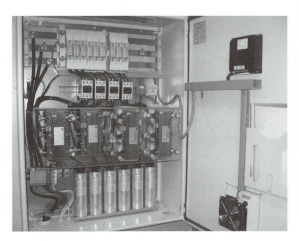

Figure 29.7 A 75 kVA power factor correction unit (PFCU). Photo courtesy Alwin Meschede

angle (i.e. the angle between the supply current and the supply voltage). If, for example, a load has an (uncorrected) phase angle of 15° lagging, the phase angle could be reduced to 0° by introducing a component that would exhibit an equal, but opposite, phase angle of 15° leading. To put this into perspective, take a look at Example 29.7.

Example 29.7 A 1.5 kW a.c. load is operated from a 220 V 50 Hz a.c. supply. If the load has a power factor of 0.6 lagging, determine the value of capacitance that must be connected in parallel with it in order to produce a unity power factor.

$$\cos\phi = 0.6, \text{ thus } \phi = \cos^{-1}(0.6) = 53.1°$$

Reactive power = True power $\times \tan\phi = 1.5\,\text{kW} \times \tan(53.1°) = 2\,\text{kVA}$

In order to increase the power factor from 0.6 to 1, the capacitor connected in parallel with the load must consume an equal reactive power. We can use this to determine the current that must flow in the capacitor and hence its reactance.

Capacitor current, $I_C = 2000/220 = 9.1\,\text{A}$

$$\text{Now } X_C = \frac{V}{I_C} = \frac{220}{9.1} = 24.12\,\Omega$$

$$\text{Since } X_C = \frac{1}{2\pi f C}$$

$$C = \frac{1}{2\pi f X_C} = \frac{1}{2\pi \times 50 \times 24.12} = \frac{1}{7578}\,\text{F}$$

Thus $C = 1.32 \times 10^{-4}\,\text{F} = 132\,\mu\text{F}$

Test your knowledge 29.7

A load has an inductance of 100 mH and a resistance of 50 Ω. Determine the value of capacitance that must be connected in parallel with the load in order to raise the power factor to unity when the load is used with a 110 V 400 Hz supply.

Key Point. Unity power factor corresponds to the condition where the supply voltage and supply current are in-phase with one another. In this condition the true power is the same as the apparent power

Key Point. It is frequently necessary to increase the power factor when there is appreciable reactance present in a load. This can be achieved by adding the opposite reactance into the circuit

SCILAB

The following SCILAB script determines the value of capacitance required for power factor correction of an inductive load (see Figures 29.8 and 29.9):

```
// Power factor correction for an inductive
load
// SCILAB source file: ex2902.sce
// Get circuit data
txt = ['Load power (W)';'Power factor';
'Supply voltage (V)';'Frequency
(Hz)'];
var = x_mdialog('Enter data for the
inductive
load',txt,['150';'0.9';'220';'50']);
pt = evstr(var(1));
pf = evstr(var(2));
vs = evstr(var(3));
f = evstr(var(4));
// Calculate the phase angle
phi = acos(pf);
// Calculate the reactive power
pr = pt * tan(phi);
// Calculate capacitor current
ic = pr/vs;
```

```
// Calculate reactance required
xc = vs/ic;
// Calculate required value of capacitance
c = 1 / (2 * %pi * f * xc);
c = c * 1E6;
// Print results
messagebox(["Capacitor current =
" + string(ic) + " A","Capacitance = " +
string(c) + " uF"],"Results");
```

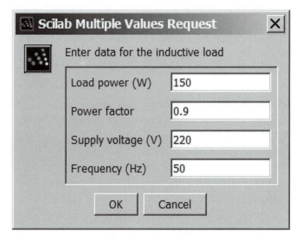

Figure 29.8 Input dialogue box for the SCILAB script

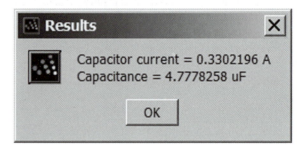

Figure 29.9 Output results box for the SCILAB script

29.4 Chapter summary

In this chapter we have introduced the concepts of power and power factor in an a.c. circuit. When an a.c. load only has resistance, the voltage and current in the load will be in phase and all of the applied power will be dissipated as heat in the resistor. This corresponds to unity power factor. Conversely, when the load consists of pure reactance the voltage and current will be 90° out of phase. In this condition, no power will be dissipated even though voltage and current will both be present.

Most a.c. loads have both resistance and reactance. In such cases the phase angle will take a value between 0° and 90° and the power factor will take a value that is somewhere between unity and zero. Note that power factor may be expressed in various ways, including the cosine of the phase angle and the ratio of true power to apparent power in the load.

Supply current can sometimes be excessive when a load has a low power factor, but the power factor can be corrected (increased to unity) by adding reactance of the correct amount but of opposite sign in parallel with the load. In the case of an inductive load an appropriate amount of capacitance must be added, whilst in the case of a capacitive load an appropriate amount of inductance must be added. In either case, the aim is that of bringing the phase angle between supply voltage and supply current to a value nearer 0°.

29.5   Review questions

1. An a.c. load draws a current of 2.5 A from a 220 V supply. If the load has a power factor of 0.7, determine the active and reactive components of the current flowing in the load.

2. An a.c. load has an effective resistance of 90 Ω connected in series with a capacitive reactance of 120 Ω. Determine the power factor of the load and the apparent power that will be supplied when the load is connected to a 415 V 50 Hz supply.

3. The phase angle between the voltage and current supplied to an a.c. load is 37°. If the supply current is 2 A and the supply voltage is 415 V, determine the true power and reactive power in the load.

4. A coil has an inductance of 0.35 H and a resistance of 65 Ω. If the coil is connected to a 220 V 50 Hz supply, determine its power factor and the value of capacitance that must be connected in parallel with it in order to raise the power factor to unity.

5. An a.c. load comprises an inductance of 1.5 H connected in series with a resistance of 85 Ω. If the load is to be used in conjunction with a 220 V 50 Hz a.c. supply, determine the power factor of the load and the supply current.

6. A power of 400 W is supplied to an a.c. load when it is connected to a 110 V 60 Hz supply. If the supply current is 6 A, determine the power factor of the load and the phase angle between the supply voltage and current.

7. A motor produces an output of 750 W at an efficiency of 60% when operated from a 220 V a.c. mains supply. Determine the power factor of the motor if the supply current is 9.5 A. If the power factor is increased to 0.9 by means of a 'power factor correcting circuit', determine the new value of supply current.

8. An a.c. load consumes a power of 800 W from a 110 V 60 Hz supply. If the load has a lagging power factor of 0.6, determine the value of parallel connected capacitor required to produce a unity power factor.

Chapter 30

Complex waveforms and Fourier analysis

In Chapter 25 we described the waveform for a sinusoidal voltage. In many practical applications the waveforms of voltage and current are not purely sinusoidal and instead comprise a fundamental component together with a number of harmonic components, each having a specific amplitude and phase relative to the fundamental. In this chapter we shall investigate these complex waveforms and describe methods that we can use to analyse them in terms of their individual component waves.

30.1   Harmonics

As mentioned in Chapter 25, the sine wave is the most fundamental of all wave shapes and all other waveforms can be synthesised from sinusoidal components. To specify a sine wave, we need to consider just three things: amplitude, frequency and phase. Since no components are present at any frequency other than that of the fundamental sine wave, such a wave is said to be a *pure tone*. All other waveforms can be reproduced by adding together sine waves of the correct amplitude, frequency and phase. The study of these techniques is called *Fourier analysis* and we will explain what this means later in this chapter.

An integer multiple of a *fundamental* frequency is known as a *harmonic*. In addition, we often specify the order of the harmonic (second, third, etc.). Thus the *second harmonic* has twice the frequency of the fundamental, the *third harmonic* has three times the frequency of the fundamental, and so on. Consider, for example, a fundamental signal at 1 kHz. The second harmonic would have a frequency of 2 kHz, the third harmonic a frequency of 3 kHz and the fourth harmonic a frequency of 4 kHz.

Note that, in musical terms, the relationship between notes that are one *octave* apart is simply that the two frequencies have a ratio of 2:1 (in other words, the higher frequency is double the lower frequency).

Complex waveforms (nothing to do with complex numbers!) comprise a fundamental component together with a number of harmonic components, each having a specific amplitude and with a specific phase relative to the fundamental. The following example shows that this is not quite so complicated as it sounds.

Consider a sinusoidal signal with an amplitude of 1 V at a frequency of 1 kHz. The waveform of this fundamental signal is shown in Figure 30.1. Now consider the second harmonic of the first waveform. Let's suppose that this has an amplitude of 0.5 V and that it is in phase with the fundamental. This 2 kHz component is shown in Figure 30.2. Finally, let's add the two waveforms together at each point in time. This produces the complex waveform shown in Figure 30.3.

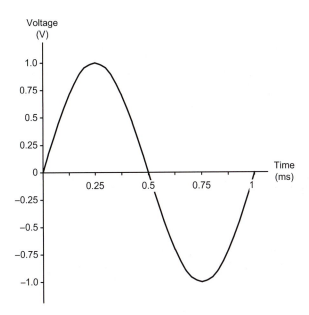

Figure 30.1 Fundamental component

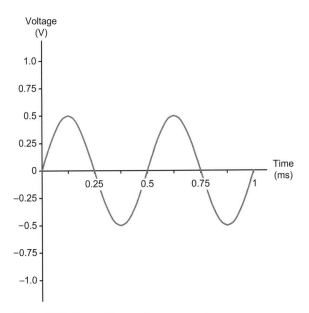

Figure 30.2 Second harmonic component

We can describe a complex wave using an equation of the form:

$$v = V_1 \sin(\omega t) + V_2 \sin(2\omega t \pm \phi_2)$$
$$+ V_3 \sin(3\omega t \pm \phi_3) + V_4 \sin(4\omega t \pm \phi_4) + \ldots$$

where v is the instantaneous voltage of the complex waveform at time, t. V_1 is the amplitude (i.e. maximum value) of the fundamental, V_2 is the amplitude of the

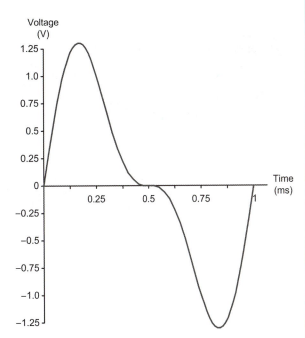

Figure 30.3 Resultant complex waveform

second harmonic, V_3 is the amplitude of the third harmonic, and so on. Similarly, ϕ_2 is the phase angle of the second harmonic (relative to the fundamental), ϕ_3 is the phase angle of the third harmonic (relative to the fundamental), and so on. The important thing to note from the foregoing equation is that *all of the individual components that make up a complex waveform have a sine wave shape*.

Key Point. A complex waveform can be described in terms of a series of a fundamental component added to a number of harmonically related components which have the same sine-wave shape but vary in amplitude and phase relative to the fundamental

Example 30.1 The complex waveform shown in Figure 30.4 is given by the equation:

$$v = 100 \sin(100\pi t) + 50 \sin(200\pi t) \text{ V}$$

Determine:

(a) the amplitude of the fundamental

(b) the frequency of the fundamental

(c) the order of any harmonic components present

(d) the amplitude of any harmonic components present

(e) the phase angle of any harmonic components present (relative to the fundamental).

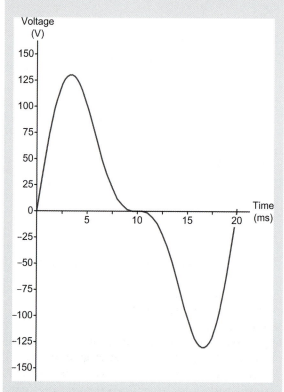

Figure 30.4 See Example 30.1

Comparing the foregoing equation with the general equation yields the following:

(a) The first term is the fundamental and this has an amplitude (V_1) of 100 V

(b) The frequency of the fundamental can be determined from:

$$\sin(\omega t) = \sin(100\pi t)$$

Thus $\omega = 100\pi$ but $\omega = 2\pi f$
thus $2\pi f = 100\pi$ and $f = 100\pi/2\pi = 50\,\text{Hz}$

(c) The frequency of the second term can similarly be determined from:

$$\sin(\omega t) = \sin(200\pi t)$$

Thus $\omega = 200\pi$ but $\omega = 2\pi f$,
thus $2\pi f = 200\pi$ and $f = 200\pi/2\pi = 100\,\text{Hz}$
Thus the second term has twice the frequency of the fundamental and so is the second harmonic.

(d) The second harmonic has an amplitude (V_2) of 50 V

(e) Finally, since there is no phase angle included within the expression for the second harmonic (i.e. there is no $\pm\phi_2$ term), the second harmonic component must be in phase with the fundamental.

Test your knowledge 30.1

An alternating voltage is given by the expression:

$$v = 180\sin(800\pi t) + 50\sin\left(2400\pi t - \frac{\pi}{4}\right)$$

Determine:

(a) the amplitude of the fundamental

(b) the frequency of the fundamental

(c) the order of any harmonic components present

(d) the amplitude of any harmonic components present

(e) the phase angle of any harmonic components present.

Test your knowledge 30.2

Calculate the instantaneous value of the voltage of the waveform in Test your knowledge 30.1 at (a) 0.5 ms and (b) 1.5 ms from the start of a cycle.

Test your knowledge 30.3

A complex waveform is described by the equation:

$$v = 100\sin(100\pi t) + 50\sin\left(200\pi t - \frac{\pi}{2}\right)$$

Determine, graphically, the shape of the waveform and estimate the peak–peak voltage.

Test your knowledge 30.4

A complex waveform is described by the equation:

$$v = 100 \sin(100\pi t) + 250 \sin\left(200\pi t - \frac{\pi}{2}\right)$$
$$+ 450 \sin(300\pi t)$$

Determine, graphically, the shape of the waveform and estimate the peak–peak voltage.

SCILAB

SCILAB provides you with an excellent tool for visualising complex waveforms. The following sample script produces a plot of the waveform in Test your knowledge 30.3 (see Figure 30.5):

```
// Resultant waveform from the addition of
two AC voltages
// SCILAB source file: ex3001.sce
// Set up the plot window
q=gda(); // Default axes
q.thickness=1; // Line thickness
q.grid=[4,4]; // Add a dashed grid in cyan
q.font_size=3; // Set the font size
t=(0:0.0001:0.02)'; // Array of values for
time, t
xdel();
// Define v1, v2 and v3
v1 = 100*sin(100*%pi*t);
v2 = 50*sin((200*%pi*t) - (%pi/2));
v3 = v1 + v2;
// Plot v1 in blue
plot2d(t,v1,rect=[0,-200,0.02,200],style=2);
// Plot v2 in red
plot2d(t,v2,rect=[0,-200,0.02,200],style=5);
// Plot v3 in green
plot2d(t,v3,rect=[0,-200,0.02,200],
style=13);
// Add legends
legend("v1","v2","v3");
```

As a further example, the following SCILAB script produces a plot of the waveform in Test your knowledge 30.4 (see Figure 30.6):

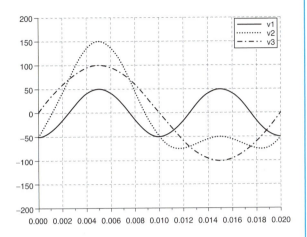

Figure 30.5 Graphical output of the SCILAB script

```
// Resultant waveform from the addition of
three AC voltages
// SCILAB source file: ex3002.sci
// Set up the plot window
q=gda(); // Default axes
q.thickness=1; // Line thickness
q.grid=[4,4]; // Add a dashed grid in cyan
q.font_size=3; // Set the font size
t=(0:0.0001:0.02)'; // Array of values for
time, t
xdel();
// Define v1, v2, v3 and v4
v1 = 100*sin(100*%pi*t);
v2 = 250*sin((200*%pi*t) - (%pi/2));
v3 = 450*sin(300*%pi*t);
v4 = v1 + v2 + v3;
// Plot v1 in blue
plot2d(t,v1,rect=[0,-800,0.02,800],style=2);
// Plot v2 in red
plot2d(t,v2,rect=[0,-800,0.02,800],style=5);
// Plot v3 in yellow
plot2d(t,v3,rect=[0,-800,0.02,800],style=6);
// Plot v4 in green
plot2d(t,v4,rect=[0,-800,0.02,800],style=13);
// Add legends
legend("v1","v2","v3","v4");
```

30.1.1   The square wave

The square wave can be created by adding a fundamental frequency to an infinite series of odd harmonics (i.e.

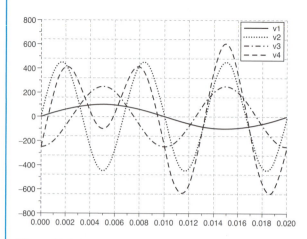

Figure 30.6 Graphical output of the SCILAB script

the third, fifth, seventh, etc.). If the fundamental is at a frequency f, then the third harmonic is at $3f$, the fifth is at $5f$, and so on.

The amplitude of the harmonics should decay in accordance with their harmonic order and they must all be in phase with the fundamental. Thus a square wave can be obtained from:

$$v = V \sin(\omega t) + \frac{V}{3} \sin(3\omega t) + \frac{V}{5} \sin(5\omega t)$$
$$+ \frac{V}{7} \sin(7\omega t) + \dots$$

where V is the amplitude of the fundamental and $\omega = 2\pi f$

Figure 30.7 shows this relationship graphically.

> **Key Point.** A square wave comprises a fundamental component together with an infinite series of odd harmonic components which decay in amplitude in accordance with their harmonic order and where all of the harmonic components are in phase with the fundamental

30.1.2   The triangular wave

The triangular wave can similarly be created by adding a fundamental frequency to an infinite series of odd harmonics. However, in this case the harmonics should decay in accordance with the square of their harmonic order and they must alternate in phase so that the third, seventh, eleventh, etc. are in antiphase whilst the fifth, ninth, thirteenth, etc. are in phase with the fundamental.

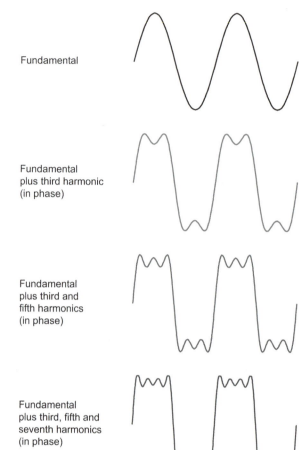

Fundamental

Fundamental plus third harmonic (in phase)

Fundamental plus third and fifth harmonics (in phase)

Fundamental plus third, fifth and seventh harmonics (in phase)

Figure 30.7 Fundamental and low-order harmonic components of a square wave

Thus a triangular wave can be obtained from:

$$v = V \sin(\omega t) + \frac{V}{9} \sin(3\omega t - \pi) + \frac{V}{25} \sin(5\omega t)$$
$$+ \frac{V}{49} \sin(7\omega t - \pi) + \dots$$

where V is the amplitude of the fundamental and $\omega = 2\pi f$

Figure 30.8 shows this relationship graphically.

> **Key Point.** A triangular wave comprises a fundamental component together with an infinite series of odd harmonic components which decay in amplitude in accordance with the square of their harmonic order and where alternate harmonic components are in antiphase with the fundamental

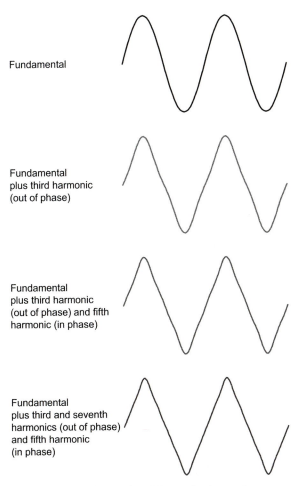

Fundamental

Fundamental plus third harmonic (out of phase)

Fundamental plus third harmonic (out of phase) and fifth harmonic (in phase)

Fundamental plus third and seventh harmonics (out of phase) and fifth harmonic (in phase)

Figure 30.8 Fundamental and low-order harmonic components of a triangular wave

30.1.3 Effect of harmonics on overall wave shape

Having considered several common types of waveform, we are now in a position to summarise the results of adding harmonics to a fundamental waveform with different phase relationships. These are listed below:

1. When odd harmonics are added to a fundamental waveform, regardless of their phase relative to the fundamental, the *positive and negative half-cycles will be similar* in shape and the resulting waveform will be symmetrical about the time axis – see Figure 30.9(a).

2. When even harmonics are added to a fundamental waveform with a phase shift of other than 180° (i.e. $\pi/2$), the *positive and negative half-cycles will be dissimilar* in shape – see Figure 30.9(b).

3. When even harmonics are added to a fundamental waveform with a phase shift of 180°, the *positive half-cycles will be the mirror image of the negative half-cycles* when reversed – see Figure 30.9(c).

4. When odd and even harmonics are added to a fundamental waveform in-phase, the *positive half-cycles will be the mirror image of the negative half-cycles* when reversed.

5. When odd and even harmonics are added to a fundamental waveform out of phase, the *positive and negative half-cycles will be dissimilar* in shape.

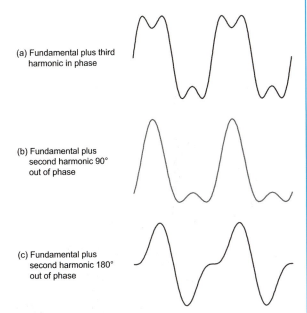

(a) Fundamental plus third harmonic in phase

(b) Fundamental plus second harmonic 90° out of phase

(c) Fundamental plus second harmonic 180° out of phase

Figure 30.9 Effects of harmonic components on wave shape

30.1.4 Generation of harmonic components

Unwanted harmonic components may be produced in any system that incorporates non-linear circuit elements (such as diodes and transistors). Figure 30.10 shows the circuit diagram of a simple half-wave rectifier. The voltage and current waveforms for this circuit are shown in Figure 30.11. The important thing to note here is the shape of the current waveforms, which comprise a series of relatively narrow pulses as the charge in the reservoir capacitor, C, is replenished at the crest of each positive-going half-cycle. These pulses of current are rich in unwanted harmonic content.

Harmonic components can also be generated in magnetic components (such as inductors and transformers) where the relationship between flux density (B)

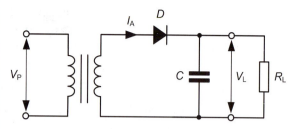

Figure 30.10 A simple half-wave rectifier

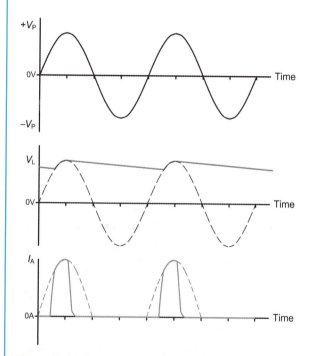

Figure 30.11 Current and voltage waveforms for the simple half-wave rectifier

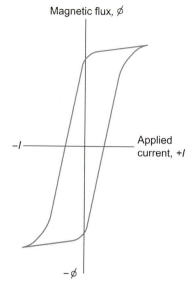

Figure 30.12 Magnetic flux plotted against applied current for a transformer

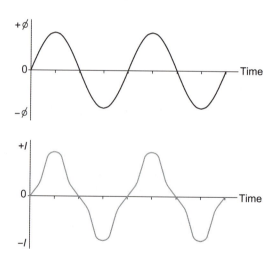

Figure 30.13 Magnetic flux and applied current waveforms for a typical transformer

and applied current is non-linear. Typical transfer characteristics and the resulting time-related input and load waveforms for a saturated transformer core are shown in Figures 30.12 and 30.13.

Alternatively, there are a number of applications in which harmonic components are actually desirable. A typical application might be the frequency multiplier stage of a transmitter in which the non-linearity of an amplifier stage (operating in Class C) can be instrumental in producing an output which is rich in harmonic content. The desired harmonic can then be selected by means of a resonant circuit tuned to the desired multiple of the input frequency – see Figure 30.14.

Test your knowledge 30.5

A sinusoidal current is specified by the equation:

$$i = 0.1 \sin\left(200\pi t + \frac{\pi}{2}\right) \text{ A}$$

Determine:

(a) the peak value of the current

(b) the frequency of the current

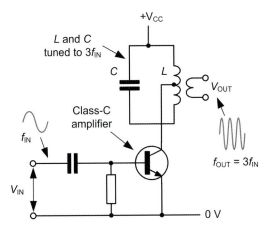

Figure 30.14 A Class C frequency multiplier stage

(c) the phase angle of the current

(d) the instantaneous value of the current at $t = 5\,\text{ms}$.

Test your knowledge 30.6

A complex waveform comprises a fundamental (sinusoidal) voltage with a peak value of 36 V and a frequency of 400 Hz, together with a third harmonic component having a peak value of 12 V leading the fundamental by 90°. Write down an expression for the instantaneous value of the complex voltage and use it to determine the value of voltage at 3 ms from the start of a cycle.

Test your knowledge 30.7

A complex waveform is given by the equation:

$$i = 10 \sin\left(100\pi t\right) + 3 \sin\left(300\pi t - \frac{\pi}{2}\right)$$

$$+ 1 \sin\left(500\pi t + \frac{\pi}{2}\right) \text{ mA}$$

Determine:

(a) the amplitude of the fundamental

(b) the frequency of the fundamental

(c) the order of any harmonic components present

(d) the amplitude of any harmonic components present

(e) the phase angle of any harmonic components present, relative to the fundamental.

30.2 Fourier analysis

Earlier in this chapter we introduced complex waveforms and showed how a complex wave could be expressed by an equation of the form:

$$v = V_1 \sin\left(\omega t\right) + V_2 \sin\left(2\omega t \pm \phi_2\right)$$

$$+ V_3 \sin\left(3\omega t \pm \phi_3\right) + V_4 \sin\left(4\omega t \pm \phi_4\right) + \dots$$

$$\text{(i)}$$

In this section we shall develop this idea further by introducing a powerful technique called Fourier analysis. Fourier analysis is based on the concept that all waveforms, whether continuous or discontinuous, can be expressed in terms of a convergent series of the form:

$$v = A_0 + A_1 \cos t + A_2 \cos 2t + A_3 \cos 3t$$

$$+ \dots + B_1 \sin t + B_2 \sin 2t + B_3 \sin 3t + \dots \quad \text{(ii)}$$

where A_0, A_1, …, B_1, B_2, … are the amplitudes of the individual components that make up the complex waveform. Essentially, this formula is the same as the simplified relationship in equation (i) but with the introduction of cosine as well as sine components. Note also that we have made the assumption that one complete cycle occurs in a time equal to 2π seconds (recall that $\omega = 2\pi f$ and one complete cycle requires 2π radians).

Another way of writing equation (ii) is:

$$v = A_0 + \sum_{n=1}^{n=\infty} \left(A_n \sin nt + B_n \cos nt\right) \quad \text{(iii)}$$

For the range $-\pi$ to $+\pi$, The value of the constant terms (the amplitudes of the individual components)

$A_0, A_1, \ldots B_1, B_2 \ldots$ etc. can be determined as follows:

$$A_0 = \frac{1}{2\pi} \int_{-\pi}^{\pi} v \, dt$$

$$= \frac{1}{2\pi} \int_{-\pi}^{\pi} \left(A_0 + \sum_{n=1}^{n=\infty} (A_n \cos nt + B_n \sin nt) \right) dt$$

(iv)

$$A_n = \frac{1}{\pi} \int_{-\pi}^{\pi} v \cos nt \, dt \quad \text{(where } n = 1, 2, 3, \ldots) \quad \text{(v)}$$

$$B_n = \frac{1}{\pi} \int_{-\pi}^{\pi} v \sin nt \, dt \quad \text{(where } n = 1, 2, 3, \ldots) \quad \text{(vi)}$$

The values $A_0, A_1, \ldots, B_1, B_2, \ldots$ etc. are called the *Fourier coefficients* of the *Fourier series* defined by equation (iii).

If you are puzzled by equations (iv) to (vi), it may help to put into words what the values of the Fourier coefficients actually represent:

- A_0 is the mean value of v over one complete cycle of the waveform (i.e. from $-\pi$ to $+\pi$, or from 0 to 2π)
- A_1 is twice the mean value of $v \cos t$ over one complete cycle
- B_1 is twice the mean value of $v \sin t$ over one complete cycle
- A_2 is twice the mean value of $v \cos 2t$ over one complete cycle
- B_2 is twice the mean value of $v \sin 2t$ over one complete cycle

and so on. We shall use this fact later when we describe an alternative method of determining the Fourier series for a particular waveform.

Just in case this is all beginning to sound a little too complex in terms of mathematics, let's take a look at some examples based on waveforms that you should immediately recognise.

Example 30.2 Determine the Fourier coefficients of the sine wave superimposed on the constant d.c. level shown in Figure 30.15.

By inspection, the equation of the wave (over the range 0 to 2π) is:

$$v = 10 + 5 \sin t$$

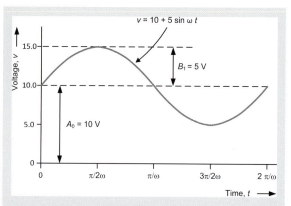

Figure 30.15 See Example 30.2

By comparison with equation (ii), we can deduce that:

$A_0 = 10$ (in other words, the mean voltage level over the range 0 to 2π is 10 V)

$B_1 = 5$ (this is simply the coefficient of $\sin t$)

There are no other terms and thus all other Fourier coefficients (A_1, A_2, B_2, etc.) are all zero.

Now let's see if we can arrive at the same results using equations (iv), (v) and (vi):

The value of A_0 is found from:

$$A_0 = \frac{1}{2\pi} \int_0^{2\pi} v \, dt = \frac{1}{2\pi} \int_0^{2\pi} 10 + 5 \sin t \, dt$$

$$A_0 = \frac{1}{2\pi} [10t]_{t=0}^{t=2\pi} + \frac{1}{2\pi} \int_0^{2\pi} 5 \sin t \, dt$$

$$A_0 = \left[\frac{20\pi}{2\pi} - 0 \right] + 0 = 10$$

The value of A_1 is found from:

$$A_1 = \frac{1}{\pi} \int_0^{2\pi} v \cos t \, dt$$

$$= \frac{1}{\pi} \int_0^{2\pi} (10 + 5 \sin t) \cos t \, dt$$

$$A_1 = \frac{1}{\pi} \int_0^{2\pi} (10 \cos t + 5 \sin t \cos t) \, dt$$

$$A_1 = \frac{1}{\pi} \int_0^{2\pi} 10 \cos t \, dt + \frac{1}{\pi} \int_0^{2\pi} 5 \sin t \cos t \, dt$$

$$A_1 = \frac{1}{\pi} \int_0^{2\pi} 10 \cos t \, dt + 0$$

$$A_1 = \frac{1}{\pi} [10 \sin t]_{t=0}^{t=2}\pi + 0 = 0$$

The value of B_1 is found from:

$$B_1 = \frac{1}{\pi} \int_0^{2\pi} v \sin t \, dt$$

$$= \frac{1}{\pi} \int_0^{2\pi} (10 + 5 \sin t) \sin t \, dt$$

$$B_1 = \frac{1}{\pi} \int_0^{2\pi} \left(10 \sin t + 5 \sin^2 t\right) dt$$

$$B_1 = \frac{1}{\pi} \int_0^{2\pi} 10 \sin t \, dt + \frac{1}{\pi} \int_0^{2\pi} 5 \sin^2 t \, dt$$

$$B_1 = 0 + \frac{1}{\pi} \left[\frac{5}{2}\left(t - \frac{1}{2}\sin 2t\right)\right]_{t=0}^{t=2\pi}$$

$$B_1 = 0 + \frac{10\pi}{2\pi} = 5$$

Example 30.3   Determine the Fourier series for the rectangular pulse shown in Figure 30.16.

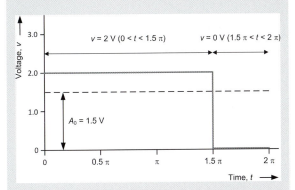

Figure 30.16 See Example 30.3

Unlike the waveform in the previous example, this voltage is discontinuous and we must therefore deal with it in two parts; that from 0 to 1.5π and that from 1.5π to 2π.

The equation of the voltage is as follows:

$$v = 2 \text{ (over the range } 0 < t < 1.5\pi)$$

$$v = 0 \text{ (over the range } 1.5\pi < t < 2\pi)$$

To find A_0 we simply need to find the mean value of the waveform over the range 0 to 2π. By considering the area under the waveform it should be obvious that this is 1.5.

Hence $A_0 = 1.5$.

Next we need to find the value of A_n for $n = 1, 2, 3, \ldots$ and so on. From equation (v):

$$A_n = \frac{1}{\pi} \int_0^{2\pi} v \cos nt \, dt$$

But since there are two distinct parts to the waveform, we need to integrate separately over the two time periods, 0 to 1.5π and 1.5π to 2π.

$$A_n = \frac{1}{\pi} \int_0^{1.5\pi} 2 \cos nt \, dt$$
$$+ \frac{1}{\pi} \int_{1.5\pi}^{2\pi} 0 \cos nt \, dt$$

$$A_n = \frac{1}{\pi} \int_0^{1.5\pi} 2 \cos nt \, dt + 0$$

$$A_n = \frac{2}{\pi n} [\sin nt]_{t=0}^{t=1.5\pi} + 0$$

$$= \frac{2}{\pi n} [\sin 1.5 n\pi - \sin 0]$$

Thus $A_n = \frac{2}{\pi n} \sin 1.5 n\pi$

Now when $n = 1$, $A_1 = \frac{2}{\pi} \sin 1.5\pi = -\frac{2}{\pi}$

when $n = 2$, $A_2 = \frac{1}{\pi} \sin 3\pi = 0$

when $n = 3$, $A_3 = \frac{2}{3\pi} \sin 4.5\pi = +\frac{2}{3\pi}$

when $n = 4$, $A_4 = \frac{1}{2\pi} \sin 6\pi = 0$

when $n = 5$, $A_5 = \frac{2}{5\pi} \sin 7.5\pi = -\frac{2}{5\pi}$

The cosine terms in the Fourier series will thus be:

$$-\frac{2}{\pi}\cos t + \frac{2}{3\pi}\cos 3t - \frac{2}{5\pi}\cos 5t \ldots$$

The sine terms can be similarly found:

$$B_n = \frac{1}{\pi} \int_0^{1.5\pi} 2 \sin nt \, dt$$

$$+ \frac{1}{\pi} \int_{1.5\pi}^{2\pi} 0 \sin nt \, dt$$

$$B_n = \frac{1}{\pi} \int_0^{1.5\pi} 2 \sin nt \, dt + 0$$

$$B_n = \frac{-2}{\pi n} [\cos nt]_{t=0}^{t=1.5\pi} + 0$$

$$= \frac{-2}{\pi n} [\cos 1.5n\pi - \cos 0]$$

Thus $B_n = \frac{-2}{\pi n} (\cos 1.5n\pi - 1)$

Now when $n = 1$, $B_1 = \frac{-2}{\pi} (\cos 1.5\pi - 1) = \frac{2}{\pi}$

when $n = 2$, $B_2 = \frac{-2}{2\pi} (\cos 3\pi - 1) = \frac{2}{\pi}$

when $n = 3$, $B_3 = \frac{-2}{3\pi} (\cos 4.5\pi - 1) = \frac{2}{3\pi}$

when $n = 4$, $B_4 = \frac{-2}{4\pi} (\cos 6\pi - 1) = 0$

when $n = 5$, $B_5 = \frac{-2}{5\pi} (\cos 7.5\pi - 1) = \frac{2}{5\pi}$

Thus the cosine terms in the Fourier series will be:

$$\frac{2}{\pi} \sin t + \frac{2}{\pi} \sin 2t + \frac{2}{5\pi} \sin 5t \ldots$$

We are now in a position to develop the expression for the pulse. This will be similar to that shown in equation (ii), but with:

$$A_0 = 1.5, A_1 = -\frac{2}{\pi}, A_2 = 0, A_3 = \frac{2}{3\pi}, \text{ and so on.}$$

The expression for the pulse is thus:

$$v = 1.5 - \frac{2}{\pi} \cos t + \frac{2}{3\pi} \cos 3t - \frac{2}{5\pi} \cos 5t \ldots$$

$$+ \frac{2}{\pi} \sin t + \frac{2}{\pi} \sin 2t + \frac{2}{3\pi} \sin 3t + \frac{2}{5\pi} \sin 5t \ldots$$

Using this function in a standard spreadsheet program (up to and including the ninth harmonic terms) produces the *synthesised* pulse shown in Figure 30.17.

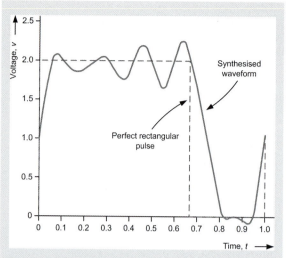

Figure 30.17 Synthesised pulse – see Example 30.3

Test your knowledge 30.8

Derive the Fourier series for each of the waveforms shown in Figure 30.18.

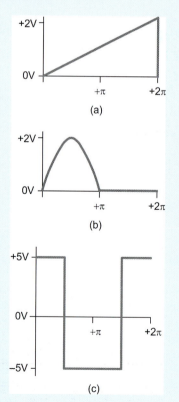

Figure 30.18 See Test your knowledge 30.8

30.2.1 An alternative method

An alternative tabular method can be used to determine Fourier coefficients. This method is based on an application of the *trapezoidal rule*. In order to determine the mean value of $y = f(t)$ over a given period, we can divide the period into a number of smaller intervals and determine the y ordinate value for each.

If the y values are $y_1, y_2, y_3, \ldots, y_m$, and there are m equal values in a time interval, t, the mean value of y, $\bar{y}$, will be given by:

$$\bar{y} = \frac{y_1 + y_2 + y_3 + \cdots y_m}{m} = \frac{\sum\limits_{n=0}^{n=m} y_n}{m}$$

We can put the trapezoidal rule to good use when a number of voltage readings are available at regular time intervals over the period of one complete cycle of a waveform. Once again, we shall show how this method works by using an example.

Example 30.4 The values of voltage over a complete cycle of a waveform (see Figure 30.19) are as follows:

Angle, θ	30	60	90	120	150	180
	210	240	270	300	330	360
Voltage, v	8.9	8.1	7.5	6.5	5.5	5.0
	4.5	3.5	2.5	1.9	1.1	5

First we need to determine the value of A_0. We can do this by calculating the mean value of voltage over the complete cycle. Applying the trapezoidal

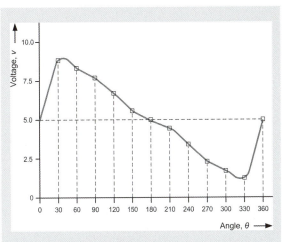

Figure 30.19 See Example 30.4

rule (i.e. adding together the values of voltage and dividing by the number of intervals) gives:

$$A_0 = (8.9 + 8.1 + 7.5 + 6.5 + \cdots + 5)/12 = 5$$

Similarly, to determine the value of A_1 we multiply each value of voltage by $\cos\theta$, before adding them together and dividing by *half* the number of intervals (recall that A_1 is *twice* the mean value of $v \cos t$ over one complete cycle). Thus:

$$A_1 = (7.707 + 4.049 - 0.002 - 3.252$$
$$+ \cdots + 5)/6 = 0$$

The same method can be used to determine the remaining Fourier coefficients. The use of a spreadsheet for this calculation is highly recommended (see Figure 30.20).

Fourier analysis - tabular method

y	θ	v	$\cos\theta$	$v\cos\theta$	$\sin\theta$	$v\sin\theta$	$\cos2\theta$	$v\cos2\theta$	$\sin2\theta$	$v\sin2\theta$	$\cos3\theta$	$v\cos3\theta$	$\sin3\theta$	$v\sin3\theta$	$\cos4\theta$	$v\cos4\theta$	$\sin4\theta$	$v\sin4\theta$
$y1$	30	8.9	0.866	7.707	0.500	4.451	0.500	4.449	0.866	7.708	0.000	-0.002	1.000	8.900	-0.500	-4.452	0.866	7.706
$y2$	60	8.1	0.500	4.049	0.866	7.015	-0.500	-4.052	0.866	7.014	-1.000	-8.100	0.000	-0.003	-0.500	-4.046	0.866	-7.017
$y3$	90	7.5	0.000	-0.002	1.000	7.500	-1.000	-7.500	0.000	-0.003	0.001	0.005	-1.000	-7.500	1.000	7.500	0.001	0.006
$y4$	120	6.5	-0.500	-3.252	0.866	5.628	-0.500	-3.247	-0.866	-5.631	1.000	6.500	0.001	0.005	-0.501	-3.256	0.865	5.626
$y5$	150	5.5	-0.866	-4.764	0.500	2.748	0.501	2.753	-0.866	-4.761	-0.001	-0.006	1.000	5.500	-0.499	-2.744	-0.867	-4.767
$y6$	180	5	-1.000	-5.000	0.000	-0.002	1.000	5.000	0.001	0.004	-1.000	-5.000	-0.001	-0.006	1.000	5.000	0.002	0.008
$y7$	210	4.5	-0.866	-3.896	-0.500	-2.252	0.499	2.246	0.867	3.899	0.001	0.006	-1.000	-4.500	-0.502	-2.257	0.865	3.893
$y8$	240	3.5	-0.500	-1.748	-0.866	-3.032	-0.501	-1.753	0.865	3.029	1.000	3.500	0.002	0.006	-0.498	-1.743	-0.867	-3.035
$y9$	270	2.5	0.001	0.002	-1.000	-2.500	-1.000	-2.500	-0.001	-0.003	-0.002	-0.005	1.000	2.500	1.000	2.500	0.002	0.006
$y10$	300	1.9	0.501	0.951	-0.866	-1.645	-0.499	-0.948	-0.867	-1.647	-1.000	-1.900	-0.002	-0.004	-0.502	-0.954	0.865	1.643
$y11$	330	1.1	0.866	0.953	-0.499	-0.549	0.501	0.551	-0.865	-0.952	0.002	0.002	-1.000	-1.100	-0.497	-0.547	-0.868	-0.954
$y12$	360	5	1.000	5.000	0.001	0.004	1.000	5.000	0.002	0.008	1.000	5.000	0.012	0.012	1.000	5.000	0.003	0.016
	Sum	60	Sum	0.001	Sum	17.367	Sum	0.000	Sum	8.666	Sum	0.001	Sum	3.810	Sum	0.000	Sum	3.131
	$A_0 =$	5	$A_1 =$	0.000	$B_1 =$	2.894	$A_2 =$	0.000	$B_2 =$	1.444	$A_3 =$	0.000	$B_3 =$	0.635	$A_4 =$	0.000	$B_4 =$	0.522

Figure 30.20 Spreadsheet table of Fourier coefficients – see Example 30.4

Having determined the Fourier coefficients, we can write down the expression for the complex waveform. In this case it is:

$$v = 5 + 2.894 \sin\theta + 1.444 \sin 2\theta$$
$$+ 0.635 \sin 3\theta + 0.522 \sin 4\theta + \cdots$$

(note that there are no terms in $\cos\theta$ in the Fourier series for this waveform).

Test your knowledge 30.9

The following voltages are taken over one complete cycle of a complex waveform:

Angle, θ	0	30	60	90	120	150	180
	210	240	270	300	330	360	
Voltage, v	0	3.5	8.7	9.1	9.5	9.5	9.1
	8.7	3.5	0	3.5	3.5	0	

Use a tabular method to determine the Fourier series for the waveform.

Earlier we discussed the effect of adding harmonics to a fundamental on the shape of the complex wave produced. Revisiting this in the light of what we now know about Fourier series allows us to relate wave shape to the corresponding Fourier series.

Figure 30.21 shows the effect of symmetry (and reflected symmetry) on the values of A_0, A_1, A_2, ..., B_1, B_2, etc. This information can allow us to take a few short-cuts when analysing a waveform (there is no point in trying to evaluate Fourier coefficients that are just not present!).

Key Point. Fourier analysis is based on the concept that all waveforms can be expressed in terms of a convergent series

SCILAB

SCILAB incorporates functions that allow you to perform Fourier transformation using techniques based on algorithms that require significantly less time for computation. As an example, the following SCILAB script uses *fast Fourier transformation* (FFT) and produces separate time and frequency domain plots

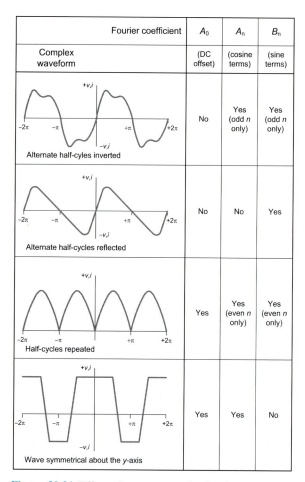

Fourier coefficient	A_0	A_n	B_n
Complex waveform	(DC offset)	(cosine terms)	(sine terms)
Alternate half-cyles inverted	No	Yes (odd n only)	Yes (odd n only)
Alternate half-cycles reflected	No	No	Yes
Half-cycles repeated	Yes	Yes (even n only)	Yes (even n only)
Wave symmetrical about the y-axis	Yes	Yes	No

Figure 30.21 Effect of symmetry and reflection on Fourier coefficients for different types of waveform

(see Figures 30.22 and 30.23, respectively) of a given complex waveform:

```
// Fast Fourier analysis of a complex
waveform
// SCILAB source file: ex3003.sce
// Set parameters
sample_rate=500; // Set sample rate
t = 0:1/sample_rate:1;
N=size(t,'*'); // Specify number of samples
// Specify a waveform with components at
20 Hz and 60 Hz
s= 0.8*sin(2*%pi*20*t) + 0.2*sin(2*%pi*60*t
- %pi/4);
// Plot the function in the time domain
set("current_figure", 0); // Time domain in
graphic window 0
clf();
```

```
plot2d(t,s,3);
title('Time domain (voltage plotted against
time)');
// Obtain the Fourier transform
y=fft(s);
// As the fft response is symmetric we need
only the first N/2 points
f=sample_rate*(0:(N/2))/N; //associated
frequency vector
n=size(f,'*');
// Clear the graphic display and plot in
the frequency domain
set("current_figure",1); // Frequency
domain in graphic window 1
clf();
ya = y * 2/N; // y-axis scale is (amplitude
* N/2)
plot2d(f,abs(ya(1:n)),5);
title('Frequency domain (amplitude plotted
against frequency)');
```

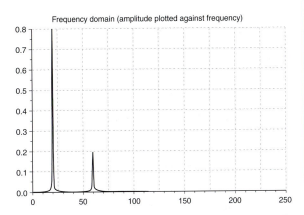

Figure 30.23 Complex waveform plotted by SCILAB in the frequency domain

30.3   Chapter summary

In this chapter we introduced some important concepts relating to complex waveforms. In particular, we explained how any complex waveform can be synthesised from a number of individual sinusoidal components each having a specific amplitude, frequency and phase. Integer multiples of a fundamental frequency are referred to as harmonics and the specific integer relates to the order of the harmonic (second, third, fourth, and so on).

Fourier analysis provides us with a powerful technique based on the concept that all waveforms, whether continuous or discontinuous, can be expressed in terms of a convergent series. Modern mathematics applications (such as MatLab and SCILAB) perform Fourier transformation using algorithms requiring significantly less time for computation. Frequency domain plots of a given complex waveform can be quickly and accurately produced using fast Fourier transformation (FFT) techniques.

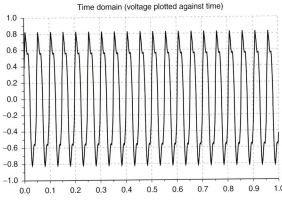

Figure 30.22 Complex waveform plotted by SCILAB in the time domain

Test your knowledge 30.10

Use SCILAB to analyse the frequency spectrum of each of the following complex waveforms:

(a) $v = 165 \sin(120\pi t) + 55 \sin(360\pi t)$
$$+ 33 \sin(600\pi t) \text{ V}$$

(b) $v = 340 \sin(100\pi t) + 20 \sin\left(300\pi t - \dfrac{\pi}{2}\right)$
$$+ 50 \sin\left(500\pi t + \dfrac{\pi}{2}\right) \text{ V}$$

30.4   Review questions

1.   A complex voltage consists of a 400 Hz fundamental with an amplitude of 110 V and a third harmonic with an amplitude of 20 V leading the fundamental by 90°. Write down an expression for the voltage and use this to determine the instantaneous value of the waveform at a time of 1 ms from the start of a cycle.

2. A current is given by the expression:

$$i = 4.5 \sin(120\pi t) + 1.5 \sin\left(360\pi t - \frac{\pi}{2}\right)$$

$$+ 0.9 \sin\left(600\pi t + \frac{\pi}{2}\right) \text{A}$$

State the amplitude and frequency of the fundamental and any harmonic components present. Also state the phase of the harmonic components relative to the fundamental.

3. Write down an expression for the instantaneous voltage of a square wave having a frequency of 100 Hz and an amplitude of 5 V (ignore any components with a harmonic order greater than seven).

4. Write down an expression for the instantaneous voltage of a triangular wave having a frequency of 500 Hz and an amplitude of 18 V (ignore any components with a harmonic order greater than seven).

5. A complex waveform is described by the equation:

$$v = 100 \sin(100\pi t) + 25 \sin(200\pi t + \pi/6)$$

$$+ 50 \sin(300\pi t + 3\pi/4) \text{V}$$

Determine:

(a) the amplitude of the fundamental

(b) the frequency of the fundamental

(c) the order of any harmonic components present

(d) the amplitude of any harmonic components present

(e) the phase angle of any harmonic components present (relative to the fundamental).

Chapter 31

Power in a complex waveform

Earlier, in Chapter 29 we introduced several important concepts relating to the power in an a.c. circuit. Throughout that chapter we assumed that the waveforms of current and voltage were purely sinusoidal, but this is often not the case. Now let's combine what we've just learned from the two previous chapters in order to determine the power and power factor when the applied waveform is complex rather than purely sinusoidal.

31.1 RMS value of a waveform

The root mean square (RMS) value of a waveform is the effective value of the current or voltage concerned. It is defined as the value of direct current or voltage that would produce the same power in a pure resistive load.

Assume that we are dealing with a voltage given by:

$$v = V_m \cos(\omega t)$$

where V_m is the amplitude, maximum or peak value of the voltage waveform (see Figure 31.1). If this voltage is applied to a pure resistance, R, the current flowing will be given by:

$$i = \frac{V_m}{R} \cos(\omega t)$$

As in Chapter 29, the instantaneous power dissipated in the resistor, p, will be given by the product of i and v.

Thus:

$$p = iv = \frac{V_m}{R} \cos(\omega t) \times V_m \cos(\omega t) = \frac{V_m^2}{R} \cos^2(\omega t)$$

From the double angle formula, $\cos^2 \theta = \dfrac{1 + \cos 2\theta}{2}$

Hence:

$$p = \frac{V_m^2}{R} \left(\frac{1 + \cos(2\omega t)}{2} \right) = \frac{V_m^2}{2R} + \frac{V_m^2 \cos(2\omega t)}{2R} \quad \text{(i)}$$

This is an important result. From equation (i) we can infer that the power waveform (i.e. the waveform of p plotted against t) will take the form of a cosine function at *twice* the frequency of the voltage (see Figure 31.1) as shown in Chapter 29. Furthermore, if we apply our recently acquired knowledge of Fourier series, we can infer that the mean value of the power waveform (over a complete cycle of the voltage or current) will be the same as its amplitude. The values of the Fourier coefficients are:

$$A_0 = \frac{V_m^2}{2R} \quad \text{(the mean power over one complete cycle}$$

of the power waveform)

$$A_2 = \frac{V_m^2}{2R} \quad \text{(the amplitude of the term in} \cos 2\omega t$$

– at twice the frequency of the voltage)

Note that no other components are present.

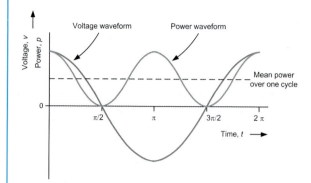

Figure 31.1 Voltage and corresponding power waveform

The mean power over one cycle, P, is given by A_0.

Thus

$$P = \frac{V_m^2}{2R}$$

from which

$$V_m = \sqrt{2PR} \qquad \text{(ii)}$$

Now, let V_{RMS} be the equivalent d.c. voltage that will produce the same power in the load.

Thus

$$P = \frac{V_{RMS}^2}{R}$$

and

$$V_{RMS} = \sqrt{PR} \qquad \text{(iii)}$$

Combining equations (ii) and (iii) gives:

$$\frac{V_m}{V_{RMS}} = \frac{\sqrt{2PR}}{\sqrt{PR}} = \sqrt{2}$$

from which

$$V_m = \sqrt{2}\, V_{RMS}$$

In other words, a waveform given by $v = 1.414$ $\cos(\omega t)\,V$ will produce the same power in a load as a direct voltage of 1 V. A similar relationship can be obtained for current:

$$I_m = \sqrt{2}\, I_{RMS}$$

where I_m is the amplitude (or peak value) of the current waveform.

If you are wondering why we've used the cosine function rather than the sine function to define our voltage waveform, this should be a little clearer in the next section where we introduce power factor (you may recall that power factor is the cosine of the phase angle, and so defining voltage this way helps to make things a little easier!).

When a wave is complex, it follows that its root mean square value can be found by adding together the values of each individual harmonic component present. Thus:

$$V_{RMS} = \sqrt{V_0^2 + \frac{V_1^2 + V_2^2 + V_3^2 + \ldots V_n^2}{2}}$$

$$= \sqrt{V_0^2 + \frac{\sum_{i=1}^{i=n} V_i^2}{2}}$$

where V_0 is the value of any d.c. component that may be present, and $V_1, V_2, V_3, \ldots, V_n$ are the amplitudes of the harmonic components present in the wave. Note that we have used the suffix i to indicate the general term in the series of voltages and n to denote the last voltage in the series.

> **Key Point.**   For a purely resistive load, the mean value of the power waveform over a complete cycle of the supply will be the same as its amplitude

SCILAB

The following SCILAB script will allow you to plot the voltage, current and power waveforms for a purely resistive load. Note that the power waveform is always positive and that it has twice the frequency of the voltage and current waveforms (see Figure 31.2):

```
// Power waveform for a resistive load
// SCILAB source file: ex3101.sce
// Clear any previous graphic window
clf();
txt = ['Input maximum value in V';
'Input frequency in Hz';'Input load
resistance in ohms'];
var = x_mdialog('Enter parameters',
txt,['10';'100';'5']);
vm = evstr(var(1));
f = evstr(var(2));
```

```
r = evstr(var(3));
t = [0:0.0001:0.02]; // set time scale
v = vm*sin(6.28*f*t);
i = v ./ r; // elementwise division
p = i .* v; // elementwise multiplication
// Plot voltage, current and power
waveforms
plot2d(t,v,style=3); // green line for
voltage
plot2d(t,i,style=4); // blue line for
current
plot2d(t,p,style=5); // red line for power
// Print grid and title
xgrid(); // add grid
xtitle('Voltage, current and power
waveforms');
hl=legend(['Voltage';'Current';'Power']);
```

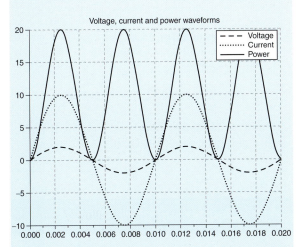

Figure 31.2 Using SCILAB to plot voltage, current and power waveforms

Test your knowledge 31.1

Write and test a SCILAB script that will plot the voltage, current and power waveforms for a purely reactive load. What does the power waveform tell you?

Key Point. For a purely resistive load the power waveform is positive at all times. For a reactive load the power waveform will be partially positive and partially negative

31.2 Power factor for a complex waveform

When harmonic components are present, the power factor can be determined by adding together the power supplied by each harmonic before dividing by the product of root mean square voltage and current. Hence:

$$\text{Power factor} = \frac{\text{Total power supplied}}{\text{True power}} = \frac{P_{\text{TOT}}}{V_{\text{RMS}}I_{\text{RMS}}}$$

where P_{TOT} is found by adding the power due to each harmonic component present.

Thus:

$$\text{Power factor} = \frac{\left(\begin{array}{l}\dfrac{V_1}{\sqrt{2}}\dfrac{I_1}{\sqrt{2}}\cos\phi_1 + \dfrac{V_2}{\sqrt{2}}\dfrac{I_2}{\sqrt{2}}\cos\phi_2 + \\[2mm] \dfrac{V_3}{\sqrt{2}}\dfrac{I_3}{\sqrt{2}}\cos\phi_3 + \cdots + \dfrac{V_n}{\sqrt{2}}\dfrac{I_n}{\sqrt{2}}\cos\phi_n\end{array}\right)}{V_{\text{RMS}}I_{\text{RMS}}}$$

$$\text{Power factor} = \frac{\left(\begin{array}{l}\dfrac{V_1 I_1\cos\phi_1}{2} + \dfrac{V_2 I_2\cos\phi_2}{2} + \\[2mm] \dfrac{V_3 I_3\cos\phi_3}{2} + \cdots + \dfrac{V_n I_n\cos\phi_n}{2}\end{array}\right)}{V_{\text{RMS}}I_{\text{RMS}}}$$

$$\text{Power factor} = \frac{\displaystyle\sum_{i=1}^{i=n}\dfrac{V_i I_i\cos\phi_i}{2}}{V_{\text{RMS}}I_{\text{RMS}}}$$

Example 31.1 Determine the effective voltage and average power in a 10 Ω resistor when the following voltage is applied to it:

$$v = 10\sin(\omega t) + 5\sin(2\omega t) + 2\sin(3\omega t)$$

Now

$$V_{\text{RMS}} = \sqrt{V_0^2 + \frac{V_1^2 + V_2^2 + V_3^2 + \cdots V_n^2}{2}}$$

In this case, $V_0 = 0$ (there is no d.c. component), $V_1 = 10$, $V_2 = 5$ and $V_3 = 2$.

Thus

$$V_{RMS} = \sqrt{\frac{10^2 + 5^2 + 2^2}{2}} = \sqrt{\frac{129}{2}} = 8.03\,\text{V}$$

$$P = \frac{8.03^2}{10} = 6.45\,\text{W}$$

Example 31.2 The voltage and current present in an a.c. circuit are described by the equations:

$$v = 50\sin\omega t + 10\sin(3\omega t + \pi/2)$$

$$i = 3.54\sin(\omega t + \pi/4) + 0.316\sin(3\omega t + 0.321)$$

Determine the total power supplied and the power factor.

The total power supplied, P_{TOT}, is given by:

$$P_{TOT} = \frac{50 \times 3.54}{2}\cos(-\pi/4)$$

$$+ \frac{10 \times 0.316}{2}\cos(\pi/2 - 0.321)$$

$$P_{TOT} = (88.5 \times 0.707) + (1.58 \times 0.32)$$

$$= 62.6 + 0.51 = 63.11\,\text{W}$$

The root mean square voltage, V_{RMS}, is given by:

$$V_{RMS} = \sqrt{\frac{50^2 + 10^2}{2}} = \sqrt{\frac{2600}{2}} = 36.1\,\text{V}$$

The root mean square current, I_{RMS}, is given by:

$$I_{RMS} = \sqrt{\frac{3.54^2 + 0.316^2}{2}} = \sqrt{\frac{12.63}{2}} = 2.51\,\text{A}$$

The true power, P, is thus:

$$P = V_{RMS}\,I_{RMS} = 36.1 \times 2.51 = 90.6\,\text{W}$$

Finally, to find the power factor we simply divide the total power, P_{TOT}, by the true power, P:

$$\text{Power factor} = \frac{P_{TOT}}{P} = \frac{63.11}{90.6} = 0.7$$

Test your knowledge 31.2

Determine the effective voltage and average power dissipated in a 50 Ω resistor when the following voltage is applied to it:

$$v = 100\sin(\omega t) + 30\sin(3\omega t)$$

Test your knowledge 31.3

A current given by:

$$i = 0.5\sin(\omega t) + 1.5\sin(2\omega t)$$

flows in a purely resistive 200 Ω load. Determine the effective current and average power dissipated in the load.

Test your knowledge 31.4

A voltage is given by:

$$v = 50 + 10\sin(\omega t) + 5\sin(3\omega t)$$

Determine the power developed in a 50 Ω load when this voltage is applied to it.

31.3   Chapter summary

This chapter has provided you with an introduction to power and power factor in circuits where waveforms are no longer purely sinusoidal. Once again, we showed how the power waveform in an a.c. circuit has twice the frequency of that of the voltage and current waveforms. We also showed how the Fourier series enables us to determine the power associated with each individual harmonic component. This brings our study of alternating current to a conclusion.

31.4   Review questions

1. A voltage is given by the expression:

$$v = 100\sin(\omega t) + 40\sin(2\omega t - \pi/2)$$

$$+ 20\sin(4\omega t - \pi/2)$$

If this voltage is applied to a 50 Ω resistor,

(a) determine the power in the resistor due to the fundamental and each harmonic component

(b) derive an expression for the current flowing in the resistor

(c) calculate the root mean square voltage and current

(d) determine the total power dissipated.

2. The voltage, v, and current, i, in a circuit are defined by the following expressions:

$$v = 30\sin(\omega t) + 7.5\sin(3\omega t + \pi/4)$$

$$i = \sin(\omega t) + 0.15\sin(3\omega t - \pi/12)$$

Determine:

(a) the root mean square voltage

(b) the root mean square current

(c) the total power

(d) the true power

(e) the power factor of the circuit.

3. A voltage given by the expression:

$$v = 100\sin(314t) + 50\sin(942t) - 40\sin(1570t)$$

is applied to an impedance of $(40 + j30)$ Ω. Determine:

(a) an expression for the instantaneous current flowing

(b) the effective (RMS) voltage

(c) the effective (RMS) current

(d) the total power supplied

(e) the true power

(f) the power factor.

Index